SolidWorks for Designers
Release 2004

Sham Tickoo
Professor
Department of Mechanical Engineering Technology
Purdue University Calumet
Hammond, Indiana
U.S.A.

CADCIM Technologies
(www.cadcim.com)
U.S.A.

CADCIM Technologies
Excellence de Technology

CADCIM Technologies

SolidWorks for Designers: Release 2004
Sham Tickoo

Published by CADCIM Technologies, 525 St Andrews Drive, Schererville, IN 46375 USA.

ISBN 0-9663537-9-X

Cover designer: *CADCIM Technologies*
Technical editors: *Gurpreet Singh, Deepak Maini, Saket Sighal*
Copy editor: *Vivek Kumar*
Cover illustration: *Created by Gurpreet Singh and Deepak Maini using SolidWorks, PhotoWorks, and Photoshop software*
Typeface: *10/12 New Baskerville Bt*

NOTICE TO THE READER

Publisher does not warrant or guarantee any of the products described in the text or perform any independent analysis in connection with any of the product information contained in the text. Publisher does not assume, and expressly disclaims, any obligation to obtain and include information other than that provided to it by the manufacturer.

The reader is expressly warned to consider and adopt all safety precautions that might be indicated by the activities herein and to avoid all potential hazards. By following the instructions contained herein, the reader willingly assumes all risks in connection with such instructions.

The Publisher makes no representation or warranties of any kind, including but not limited to, the warranties of fitness for particular purpose or merchantability, nor are any such representations implied with respect to the material set forth herein, and the publisher takes no responsibility with respect to such material. The publisher shall not be liable for any special, consequential, or exemplary damages resulting, in whole or part, from the reader's use of, or reliance upon, this material.

www.cadcim.com

DEDICATION

To teachers, who make it possible to disseminate knowledge
to enlighten the young and curious minds
of our future generations

To students, who are dedicated to learning new technologies
and making the world a better place to live

THANKS

To the faculty and students of the MET department of
Purdue University Calumet for their cooperation

To Santosh Tickoo, Gurpreet Singh, Deepak Maini,
Saket Singhal, and Abhinav Rai
of CADCIM Technologies for their valuable help

Free Teaching Aids for Faculty

The following teaching aids are available for free to faculty.

1. *All Part files (.sldprt), Assembly files (.sldasm), and Drawing files (.slddrw) used for tutorials and exercises in this book.*
2. *PowerPoint presentations for every chapter of the book.*
3. *Instructor's Guide with answers to review questions and solution to exercises.*
4. *Course outlines.*
5. *Students projects.*
6. *Free online technical support by contacting* **techsupport@cadcim.com**.

To access the web site that contains these teaching aids, please contact the author, Prof. Sham Tickoo, at the following address.

stickoo@calumet.purdue.edu
or
tickoo@cadcim.com

For Students: *You can download solid modeling exercises, tutorials, and special topics by accessing the author's web site at the following address.*

www.cadcim.com
or
http://technology.calumet.purdue.edu/met/tickoo/students/students.htm.

For more information, please visit www.cadcim.com

Table of Contents

Dedication iii
Preface xv
Introduction xix

Chapter 1: Drawing Sketches for the Solid Models

The Sketching Environment 1-2
Starting a New Document in SolidWorks 2004 1-3
Setting Up the Document Options 1-7
Modifying the Dimensioning Standards 1-8
Modifying the Linear and Angular Units 1-8
Modifying Snap and Grid Settings 1-9
Learning About Sketcher Terms 1-11
Origin 1-12
Inferencing Lines 1-12
Select Tool 1-13
Drawing Lines 1-13
Drawing Continuous Lines 1-14
Drawing Individual Lines 1-14
Drawing Tangent or Normal Arcs Using the Line Tool 1-17
Drawing Construction Lines or Centerlines 1-18
Drawing Circles 1-18
Drawing a Construction Circle 1-19
Drawing Arcs 1-20
Drawing Tangent/Normal Arcs 1-20
Drawing Centerpoint Arcs 1-22
Drawing 3 Point Arcs 1-23
Drawing Rectangles 1-25
Drawing Parallelogram 1-26
Drawing a Rectangle at an Angle 1-26
Drawing Parallelograms 1-26
Drawing Polygons 1-27
Drawing Splines 1-29
Drawing Points 1-30
Drawing Ellipses 1-31
Drawing Elliptical Arcs 1-32
Drawing Parabolic Curve 1-33
Drawing Display Tools 1-34
Zoom to Fit 1-35
Zoom Area 1-35
Zoom In/Out 1-35
Zoom to Selection 1-35

Pan	1-36
Redraw	1-36
Deleting the Sketched Entities	1-36
Tutorial 1	1-36
Tutorial 2	1-47
Tutorial 3	1-51
Tutorial 4	1-56
Self-Evaluation Test	1-61
Review Questions	1-61
Exercise 1	1-62
Exercise 2	1-63

Chapter 2: Editing and Modifying the Sketches

Editing the Sketched Entities	2-2
Trimming the Sketched Entities	2-2
Extending the Sketched Entities	2-3
Filleting the Sketched Entities	2-4
Chamfering the Sketched Entities	2-5
Offsetting the Sketched Entities	2-7
Mirroring the Sketched Entities	2-9
Moving, Rotating, and Scaling the Sketched Entities	2-11
Creating Patterns	2-15
Creating Linear Pattern of the Sketched Entities	2-16
Creating Circular Pattern of the Sketched Entities	2-21
Editing Patterns	2-25
Writing Text in the Sketching Environment	2-26
Modifying Sketched Entities	2-27
Modifying a Sketched Line	2-27
Modifying a Sketched Circle	2-27
Modifying a Sketched Arc	2-27
Modifying a Sketched Polygon	2-28
Modifying a Spline	2-28
Modifying a Sketched Point	2-29
Modifying an Ellipse or an Elliptical Arc	2-29
Modifying a Parabola	2-29
Dynamically Modifying and Coping the Sketched Entities	2-29
Splitting the Sketched Entity	2-30
Tutorial 1	2-30
Tutorial 2	2-40
Tutorial 3	2-46
Self-Evaluation Test	2-52
Review Questions	2-52
Exercise 1	2-53
Exercise 2	2-54
Exercise 3	2-55
Exercise 4	2-56

Chapter 3: Adding Relations and Dimensions to the Sketches

Adding Geometric Relations to the Sketch 3-2
Adding Relations Using the Add Relations PropertyManager 3-2
Automatic Relations 3-7
Dimensioning the Sketch 3-8
Horizontal/Vertical Dimensioning 3-10
Aligned Dimensioning 3-20
Angular Dimensioning 3-20
Diameter Dimensioning 3-22
Radius Dimensioning 3-23
Linear Diameter Dimensioning 3-23
Ordinate Dimensioning 3-24
Additional Dimensioning Options 3-26
Concept of Fully Defined Sketch 3-27
Fully Defined 3-27
Over Defined 3-28
Under Defined 3-28
Dangling 3-28
No Solution Found 3-28
Invalid Solution Found 3-29
Sketch Dimension or Relation Status 3-29
Deleting the Over Defining Dimensions 3-30
Displaying and Deleting Relations 3-31
Viewing and Examining Relations 3-35
Opening an Existing File 3-36
Tutorial 1 3-38
Tutorial 2 3-48
Tutorial 3 3-55
Self-Evaluation Test 3-61
Review Questions 3-62
Exercise 1 3-63
Exercise 2 3-63
Exercise 3 3-64

Chapter 4: Advanced Dimensioning and Base Feature Options

Advanced Dimensioning Techniques 4-2
Autodimension the Sketches 4-2
Dimensioning of the True Length of an Arc 4-5
Measuring Distances and Viewing Section Properties 4-6
Measuring Distances 4-6
Determining Section Properties of Closed Sketches 4-10
Creating the Base Feature by Extruding the Sketches 4-11
Creating the Solid Extruded Features 4-12
Creating the Thin Extruded Features 4-17
Creating the Base Features by Revolving the Sketches 4-20

Creating the Solid Revolve Features	4-21
Creating the Thin Revolved Features	4-24
Dynamically Rotating the View of the Model	4-26
Rotating the View Freely in 3D Space	4-26
Rotating the View Around a Selected Vertex, Edge, or Face	4-26
Modifying the View Orientation	4-27
Restoring Previous View	4-29
Display Modes of the Model	4-29
Wireframe	4-29
Hidden Lines Visible	4-29
Hidden Lines Removed	4-30
Shaded With Edges	4-30
Shaded	4-30
Shadows In Shaded Mode	4-30
Perspective	4-30
Assigning Material and Texture to the Model	4-32
Assigning Material to a Model New	4-32
Assigning Textures to a Model New	4-33
Tutorial 1	4-34
Tutorial 2	4-41
Tutorial 3	4-46
Self-Evaluation Test	4-52
Review Questions	4-53
Exercise 1	4-54
Exercise 2	4-55

Chapter 5: Creating Reference Geometries

Importance of the Sketching Planes	5-2
Reference Geometry	5-2
Creating New Planes	5-5
Creating Reference Axis	5-10
Creating Reference Points New	5-14
Creating Reference Coordinate System	5-16
Other Boss/Base Options	5-17
End Condition	5-17
Direction of Extrusion New	5-21
Modeling using the Contour Selection Method Enhanced	5-21
Creating the Cut Features	5-25
Creating Extruded Cuts	5-25
Handling Multiple Bodies in Cut Feature	5-28
Creating Revolved Cuts	5-30
Concept of the Feature Scope	5-31
Tutorial 1	5-32
Tutorial 2	5-38
Tutorial 3	5-45
Self-Evaluation Test	5-54

Review Questions 5-55
Exercise 1 5-56
Exercise 2 5-57

Chapter 6: Advanced Modeling Tools-I

Advanced Modeling Tools 6-2
Creating Simple Hole 6-2
Creating the Standard Holes Using Hole Wizard Enhanced 6-3
Creating Fillets Enhanced 6-15
Selection Methods 6-31
Creating Chamfer Enhanced 6-33
Creating the Shell Feature Enhanced 6-37
Creating Wrap Feature New 6-41
Tutorial 1 6-43
Tutorial 2 6-53
Self-Evaluation Test 6-62
Review Questions 6-63
Exercise 1 6-64
Exercise 2 6-65

Chapter 7: Advanced Modeling Tools-II

Advanced Modeling Tools 7-2
Creating Mirror Feature 7-2
Creating Linear Pattern Feature 7-8
Creating Circular Pattern Feature 7-15
Creating Sketch-driven Pattern 7-19
Creating Curve-driven Pattern 7-21
Creating Table-driven Pattern 7-24
Creating the Rib Feature 7-26
Creating the Dome Feature Enhanced 7-31
Creating the Shape Feature 7-35
Displaying the Section View of the Model Enhanced 7-37
Tutorial 1 7-39
Tutorial 2 7-48
Tutorial 3 7-54
Self-Evaluation Test 7-65
Review Questions 7-65
Exercise 1 7-66
Exercise 2 7-68
Exercise 3 7-69

Chapter 8: Editing Features

Editing the Features of the Model 8-2
Editing Using Edit Feature Option 8-2

Editing the Sketches of the Sketched Features 8-4
Changing the Sketch Plane of the Sketches 8-4
Editing by Double-clicking the Entities and Features 8-4
Editing using the Move/Size Features 8-6
Editing the Sketches with the Move/Size Features Tool Active 8-9
Editing the Features and Sketches by Cut, Copy, and Paste 8-10
Cut, Copy, and Paste Features and Sketches from One Document to the other 8-11
Coping Features Using Drag and Drop 8-11
Deleting the Features 8-13
Deleting the Bodies 8-13
Suppressing the Features 8-14
Unsuppressing the Features 8-15
Unsuppressing the Features with Dependents 8-15
Hiding the Bodies 8-15
Move/Copy Bodies 8-16
Reordering the Features 8-18
Rolling Back the Model 8-20
Renaming the Features 8-21
Creating Folders in the FeatureManager Design Tree 8-21
What's Wrong? Functionality 8-22
Tutorial 1 8-23
Tutorial 2 8-31
Tutorial 3 8-38
Self-Evaluation Test 8-46
Review Questions 8-47
Exercise 1 8-48
Exercise 2 8-49

Chapter 9: Advanced Modeling Tools-III

Advanced Modeling Tools 9-2
Creating the Sweep Feature Enhanced 9-2
Creating Cut Sweep Features Enhanced 9-14
Creating Loft Feature Enhanced 9-15
Creating 3D Sketches 9-25
Editing 3D Sketches 9-27
Creating Curves 9-27
Extruding a 3D Sketch New 9-40
Creating Draft Features 9-40
Creating the Deform Feature New 9-42
Tutorial 1 9-47
Tutorial 2 9-51
Tutorial 3 9-55
Self-Evaluation Test 9-62
Review Questions 9-62
Exercise 1 9-63
Exercise 2 9-65

Chapter 10: Assembly Modeling-I

Assembly Modeling 10-2
    Types of Assembly Design Approach 10-2
Creating Bottom-Up Assemblies 10-4
    Placing Components in the Assembly File 10-4
    Assembling the Components 10-9
Creating Top-down Assemblies 10-26
    Creating the Component in the Top-down Assembly 10-27
Moving the Individual Components 10-28
    Moving the Individual Components by Dragging New 10-28
    Moving the Individual Components using the Move Component Tool 10-28
Rotating the Individual Components 10-29
    Rotating the Individual Components by Dragging New 10-29
    Rotating the Individual Components using the Rotate Component Tool 10-30
Moving and Rotating an Individual Components using the Triad New 10-30
Tutorial 1 10-32
Tutorial 2 10-47
Self-Evaluation Test 10-52
Review Questions 10-53
Exercise 1 10-54

Chapter 11: Assembly Modeling-II

Advanced Assembly Mates 11-2
    Applying the Symmetric Mate 11-2
    Applying the Cam Mate 11-2
    Applying Gear Mate New 11-4
    Applying Limit Mate New 11-4
Creating Subassemblies 11-5
    Bottom-up Subassembly Design 11-5
    Top-down Subassembly Design 11-5
Deleting Components and Subassemblies 11-6
Editing Assembly Mates 11-7
    Replacing the Mated Entities 11-8
Editing the Components 11-9
Editing the Subassemblies 11-10
Dissolving the Subassembly 11-11
Replacing the Components 11-11
Creating Patterns of the Components in an Assembly 11-14
    Feature-driven Pattern 11-14
    Local Pattern 11-15
Mirroring the Components 11-16
Simplifying the Assemblies using Visibility Options 11-18
    Hiding the Components 11-19
    Suppressing the Components 11-19
    Changing the Transparency Conditions 11-20

Checking the Interferences in the Assembly 11-21
Creating the Assemblies for Mechanism 11-22
    Analyzing the Collisions using the Collision Detection Tool 11-23
Creating the Explode State of the Assembly 11-25
    Creating Systematic Explode State 11-26
    Creating the Explode Line Sketch 11-28
Tutorial 1 11-30
Tutorial 2 11-41
Self-Evaluation Test 11-46
Review Questions 11-47
Exercise 1 11-48

Chapter 12: Working With Drawing Views-I

The Drawing Mode 12-2
Starting a Drawing Document 12-2
    Starting a New Drawing Document Using the New SolidWorks Document Dialog box 12-2
    Starting a New Drawing Document From Within the Part/Assembly Document 12-4
Type of Views 12-5
Generating Standard Drawing Views 12-6
    Generating Model Views 12-6
    Generating the Three Standard Views 12-9
    Generating Standard Views Using the Relative View Tool 12-11
    Generating Standard Views Using the Predefined View Tool 12-12
Generating the Derived Views 12-15
    Generating Projected Views 12-15
    Generating Section Views 12-16
    Generating Aligned Section Views 12-21
    Generating Broken-out Section Views 12-23
    Generating Auxiliary Views 12-25
    Generating Detail Views 12-27
    Cropping Drawing Views 12-29
    Generating Broken View 12-30
    Alternate Position View 12-32
    Creating the Drawing view of the Exploded State of the Assembly 12-34
Working with Interactive Drafting in SolidWorks 12-35
Editing and Modifying the Drawing Views 12-35
    Changing the View Orientation 12-36
    Changing the Scale of the Drawing Views 12-36
    Deleting the Views 12-36
    Rotating the Drawing View 12-36
Modifying the Hatch Pattern of the Section View 12-37
Tutorial 1 12-39
Tutorial 2 12-45
Self-Evaluation Test 12-50

Review Questions 12-51
Exercise 1 12-52

Chapter 13: Working With Drawing Views-II

Adding Annotations to the Drawing Views 13-2
Generating the Annotations using the Model Items Tool 13-2
Adding the Reference Annotations 13-5
Editing the Annotations 13-21
Adding the BOM in the Drawing Enhanced 13-21
Adding Balloons to the Drawing Views 13-23
Adding Balloons using the Balloon Tool 13-23
Adding Balloons using the AutoBalloon Tool New 13-25
Adding New Sheets to the Drawing Views 13-26
Editing the Sheet Format 13-27
Creating a User-Defined Sheet Format 13-27
Tutorial 1 13-28
Tutorial 2 13-34
Self-Evaluation Test 13-39
Review Questions 13-40
Exercise 1 13-41

Student Project 1

Index 1

About the Author

Sham Tickoo is a professor of Mechanical Engineering Technology at Purdue University Calumet. He has been in the education and training since 1981 and joined Purdue University in the year 1987. Since then he has been teaching drafting and design, AutoCAD, AutoLISP, 3ds max, 3ds viz, Mechanical Desktop, Pro/Engineer, Autodesk Inventor, SolidWorks etc.

Prof. Tickoo is also one of the best known authors of CAD/CAM books in the world market. He has authored a number of books on the latest CAD/CAM technology. He is one of the very few authors in the world whose books are translated and sold in the countries like Italy, China, Philippines, and Russia. The list of some of his books is given below:

SolidWorks for Engineers and Designers Release 2004, dreamtech PRESS, India
SolidWorks for Designers Release 2003, CADCIM Technologies, USA
SolidWorks for Engineers and Designers Release 2003, dreamtech PRESS, India
Pro/ENGINEER Wildfire for Designers, CADCIM Technologies, USA
Pro/ENGINEER Wildfire for Engineers & Designers, dreamtech PRESS, India
Pro/ENGINEER for Designers Release 2001, CADCIM Technologies, USA
Autodesk Inventor for Designers Release 6, CADCIM Technologies
Autodesk Inventor for Engineers & Designers Release 6, dreamtech PRESS, India
Autodesk Inventor for Designers Release 5, CADCIM Technologies
AutoCAD: A Problem Solving Approach, Delmar Publishers, USA
AutoCAD LT: A Problem Solving Approach, Delmar Publishers, USA
Customizing AutoCAD, Delmar Publishers, USA
3D Studio VIZ, Goodheart-Willcox Publishers, USA
3D Studio MAX, Goodheart-Willcox Publishers, USA
Mechanical Desktop Instructor, McGraw-Hill, USA
AutoCAD 2000 Tecniche Avanzate, APOGeO education, Italy
AutoCAD 2000 Fondamenti, APOGeO education, Italy
AutoCAD 2000: A Problem Solving Approach, China Machine Press, China
Customizing AutoCAD 2000, China Machine Press, China
AutoCAD 2002, Russian Edition

In addition to his academic career, he has been employed as a design engineer, quality control engineer, software developer, and has also worked as a consultant for a number of companies in US and Canada.

Prof. Tickoo has developed a software package called "SMLayout" that has been registered by Autodesk Inc. as a third party software product. This software package is currently being used by some companies in Canada, United States, and Columbia. SMLayout is written in the AutoLISP programming language. It generates flat layout drawing of some complicated geometric shapes like: transitions, cones, cone-cylinder intersections, sphere-cylinder intersections, and cylinder-cylinder intersections. This software package addresses the needs of steel fabricators and manufacturers of sheet metal products. The programs also provide a dimensioning option for automatic dimensioning of the drawing.

In March 2000, Prof. Tickoo was issued a US patent for a product called "Self Adjusting Cargo Organizer" that he designed for vehicles.

Preface

SolidWorks 2004

SolidWorks, developed by SolidWorks Corporation, is one of the world's fastest growing solid modeling softwares. It is a parametric feature-based solid modeling tool and it not only unites the three-dimensional (3D) parametric features with two-dimensional (2D) tools, but also addresses every design-through-manufacturing process. SolidWorks is also known as standard in 3D. SolidWorks 2004 is the latest in the family of SolidWorks, and includes about 250 customer requested enhancements. This implies that SolidWorks is purely based on the customer's needs. Based mainly on the solid modeling users feedback, this solid modeling tool is remarkably user-friendly and it allows you to be productive from day one.

The 2D drawing views of the components are easily generated in the **Drawing** mode. The drawing views that can be generated include detailed, orthographic, isometric, auxiliary, section, and so on. You can use any predefined standard drawing document to generate the drawing views. You can display the model dimensions in the drawing views or add reference dimensions and other annotations whenever you want. You can also add the parametric Bill Of Materials (BOM) and balloons in the drawing view. If a component in the assembly is replaced, removed, or a new component is assembled, the modification will be automatically reflected in the BOM placed in the drawing document. The bidirectional associative nature of this software ensures that any modification made in the model is automatically reflected in the drawing views and any modification made in the dimensions in drawing views automatically updates the model.

SolidWorks for Designers Release 2004 is a text book written with an intent of helping the users who are into 3D design. This book is written with the tutorial point of view with learn-by-doing as the theme. Real world mechanical engineering industry examples and tutorials are used in this book to ensure that the user can relate his knowledge of this book with the actual mechanical industry designs. The main features of the book are as follows.

- **Tutorial Approach**

  The author has adopted the tutorial point-of-view with learn-by-doing as the theme throughout the book. This approach will guide the users through the process of creating the model in the tutorial.

- **Real-world Mechanical Engineering Projects as Tutorials**
  The author has used the real-world mechanical engineering projects as tutorials in this book so that the reader can correlate the tutorials in this book with the real-time models in the mechanical engineering industry.

- **Coverage of all SolidWorks Modes**
  All the modes of SolidWorks are covered in this book, which include the **Part** mode, the **Assembly** mode, and the **Drawing** mode.

- **Tips and Notes**
  Additional information related to the topics is provided to the users in the form of tips and notes.

- **Learning Objectives**
  The first page of every chapter provides in brief the topics that will be covered in that chapter. This will help the users to easily refer to a topic.

- **Tools Section**
  Every chapter begins with the tools section that provides the detailed explanation of the SolidWorks tools and commands.

- **Self-Evaluation Test, Review Questions, and Exercises**
  Every chapter ends with a Self-Evaluation Test so that the users can assess their knowledge of the chapter. The author has given the answers of the Self-Evaluation Tests so that the users can compare their answers with the correct answers. The Review Questions and Exercises are also given at the end of each chapter and can be used by the Instructors as test questions and exercises in the classroom.

- **Heavily Illustrated Text**
  The text in this book is heavily illustrated with the help of around 700 line diagrams and 800 photos that support the tools sections and tutorials.

Features of the Text

Learning Objectives

Each chapter begins with Learning Objectives. These are the points that in brief explain that tools and options that the reader will be able to work with after completing the chapter.

Tutorials

These are real-world mechanical, automobile engineering projects that are explained in a step-by-step method.

Self-Evaluation Test

These are the questions given at the end of each chapter. Answer to these questions are also given at the end of the chapter so that the user can check his answer with the correct answers.

Review Questions

These are the questions that are also given at the end of the chapter. However, the answer to these questions are not given in the book and so the instructor can use these questions as the test questions. The teaching faculty will get answers to these questions in the Instructor's Guide that is provided by the author.

Note

The author have provided a lot of additional information to the users about the topic being discussed in the form of notes.

Tip

Special information and techniques are provided in the form of tip that allow the users to increase their efficiency.

New

This indicates the new commands or tools introduced in SolidWorks 2004.

Enhanced

This indicates the existing commands and tools that are enhanced in SolidWorks 2004.

...... advanced chapter..... For some advanced topics in this textbook, the reader is asked to refer to advanced chapters in part 2 of this book. The part 2 of this book is the Advanced SolidWorks for Designers Release 2004 textbook by Sham Tickoo, published by CADCIM Technologies. This textbook covers advanced chapters and topics of SolidWorks 2004.

Introduction

SolidWorks 2004

Welcome to the entire new world of Computer Aided Designing (CAD) with SolidWorks. If you are a new user of this software package, you are going to join hands with thousands of users of this parametric feature-based and one of the most easy to learn software package. If you are a user of the previous releases of this software, you are going to upgrade your designing skills with the tremendous improvement in this latest release.

SolidWorks, developed by SolidWorks Corporation, USA, is a feature-based, parametric solid-modeling mechanical design and automation software. SolidWorks is also known as standard in 3D. SolidWorks is the first CAD package to use the Microsoft Windows graphic user interface. SolidWorks uses the drag-drop (DD) functionality of Windows, which makes this CAD package extremely easy to learn. The Windows graphic user interface makes it possible for the mechanical design engineers to innovate their ideas and implement them in the form of virtual prototypes or solid model, large assemblies, subassemblies, and detailing and drafting.

SolidWorks is one of the products of SolidWorks Corporation, which is a part of Dassault Systemes. SolidWorks also works as the platform software for a number of softwares. This means you can also use other compatible softwares within the SolidWorks window. There are a number of softwares provided by the SolidWorks Corporation that are used as Add-Ins with SolidWorks. Some of the softwares that can be used on SolidWorks's working platform are listed below.

1. SolidWorks Animator
2. PhotoWorks
3. FeatureWorks
4. COSMOS/Works
5. COSMOS/Motion
6. COSMOS/Flow
7. eDrawings
8. SolidWorks Piping
9. CAMWorks
10. Toolbox
11. Mold Base

As mentioned earlier, SolidWorks is a parametric, feature-based and easy to use mechanical design automation software. It allows you to convert the basic 2D sketch into a solid model using very simple but highly effective modeling tools. SolidWorks does not restrict you only to 3D solid output, but it also extends them to the bidirectional associative generative drafting. It also enables you to create the virtual prototype of a sheet metal component and its flat pattern, which helps you in complete process planning for designing and creating a press tool. SolidWorks also helps you in the extraction of core and cavity of a model that has to be molded or casted. With SolidWorks, you can also create complex parametric shapes as surfaces. Some of the important modes of SolidWorks are as follows.

Part Mode

The **Part** mode of SolidWorks is a feature-based parametric environment. This environment allows you to create the solid models. You are provided with three default planes named as Front, Top, and Right. First, you have to select the sketching plane on which you have to create the sketch for the base feature. After selecting the sketching plane, you enter the sketching environment. The sketches for the model are drawn in the sketching environment using very easy to use tools. After drawing the sketches, you will dimension the sketch and apply the required relations in the same sketching environment. The design intent is captured very easily by adding relations and equations, and by using design table in the design. In the **Part** mode, you are provided with the standard hole library known as HOLE WIZARD. You can create simple holes, tapped holes, counterbore, countersink, and so on. The holes can be of any standard such as ISO, ANSI, JIS, and so on. You can also create complicated surfaces using the surface modeling available within the **Part** mode. Annotations such as weld symbols, geometric tolerance, datum references, and surface finish symbols can be added to the model within the **Part** mode. The standard features that are to be used frequently can be saved as library features and can be retrieved when needed. The palette feature library is also provided in SolidWorks. It contains a number of standard mechanical parts and features. You can also create the sheet metal components in the **Part** mode of SolidWorks using the related tools. In this release of SolidWorks you can also analyze the part model for various stresses applied to the model in the real physical conditions. COSMOSXpress is a very easy and user-friendly tool for stress analysis in SolidWorks. COSMOSXpress helps you to reduce the cost and time in testing your design in real physical testing conditions (destructive tests); you can analyze the component during modeling in the SolidWorks windows. You can also work with the weld modeling within the **Part** mode of SolidWorks by creating the steel structures and adding the weld beads. All the standard weld types and welding conditions are available for your reference. You can extract the core and cavity in the **Part** mode by using the mould design tools.

Assembly Mode

In the **Assembly** mode, assemble the components of the assembly with the help of the required tools. There are two types of methods of assembling the components:

1. Bottom-up assembly
2. Top-down assembly

In the bottom-up approach of assembly, the assembly is created by assembling the previously created components and maintaining their design intent. In the top-down approach of assembly the components are created in the assembly mode. You may begin with some ready-made parts and then you create other components in the context of the assembly. You can refer to the features of some components of the assembly to drive the dimensions of the other components. The **SmartMates** option allows you to assemble all the components using a single button. While assembling the components of an assembly in SolidWorks, you can also animate the assembly by dragging the assembly; with this you can also check the working of your assembly. Collision detection is one major feature of the SolidWorks assembly. Using this feature you can rotate and move the components and detect the collision between the components that you assembled. With the use of physical dynamics you can see the realistic motion of the assembly. Physical simulation is used to simulate the assembly with the effects of motors, springs, and gravity on the assemblies.

Drawing Mode

The **Drawing** mode is used for the documentation of the parts or the assemblies created earlier in the form of drawing views and the detailing of the drawing views. There are two types of drafting done in SolidWorks:

1. Generative drafting
2. Interactive drafting

Generative drafting is done by generating the drawing views of the part or assembly created earlier. In generative drafting, the parametric dimensions and annotations added to the component in the part mode are generated in the drawing views. Generative drafting is bidirectional associative in nature. Automatic BOMs and balloons can be added while generating the drawing views of an assembly.

In interactive drafting you have to create the drawing views by sketching them using the normal sketching tools and then you have to add the dimensions to them.

SYSTEM REQUIREMENTS

The system requirements to ensure smooth running of SolidWorks on your system are as follows:

- Microsoft Windows XP Professional or Windows 2000 recommended; also runs on Windows NT (SP 6 or higher recommended).
- Intel Pentium, or AMD Athlon-based computer.
- 128 MB RAM minimum (512 MB to 1 GB or higher recommended).
- Graphic card OpenGL (recommended).
- Mouse or other compatible pointing device.
- Microsoft Office XP or Microsoft Office 2000.
- Internet Explorer version 5.5 or higher is recommended.

GETTING STARTED WITH SolidWorks

Install SolidWorks on your system and then start it by choosing the **Start** button available on the lower left corner of the screen. Choose **Programs** to display the **Program** menu. In the

Program menu, choose **SolidWorks 2004** to display the cascade menu. In this cascade menu choose **SolidWorks 2004** as shown in Figure I-1.

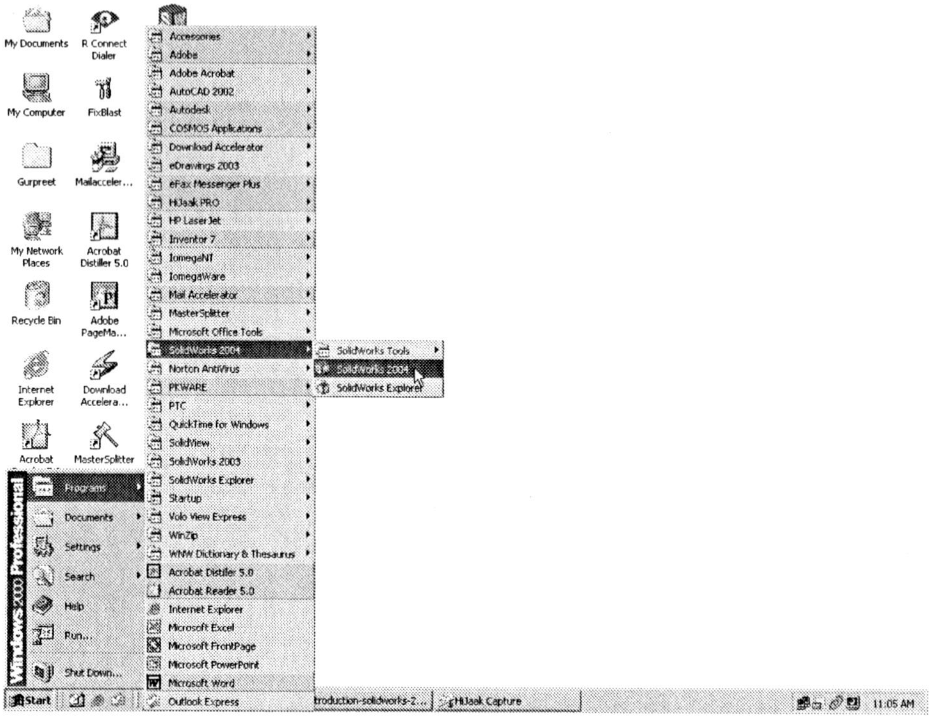

Figure I-1 Starting SolidWorks using the taskbar shortcut

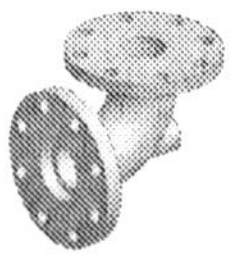

Tip. *You can also start* ***SolidWorks 2004*** *by double-clicking the* ***SolidWorks 2004*** *shortcut icon from the desktop of your computer. If the* ***SolidWorks 2004*** *shortcut icon is not created by default then you need to create this shortcut icon. The shortcut icon of* ***SolidWorks 2004*** *is created by choosing the* ***Start*** *button available on the lower left corner of the screen. Choose* ***Programs*** *to display the* ***Program*** *menu. In the* ***Program*** *menu, choose* ***SolidWorks 2004*** *to display the cascade menu. Move the cursor on* ***SolidWorks 2004*** *and right-click to display the shortcut menu. When you move the cursor to the* ***Send To*** *option the cascade menu is displayed. Choose the* ***Desktop (create shortcut)*** *option from the cascade menu. The* ***SolidWorks 2004*** *shortcut icon is placed on the desktop of your computer.*

The system will now prepare to start SolidWorks. The SolidWorks window will be displayed on the screen. If you are opening SolidWorks for the first time, the **SolidWorks License Agreement** dialog box will be displayed as shown in Figure I-2; choose the **Accept** button from this dialog box.

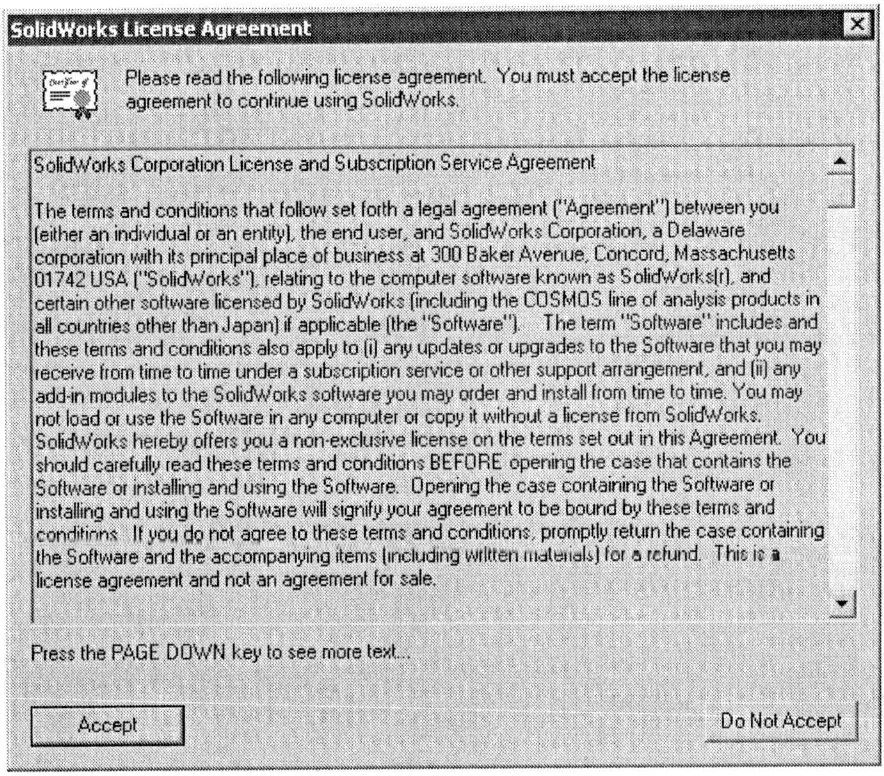

Figure I-2 **SolidWorks License Agreement** dialog box

The **Welcome to SolidWorks 2004** window is displayed as shown in Figure I-3. This window can be used to open a new file, open an existing file, and use online tutorials. This window is also used to visit the Web site of the SolidWorks partners.

Choose the **New Document** button from the **Welcome to SolidWorks 2004** window to open a new file. The **New SolidWorks Document** dialog box is displayed as shown in Figure I-4.

Tip. *In the later releases of SolidWorks the **Tip of the Day** dialog box is displayed by default when you start a SolidWorks session. Now, you have to invoke the **Tip of the Day** dialog box by choosing **Help > Tip of the Day** from the menu bar. The **Tip of the Day** dialog box is displayed. Select the **Show tip at startup** check box to display tips every time you start the SolidWorks session. The tips displayed in this dialog box are very valuable tips that help a fresher as well as an advanced user for the maximum utilization of the CAD package.*

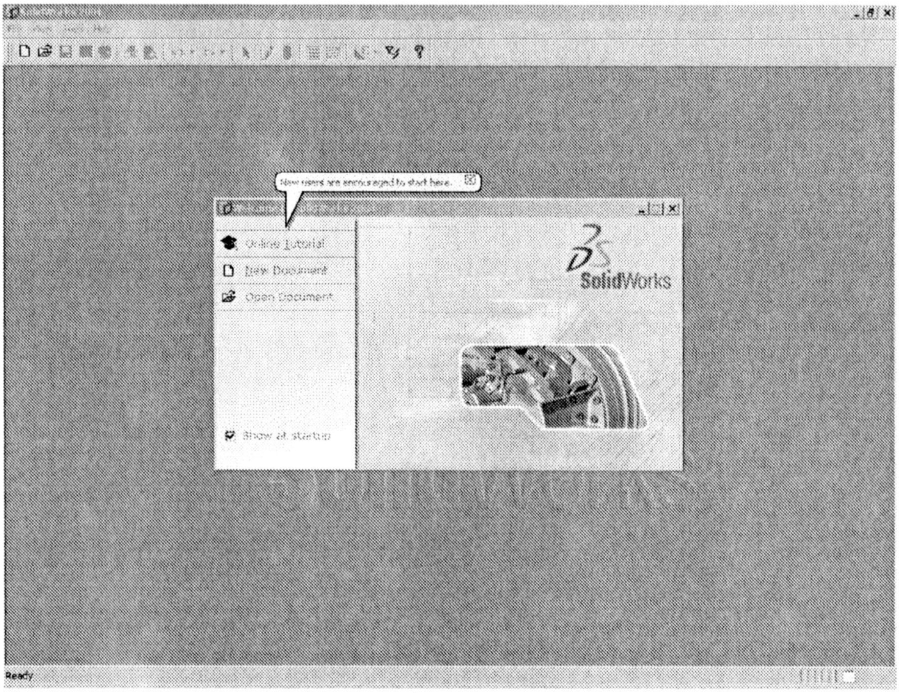

Figure I-3 *SolidWorks window and the* ***Welcome to SolidWorks 2004*** *window*

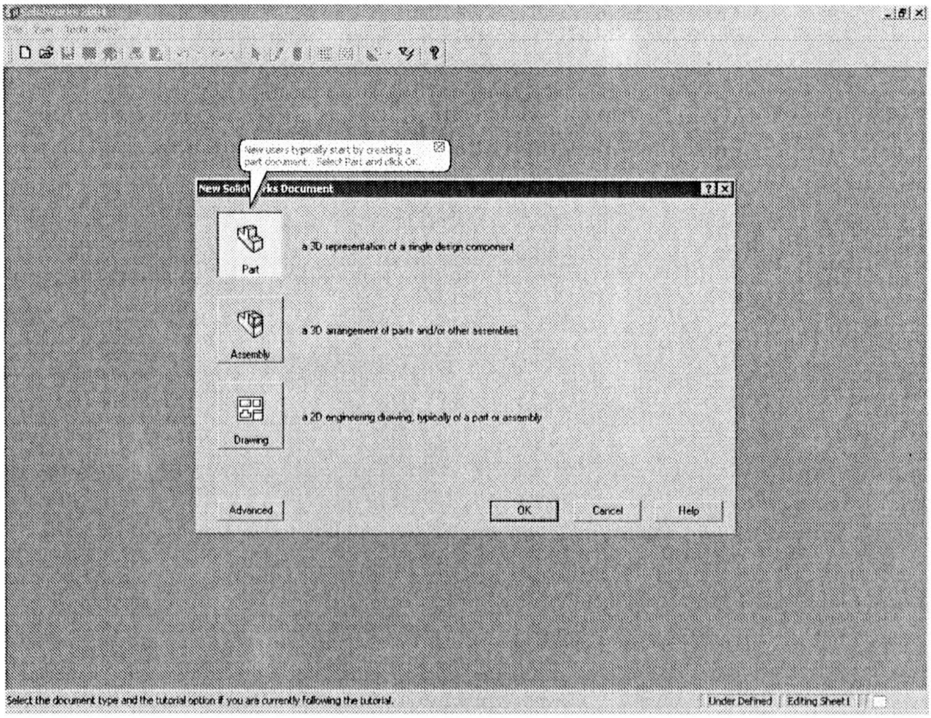

Figure I-4 *The* ***New SolidWorks Document*** *dialog box*

Choose the **Part** button for creating a part model and choose **OK** from the **New SolidWorks Document** dialog box to enter the **Part** mode of SolidWorks. The initial screen of the **Part** mode of SolidWorks along with its components is shown in Figure I-5. It is evident from the screen that SolidWorks is a very user-friendly solid modeling tool. Apart from the default CommandManager shown in Figure I-5 you can also invoke a number of other CommandManagers by moving the cursor on the CommandManager and right-clicking to display the shortcut menu. From this shortcut menu, choose the **Customize CommandManager**. A cascading menu is displayed; using this cascading menu you can select any CommandManager that you need to display.

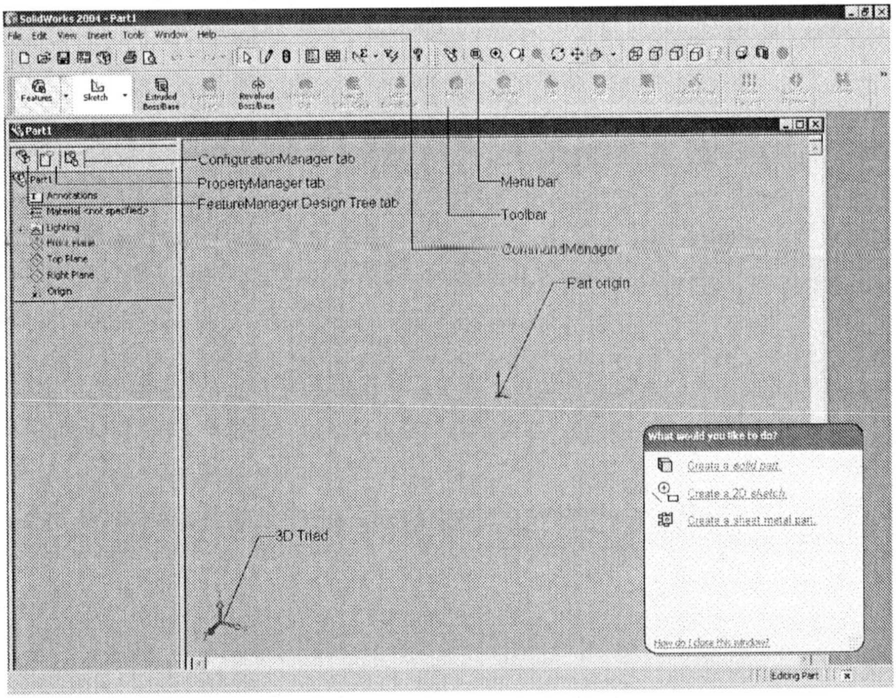

Figure I-5 *Components of new part document*

COMMANDMANAGER

In SolidWorks there are three options to invoke any tool. The first option is the use of CommandManager, the second option is the use of menu bars available on the top of the screen, and the third option is the use of shortcut menu. The menu bar and the shortcut menu are discussed later. CommandManager is the latest enhancement in SolidWorks 2004. While working with CommandManager you will able to know that this is the most convenient method to invoke a tool. There are different types of CommandManagers used for different design environments. The CommandManagers available in various modes of SolidWorks are discussed next.

Part Mode CommandManagers

There are a number of CommandManagers that can be invoked in the **Part** mode. The CommandManagers that are extensively used during the designing process in this environment are described below.

Sketch CommandManager

This CommandManager is used to enter and exit the 2D sketching environment and the 3D sketching environment. The tools available in the CommandManager are used to draw sketches for the features. This CommandManager is also used to add relations and smart dimensions to the sketched entities. The **Sketch CommandManager** is shown in Figure I-6.

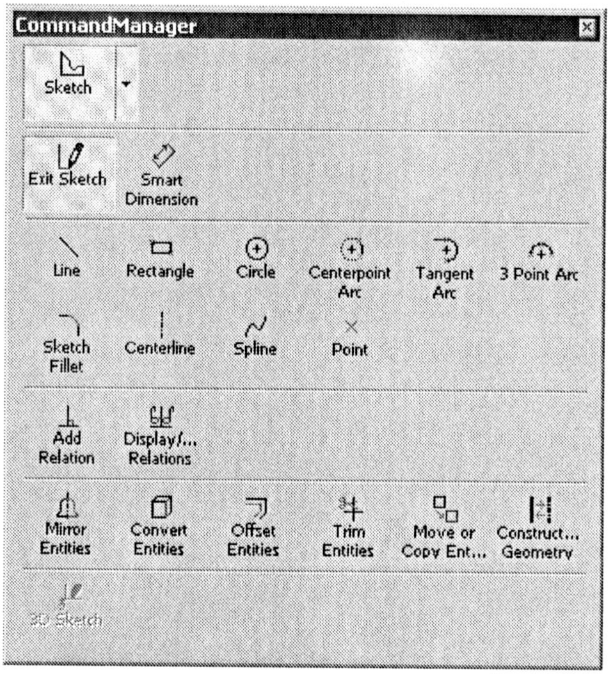

Figure I-6 Sketch CommandManager

Dimension/Relations CommandManager

This CommandManager is used to add various dimensions such as horizontal dimension, vertical dimension, ordinate dimension, and so on. The **Dimension/Relation CommandManager** is displayed in Figure I-7.

Features CommandManager

This is one of the most important CommandManagers provided in the **Part** mode. Once the sketch is completed, you need to convert the sketch into a feature using the modeling tools. This CommandManager provides all the modeling options that are used for feature-based solid modeling. The **Features CommandManager** is shown in Figure I-8.

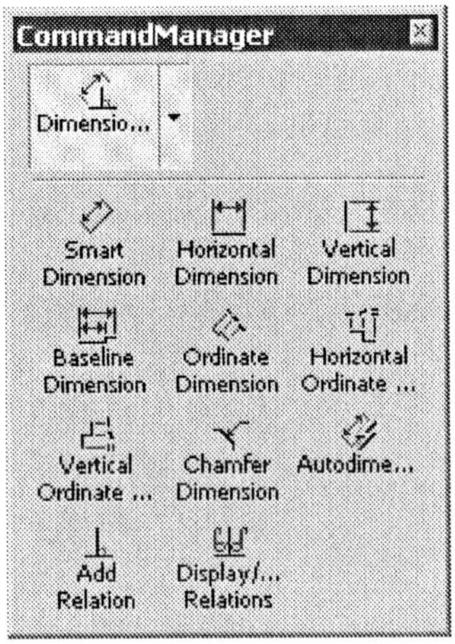

Figure I-7 Dimension/Relations CommandManager

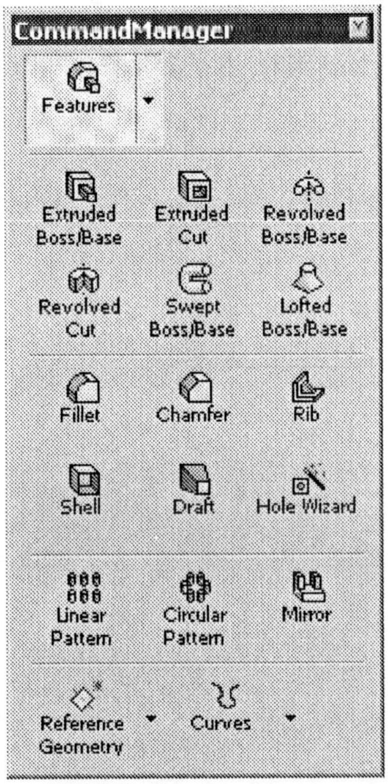

Figure I-8 Features CommandManager

Reference Geometry CommandManager

This CommandManager is a an important toolbar used in the **Part** mode. This toolbar is used to create the reference geometry features like planes, axes, coordinate systems, points, and mate references. The **Reference Geometry CommandManager** along with all its options is shown in Figure I-9.

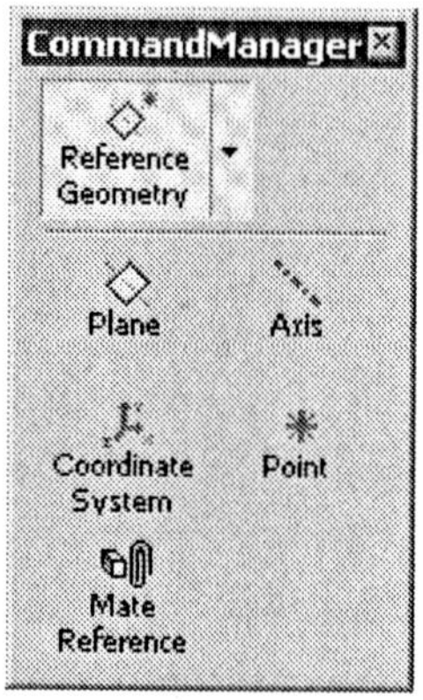

Figure I-9 Reference Geometry CommandManager

Sheet Metal CommandManager

This CommandManager provides you with the tools that are used in creating the sheet metal parts. As mentioned earlier in SolidWorks, you can create sheet metal part while working in the **Part** mode. This is done with the help of the **Sheet Metal CommandManager** shown in Figure I-10.

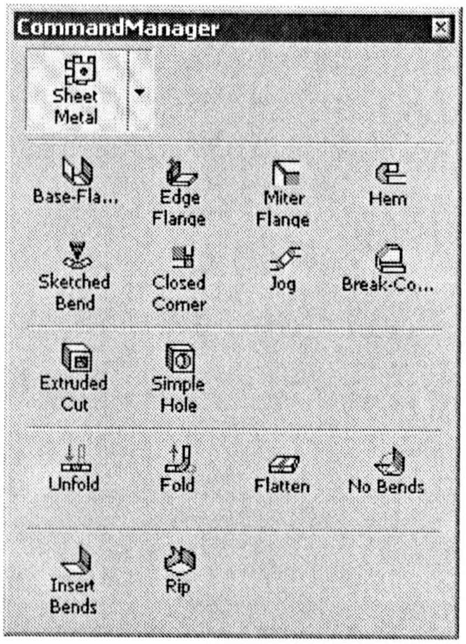

Figure I-10 Sheet Metal CommandManager

Surfaces CommandManager

The options available in this CommandManager are used to create complicated surface features. These surface features are also converted into the solid feature. The **Surfaces CommandManager** is shown in Figure I-11.

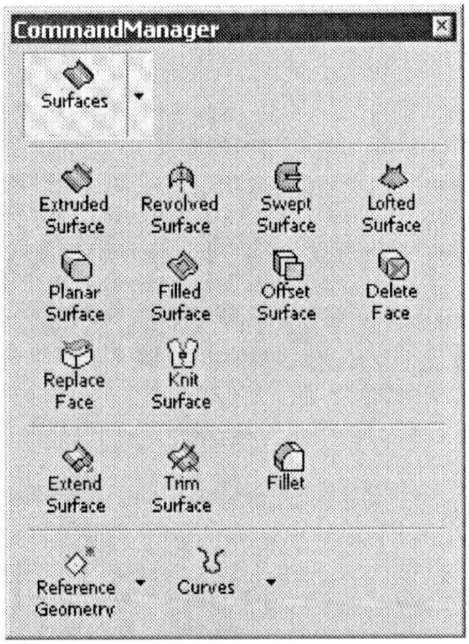

Figure I-11 Surfaces CommandManager

Curves CommandManager

The options available in this CommandManager are used to create the curves. The curves include the projected curve, split line, curve through points, helical curve, and so on. The **Curves CommandManager** is shown in Figure I-12.

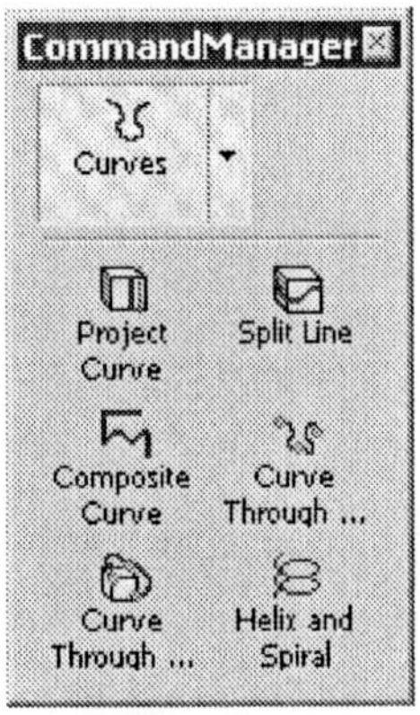

Figure I-12 Curves CommandManager

Molds CommandManager

This CommandManager is used to design the mold in SolidWorks. You can extract the core and cavity using the options available in this CommandManager. The **Molds CommandManager** is shown in Figure I-13.

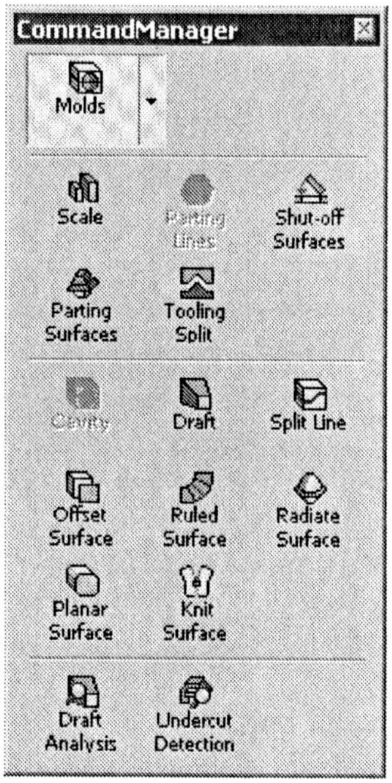

Figure I-13 Molds CommandManager

View CommandManager

This CommandManager is used to zoom, pan, and orient the model. The shading mode of the solid model is also changed using this toolbar. You can also create the section view of a model using the option available in this CommandManager. The **View CommandManager** along with all its buttons is shown in Figure I-14.

Standard Views CommandManager

This CommandManager is used to orient the model using the standard view options. The **Standard Views CommandManager** is shown in Figure I-15.

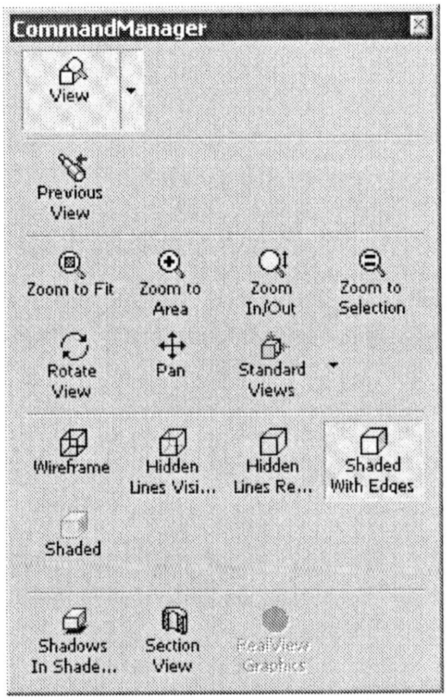

Figure I-14 View CommandManager

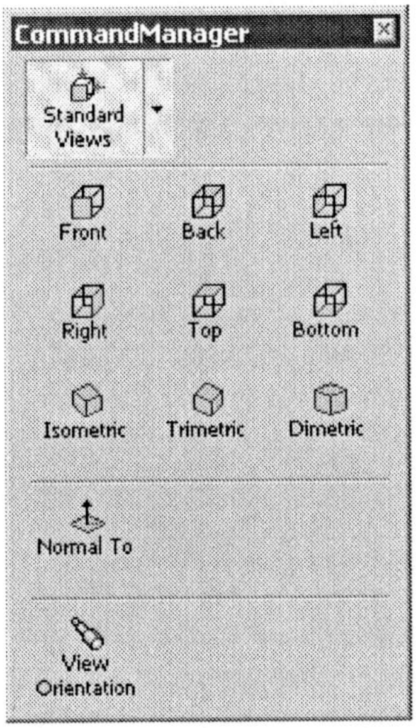

Figure I-15 Standard Views CommandManager

Tools CommandManager

This CommandManager is used to measure the distance between two entities, to add the equations in the design, to calculate the mass properties of a solid model, and so on. The **Tools CommandManager** is shown in Figure I-16.

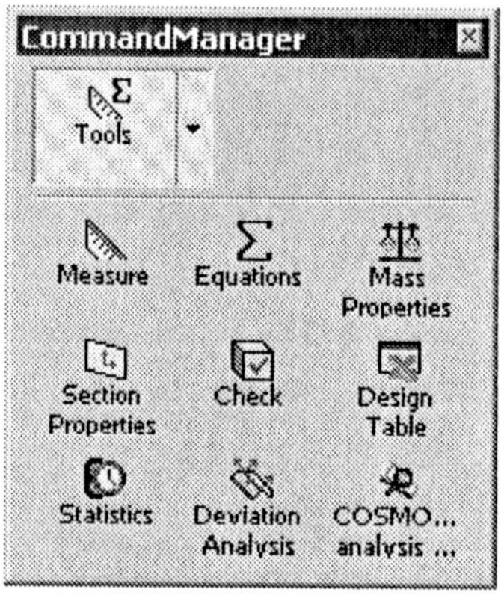

Figure I-16 Tools CommandManager

Assembly Mode CommandManagers

The CommandManager in the **Assembly** mode is used to assemble the components, create an explode line sketch, and simulate the assembly. The CommandManager that is used during the designing process in the **Assembly** mode is discussed next.

Assemblies CommandManager

This CommandManager is used to apply the various types of mates to the assembly. Mates are the constraints that can be applied to the components to restrict their degree of freedom. You can also move and rotate the component in the assembly, change the hidden and the suppression state of the assembly and individual component, edit the component of assembly, and so on. The **Assemblies CommandManager** is shown in Figure I-17.

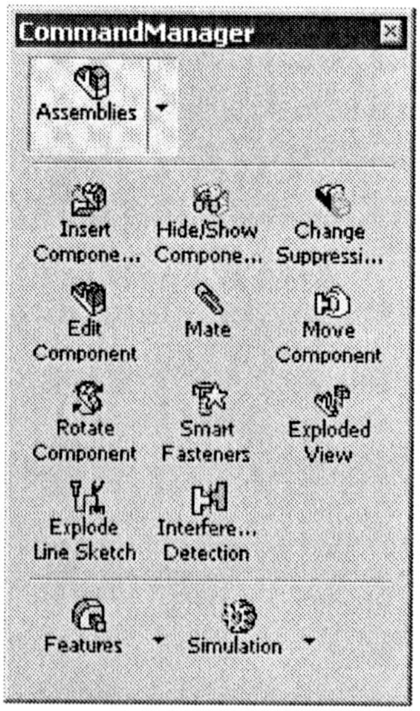

Figure I-17 Assemblies CommandManager

Drawing Mode CommandManagers

There are a number of CommandManagers that can be invoked in the **Drawing** mode. The CommandManagers that are extensively used during the designing process in this mode are described below.

Drawings CommandManager

This CommandManager is used to generate the drawing views from an existing model or an assembly. The views that can be generated using this CommandManager are model view, three standard views, projected view, section view, aligned section view, detail view, crop view, relative view, auxiliary view, and so on. The **Drawings CommandManager** is shown in Figure I-18.

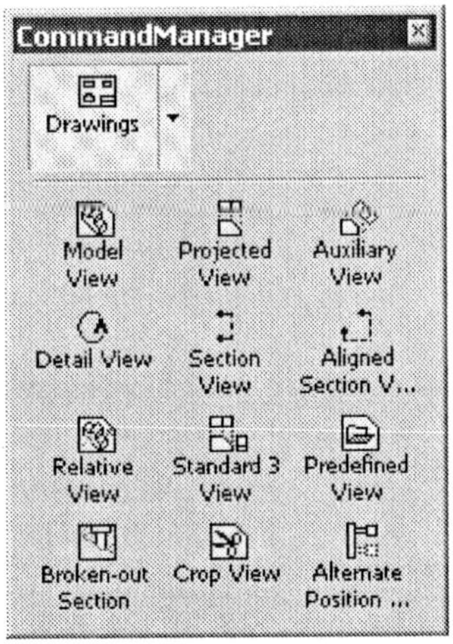

Figure I-18 Drawings CommandManager

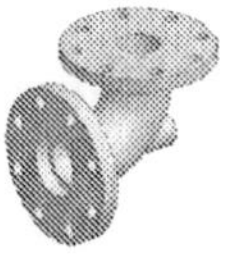

Tip. *You can also create 2D drawings and drawing views using the normal sketching tools available in the **Sketch Tools** toolbar. The **Sketch Tools** toolbar is also available in the **Drawing** mode.*

Annotations CommandManager

The **Annotations CommandManager** is used to generate the model items and to add the notes, balloons, geometric tolerance, surface finish symbols, and so on to the drawing views. The **Annotations CommandManager** is shown in Figure I-19.

DIMENSIONING STANDARD AND UNITS

While installing SolidWorks on your system, you can specify the units and the dimensioning standard that you will use for dimensioning the model. You are provided with many dimensioning standards such as ANSI, ISO, DIN, JIS, BSI, and GOST. You are also provided with various units such as millimeters, centimeters, inches, and so on. This book follows millimeters as units and ISO as the dimensioning standard. Therefore, it is recommended that you install SolidWorks with ISO as dimensioning standard and millimeters as units.

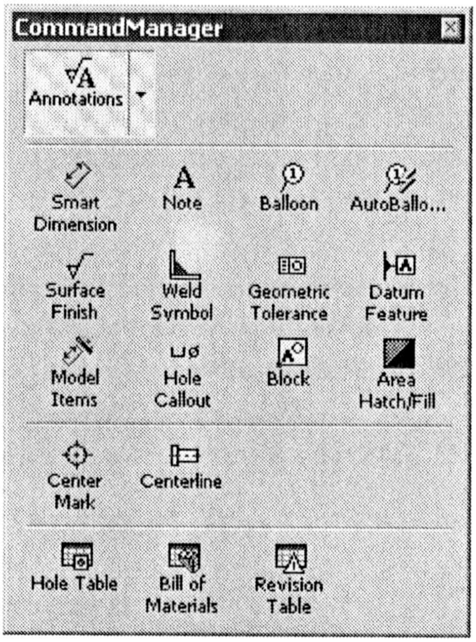

Figure I-19 Annotations CommandManager

IMPORTANT TERMS AND THEIR DEFINITIONS

Before you proceed further in SolidWorks, it is very important for you to understand the following terms. These terms will be widely used in this book.

Feature-based Modeling

Feature is defined as the smallest building block that can be modified individually. In SolidWorks, the solid models are created by integrating a number of these building blocks. A model created in SolidWorks is a combination of a number of individual features and each feature is related to the other feature directly or indirectly. These features understand their fit and function properly, and therefore can be modified any time during the design process. If the proper design intent is maintained while creating the model then these features automatically adjust their values if there is any change in their surrounding. This provides greater flexibility to the design.

Parametric Modeling

The parametric nature of a software package is defined as its ability to use the standard properties or parameters in defining the shape and size of a geometry. The main function of this property is to derive the selected geometry to new size or shape without considering its original size or shape. You can change or modify the shape and size of any feature at any stage of the design process. This property makes the designing process very easy. For example, consider the design of the body of a pipe housing as shown in Figure 1-20.

If you have to change the design by modifying the diameter of the holes and the number of holes on the front, top, and the bottom face, you have to simply select the feature and change the diameter and the number of instances in the pattern. The modified design is shown in Figure 1-21.

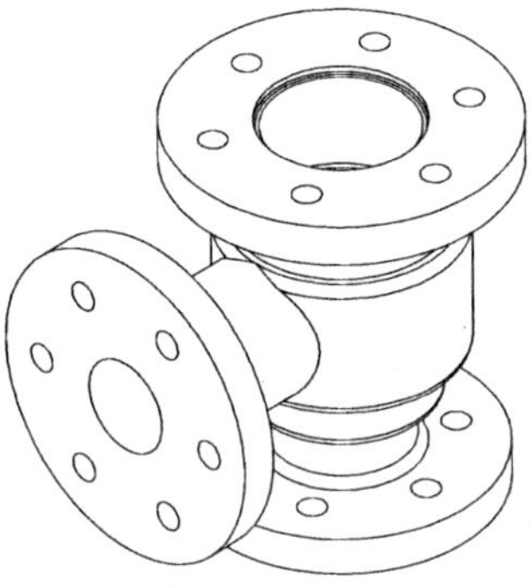

Figure I-20 *Body of pipe housing*

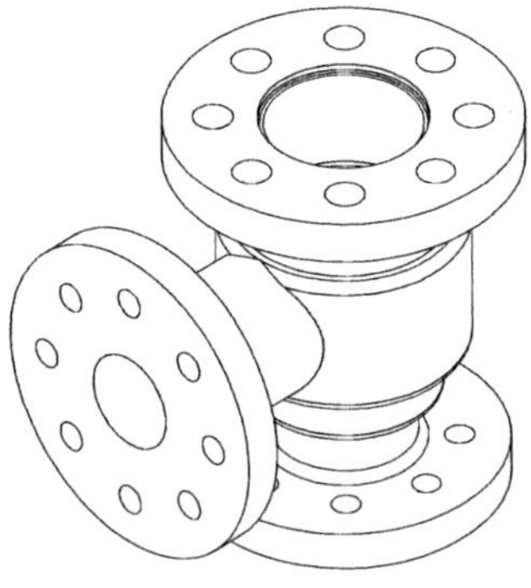

Figure I-21 *Design after modifications*

Bidirectional Associativity

As mentioned earlier, SolidWorks has different modes such as **Part** mode, **Assembly** mode, and **Drawing** mode. There exists a bidirectional associativity between all these modes. The bidirectional associativity ensures that if any modification is made in the model in any one of the modes of SolidWorks, it is automatically reflected in the other mode immediately. For example, if you modify the dimension of a part in the **Part** mode, the change will be reflected in the **Assembly** and the **Drawing** modes. Similarly if you modify the dimensions of a part in the drawing views generated in the **Drawing** mode, the changes will be reflected in the **Part** and **Assembly** modes. Consider the drawing views shown in Figure I-20. These are the drawing views of the body of the pipe housing shown in Figure I-22. Now, when you modify the model of the body of pipe housing in the **Part** mode, the changes are reflected in the **Drawing** mode automatically. Figure I-23 shows the drawing views of the pipe housing after increasing the diameter and the number of holes.

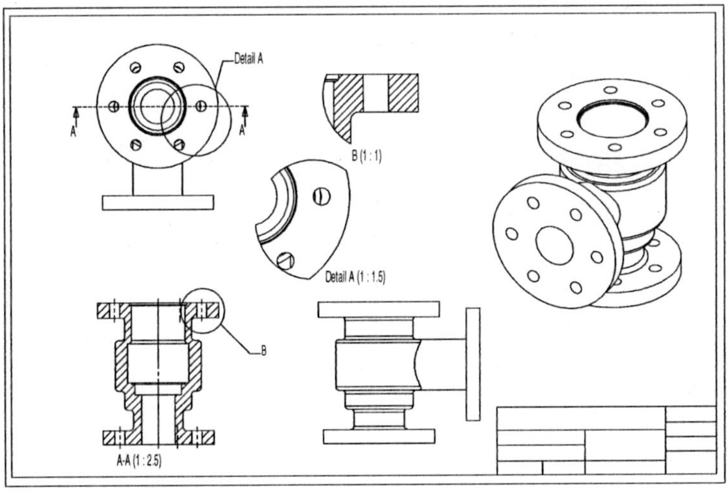

Figure I-22 *Drawing views of the body part before making the modifications*

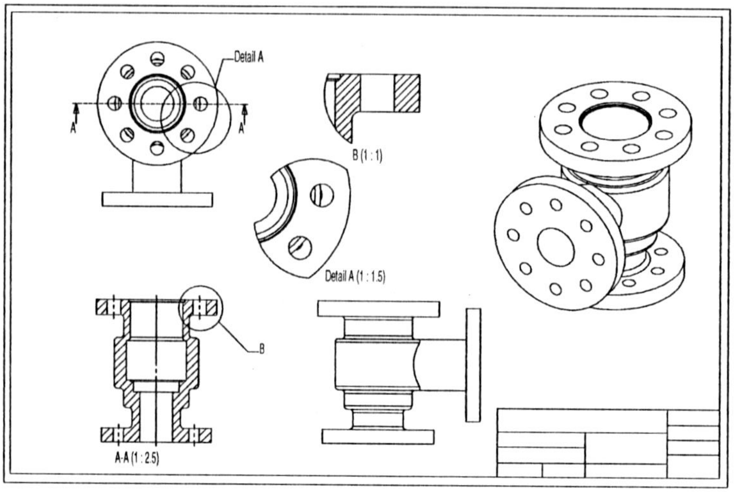

Figure I-23 *Drawing views after modifications*

Windows functionality

SolidWorks is the first Windows-based 3D CAD package. SolidWorks uses the graphical user interface of Windows, drag and drop functionality, and copy paste functionality of Window. For example, consider a case where you had created a hole feature on the front planar surface of a model. Now, to create another hole feature on the top planar surface of the same model, just select the hole feature and press CTRL+C (copy) on the keyboard. Next, select the top planar surface of the base feature and press CTRL+V (paste). You can also drag and drop the standard features from the feature pallet library on the face of the model on which the feature has to be added. Therefore, SolidWorks is the most easy to learn 3D CAD package for mechanical engineers.

Geometric Relations

Geometric relations are the logical operations that are performed to add a relationship (like tangent or perpendicular) between the sketched entities, planes, axes, edges, or vertices. When adding the relations, one entity should be a sketched entity and the other entity can be a sketched entity, or an edge, face, vertex, origin, plane, and so on. There are two methods to create the geometric relations:

1. Automatic Relations
2. Add Relations

Automatic Relations

The sketching environment of SolidWorks has been provided with the facility of autorelations. This facility ensures that the geometric relations are applied to the sketch automatically while creating the sketch. Automatic relations are also provided in the drawing mode while working with interactive drafting.

Add Relations

Add relations is used to add the geometric relations manually to the sketch. There are 16 types of geometric relations that can be manually applied to the sketch. They are as follows:

Horizontal

This relation forces the selected line segment to become a horizontal line. You can also select two points and force them to be aligned horizontally.

Vertical

This relation forces the selected line segment to become a vertical line. You can also select two points and force them to be aligned vertically.

Collinear

This relation forces two selected entities to be placed in the same line.

Coradial

This relation is applied to two selected arcs, circles, or an arc and a circle to force them to become equi-radius and at the same time share the same centerpoint.

Perpendicular

This relation is used to make the selected line segment perpendicular to another selected segment.

Parallel

This relation is used to make the selected line segment parallel to another selected segment.

Tangent

This relation is used to make the selected line segment, arc, spline, circle, or ellipse tangent to another arc, circle, spline, or ellipse.

Note
In case of splines, the relations are applied to their control points.

Concentric

This relation forces two selected arcs, circles, a point and an arc, a point and a circle, or an arc and a circle to share the same centerpoint.

Midpoint

This relation forces a selected point to be placed on the midpoint of a line.

Intersection

This relation forces a selected point to be placed at the intersection of two selected entities.

Coincident

This relation is used to make two points or a point and a line or a point and an arc coincident.

Equal
The equal relation forces the two selected lines to become equal in length. This relation is also used to force two arcs or two circles or an arc and a circle to have equal radii.

Symmetric
The symmetric relation is used to force the selected entities to become symmetrical about a selected center line, so that the selected entities remain equidistant from the center line.

Fix
This relation is used to fix the selected entity to a particular location with respect to the coordinate system of the current sketch. The endpoints of the fixed line, arc, spline, or an elliptical segment are free to move along the line.

Pierce
This relation forces the sketched point to be coincident to the selected axis, edge, or curve where it pierce the sketch plane. The sketched point in this relation can be the endpoint of the sketched entity.

Merge
This relation is used to merge two sketched points or endpoints.

Library Feature
Generally, in a mechanical design, some of the features are used frequently. In most of the other solid modeling tools, you have to create these features whenever you need them. However, SolidWorks allows you to save this type of feature in a library so that you can retrieve them whenever you want. This saves lot of designing time and effort of a designer.

Feature Palette
The feature palette is also a very unique feature of SolidWorks. Using the feature palette library you can access some ready-to-use drop-down standard mechanical parts such as shafts, keyways, forming parts for sheet metal, and so on.

Design Table
Design tables are used to create a multi-instance parametric component. For example, you have some components in your organization that have the same geometry but different dimensions. Instead of creating each component of the same geometry with different size, you can create one component and then using the design table, create different instances by changing the dimension as per your requirement. Using the design table you can access all those components with the same geometry having different sizes in a single part file.

Equations
Equations are the analytical and numerical formulas applied to the dimensions during the sketching of the feature sketch or after sketching the feature sketch. The equations can also be applied to the placed features.

Collision Detection

Collision detection is used to detect the interference and collision between the parts of an assembly when the assembly is in motion. While creating the assembly in SolidWorks, you can detect the collision between different parts of the assembly by moving and rotating the components of the assembly.

What's Wrong? functionality

While creating a feature of the model or after editing a feature created earlier, if the geometry of the feature is not compatible and the system is not able to construct that feature then the **What's Wrong?** functionality is used to detect the possible error that occurs while creating the feature.

2D Command Line Emulator

The 2D command emulator is an Add-In of SolidWorks. You can activate this by choosing **Tools > Add-Ins** from the menu bar. When you choose this option, the **Add-Ins** dialog box is displayed. Select the **SolidWorks 2D Emulator** check box and choose **OK** from the **Add-Ins** dialog box. A command section is displayed at the bottom of the graphics area. This 2D Command line emulator is useful for invoking the commands by typing them. You can type the commands in the 2D Command line emulator.

COSMOSXpress

In SolidWorks 2004, you are provided with COSMOSXpress, which is an analysis tool in SolidWorks to execute the static or stress analysis. In COSMOSXpress you can only execute the linear static analysis. Using the linear static analysis you can calculate the displacement, strain, and stresses applied on a component with the effect of material, various loading conditions, and restraint conditions applied on a model. A component fails when the stress applied to that component reaches a certain permissible limit. The Static Nodal stress plot of the crane hook designed in SolidWorks and analyzed using COSMOSXpress is shown in Figure I-24.

Physical Dynamics

Physical Dynamics is used to observe the motion of the assembly. With this option selected when you drag a component in the assembly, the component applies a force to the component that it touches. With the application of that force the other component is moved or rotated within the allowable degrees of freedom of that component.

Physical Simulation

Using **Physical Simulation** you can simulate the assemblies created in the assembly environment of SolidWorks. You can assign and simulate the effect of different simulation elements such as linear, rotary motors, and gravity to the assemblies. After creating a simulating assembly you can record and replay simulation.

Seed Feature and Seed Component

The original feature that is used as the parent feature to create any type of pattern or mirror

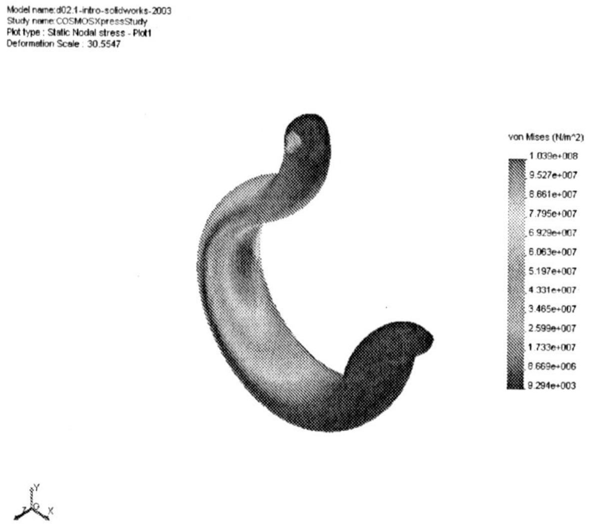

Figure I-24 *The crane hook analyzed using COSMOSXpress*

feature is known as seed feature. You can edit or modify only a seed feature; you cannot edit the instances of the pattern feature.

The original component that is used to create a derived pattern, local pattern, or a mirrored component is known as the seed component.

FeatureManager Design Tree

The **FeatureManager Design Tree** is one the most important component of SolidWorks screen. It contains the information about the default planes, material, lights, and all the features that are added to the model. When you add features to the model using various modeling tools, the same are also displayed in the **FeatureManager Design Tree**. You can easily select and edit the features using the **FeatureManager Design Tree**. When you invoke any tool to create a feature, the **FeatureManager Design Tree** is replaced the by respective **PropertyManager**. At this stage the **FeatureManager Design Tree** is displayed in the drawing area.

Absorbed Features

Features that are directly involved in creating other features are known as absorbed features. For example, sketch of an extruded feature is an absorbed feature of the extrude feature.

Child Features

The features that are dependent on their parent and whose existence is not possible without their parent features are known as child features. For example, consider a cube with filleted edges. If you delete the cube, the fillet feature will also be deleted because its existence is not possible without its parent feature.

Dependent Features

Dependent features are those that depend on their parent feature but can still exist without the parent feature with some minor modifications. If the parent feature is deleted, then by specifying other references and modifying the feature you can retain the same feature.

AUTO-BACKUP OPTION

SolidWorks also allows you to set the option to save the SolidWorks document automatically after a regular interval of time. While working on a design project, if the system crashes you may loose the unsaved design data. If the auto-backup option is turned on, your data is saved automatically after regular intervals. To set this option on choose **Tools > Options** from the menu bar; the **System Options - General** dialog box is displayed. Select the **Backup** option from the display area provided on the left of this dialog box. Now, choose the **Save auto recover info every** check box. The spinner provided on the right of the check box is invoked. Using this spinner you can set the number of changes after which the document would be saved automatically. By default, the backup files are saved in *X:\Documents and Settings\Admisitrator <name of your machine>\Local Settings\TempSWBackupDirectory* location (where *X* is the drive in which you have installed SolidWorks 2004 and the *Local Settings* folder is a hidden folder). You can also change the path of this location. To change this path select the button provided on the right of the edit box. The **Browse For Folder** dialog box is displayed; using this you can specify the location of the folder in which you need to save the backup files. If you need to save the backup files in the current folder in which you are working, then select the **Save backup files in the same location as the original** check box. Using the **Number of backup copies per document** spinner you can set the number of backup files that you need to save. After setting all the options choose the **OK** button from the **System Options - Backup** dialog box.

COLOR SCHEME

SolidWorks allows you to use various color schemes as the background color for the screen, color and display style of **FeatureManager Design Tree**, and for displaying the entities in the screen. Note that the color scheme used in this book is neither the default color scheme nor the predefined color schemes. To set this color scheme, choose **Tools > Options** from the menu bar. The **System Options - General** dialog box is displayed. Select the **Color** option from the left of this dialog box. The option related to the color scheme is displayed in the dialog box and the name of the dialog box is changed to **System Options - Color** dialog box. In the **System colors** area the **Viewport Background** option available in the scroll down area is selected by default; choose the **Edit** button. Select white color from the **Color** dialog box and choose the **OK** button. Now, select the **Blue** option from the **PropertyManager Color** drop-down list available in the **System colors** area. Now choose the **None** option from the **PropertyManager Skin** drop-down list. Now, clear the **Match graphics area and FeatureManager backgrounds** check box. Next, you need to select the **Reset All To Defaults** button to complete the settings of the color scheme. After setting the color scheme, you need to save the color scheme so that next time if you need to set this color scheme, you do not need to apply all the settings. You just need to select the name of the saved color scheme from the **Current Color Scheme** drop-down list. Choose the **Save As Scheme** button; the **Color Scheme Name** dialog box is displayed. Enter the name of the color scheme as

SolidWorks 2004 in the edit box provided in the **Color Scheme Name** dialog box and choose the **OK** button. Now, choose the **OK** button from the **System Options - Color** dialog box.

Note that the colors of the entities that are mentioned in the description or tutorials in this book are those that are displayed if the operating system is Windows 2000 Professional. However, if you use Windows XP as the operating system, the colors of the entities that you get on the screen may be different from those mentioned in this book.

Chapter 1

Drawing Sketches for the Solid Models

Learning Objectives

After completing this chapter you will be able to:

- *Understand the requirement of the sketching environment.*
- *Open a new part document.*
- *Understand the various terms used in sketching environment.*
- *Work with various sketching tools.*
- *Use the drawing display tools.*
- *Delete the sketched entities.*

THE SKETCHING ENVIRONMENT

Most of the products designed using SolidWorks are a combination of sketched features, placed features, and derived features. The placed and derived features are created without creating a sketch, but the sketched features require a sketch to be created first. Generally, the base feature of any design is a sketched feature and is created using a sketch. Therefore, while creating any design, the first and foremost point is to draw the sketch for the base feature. Once you have drawn the sketch of the base feature, you can convert it into the base feature and then add the other sketched, placed, and derived features to complete the design. In this chapter, you will learn to create the sketch for the base feature using the various sketcher entities.

In general terms, a sketch is defined as the basic contour for the feature. For example, consider the spanner shown in Figure 1-1.

Figure 1-1 *Solid model of a spanner*

This spanner consists of a base feature, a cut feature, a mirror feature (cut on the back face), fillets, and an extruded text feature. The base feature of this spanner is shown in Figure 1-2. This base feature is created using a single sketch drawn on the **Front** plane, as shown in Figure 1-3. This sketch is drawn in the sketching environment using the various sketching tools. Therefore, to draw the sketch of the base feature, you first need to invoke the sketching environment where you will draw the sketch.

Note

Once you are conversant with the various options of SolidWorks, you can also use a derived feature or a derived part as the base feature.

Figure 1-2 Base feature of the spanner

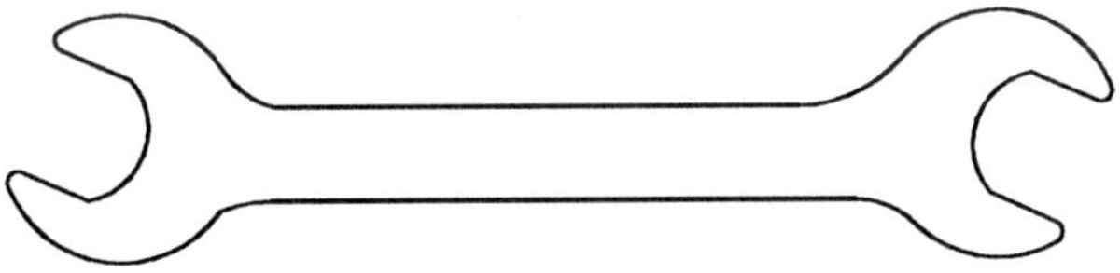

Figure 1-3 Sketch for the base feature of the spanner

The sketching environment of SolidWorks can be invoked any time in the **Part** mode, **Assembly** mode, and **Drawing** mode. You just have to specify that you want to create the sketch of a feature and select the plane on which you want to create the sketch.

Note

You will learn to invoke the sketching environment in SolidWorks 2004 later in this chapter.

STARTING A NEW DOCUMENT IN SolidWorks 2004

If the proper settings are configured, the **Tip of the day** dialog box is displayed whenever you start a new session of SolidWorks 2004, as shown in Figure 1-4.

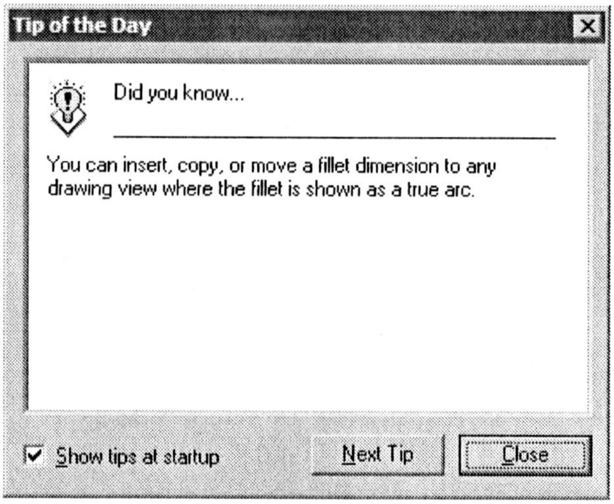

***Figure 1-4 Tip of the Day** dialog box*

Tip. *If the **Tip of the Day** dialog box is not displayed when you start the SolidWorks session, choose **Help > Tip of the Day** from the menu bar. The **Tip of the Day** dialog box is displayed. Select the **Show tips at startup** option from this dialog box. By selecting this option, the **Tip of the Day** dialog box will be displayed every time you start a new SolidWorks session. The tips displayed in this dialog box are helpful in the full utilization of this solid modeling package.*

Choose the **Close** button from this dialog box. You will notice that the **Welcome to SolidWorks 2004** window is displayed, as shown in Figure 1-5. This window can be used to start a new document, open an existing document, and start the various types of online tutorials available in SolidWorks 2004.

Choose **New Document** from the **Welcome to SolidWorks 2004** window. The **New SolidWorks Document** dialog box is displayed, as shown in Figure 1-6. The dialog displays three options to start a new part, assembly, or drawing document. These three options are discussed next.

Part

Choose the **Part** button and choose **OK** from the **New SolidWorks Document** dialog box to start a new part document to create solid models or sheet metal components. When you start a new part document, you will enter the **Part** mode. You can invoke the sketching environment using the methods discussed earlier to draw the sketch of the base feature.

Tip. *When you choose any option from the **Welcome to SolidWorks 2004** window, it is not closed. It is minimized on the screen and you can restore this window to open an existing or new file, or use any other option.*

Assembly

Choose the **Assembly** button and choose **OK** from the **New SolidWorks Document** dialog box to start a new assembly document. In an assembly document, you can assemble the

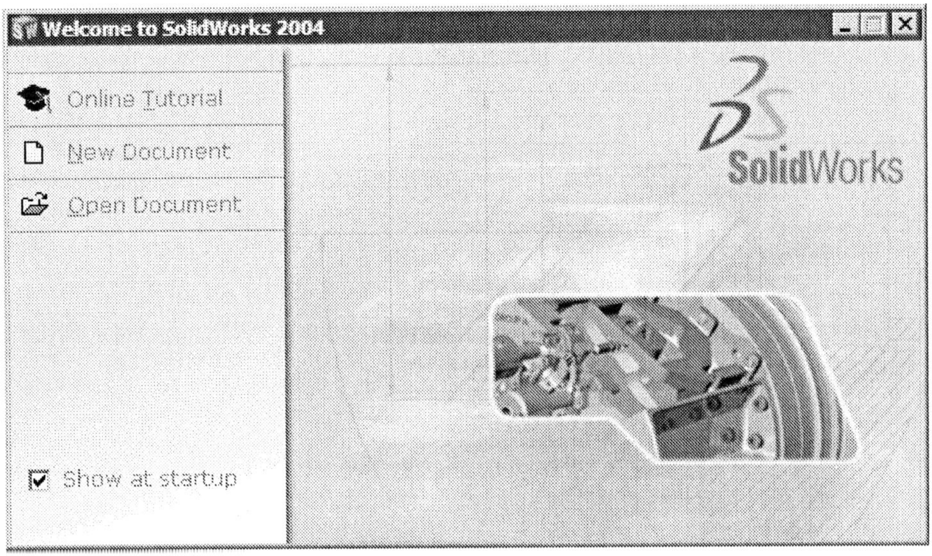

Figure 1-5 ***Welcome to SolidWorks 2004*** *window*

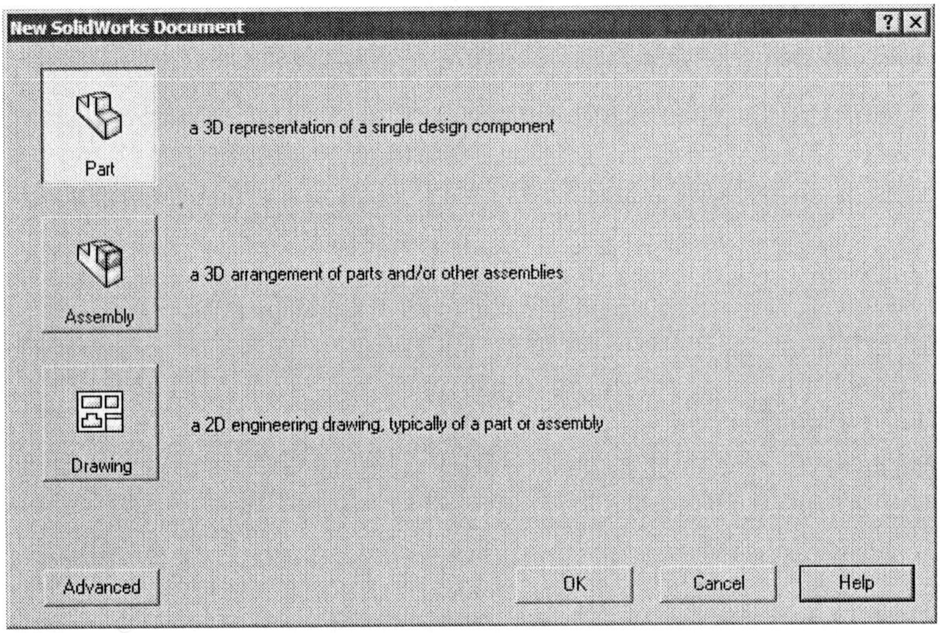

Figure 1-6 ***New SolidWorks Document*** *dialog box*

components created in the part documents. You can also create components in the assembly document.

Drawing

Choose the **Drawing** button and choose **OK** from the **New SolidWorks Document** dialog box to open a new drawing document. In a drawing document, you can generate or create the drawing views of the parts created in the part documents or the assemblies created in the assembly documents.

Note

*The **Advanced** button available in the **New SolidWorks Document** dialog box is used to display the advanced version of this dialog box. Once you are familiar with SolidWorks 2004, you can use the advanced version of the **New SolidWorks Document** dialog box to start a new document.*

Tip. *When you open a new SolidWorks document, it is not maximized in the SolidWorks window. This is the reason the area of the new document is not maximum. To maximize the document, choose the **Maximize** button provided on the upper right corner of the document. You can also maximize the document window by double-clicking the blue bar on top of the document window.*

THE SKETCHING ENVIRONMENT

Whenever you start a new part document, by default you are in the part modeling environment. However, as mentioned earlier, you need to start the design by first creating the sketch of the base feature in the sketching environment. You can invoke the sketching environment using the **Sketch** tool available in the **Standard** toolbar or by using the **Sketch** tool available in the **CommandManager**, see Figure 1-7.

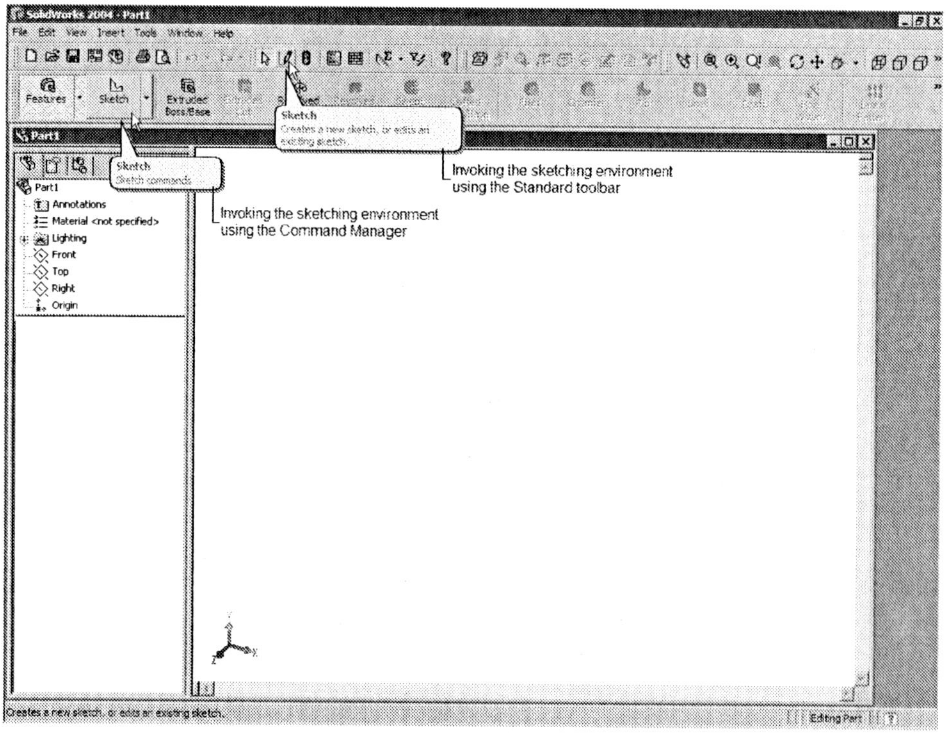

***Figure 1-7** Different methods of invoking the sketching environment in SolidWorks 2004*

When you choose the **Sketch** tool from the **Standard** toolbar, the **Edit Sketch PropertyManager** is displayed and you are prompted to select the plane on which you want to create the sketch, see Figure 1-8. Also, the three default planes available in SolidWorks 2004 (Front, Right, and Top) are temporarily displayed on the screen, as evident in Figure 1-8.

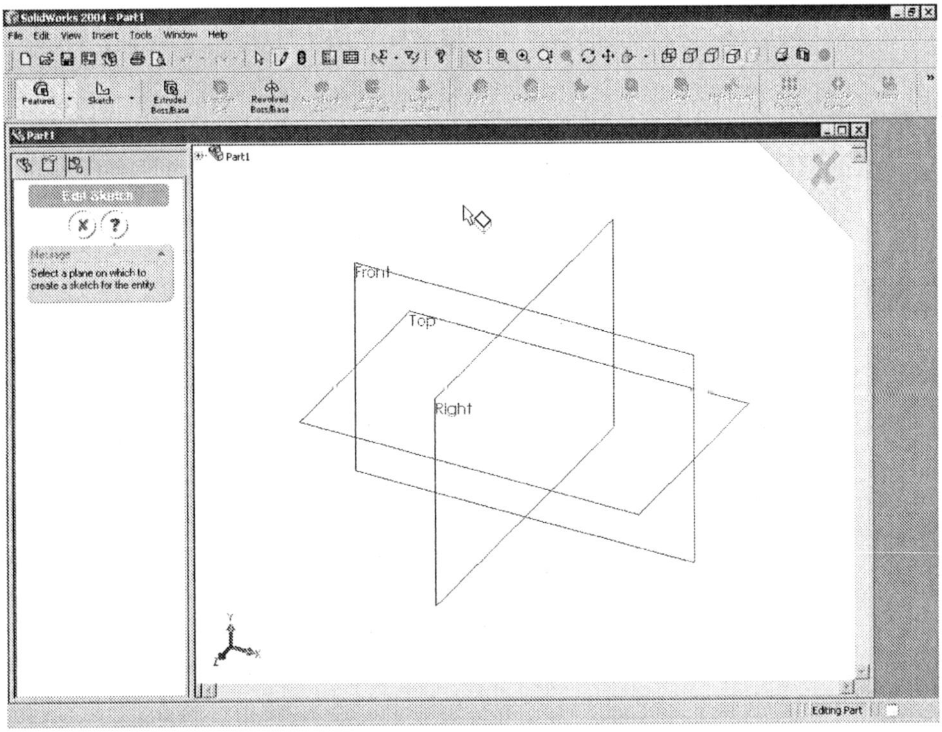

***Figure 1-8** Selecting the plane to create the sketch of the base feature*

Depending on the requirement of the design, you can select any plane to create the sketch of the base feature. The selected plane is automatically oriented normal to the view so that you can easily create the sketch. Also, the **CommandManager** now displays the various sketching tools to create the sketch.

When you choose the **Sketch** button from the **CommandManager**, various sketching tools are displayed. Now, if you choose any sketching tool, the **Edit Sketch PropertyManager** is displayed and you are prompted to select the plane on which the sketch will be created.

The default screen appearance of a SolidWorks part document in the sketching environment is shown in Figure 1-9.

SETTING UP THE DOCUMENT OPTIONS

When you install SolidWorks on your system, you are prompted to specify the dimensioning standards and units for measuring linear distances. You can specify these options at that time. The settings specified at that time are made the default settings and whenever you start

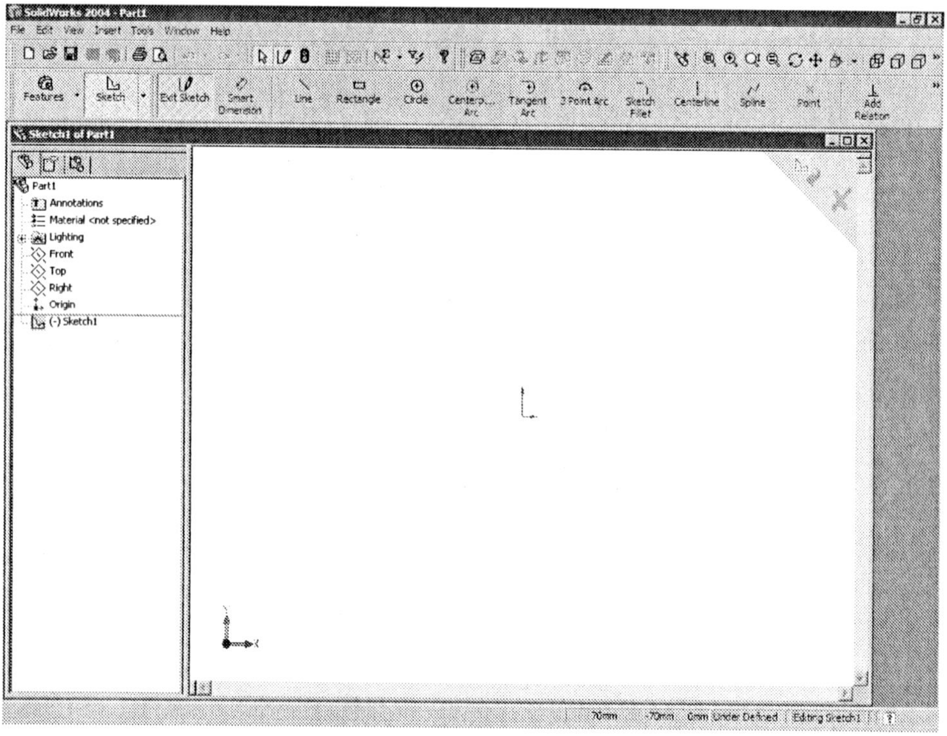

Figure 1-9 Default screen display of a part document in the sketcher environment

a new SolidWorks document, it will use those settings. However, if you want to modify these settings for a particular document, you can easily do so using the **Document Properties** dialog box. To invoke this dialog box, choose **Tools > Options**. When you invoke this option, the **System Options** dialog box will be displayed. In this dialog box, choose the **Document Properties** tab. The name of this dialog box is changed to the **Document Properties** dialog box.

Modifying the Dimensioning Standards

To modify the dimensioning standards, invoke the **System Options** dialog box and then choose the **Document Properties** tab. You will notice that the **Detailing** option is selected by default from the area on the left to display the detailing options, as shown in Figure 1-10.

The default dimensioning standard that was selected while installing SolidWorks will be selected in the drop-down list provided in the **Dimensioning standard** area. You can select the required dimensioning standard from this drop-down list. The standards that are available are ANSI, ISO, DIN, JIS, BSI, GOST, and GB. You can select any one of these dimensioning standards for the current document.

Modifying the Linear and Angular Units

To modify the linear and angular units, invoke the **System Options** dialog box and then choose the **Document Properties** tab. In this tab, select the **Units** option from the area on

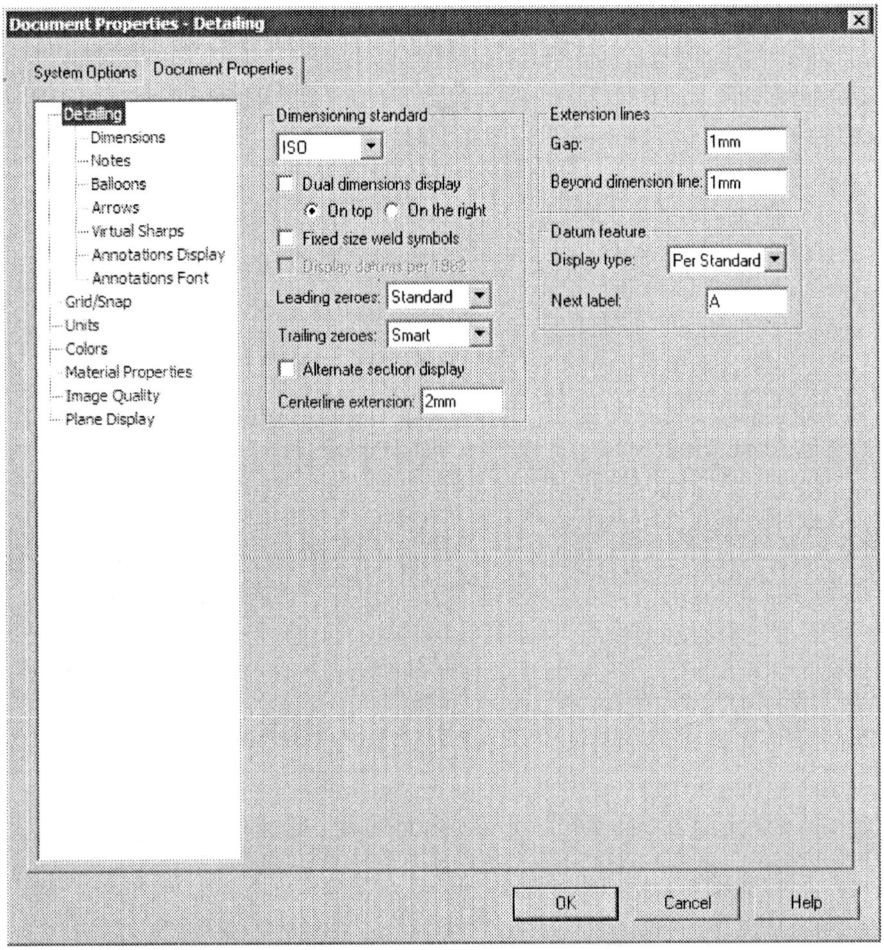

Figure 1-10 *Setting the dimensioning standards*

the left to display the options related to linear and angular units, as shown in Figure 1-11. The default option for measuring the linear distances that was selected while installing SolidWorks will be provided in the drop-down list provided in the **Length units** area. The various types of units that you can select are angstroms, nanometers, microns, millimeters, centimeters, meters, microinches, mils, inches, feet, and feet & inches. You can also change the units for angular dimensions by selecting it from the drop-down list in the **Angular units** area. The various types of angular units that can be selected are degrees, deg/min, deg/min/sec, and radians.

Modifying the Snap and Grid Settings

In the sketching environment of SolidWorks, you can force the cursor to jump through a specified distance while creating the sketch. Therefore, if you draw a sketch entity, its length will change in the specified increment. For example, you can force the cursor to jump through a distance of 10 mm. As a result, if you are drawing a line, its length will be incremented through a distance of 10 mm while sketching. To do this, choose **Tools > Options** from the

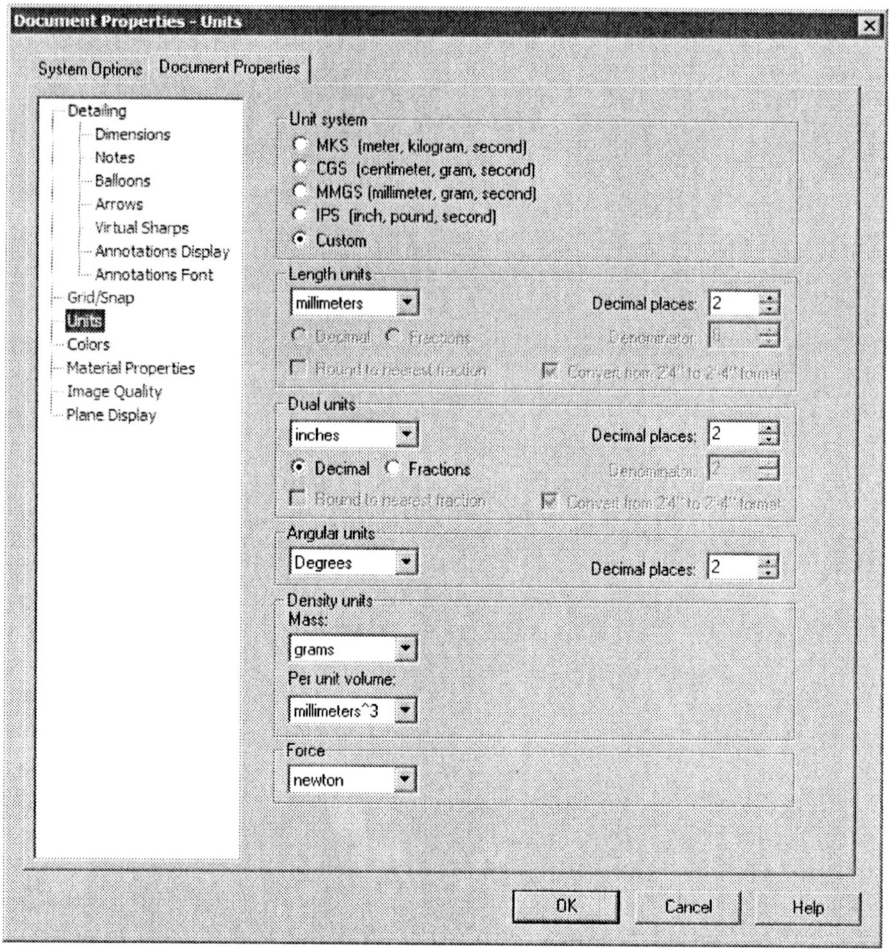

Figure 1-11 *Setting the dimensioning units*

menu bar to display the **System Options** dialog box. Choose the **Document Properties** tab to invoke the **Document Properties** dialog box. Now, select the **Grid/Snap** option to display the related options, see Figure 1-12.

Select the **Snap to points** check box in the **Snap** area to ensure that the cursor jumps through the specified distance. Now, the distance through which the cursor will jump is dependent on the ratio between the values in the **Major grid spacing** and **Minor-lines per major** spinners available in the **Grid** area. For example, if you want that the coordinates should change in the increment of 10 mm, you will have to make the ratio of major and minor lines to 10. This can be done by setting the value of the **Major grid spacing** spinner to **100** and that of the **Minor-lines per major** spinner to **10**. Similarly, to make the cursor jump through a distance of 5 mm, set the value of the **Major grid spacing** spinner to **50** and that of the **Minor-lines per major** spinner to **10**.

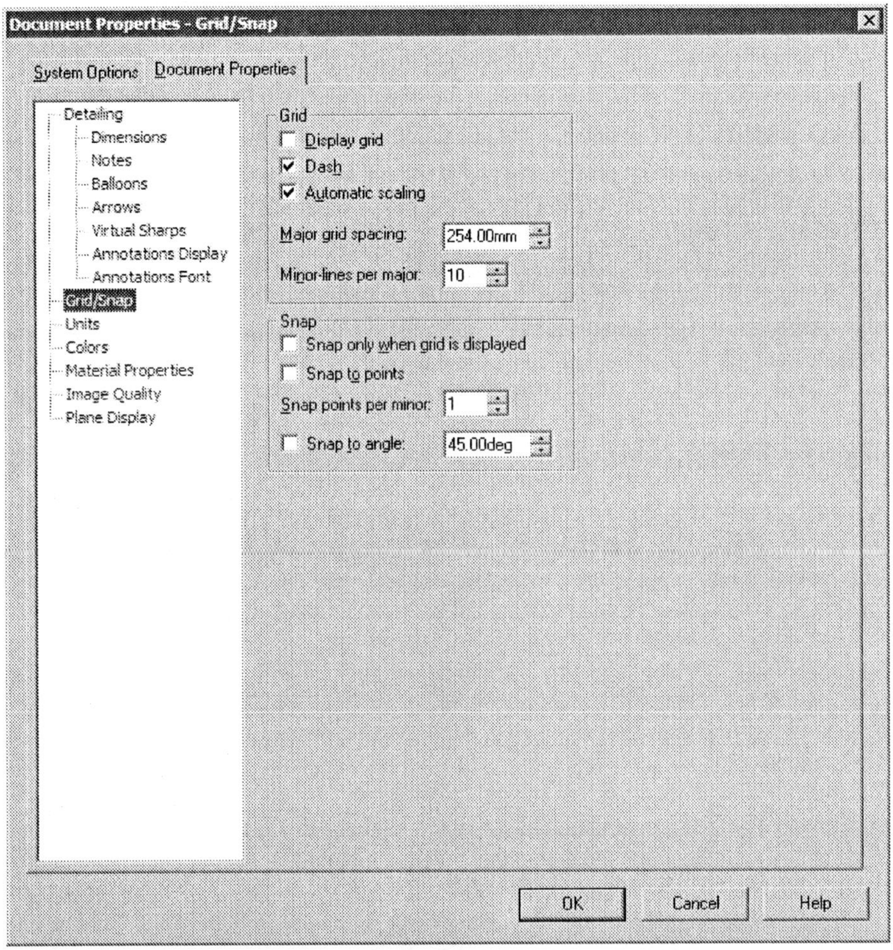

Figure 1-12 *Modifying the grid and snap settings*

Note

Remember that this setting will only be for the current documents. When you open another document, it will again have the settings that were defined while installing SolidWorks.

Tip. *If you want to display the grid in the sketching environment, select the* ***Display grid*** *check box from the* ***Grid*** *area of the* ***Document Properties - Grid/Snap*** *dialog box. If you do not want the cursor to snap to a point, clear the* ***Snap to points*** *check box from the* ***Snap*** *area.*

LEARNING ABOUT SKETCHER TERMS

Before you learn about the various sketching tools, it is important for you to understand some terms that are used in the sketching environment. These terms are discussed next.

Origin

The origin is a red color icon that is displayed in the center of the sketching environment screen. This icon consists of two arrows displaying the X axis and the Y axis direction. The point of intersection of these two axes is the origin point and the coordinates of this point are 0,0.

Inferencing Lines

The inferencing lines are the temporary lines that are used to track a particular point on the screen. These lines are dashed lines and are automatically displayed when you select a sketching tool in the sketcher environment. These lines are created from the endpoint of a sketched entity or from the origin. For example, if you want to draw a line from the point where two imaginary lines intersect, you can use the inferencing lines to locate the point and then draw the line from that point. Figure 1-13 shows the use of inferencing lines to locate the point of intersection of two imaginary lines.

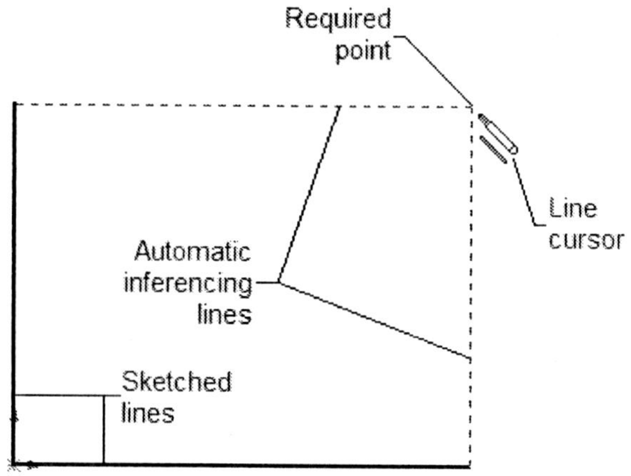

Figure 1-13 *Using inferencing lines to locate a point*

Figure 1-14 shows the use of inferencing lines to locate the center of an arc. Notice that the inferencing lines are created from the endpoint of the line and from the origin.

Note

The inferencing lines that are displayed on the screen will be either blue or brown in color. The blue inferencing lines suggest that the relations are not added to the sketched entity and the brown inferencing lines suggest that the relations are added to the sketched entity. You will learn about various relations in later chapters.

Tip. *You can also turn off the automatic inferencing lines by choosing* ***Tools > Sketch Settings > Automatic Inferencing Lines*** *from the menu bar.*

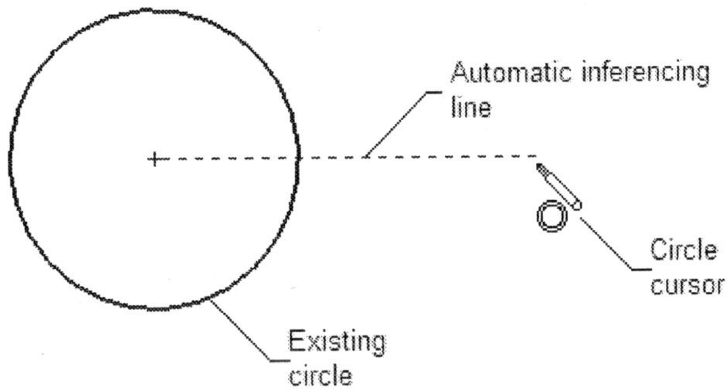

Figure 1-14 *Using inferencing lines to locate the center of a circle*

Select Tool

Toolbar:	Standard > Select

The **Select** tool is used to select a sketched entity or exit any sketching tool that is active. You can select the sketched entities by picking them one by one using the left mouse button. You can also define a window by holding the left mouse button down and dragging the cursor around the sketched entities to select them. Remember that only those entities will be selected that lie completely inside the window that you have defined.

Note
You can also invoke the ***Select*** *tool or exit a sketching tool by pressing the ESC key.*

Now that you are familiar with these terms, you will learn about various sketcher tools available in SolidWorks. The various sketcher tools are discussed next.

DRAWING LINES

CommandManager:	Sketch > Line
Menu:	Tools > Sketch Entities > Line
Toolbar:	Sketch > Line

The lines are one of the basic sketching tools available in SolidWorks. In general terms, a line is defined as the shortest distance between two points. As mentioned earlier, SolidWorks is a parametric solid modeling tool. This property allows you to draw a line of any length and at any angle and later force it to the desired length and angle. To draw a line in the sketcher environment of SolidWorks, choose the **Line** tool. You will notice that the cursor that was an arrow earlier is replaced by the line cursor. The line cursor is actually a pencil-like cursor with a small inclined line below the pencil.

In SolidWorks, there are two methods to draw lines. The first method is to draw continuous lines and the second method is to draw individual lines. Both these methods are discussed next.

Drawing Continuous Lines

This is the default method of drawing lines. In this method you just have to specify the start point and the endpoint of the line using the left mouse button. As soon as you specify the start point of the line, the **Line PropertyManager** will be displayed on the left of the screen. However, when you are drawing continuous lines, the options in the **Line PropertyManager** will not be available. Now, move the cursor away from the start point.

Next, specify the endpoint of the line using the left mouse button. A line will be drawn between the two points. You will notice that the line is green in color and has square boxes at the two ends. The line is displayed in green color because it is still selected. As soon as you draw another line, choose the **Select** tool, or choose another tool, the line turns blue in color and the new line is displayed in green color.

You will notice that after you have drawn the first line, another line is displayed on the screen. The start point of this line is the endpoint of the last line and the length of this line will increase or decrease as you move the mouse. This line is called a rubber-band line because this line will stretch like a rubber-band as you move the cursor. The point that you specify next on the screen will be taken as the endpoint of the second line and a line will be drawn such that the endpoint of the first line is taken as the start point of the new line and the point you specify is taken as the endpoint of the new line. Now, a new rubber-band line is displayed starting from the endpoint of the last line. This is a continuous process and you can draw as many continuous lines as you need by specifying the points on the screen using the left mouse button.

You can exit the continuous line drawing process by pressing the ESC key from the keyboard, by choosing the **Select** tool, or by double-clicking the screen. You can also right-click to display the shortcut menu and choose the **End Chain** option from the shortcut menu.

Figure 1-15 shows a sketch drawn using continuous lines. This sketch is started from the lower left corner and the horizontal line is drawn first. To close the loop using the last line, as soon as you move the cursor close to the start point of the first line, you will notice that a red circle is displayed at the start point and the line cursor turns yellow in color. If you specify the endpoint of the line at this stage, the loop will be closed and no rubber-band line will be displayed now. This is because the loop is already closed and you may not need another continuous line now. However, the **Line** tool is still active and you can create other lines.

Note

When you exit the line drawing process by double-clicking the screen or by choosing ***End chain*** *from the shortcut menu, the current chain is ended but the* ***Line*** *tool is still active and you can draw additional lines.*

Drawing Individual Lines

This is the second method of drawing lines. Using this method you can draw individual lines and the start point of the next line will not necessarily be the endpoint of the last line. To draw individual lines, you need to press and hold down the left mouse button and drag the cursor from the start point of the line to the endpoint. Once you have dragged the cursor to

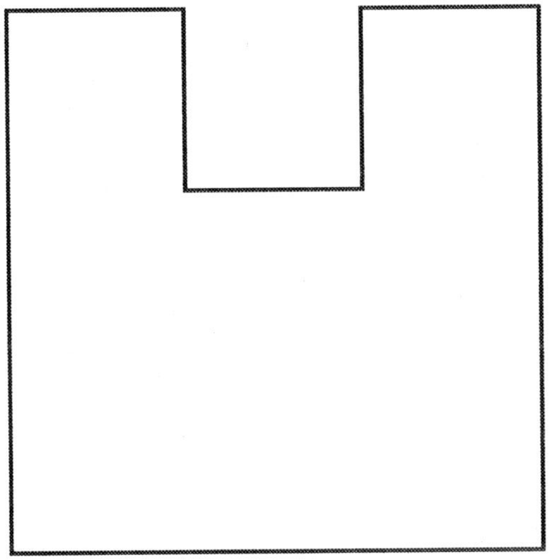

Figure 1-15 *Sketch drawn with the help of continuous lines*

the endpoint, release the left mouse button. A line will be drawn between the point from where you started dragging the mouse and the point where you released the mouse.

To make the process of sketching in SolidWorks easy, you are provided with the **PropertyManager**. The **PropertyManager** is a table that is displayed on the left of the screen as soon as you select the first point of any sketch entity. The **PropertyManager** has all the parameters related to the sketched entity such as the start point, endpoint, angle, length, and so on. You will notice that as you start dragging the mouse, the **Line PropertyManager** is displayed on the left of the drawing area. All the options in the **Line PropertyManager** will be available when you release the left mouse button. Figure 1-16 shows a partial display of the **Line PropertyManager**.

Note
*The **Line PropertyManager** will also display addition options about relations. You will learn more about relations in later chapters.*

After you have drawn the line, modify the parameters in the **Line PropertyManager** to force the line to the desired length and angle. You can also dynamically modify the line by holding the square boxes at the endpoints of the line and dragging them.

Line Cursor Parameters

When you draw lines in the sketcher environment of SolidWorks, you will notice that a numeric value is displayed above the line cursor, see Figure 1-17. This numeric value indicates the length of the line that you draw. This value is the same as that in the **Length** spinner of the **Line PropertyManager**. The only difference is that in the **Line PropertyManager**, the value will be displayed with more precision.

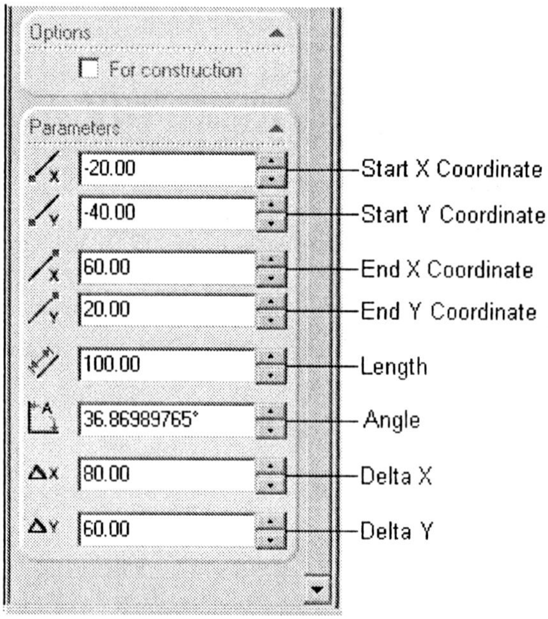

Figure 1-16** Partial display of the **Line PropertyManager

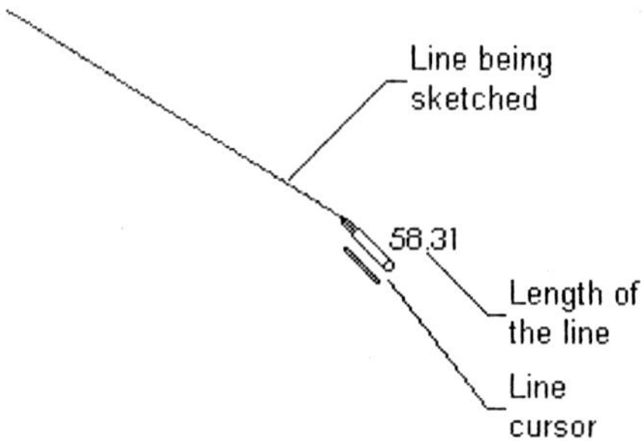

***Figure 1-17** The length of the line displayed on the screen while drawing the line*

The other thing that you will notice while sketching is that sometimes when you are drawing vertical or horizontal lines, a **V** or an **H** symbol is displayed below the line cursor. These are the symbols of the **Vertical** and **Horizontal** relations. SolidWorks automatically applies these relations to the lines. These relations ensure that the lines that you draw are vertical or horizontal and not inclined. Figure 1-18 shows the symbol of the **Vertical** relation on a line and Figure 1-19 shows the symbol of the **Horizontal** relation on a line.

Note

*In addition to the **Horizontal** and **Vertical** relations, you can also apply a number of other relations such as **Tangent**, **Concentric**, **Perpendicular**, **Parallel**, and so on. You will learn about all these relations in later chapters.*

*The other options of the **Line PropertyManager** will be discussed in later chapters.*

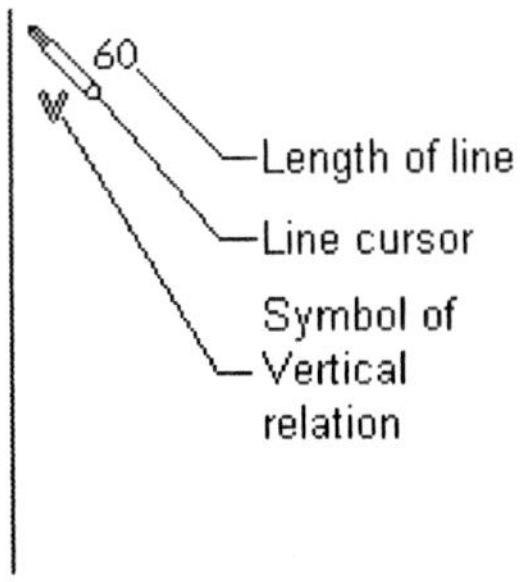

Figure 1-18 Symbol of the ***Vertical*** *relation*

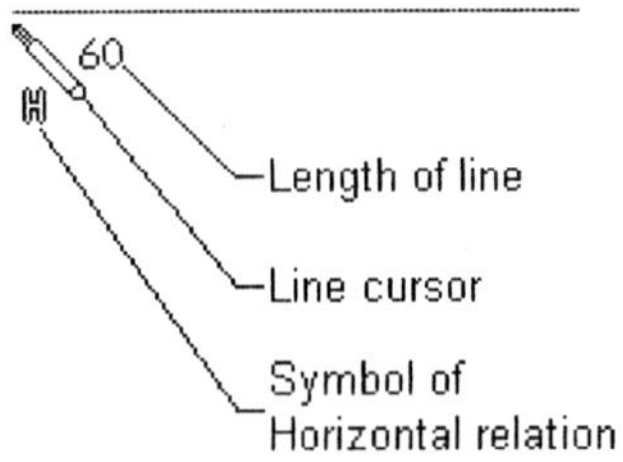

Figure 1-19 Symbol of the ***Horizontal*** *relation*

Drawing Tangent or Normal Arcs Using the Line Tool

SolidWorks allows you to draw tangent or normal arcs originating from the endpoint of the line while drawing continuous lines. Note that these arcs can be drawn only if you have drawn at least one line, arc, or spline. To draw such arcs, draw a line by specifying the start point and the endpoint. Move the cursor away from the endpoint of the last line to display the rubber-band line. Now, when you move the cursor back to the endpoint of the last line, the arc mode is invoked and the line cursor is replaced by the arc cursor. The angle and the radius of the arc is displayed above the arc cursor. You can also invoke the arc mode by right-clicking and choosing **Switch to arc** from the shortcut menu or pressing the A key from the keyboard.

To draw a tangent arc, invoke the arc mode by moving the cursor back to the endpoint of the last line. Now, move the cursor through a small distance along the tangent direction of the line; a dotted line will be drawn. Next, move the cursor in the direction in which the arc should be drawn. You will notice that a tangent arc is drawn. Specify the endpoint of the tangent arc using the left mouse button. Figure 1-20 shows an arc tangent to an existing line.

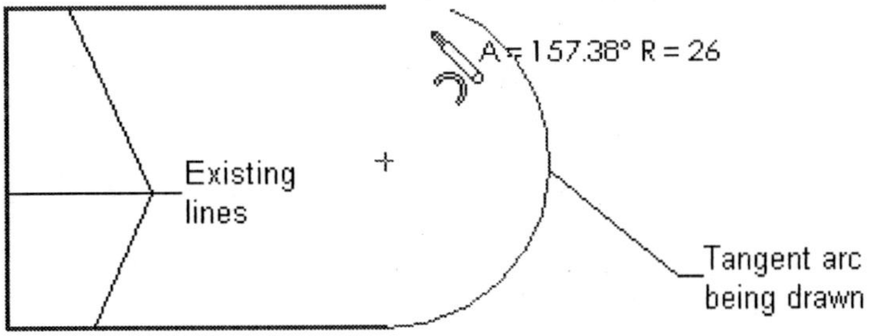

Figure 1-20 *Drawing a tangent arc using the* ***Line*** *tool*

To draw a normal arc, invoke the arc mode. Now, move the cursor through a small distance in the direction normal to the line and then move it in the direction of the endpoint of the arc. The normal arc will be drawn, as shown in Figure 1-21.

As soon as the endpoint of the tangent or normal arc is defined, the line mode is invoked again. You can continue drawing lines using the line mode or move the cursor back to the endpoint of the arc to invoke the arc mode.

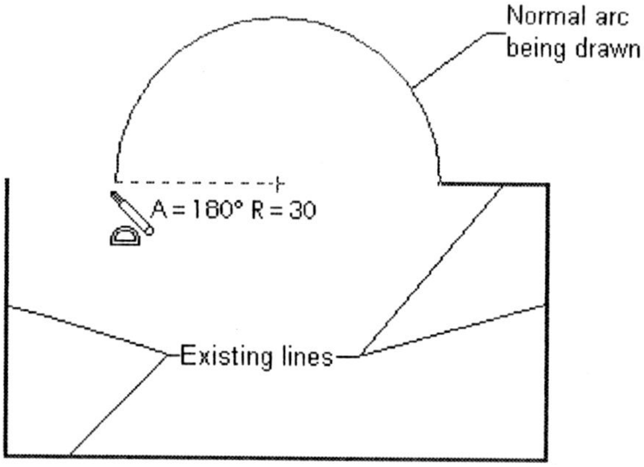

***Figure 1-21** Drawing a normal arc using the **Line** tool*

Note

*If the arc mode is invoked by mistake while drawing lines, you can cancel the arc mode and invoke the line mode again by pressing the A key from the keyboard. Alternatively, you can right-click and choose **Switch to Line** from the shortcut menu or move the cursor back to the endpoint and press the left mouse button to invoke the line mode.*

Drawing Construction Lines or Centerlines

CommandManager:	Sketch > Centerline
Menu:	Tools > Sketch Entities > Centerline
Toolbar:	Sketch > Centerline

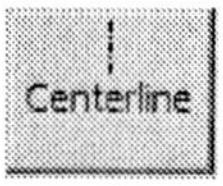

Construction lines or centerlines are the ones that are drawn only for the aid of sketching. These lines are not considered while converting the sketches into features. You can draw a construction line similar to the sketched line by using the **Centerline** tool. You will notice that when you draw a construction line, the **For Construction** check box in the **Line PropertyManager** is selected. You can also draw a construction line by sketching the line using the **Line** tool and then selecting the **For Construction** check box available in the **Line PropertyManager**. When you select this check box, the line is turned into a centerline.

DRAWING CIRCLES

CommandManager:	Sketch > Circle
Menu:	Tools > Sketch Entities > Circle
Toolbar:	Sketch > Circle

In SolidWorks, the circles are drawn by specifying the center point of the circle using the left mouse button and then moving the mouse on the screen to define the radius of the circle. Similar to the lines, as soon as you specify the center of the circle, the **Circle PropertyManager** is displayed. However, note that the options in the **Circle PropertyManager** will be available only after you have defined the radius of the circle.

To draw the circle, choose the **Circle** button from the **CommandManager**. You will notice that the arrow cursor is replaced by the circle cursor. The circle cursor consists of a pencil and two concentric circles below the pencil. Specify the center point of the circle and then move the cursor to define the radius of the circle. The current radius of the circle is displayed above the circle cursor. This radius will change as you move the cursor. You can define any arbitrary radius of the circle and then modify it to the desired value by using the **Circle PropertyManager**. Figure 1-22 shows a circle being drawn using the **Circle** tool.

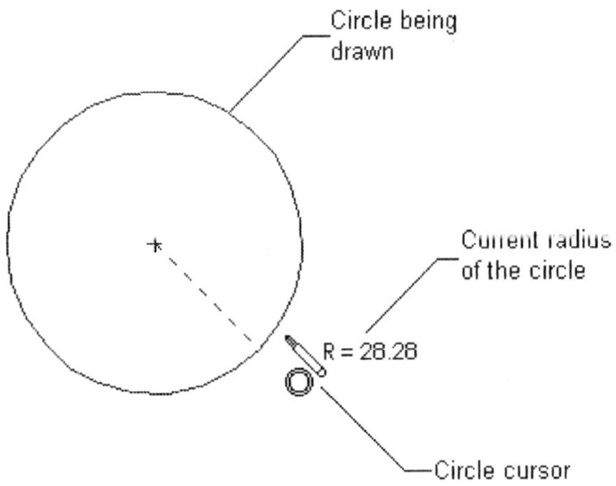

Figure 1-22 *Sketching a circle*

Figure 1-23 shows a partial display of the **Circle PropertyManager**.

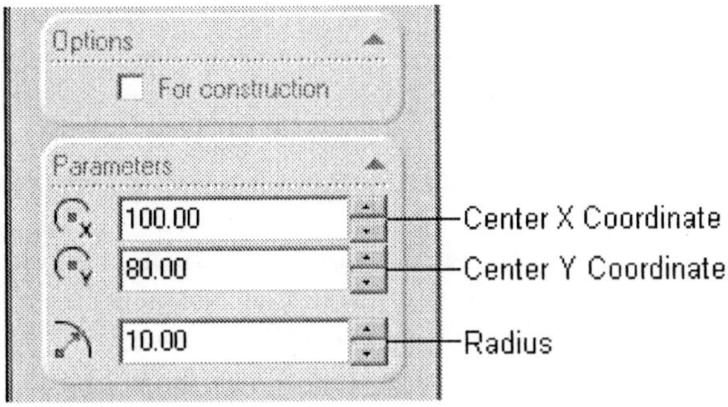

Figure 1-23 *Partial display of the* ***Circle PropertyManager***

Drawing a Construction Circle

If you want to sketch a construction circle, draw the circle using the **Circle** tool and then select the **For Construction** check box available in the **Circle PropertyManager**.

Tip. *To convert a construction entity back to the sketched entity, invoke the select tool and then select the construction entity. The entity will turn green in color and the **PropertyManager** will be displayed on the left of the drawing area. From the **PropertyManager**, clear the **For Construction** check box. The construction entity will again be changed into a sketched entity and will be displayed with continuous line.*

DRAWING ARCS

In SolidWorks, you can draw the arcs using three methods: **Tangent/Normal Arc**, **Centerpoint Arc**, and **3 Point Arc**. All these methods can be invoked separately by choosing their respective buttons from the **Sketch Tools** toolbar. All these three methods used to draw arcs are discussed next.

Drawing Tangent/Normal Arcs

CommandManager:	Sketch > Tangent Arc
Menu:	Tools > Sketch Entities > Tangent Arc
Toolbar:	Sketch > Tangent Arc

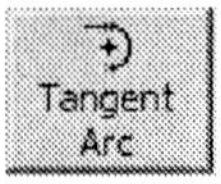

The tangent arcs are the ones that are drawn tangent to an existing sketched entity. The existing sketched entities include sketched and construction lines, arcs, and splines. The normal arcs are those that are drawn normal to an existing line segment. You can draw tangent and normal arcs using the **Tangent Arc** tool.

To draw a tangent arc, invoke the **Tangent Arc** tool. The arrow cursor is replaced by the arc cursor. Move the arc cursor close to the endpoint of the entity that you want to select as the tangent entity. You will notice that a red circle is displayed at the endpoint and the entity selected as the tangent entity is turned green in color. Also, the color of the pencil in the cursor is changed to yellow and an orange-colored box is displayed below the pencil. Now, press the left mouse button once and move the cursor along the tangent direction through a small distance and then move the cursor to size the arc. The arc will start from the endpoint of the tangent entity and its size will change as you move the cursor. Note that the angle and the radius of the tangent arc are displayed above the arc cursor, see Figure 1-24.

To draw a normal arc, invoke the **Tangent Arc** tool. Move the arc cursor close to the endpoint of the entity that you want to select as the normal entity. A red circle is displayed at the endpoint and the entity selected as the tangent entity is turned green in color. Also, the color of the pencil in the cursor is changed to yellow and an orange-colored box is displayed below the pencil. Now, press the left mouse button once and move the cursor along the normal direction through a small distance and then move the cursor to size the arc, refer to Figure 1-25.

While drawing the arc, as soon as you start moving the cursor after specifying the start point, the **Arc PropertyManager** is displayed. However, the options in the **Arc PropertyManager** are not enabled at this stage. These options are enabled only after you have completed drawing the tangent or normal arc.

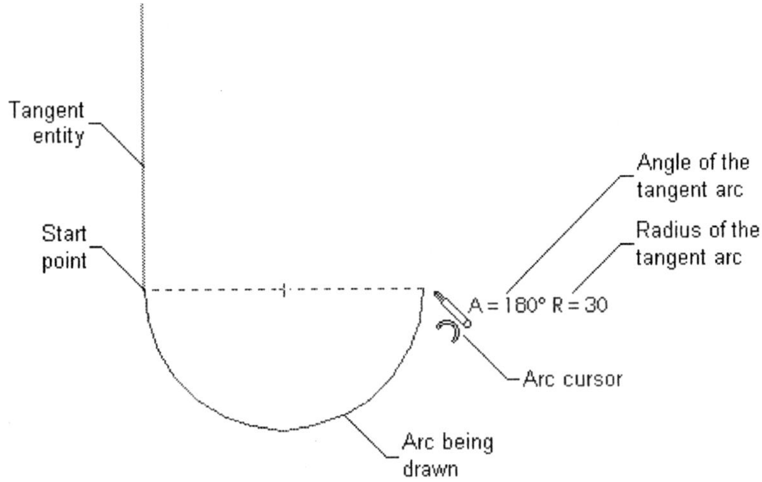

Figure 1-24 *Drawing tangent arc*

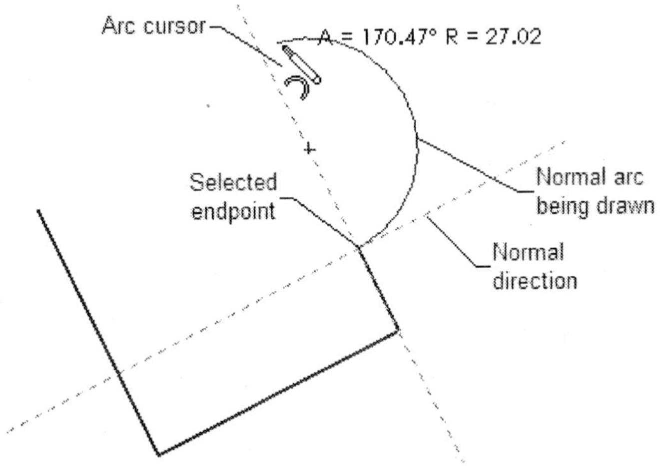

Figure 1-25 *Drawing normal arc*

You can draw an arbitrary arc and then modify its value using the **Arc PropertyManager**. Figure 1-26 shows a partial view of the **Arc PropertyManager**.

Note

*When you select a tangent entity to draw a tangent arc, the **Tangent** relation is applied between the start point of the arc and the tangent entity. Therefore, if you change the coordinates of the start point of the arc, the tangent entity will also be modified accordingly.*

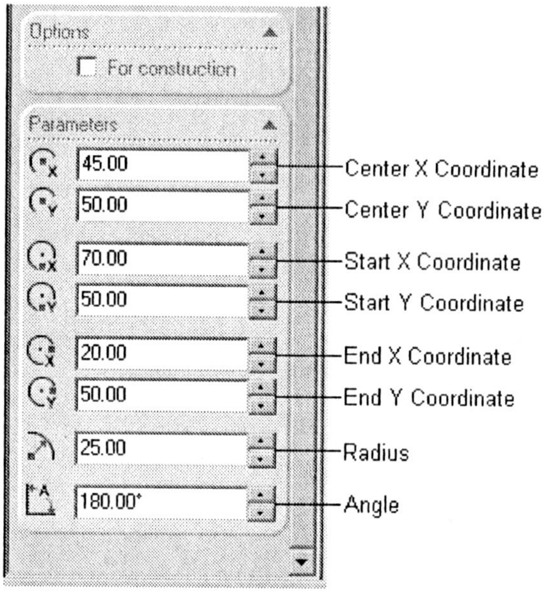

Figure 1-26** Partial display of the **Arc PropertyManager

Drawing Centerpoint Arcs

CommandManager:	Sketch > Centerpoint Arc
Menu:	Tools > Sketch Entity > Centerpoint Arc
Toolbar:	Sketch > Centerpoint Arc

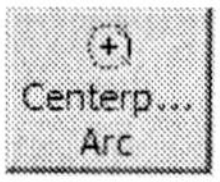

The centerpoint arcs are the ones that are drawn by defining the centerpoint, start point, and endpoint of the arc. When you invoke this tool, the arrow cursor is replaced by the arc cursor. As mentioned earlier, an arc cursor consists of a pencil and an arc below the pencil.

To draw a centerpoint arc, invoke the **Centerpoint Arc** tool and then move the arc cursor to the point that you want to specify as the centerpoint of the arc. Press the left mouse button once at the location of the centerpoint and then move the cursor to the point from where you want to start the arc. You will notice that a dotted circle is displayed on the screen. The size of this circle will modify as you move the mouse. This circle is drawn for your reference and the centerpoint of this circle lies at the point that you specified as the center of the arc. Press the left mouse button once at the point that you want to select as the start point of the arc. Next, move the mouse to specify the endpoint of the arc. You will notice that the reference circle is no more displayed and an arc is being drawn with the start point as the point that you specified after specifying the centerpoint. Also, the **Arc PropertyManager**, similar to the one that is shown in the tangent arc, is displayed on the left of the drawing area. Note that the options in the **Arc PropertyManager** will not be available at this stage.

If you move the cursor in the clockwise direction, the resultant arc will be drawn in the clockwise direction. However, if you move the cursor in the counterclockwise direction, the resultant arc will be drawn in the counterclockwise direction. Specify the endpoint of the arc using the left mouse button. Figure 1-27 shows the reference circle that is drawn when you move the mouse button after specifying the centerpoint of the arc and Figure 1-28 shows the resultant centerpoint arc.

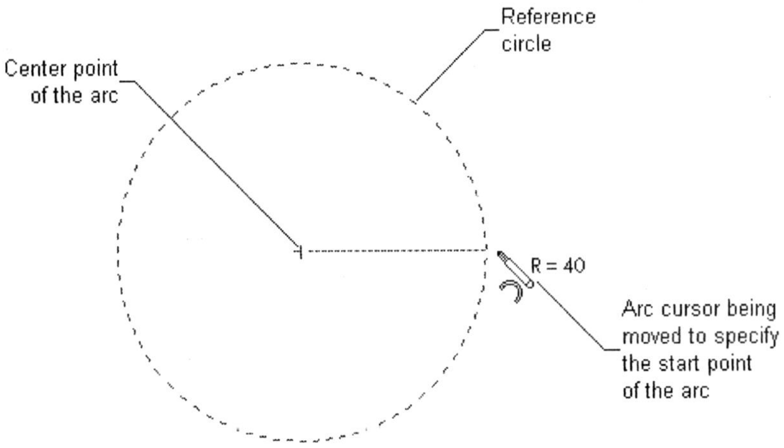

Figure 1-27 *Specifying the centerpoint and the start point of the centerpoint arc*

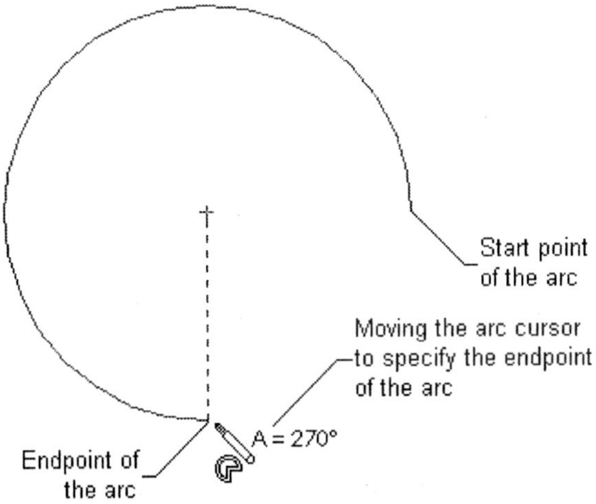

Figure 1-28 *Moving the cursor to specify the start point and the endpoint of the arc*

Drawing 3 Point Arcs

CommandManager:	Sketch > 3 Pt Arc
Menu:	Tools > Sketch Entities > 3 Point Arc
Toolbar:	Sketch > 3 Pt Arc

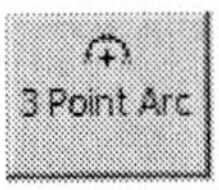

The 3 point arcs are the ones that are drawn by defining the start point and endpoint of the arc, and a point somewhere on the arc. When you invoke this tool, the arrow cursor is replaced by the arc cursor.

To draw a 3 point arc, invoke the **3 Pt Arc** tool and then move the arc cursor to the point that you want to specify as the start point of the arc. Press the left mouse button once at the location of the start point and then move the cursor to the point that you want specify as the endpoint of the arc. As soon as you start moving the cursor after specifying the start point, a reference arc will be drawn and the **Arc PropertyManager** will be displayed. However, the options in the **Arc PropertyManager** will not be available at this stage.

Using the left mouse button, specify the endpoint of the arc. You will notice that the reference arc is no more displayed. Instead, a solid arc is displayed and the cursor is attached to the arc. As you move the cursor, the arc will also be modified dynamically. Using the left mouse button, specify a point on the screen to create the arc. The last point that you specify will determine the direction of the arc. The options in the **Arc PropertyManager** will be displayed after you have drawn the arc. You can modify the properties of the arc using the **Arc PropertyManager**. Figure 1-29 shows the reference arc that is drawn by specifying the start point and the endpoint of the arc and Figure 1-30 shows the resultant 3 point arc.

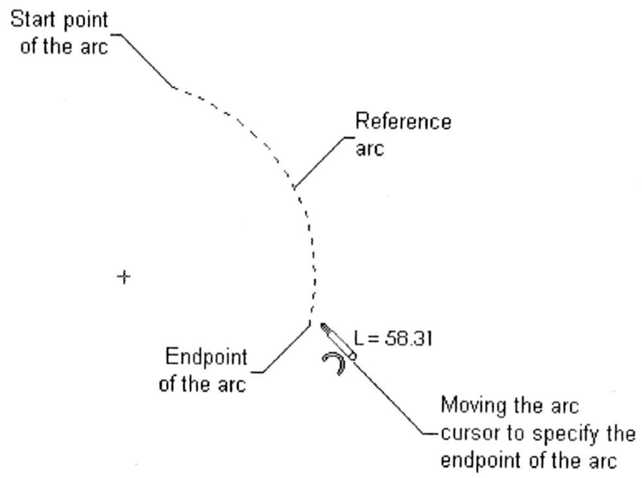

Figure 1-29 *Specifying the start point and the endpoint of the arc*

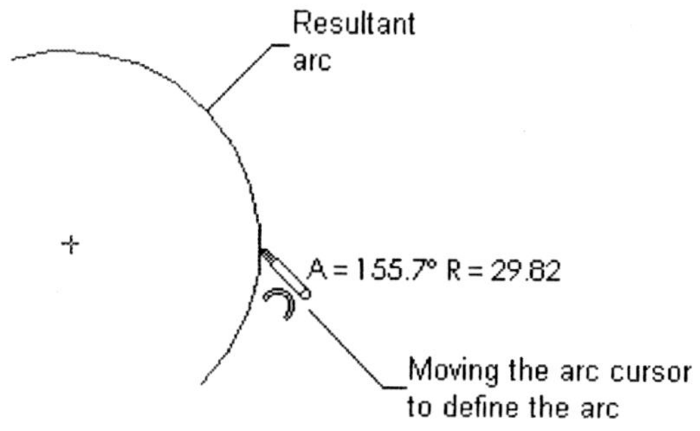

Figure 1-30 *Specifying a point on the arc to draw it*

DRAWING RECTANGLES

CommandManager:	Sketch > Rectangle
Menu:	Tools > Sketch Entities > Rectangle
Toolbar:	Sketch > Rectangle

In SolidWorks, the rectangles are drawn by specifying two opposite corners of the rectangle. To draw a rectangle, invoke the **Rectangle** tool. The arrow cursor will be replaced by the rectangle cursor. Move the cursor to the point that you want to specify as the first corner of the rectangle. Press the left mouse button once at the first corner and then move the cursor and specify the other corner of the rectangle using the left mouse button. You will notice that the length and width of the rectangle are displayed above the rectangle cursor. The length is measured along the X axis and the width is measured along the Y axis. Figure 1-31 shows a rectangle being drawn by specifying two opposite corners.

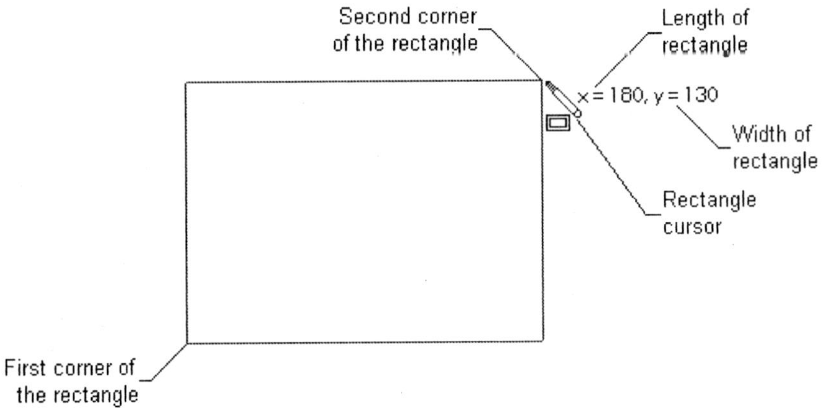

Figure 1-31 *Drawing a rectangle by specifying two opposite corners*

Note

*When you draw a rectangle, the **PropertyManager** will not be displayed. This is because a rectangle is considered as a combination of four individual lines. Therefore, after drawing the rectangle, if you select one of the lines of the rectangle using the **Select** tool, the **Line PropertyManager** will be displayed. You can modify the parameters of the selected line using the **Line PropertyManager**.*

*Remember that as the relations are applied to all the four corners of the rectangle, if you modify the parameters of one of the lines using the **Line PropertyManager**, the other three lines will also be modified accordingly.*

*You can convert a rectangle into a construction rectangle by selecting all the lines together using a window and then selecting the **For Construction** check box from the **PropertyManager**.*

DRAWING PARALLELOGRAM

Menu: Tools > Sketch Entities > Parallelogram

In SolidWorks, the **Parallelogram** tool can be used to draw a parallelogram and also to draw a rectangle at an angle. The methods used to draw both these entities are discussed next.

Drawing a Rectangle at an Angle

To draw a rectangle at an angle, invoke the **Parallelogram** tool from the menu bar. The cursor will be replaced by the parallelogram cursor. Move the cursor to the point that you want to select as the start point of one of the edges of the rectangle. Press the left mouse button once at this point and move the cursor to size the edge. You will notice that a reference line is being drawn. Based on the current position of the cursor, the reference line will be horizontal, vertical, or inclined at some angle. The current length of the edge and its angle will be displayed above the parallelogram cursor. Using the left mouse button, specify the endpoint of the edge such that the resultant reference line is at an angle.

Next, move the cursor to specify the width of the rectangle. You will notice that a reference rectangle is drawn at an angle. Also, irrespective of the current position of the cursor, the width will be specified normal to the first edge, either above or below. Using the left mouse button, specify a point on the screen to define the width of the rectangle. The reference rectangle will be converted into a sketched rectangle. Figure 1-32 shows a rectangle drawn at an angle.

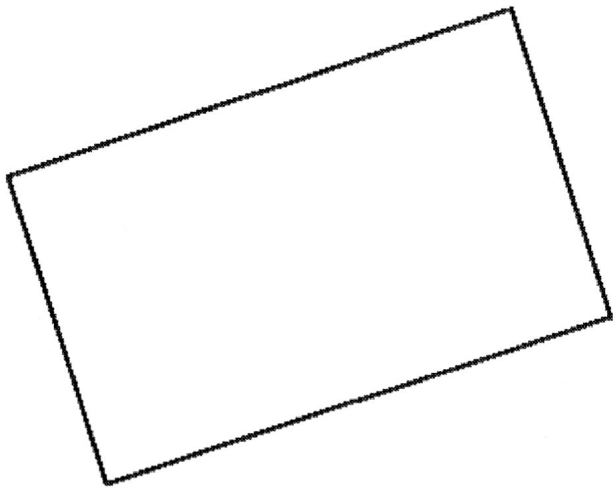

Figure 1-32 *Rectangle at an angle*

Drawing Parallelograms

To draw a parallelogram, invoke the **Parallelogram** tool from the menu bar. The cursor will be replaced by the parallelogram cursor. Specify two points on the screen to define one edge of the parallelogram. Next, press the CTRL key from the keyboard once and then move the mouse to define the width of the parallelogram. You will notice that the width is no more added normal to the first edge. As you move the mouse, a reference parallelogram will be drawn. The size and the shape of the reference parallelogram will depend upon the current location of the cursor.

Specify a point on the screen to define the parallelogram. Figure 1-33 shows a parallelogram drawn at an angle.

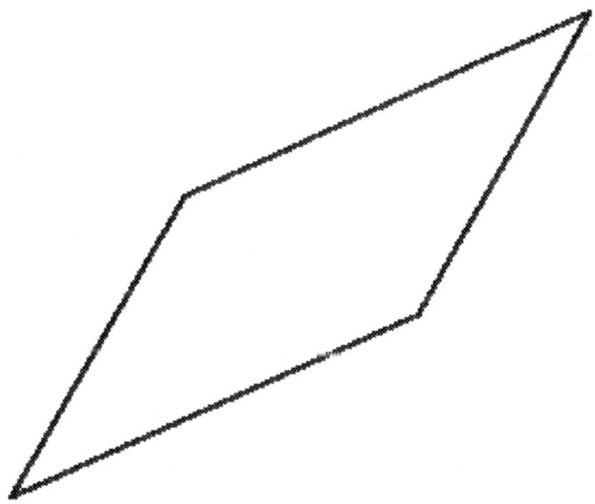

Figure 1-33 *Parallelogram at an angle*

Note
Similar to the rectangles, each edge of a parallelogram is considered as a separate line. Also, in case of parallelograms, the ***PropertyManager*** *is not displayed while you are drawing it.*

Tip. *You can also invoke the tools to draw lines, arcs, circles, or rectangles using the shortcut menu that is displayed when you right-click in the drawing area.*

DRAWING POLYGONS

Menu:	Tools > Sketch Entities > Polygon

A polygon is defined as a multisided geometric figure in which the length of all the sides and the angle between all the sides are the same. In SolidWorks, you can draw a polygon with the number of sides ranging from 3 to 40. The dimensions of a polygon are controlled using the diameter of a construction circle that is either inscribed inside the polygon or circumscribed about the polygon. If the construction circle is inscribed inside the polygon, the diameter of the construction circle is taken from the edges of the polygon. If the construction circle is circumscribed about the polygon, the diameter of the construction circle is taken from the vertices of the polygon.

To draw a polygon, invoke the **Polygon** tool; the **Polygon PropertyManager** will be displayed, as shown in Figure 1-34.

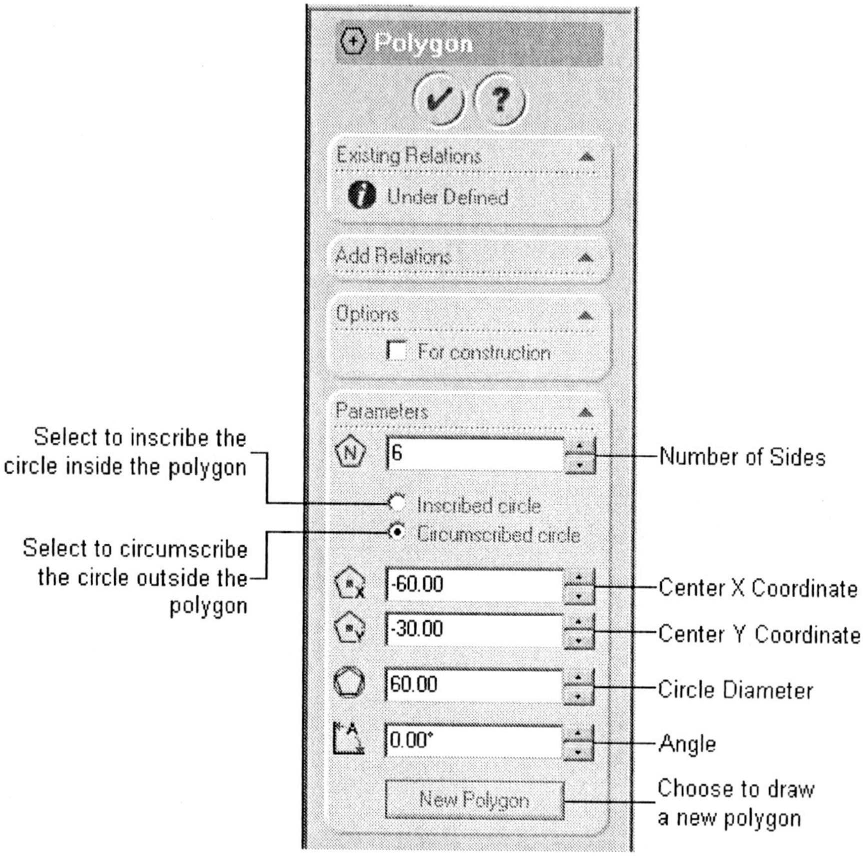

Figure 1-34 Polygon PropertyManager

Set the parameters such as the number of sides, inscribed or circumscribed circle, and so on in the **Polygon PropertyManager**. You can also modify these parameters after drawing the polygon. When you invoke this tool, the arrow cursor will be replaced by the polygon cursor. Press the left mouse button at the point that you want to select as the centerpoint of the polygon and then move the cursor to size the polygon. The length of each side and the rotation angle of the polygon will be displayed above the polygon cursor as you drag it. Using the left mouse button, specify a point on the screen after you get the desired length and rotation angle of the polygon. You will notice that based on whether you selected the **Inscribed circle** or the **Circumscribed circle** radio button in the **Polygon PropertyManager**, a construction circle will be drawn inside or outside the polygon. After you have drawn the polygon, you can modify the parameters such as the centerpoint of the polygon, the diameter of the construction circle, the angle of rotation of the polygon, and so on using the **Polygon PropertyManager**. If you want to draw another polygon, choose the **New polygon** button provided below the **Angle** spinner in the **Polygon PropertyManager**.

Figure 1-35 shows a six-sided polygon with the construction circle inscribed inside the polygon and Figure 1-36 shows a five-sided polygon with the construction circle circumscribed outside the polygon. Notice that the reference circle is retained with the polygon. Remember that this circle will not be considered while converting the polygon into a feature.

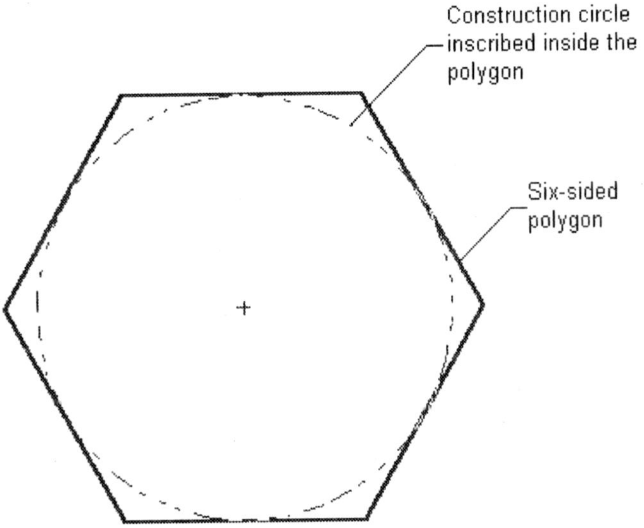

Figure 1-35 Six-sided polygon with construction circle inscribed inside the polygon

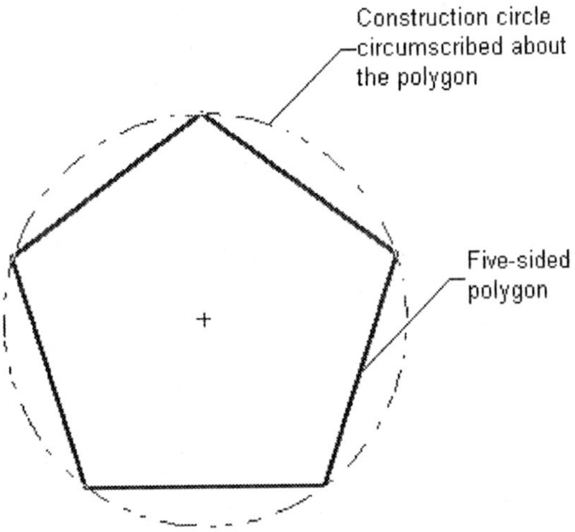

Figure 1-36 Five-sided polygon with construction circle circumscribed outside the polygon

DRAWING SPLINES

CommandManager:	Sketch > Spline
Menu:	Tools > Sketch Entities > Spline
Toolbar:	Sketch > Spline

In SolidWorks, you can draw a spline by continuously specifying the endpoints of the spline segments using the left mouse button. This method of drawing splines is similar to the method of drawing continuous lines. After specifying all the points of the spline, right-click and choose **Select** to exit the **Spline** tool. Choosing **End Spline** will exit the current spline but the **Spline** tool will still be active and you can draw another spline. Figure 1-37 shows a spline drawn with start point at the origin.

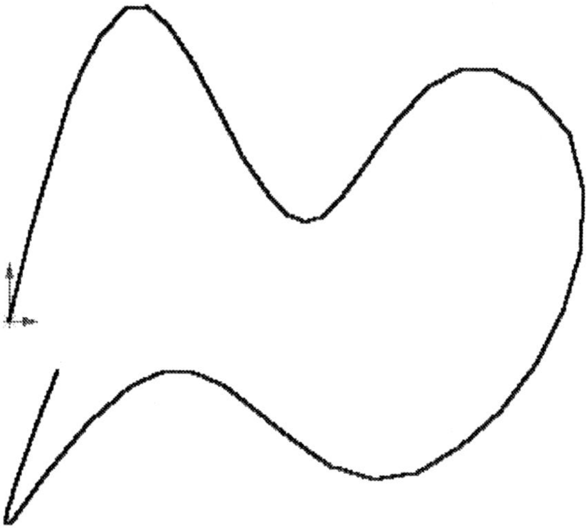

Figure 1-37 *Sketched spline with start point at the origin*

Note

*The number of points that were used to draw a spline are taken as its handles. These handles are displayed when you select the spline using the **Select** tool.*

Similar to drawing individual lines, you can also create individual spline segments by specifying the start point and then dragging the mouse to specify the endpoint.

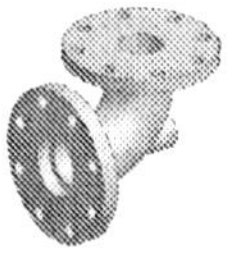

Tip. *After creating the spline, when you select it using the **Select** tool, the **Spline PropertyManager** is displayed. The current handle will be displayed with a filled square and its number and the corresponding X and Y coordinates will be displayed in the **Spline PropertyManager**. You can modify these coordinates to modify the selected spline.*

DRAWING POINTS

CommandManager:	Sketch > Point
Menu:	Tools > Sketch Entities > Point
Toolbar:	Sketch > Point

To draw a point, choose the **Point** tool from the **CommandManager** and then specify the point on the screen where you want to place the point. The **Point PropertyManager** will be displayed with the X and Y coordinates of the current point. You can modify the location of the point by modifying its X and Y coordinates in the **Point PropertyManager**.

DRAWING ELLIPSES

Menu: Tools > Sketch Entities > Ellipse

In SolidWorks, an ellipse is drawn by specifying the centerpoint of the ellipse and then specifying the two ellipse axes by moving the mouse. To draw an ellipse, invoke the **Ellipse** tool from the menu bar. The arrow cursor will be replaced by the ellipse cursor. Move the cursor to the point that you want to select as the centerpoint of the ellipse. Press the left mouse button once at the centerpoint of the ellipse and then move the cursor to specify one of the ellipse axis. You will notice that a reference circle is drawn and two values are displayed above the ellipse cursor, see Figure 1-38. The first value that shows R = * is the radius of the first axis that you are defining and the second value that shows r = * is the radius of the other axis. While you are defining the first axis, the second axis is taken equal to the first axis. This is the reason a reference circle is drawn and not a reference ellipse.

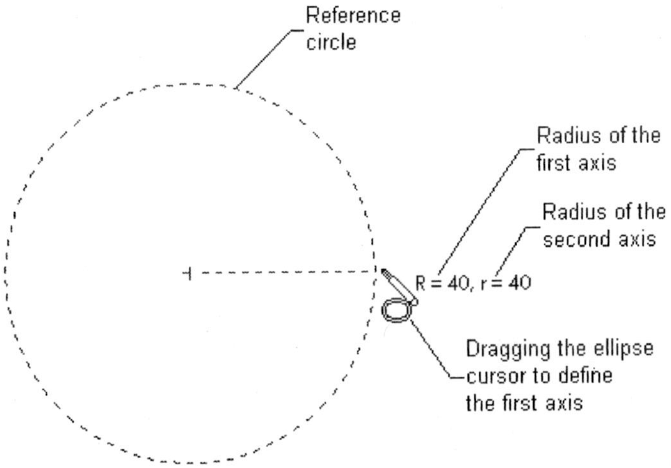

***Figure 1-38** Dragging the cursor to define the first axis*

Specify a point on the screen to define the first axis. Next, move the cursor to size the other ellipse axis. You will notice that the **Ellipse PropertyManager** is displayed. As you move the cursor, the second value above the ellipse cursor that shows r = * will change dynamically as you move the cursor on the screen. Using the left mouse button, specify a point on the screen to define the second axis of the ellipse, see Figure 1-39.

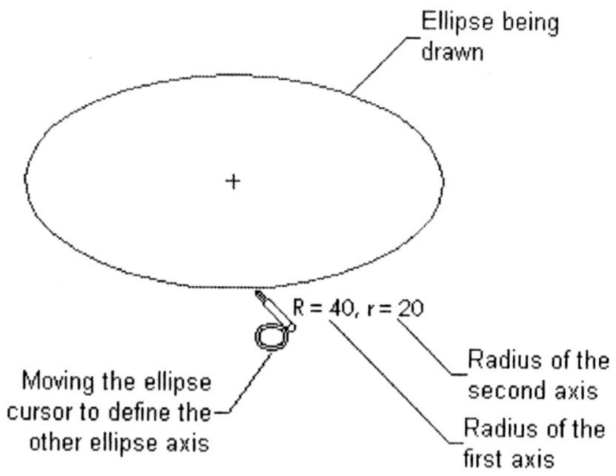

***Figure 1-39** Defining the second axis of the ellipse*

DRAWING ELLIPTICAL ARCS

Menu:	Tools > Sketch Entities > Partial Ellipse

In SolidWorks, the process of drawing an elliptical arc is similar to that of drawing an ellipse. You will follow the same process of defining the ellipse first. The point that you specify on the screen to define the other axis of the ellipse is taken as the start point of the elliptical arc. You can define the endpoint of the elliptical arc by specifying a point on the screen as shown in Figure 1-40.

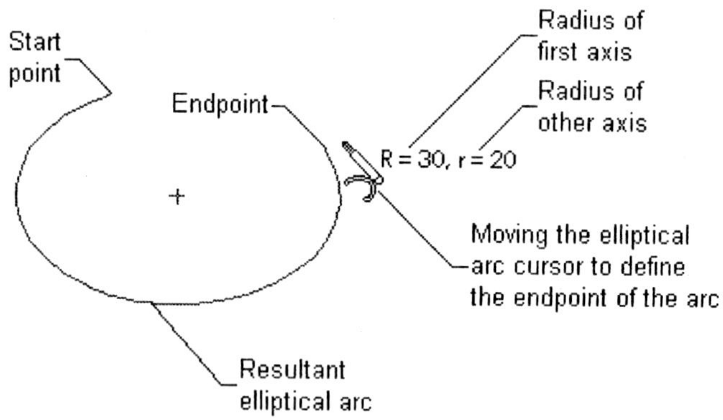

***Figure 1-40** Drawing the elliptical arc*

You can also set the parameters of the elliptical arc in the **Ellipse PropertyManager** shown in Figure 1-41.

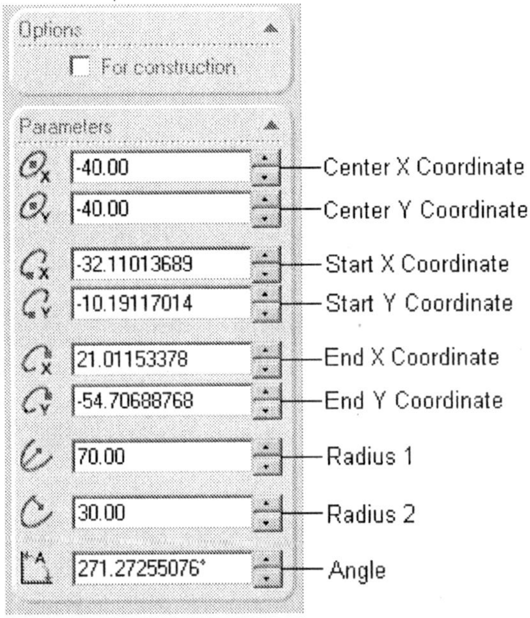

Figure 1-41 *Partial view of the* ***Ellipse PropertyManager***

DRAWING A PARABOLIC CURVE

Menu: Tools > Sketch Entities > Parabola

In SolidWorks, you will draw a parabolic curve by specifying the focal point of the parabola and then specifying two points on the guide of the parabolic curve. To draw a parabolic curve, invoke the **Parabola** tool from the menu bar. The cursor will be replaced by the parabola cursor. Move the cursor to the point that you want to specify as the focal point of the parabola. Press the left mouse button once at the focal point and then move the cursor to define the apex point and size the parabola. You will notice that a reference parabolic arc is displayed. As you move the cursor away from the focal point, the parabola is flattened. After you get the basic shape of the parabolic curve, specify a point on the screen using the left mouse button. This point is taken as the apex of the parabolic curve. Next, specify two point on the screen with respect to the reference parabola to define the guide of the parabolic curve, see Figure 1-42.

As you move the mouse after specifying the focal point of the parabola, the **Parabola PropertyManager** will be displayed. But the options in the **Parabola PropertyManager** will not be available. These options will be available only after you have drawn the parabola. Figure 1-43 shows a partial view of the **Parabola PropertyManager**.

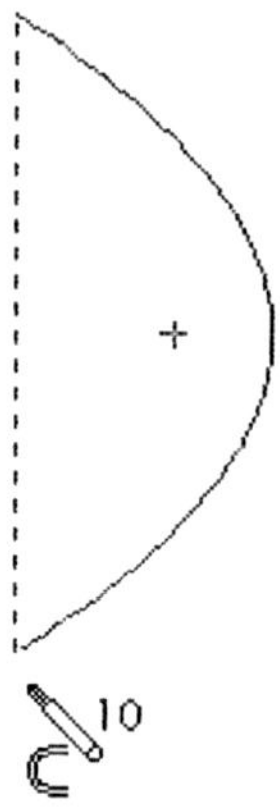

Figure 1-42 *Drawing the parabola*

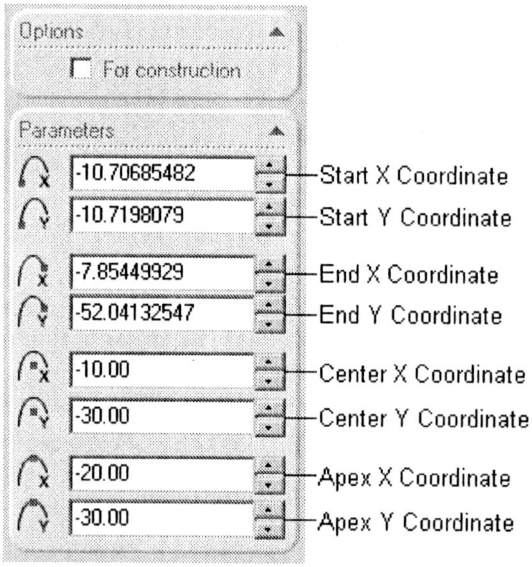

Figure 1-43 *Partial view of the* ***Parabola PropertyManager***

DRAWING DISPLAY TOOLS

The drawing display tools are one of the most important tools provided in any of the solid modeling software. These tools allow you to modify the display of the drawing by zooming or panning the drawing. Some of the drawing display tools that are available in SolidWorks are discussed in this chapter. The remaining tools are discussed in the later chapters.

Zoom to Fit

Menu:	View > Modify > Zoom to Fit
Toolbar:	View > Zoom to Fit

The **Zoom to Fit** tool is used to increase or decrease the drawing display area so that all the sketched entities or dimensions are fitted inside the current view.

Zoom to Area

Menu:	View > Modify > Zoom to Area
Toolbar:	View > Zoom to Area

The **Zoom to Area** tool is used to magnify a specified area so that the part of the drawing inside the magnified area can be viewed in the current window. The area is defined by a window that is created by dragging the cursor and specifying two opposite corners of the window. When you choose this tool, the cursor is replaced by a magnifying glass cursor. Press and hold the left mouse button down and drag the cursor to specify two opposite corners of the window. The area enclosed inside the window will be magnified.

Zoom In/Out

Menu:	View > Modify > Zoom In/Out
Toolbar:	View > Zoom In/Out

The **Zoom In/Out** tool is used to dynamically zoom in or out of the drawing. When you invoke this tool, the cursor is replaced by the zoom cursor. To zoom out of a drawing, press and hold the left mouse button down and drag the cursor in the downward direction. Similarly, to zoom in a drawing, press and hold the left mouse button down and drag the cursor in the upward direction. As you drag the cursor, the drawing display will be modified dynamically. After you get the desired view, exit this tool by choosing the **Select** tool from the **Sketch** toolbar. You can also exit this tool by right-clicking and choosing **Select** from the shortcut menu or by pressing the ESC key.

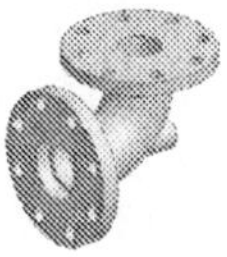

Tip. *You can also use the keyboard shortcuts to invoke some of the drawing display tools. For example, to invoke the* ***Zoom to Fit*** *tool, press the F key. Similarly, to zoom out of a drawing, press the Z key and to zoom in, press the SHIFT+Z keys.*

Zoom to Selection

Menu:	View > Modify > Zoom to Selection
Toolbar:	View > Zoom to Selection

The **Zoom to Selection** tool is used to modify the drawing display area such that the selected entity is fitted inside the current display. After selecting the entity, choose the **Zoom to Selection** button. The drawing display area will be modified such that the selected entity fits inside the current view.

Pan

Menu:	View > Modify > Pan
Toolbar:	View > Pan

The **Pan** tool is used to drag the view in the current display. This process is similar to changing the view by using the scroll bars available in the drawing area.

Tip. *You can also invoke the* ***Pan*** *tool using the CTRL key and the arrow keys on the keyboard. For example, to pan toward the right, press the CTRL key and then press the right arrow key a few times. Similarly, to pan upward, press the CTRL key and then press the up arrow key a few times.*

Redraw

Menu:	View > Redraw

The **Redraw** tool is used to refresh the screen. Sometimes when you draw a sketched entity, some unwanted elements remain on the screen. To remove these unwanted elements from the screen, choose this tool. The screen will be refreshed and all the unwanted elements will be removed. You can also invoke this tool by pressing the CTRL+R keys from the keyboard.

DELETING THE SKETCHED ENTITIES

You can delete the sketched entities by selecting them using the **Select** tool and then pressing the DELETE key from the keyboard. You can select the entities by picking them individually or select more than one entity by defining a window around the entities. When you select the entities, they turn green in color. When they turn green, press the DELETE key from the keyboard. You can also delete the sketched entities by selecting them and choosing the **Delete** option from the shortcut menu that is displayed upon right-clicking.

TUTORIALS

Tutorial 1

In this tutorial, you will draw the sketch of the model shown in Figure 1-44. The sketch is shown in Figure 1-45. You will not dimension the sketch. The solid model and the dimensions are given only for your reference. **(Expected time: 30 min)**

The steps that will be followed to complete this tutorial are listed below:

a. Start SolidWorks and then start a new part document.
b. Maximize the part file document and then switch to the sketching environment.
c. Draw the sketch of the model using the **Line** and the **Circle** tools, refer to Figures 1-48 through 1-50.
d. Save the sketch and then close the document.

Figure 1-44 Solid model for Tutorial 1

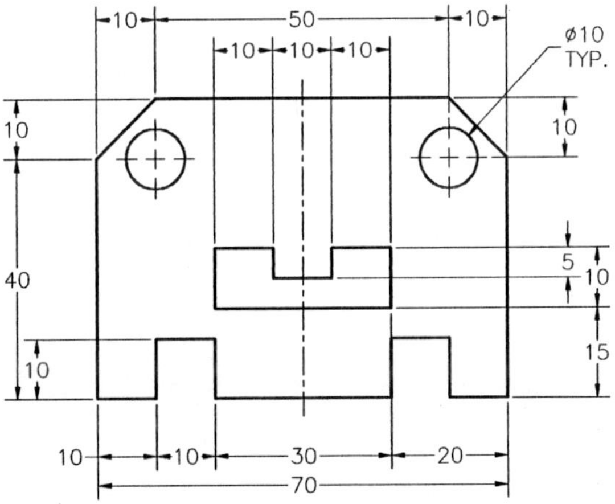

Figure 1-45 Sketch of the model

Starting SolidWorks and Starting a New Part Document

1. Start SolidWorks by choosing **Start > Programs > SolidWorks 2004 > SolidWorks 2004** or by double-clicking the shortcut icon of SolidWorks available on the desktop of your computer.

   The **Tip of the Day** dialog box will be displayed. If the **Tip of the Day** dialog box is not displayed when you start the SolidWorks session, you need to set the option to display the **Tip of the Day** dialog box when you start the new SolidWorks session.

2. Choose **Help > Tip of the Day** from the menu bar. The **Tip of the Day** dialog box is displayed. Select the **Show tips at startup** check box from this dialog box.

Tip. *If the shortcut icon of SolidWorks is not created automatically on the desktop of your system when you install SolidWorks, you can create it manually. To create the shortcut icon on the desktop, choose* ***Start > Programs > SolidWorks 2004*** *to display the SolidWorks cascading menu. Right-click* ***SolidWorks 2004*** *in the cascading menu and then choose* ***Send To > Desktop (create shortcut)*** *from the shortcut menu.*

By selecting the **Show tips at startup** check box, the **Tip of the Day** dialog box will be displayed every time you start the SolidWorks session. You can get many valuable tips from the **Tip of the Day** dialog box. The tips displayed in this dialog box are helpful to make the full utilization of this CAD package.

3. Close the **Tip of the Day** dialog box by choosing the **Close** button. The **Welcome to SolidWorks 2004** window is displayed. Choose the **New Document** option from this window.

   The **New SolidWorks Document** dialog box is displayed.

4. The **Part** button is chosen by default. Choose the **OK** button from the **New SolidWorks Document** dialog box as shown in Figure 1-46.

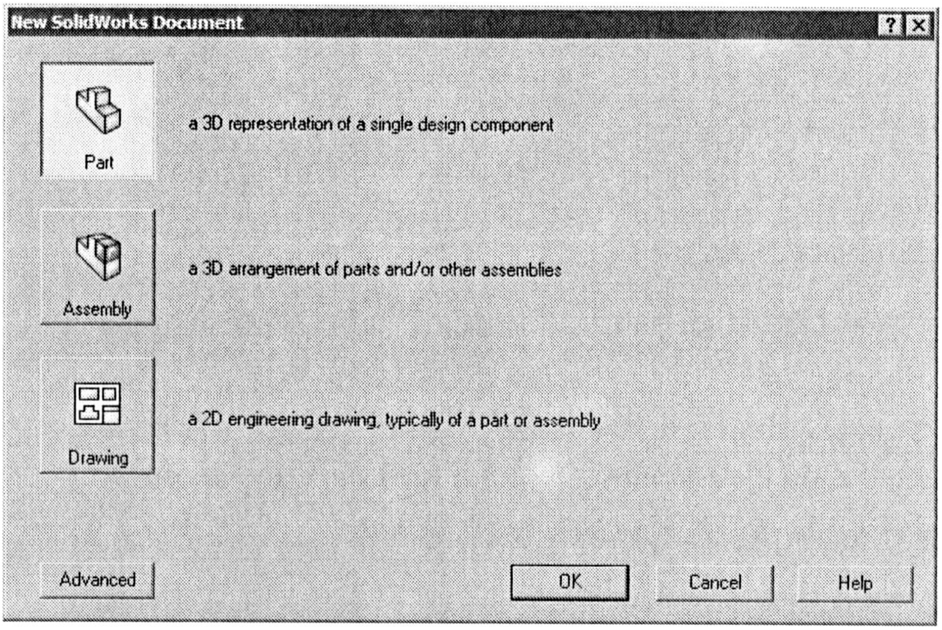

Figure 1-46 New SolidWorks Document *dialog box*

A new SolidWorks part document will be opened. But the part document window will not be maximized in the SolidWorks window.

5. Choose the **Maximize** button available on the upper right corner of the part document window to maximize the document window.

   When you open a new part document, the part modeling environment is active by default. As you first need to draw the sketch of the feature, you need to invoke the sketching environment.

6. Choose the **Sketch** button from the **Standard** toolbar. The **Edit Sketch PropertyManager** is displayed and you are prompted to select a plane on which you want to create the sketch.

7. Select the **Front** plane. The sketching environment is invoked and the plane is oriented normal to the view. You will notice that a red color origin is displayed in the center of the screen, indicating the sketching environment. The default screen appearance of the sketching environment of SolidWorks is shown in Figure 1-47.

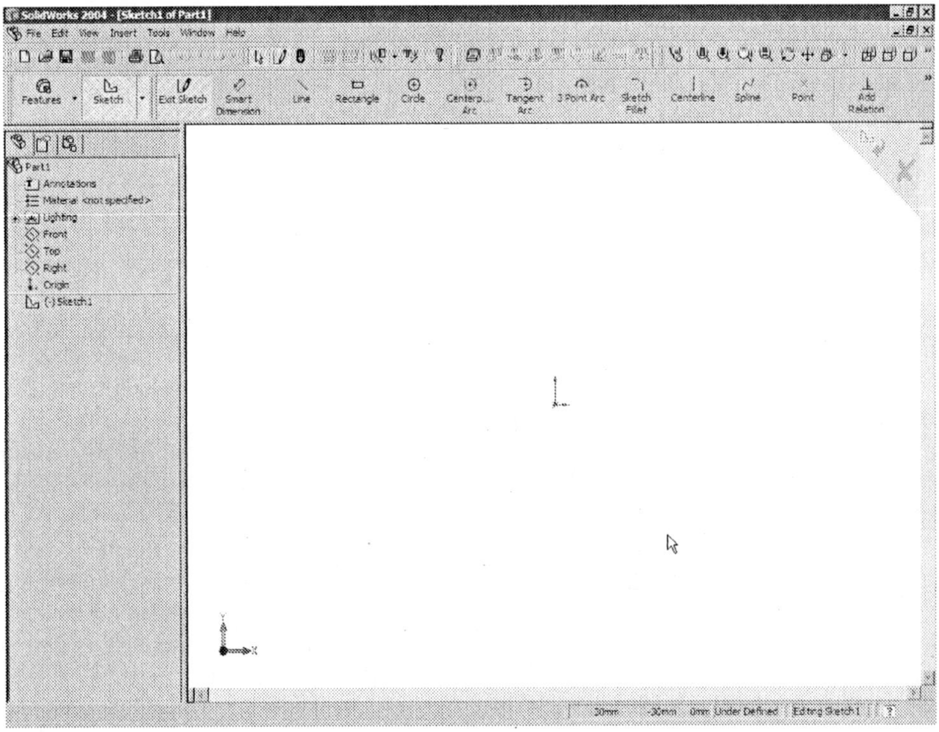

Figure 1-47 Screen display in the sketcher environment

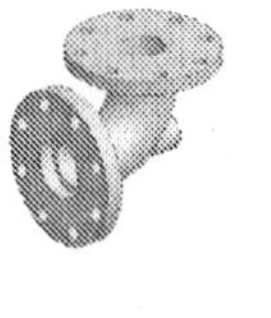

Tip. *If the grid is displayed on the screen when you invoke the sketching environment for the first time, then you can set the option to turn off the grid display. Right-click in the drawing area to display the shortcut menu. The* ***Display Grid*** *option has a check mark on its left, suggesting that this option is chosen. Select this option again to turn the grids off.*

Setting the Units and Grid

It is assumed that while installing SolidWorks, you selected the option of measuring the length in millimeters. This is the reason the length will be measured in millimeter in the current file. But if you selected some other unit, you need to make some initial settings of changing the linear and angular units before you proceed with drawing the sketch.

1. Choose **Tools > Options** from the menu bar to invoke the **System Options - General** dialog box.

2. Choose the **Document Properties** tab. The name of the dialog box is changed to the **Document Properties - Detailing** dialog box.

3. Select the **Units** option from the area on the left to display the options related to linear and angular units.

4. Select **millimeters** from the drop-down list available in the **Length units** area. Also, select the **Degrees** option from the drop-down list provided in the **Angular units** area.

Note

*If you selected **Millimeters** as the unit while installing SolidWorks, you can skip the points discussed earlier in this section.*

5. Select **Grid/Snap** from the area on the left. Set the value of the **Major grid spacing** spinner to **100** and the value of the **Minor-lines per major** spinner to **10**.

6. Select the **Snap to points** check box in the **Snap** area, if it is cleared. Choose **OK** to exit the dialog box.

Drawing the Outer Loop of the Sketch

It is a good practice to draw the sketch on one side of the origin, preferably in the first quadrant. This is because while generating the part program for manufacturing the part, you will have a reference for work origin in advance.

The sketch of the model consists of an outer loop, two circles inside the outer loop, and a cavity. Therefore, it will be drawn using the **Line** and the **Circle** tools. You will first draw the outer loop and then the inner entities. Note that in the sketcher environment, the lower right corner of the SolidWorks window displays three areas. The first area displays the X, Y, and Z coordinates of the current location of the cursor. These coordinates will be modified as you move the cursor around the drawing area. You will use the coordinate display to draw the sketch of the model.

You will start drawing the sketch from the lower left corner of the sketch and the outer loop will be drawn using the continuous lines.

1. Choose the **Line** button from the **Sketch CommandManager** to invoke the **Line** tool. The arrow cursor will be replaced by the line cursor.

2. Move the cursor in the first quadrant close to the origin. The coordinates of the point are displayed close to the lower right corner of the screen.

3. Press the left mouse button at the point whose coordinates are 10 mm 10 mm 0 mm and then move the cursor horizontally toward the right.

   You will notice that the symbol of **Horizontal** relation is displayed below the line cursor and the length of the line is displayed above the line cursor.

   As the length of the first horizontal line at the lower left corner is 10 mm, you will move the mouse until the length of the line is shown as 10 above the line cursor.

4. Press the left mouse button when the length of the line that is displayed above the line cursor shows a value of 10.

   The first horizontal line is drawn. Because you are drawing continuous lines, the endpoint of the last line is automatically selected as the start point of the next line.

5. Move the line cursor vertically upward. The symbol of **Vertical** relation is displayed below the line cursor and the length of the line is displayed above the line cursor.

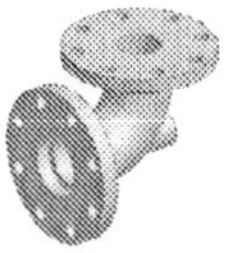

Tip. *If by mistake you invoke the arc mode while drawing lines, move the cursor back to the endpoint of the previous line and press the left mouse button. The line mode will be invoked again.*

6. Press the left mouse button when the length of the line displayed above the line cursor shows a value of 10.

   A vertical line of length 10 mm will be drawn and will be displayed in green color. Also, as this is the line that is selected now, the previous line will no more be highlighted and therefore will be displayed in blue color.

7. Move the line cursor horizontally toward the right. Press the left mouse button when the length of the line above the line cursor shows a value of 10. This draws the next horizontal line of 10 mm length.

8. Move the line cursor vertically downward and press the left mouse button when the length of the line on the line cursor shows a value of 10.

9. Move the line cursor horizontally toward the right and press the left mouse button when the length of the line on the line cursor shows a value of 30.

10. Move the line cursor vertically upward and press the left mouse button when the length of the line on the line cursor shows a value of 10.

11. Move the line cursor horizontally toward the right and press the left mouse button when the length of the line on the line cursor shows a value of 10.

12. Move the line cursor vertically downward and press the left mouse button when the length of the line on the line cursor shows a value of 10.

13. Move the line cursor horizontally toward the right and press the left mouse button when the length of the line on the line cursor shows a value of 10.

14. Move the line cursor vertically upward and press the left mouse button when the length of the line on the line cursor displays a value of 40.

    The next line that you will draw is an inclined line that makes an angle of 135-degree. To draw this line, you will move the cursor in a direction that makes an angle of 135-degree.

    The aligned length of the line is not given for this line. Instead, the delta X and delta Y values are given. Therefore, you will draw a line at this point such that the delta X and delta Y value of this line is 10 mm each. These values can be viewed in the **Line PropertyManager**. The last spinners in the **Line PropertyManager** are for the delta values.

15. Move the line cursor such that the line is drawn at an angle of 135-degree. The current angle will be displayed in the spinner above the delta spinners in the **Line PropertyManager**.

16. Press the left mouse button when the delta X value in the **Delta X** spinner shows a value of 10 and the value of delta Y in the **Delta Y** spinner displays a value of 10 in the **Line PropertyManager**. At this stage, the length of the line will be displayed as 14.14 above the line cursor and also in the **Length** spinner in the **Line PropertyManager**.

17. Move the line cursor horizontally toward the left and press the left mouse button when the length of the line on the line cursor displays a value of 50.

    You will notice that some blue and brown inferencing lines are displayed when you move the cursor.

18. Move the line cursor in the direction of 225-degree angle.

    You will notice that two blue inferencing lines are displayed. The first one originates from the start point of the first inclined line and the other one from the start point of the first line of the sketch.

19. Press the left mouse button at the point where the two inferencing lines intersect.

    You will notice that at this point, the length of the line on the line cursor shows a value of 14.14. Also, the delta X and delta Y values in the **Line PropertyManager** are displayed as 10 each and the angle is shown as 225-degree.

20. Move the cursor vertically downward to the start point of the first line.

You will notice that when you move the cursor close to the start point of the first line, a red circle is displayed. Also, the line cursor turns yellow in color and an orange-colored box is displayed below the line cursor. The length of the line shows a value of 40.

21. Press the left mouse button when the red circle is displayed. Right-click to display the shortcut menu and choose the **Select** option to exit the **Line** tool.

    This completes the sketch of the outer loop. Note that the display of the sketch is small. Therefore, you need to modify the drawing display area such that the sketch fits the screen. This is done using the **Zoom to Fit** tool.

22. Choose the **Zoom to Fit** button from the **View** toolbar to fit the current sketch on the screen.

    The outer loop of the sketch is completed and is shown in Figure 1-48.

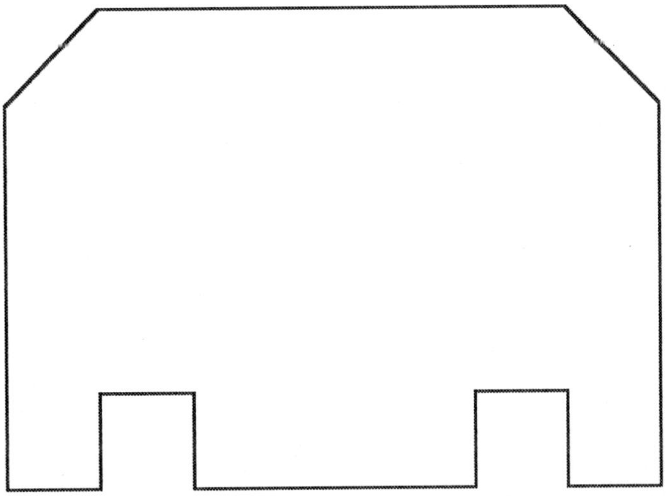

***Figure 1-48** Outer loop of the sketch*

Drawing Circles

The circles will be drawn using the **Circle** tool. You will use the inferencing line originating from the start points and endpoints of the inclined lines to specify the center point of the circles. As mentioned earlier, the cursor in the sketching environment jumps through a distance of 10 mm by default. Therefore, when you move the cursor to define the radius of the circle, the minimum value that can be set is 10 mm. But the radius of the circle is 5 mm and to specify this value, the cursor needs to jump only through a distance of 5 mm. This is the reason you need to modify the document settings so that the cursor jumps through a distance of 5 mm.

1. Choose **Tools > Options** from the menu bar to invoke the **System Options - General** dialog box. Choose the **Document Properties** tab.

2. Select the **Grid/Snap** option from the area on the left to display the options related to linear and angular units. Set the value of the **Major grid spacing** spinner to **50** and choose **OK**.

   Setting this value to 50 will force the cursor to jump through a distance of 5 mm instead of 10 mm. Therefore, when you move the cursor now, the values shown on the cursor and the coordinates of the points shown close to the lower right corner of the SolidWorks window will be in the increment of 5 mm.

3. Choose the **Circle** button from the **Sketch CommandManager** to invoke the **Circle** tool.

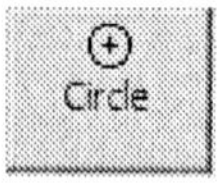

   As the **Select** tool was active earlier, the cursor earlier was the arrow cursor. But when you invoke the **Circle** tool, the arrow cursor will be replaced by the circle cursor.

4. Move the circle cursor close to the lower endpoint of the right inclined line and then move it toward the left. Remember that you will not press the left mouse button at this moment.

   An inferencing line is displayed generating from the lower endpoint of the right inclined line. When you move the cursor toward the left, you will notice that at the point where the cursor is vertically in line with the upper endpoint of the right inclined line, another inferencing line is generated from the upper endpoint of the right inclined line. This inferencing line will intersect the inferencing line generated from the lower endpoint of the inclined line.

5. Press the left mouse button at the point where the inferencing lines from both the endpoints of the inclined lines intersect. Now, move the circle cursor toward the left to define a circle.

6. Press the left mouse button when the radius of the circle displayed above the circle cursor shows a value of 5.

7. Similarly, draw the circle on the left using the inferencing lines generating from the endpoints of the left inclined line. The sketch after drawing the two circles inside the outer loop is shown in Figure 1-49.

8. Right-click in the drawing area and choose **Select** to exit the **Circle** tool.

Drawing the Sketch of the Inner Cavity

Next, you will draw the sketch of the inner cavity. To draw the sketch of the inner cavity, you will start drawing with the lower horizontal line.

1. Choose the **Line** button from the **Sketch CommandManager**. The arrow cursor is replaced by the line cursor.

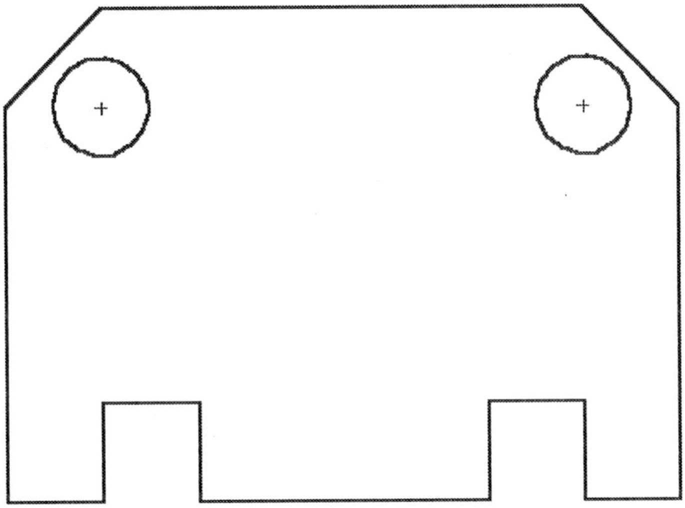

Figure 1-49 Sketch after drawing the two inner circles

2. Move the line cursor to a location whose coordinates are 30 mm 25 mm 0 mm.

3. Press the left mouse button at this point and move the cursor horizontally toward the right. Press the left mouse button when the length of the line above the line cursor shows a value of 30.

4. Move the line cursor vertically upward and press the left mouse button when the length of the line on the line cursor displays a value of 10.

5. Move the line cursor horizontally toward the left and press the left mouse button when the length of the line on the line cursor displays a value of 10.

6. Move the line cursor vertically downward and press the left mouse button when the length of the line on the line cursor displays a value of 5.

7. Move the line cursor horizontally toward the left and press the left mouse button when the length of the line on the line cursor displays a value of 10.

8. Move the line cursor vertically upward and press the left mouse button when the length of the line on the line cursor displays a value of 5.

9. Move the line horizontally toward the left and press the left mouse button when the length of the line on the line cursor displays a value of 10.

10. Move the line cursor vertically downward to the start point of the first line. Press the left mouse button when the red circle is displayed. The length of the line at this point will show a value of 10.

11. Right-click and choose **Select** from the shortcut menu. This completes the sketch for Tutorial 1. The final completed sketch for Tutorial 1 is shown in Figure 1-50.

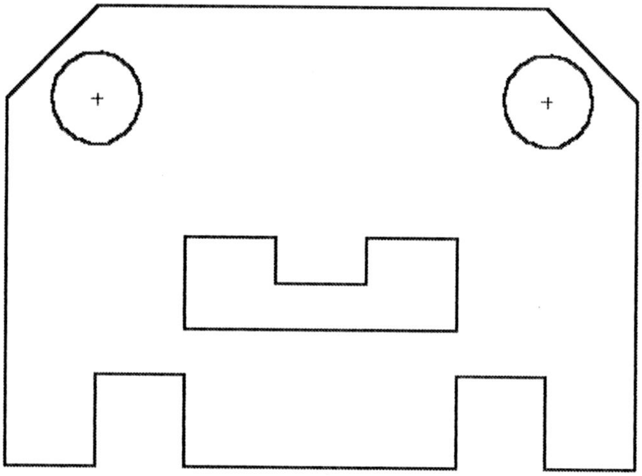

Figure 1-50 *Final sketch for Tutorial 1*

Note

By default, the entity points are displayed at the endpoints when you sketch an entity. The entity points are dots at the endpoints of lines, arcs, splines, and so on. If you do not want the entity points to be displayed then choose ***Tools*** *>* ***Options*** *from the menu bar to invoke the* ***System Options - General*** *dialog box. Select the* ***Sketch*** *option from the left of this dialog box to display the option related to sketch. Select the* ***Display entity points in part/assembly sketches*** *check box and choose the* ***OK*** *button from this dialog box.*

Saving the Sketch

It is recommended that you create a separate folder for saving the tutorial files of this book. When you invoke the option to save a document, the default folder that is displayed is */My Documents*. You will create a folder with the name **SolidWorks** in the *My Documents* folder and then create the folder of each chapter inside the *SolidWorks* folder. As a result, you can save the tutorials of a chapter in the folder of that chapter.

1. Choose the **Save** button from the **Standard** toolbar to invoke the **Save As** dialog box. Create the *SolidWorks* folder inside the */My Documents* folder and then create the *c01* folder inside the *SolidWorks* folder.

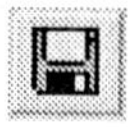

2. Enter the name of the document as *c01-tut01.sldprt* in the **File name** edit box and choose the **Save** button. The document will be saved in the */My Documents/SolidWorks/c01* folder.

3. Close the document by choosing **File > Close** from the menu bar.

Tip. *If you open a document that was saved in the sketching environment, it will be opened in the sketching environment only and not in the part modeling environment.*

Tutorial 2

In this tutorial, you will draw the basic sketch for the revolved solid model shown in Figure 1-51. The sketch for the revolved solid model is shown in Figure 1-52. Do not dimension the sketch as the solid model and the dimensions are given only for your reference.

(Expected time: 30 min)

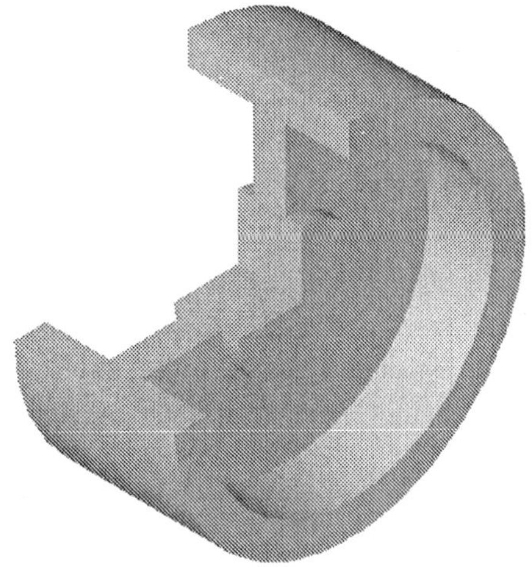

Figure 1-51 *Revolved model for Tutorial 2*

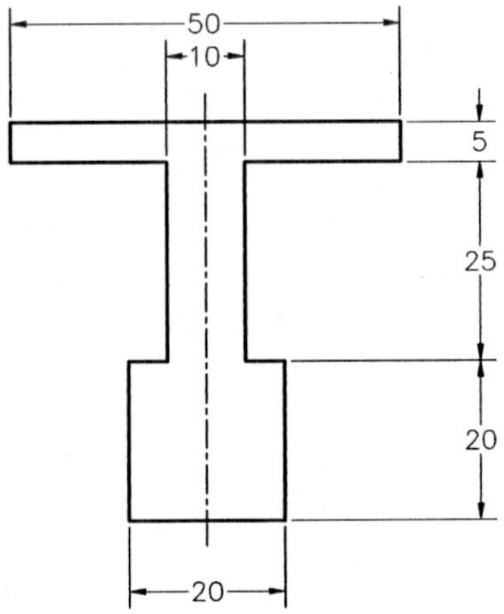

Figure 1-52 *Sketch for the revolved model*

The steps that will be followed to complete this tutorial are listed below.

a. Start a new part document.
b. Maximize the part document and then switch to the sketching environment.
c. Modify the settings of the snap and grid so that the cursor jumps through a distance of 5 mm instead of 10 mm.
d. Draw the sketch of the model using the **Line** tool, refer to Figure 1-53.
e. Save the sketch and then close the document.

Starting a New Document

1. Choose the **New** button from the **Standard** toolbar to invoke the **New SolidWorks Document** dialog box.

2. The **Part** button is chosen by default in the **New SolidWorks Document** dialog box. Choose **OK**.

   A new SolidWorks part document is opened. But the part document window is not maximized in the SolidWorks window.

3. Choose the **Maximize** button available on the upper right corner of the part document window to maximize the document window.

   As mentioned earlier, when you open a new part document, the part modeling environment is active by default. But because you first need to draw the sketch of the revolved model, you need to invoke the sketching environment.

4. Choose the **Sketch** button from the **Standard** toolbar. The **Edit Sketch PropertyManager** is displayed. Select the **Front** plane.

   A red color origin is displayed and the **Sketch CommandManager** is displayed. Also, the confirmation corner is displayed with the **Exit Sketch** and the **Delete Sketch** options on the upper right corner of the drawing area. This suggests that the sketching environment is activated.

Modifying the Snap and Grid Settings and the Dimensioning Units

Before you proceed with drawing the sketch, you need to modify the grid and snap settings so that you can make the cursor jump through a distance of 5 mm instead of 10 mm, which is the default value.

1. Choose **Tools > Options** from the menu bar to invoke the **System Options - General** dialog box. Choose the **Document Properties** tab.

2. Select the **Grid/Snap** option from the area on the left to display the options related to linear and angular units. Set the value of the **Major grid spacing** spinner to **50**. Make sure the value of the **Minor-lines per major** spinner is **10**. Choose **OK** to close the dialog box.

The coordinates close to the lower left corner of the SolidWorks window will show an increment of 5 mm instead of the default increment of 10 mm when you exit the dialog box.

3. Make sure the **Snap to points** check box in the **Snap** area is selected.

   If you selected a unit other than millimeter to measure the length while installing SolidWorks, you need to change the unit for the current drawing.

4. Select the **Units** option from the area on the left of the **Document Properties - Grid/ Snap** dialog box.

5. Select **millimeters** from the drop-down list available in the **Length units** area and **Degrees** from the drop-down list available in the **Angular units** area. Choose the **OK** button after making the necessary settings.

Drawing the Sketch

As evident from Figure 1-52, the sketch will be drawn using the **Line** tool. You will start drawing the sketch from the lower left corner of the sketch.

1. Choose the **Line** button from the **Sketch CommandManager**. The arrow cursor will be replaced by the line cursor.

2. Move the line cursor to a location whose coordinates are 40 mm 0 mm 0 mm.

   An inferencing line originating from the origin is displayed.

3. Press the left mouse button down at this point and move the cursor horizontally toward the right. Press the left mouse button again when the length of the line above the line cursor shows a value of 20.

4. Move the line cursor vertically upward and press the left mouse button when the length of the line on the line cursor displays a value of 20.

5. Move the cursor horizontally toward the left and press the left mouse button when the length of the line on the line cursor displays a value of 5.

6. Move the line cursor vertically upward and press the left mouse button when the length of the line on the line cursor displays a value of 25.

7. Move the line cursor horizontally toward the right and press the left mouse button when the length of the line on the line cursor displays a value of 20.

8. Move the line cursor vertically upward and press the left mouse button when the length of the line on the line cursor displays a value of 5.

9. Move the line cursor horizontally toward the left and press the left mouse button when the length of the line on the line cursor displays a value of 50.

10. Move the line cursor vertically downward and press the left mouse button when the length of the line on the line cursor displays a value of 5.

11. Move the line cursor horizontally toward the right and press the left mouse button when the length of the line on the line cursor displays a value of 20.

12. Move the line cursor vertically downward and press the left mouse button when the length of the line on the line cursor displays a value of 25.

13. Move the line cursor horizontally toward the left and press the left mouse button when the length of the line on the line cursor displays a value of 5.

14. Move the line cursor vertically downward to the start point of the first line. Press the left mouse button when the red circle is displayed.

    The length of the line at this point will be 20 mm

15. Right-click and choose **Select** from the shortcut menu.

    The sketch is completed but does not fit the screen. Therefore, you need to modify the display area such that the sketch fits the screen.

16. Choose the **Zoom to Fit** button from the **View** toolbar to fit the sketch on the screen. The completed sketch for Tutorial 2 is shown in Figure 1-53.

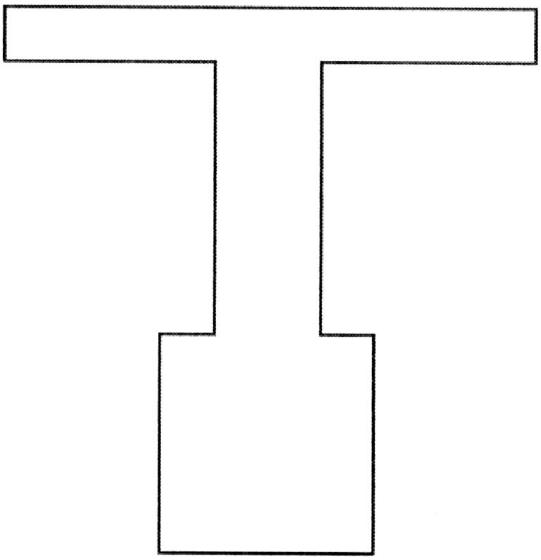

***Figure 1-53** Final sketch for Tutorial 2*

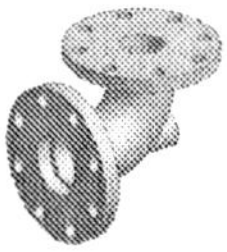

Tip. *You will notice that the bottom horizontal line in the sketch is black in color and the remaining lines are blue in color. In the next chapter, you will learn about the reason why some entities in the sketch have a different color.*

Saving the Sketch

1. Choose the **Save** button from the **Standard** toolbar to invoke the **Save As** dialog box. 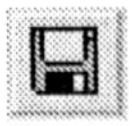

2. Enter the name of the document as *c01-tut02.sldprt* in the **File name** edit box and choose the **Save** button.

   The document will be saved in the */My Documents/SolidWorks/c01* folder.

3. Close the document by choosing **File > Close** from the menu bar.

Tutorial 3

In this tutorial you will draw the basic sketch of the model shown in Figure 1-54. The sketch to be drawn is shown in Figure 1-55. Do not dimension the sketch as the solid model and the dimensions are given only for your reference. **(Expected time: 30 min)**

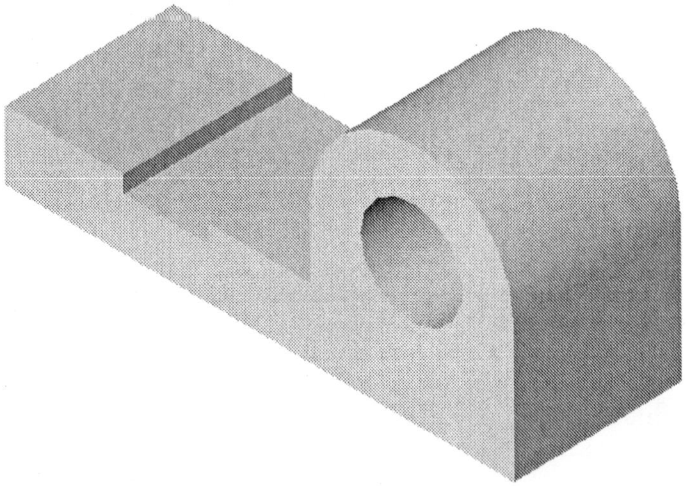

Figure 1-54 *Solid model for Tutorial 3*

The steps that will be followed to complete this tutorial are listed below.

a. Start SolidWorks and then start a new part file.
b. Maximize the part file document and then switch to the sketching environment.
c. Modify the settings of the snap and grid so that the cursor jumps through a distance of 5 mm instead of 10 mm.
d. Draw the outer loop of the sketch using the **Line** tool, refer to Figure 1-56.
e. Draw the inner circle using the **Circle** tool, refer to Figure 1-57
f. Save the sketch and then close the file.

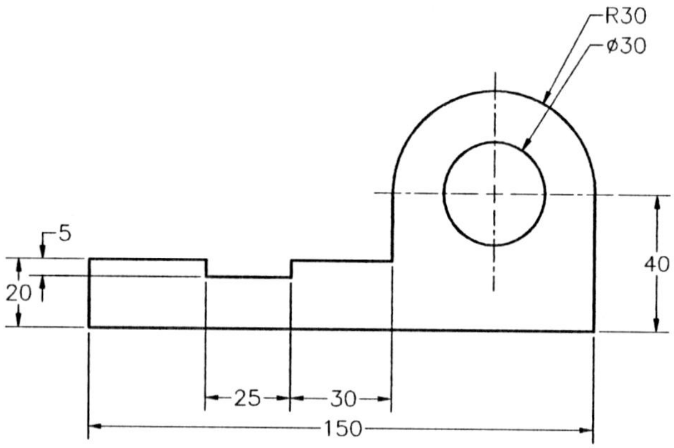

Figure 1-55 Sketch for Tutorial 3

Starting a New File

1. Choose the **New** button from the **Standard** toolbar to invoke the **New SolidWorks Document** dialog box.

2. The **Part** button is chosen by default in the **New SolidWorks Document** dialog box. Choose the **OK** button.

   A new SolidWorks part document will be opened. However, as mentioned earlier, the part document window will not be maximized in the SolidWorks window.

3. Choose the **Maximize** button available on the upper right corner of the part document window to maximize the document window.

   As you first need to draw the sketch of the revolved model, you need to invoke the sketching environment.

4. Choose the **Sketch** button from the **Standard** toolbar. The **Edit Sketch PropertyManager** is displayed. Select the **Front** plane.

   A red color origin is displayed and the **Sketch CommandManager** is displayed above the drawing area. Also, the confirmation corner with the **Exit Sketch** and **Delete Sketch** options is displayed on the upper right corner of the drawing area. This suggests that the sketching environment is activated.

Modifying the Snap and Grid Settings and Dimensioning Units

As the dimensions in the sketch are in multiples of 5 mm, you need to modify the grid and snap settings so that you can make the cursor jump through a distance of 5 mm instead of 10 mm.

1. Choose **Tools > Options** from the menu bar to invoke the **System Options - General** dialog box. Choose the **Document Properties** tab.

2. Select the **Grid/Snap** option from the area on the left to display the options related to linear and angular units. Set the value of the **Major grid spacing** spinner to **50** and choose **OK**. Make sure the value of the **Minor-lines per major** spinner is **10**.

   The coordinates displayed close to the lower left corner of the SolidWorks window will show an increment of 5 mm instead of the default increment of 10 mm when you close the dialog box.

3. If you selected a unit other than millimeter to measure the length while installing SolidWorks, select the **Units** option from the area on the left of the **Document Properties - Grid/Snap** dialog box. Select **millimeters** from the drop-down list available in the **Length units** area and **Degrees** from the drop-down list available in the **Angular units** area.

4. Choose **OK** to close the dialog box.

Drawing the Outer Loop

As evident from Figure 1-55, the sketch consists of an outer loop and an inner circle. Therefore, this sketch will be drawn using the **Line** and the **Circle** tools. You will start drawing from the lower left corner of the sketch. As the length of the lower horizontal line is 150 mm, you need to modify the drawing display area such that the drawing area in the first quadrant is increased. This can be done using the **Pan** tool.

1. Choose the **Pan** button from the **View** toolbar. The arrow cursor will be replaced by the pan cursor.

2. Press and hold the left mouse button down and drag the cursor toward the bottom left corner of the screen.

   You will notice that the origin also moves toward the bottom left corner of the screen, thus increasing the drawing area in the first quadrant.

3. After dragging the origin close to the lower left corner, release the left mouse button.

4. Choose the **Line** button from the **Sketch CommandManager**.

   The pan cursor will be replaced by the line cursor.

5. Move the line cursor to a location whose coordinates are 40 mm 0 mm 0 mm.

   An inferencing line originating from the origin will be displayed.

6. Press the left mouse button down at this point and move the cursor horizontally toward the right. Press the left mouse button again when the length of the line above the line cursor shows a value of 150.

7. Move the line cursor vertically upward and press the left mouse button when the length of the line on the line cursor displays a value of 40.

   The next entity that has to be drawn is a tangent arc. As mentioned earlier, you can draw the tangent arc using the **Line** tool also. Drawing arcs from within the **Line** tool is a recommended method when you need to draw a sketch that is a combination of lines and arcs. This increases the productivity by reducing the time taken in invoking the tools for drawing an arc and then invoking the **Line** tool to draw lines.

8. Move the line cursor away from the endpoint of the last line and then move it back to the endpoint.

   The arc mode is invoked and the line cursor is replaced by the arc cursor. Also, the **Arc PropertyManager** is displayed in place of the **Line PropertyManager**.

9. Move the arc cursor vertically upward to a small distance. A dotted reference line is displayed.

10. When the dotted line is displayed, move the cursor toward the left.

    You will notice that a tangent arc is being drawn. The angle of the tangent arc and its radius are displayed above the arc cursor.

11. Press the left mouse button when the angle value above the arc cursor shows 180 and the radius shows a value of 30 to complete the arc.

    The required tangent arc is drawn. As mentioned earlier, the line mode is automatically invoked after you have drawn the arc using the **Line** tool. Therefore, the arc cursor will be replaced by the line cursor and the **Arc PropertyManager** will be replaced by the **Line PropertyManager**.

12. Move the line cursor vertically downward and press the left mouse button when the length of the line on the line cursor displays a value of 20.

13. Move the line cursor horizontally toward the left and press the left mouse button when the length of the line on the line cursor displays a value of 30.

14. Move the line cursor vertically downward and press the left mouse button when the length of the line on the line cursor displays a value of 5.

15. Move the line cursor horizontally toward the left and press the left mouse button when the length of the line on the line cursor displays a value of 25.

16. Move the line cursor vertically upward and press the left mouse button when the length of the line on the line cursor displays a value of 5.

17. Move the line cursor horizontally toward the left and press the left mouse button when the length of the line on the line cursor displays a value of 35.

18. Move the line cursor to the start point of the first line. Press the left mouse button when the red circle is displayed.

    The length of the line at this point will be 20 mm

19. Right-click and choose **Select** from the shortcut menu to exit the **Line** tool.

20. Choose the **Zoom to Fit** button to fit the sketch on the screen. This completes the outer loop of the sketch. The sketch after drawing the outer loop is shown in Figure 1-56.

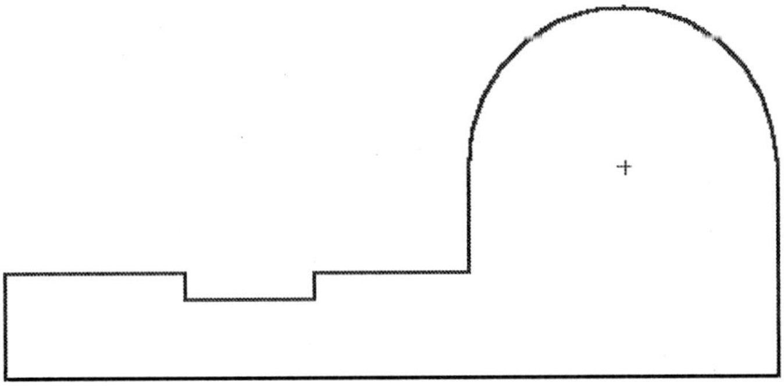

Figure 1-56 *Sketch after drawing the outer loop*

Drawing the Circle

The circle in the sketch will be drawn using the **Circle** tool. The centerpoint of the circle will be the centerpoint of the arc, which is displayed by a plus sign. This plus sign is automatically drawn when you draw the arc. You can select this center point to draw the circle.

1. Choose the **Circle** button from the **Sketch CommandManager** to invoke the **Circle** tool.

   The arrow cursor will be replaced by the circle cursor.

2. Move the circle cursor close to the centerpoint of the arc and press the left mouse button when the red circle is displayed.

3. Move the cursor toward the left and when the radius of the circle above the circle cursor shows a value of 15, press the left mouse button. A circle of 15 mm radius is drawn.

4. This completes the sketch for Tutorial 3. Right-click and choose the **Select** option from the shortcut menu to exit the **Circle** tool.

   The final sketch for Tutorial 3 is shown in Figure 1-57.

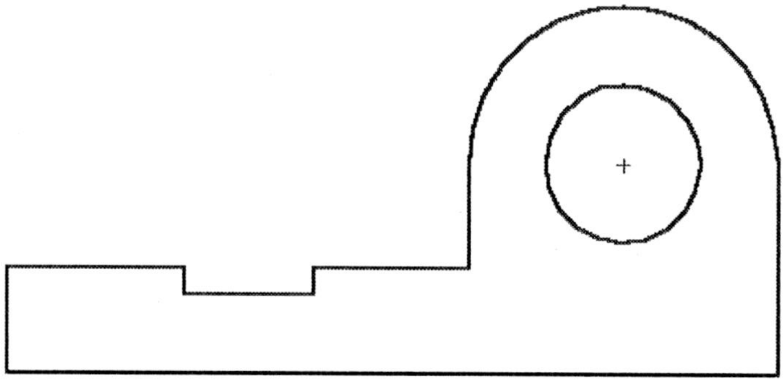

Figure 1-57 *Final sketch for Tutorial 3*

Saving the Sketch

1. Choose the **Save** button from the **Standard** toolbar to invoke the **Save As** dialog box.

2. Enter the name of the document as *c01-tut03.sldprt* in the **File name** edit box and choose the **Save** button.

3. Close the document by choosing **File > Close** from the menu bar.

Tutorial 4

In this tutorial, you will draw the sketch of the model shown in Figure 1-58. The sketch for the model is shown in Figure 1-59. Do not dimension the sketch as the dimensions and the solid model are given only for your reference. **(Expected time: 30 min)**

The steps that will be followed to complete this tutorial are listed below.

a. Start SolidWorks and then start a new part document.
b. Maximize the part document and then switch to the sketching environment.
c. Modify the settings of the snap and grid so that the cursor jumps through a distance of 5 mm instead of 10 mm.
d. Draw the sketch using the **Line** tool, refer to Figure 1-60.
e. Save the sketch and then close the file.

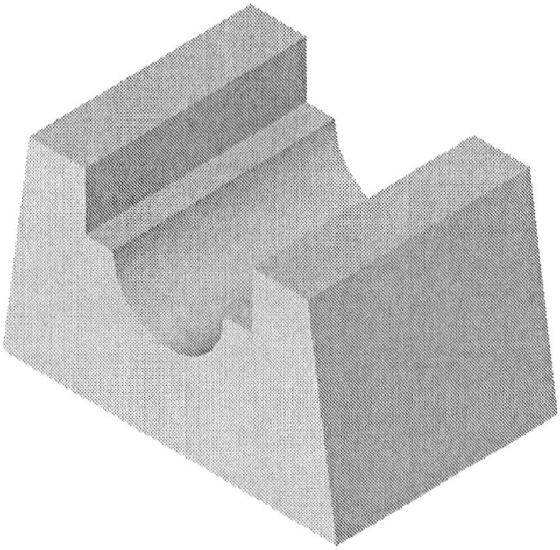

Figure 1-58 Model for Tutorial 4

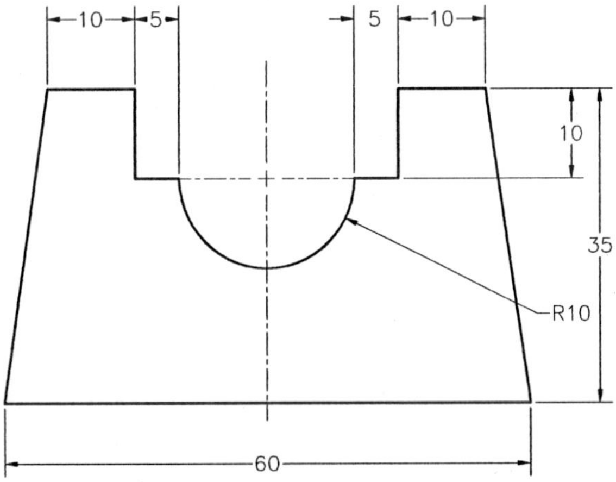

Figure 1-59 Sketch for Tutorial 4

Opening a New File

1. Choose the **New** button from the **Standard** toolbar to invoke the **New SolidWorks Document** dialog box.

2. The **Part** button is chosen by default in the **New SolidWorks Document** dialog box. Choose the **OK** button.

3. Choose the **Maximize** button available on the upper right corner of the part document window to maximize the document window.

As you first need to draw the sketch of the model, you need to invoke the sketching environment.

4. Choose the **Sketch** button from the **Standard** toolbar and select the **Front** plane to invoke the sketching environment.

Modifying the Snap and Grid Settings and Dimensioning Units

As evident in Figure 1-59, the dimensions in the sketch are in the multiples of 5 mm. Therefore, you need to modify the grid and snap settings so that the cursor jumps through a distance of 5 mm instead of 10 mm.

1. Choose **Tools > Options** from the menu bar to invoke the **System Options - General** dialog box. Choose the **Document Properties** tab.

2. Select the **Grid/Snap** option from the area on the left to display the options related to linear and angular units. Set the value of the **Major grid spacing** spinner to **50** and choose **OK**. Make sure the value of the **Minor-lines per major** spinner is **10**.

   The coordinates displayed close to the lower left corner of the SolidWorks window shows an increment of 5 mm when you close this dialog box.

3. If you selected a unit other than millimeter to measure the length while installing SolidWorks, select the **Units** option from the area on the left of the **Document Properties - Grid/Snap** dialog box. Select **millimeters** from the drop-down list available in the **Length units** area and **Degrees** from the drop-down list available in the **Angular units** area.

4. Choose **OK** to close the dialog box.

Drawing the Sketch

The sketch will be drawn using the **Line** tool. The arc in the sketch will also be drawn using the same tool. You will start drawing from the lower left corner of the sketch.

1. Choose the **Line** button from the **Sketch CommandManager**.

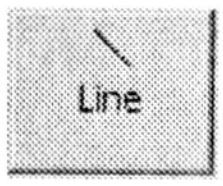

   The arrow cursor is replaced by the line cursor.

2. Move the line cursor to a location whose coordinates are 30 mm 0 mm 0 mm.

   An inferencing line originating from the origin is displayed.

3. Press the left mouse button at this point and move the cursor horizontally toward the right. Press the left mouse button again when the length of the line above the line cursor shows a value of 60.

   The bottom horizontal line of 60 mm length is drawn.

4. Choose the **Zoom to Fit** button from the **View** toolbar to increase the display of the line that is drawn.

As mentioned earlier, you can also invoke the drawing display tools while you are inside some other tool. After modifying the drawing display, the tool that was active before invoking the drawing display tool will be restored and you can continue using that tool. Therefore, after the drawing display area is modified, the **Line** tool will be restored and you can continue drawing lines.

5. Move the line cursor along a direction that makes an angle close to 98-degree with the positive X axis direction. The angle can be checked from the spinner above the **Delta X** spinner in the **Line PropertyManager**.

6. When the **Delta X** spinner shows a value of **5** and **Delta Y** spinner shows a value of **35**, press the left mouse button.

The length of the line displayed above the line cursor at this point will be 35.36.

7. Move the line cursor horizontally toward the left and press the left mouse button when the length of the line above the line cursor shows a value of 10.

8. Move the line cursor vertically downward and press the left mouse button when the length of the line above the line cursor shows a value of 10.

9 Move the line cursor horizontally toward the left and press the left mouse button when the length of the line above the line cursor shows a value of 5.

Next, you need to draw the arc that is normal to the last line.

10. Move the line cursor away from the endpoint of the last line and then move it back close to the endpoint.

The arc mode will be invoked and the line cursor will be replaced by the arc cursor. Also, the **Line PropertyManager** will be replaced by the **Arc PropertyManager**.

11. Move the arc cursor vertically downward through a small distance.

A dotted line will be displayed.

12. When the dotted line is displayed, move the arc cursor toward the left.

You will notice that a normal arc is being drawn and the angle and radius of the arc is displayed above the arc cursor.

13. Press the left mouse button when the angle value on the arc cursor is 180 and the radius value is 10. An arc normal to the last line will be drawn and the line mode will be activated.

14. Move the line cursor horizontally toward the left and press the left mouse button when the length of the line on the line cursor shows a value of 5.

15. Move the line cursor vertically upward and press the left mouse button when the length of the line on the line cursor shows a value of 10.

16. Move the line cursor horizontally toward the left and press the left mouse button when the length of the line on the line cursor shows a value of 10.

17. Move the line cursor to the start point of the first line and press the left mouse button when the red circle is displayed.

18. Right-click to display the shortcut menu and choose **Select** to exit the **Line** tool.

    This completes the sketch. However, you need to modify the drawing display area such that the sketch fits the screen.

19. Choose the **Zoom to Fit** button from the **View** toolbar to modify the drawing display area. The final sketch for Tutorial 4 is shown in Figure 1-60.

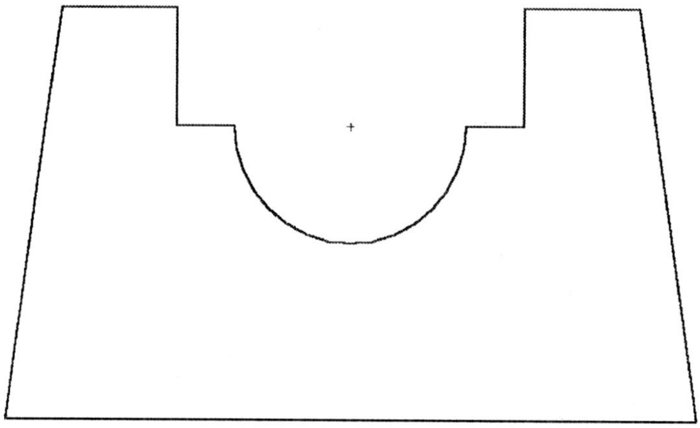

Figure 1-60 *Final sketch for Tutorial 4*

Saving the Sketch

1. Choose the **Save** button from the **Standard** toolbar to invoke the **Save As** dialog box.

2. Enter the name of the document as *c01-tut04.sldprt* in the **File name** edit box and choose the **Save** button.

3. Close the document by choosing **File > Close** from the menu bar.

SELF-EVALUATION TEST

Answer the following questions and then compare your answers with the answers given at the end of this chapter.

1. The base feature of any design is a sketched feature and is created by drawing the sketch. (T/F)

2. You can also invoke the **3Pt Arc** tool from within the **Line** tool. (T/F)

3. By default, the cursor jumps through a distance of 5 mm. (T/F)

4. When you save a file in the sketching environment, it is opened in the part modeling environment when you open it for the next time. (T/F)

5. You can convert a sketched entity into a construction entity by selecting the __________ check box provided in the **PropertyManager**.

6. To draw a rectangle at an angle, you need to use the __________ tool.

7. The __________ are the temporary lines that are used to track a particular point on the screen.

8. You can also invoke the __________ tool or exit a sketching tool by pressing the ESC key.

9. When you select a tangent entity to draw a tangent arc, the __________ relation is applied between the start point of the arc and the tangent entity.

10. The rectangle is considered as a combination of individual __________.

REVIEW QUESTIONS

Answer the following questions.

1. The 3 point arcs are the ones that are drawn by defining the start point of the arc, the endpoint of the arc, and a point on the arc. (T/F)

2. You can also delete the sketched entities by selecting them and choosing the **Delete** option from the shortcut menu that is displayed upon right-clicking. (T/F)

3. The origin is a blue-colored icon that is displayed in the middle of the sketcher screen. (T/F)

4. In SolidWorks, circles are drawn by specifying the centerpoint of the circle and then entering the radius of the circle in the dialog box that is displayed. (T/F)

5. When you open a new SolidWorks document, it is not maximized in the SolidWorks window. (T/F)

6. In SolidWorks, a polygon is considered as a combination of which of the following entities.

   (a) Lines (b) Arcs
   (c) Splines (d) None

7. Which one of the following options is not displayed in the **New SolidWorks Document** dialog box?

   (a) **Part** (b) **Assembly**
   (c) **Drawing** (d) **Sketch**

8. Which one of the following entities will not be considered while converting a sketch into a feature?

   (a) Sketched circles (b) Sketched lines
   (c) Construction lines (d) None

9. When you select a line of the rectangle, which one of the following **PropertyManager** will be displayed?

   (a) **Line PropertyManager** (b) **Line/Rectangle PropertyManager**
   (c) **Rectangle PropertyManager** (d) None

10. While drawing an elliptical arc, which one of the following **PropertyManager** will be displayed?

    (a) **Arc PropertyManager** (b) **Ellipse PropertyManager**
    (c) **Elliptical Arc PropertyManager** (d) None

EXERCISES

Exercise 1

Draw the sketch of the model shown in Figure 1-61. The sketch to be drawn is shown in Figure 1-62. Do not dimension the sketch as the solid model and the dimensions are given only for your reference. **(Expected time: 30 min)**

Figure 1-61 Solid model for Exercise 1

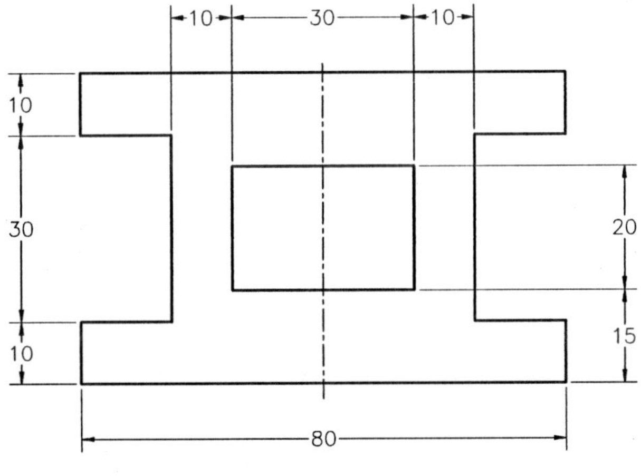

Figure 1-62 Sketch for Exercise 1

Exercise 2

Draw the sketch of the model shown in Figure 1-63. The sketch to be drawn is shown in Figure 1-64. Do not dimension the sketch as the solid model and the dimensions are given only for your reference. **(Expected time: 30 min)**

Figure 1-63 Solid model for Exercise 2

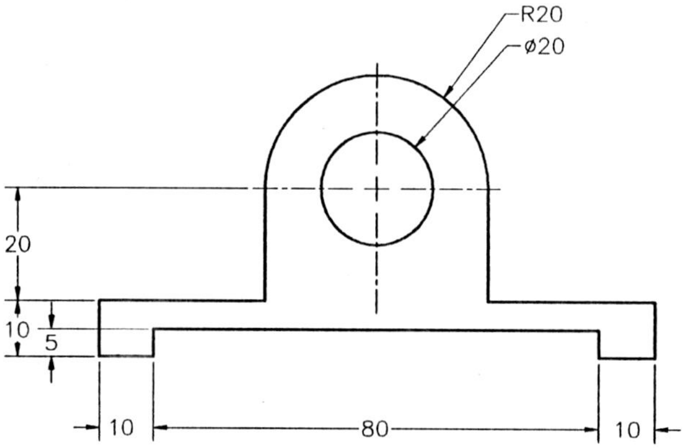

Figure 1-64 Sketch for Exercise 2

Answers to Self-Evaluation Test

1. T, **2.** T, **3.** F, **4.** F, **5. For Construction**, **6. Parallelogram**, **7.** inferencing lines, **8. Select**, **9. Tangent**, **10.** lines

Chapter 2

Editing and Modifying the Sketches

Learning Objectives

After completing this chapter you will be able to:

- *Edit sketches using various editing tools.*
- *Create rectangular patterns of sketched entities.*
- *Create circular patterns of sketched entities.*
- *Write text in the sketching environment.*
- *Modify sketch entities.*
- *Modify sketches by dynamically dragging the sketched entities.*

EDITING THE SKETCHED ENTITIES

SolidWorks provides you with a number of tools that can be used to edit the sketched entities. These options include trimming the sketched entities, extending the sketched entities, offsetting the sketched entities, mirroring the sketched entities, and so on.

Trimming the Sketched Entities

CommandManager:	Sketch > Trim Entities
Menu:	Tools > Sketch Tools > Trim
Toolbar:	Sketch > Trim Entities

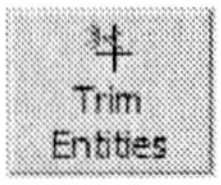

The **Trim Entities** tool is used to trim the unwanted entities in a sketch. You can use this tool to trim a line, arc, ellipse, parabola, circle, spline, or centerline that is intersecting a line, arc, circle, ellipse, parabola, spline, or centerline. To use the trim option, choose the **Trim Entities** button from the **Sketch CommandManager**. The cursor will be replaced by the trim cursor; move the trim cursor near to the portion of the sketched entity to be removed. If the entity will be completely deleted using the trim option, then that entity will be highlighted in red. But if a portion of the entity will be removed, then that portion will be highlighted in black. Press the left mouse button to remove the highlighted entity. Figures 2-1 and 2-2 show the sketch before and after trimming the entities.

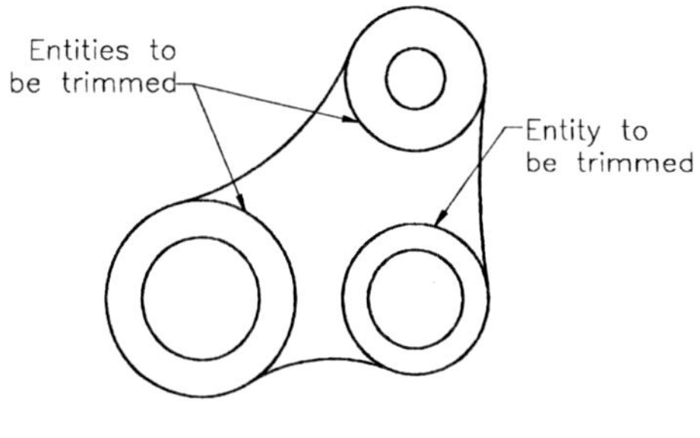

***Figure 2-1** Entities to be selected for trimming*

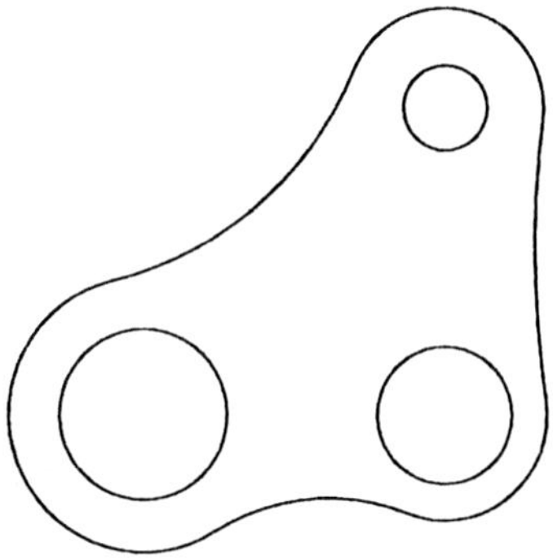

***Figure 2-2** Sketch after trimming the entities*

Using the Trim Option to Extend a Sketched Entity

You can also use the **Trim Entities** tool to extend the sketched entities. To do this, choose the **Trim Entities** button from the **Sketch CommandManager** and move the trim cursor to the entity to be extended. When the sketched entity turns red or black in color, drag the cursor to the entity up to which the selected entity is to be extended. You can notice the preview of the extended entity in yellow color. Release the left mouse button when the preview of the extended entity appears.

Tip. *In SolidWorks all the buttons are not displayed by default in the toolbars or* ***CommandManagers****. Therefore, you have to customize them according to your need and specifications. The next topic is Extending the Sketched Entities. The tool button for the* ***Extend*** *option is not available in the* ***Sketch Tools*** *toolbar. To insert the* ***Extend Entities*** *button in the* ***Sketch CommandManager****, you need to follow the procedure given as under:*

1. Choose ***Tools > Customize*** *from the menu bar to display the* ***Customize*** *dialog box.*
2. Choose the ***Commands*** *tab from the* ***Customize*** *dialog box.*
3. Choose ***Sketch*** *from the* ***Categories*** *area of the* ***Customize*** *dialog box.*
4. Press and hold down the left mouse button to select the ***Extend Entities*** *button from the* ***Buttons*** *area of the* ***Customize*** *dialog box. You can view the description of the* ***Extend Entities*** *button in the* ***Description*** *area.*
5. Drag the mouse to the ***Sketch CommandManager*** *and release the left mouse button to place the* ***Extend Entities*** *button on this* ***CommandManager****.*
6. Choose ***OK*** *from the* ***Customize*** *dialog box.*

You can insert the other tool buttons in the ***Sketch CommandManager*** *or any other CommandManager by following the same procedure. To remove the tool button from any of the CommandManager, invoke the* ***Customize*** *dialog box and using the left mouse button select the tool button that you want to remove and drag it in the graphics area.*

When you customize and add a tool to the ***CommandManager****, it is automatically added to the toolbar also.*

Extending the Sketched Entities

CommandManager:	Sketch > Extend Entities	*(Customize to Add)*
Menu:	Tools > Sketch Tools > Extend	
Toolbar:	Sketch > Extend Entities	*(Customize to Add)*

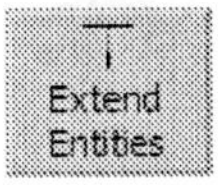

The **Extend Entities** tool is used to extend the sketched entity to intersect the next available entity. The tool is used to extend a line, arc, ellipse, parabola, circle, spline, or center line to intersect a line, arc, circle, ellipse, parabola, spline, or center line. The sketched entity is extended up to its intersection with another sketched entity or a model edge. Choose the **Extend Entities** button from the **Sketch CommandManager** and move the extend cursor close to the portion of the sketched entity to be extended. The entity to be extended is displayed in black and the preview of the extended portion of the entity is displayed in yellow. Press the left mouse button to complete the extend operation. Figure 2-3 shows the sketched entities before extending and Figure 2-4 shows the sketched entities after extending.

Tip. *If the preview of the sketched entity to be extended is shown in the wrong direction, move the extend cursor to a position on the other half of the entity and observe the new preview.*

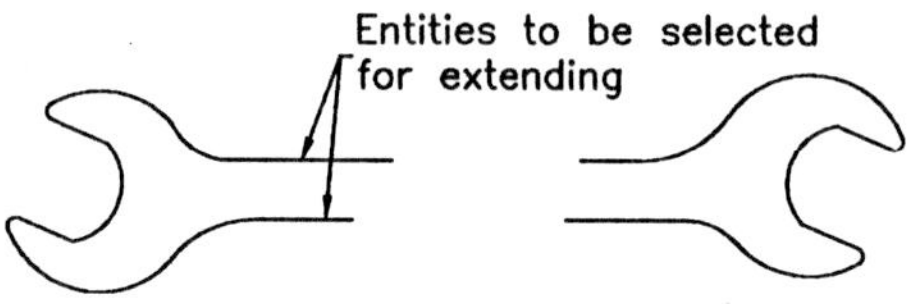

Figure 2-3 Sketched entities before extending

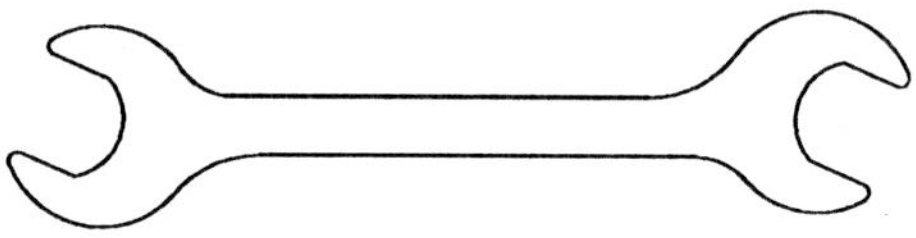

Figure 2-4 Sketched entities after extending

Tip. *You can toggle between the **Trim Entities** and **Extend Entities** tools using the shortcut menu that is displayed upon right-clicking when one of these tools is active.*

Filleting the Sketched Entities

CommandManager:	Sketch > Sketch Fillet
Menu:	Tools > Sketch Tools > Fillet
Toolbar:	Sketch > Sketch Fillet

Fillet creates a tangent arc at the intersection of two sketched entities. It trims or extends the entities to be filleted, depending on the geometry of the sketched entity. You can apply fillet between two nonparallel lines, between two arcs, between two splines, between an arc and a line, between a spline and a line, and between a spline and an arc. The fillet between two arcs, or between an arc and a line depends on the compatibility of the geometry to be extended or filleted along a given radius. You can invoke the **Sketch Fillet** tool first and then select the entities to be filleted. You can also hold down the CTRL key and first select two entities to be filleted and then invoke this tool. When you invoke this tool, the **Sketch Fillet PropertyManager** is displayed, as shown in Figure 2-5. Set the value of the **Fillet Radius** spinner and press ENTER or choose the **Apply** button from the **Sketch Fillet** dialog box. You can also select the sketched entities after invoking the **Sketch Fillet PropertyManager**.

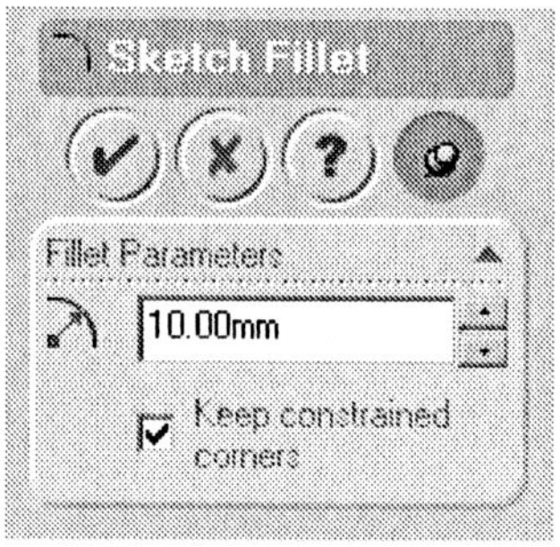

Figure 2-5 Sketch Fillet PropertyManager

You can also select the nonintersecting entities for creating fillet. While filleting the nonintersecting entities the selected entities will be extended to form a fillet. If the **Keep constrained corners** check box is selected, the dimension and geometric relations applied to the sketch will not be deleted. If you clear the **Keep constrained corners** check box, you will be prompted to delete the relations applied to the corners of the sketched entities to be filleted. The **Undo** button is displayed in the **Sketch Fillet PropertyManager** when you create at least one sketch fillet. Figures 2-6 and 2-7 show the sketched entities before and after applying the radius.

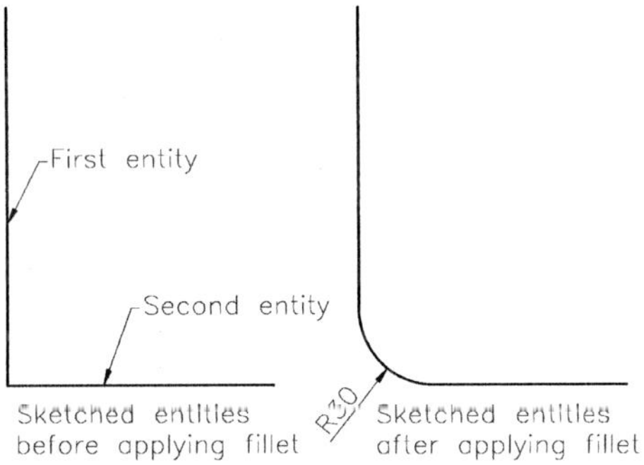

Figure 2-6 *Sketched entities before and after creating fillet.*

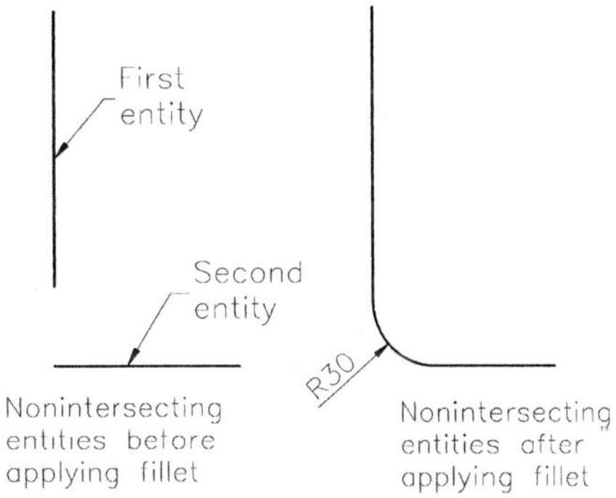

Figure 2-7 *Sketched entities before and after creating fillet.*

Note

The consecutive fillets with the same radius are not dimensioned individually; an automatic equal radii relation is applied to all the fillets.

The fillet creation between two splines, between a spline and a line, and between a spline and an arc depends on the compatibility of the spline to be trimmed or extended.

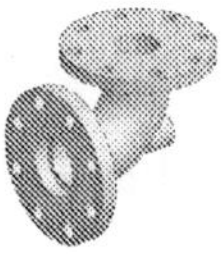

Tip. *The other method of creating a sketched fillet between two entities is to drag a window around the entities to be filleted. To fillet the entities using this method, invoke the* ***Sketch Fillet PropertyManager****. Set the fillet radius spinner to the required value. Now, drag the cursor to create a window around the two entities to be filleted such that they are enclosed inside the window. As soon as you release the left mouse button, the fillet will be created between the two selected entities.*

You can also select a vertex to fillet the two entities that form that vertex.

Chamfering the Sketched Entities

CommandManager:	Sketch > Sketch Chamfer	*(Customize to Add)*
Menu:	Tools > Sketch Tools > Chamfer	
Toolbar:	Sketch > Sketch Chamfer	*(Customize to Add)*

The **Sketch Chamfer** tool is used to apply a chamfer to adjacent sketch entities. The chamfer can be specified by angle-distance or distance-distance options. You can apply chamfer between two nonparallel lines; the lines may be intersecting lines or nonintersecting lines. The creation of chamfer between two nonintersecting lines depends on the length of the lines and the chamfer distance. To create chamfer, choose the **Sketch Chamfer** button from the **Sketch CommandManager** and select two entities to be chamfered. You can also select the two entities before invoking the **Chamfer** tool. When you invoke this tool, the **Sketch Chamfer PropertyManager** is displayed, shown in Figure 2-8. The options available in the **Sketch Chamfer PropertyManager** are discussed next.

Figure 2-8 Sketch Chamfer PropertyManager

Angle-distance

The **Angle-distance** radio button is selected to create the chamfer by specifying the angle and the distance, refer to Figure 2-9. When you select this radio button, the **Direction 1 Angle** spinner is displayed below the **Distance 1** spinner. The **Distance 1** spinner is used to specify the distance and the **Direction 1 Angle** spinner is used to specify the angle. Note that the angle will be measured from the first entity that you selected.

Distance-distance

The **Distance-distance** radio button is selected to create the chamfer by specifying the distances along both the selected lines, refer to Figure 2-9. When you invoke the **Sketch Chamfer PropertyManager**, the **Distance-distance** radio button and the **Equal distance** check box are selected by default. Therefore, you can create an equal distance chamfer between the selected entities. The distance value can be specified in the **Distance 1** spinner.

Equal distance

The **Equal distance** check box is selected to specify an equal distance in both the directions for creating the sketched chamfer. This check box is selected by default. If you clear this check box, you can specify two different distances for creating the chamfer, refer to Figure 2-9. When you clear this check box the **Distance 2** spinner is displayed below the **Distance 1** spinner to set the value of distance in the second direction. Note that the distance 1 value will be measured along the first entity that you selected and distance 2 is measured along the second entity that you selected.

You can also use the shortcut menu that is invoked to use the various methods of creating chamfer.

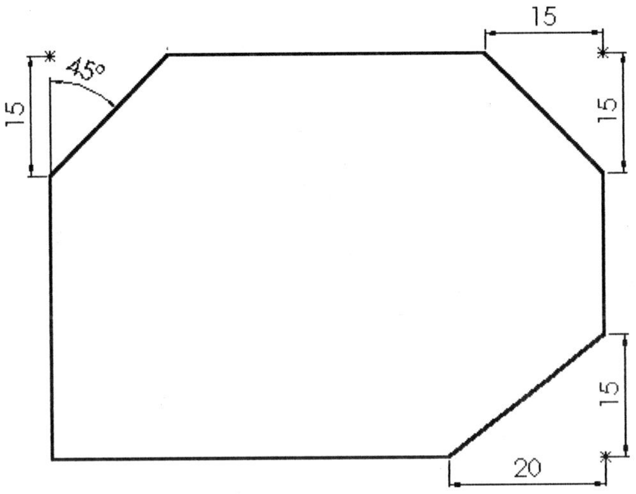

Figure 2-9 Creating various types of chamfers

Note

If you apply a sketch chamfer to the entities that are constrained using some relations or dimensions, the ***SolidWorks*** *warning box will be displayed. This window will inform you* ***At least one sketch constraint is about to be lost, Chamfer anyway?*** *Choose the* ***Yes*** *button from this warning box. You will learn more about the relations and dimensions that constrain the sketch in later chapters.*

Offsetting the Sketched Entities

CommandManager:	Sketch > Offset Entities
Menu:	Tools > Sketch Tools > Offset Entities
Toolbar:	Sketch > Offset Entities

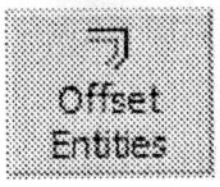

Offsetting is one of the easiest methods of drawing parallel lines or concentric arcs, and circles. You can select the entire chain of the entities as a single entity or select an individual entity to offset. You can offset selected sketched entities, edges, loops, faces, curves. When you choose the **Offset Entities** button from the **Sketch Tools** toolbar the **Offset Entities PropertyManager** is displayed as shown in Figure 2-10. The various options available in the **Offset Entities PropertyManager** are discussed next.

Note

While performing any kind of editing operation, if you want to clear the current selection set, right-click in the drawing area and choose ***Clear Selections*** *from the shortcut menu.*

Offset Distance

The **Offset Distance** spinner is used to set the distance through which the selected entity needs to be offset. You can set the value of the offset distance using this spinner or by dragging the offset entity in the drawing area.

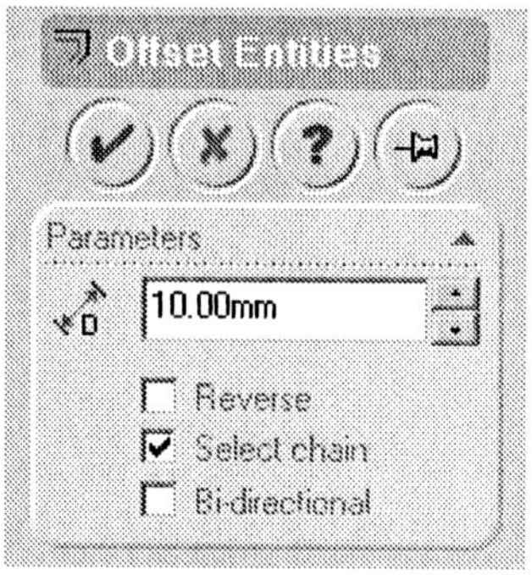

Figure 2-10 Offset Entities PropertyManager

Reverse

The **Reverse** check box is selected to change the direction of offset. Note that while offsetting the entities by dragging, you do not need this check box. This is because you can define the direction of offset by dragging the entities in that direction.

Select Chain

The **Select Chain** check box is selected to select the entire chain of continuous sketched entities that are in contact with the selected entity. When you invoke the offset tool, the **Select Chain** check box is selected by default. If you clear this check box, then only the selected sketched entity will be offset.

Bi-directional

The **Bi-directional** radio button is selected to create offset entities in two directions. If the **Bi-directional** check box is selected, then the **Reverse** check box will not be available in the **Parameters** area of the **Offset Entities PropertyManager**.

Figure 2-11 shows a new chain of entities created by offsetting the chain of entities. Figure 2-12 shows the offsetting of a single entity.

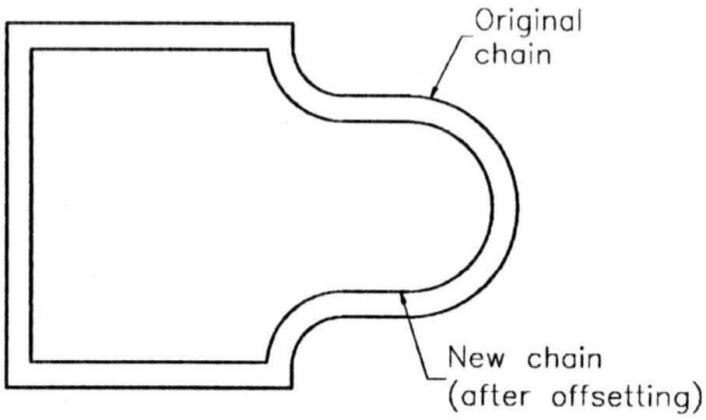

Figure 2-11 Offsetting a chain of entities

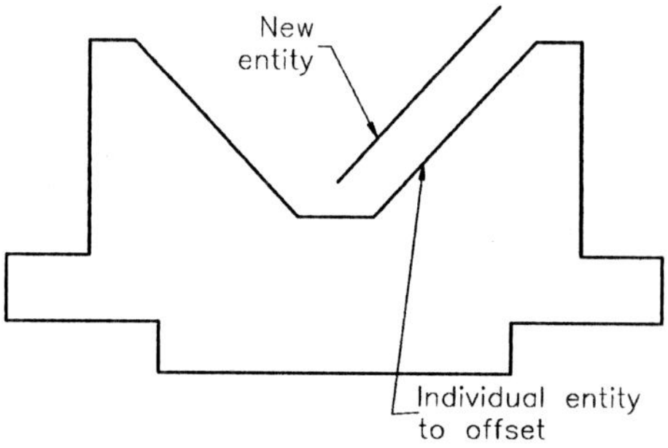

Figure 2-12 Offsetting a single entity

Note

Whenever you offset an entity, the offset distance between the old and the new entities is displayed on the screen in the form of a dimension.

Mirroring the Sketched Entities

CommandManager:	Sketch > Mirror Entities
Menu:	Tools > Sketch Tools > Mirror
Toolbar:	Sketch > Mirror Entities

The **Mirror Entities** tool is used to create the mirror image of the selected entities. The entities are mirrored about a center line. When you create the mirrored entity, SolidWorks applies the symmetric relation between the sketched entities. If you change the entity, its mirror image will also change. There are two methods to create the mirrored image of a sketched entity. They are discussed next.

Mirroring the Existing Items

Create the sketch of the entities to be mirrored using the normal sketch tools in the sketching environment. Choose the **Centerline** button from the **Sketch CommandManager** and create a centerline about which the sketched entities will be mirrored. Now, hold down the CTRL key and using the left mouse button, select the entities to be mirrored and the centerline about which the entities will be mirrored. You can also select the entities by dragging a window around the sketched entities and the centerline. After selecting the entities, invoke the **Mirror Entities** tool from the **Sketch CommandManager**. The sketched entities are mirrored about the centerline. If you modify the sketch of the sketched entity, the same will be reflected on the mirror image. Figure 2-13 shows the sketched entities with the centerline and Figure 2-14 shows the resultant mirror image of the sketched entities.

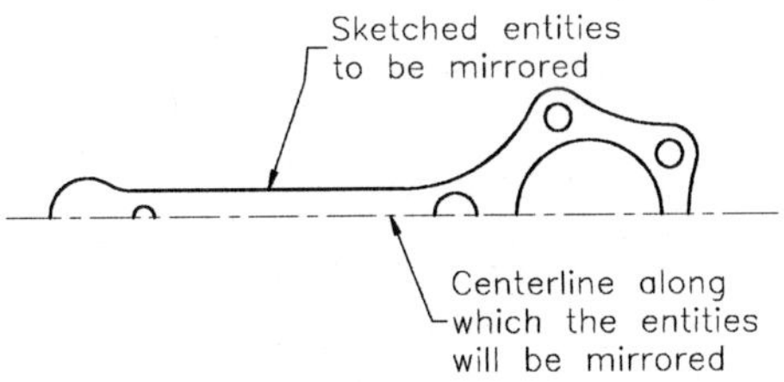

***Figure 2-13** Selecting the sketched entities and the centerline*

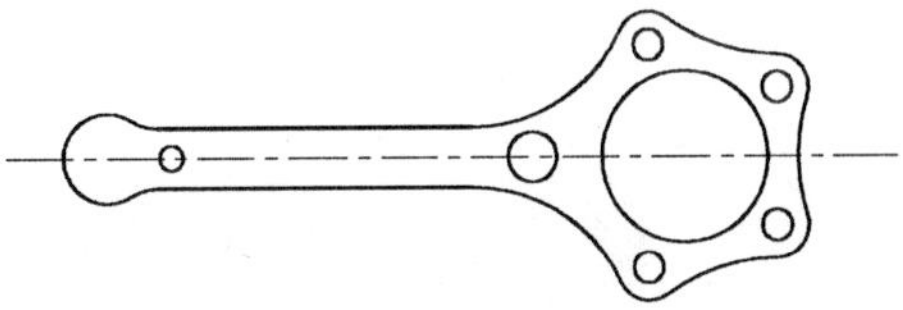

***Figure 2-14** Sketch after mirroring the geometry*

Mirroring Items While Sketching (Automatic Mirroring)

This method is recommended when you are creating symmetric sketches. To use this method, create a centerline along which the sketched entities will be mirrored. Select the centerline and choose the **Mirror Entities** button from the **Sketch CommandManager**. The symmetry symbols appear at both ends of the centerline to indicate that automatic mirroring is activated, as shown in Figure 2-15. Now, start creating the sketch. The sketched entity that you create on one side of the centerline will automatically be created on the other side of the mirror line (centerline).

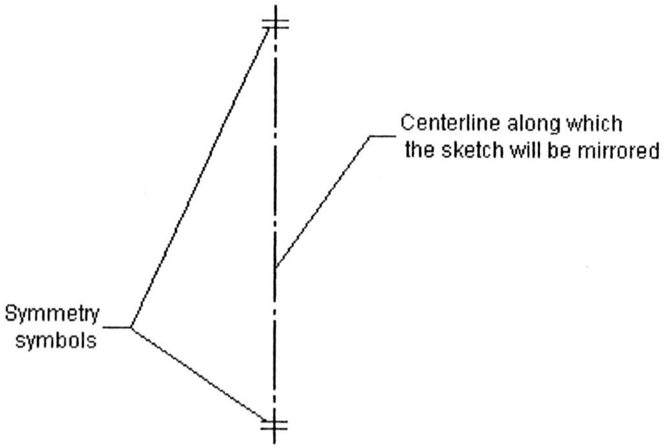

Figure 2-15 *The centerline and the symmetry symbols that indicate that automatic mirroring is activated*

As evident in Figure 2-16, the entities are mirrored automatically while sketching.

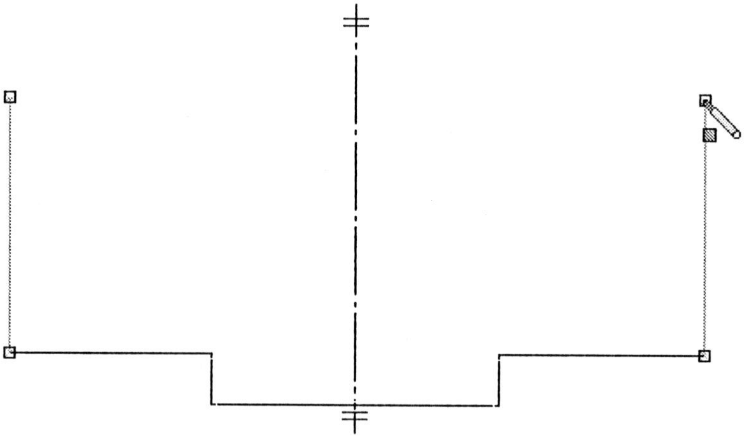

Figure 2-16 *Sketching using automatic mirroring active*

Figure 2-17 shows the complete sketch with automatic mirroring.

Note

After completing the sketch using the automatic mirroring option, you can clear this option by choosing the ***Mirror Entities*** *button from the* ***Sketch CommandManager****.*

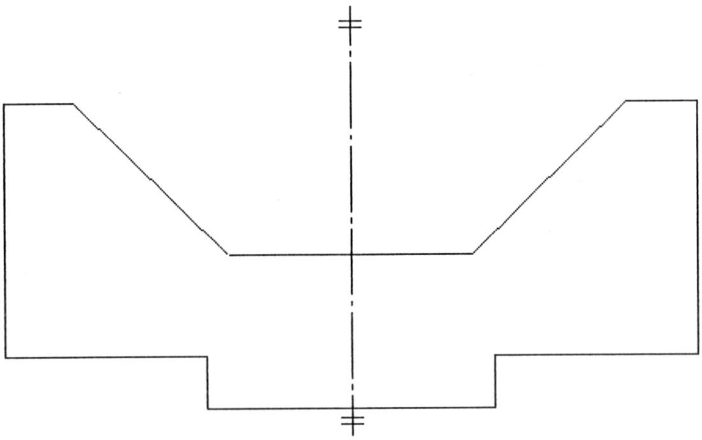

Figure 2-17 *Sketch drawn with automatic mirroring*

MOVING, ROTATING, AND SCALING THE SKETCHED ENTITIES

One of the latest enhancements in SolidWorks 2004 is the tool that allows you to move, rotate, or scale the selected sketched entities. All these operations can be performed using the **Move or Copy Entities** tool. You can also make a copy of the sketched entities while you are moving, rotating, or scaling the entities. When you invoke this tool, the **Move or Copy PropertyManager** is displayed. You can select the option to move, rotate, or scale the selected entities in this **PropertyManager**. However, the menu bar provides you separate options to invoke any of these three operations. You can choose any of these operations from the **Tools > Sketch Tools** menu. You can also invoke any of these operations using the shortcut menu that is displayed when you select the entities and right-click.

The various operations that can be performed using the **Move or Copy Entities** tool are discussed next.

Moving or Copying the Sketched Entities

Command Manager:	Sketch > Move or Copy
Menu:	Tools > Sketch Tools > Move or Copy
Toolbar:	Sketch > Move or Copy

To move or copy the sketched entities, invoke this tool. Remember that this tool will be available only when at least one sketched entity is drawn. When you invoke this tool, the **Move or Copy PropertyManager** is displayed, as shown in Figure 2-18 and you are prompted to select sketch items or annotations. The options available in this **PropertyManager** are discussed next.

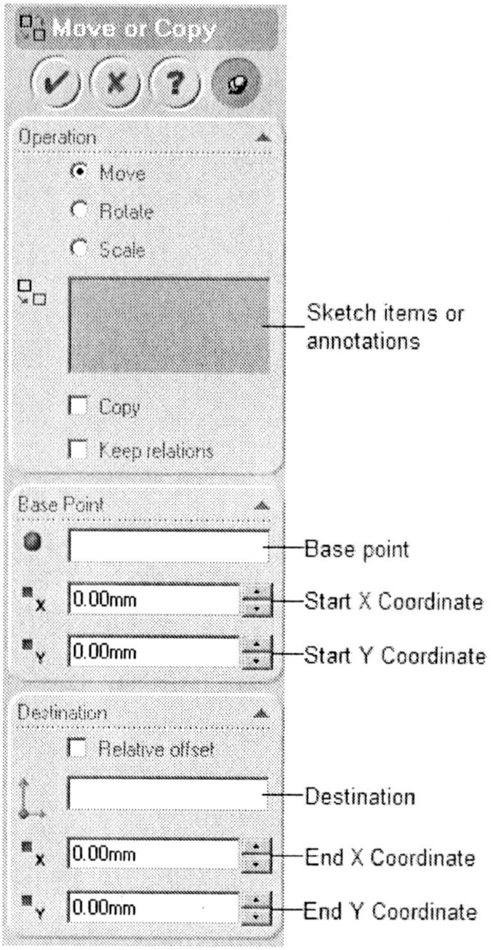

Figure 2-18 Move or Copy PropertyManager

Operation Rollout

The options in the **Operation** rollout are used to select the operation to be performed and to select the entities. You can first invoke the **Move or Copy PropertyManager** and then select the entities to be moved or copied. The selected entities are displayed in the **Sketch items or annotations** selection area. After selecting all the entities, right-click to accept the selection set.

You can select the **Copy** check box if you want to copy the selected entities. When you select this check box, the **Number of Copies** spinner is displayed below this check box. You can specify the number of copies to be created using this spinner. You can select the **Keep relations** check box from this area to retain the relations applied to the selected entities even after performing this operation. This check box is not available if you select the **Copy** check box.

Base Point Rollout

The options in this rollout are used to specify the base point for moving or copying the selected entities. After selecting the entities to be moved or copied, as soon as you right-click, you are prompted to specify the base point of the move. Also, the base point icon is attached to the arrow cursor. You can define the base point by specifying a point on the screen or by entering the coordinates of the base point in the **Start X Coordinate** and **Start Y Coordinate** spinners available in this area. If you want to use the spinners, set the values in the spinners and then press the TAB key to accept the values. The base point icon is automatically fixed at the base point that you define.

If you want to redefine the base point, right-click in the drawing area and choose **Define Base Point**. The previous selection is cleared and you are prompted to specify the base point again.

Destination Rollout

The options in this rollout are used to specify the destination where the entities will be placed after moving or copying. As soon as you define the base point, this rollout is activated and you are prompted to specify the destination of the move. Also, the destination icon is attached to the arrow cursor. As you move the cursor on the screen, the selected entities also move dynamically. You can define the destination point by specifying a point on the screen or use the **End X Coordinate** and **End Y Coordinate** spinners to enter the values.

If you select the **Relative offset** check box, the **Base Point** rollout will no more be displayed and the **Destination** rollout also shows **Delta X** and **Delta Y** spinners. By selecting this check box, you can define the displacement along X and Y directions. All the selected entities will be moved in the X and Y directions through the distances specified in the two spinners.

If you selected the **Copy** check box in the **Operation** rollout, the preview of the specified number of copies will be displayed and the gap between the copies is defined by the difference between the base point and the destination point.

Figure 2-19 shows the selected entities being moved using the **Move or Copy** tool. In this figure, the upper and the right edge of a rectangle is selected to be moved.

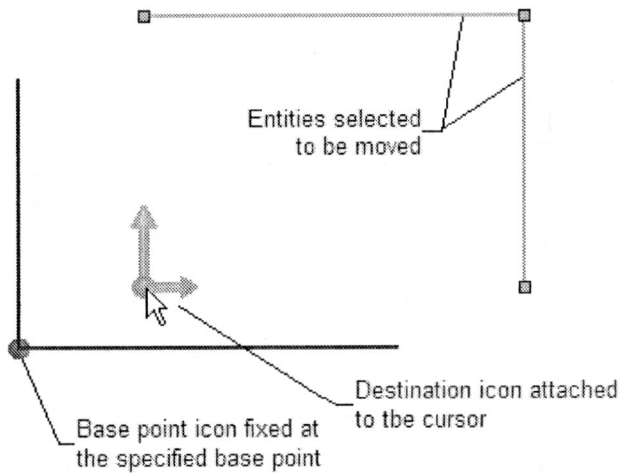

***Figure 2-19** Moving the selected entities*

Rotating or Copying the Sketched Entities

Command Manager:	Sketch > Rotate or Copy *(Customize to Add)*
Menu:	Tools > Sketch Tools > Rotate or Copy
Toolbar:	Sketch > Rotate or Copy *(Customize to Add)*

To rotate or copy the sketched entities, invoke the **Rotate or Copy** tool. You can also select the entities and right-click to display the shortcut menu. In this menu, choose **Rotate or Copy** to invoke this tool. When you invoke this tool, the **Move or Copy PropertyManager** is displayed, as shown in Figure 2-20, and you are prompted to select sketch items or annotations.

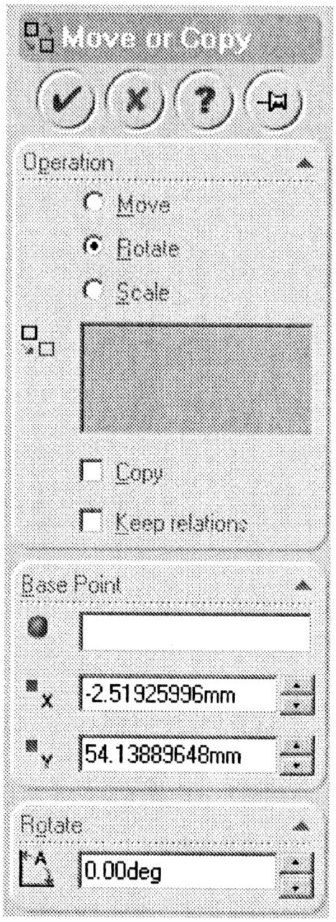

***Figure 2-20 Move or Copy PropertyManager** for the **Rotate** tool*

The entities that you select to rotate are displayed in the **Sketch items or annotations** selection area of the **Operation** rollout. After selecting the entities to rotate, right-click. You will be prompted to specify the base point about which to rotate. You can use the options in the **Base Point** rollout. These options are similar to those discussed while moving the entities.

As soon as you specify the base point, the **Angle** spinner in the **Rotate** rollout is highlighted. You can specify the angle of rotation using this spinner. You can also drag the mouse on the screen to define the angle of rotation.

Figure 2-21 shows a rectangle being rotated by dragging the cursor to define the angle of rotation. The lower right vertex of the rectangle is taken as the base point of rotation.

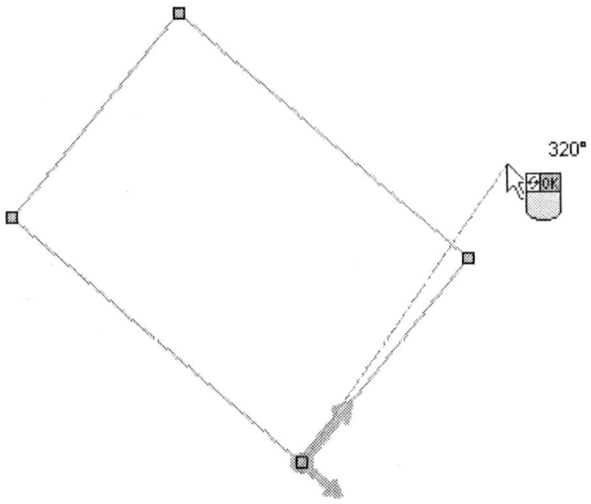

Figure 2-21 *Defining the rotation angle for the rectangle by dragging the mouse*

Scaling or Copying the Sketched Entities

Command Manager:	Sketch > Scale or Copy *(Customize to Add)*
Menu:	Tools > Sketch Tools > Scale or Copy
Toolbar:	Sketch > Scale or Copy *(Customize to Add)*

To scale or copy the sketched entities, invoke the **Scale or Copy** tool using the menu bar or using the shortcut menu. The **PropertyManager** that is displayed is similar to the one that is displayed for rotating. The only difference is that the **PropertyManager** for scaling has the **Scale Factor** spinner in the **Operation** rollout. This spinner is used to specify the scale factor for scaling the selected entities. Note that in this case, the **Keep relations** check box is not available.

After you select the entities to be scaled, define the base point. Next, specify the scale factor using the spinner and then right-click to complete this operation.

CREATING PATTERNS

While sketching the base feature of a model, sometimes you may need to place the sketched entities in a particular arrangement such as along linear edges or around a circle. For example, refer to Figures 2-22 and 2-23. These figures show the base features with the slots. These slots are

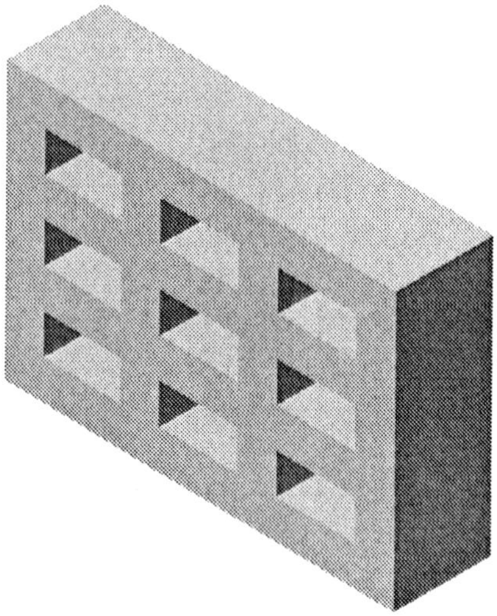

Figure 2-22 *Base feature with slots created about linear edges*

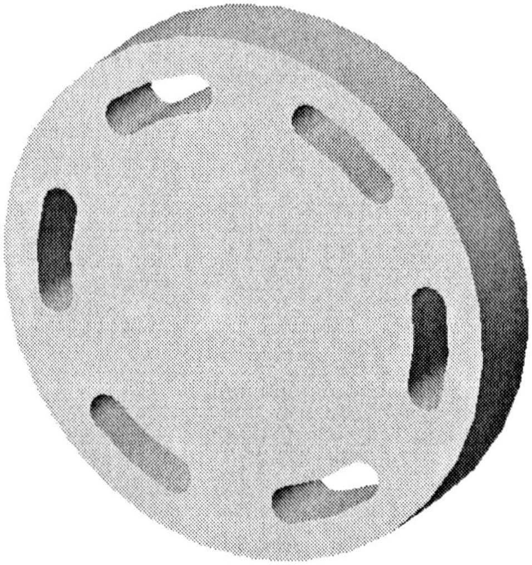

Figure 2-23 *Base feature with slots created around a circle*

created with the help of linear and circular patterns of sketched entities. The tools that are used to create linear and circular patterns of sketched entities are discussed next.

Creating Linear Pattern of the Sketched Entities

CommandManager:	Sketch > Linear Sketch Step and Repeat *(Customize to Add)*
Menu:	Tools > Sketch Tools > Linear Sketch Step and Repeat
Toolbar:	Sketch > Linear Sketch Step and Repeat *(Customize to Add)*

In SolidWorks, the linear pattern of the sketched entities is created using the **Linear Sketch Step and Repeat** tool. To create the linear pattern, select the sketched entities using the **Select** tool and then choose this button from the **Sketch CommandManager**. The **Linear Sketch Step and Repeat** dialog box will be displayed as shown in Figure 2-24 and the preview of the linear pattern will be shown on the screen in the background. Also, the arrow cursor is replaced by the linear pattern cursor. Note that if you have not selected the sketched entities to pattern before invoking this tool, you will have to select them one by one using the linear pattern cursor. You cannot define a window to select more than one entity using the linear pattern cursor.

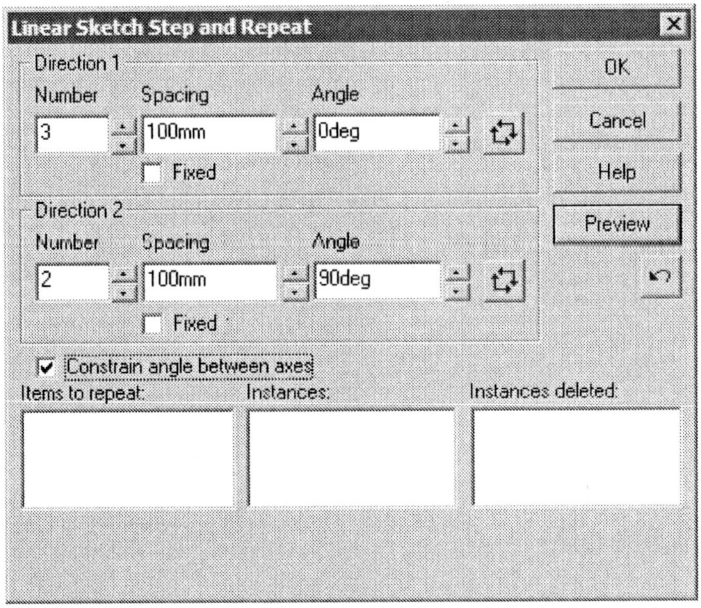

***Figure 2-24** **Linear Sketch Step and Repeat** dialog box*

The options available in this dialog box are discussed next.

Direction 1 Area

By default, the first direction of a linear pattern is taken along the direction of the positive X axis and is indicated in the preview of the pattern along direction 1. The options under this area are used to set the number of instances, spacing, and the angle along the first direction. These options are discussed next.

Number

The **Number** spinner is used to specify the number of instances in the linear pattern along the first direction. As you increase or decrease the value of this spinner, the number of instances in the preview of the pattern are also modified. The default value of this spinner is 3.

Spacing

The **Spacing** spinner is used to specify the spacing between each instance along the first direction. Remember that the spacing is measured from the center of the entities selected

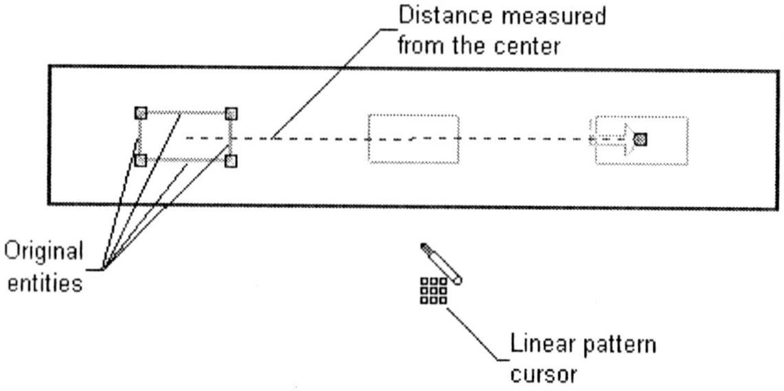

Figure 2-25 Preview of the linear pattern with 3 instances along direction 1 and 1 along direction 2

for creating the pattern. Figure 2-25 shows the preview of the linear pattern that is displayed in the background. As evident in the figure, the number of instances along direction 1 is three and the distance is measured between the center of two consecutive entities.

Angle

The **Angle** spinner is used to specify the angle of the linear pattern along the first direction. Remember that the orientation of the entities selected to be patterned will not change as you enter an angle value for the first direction. The default value of the angle is 0-degree. Figure 2-26 shows a linear pattern at an angle of 30-degree.

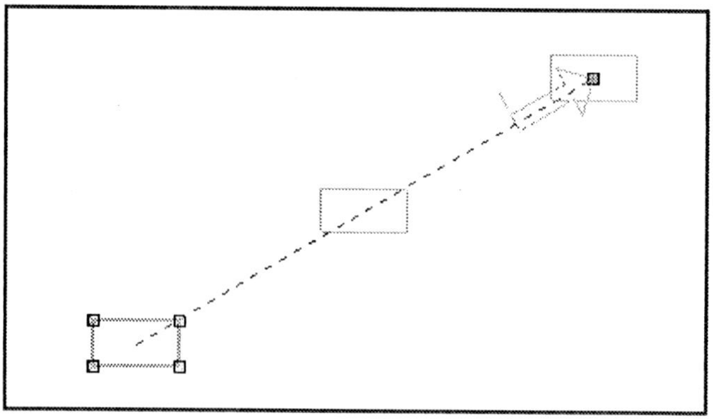

Figure 2-26 Preview of the linear pattern at an angle of 30-degree along direction 1

Reverse Direction 1

The **Reverse Direction 1** button is available on the right of the **Angle** spinner. This button is chosen to reverse direction 1.

Fixed

The **Fixed** check box is available below the **Spacing** spinner. If this check box is selected, the incremental spacing between the instances along direction 1 will be placed as a dimension between the first two instances after you create the pattern. Also, a center line will be drawn between the first two instances. Note that this dimension will not be placed from center to center. It will be displayed from the first entity in the first instance in the pattern to the first entity in the second instance.

Direction 2 Area

By default, the second direction of a linear pattern is taken along the direction of the positive Y axis. The options under this area are used to set the number of instances, spacing, and the angle along the second direction. These options are discussed next.

Number

The **Number** spinner is used to specify the number of instances in the linear pattern along the second direction. The default value of this spinner is 1. This is the reason the other options in the **Direction 2** area are not available. As soon as you set the value of this spinner to more than 1, the other options in this area will be available.

Spacing

The **Spacing** spinner is used to specify the spacing between each instance along the second direction. As mentioned earlier, the spacing is measured from the center of the entities selected for creating the pattern.

Angle

The **Angle** spinner is used to specify the angle of the linear pattern along the second direction. This value will always be measured from the positive X axis direction. As mentioned earlier, the orientation of the entities selected to be patterned will not change as you enter an angle value for the pattern. The default value of the angle is 90-degree. Figure 2-27 shows a linear pattern with the angle of direction 1 as 0-degree and the angle of direction 2 as 60-degree.

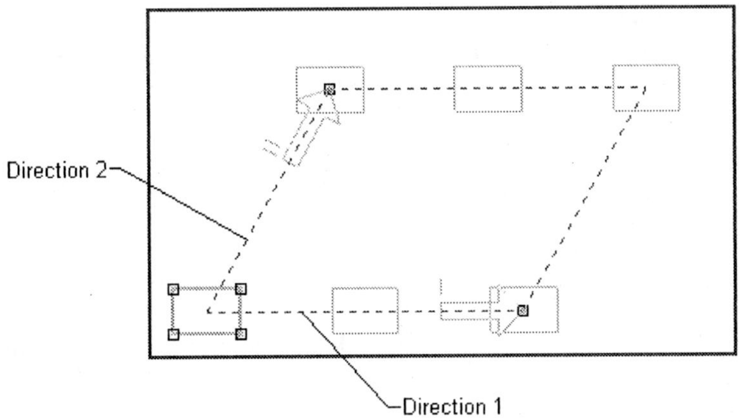

Figure 2-27 Preview of the linear pattern at an angle of 0-degree along direction 1 and 60-degree along direction 2

> **Tip**. *You can also specify the spacing and angle value dynamically in the preview of the linear pattern. To do this, press the left mouse button on the control points displayed at the end of the arrow in the pattern preview and drag it. After placing the arrow at the desired location, release the left mouse button. The new spacing and angle values will be displayed in their respective spinners.*

Reverse Direction 2

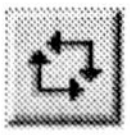

The **Reverse Direction 2** button is available on the right of the **Angle** spinner. This button is chosen to reverse direction 2.

Fixed

The **Fixed** check box is available below the **Spacing** spinner. If this check box is selected, the incremental spacing between the instances along direction 2 will be placed as a dimension between the first two instances after you create the pattern. Also, a center line will be drawn between the first two instances.

Constrain angle between axes

The **Constrain angle between axes** check box will be available only if you set the value of the **Number** spinner in the **Direction 2** area to more than 1. This check box is selected by default. As a result, the angle between the two directions of the linear pattern will be constrained and will be placed as the dimension value when you create the pattern. However, if this check box is cleared, the angle between the two directions will not be displayed and will not be related to each other.

Items to repeat

The **Items to repeat** list box displays all the entities selected to pattern. If you want to remove an entity from the selection set, select that entity from this box. The background color of the box will be changed to a reddish color and the entity will be highlighted in blue. Now, press the DELETE key from the keyboard. The removed entity will no more be displayed in the preview of the pattern and will not be selected in the original entities.

Instances

The **Instances** list box displays all the instances of the linear pattern, except the original instance. The instances are displayed in a matrix format with the first number indicating the instances along direction 1 and the second number indicating the instances along direction 2. Note that the matrix [1,1] is not displayed because it is the original instance. If you want to remove an instance from the pattern, select it in this list box and then press the DELETE key from the keyboard. You will notice that the deleted instance is no more displayed in the preview of the pattern. Remember that this instance will not be deleted from the memory of the pattern. It will be displayed in the **Instances deleted** list box.

Instances deleted

The **Instances deleted** list box displays all the instances that are deleted from the linear pattern. As mentioned earlier, the deleted instances are not removed from the memory of the pattern. They are stored in this list box. If you want to restore the deleted instance, select it in this box

Tip. *You can also add entities in the current selection set or remove them from the current selection set by selecting them using the linear pattern cursor. As you add or remove the instances, the effect can be seen dynamically in the preview of the pattern.*

and then press the DELETE key. You will notice that the selected instance is again displayed in the preview and also in the **Instances** list box.

Undo

The **Undo** button is chosen to undo the operations performed in the **Linear Step Sketch and Repeat** dialog box. All the operations will be undone one by one if you keep choosing this button and the original parameters will be restored. After everything is undone, this button will not be available.

Creating Circular Pattern of the Sketched Entities

CommandManager:	Sketch > Circular Sketch Step and Repeat *(Customize to Add)*
Menu:	Tools > Sketch Tools > Circular Sketch Step and Repeat
Toolbar:	Sketch > Circular Sketch Step and Repeat *(Customize to Add)*

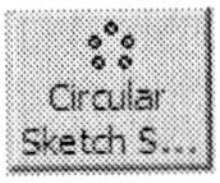

In SolidWorks, the circular pattern of the sketched entities is created using the **Circular Sketch Step and Repeat** tool. To create the circular pattern, select the sketched entities using the **Select** tool and then choose this button from the **Sketch CommandManager**. The **Circular Sketch Step and Repeat** dialog box will be displayed, as shown in Figure 2-28, and the preview of the circular pattern will be shown on the screen in the background. Also, the arrow cursor will be replaced by the circular pattern cursor.

The options provided in the **Circular Sketch Step and Repeat** dialog box are discussed next.

Arc Area

The options under the **Arc** area are used to specify the radius and angle of the imaginary circle, around the circumference of which the instances will be arranged. These options are discussed next.

Radius

The **Radius** spinner is used to specify the radius of the imaginary circle around the circumference of which the instances of the circular pattern are arranged. As you modify this value, the center point of the imaginary circle will also change.

Angle

As you invoke the **Circular Sketch Step and Repeat** tool, a reference circle is drawn with dashed lines around which the instances of the circular pattern will be arranged. Also, a reference line is drawn from the center of the entities selected to pattern and the center of the reference circle. The **Angle** spinner is used to modify the angle of this reference line drawn between the center of entities and the center of the reference circle. As you modify the value of this spinner, the center of the reference circle will also modify accordingly.

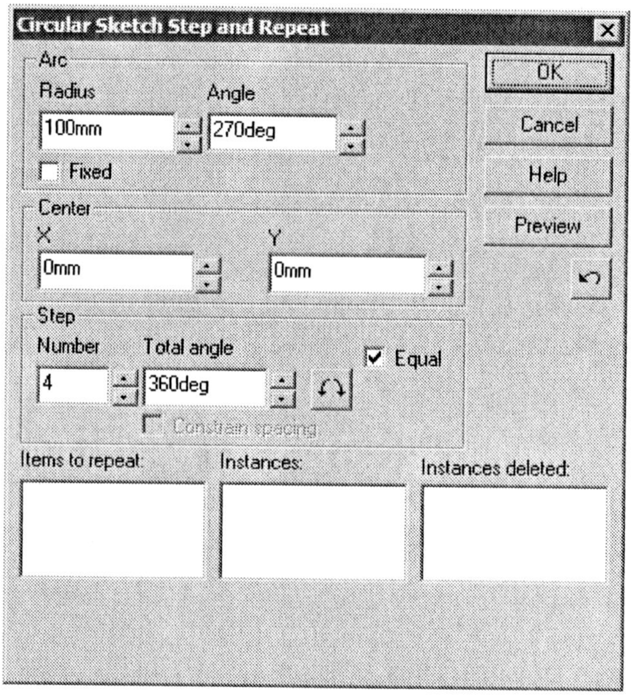

Figure 2-28 ***Circular Sketch Step and Repeat*** *dialog box*

Fixed

If you select the **Fixed** check box, the radius of the reference circle will be placed as a linear dimension value after you create the circular pattern.

Center Area

There are two spinners that are available in the **Center** area. They are the **X** spinner and the **Y** spinner. These two spinners are used to specify the X and Y coordinates of the center point of the reference circle around which the instances of the circular pattern will be arranged. By default, the center of the reference circle is coincident with the origin. Therefore, the value of both these spinners will be 0 mm, 0 mm. You will notice that an arrow is displayed at the center point of the reference circle and the tip of the arrow coincides with the origin. You can define the new coordinates of the center of the reference circle in these spinners. You can also define the center of the reference circle by pressing the left mouse button on the control point provided on the tip of the arrow and dragging the arrow to the desired location. The values of the **Radius** and the **Angle** spinner in the **Arc** area and the **X** and **Y** spinners in the **Center** area will modify accordingly.

Note

If you know the location of the center point of the circular pattern, dragging the arrow at the center of the reference circle is the recommended method of defining the center of the circular pattern.

Step Area

The options in the **Step** area are used to specify the parameters such as the number of instances in the pattern, the angle between the instances, and so on. These options are discussed next.

Number

The **Number** spinner is used to specify the number of instances in the circular pattern. The default value of this spinner is **4**.

Total angle

The **Total angle** spinner is used to specify the total angle between all the instances of the circular pattern. The default value of this spinner is 360-degree. You can enter the desired value of the angle in this spinner.

Reverse Rotation

The **Reverse Rotation** button is chosen to reverse the direction of the pattern. By default, the instances of the circular pattern are arranged in the clockwise direction. When you choose this button, the direction is reversed to counterclockwise direction. To restore the counterclockwise direction, choose this button again. Figure 2-29 shows the preview of the circular pattern with 4 instances arranged in the counterclockwise direction.

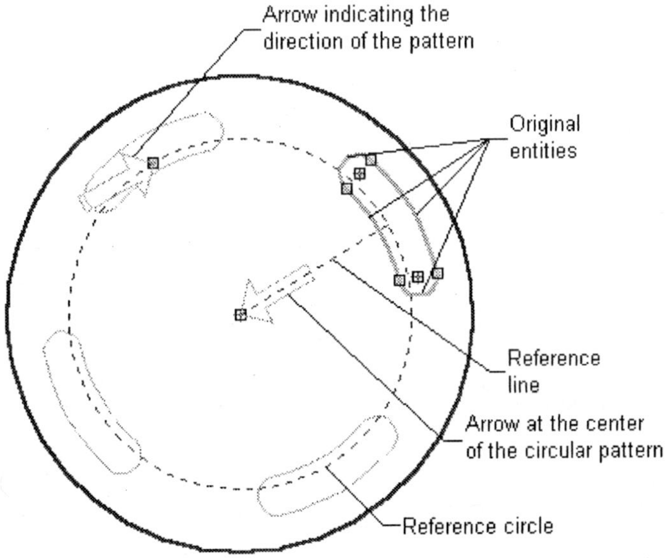

Figure 2-29 Parameters associated with the circular pattern

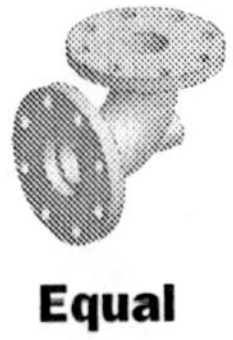

Tip. *You can modify the total angle between the instances by pressing the left mouse button on the tip of the direction arrow and dragging it.*

Equal

The **Equal** check box is selected to specify the total angle between all the instances of the circular pattern. If you clear this check box, the **Total angle** spinner is replaced by the

Spacing spinner. You can specify the incremental spacing between two successive instances of the circular pattern in this spinner. Figure 2-30 shows the preview of the circular pattern with 70-degree incremental spacing between two successive instances.

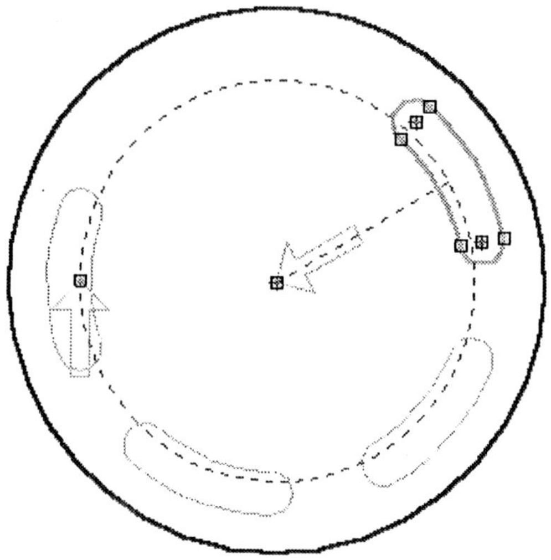

Figure 2-30 *Creating circular pattern by defining incremental angle between individual instances*

Constrain spacing

The **Constrain spacing** check box is selected to constrain the incremental spacing between the individual instances of the circular pattern. If this check box is selected, the incremental spacing will be placed as an angular dimension when you create the pattern. Figure 2-31 shows a circular pattern with the angle and the radius values placed in the pattern.

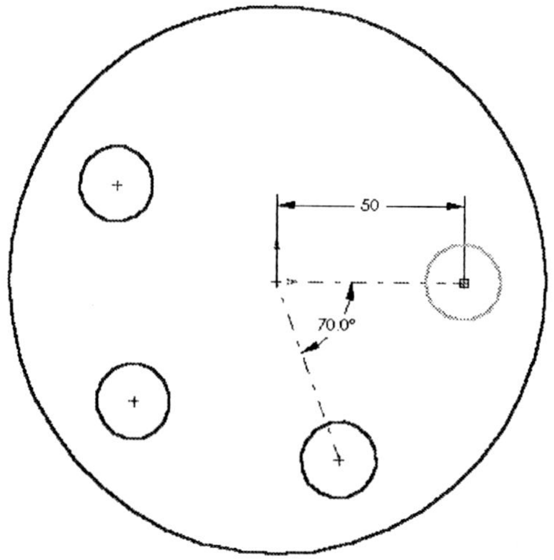

Figure 2-31 *Radius and angle values placed in the circular pattern*

Note

The other options in this dialog box are similar to those discussed in the linear pattern.

EDITING PATTERNS

You can edit the patterns of the sketched entities by using the shortcut menu that is displayed when you right-click on any instance of the pattern. Based on whether you right-clicked on the instance of the linear or the circular pattern, the **Edit Linear Step and Repeat** or the **Edit Circular Step and Repeat** option will be available in the shortcut menu. Figure 2-32 shows the shortcut menu that is displayed when you right-click one of the instances of a circular pattern.

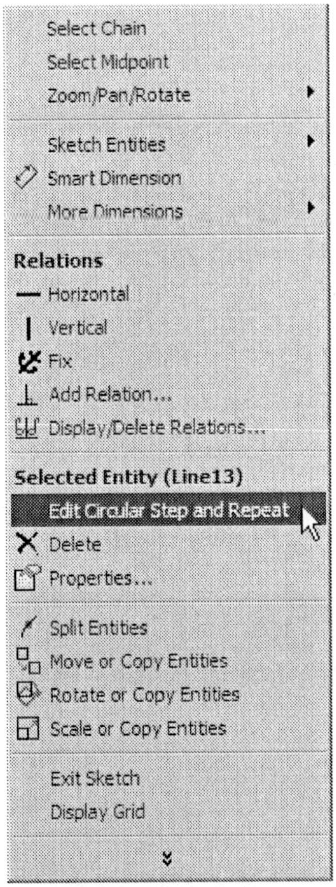

Figure 2-32 *Shortcut menu displayed upon right-clicking an instance of a linear pattern*

Depending on whether you choose the option to edit a linear pattern or a circular pattern, the **Linear Sketch Step and Repeat** or the **Circular Sketch Step and Repeat** dialog box will be displayed. Note that only those options that can be edited will be available in these dialog boxes.

WRITING TEXT IN THE SKETCHING ENVIRONMENT

Command Manager:	Sketch > Text *(Customize to Add)*
Menu:	Tools > Sketch Entities > Text
Toolbar:	Sketch > Text *(Customize to Add)*

You can also create text in the sketching environment of SolidWorks, that can be later used to create join or cut features. To create the text, choose the **Text** button from the **Sketch CommandManager**. The **Sketch Text PropertyManager** is displayed, as shown in Figure 2-33. Enter the text in the box available in the **Text** rollout. You will notice that by default, the text will be started from the sketch origin. However, the text will actually be placed at the location that you specify on the screen before or after writing the text. You can also change the format, font, justification, and so on using the options available in the **Text** rollout.

Figure 2-33 Sketch Text PropertyManager

Figure 2-34 shows the text created using the **Text** tool.

***Figure 2-34** Sketch created using the **Text** tool*

You can also create the text along a curve. To create the text along a curve, you first need to create the curve. The curve can be an arc, spline, a line, or a combination of line, arc, and spline. Next, invoke the **Sketch Text PropertyManager** and select the curve or curves along which you need to create the text. Now, enter the text in the **Text** edit box. You will observe that the text will be created along the arc. You can use the **Flip Horizontal** and **Flip Vertical** options to modify the display of the text. Figure 2-35 shows the text created along an arc.

***Figure 2-35** Text created along the arc*

MODIFYING SKETCHED ENTITIES

Most of the sketches require modification at some stage of the design. Therefore, it is important for any designer to understand the process of modification in SolidWorks. Modification of various sketched entities is discussed next.

Modifying a Sketched Line

You can modify the sketched lines by using the **Line PropertyManager** that is displayed when you select it using the **Select** tool. Note that if the selected line is a part of a rectangle, polygon, or a parallelogram, the entire object will be modified as you modify the line. This is because the relations are applied to all the lines of a rectangle, polygon, and a parallelogram.

Similarly, you can also modify the centerlines using the **Line PropertyManager** that is displayed when you select a centerline.

Modifying a Sketched Circle

To modify a sketched circle, select it using the **Select** tool to display the **Sketched Circle PropertyManager**. The coordinate values of the center point of the circle and the value of the radius will be displayed. You can modify the values that you want.

Modifying a Sketched Arc

To modify a sketched arc, select it using the **Select** tool. The **Arc PropertyManager** will be displayed with the coordinate values of the center point, the start point, and the endpoint. The

Tip. *The status of the sketched entity that you select for modification is displayed in the **Existing Relations** rollout of the **PropertyManager** that is displayed. For example, if the selected sketched entity is fully defined, it will be displayed in the **PropertyManager** and if the entity is underdefined, the **PropertyManager** will display that the entity is underdefined.*

values of the radius and the included angle will also be displayed. You can modify the values that you want.

Modifying a Sketched Polygon

To modify a sketched polygon, right-click on any one edge of the polygon to display the shortcut menu. Choose the **Edit polygon** option from the shortcut menu. The **Polygon PropertyManager** will be displayed. You can modify the selected polygon using the options in the **Polygon PropertyManager** or draw a new polygon.

Note

*If you right-click the reference circle that is automatically drawn when you draw a polygon, the **Edit polygon** option will not be available in the shortcut menu.*

Modifying a Spline

You can perform four types of modifications on a spline. The first one is the modification of the coordinates of the selected control point. The second one is adding more control points on the spline. The third one is modifying the spline using the moving frame. The fourth one is adding more moving frames. These modifications are discussed next.

Modifying the Coordinates of Control Points

To modify the control points of a spline, select it using the **Select** tool. The **Spline PropertyManager** will be displayed. All the control points of the spline will be displayed and the current control point will be displayed with a filled square. The number and the coordinates of the current handle will be displayed in the **Spline PropertyManager**. You can modify the coordinates using the spinner. You can select any other handle by using the **Spline Point Number** spinner.

Adding Control Points

To add control points to the spline, choose **Tools > Spline Tools > Insert Splint Point** from the menu bar. You can also select the spline and right-click to invoke the shortcut menu. In this menu, choose the **Spline Point** option. Now, select the points on the spline where you want to add the control point. A circle will be displayed at the point specified on the spline. Similarly, you can add as many control points as you want to the spline. After adding the required number of control points, invoke the **Select** tool and then select the spline again. You will notice that the boxes of control points will be displayed on all the points you specified on the spline. You can display the coordinates of a control point using the **Spline Point Number** spinner available in the **Spline PropertyManager**.

Modifying the Spline Using the Moving Frames

The moving frame is a combination of three mutually perpendicular lines (called handles) and is displayed when you select a spline control point. The moving frame is used to modify the curvature and tangency of the spline at the selected control point. To modify a spline using the moving frame, select a particular control point of the spline using the **Select** tool. The moving frame will be displayed at the selected point. Three handles are provided on the moving frame to modify the spline. Select one of the handles and drag the mouse. The spline will be modified dynamically as you drag the cursor. You can also move the moving frame by dragging the frame point along the spline.

Adding the Moving Frames

To add the moving frame on the spline, select the spline where you want to add the moving frame using the **Select** tool and right-click to invoke the shortcut menu. Choose the **Moving Frame** option from the shortcut menu. A moving frame will be displayed at the selected point. You can add as many moving frames as you want to the spline. When a moving frame is not selected, it is displayed as a triangle. As you select a moving frame, the **Frame Point PropertyManager** is displayed. The **Frame Point PropertyManager** is used to display the values of **X Coordinate**, **Y Coordinate**, **Curvature**, and **Tangency** of the selected moving frame. Choose the moving frame and press the DELETE key from the keyboard to delete the moving frame.

Note

You can also simplify the splines and convert a sketched entity into a spline. These options will be discussed in advanced chapters.

Modifying a Sketched Point

To modify a sketched point, select it using the **Select** tool. The **Sketched Point PropertyManager** will be displayed. You can modify the coordinates of the sketched point using this **PropertyManager**.

Modifying an Ellipse or an Elliptical Arc

To modify an ellipse or an elliptical arc, select it using the **Select** tool. The **Ellipse PropertyManager** will be displayed. You can modify the required parameters using the options available in this **PropertyManager**.

Modifying a Parabola

To modify a parabola, select it using the **Select** tool. The **Parabola PropertyManager** will be displayed. You can modify the parameters of a parabola from this **PropertyManager**.

Dynamically Modifying and Copying the Sketched Entities

In the sketcher environment of SolidWorks, you can modify or copy the sketched entities by dynamically dragging them using the left mouse button. For example, consider a case where you create a sketch of a rectangle and you want to increase the size of the rectangle. You simply have to select any of the lines of the rectangle or select any of the vertices and hold down the left mouse button to drag the mouse. Drag the sketch according to your requirement and then release the left mouse button. If you choose **Tools > Sketch Setting > Detach Segment on Drag** from the menu bar and select a line of a rectangle to drag, the line segment will be detached from the rectangle. This option is not activated by default. As a result, the segments are not detached upon dragging.

You can also copy the sketched entities dynamically. Select the sketched entity or entities to be copied using the CTRL key or by dragging a window around them. Press and hold down the CTRL key on the keyboard. Now, press and hold the left mouse button down and drag the selected entity or entities. The preview of the copied entities will be displayed. Release the left mouse button at the location where you want to place the new entities. You can make multiple copies of the selected entities by repeating this procedure.

Splitting the Sketched Entity

CommandManager: Sketch > Split Entities *(Customize to Add)*
Menu: Tools > Sketch Tools > Split Curve
Toolbar: Sketch > Split Curve *(Customize to Add)*

Using the **Split Entities** option you can split a sketched entity into two or more than two entities by specifying the split points. To split an entity, choose the **Split Entities** button from the **Sketch CommandManager**; the current cursor will be replaced by the split entities cursor. Move the cursor to an appropriate location where you want to split the sketched entity. When the cursor snaps to the entity, press the left mouse button to add a split point. Now, right click to display the shortcut menu and choose the select option from the shortcut menu. Using the select cursor, select the sketched entity. You will notice that the sketched entity is divided in two entities and a split point is added between the two sketched entities. You can add as many split points as you need. Remember that to split a circle, a full ellipse, or a closed spline, you need to split them at least at two points.

You can also delete the split points to convert the split entity into a single entity. To delete the split point, select the split point and press the DELETE key from the keyboard. You can also right-click on the split point to display the shortcut menu and choose the **Delete** option from the shortcut menu.

NOTE
*You can use the **Modify Sketch** tool to move, scale, or rotate the entire sketch. This will be discussed in advanced chapters.*

TUTORIALS

Tutorial 1

In this tutorial, you will create the base sketch of the model shown in Figure 2-36. The sketch of the model is shown in Figure 2-37. You will create the sketch of the base feature using the normal sketch tools and then modify and edit the sketch using various modifying options.

(Expected time: 30 min)

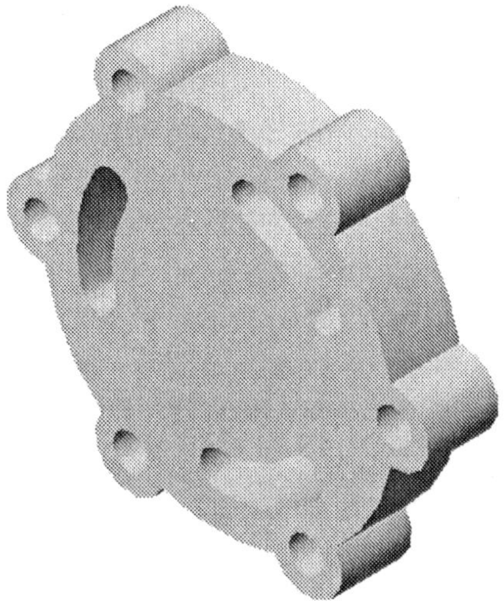

Figure 2-36 *Solid Model for Tutorial 1*

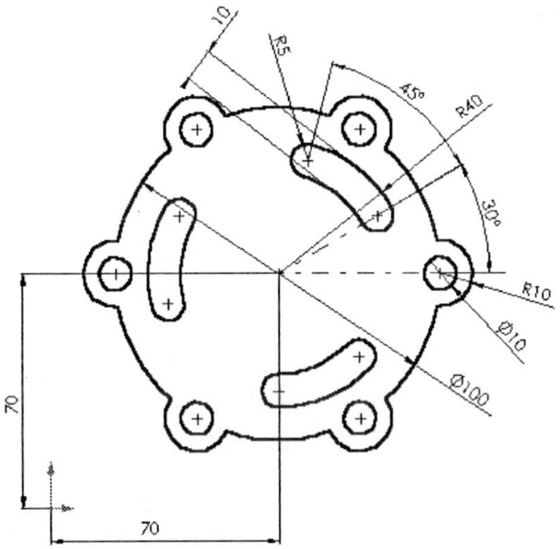

Figure 2-37 *The sketch of the base feature*

The steps that will be followed to complete this tutorial are listed next.

a. Start SolidWorks and then start a new part document.
b. Maximize the part file document and then switch to the sketching environment.
c. Draw the outer loop of the sketch of the given model, refer to Figures 2-38 and 2-39.
d. Create the inner cavity using the **Centerpoint Arc** and the **Tangent arc** tools, refer to Figure 2-40.
e. Use the **Circular Sketch Step and Repeat** tool to create a circular pattern of the inner cavity, refer to Figure 2-41.
f. Complete the sketch by creating the circles that define the hole in the outer loop.

g. Using the **Circular Sketch Step and Repeat** tools, create the circles for the remaining holes in the outer loop, refer to Figure 2-42.

Starting SolidWorks and Starting a New SolidWorks Document

1. Start SolidWorks by choosing **Start > Programs > SolidWorks 2004 > SolidWorks 2004** or by double-clicking the shortcut icon of SolidWorks available on the desktop of your computer.

   The **Welcome to SolidWorks 2004** window and **Tip of the Day** dialog box are displayed. The **Tip of the Day** dialog box is overlapping this window. As mentioned earlier, the tips that are displayed in this dialog box are very useful in making the best use of SolidWorks.

2. Close the **Tip of the Day** dialog box by choosing the **Close** button.

3. Choose the **New Document** option from the **Welcome to SolidWorks 2004** window.

   The **New SolidWorks Document** dialog box will be displayed.

4. The **Part** button is chosen by default. Choose the **OK** button from this dialog box.

   A new SolidWorks part document is started. But the part document window is not maximized in the SolidWorks window.

5. Choose the **Maximize** button available on the upper right corner of the part document window to maximize the document window.

   By default, when you open a new part document, the part modeling environment is active. But because you need to draw the sketch of the feature, you need to invoke the sketching environment.

6. Choose the **Sketch** button from the **Standard** toolbar. The **Edit Sketch PropertyManager** is invoked and you are prompted to select the plane to create the sketch.

7. Select the **Front** plane; the sketching environment is invoked and the plane is oriented normal to the view.

   You will notice that red color sketch origin is displayed along with the confirmation corner.

Modifying the Snap and Grid Settings and the Dimensioning Units

Before you proceed with drawing the sketch, you need to modify the grid and snap settings so that you can make the cursor jump through a distance of 10 mm.

1. Choose **Tools > Options** from menu bar and then choose the **Document Properties** tab to invoke the **Document Properties - Detailing** dialog box.

2. Choose the **Grid/Snap** option from the area on the left. Set the value of the **Minor-lines per major** spinner to **10**, and **Major grid spacing** spinner to **100**.

3. Select the **Snap to points** check box from the **Snap** area, if it is not selected.

   The coordinates displayed close to the lower left corner of the SolidWorks window will show an increment of 10 mm when you move the cursor in the drawing area after exiting the dialog box.

4. If by default the grid is displayed when you invoke the sketching environment, you can remove the display of grid. To remove the grid clear the **Display grid** check box from the **Grid** area.

   If you selected a unit other than millimeter to measure the length while installing SolidWorks, you need to select the units for the current drawing.

5. Select the **Units** option from the area on the left of the **Document Properties - Grid/Snap** dialog box.

6. Select **millimeters** from the drop-down list available in the **Length units** area and **Degrees** from the drop-down list available in the **Angular units** area.

7. Choose the **OK** button after making the necessary settings.

Creating the Outer Loop of the Sketch

As evident from Figure 2-37, the sketch consists of the outer loop and inner cavities. It is recommended that for complex sketches, you should first create the outer loop of the sketch. Therefore, you will create the outer loop of the sketch first. Then you can proceed with the inner cavities.

The origin of the sketcher environment is placed in the middle of the drawing area and you have to create the sketch in the first quadrant. Therefore, it is recommended that you modify the drawing area. This can be done using the **Pan** tool.

1. Choose the **Pan** button from the **View** toolbar to invoke the pan tool. The select cursor is replaced by the pan cursor.

2. Press and hold down the left mouse button near the sketch origin and drag the cursor to the lower left corner of the drawing area. The sketch origin moves near the lower left corner of the drawing area.

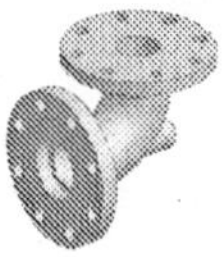

Tip. *You can also pan using the CTRL+middle mouse button. To use this option, hold down the CTRL key and then press the middle mouse button and drag the cursor.*

3. Choose the **Circle** button from the **Sketch CommandManager**. The pan cursor will be replaced by the circle cursor.

4. Move the cursor to a location where the value of the coordinates is 70 mm 70 mm 0 mm.

5. Specify the centerpoint of the circle at this location and move the cursor horizontally toward the right. When the radius above the circle cursor shows a value of 50, press the left mouse button.

6. Choose the **Zoom to Area** button from the **View** toolbar. Using the left mouse button, drag a window such that the sketched circle and the sketch origin are placed in the window.

   When you release the left mouse button the display area of the sketch will be increased.

7. Choose the **Centerline** button from the **Sketch CommandManager** and draw a horizontal center line from the center of the circle toward the right, refer to Figure 2-38.

8. Choose the **Circle** button from the **Sketch CommandManager** and move the cursor at the intersection of the centerline and the circle. At the intersection point, a cross will be displayed below the cursor.

9. Using the left mouse button, specify the centerpoint of the circle at this location and move the cursor horizontally toward the right. When the value of the radius above the circle cursor shows a value of 10, press the left mouse button.

10. Choose the **Trim Entities** button from the **Sketch CommandManager**.

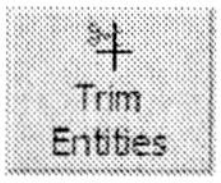

11. Trim the part of the sketch so that the sketch looks similar to the one shown in Figure 2-38.

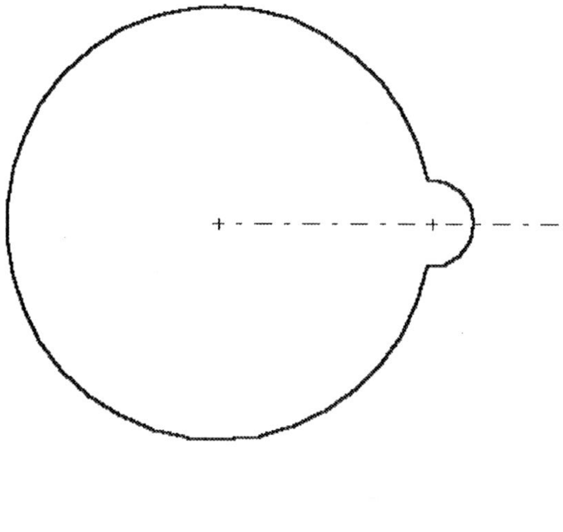

Figure 2-38 *Sketch after trimming the unwanted entities*

12. Choose the **Circular Sketch Step and Repeat** button from the **Sketch CommandManager**. You may have to customize this button to add in the **Sketch CommandManager**.

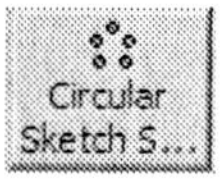

The **Circular Sketch Step and Repeat** dialog box is displayed and the cursor is replaced by the circular pattern cursor.

13. Using the circular pattern cursor, select the smaller trimmed circle. The preview of the circular pattern with default setting is displayed in the drawing area.

    You will notice that the center of the circular pattern, which is displayed by an arrow, is placed at the origin. But because the origin is not the actual center of the circular pattern, you need to modify the centerpoint of the circular pattern. You can modify the center of the circular pattern by entering the coordinates of the point in the **X** and **Y** spinners available in the **Center** area of this dialog box. But the recommended method of modifying the center of the circular pattern is by dragging the arrow displayed at the center of the pattern.

14. Move the circular pattern cursor at the control point available at the end of the arrow that is displayed at the origin. The control point turns red in color.

15. Press and hold the left mouse button down at the control point and drag it to the center of the outer trimmed circle in the sketch. Release the left mouse button when a red circle is displayed at the centerpoint of the outer trimmed circle.

    You will notice that both **X** and **Y** spinners in the **Center** area show 70 mm as the values. This is because the center of the outer trimmed circle is located at a distance of 70 mm along the X and Y axis directions.

16. Set the value of the **Number** spinner in the **Step** area to **6**. Accept all the other default values and choose the **OK** button to create the pattern.

17. Trim the unwanted portion of the outer trimmed circle using the **Trim Entities** tool. After trimming, the sketch should look similar to that shown in Figure 2-39.

***Figure 2-39** Outer loop of the sketch*

Drawing the Sketch of the Inner Cavity

Next, you need to draw the sketch of the inner cavities. You will draw the sketch of one of the cavities and then create a circular pattern of this cavity. The number of items in the circular pattern will be 3.

Before creating the sketch of the inner cavity, you need to create a centerline that will act as a reference for creating the sketch of the inner cavity.

1. Invoke the **Document Properties - Grid/Snap** dialog box. Set the value of the spinner available on the right of the **Snap to angle** check box in the **Snap** area to **30**. Also, select the **Snap to angle** check box.

2. Choose the **Centerline** button from the **Sketch CommandManager**. Move the cursor close to the center of the outer circle. When the red circle is displayed, specify the start point of the centerline. Move the cursor to a location where the angle of the line is shown as exactly 30-degree and the length is close to 35. At this point, specify the endpoint of the line.

3. Exit the **Centerline** tool and select the centerline. The **Line PropertyManager** is displayed. Set the value of the **Length** spinner to **35**.

4. Choose the **Circle** button from the **Sketch CommandManager**.

5. Move the cursor to the upper endpoint of the inclined centerline. When the red circle is displayed, specify the centerpoint. Move the cursor horizontally toward the right and create a circle of radius close to 5 mm.

6. Set the value of the radius of the circle to **5** in the **Radius** spinner of the **Circle PropertyManager**.

7. Choose the **Circular Sketch Step and Repeat** button from the **Sketch CommandManager**.

The **Circular Sketch Step and Repeat** dialog box is displayed and the circle is automatically selected to be patterned. This is because the circle was selected when you invoked this tool.

Note that the center of the circular pattern is placed at the origin. You need to change the center of the circular pattern to the center of the outer trimmed circle.

8. Move the circular pattern cursor at the control point available at the end of the arrow that is displayed at the origin.

9. When the red circle is displayed, press and hold down the left mouse button and drag the cursor to the centerpoint of the outer trimmed circle. Release the left mouse when the red circle is displayed.

10. Set the value of the **Number** spinner to **2**.

11. Set the value of the **Total angle** spinner to 45-degree and choose the **Reverse Rotation** button.

12. Choose **OK** from the **Circular Sketch Step and Repeat** dialog box.

13. Choose the **Centerpoint Arc** button from the **Sketch CommandManager**.

14. Move the cursor to the centerpoint of the outer trimmed circle and specify the centerpoint of the arc. Now, move the cursor to the intersection of the centerline and the circle. Specify the start point of the arc when the intersection symbol appears. Exit this tool.

15. Move the cursor in the counterclockwise direction. When the value of the angle above the arc cursor shows a value of 45-degree, press the left mouse button to specify the endpoint of the arc.

    Next, you will offset the arc created in the previous step through an offset distance of 10.

    The arc created earlier is displayed in green, suggesting that it is already selected. If the arc is not selected, select the arc using the left mouse button before invoking the offset tool.

16. Choose the **Offset Entities** button from the **Sketch CommandManager**. The **Offset Entities PropertyManager** is displayed.

17. The default value of the **Offset Distance** spinner is **10**. Clear the **Reverse** check box, if selected. Choose the **OK** button from the **Offset Entities PropertyManager**.

18. Choose the **Trim Entities** button from the **Sketch CommandManager**. Trim the unwanted portion of the inner cavity and then exit this tool. The sketch after completing the inner cavity is shown in Figure 2-40.

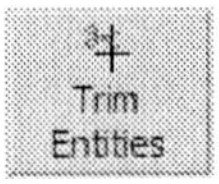

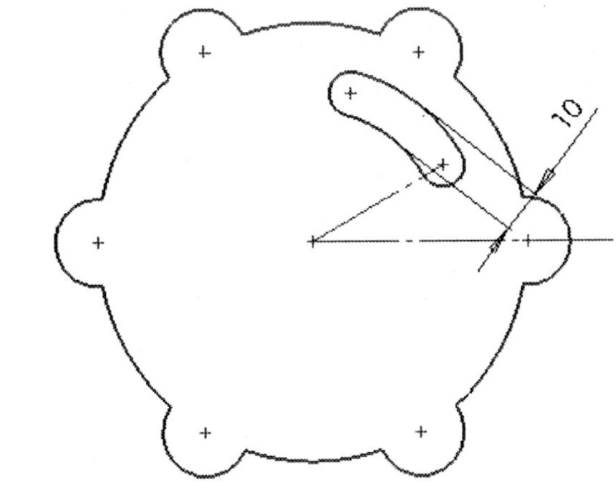

Figure 2-40 *The sketch of the outer loop and the inner cavity*

Creating the Pattern of the Inner Cavity

Next, you will create the pattern of the inner cavity using the **Circular Sketch Step and Repeat** tool. The pattern of the inner cavity consists of three instances. The centerpoint of the pattern lies at the centerpoint of the outer trimmed circle.

Before invoking this option, press the CTRL key from the keyboard and using the left mouse button, select all the entities of the inner cavity.

1. Choose the **Circular Sketch Step and Repeat** button from the **Sketch CommandManager** toolbar to invoke the **Circular Sketch Step and Repeat** dialog box.

2. Press and hold the left mouse button down at the control point of the arrow that is displayed at the origin. Drag it to the center of the outer trimmed circle in the sketch. Release the left mouse button when the red circle is displayed.

   You will notice that the **X** and **Y** spinners in the **Center** area show 70 mm as the values.

   The default value of the number of items in the pattern is 4. But because you need 3 items in the pattern, you need to modify this value.

3. Set the value of the **Number** spinner in the **Step** area to **3**. Accept all the other default values and choose the **OK** button to create the pattern. The sketch after creating the pattern of the inner cavity is shown in Figure 2-41.

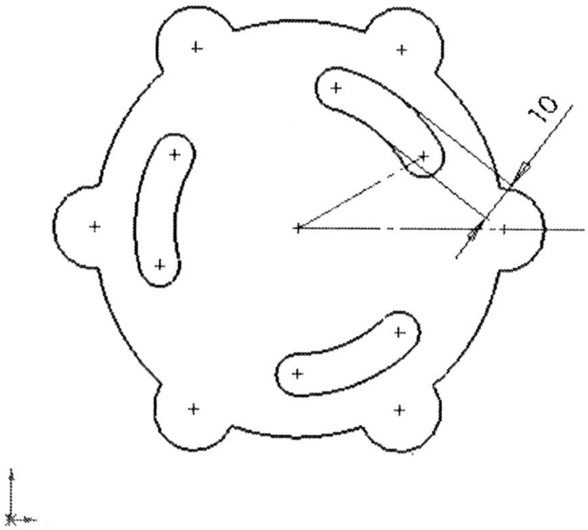

Figure 2-41 *Sketch after creating the pattern of the inner cavity*

Sketching the Holes

Next, you need to create the sketch of the holes. As evident from Figure 2-37, you need to create a total of six circles. After creating the first circle, you will create the other five circles by creating a circular pattern of the parent circle.

1. Choose the **Circle** button from the **Sketch CommandManager**.

2. Move the cursor to the centerpoint of the right trimmed circle. Taking the centerpoint of this as the center, draw a circle of radius close to 5. You can press the CTRL key to make sure the cursor is not snapping to the points.

3. Set the value of the **Radius** spinner to **5** in the **Circle PropertyManager**.

4. Choose the **Circular Sketch Step and Repeat** button from the **Sketch CommandManager** to invoke the **Circular Sketch Step and Repeat** dialog box.

5. Move the center of the circular pattern to the centerpoint of the outer trimmed circle.

6. Set the value of the **Number** spinner to **6** in the **Step** area of this dialog box. Accept all the other default values and choose **OK** in this dialog box.

   The final sketch of Tutorial 1 is shown in Figure 2-42.

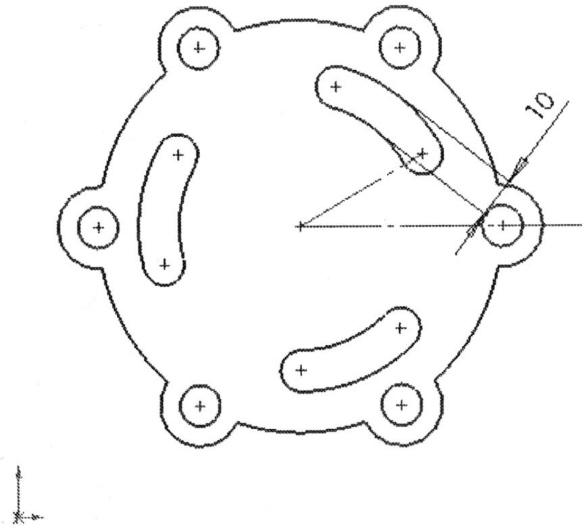

Figure 2-42 *Final sketch for Tutorial 1*

Saving the Sketch

As mentioned earlier, it is recommended that you create a separate folder for saving the tutorial files of each chapter. When you invoke the option to save the document, the default folder that is displayed is the one in which you saved the last document. Browse to the *My Documents/SolidWorks* folder. Now, you will create another folder inside the *SolidWorks* folder and save the document.

1. Choose the **Save** button from the **Standard** toolbar to invoke the **Save As** dialog box.

2. Choose the **Create New Folder** button from the **Save As** dialog box. Enter the name of the folder as *c02* and press ENTER.

3. Enter the name of the document as *c02-tut01.sldprt* in the **File name** edit box and choose the **Save** button.

   The document will be saved in the */My Documents/SolidWorks/c02* folder.

4. Close the file by choosing **File** > **Close** from the menu bar.

Tutorial 2

In this tutorial you will create the base sketch of the model shown in Figure 2-43. The sketch of the model is shown in Figure 2-44. You will create the sketch with mirror line and mirror tool. After creating the sketch, you will modify it by dragging the sketched entities.

(Expected time: 30 min)

Figure 2-43 *Solid model for Tutorial 2*

The steps that will be followed to complete this tutorial are listed below.

a. Start a new part document.
b. Maximize the part document and then switch to the sketching environment.
c. Create the centerlines and create a mirror line using one of the centerline and the **Mirror Entities** tool.
d. Create the sketch in the third quadrant; the sketch will automatically be mirrored on other side, refer to Figure 2-45
e. Mirror the entire sketch along the second mirror line, refer to Figure 2-46.
f. Modify the sketch by dragging the sketched entities, refer to Figure 2-47.

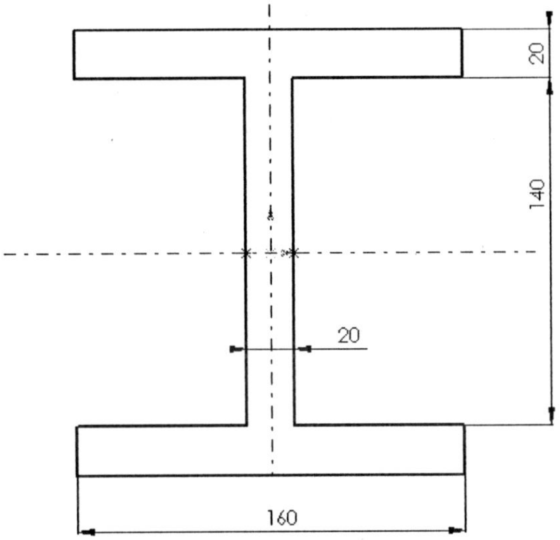

Figure 2-44 Sketch of the model

Starting a New Document

1. Choose the **New** button from the **Standard** toolbar to invoke the **New SolidWorks Document** dialog box.

2. The **Part** button is chosen by default. Choose the **OK** button from the **New SolidWorks Document** dialog box. A new SolidWorks part document is started.

3. Choose the **Maximize** button available on the upper right corner of the part document window to maximize the document window.

   Next, you need to invoke the sketching environment.

4. Choose the **Sketch** button from the **Standard** toolbar and select the **Front** plane to invoke the sketching environment.

   Next, you need to modify the snap and grid settings.

5. Invoke the **Document Properties - Grid/Snap** dialog box. Set the value of the **Major grid spacing** spinner to **100** and **Minor-lines per Major** spinner to **10**. Select the **Snap to points** check box to activate the snap option. Make sure the **Snap to angle** check box is cleared.

   It is assumed that you have selected the unit of measurement as millimeter while installing SolidWorks. However, if you have selected any other unit while installation then you need to change the unit.

6. Select the **Units** option available below the **Grid/Snap** option. Select the **millimeters** option from the drop-down list available in the **Length units** area. Select the **Degrees** option from the drop-down list available in the **Angular units** area. Choose **OK** to exit the dialog box.

Creating the Center Lines and a Mirror Line

In this tutorial, you will create the sketch of the given model with the help of a mirror line. The sketches that are symmetrical along any axis are recommended to be drawn using the mirror line. The mirror line is created using the **Centerline** and **Mirror** tools. After creating a mirror line, when you draw the entities on one side of the mirror line, the same entities are also automatically created on the other side of the mirror line. This is because the symmetrical relation is applied to the entities on both sides of the mirror line. Therefore, if you modify an entity on one side of the mirror line, the same modifications are applied to the mirrored entity, and vice versa. First, you need to create a mirror line.

1. Choose the **Centerline** button from the **Sketch CommandManager**.

2. Move the line cursor to a location where the value of the coordinates is 0 mm 100 mm 0 mm. You may have to pan the drawing to reach that point.

3. Specify the start point of the centerline at this point and move the cursor vertically downward and create a line of 200 mm length. You may have to zoom and pan the drawing to draw a line of this length.

4. Now, double-click anywhere on the screen to end the current chain. You can also press the right mouse button anywhere on the drawing area to invoke the shortcut menu. Choose the **End chain** option from the shortcut menu.

5. Move the line cursor to a location where the value of the coordinates is -100 mm 0 mm 0 mm.

6. Specify the start point of the centerline and move the cursor horizontally toward the right to draw a line of 200 mm length. Pan the drawing, if required.

7. Zoom to fit the drawing and then right-click to invoke the **Select** tool. The line cursor will be replaced by the select cursor.

8. Select the vertical centerline; it will be displayed in green.

9. Choose **Mirror Entities** from the **Sketch CommandManager** to create the mirror line and activate the automatic mirror option.

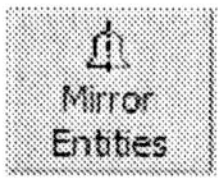

   You can confirm the creation of the mirror line and the activation of the automatic mirror option by observing the symmetrical symbol applied to both the ends of the centerline.

Drawing the Sketch

Next, you will create the sketch of the base feature. You will create the sketch in the third quadrant. The same sketch will be created automatically on the other side of the mirror line. The symmetrical relation is applied between the parent entity and the mirrored entity.

1. Choose the **Line** button from the **Sketch CommandManager** to invoke the line tool.

2. Move the line cursor to a location where the value of the coordinates is 0 mm -100 mm 0 mm.

3. Specify the start point of the line at this location and move the cursor horizontally toward the left.

4. Specify the endpoint of the line when the length of the line above the line cursor shows a value of 90.

   You will notice that as soon as you specify the endpoint of the line, a mirror image is automatically created on the other side of the mirror line. This line created as the mirrored entity is merged with the line drawn on the left. Therefore, the entire line becomes a single entity. Remember that the lines will merge only if one of the endpoints of the line you draw is coincident with the mirror line.

5. Move the cursor vertically upwards and specify the endpoint of the line when the length of the line above the line cursor shows a value of 30.

   You will notice that as soon as you specify the endpoint of a line, a mirror image is automatically created on the other side of the mirror line.

6. Move the line cursor toward the right and specify the endpoint of the line when the length of the line above the line cursor shows a value of 30. A mirror image is automatically created on the other side of the mirror line.

7. Move the line cursor vertically upward and specify the endpoint of the line when the line cursor snaps the horizontal center line. Exit the **Line** tool. The sketch after creating the lines is shown in Figure 2-45.

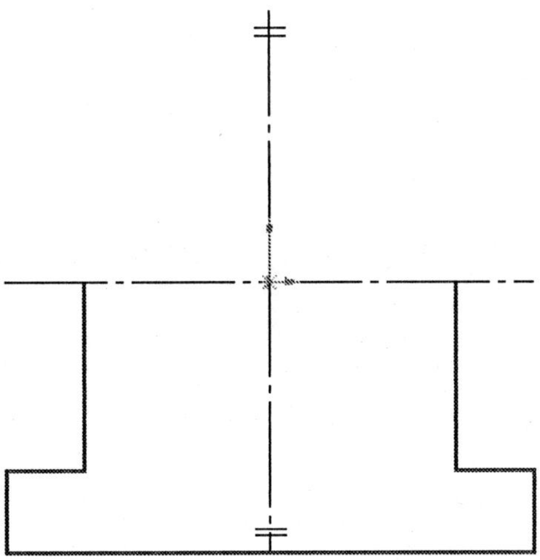

Figure 2-45 *Sketch after creating the lines*

Mirroring the Entire Sketch

After creating one half of the sketch, you need to mirror the entire sketch about the horizontal centerline. But first, you need to disable automatic mirroring.

1. As evident from the symmetric relation symbol on the vertical centerline, automatic mirroring is activated. Choose the **Mirror Entities** button again to disable automatic mirroring.

   Now, the vertical centerline will not work as a mirror line.

2. Use the CTRL key from the keyboard and the left mouse button to select all the lines sketched earlier and the horizontal centerline.

3. Choose the **Mirror Entities** button from the **Sketch CommandManager**. The entire sketch is mirrored about the horizontal centerline. The sketch after mirroring is shown in Figure 2-46.

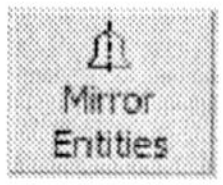

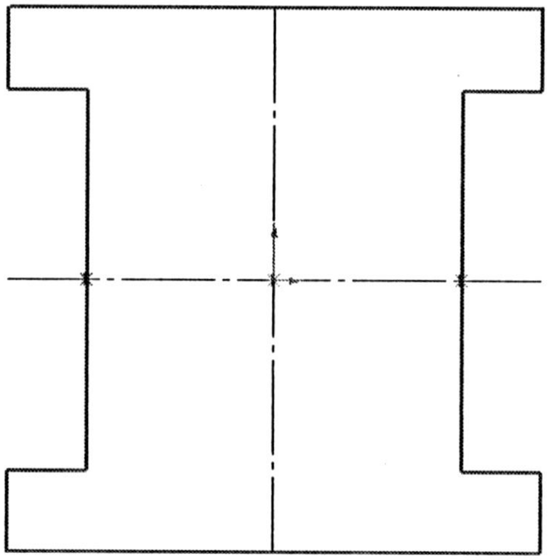

***Figure 2-46** Sketch after mirroring the sketched entities*

Modifying the Sketch by Dragging

Next, you will modify the sketch by dragging. While dragging the entities, you will observe that the corresponding mirrored entity will also be modified.

1. Select the lower right vertical line. The **Line PropertyManager** will be displayed on the left of the drawing area.

   You will notice that the value of the length in the **Length** spinner shows a value of **30**. But the required length of this line is 20. Therefore, you need to edit the length of this line. In this tutorial, you will edit the sketch by dragging. You will observe that when you drag one line, all the dependent lines are also modified. This is because they are created as mirror images.

2. Select the top horizontal line and drag the cursor vertically downward. Release the left mouse button when the value of the **Start Y Coordinate** spinner and the **End Y Coordinate** spinner decreases by 10.

   You will observe that all the sketched entities that are related to the dragged entity are also modified.

3. Select the lower left vertical line and drag the cursor horizontally toward the right. Release the left mouse button when the value of the **Start X Coordinate** spinner and the **End Y Coordinate** spinner decreases by 10.

4. Select the middle left vertical line and drag the cursor horizontally toward the right. Release the left mouse button when the value of the **Start X Coordinate** spinner and the **End X Coordinate** spinner shows a value of -10.

The final sketch after modifying the sketched entities by dragging is shown in Figure 2-47.

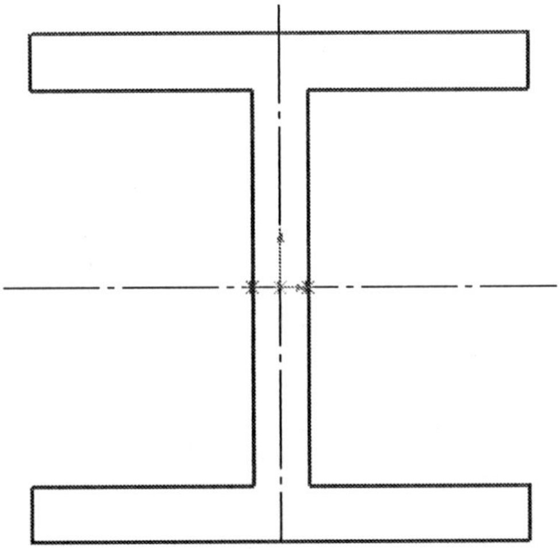

Figure 2-47 *Final sketch for Tutorial 2*

Saving the Sketch

1. Choose the **Save** button from the **Standard** toolbar to invoke the **Save As** dialog box.

2. Browse to the *SolidWorks/c02* folder if it is not displayed by default. Enter the name of the document as *c02-tut02.sldprt* in the **File name** edit box and choose the **Save** button.

   The document will be saved in the */My Documents/SolidWorks/c02* folder.

3. Close the document by choosing **File > Close** from the menu bar.

Tutorial 3

In this tutorial you will create the basic sketch of the model shown in Figure 2-48. The sketch is shown in Figure 2-49. Do not dimension the sketch as the solid model and the dimensions are given only for your reference. **(Expected Time: 30 min.)**

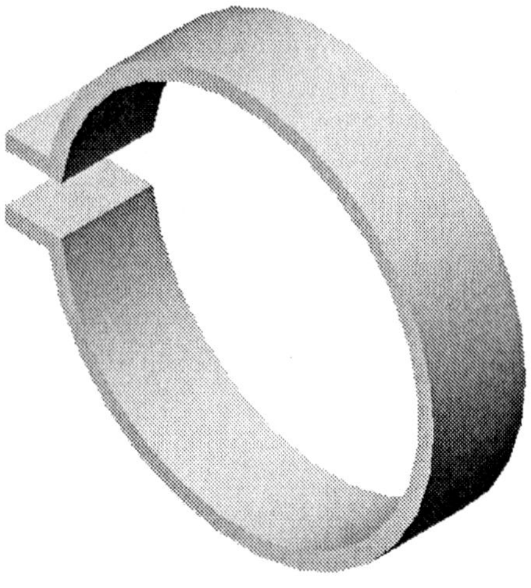

Figure 2-48 *Solid model for Tutorial 3*

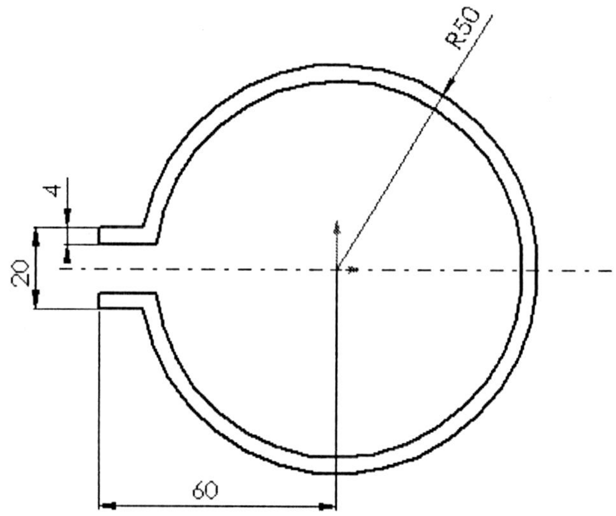

Figure 2-49 *Sketch for Tutorial 3*

The steps that will be followed to complete this tutorial are listed below:

a. Start a new part document.
b. Maximize the part file document and then switch to the sketching environment.
c. Create a centerline that will be used for reference.

d. Draw and edit the sketch using the **Mirror Entities** and **Trim Entities** tools, refer to Figure 2-50.
e. Offset the entire sketch, refer to Figure 2-51.
f. Complete the final editing of the sketch using the sketch extend and sketch trim tools, refer to Figures 2-52 and 2-53.

Starting a New Document

1. Choose the **New** button from the **Standard** toolbar to invoke the **New SolidWorks Document** dialog box.

2. The **Part** button is chosen by default. Choose the **OK** button from the **New SolidWorks Document** dialog box.

3. Choose the **Maximize** button available on the upper right corner of the part document window to maximize the document window.

4. Choose the **Sketch** button from the **Standard** toolbar and select the **Front** plane to invoke the sketching environment.

5. Using the **Document Properties - Grid/Snap** dialog box, set the value of the **Major grid spacing** to **100** and **Minor-lines per major** spinner to **10**. Also, select the **Snap to point** check box.

Creating the Centerline

Before creating the sketch, you need to create a centerline that will act as a reference for the other sketch entities. This centerline will also be used for mirroring.

1. Choose the **Centerline** button from the **Sketch CommandManager**.

2. Move the cursor to a location where the value of the coordinates is -70 mm 0 mm 0 mm.

3. Specify the start point at this location and move the cursor horizontally toward the right.

4. Specify the endpoint of the centerline when the value of the length of the centerline above the line cursor shows a value of 140.

5. Double-click anywhere in the drawing area to end line creation.

Drawing the Outer Loop of the Sketch

Next, you need to create the outer loop of the sketch using the sketch tools. As evident from Figure 2-49, the sketch is drawn using the **Circle** and the **Line** tools.

1. Choose the **Circle** button from the **Sketch CommandManager**. The line cursor will be replaced by the circle cursor.

2. Move the circle cursor to the origin and when the red circle is displayed, press the left mouse button to specify the centerpoint of the circle.

3. Move the cursor horizontally toward the right and draw a circle of 50 mm radius.

4. Choose the **Zoom to Fit** button from the **View** toolbar to increase the display of the sketch.

5. Choose the **Line** button from the **Sketch CommandManager** and move the cursor to a location where the value of the coordinates is -60 mm 10 mm 0 mm. Specify the start point of the line at this location.

6. Move the cursor horizontally toward the right and press the left mouse button to specify the endpoint of the line when the line cursor snaps the circle. Exit the **Line** tool.

7. Press the CTRL key from the keyboard and using the left mouse button, select the horizontal line created in the last step and the horizontal centerline. Both the selected entities will be displayed in green color.

8. Choose the **Mirror Entities** button from the **Sketch CommandManager**. The mirror image of the lower horizontal line will be created on the other side of the centerline.

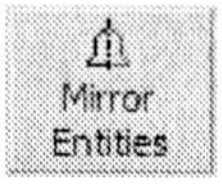

9. Choose the **Line** button from the **Sketch CommandManager** and move the cursor to the left endpoint of the upper horizontal line. Specify the start point of the line when the red circle is displayed.

10. Move the cursor vertically downwards. Specify the endpoint of the line when the cursor snaps to the left endpoint of the lower horizontal line.

11. Choose the **Trim Entities** button from the **Sketch CommandManager**. The line cursor will be replaced by the trim cursor. Trim the unwanted portion of the circle, as shown in Figure 2-50.

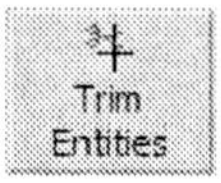

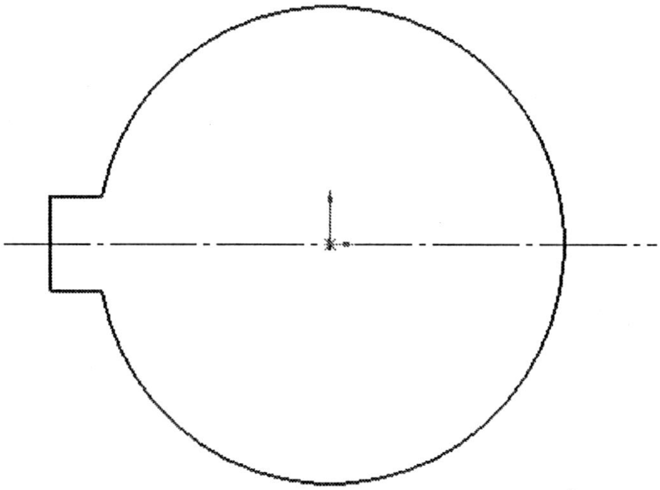

Figure 2-50 *Sketch of the outer loop*

Offsetting the Entities

After creating the outer loop of the sketch, you will create the inner cavity. The first step in creating the inner cavity of the sketch will be offsetting the entire sketch inwards.

1. Choose the **Offset Entities** button from the **Sketch CommandManager**. The **Offset Entities PropertyManager** is displayed on the left of the drawing area.

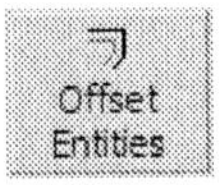

2. Set the value of the **Offset Distance** spinner to **4**. Select any one entity of the sketch; the entire sketch will be selected.

   When you select the sketch, the preview of the offset sketch is displayed in the drawing area. But the direction of the offset is outside the sketch. The required direction of offset is inside the sketch. Therefore, you need to flip the direction.

3. Move the cursor inside the sketch and press the left mouse button to create the offset sketch with reverse direction. A dimension with a value of 4 is displayed with the sketch.

   The sketch after offsetting the outer loop is shown in Figure 2-51.

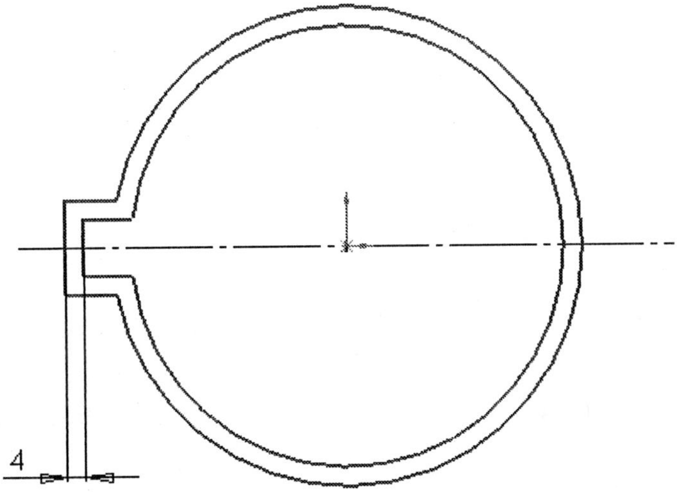

Figure 2-51 *Sketch after offsetting the outer loop*

4. Choose the **Extend Entities** button from the **Sketch CommandManager**. The select cursor will be replaced by the extend cursor.

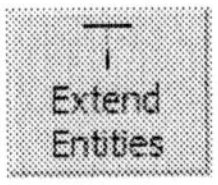

   If the **Extend Entities** button is not available in the **Sketch CommandManager**, you can invoke the extend option from **Tools > Sketch Tools > Extend**. As discussed earlier, you can insert the **Extend Entities** button in the **Sketch CommandManager** using the **Customize** dialog box.

5. Move the cursor close to the left end of the lower horizontal line of the inner sketch. You can preview the extended line in magenta color. (The preview of the extended entity will be displayed in magenta if you are working on Windows 2000 platform. If you are working on Windows XP, the preview will be displayed in yellow.) Move the cursor a little bit toward the left if the preview of the extended line appears to the right. Press the left mouse button to extend the line.

6. Similarly, extend the upper horizontal line of the inner sketch. The sketch after extending the lines is shown in Figure 2-52.

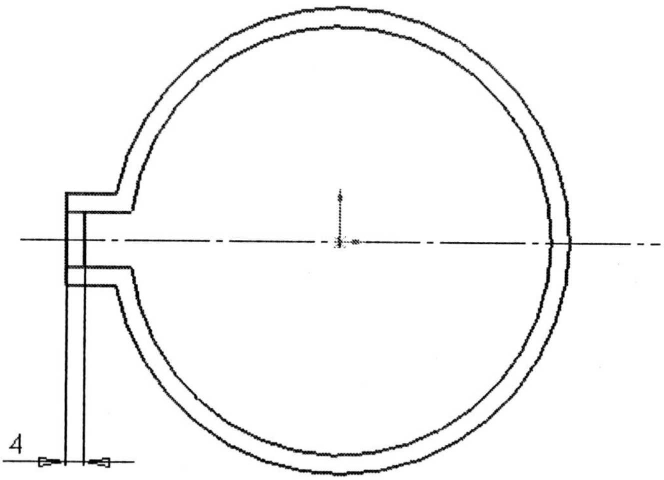

Figure 2-52 *Sketch after extending the lines*

7. Choose the **Trim Entities** button from the **Sketch CommandManager** and the extend cursor will be replaced by the trim cursor.

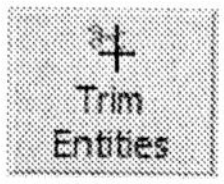

8. Using the left mouse button trim the unwanted entities. The final sketch is shown in Figure 2-53.

Saving the Sketch

1. Choose the **Save** button from the **Standard** toolbar to invoke the **Save As** dialog box.

2. Enter the name of the document as *c02-tut03.sldprt* in the **File name** edit box and choose the **Save** button. Close the file by choosing **File** > **Close** from the menu bar.

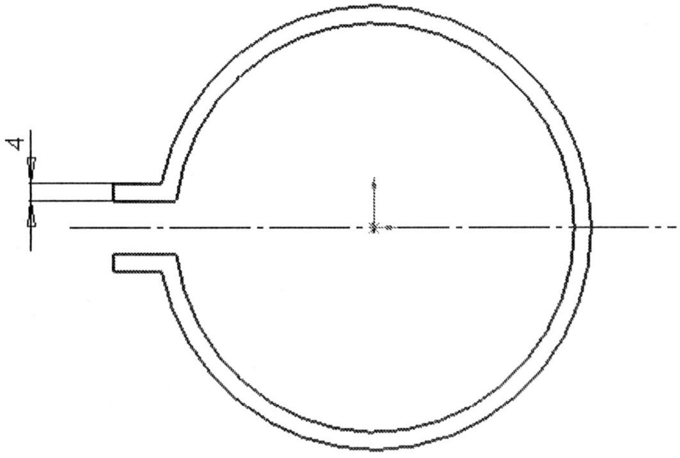

Figure 2-53 *Final sketch for Tutorial 3*

Tip. *Setting the snap and grid parameters for each document is a very tedious and time-consuming process. To overcome this problem you can create a template in which all the snap, grid, and units are set. Whenever you have to create a part document you just have to choose that template from the* ***Template*** *tab of the* ***New SolidWorks Document*** *dialog box. To create and save a template, create a new document in the Part mode and set the parameters of units, snap, and grid from the* ***Document Properties*** *dialog box. Choose the* ***Save*** *button from the* ***Standard*** *toolbar or choose* ***File > Save*** *from the menu bar. The* ***Save As*** *dialog box will be displayed. Browse* ***X:/Program Files/SolidWorks/data/templates*** *location to save the file.* ***X:*** *is the drive in which you have installed SolidWorks. Select the* ***Part Templates (*.prtdot)*** *option from the* ***Save as type*** *drop-down list. Enter the name of the template file in the* ***File name*** *edit box. Choose the* ***Save*** *button from the* ***Save As*** *dialog box.*

Next time when you invoke the ***New SolidWorks Document*** *dialog box to start a new part document, choose the* ***Advanced*** *button to display the advanced version of this dialog box. The advanced version of the* ***New SolidWorks Document*** *dialog box will have the* ***Tutorial*** *tab that lists the template that you created, along with the other default templates. Using the same procedure, you can also create the user-defined templates for drawing and assembly documents. You will learn more about drawing and assembly documents in later chapters.*

Next time when you invoke the ***New SolidWorks Document*** *dialog box, the advanced version of this dialog box will be displayed. To restore the original version, choose the* ***Novice*** *button.*

SELF-EVALUATION TEST

Answer the following questions and then compare your answers with the answers given at the end of this chapter.

1. The **Trim** option is also used to extend the sketched entities. (T/F)

2. In the sketching environment, you can apply fillets to two parallel lines. (T/F)

3. You can apply a fillet to two nonparallel and nonintersecting entities. (T/F)

4. You cannot offset a single entity; you have to select a chain of entities to create an entity using the **Offset Entities** tool. (T/F)

5. You can choose **Insert** > **Customize** from the menu bar to display the **Customize** dialog box. (T/F)

6. The design intent is not captured in the sketch created using the mirror line. (T/F)

7. The __________ tool is used to create a linear pattern in the sketcher environment of SolidWorks.

8. The __________ tool is used to create a circular pattern in the sketcher environment of SolidWorks.

9. To modify a sketched circle, select it using the __________ tool to display the __________ **PropertyManager**.

10. The __________ tool is used to invoke the automatic mirroring option.

REVIEW QUESTIONS

Answer the following questions.

1. You cannot extend the sketched entity using the trim tool. (T/F)

2. The preview of the entity to be extended is displayed in red color. (T/F)

3. You cannot apply a sketch fillet to two nonintersecting entities. (T/F)

4. The sketched entities can be mirrored without using a centerline. (T/F)

5. The **Fixed** check box in the **Linear Sketch Step and Repeat** dialog box is selected to specify the dimension between the first two instances along a specified direction after you create the pattern. (T/F)

6. Which **PropertyManager** is displayed when you choose the **Sketch Fillet** button from the **Sketch CommandManager**?

   (a) **Sketch Fillet** (b) **Fillet**
   (c) **Surface Fillet** (d) **Sketching Fillet**

7. When you choose **Tools > Sketch Tools > Chamfer** from the menu bar to invoke the chamfer tools, which **PropertyManager** is displayed on the left of the drawing area.

   (a) **Sketch Chamfer** (b) **Sketcher Chamfer**
   (c) **Sketching Chamfer** (d) **Chamfer**

8. Which tool is used to create automatic mirror line?

   (a) **Mirror Entities** (b) **Mirror**
   (c) **Automatic Mirror** (d) None

9. Which tool is used to break a sketched entity in two or more than two entities?

   (a) **Split Entities** (b) **Trim Sketch**
   (c) **Break Curve** (d) **Trim Curve**

10. Which tool is used to create a circular pattern in SolidWorks?

    (a) **Circular Pattern** (b) **Circular Sketch Step and Repeat**
    (c) **Array** (d) None

EXERCISES

Exercise 1

Create the sketch of the model shown in Figure 2-54. The sketch is shown in Figure 2-55. The solid model and dimensions are given only for reference. Create the sketch on one side and then mirror it on the other side. Make sure you do not use the mirror line option to draw this sketch. This is because if you draw the sketch using the mirror line, some relations are applied to the sketch. These relations interfere while creating fillets. **(Expected time: 30 min)**

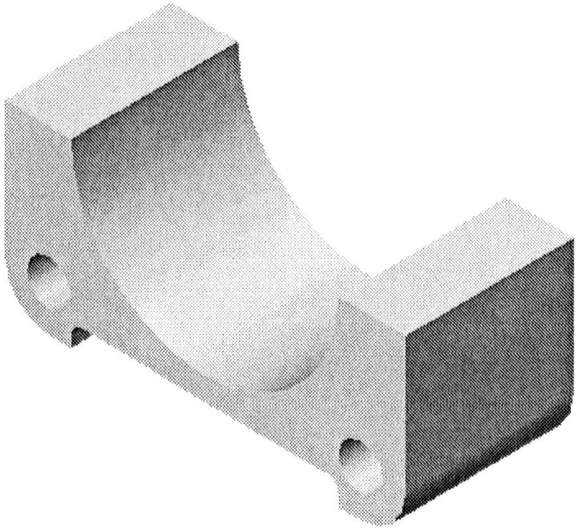

Figure 2-54 Solid model for Exercise 1

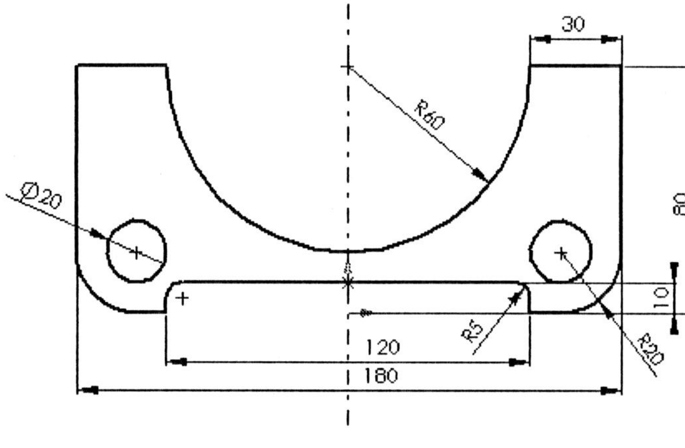

Figure 2-55 Sketch for Exercise 1

Exercise 2

Create the sketch of the model shown in Figure 2-56. The sketch is shown in Figure 2-57. The solid model and dimensions are given only for reference. Create the sketch using the sketching tools and then edit the sketch using the circular pattern tool and the trim tool.

(Expected time: 30 min)

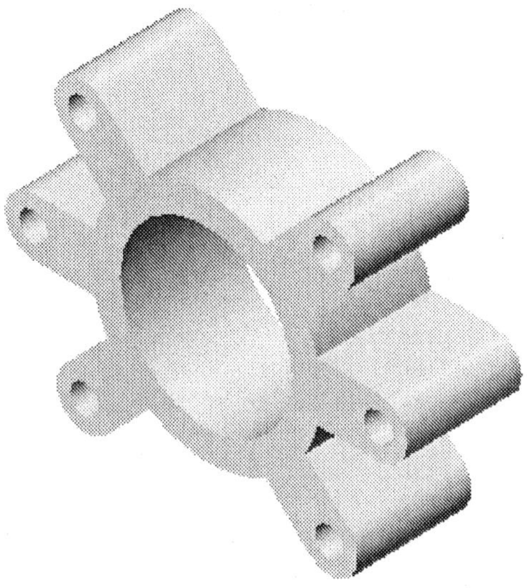

Figure 2-56 Solid model for Exercise 2

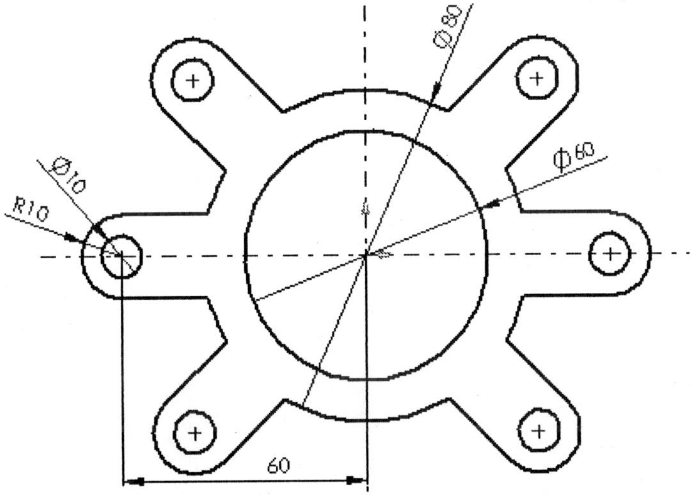

Figure 2-57 Sketch for Exercise 2

Exercise 3

Create the sketch of the model shown in Figure 2-58. The sketch is shown in Figure 2-59. This model is created using a revolved feature; therefore, you will create the sketch on one side of the centerline. The solid model and dimensions are given only for reference.

(Expected time: 30 min)

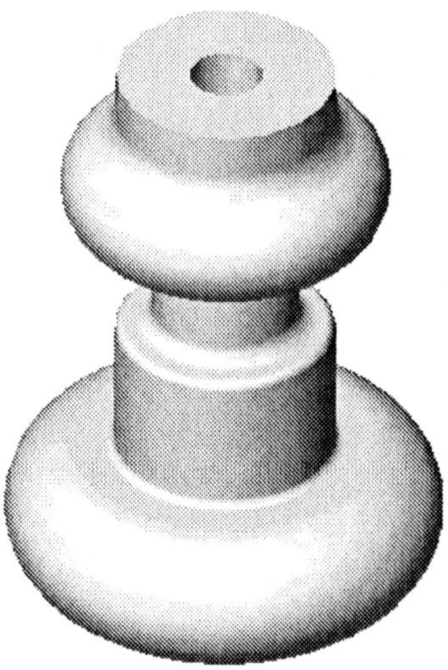

Figure 2-58 Solid model for Exercise 3

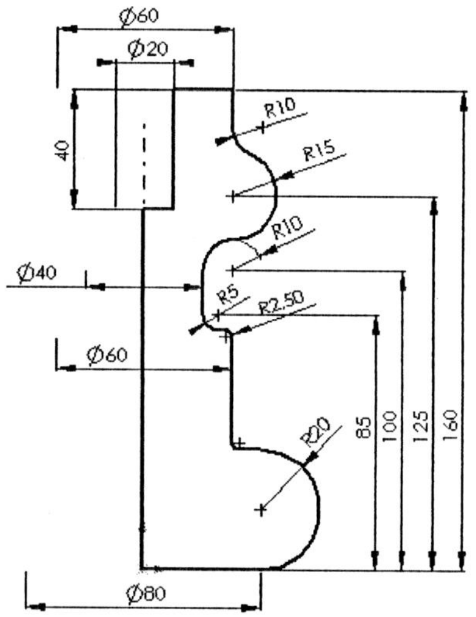

Figure 2-59 Sketch for Exercise 3

Exercise 4

Create the sketch of the model shown in Figure 2-60. The sketch is shown in Figure 2-61. This model is created using a revolved feature; therefore, you will create the sketch on one side of the centerline. The solid model and dimensions are given only for reference.

(Expected time: 30 min)

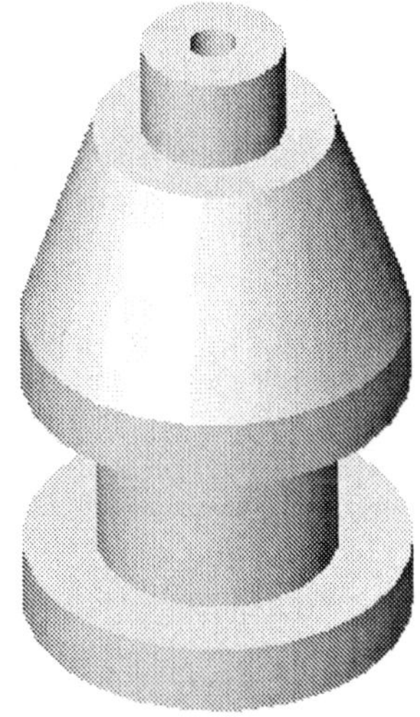

Figure 2-60 *Solid model for Exercise 4*

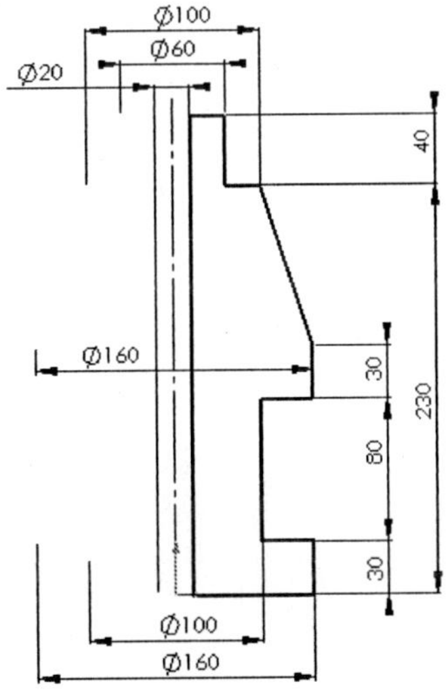

Figure 2-61 *Sketch for Exercise 4*

Answers to Self-Evaluation Test

1. T, **2.** F, **3.** T, **4.** F, **5.** F, **6.** F, **7. Linear Sketch Step and Repeat**, **8. Circular Sketch Step and Repeat**, **9. Select**, **Circle**, **10. Mirror Entities**

Chapter 3

Adding Relations and Dimensions to the Sketches

Learning Objectives

After completing this chapter you will be able to:

- *Add geometric relations to the sketch.*
- *Dimension the sketches.*
- *Modify the dimensions of the sketch.*
- *Understand the concept of fully defined sketch.*
- *View and examine the relations applied to the sketches.*
- *Open an existing file.*

ADDING GEOMETRIC RELATIONS TO THE SKETCH

Geometric relations are the logical operations that are performed to add a relationship (such as tangent or perpendicular) between the sketched entities, planes, axes, edges, or vertices. The relations applied to the sketched entities are used to capture the design intent. Geometric relations constrain the degree of freedom of the sketched entities. There are two methods to apply the relations to the sketch. These two methods are

1. Add Relations PropertyManager
2. Automatic Relations

Adding Relations Using the Add Relations PropertyManager

CommandManager:	Sketch > Add Relation
Menu:	Tools > Relations > Add
Toolbar:	Sketch > Add Relation

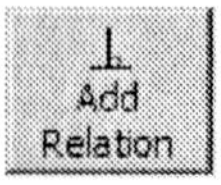

The **Add Relations PropertyManager** is widely used to apply relation to the sketch in the sketcher environment of SolidWorks. The **Add Relations PropertyManager** is invoked using the **Add Relation** button from the **Sketch CommandManager**. You can also invoke the **Add Relations PropertyManager** by right-clicking in the drawing area and choosing the **Add Relation** option from the shortcut menu. The **Add Relations PropertyManager** is shown in Figure 3-1. The confirmation corner is also displayed at the top right corner of the drawing area. The various options available in the **Add Relations PropertyManager** are discussed next.

*Figure 3-1 The **Add Relations PropertyManager***

Selected Entities

The **Selected Entities** rollout displays the name of the entities that are selected to apply the relations. The entities that you select are displayed in green color and are added in the area under the **Selected Entities** rollout. You can remove the selected entity from the selection set by selecting the same entity again in the drawing area.

> **Tip**. *You can also remove the selected entity from the selection set by selecting that entity in the **Selected Entities** rollout and right-clicking to invoke the shortcut menu. Choose the **Delete** option from this menu to remove the entity from the selection set. If you choose the **Clear Selections** option from the shortcut menu, all the entities will be removed from the selection set.*

Existing Relations

The **Existing Relations** rollout displays the relations that are already applied to the selected sketch entities. It also shows the status of the sketch entities. You can delete the already existing relation from this rollout. Select the already existing relation from the selection area and right-click to display the shortcut menu. Choose the **Delete** option from this shortcut menu to delete the selected relation. If you choose the **Delete All** option from the shortcut menu, all the relations displayed in the selection area of the **Existing Relations** rollout will be deleted.

Add Relations

The **Add Relations** rollout is used to apply the relations to the selected entity. The list of relations that can be applied to the selected entity or entities is shown in the **Add Relations** rollout. The most appropriate relation for the selected entities appears in bold letters.

> **Tip**. *You can apply a relation to a single entity or a relation between two or more than two entities. To apply a relation between two or more than two entities, at least one entity should be a sketched entity. The other entity or entities can be sketch entities, edges, faces, vertices, origins, plane, or axes. The sketch curves from other sketches that form lines or arcs when projected on the sketch plane can also be included in the relation.*

All the relations that can be applied to the sketches using the **Add Relations** rollout are discussed next.

Horizontal

The **Horizontal** relation forces one or more selected lines or centerlines to become horizontal. You can also select an external entity such as an edge, plane, axis, or sketch curve on an external sketch that projects as a line in the sketch to apply this relation. Using the **Horizontal** relation you can also force two or more points to become horizontal. A point can be a sketch point, a center point, an endpoint, a control point of a spline, or an external entity such as origin, vertex, axis, or point in an external sketch that projects as a point. To use this relation, invoke the **Add Relations PropertyManager**. Select the entity or entities to apply the **Horizontal** relation. Choose the **Horizontal** button from the **Add Relations** rollout provided in the **Add Relations**

PropertyManager. You will notice that the name of the horizontal relation is displayed in the **Existing Relations** rollout.

Vertical

The **Vertical** relation forces one or more selected lines or centerlines to become vertical. Using the **Vertical** relation, you can also force two or more points to become vertical. To use this relation, invoke the **Add Relations PropertyManager** and select the entity or entities to apply the **Vertical** relation. Choose the **Vertical** button from the **Add Relations** rollout. You will notice that the name of the vertical relation is displayed in the **Existing Relations** rollout.

Collinear

The **Collinear** relation forces the selected lines to lie on the same infinite line. To use this relation, select the lines to apply the **Collinear** relation. Choose the **Collinear** button from the **Add Relations** rollout.

Coradial

The **Coradial** relation forces the selected arcs or circles to share the same radius and the same center point. You can also select an external entity that projects as an arc or a circle in the sketch to apply this relation. To use this relation, invoke the **Add Relations PropertyManager**. Select two arcs or circles, or an arc and a circle to apply the **Coradial** relation. Choose the **Coradial** button from the **Add Relations** rollout.

Perpendicular

The **Perpendicular** relation forces the selected lines to become perpendicular to each other. To use this relation, invoke the **Add Relations PropertyManager**. Select two lines and choose the **Perpendicular** button from the **Add Relations** rollout. Figure 3-2 shows two lines before and after applying the perpendicular relation.

Parallel

The **Parallel** relation forces the selected lines to become parallel to each other. To use this relation, invoke the **Add Relations PropertyManager**. Select two lines and choose the **Parallel** button from the **Add Relations** rollout. Figure 3-3 shows two lines before and after applying this relation.

ParallelYZ

The **ParallelYZ** relation forces a line in the 3D sketch to become parallel to the YZ plane with respect to the selected plane. To use this relation, invoke the **Add Relations PropertyManager**. Select the line in the 3D sketch and a plane and choose the **ParallelYZ** button from the **Add Relations** rollout.

Note

You will learn more about the 3D curves in later chapters.

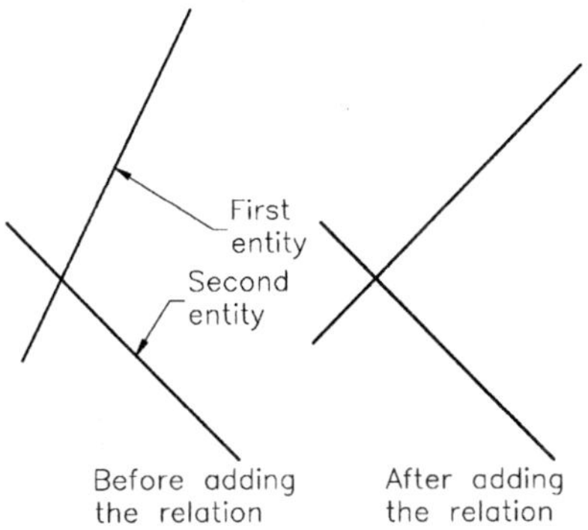

Figure 3-2 Entities before and after applying the ***Perpendicular*** *relation*

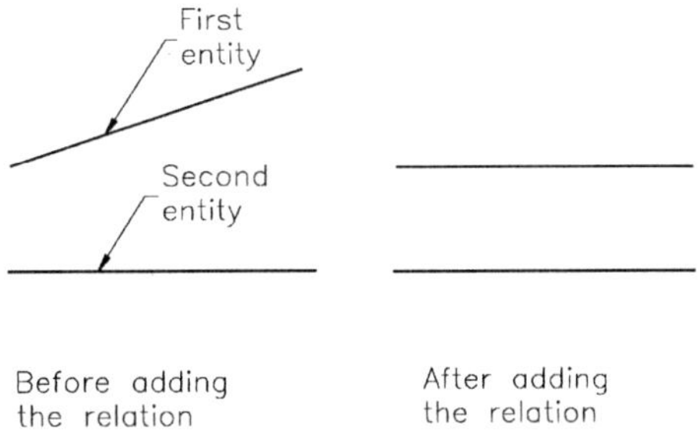

Figure 3-3 Entities before and after applying the ***Parallel*** *relation*

ParallelZX

The **ParallelZX** relation forces a line in the 3D sketch to become parallel to the ZX plane with respect to the selected plane. To use this relation, invoke the **Add Relations PropertyManager**. Select the line in the 3D sketch and a plane and choose the **ParallelZX** button from the **Add Relations** rollout.

AlongZ

The **AlongZ** relation forces a line in the 3D sketch to become normal to the selected plane. To use this relation, invoke the **Add Relations PropertyManager**. Select the line in the 3D sketch and a plane and choose the **AlongZ** button from the **Add Relations** rollout.

Tangent

The **Tangent** relation forces the selected arc, circle, spline, or ellipse to become tangent to other arc, circle, spline, ellipse, line, or edge. To use this relation, invoke the **Add Relations PropertyManager**. Select two entities and choose the **Tangent** button from the **Add Relations** rollout. Figures 3-4 and 3-5 show the entities before and after applying this relation.

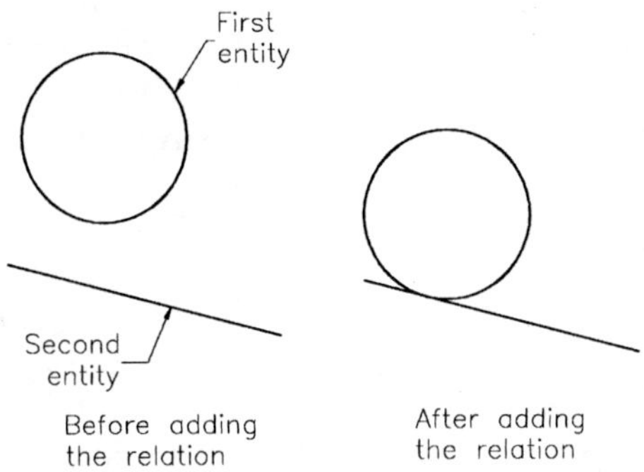

Figure 3-4 Applying the ***Tangent*** *relation to a line and a circle*

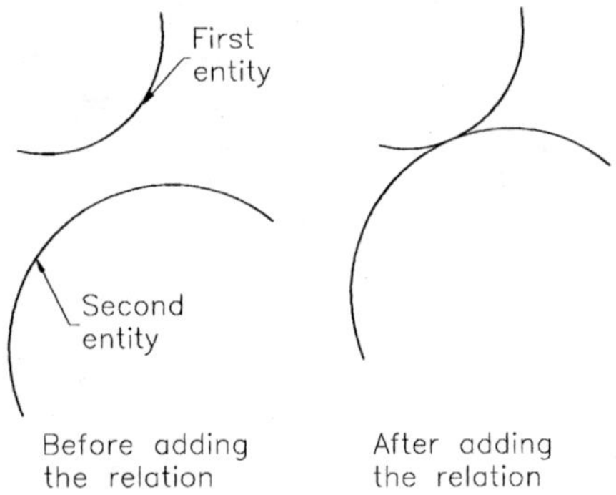

Figure 3-5 Applying the ***Tangent*** *relation to two arcs*

Concentric

The **Concentric** relation forces the selected arc or circle to share the same center point with other arc, circle, point, vertex, or a circular edge. To use this relation, invoke the **Add Relations PropertyManager**. Select the required entity to apply the **Concentric** relation and then choose the **Concentric** button from the **Add Relations** rollout.

Midpoint

The **Midpoint** relation forces the selected point to move to the midpoint of a selected line. To use this relation, invoke the **Add Relations PropertyManager**. Select the point to be moved and the line to the midpoint of which the point will be moved. Choose the **Midpoint** button from the **Add Relations** rollout.

Intersection

The **Intersection** relation forces the selected point to move at the intersection of two selected lines. To use this relation, invoke the **Add Relations PropertyManager**. Select the required entity to apply the **Intersection** relation. Choose the **Intersection** button from the **Add Relations** rollout.

Coincident

The **Coincident** relation forces the selected point to be coincident with the selected line, arc, circle, or ellipse. To use this relation, invoke the **Add Relations PropertyManager**. Select the required entity to apply the **Coincident** relation. Choose the **Coincident** button from the **Add Relations** rollout.

Equal

The **Equal** relation forces the selected lines to have equal length and selected arcs, circles, or an arc and a circle to have equal radii. To use this relation, invoke the **Add Relations PropertyManager**. Select the required entity to apply the **Equal** relation and choose the **Equal** button.

Symmetric

The **Symmetric** relation forces two selected lines, arcs, points, and ellipses to remain equidistant from a centerline. This relation also forces the entities to have the same size and orientation. To use this relation, invoke the **Add Relations PropertyManager**. Select the required entity to apply the **Symmetric** relation and select a centerline. Choose the **Symmetric** button from the **Add Relations** rollout.

Fix

The **Fix** relation forces the selected entity to be fixed at the specified position. If you apply this relation to a line or an arc, its location is fixed but you can change its size by dragging the endpoints. To use this relation, invoke the **Add Relations PropertyManager**. Select the required entity to apply the **Fix** relation and choose the **Fix** button.

Pierce

The **Pierce** relation forces a sketch point or an endpoint of an entity to be coincident with an entity of another sketch. To use this relation, invoke the **Add Relations PropertyManager**. Select the required entities to apply the **Pierce** relation and choose the **Pierce** button from the **Add Relations** rollout.

Merge Points

The **Merge Points** relation forces two sketch points or endpoints to merge in a single point. To use this relation, invoke the **Add Relations PropertyManager**. Select the required entities to apply the **Merge Points** relation and choose the **Merge Points** button from the **Add Relations** rollout.

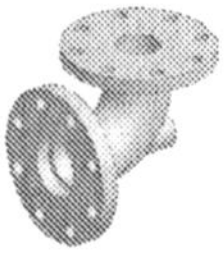

Tip. *You can also apply the relations using the* ***Properties PropertyManager****. This* ***PropertyManager*** *is automatically invoked if you select more than one entity from the drawing area. The possible relations for the selected geometry will be displayed in the* ***Add Relations*** *rollout. Choose the relation that you want to apply to the selected geometry.*

Another method of applying relations to the sketches is by selecting the entity or entities to which you have to apply the relation and then right-clicking to display the shortcut menu. The relation that can be applied to the selected entities will be displayed in the shortcut menu under the ***Relations*** *section. Choose the relation from the shortcut menu.*

Automatic Relations

Automatic relations are applied automatically to the sketch while drawing. For example, you will notice that when you specify the start point of the line and move the cursor horizontally toward the right or left, the symbol **H** is displayed below the line cursor. This is the symbol of the Horizontal relation that is applied to the line while drawing. If you move the cursor vertically downward or upward, the **V** symbol for Vertical relation is displayed below the line cursor. If you move the cursor to the intersection of two or more than two sketched entities, the intersection symbol appears below the cursor. Similarly, other relations are also automatically applied to the sketch while creating the sketch.

You can activate the automatic relations option if it is not available. Invoke the **System Options - Sketch** dialog box by choosing **Tools** > **Options** > to invoke the **System Options - General** dialog box and then select the **Sketch** option from the area on the left. Select the **Automatic Relations** check box from the **System Options - Sketch** dialog box and choose the **OK** button. The relations that are applied automatically are listed below.

1. Horizontal
2. Vertical
3. Coincident
4. Midpoint
5. Intersection
6. Tangent
7. Perpendicular

Tip. *You will observe that while sketching, two types of inferencing lines are displayed. One is displayed in blue and the second in brown. The brown inferencing line indicates that relation is applied automatically to the sketch. The blue inferencing line indicates that no automatic relation is applied.*

DIMENSIONING THE SKETCH

After drawing the sketches and adding the relations, dimensioning is the most important step in creating a design. As mentioned earlier, SolidWorks is a parametric software. The parametric nature of SolidWorks ensures that irrespective of the original size, the selected entity is driven by the dimension value that you specify. Therefore, when you apply and modify a dimension on an entity, it is forced to change its size in accordance with the specified dimension value.

In SolidWorks, you can use the **Smart Dimension** tool to dimension any kind of entity. You can also select the individual type of dimensioning tool from the **Dimension/Relations CommandManager**. This **CommandManager** is not available by default. To invoke it, right-click on the **Sketch CommandManager** and choose **Customize Command Manager** from the shortcut menu. The **Customize CommandManager** flyout is displayed. In this flyout, choose the **Dimension/Relations** option and then click anywhere on the screen. Again, right-click on the **Sketch CommandManager** and choose **Dimension/Relations** to display the tools available on this **CommandManager**. Figure 3-6 shows the default dimension types available in the **Dimension/Relations CommandManager**.

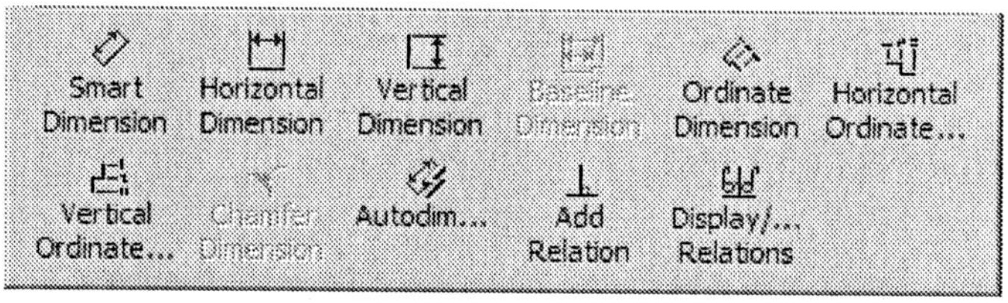

Figure 3-6 Dimension/Relations CommandManager

If you use the **Smart Dimension** tool, the type of dimension that will be applied will depend on the type of entity selected. For example, if you select a line, horizontal, vertical, or aligned dimension will be applied. If you select a circle, diametric dimension is applied. Similarly, if you select an arc, a radial dimension is applied. However, if you want to apply a particular type of dimension, choose its tool from the **Dimension/Relations CommandManager** and then apply the dimension.

In SolidWorks 2004, as soon as you place the dimension, the **Modify** dialog box is displayed, as shown in Figure 3-7. You can enter a new value in this box to modify the dimension. You can modify the default dimension value using the spinner or by entering a new value in the edit box available in the **Modify** dialog box.

The buttons available in the **Modify** dialog box are discussed next.

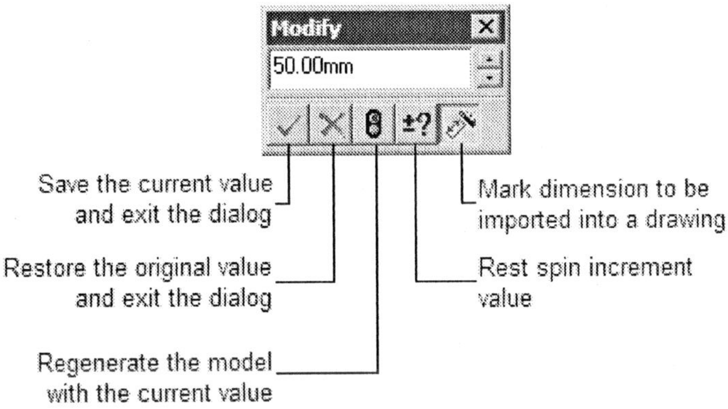

Figure 3-7 ***Modify*** *dialog box*

Tip. *If the* ***Modify*** *dialog box is not displayed when you place the dimension, you need to set its preference manually. To do this, invoke the* ***System Options - General*** *dialog box and choose the* ***Input dimension value*** *check box.*

The **Save the current value and exit the dialog** button is used to accept the current value and exit the dialog box. The **Restore the original value and exit the dialog** button is used to restore the last dimensional value applied to the sketch and exit the dialog box. The **Regenerate the model with the current value** button is used to preview the geometry of the sketch with the new modified dimensional value. The **Reset spin increment value** button is used to enter a new spin increment value. This is the value that is added or subtracted from the current value when you click once on the spinner arrow. When you choose this button the **Increment** dialog box is displayed. The **Mark dimensions to be imported into a drawing** button is chosen to make sure the selected dimension is generated as model annotation in the drawing views. If this button is not chosen, the dimensions will not be generated.

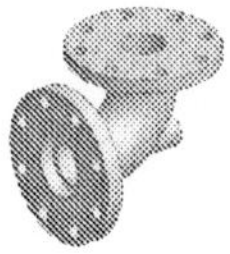

Tip. *You can also enter the arithmetic symbols directly into the edit box of the* ***Modify*** *toolbar to calculate the dimension. For example, if you have a dimension as a complex arithmetic function such as (220*12.5)-3+150, which is equal to 1247, you do not need to calculate this using the calculator. Just enter the statement in the edit box and press ENTER. SolidWorks will automatically solve the function to get the value of the dimension.*

The various types of dimensions that can be applied to the sketches in the sketching environment of SolidWorks 2004 are discussed next.

Horizontal/Vertical Dimensioning

CommandManager:	Dimension/Relations > Horizontal/Vertical Dimension
Menu:	Tools > Dimension > Horizontal/Vertical
Toolbar:	Dimension/Relations > Horizontal/Vertical Dimension

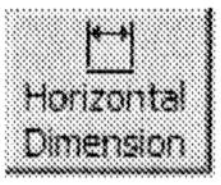

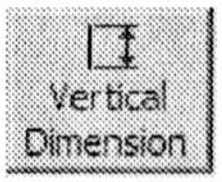

These dimensions are used to define horizontal or vertical dimensions of a selected line or between two points. The points can be the endpoints of lines or arcs, or the center points of circles, arcs, ellipses, or parabolas. You can dimension a vertical or a horizontal line by directly selecting it. Choose the **Horizontal/Vertical Dimension** button from the **Dimension/Relations CommandManager**. You can also right-click in the drawing area and choose the **More Dimensions > Horizontal/Vertical** option from the shortcut menu to activate these tools. When you move the cursor on the line, it is highlighted and turns red in color. As soon as you select the line, it turns to green, and the dimension is attached to the cursor. Move the cursor and place the dimension at an appropriate place using the left mouse button. As the priority of editing the dimension as it is placed is set by default, the **Modify** dialog box is displayed with the default value in it. Enter the new value of dimension in the **Modify** dialog box and press ENTER. Figure 3-8 shows horizontal and vertical dimensioning of lines.

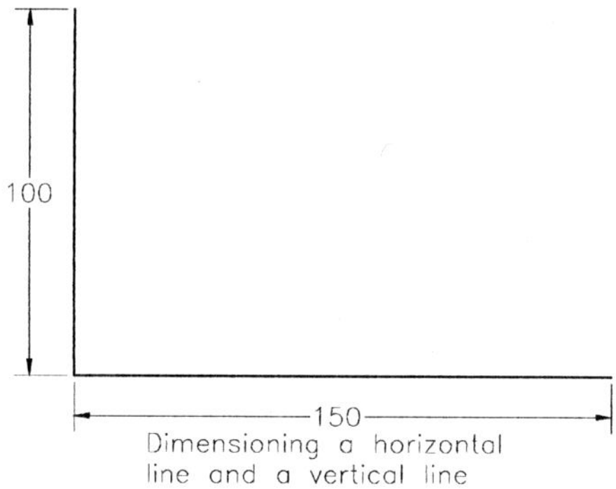

***Figure 3-8** Linear dimensioning of lines*

If the dimension is selected in the drawing area, the **Dimension PropertyManager** is displayed at the left of the drawing area as shown in Figure 3-9. The various options available in the **Dimension PropertyManager** are discussed next.

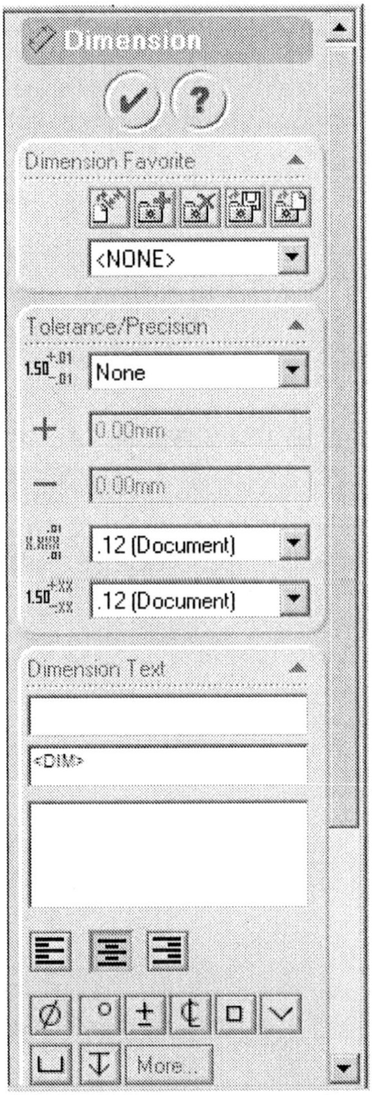

*Figure 3-9 The **Dimension PropertyManager***

Dimension Favorite

The **Dimension Favorite** rollout shown in Figure 3-10 is used to create, save, delete, and retrieve the dimension style in the current document. You can also retrieve the dimensions styles saved in other documents using this rollout. The options available in this rollout are discussed next.

Apply the default attributes to selected dimensions

The **Apply the default attributes to selected dimensions** button is used to apply the default attributes to the selected dimension or dimensions. The attributes include the tolerance, precision, arrow style, dimension text, and so on. This option is generally used when you have modified the settings applied to a dimension and then you want to restore the default settings on that dimension.

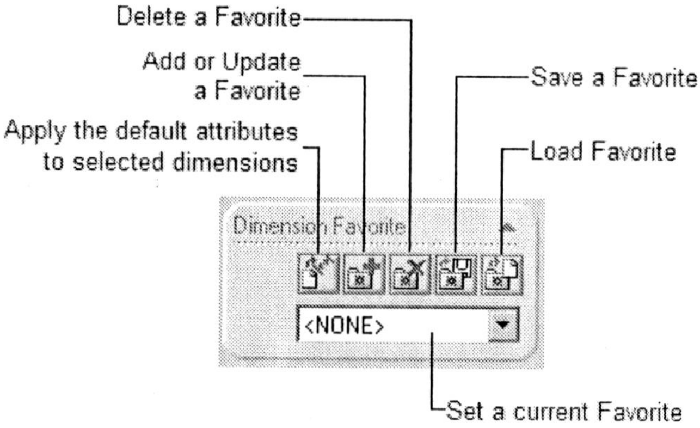

*Figure 3-10 The **Dimension Favorite** rollout*

Add or Update a Favorite

The **Add or Update a Favorite** button is used to add a dimension style to the current document for a selected dimension. After invoking the **Dimension PropertyManager**, set the attributes using various options provided in this **PropertyManager**. Now choose the **Add or Update a Favorite** button. The **Add or Update a Favorite** dialog box is displayed, as shown in Figure 3-11. Enter the name of the dimension style in the edit box and press ENTER or choose the **OK** button from this dialog box. The dimension style will be added to the current document.

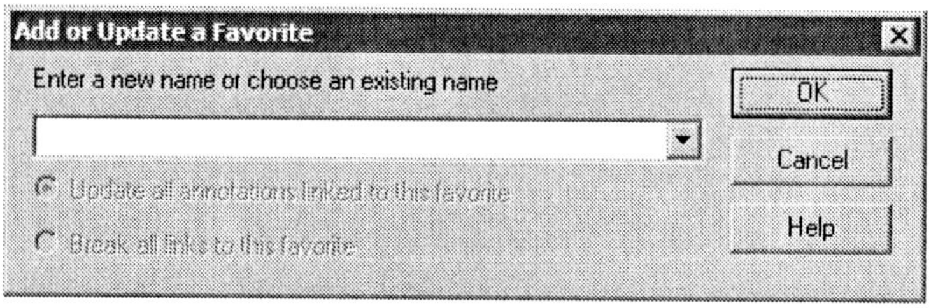

*Figure 3-11 The **Add or Update a Favorite** dialog box*

You can apply the new dimension style to the selected dimension by selecting the dimension style from the drop-down list in the **Dimension Favorite** rollout. You can also update the dimension style. To update a dimension style, select the dimension and set the options of the dimension style according to your need and then choose the **Add or Update a Favorite** button to invoke the **Add or Update a Favorite** dialog box. Select the dimension style to update from the drop-down list provided in the dialog box. The two radio buttons in this dialog box are enabled. Select the **Update all the annotations linked to this favorite** radio button and choose the **OK** button to update all the dimensions linked with the selected favorite. If you select the **Break all links to this favorite** radio button and choose the **OK**

Tip. *You can load more than one favorite by pressing the CTRL key and selecting the favorites from the* ***Open*** *dialog box. All the favorites will be displayed in the* ***Set a current Favorite*** *drop-down list.*

button, then the link between the other dimensions having the same favorite and the selected favorite will be broken.

Delete a Favorite

The **Delete a Favorite** button is used to delete a dimension style. Select the dimension style or favorite from the **Set a current Favorite** drop-down list and choose the **Delete a Favorite** button. Note that even after you delete the dimension style, the properties of the dimensions will be the same as with the deleted favorite. You can set the properties of a dimension to the default settings using the **Apply the default attributes to the selected dimension** button.

Save a Favorite

The **Save a Favorite** button is used to save a dimension style so that it can be retrieved in some other document. Select the dimension style or favorite from the **Set a current Favorite** drop-down list and choose the **Save a Favorite** button. The **Save As** dialog box will be displayed. Browse the folder in which you want to save the favorite and enter the name of the favorite in the **File name** edit box. Choose the **Save** button from the **Save As** dialog box. The extension for the file in which the favorite is saved is *.sldfvt*.

Load Favorite

The **Load Favorite** button is used to open a saved favorite in the current document. The properties of that favorite will be applied to the selected dimension. To load a favorite, choose the **Load favorite** button to invoke the **Open** dialog box. Browse the folder in which the favorite is saved. Now, select the file with the extension *.sldfvt* and choose the **Open** button. The **Add or Update a Favorite** dialog box is displayed. Choose the **OK** button from this dialog box.

Tolerance/Precision Rollout

The **Tolerance/Precision** rollout shown in Figure 3-12 is used to specify the tolerance and precision in the dimensions. The various options available in this rollout are discussed next.

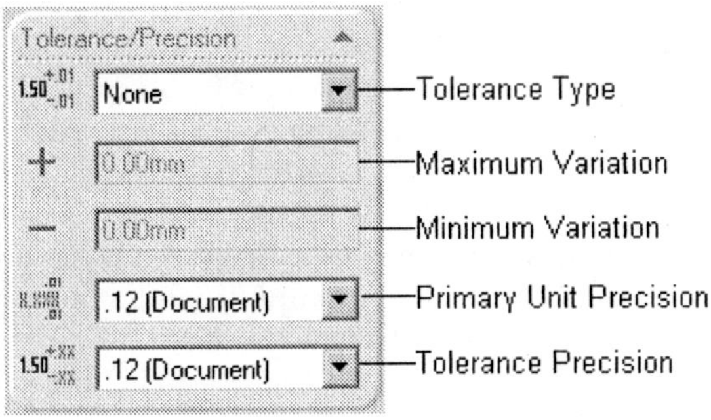

*Figure 3-12 The **Tolerance/Precision** rollout*

Tolerance Type

The **Tolerance Type** drop-down list is used to apply the tolerance to the dimension. By default, the **None** option is selected. Therefore, no tolerance is applied to the dimensions. The other tolerance methods available in this drop-down list are discussed next.

Basic

The basic dimension is the one that is enclosed in a rectangle. To display a basic dimension, select a dimension that you want to display as a basic dimension and then select the **Basic** option from the **Tolerance Display** drop-down list. You will notice that the dimension is enclosed in a rectangle, suggesting that it is a basic dimension. The basic dimension is shown in Figure 3-13.

Bilateral

The bilateral tolerance provides the maximum and the minimum variation in the value of the dimension that is acceptable in the design. To apply the bilateral tolerance, select the dimension and then select the **Bilateral** option from the **Tolerance Type** drop-down list. The **Maximum Variation** and the **Minimum Variation** edit boxes are enabled. These edit boxes are used to apply the value of the maximum and minimum variation to a dimension. If you select the **Show parentheses** check box, the bilateral tolerance will be displayed with parentheses. This check box is available when you apply the bilateral tolerance. The dimension with bilateral tolerance is shown in Figure 3-14.

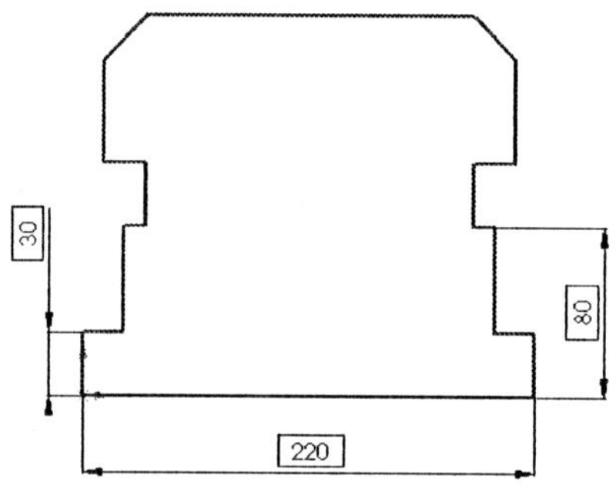

Figure 3-13 Basic dimension

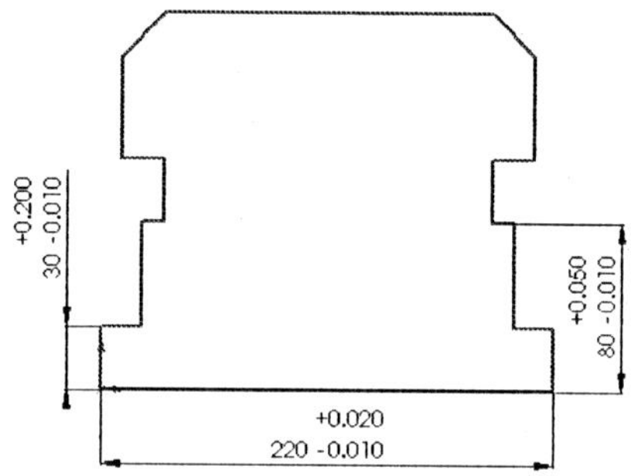

Figure 3-14 Bilateral tolerance

Note
The dimension standard used in the sketches in this book is the ISO standard.

Limit

In the limit dimension, the dimension is displayed as its maximum and minimum values that are allowed in the design. To apply this tolerance type, select the dimension to display as limit dimension and select the **Limit** option. The **Maximum Variation** and the **Minimum Variation** edit boxes are enabled to enter the value of maximum and minimum variation. The dimension along with limit tolerance is shown in Figure 3-15.

Symmetric

The symmetric tolerance is displayed with plus and minus signs. To use this tolerance, select the dimension and select the **Symmetric** option. The **Maximum Variation** edit box will be displayed to enter the value of tolerance. You can select the **Show parentheses** check box to show the tolerance in parentheses. The dimension along with the symmetric tolerance is shown in Figure 3-16.

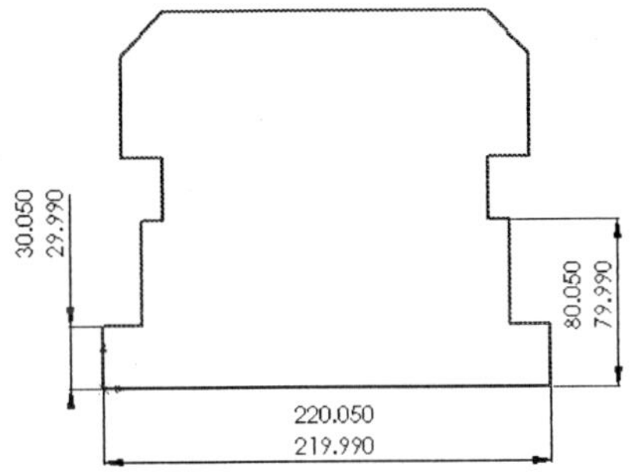

Figure 3-15 Limit tolerance

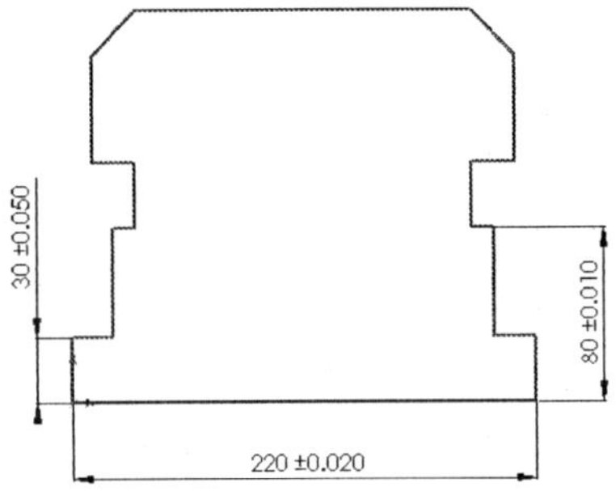

Figure 3-16 Symmetric tolerance

MIN

In this type of dimensional tolerance, the **min.** symbol is added to the dimension as a suffix. This implies that the dimensional value is the minimum value that is allowed in the design. To display this dimensional tolerance, select the dimension and select the **MIN** option from the **Tolerance Type** drop-down list. The dimension along with the minimum tolerance is shown in Figure 3-17.

MAX

In this type of dimensional tolerance, the **max.** symbol is added to the dimension as suffix. This implies that the dimensional value is the maximum value that is allowed in the design. To display this dimensional tolerance, select the dimension and select the **MAX** option from the **Tolerance Type** drop-down list. The dimension along with the maximum tolerance is shown in Figure 3-18.

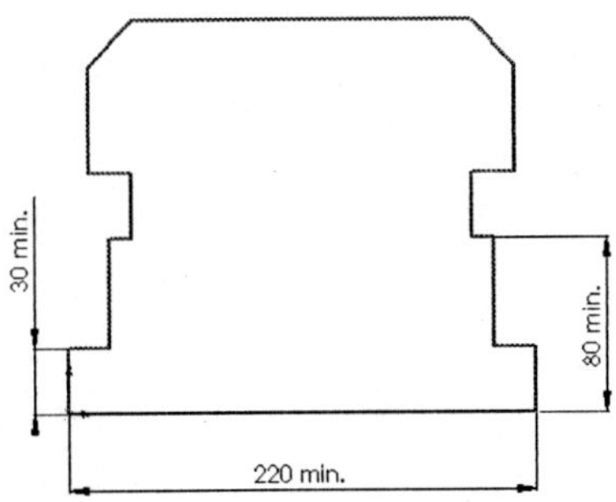

Figure 3-17 Minimum tolerance

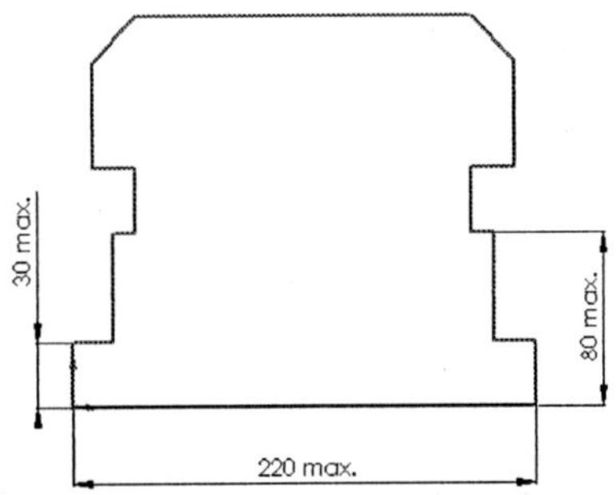

Figure 3-18 Maximum tolerance

Fit

The **Fit** option is used to apply the fit according to the **Hole Fit** and the **Shaft Fit** system. The **Tolerance/Precision** rollout with the **Fit** option selected is shown in Figure 3-19. Specify the type of fit from the **Classification** drop-down list. The **Classification** drop-down list is used to define **User Defined** fit, **Clearance** fit, **Transitional** fit, and **Press** fit. To apply the fit using the hole fit system or the shaft fit system, select the dimension and select the **Fit** option from the **Tolerance Type** drop-down list. The **Classification** drop-down list, the **Hole Fit** drop-down list, and the **Shaft Fit** edit drop-down list are displayed below the **Tolerance Type** drop-down list. Choose the type of fit from the **Classification** drop-down list and select the standard of fit from the **Hole Fit** drop-down list or the **Shaft Fit** drop-down list. If you choose the **Clearance**, **Transitional**, or **Press** option from the **Classification** drop-down list and you choose the standard of fit from the **Hole Fit** drop-down list then only the standards that match with the selected hole fit will be displayed in the **Shaft Fit** drop-down list and vice-versa. If you choose the **User Defined** option from the **Classification** drop-down list then you can choose any standard from the **Hole Fit** and the **Shaft Fit** drop-down lists.

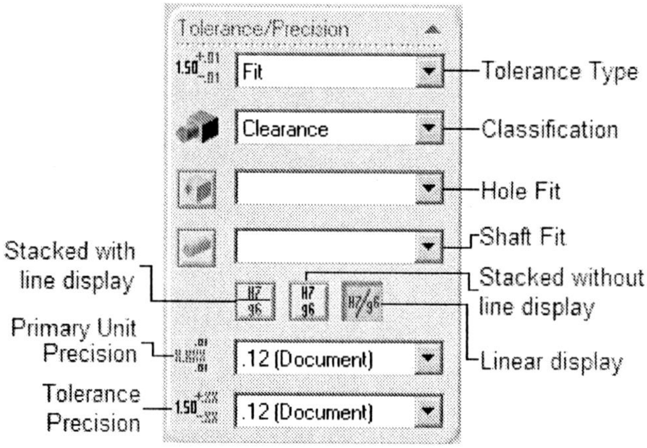

*Figure 3-19 The **Tolerance/Precision** rollout with the **Fit** option selected from the **Tolerance Type** drop-down list.*

The **Stacked with line display** button provided under the **Shaft Fit** drop-down list is chosen to display stacked tolerance with a line. You can also display the tolerance as stacked without a line using the **Stacked without line display** button. If you choose the **Linear display** button, the tolerance will be displayed in linear form. The dimension along with the hole fit and the shaft fit is shown in Figure 3-20.

Fit with tolerance

The **Fit with tolerance** option in the **Tolerance Type** drop-down list is used to display the tolerance along with the hole fit and shaft fit in a dimension. To apply the fit with the tolerance, select the dimension and select the **Fit with tolerance** option from the **Tolerance Type** drop-down list. Select the type of fit from the **Classification** drop-down list. Now, select the fit standard from the **Hole Fit** drop-down list or the **Shaft Fit** drop-down list. The tolerance will be displayed with the fit standard only if you select only one fit system either from the hole drop-down list or from the shaft drop-down list. The tolerance will be displayed along with the fit standard in the drawing area. In this release of SolidWorks the tolerance is calculated automatically depending on the type of fit selected and the standard of fit selected. The **Show parentheses** check box can be selected to show the tolerance in parentheses. The dimension along with fit and tolerance is shown in Figure 3-21.

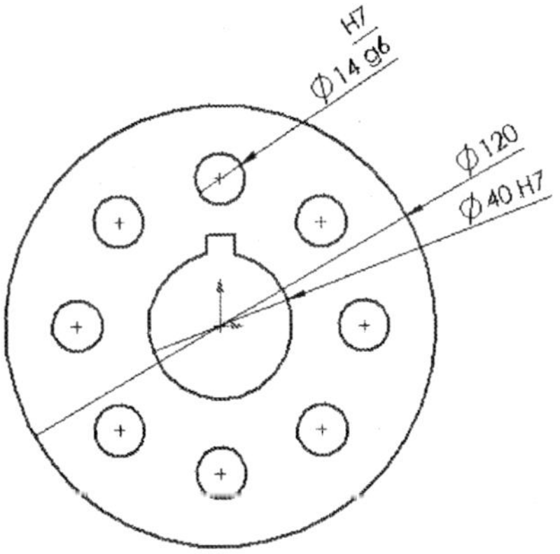

Figure 3-20 *Hole fit and shaft fit*

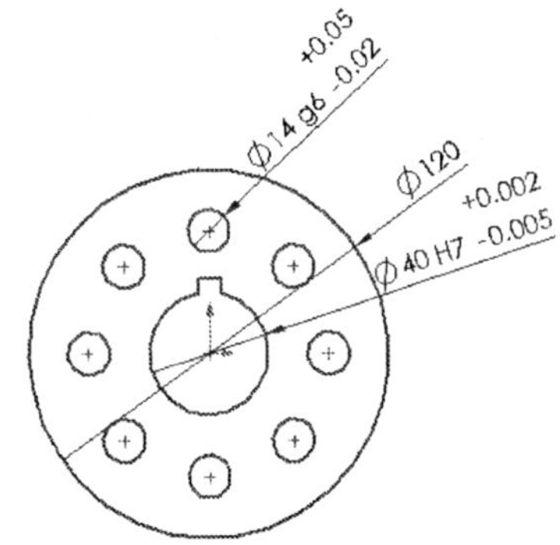

Figure 3-21 *Fit with tolerance*

Note
The diametrical dimension will be discussed later in this chapter.

Primary Unit Precision

The **Primary Unit Precision** drop-down list is used to specify the precision of the number of places after the decimal for dimensions. By default, the selected precision is two places after the decimal.

Tolerance Precision

The **Tolerance Precision** drop-down list is used to specify the precision of the number of places after the decimal for tolerance. By default, the selected precision is two places after the decimal.

Dimension Text

The **Dimension Text** rollout displayed in Figure 3-22 is used to add the text and symbols to the dimension. Three text boxes are provided in this rollout. These three text boxes are the three lines in which the text or symbols can be added. The first text box in this rollout is used to enter the text or symbols in the first line, which is above the dimension value. The second text box is used to enter the text or symbols in the second line, in which the dimension is also displayed. The third text box is used to enter the text or symbols in the third line, which is below the dimension value.

This rollout also provides buttons to add the symbols such as **Diameter**, **Degree**, **Plus/Minus**, **Centerline**, and so on to the dimension text. You can invoke the **Symbols** dialog box to add

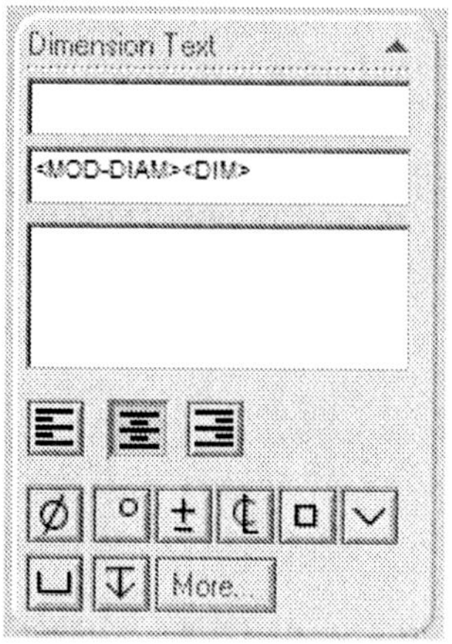

***Figure 3-22 Dimension Text** rollout*

more symbols by choosing the **More Symbols** button from the **Dimension Text** rollout. When you choose this button, the **Symbols** dialog box is displayed as shown in Figure 3-23.

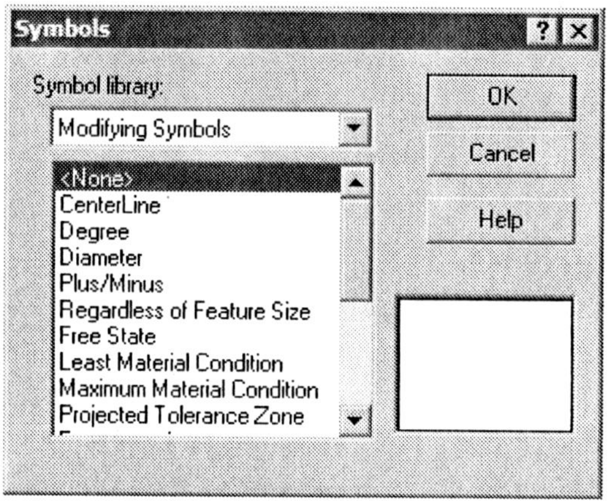

***Figure 3-23** The **Symbols** dialog box*

Arrows

The **Arrows** rollout shown in Figure 3-24 is used to specify the arrow style in the dimensions. The various options available in this rollout are discussed next.

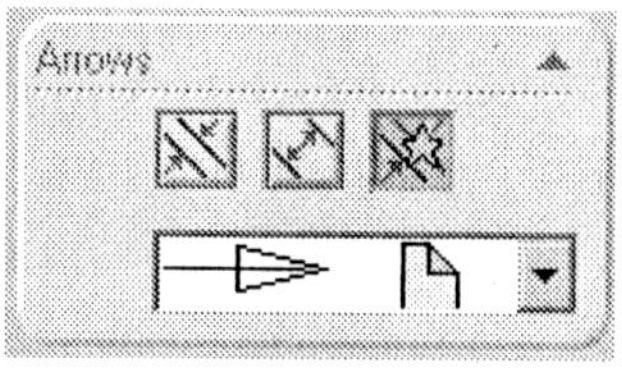

***Figure 3-24** The **Arrows** rollout*

Outside

The **Outside** button is used to display the arrows outside the dimension line. Select a dimension from the drawing area and choose the **Outside** button from the **Arrows** rollout.

Inside

The **Inside** button is used to display the arrows inside the dimension line. Select a dimension from the drawing area and choose the **Inside** button from the **Arrows** rollout.

Note

You can also click on the control point displayed on the arrowhead to reverse the direction of the arrowheads.

Smart

The **Smart** button is used to display the dimension inside or outside the dimension line, depending on the surrounding geometry. The **Smart** button is chosen by default in the **Arrows** rollout.

Style

The **Style** drop-down list is used to select the arrowhead style. The unfilled triangular arrow is select by default. You can select any arrowhead style for a particular dimension or dimension style. To change the arrow style, select the dimension from the drawing area and then select the arrowhead style from the **Style** drop-down list.

The **More Properties** button is used to invoke the **Dimension Properties** dialog box to modify the properties of the dimension. All the options available in the **Dimension PropertyManager** are available in the **Dimension Properties** dialog box with some additional options. You can modify the dimension properties from the **Dimension PropertyManager** or from the **Dimension Properties** dialog box.

Horizontal/Vertical Dimensioning Between Points

As mentioned earlier, you can add a horizontal or a vertical dimension between two points. To add horizontal or vertical dimension between two points, choose the required button from the **Dimension/Relations CommandManager**. Select the first point, and then select the second point. Specify a point to place the dimension. The **Modify** dialog box will be displayed. Enter a new dimension value in this dialog box and press ENTER. Figures 3-25 and 3-26 show the horizontal and vertical linear dimensioning between points.

You can also apply horizontal or vertical (linear) dimensioning to the circle. However, note that these type of dimensions can be applied to the circle using only the **Smart Dimension** tool. To apply this type of dimension to the circle, choose the **Smart Dimension** button and select the circle. The dimension will be attached to the cursor. If you want to create the vertical dimension, move the cursor to the right or left of the sketch. If you want to create the horizontal dimension, move the cursor to the top or bottom of the sketch. Using the left mouse button,

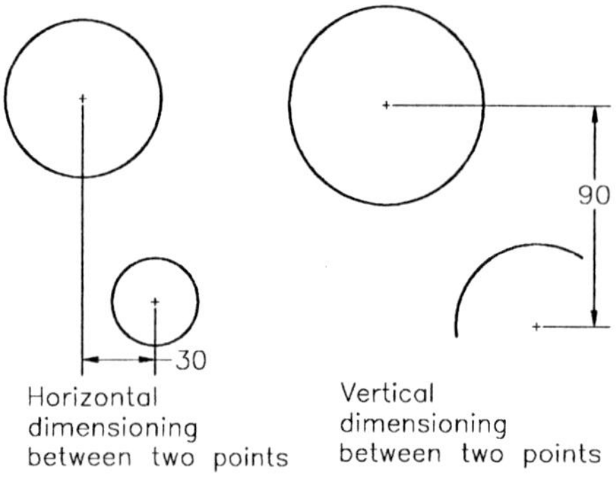

Figure 3-25 Applying horizontal and vertical dimensions to two points

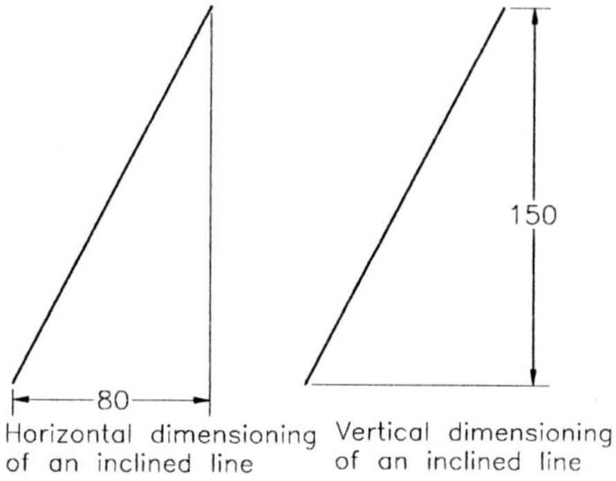

Figure 3-26 Applying horizontal and vertical dimensions to inclined lines

place the dimension and enter a new value in the **Modify** dialog box. The linear dimensioning of the circle is shown in Figure 3-27.

Aligned Dimensioning

CommandManager:	Sketch > Smart Dimension
Menu:	Tools > Dimension > Smart
Toolbar:	Sketch > Smart Dimension

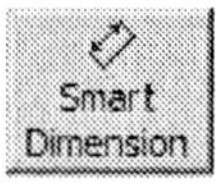

Aligned dimensions are used to dimension lines that are at an angle with respect to the X axis and the Y axis. These type of dimensions measure the actual distance of the inclined lines. You can directly select the inclined line to apply this dimension or select two points. The points that can be used to apply aligned dimensions include the endpoints of line, arc, parabolic arc, or spline and the center points of arcs, circles, ellipse, or parabolic arc. To apply an aligned dimension to an inclined line, choose the **Smart Dimension** button from the **Sketch CommandManager** and select the line. Move the cursor at an angle such that the dimension line is parallel to the inclined line. Place the dimension at an appropriate place and enter a new value in the **Modify** dialog box.

To apply an aligned dimension between two points, choose the **Smart Dimension** button and select the first point to which you have to apply the dimension. Next, select the second point to apply the dimension. The dimension will be attached to the cursor. Move the cursor such that the dimension line is parallel to the imaginary line that joins two points. Now, place the dimension at an appropriate location. Enter a new value in the **Modify** dialog box and press ENTER. Figure 3-28 shows aligned dimensioning of a line and an aligned dimension between two points.

Angular Dimensioning

CommandManager:	Sketch > Smart Dimension
Menu:	Tools > Dimension > Smart
Toolbar:	Sketch > Smart Dimension

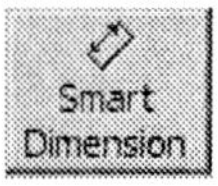

Angular dimensions are used to dimension angles. You can select two line segments to apply the angular dimensions or use three points to apply the angular dimensions.

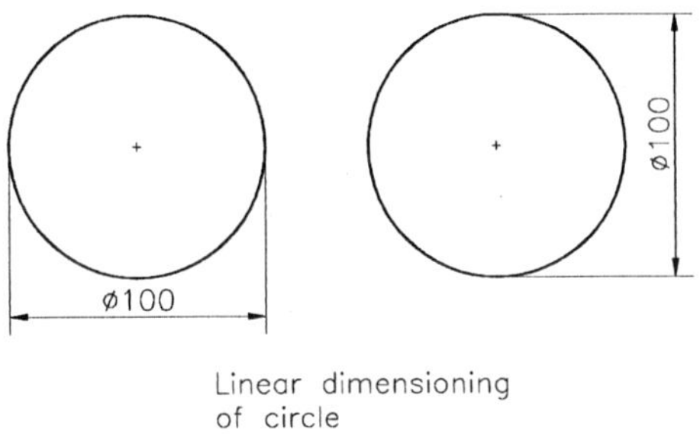

Figure 3-27 Linear dimensioning of circle

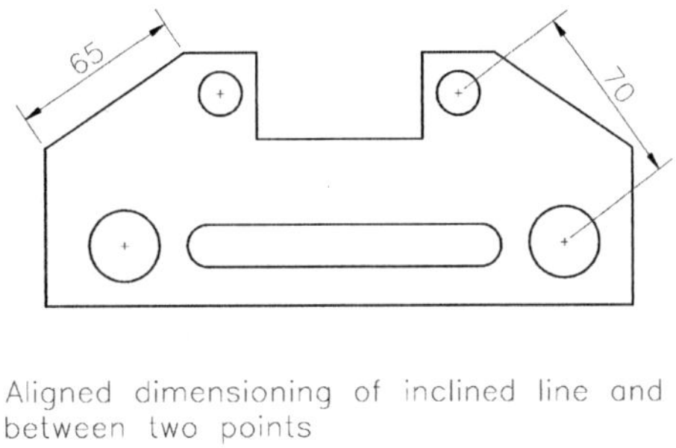

Figure 3-28 Aligned dimensioning

You can also use angular dimensioning to dimension an arc. All these options of angular dimensioning are discussed next.

Angular Dimensioning Using Two Line Segments

To apply angular dimensions to two lines, choose the **Smart Dimension** button from the **Sketch CommandManager** and select the first line segment. A dimension will be attached to the cursor. Now, select the second line segment. An angular dimension will be attached to the cursor. Place the angular dimension and enter the new value of angular dimension in the **Modify** dialog box. You have to be very careful while placing the angular dimension. This is because depending on the location of the dimension placement, the interior angle, exterior angle, major angle, or minor angles are displayed. Figures 3-29 through 3-32 illustrate various angular dimensions depending upon the dimension placement point.

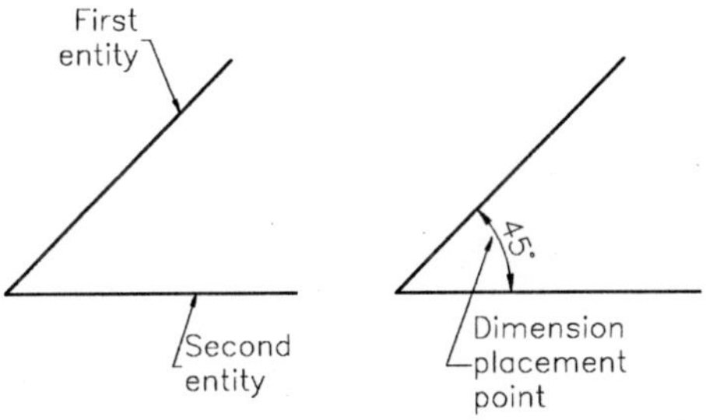

Figure 3-29 Angular dimension displayed according to the dimension placement point

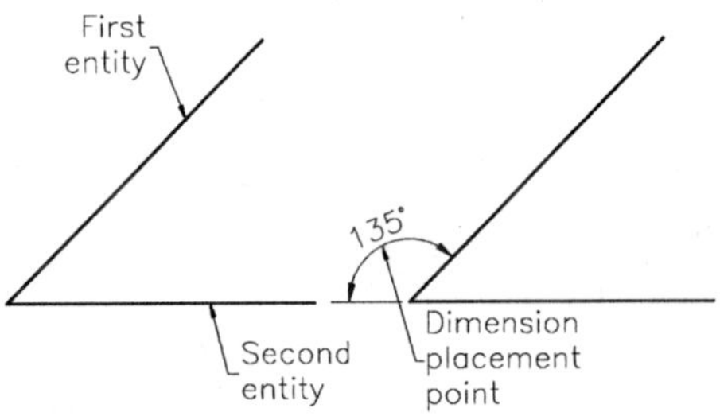

Figure 3-30 Angular dimension displayed according to the dimension placement point

Angular Dimensioning Using Three Points

While adding angular dimensions to three points, you need to be extremely careful in selecting the points. Remember that the first point that you select is the angle vertex point. To apply angular dimensions using three points, choose the **Smart Dimension** button from the **Sketch CommandManager**. Select the first point using the left mouse button. This is the angle vertex point. Select the second point. A linear dimension is attached to the cursor. Next, select the third point; an angular dimension is attached to the cursor. Place the angular dimension and

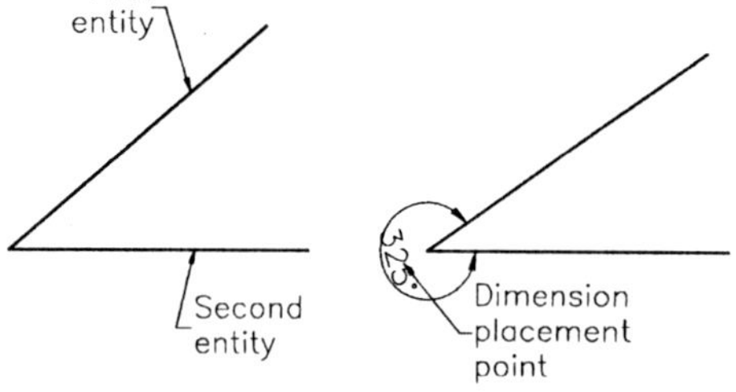

Figure 3-31 Angular dimension displayed according to the dimension placement point

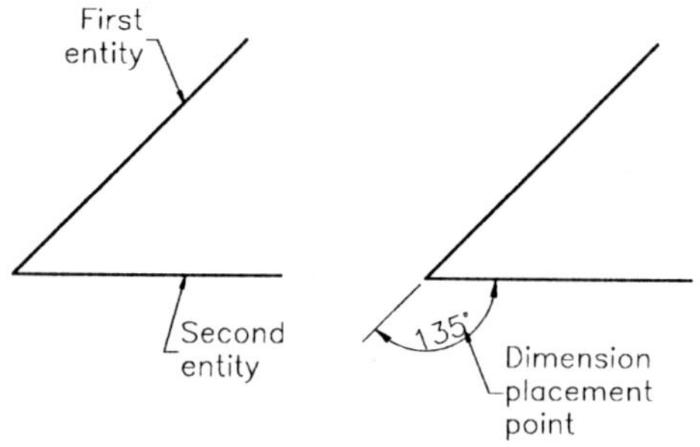

Figure 3-32 Angular dimension displayed according to the dimension placement point

enter a new value of angular dimension in the **Modify** dialog box. Figure 3-33 shows the angular dimensioning using three points.

Angular Dimensioning of an Arc

You can use angular dimensions to dimension an arc. In case of arcs, the three points that should be used are the endpoints of the arc and the center point of the arc. Figure 3-34 shows the angular dimensioning of an arc.

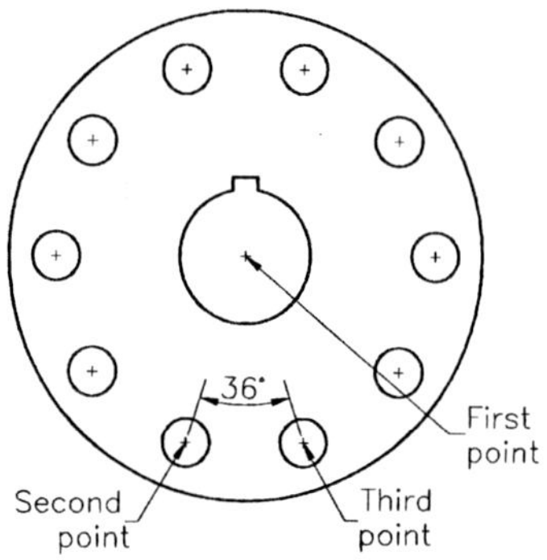

Figure 3-33 Angular dimension displayed according to the dimension placement point

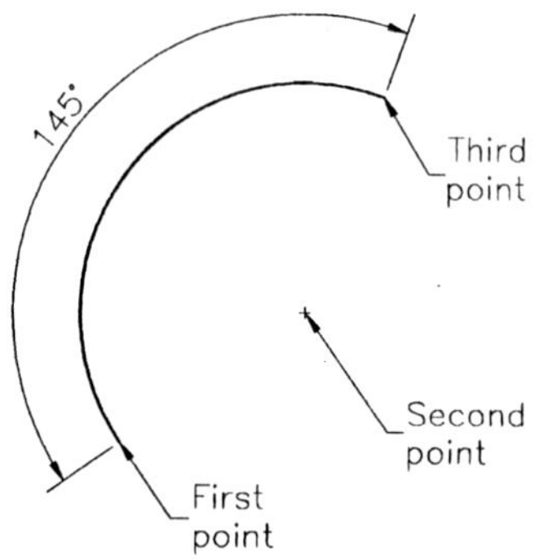

Figure 3-34 Angular dimension displayed according to the dimension placement point

Diameter Dimensioning

CommandManager:	Sketch > Smart Dimension
Menu:	Tools > Dimension > Smart
Toolbar:	Sketch > Smart Dimension

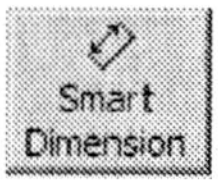

Diameter dimensions are applied to dimension a circle or an arc in terms of its diameter. To apply the diameter dimension, choose the **Smart Dimension** button from the **Sketch CommandManager**. Select the circle or the arc and place the dimension. In SolidWorks, when you select a circle to dimension, the diameter dimension is applied to it by default. However, when you select an arc, the radius dimension is displayed to an arc. To apply the diameter dimension to an arc, select the arc and place the radius dimension. Next right-click to display the shortcut menu and choose the **Properties** option from the shortcut menu. The **Dimension Properties** dialog box is displayed. Select the **Diameter dimension** check box and choose **OK**. Figure 3-35 shows a circle and an arc with the diameter dimension.

Radius Dimensioning

CommandManager:	Sketch > Smart Dimension
Menu:	Tools > Dimension > Smart
Toolbar:	Sketch > Smart Dimension

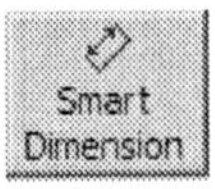

Radius dimensions are applied to dimension a circle or an arc in terms of its radius. As mentioned earlier, by default the dimension applied to a circle is in the diameter form and the dimension applied to an arc is a radius dimension. To apply radius dimension, choose the **Dimension** button from the **Sketch** toolbar and select the arc. A radius dimension will be attached to the cursor. Using the left mouse button place the dimension at an appropriate place. To convert the radius dimension to the diameter dimension you need to select the **Diameter dimension** check box from the **Dimension Properties** dialog box. Figure 3-36 illustrates the radial dimensioning of a circle and an arc.

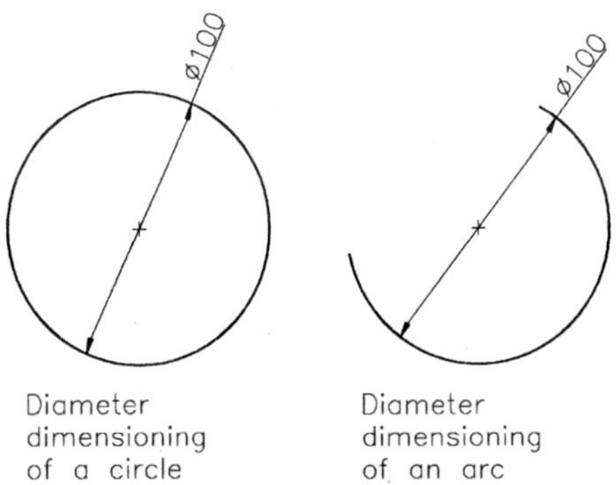

Figure 3-35 Diameter dimensioning of a circle and an arc

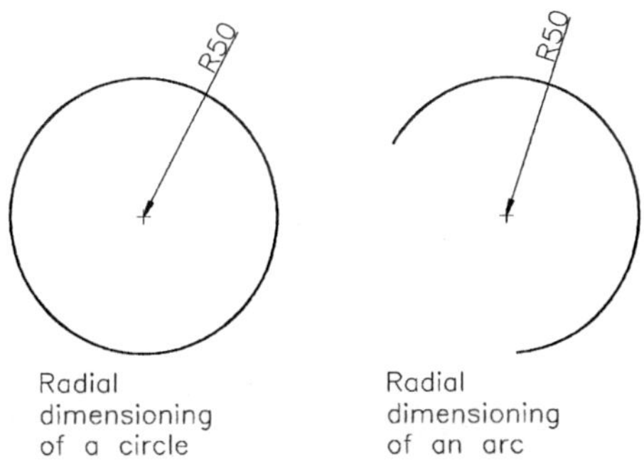

Figure 3-36 Radial dimensioning of a circle and an arc

Linear Diameter Dimensioning

CommandManager:	Sketch > Smart Dimension
Menu:	Tools > Dimension > Smart
Toolbar:	Sketch > Smart Dimension

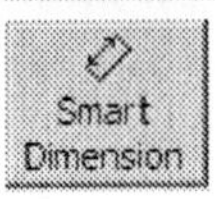

Linear diameter dimensioning is used to dimension the sketch of a revolved component. An example of a revolved component is shown in Figure 3-37. The sketch for a revolved component is drawn using simple sketcher entities as shown in Figure 3-38. If you dimension the sketch of the base feature of the given model using the linear dimensioning method then the same dimensions will be generated in drawing views. This may be confusing because in the shop floor drawing, you need diameter dimension of a revolved model. To overcome this problem, it is recommended that you create a linear diameter dimension as shown in Figure 3-38. To create the linear diameter dimension, choose the **Smart Dimension** button from the **Sketch CommandManager**. Select the entity to be dimensioned and then select the centerline around which the sketch will be revolved. Move the cursor to the other side of the centerline. A linear diameter dimension is displayed. Place the dimension and enter a new value in the **Modify** dialog box.

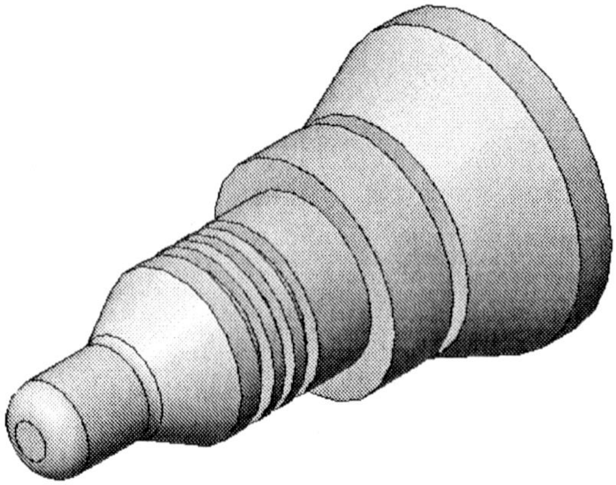

Figure 3-37 *A revolved component*

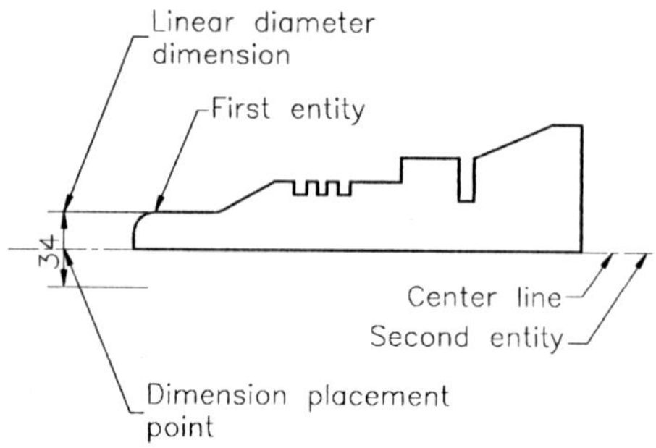

Figure 3-38 *Sketch for the revolved feature and linear diameter dimension*

Ordinate Dimensioning

The ordinate dimensions are used to dimension the sketch with respect to a specified datum. Depending on the requirement of the design, the datum can be an entity in the sketch or can be the origin. The ordinate dimensions are of two types, horizontal and vertical. The method of creating both these types of ordinate dimensions is discussed next.

Horizontal Ordinate Dimensioning

CommandManager:	Dimension/Relations > Horizontal Ordinate Dimension
Menu:	Tools > Dimension > Horizontal Ordinate
Toolbar:	Dimension/Relations > Horizontal Ordinate Dimension

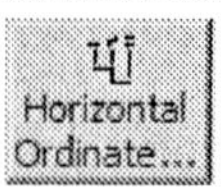

Horizontal ordinate dimensions are used to dimension the horizontal distances of the selected entities from the specified datum, see Figure 3-39. Note that when you create the ordinate dimensions, the **Modify** dialog box is not displayed to modify the dimension value. After placing all the ordinate dimensions, you need to exit the tool and then double-click on the dimensions to modify their values.

To add horizontal ordinate dimension, choose the **Horizontal Ordinate Dimension** button from the **Dimension/Relations CommandManager**. You are prompted to select an edge or a vertex. Note that the first entity that you select is taken as the datum entity from where the remaining entities will be measured. Select the first entity and place the dimension above or below it. You will notice that the dimension shows a value of 0, refer to the dimension of the left vertical line in Figure 3-39.

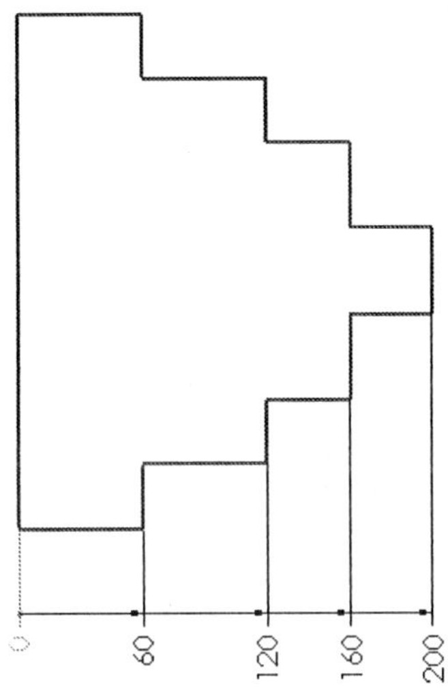

Figure 3-39 *Horizontal ordinate dimensions*

After placing the first dimension, you are again prompted to select an edge or a vertex. Select the edge that you need to dimension using the first selected edge as datum. As soon as you select the edge, a horizontal dimension between the datum and this entity will be placed. Similarly, continue placing the dimensions to create multiple horizontal ordinate dimensions, refer to Figure 3-39.

Vertical Ordinate Dimensioning

CommandManager:	Dimension/Relations > Vertical Ordinate Dimension
Menu:	Tools > Dimension > Vertical Ordinate
Toolbar:	Dimension/Relations > Vertical Ordinate Dimension

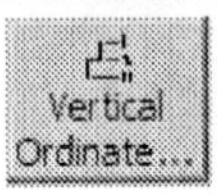

Vertical ordinate dimensions are used to dimension the vertical distances of the selected entities from the specified datum, see Figure 3-40. To add vertical ordinate dimension, choose the **Vertical Ordinate Dimension** button from the **Dimension/ Relations CommandManager**. You are prompted to select an edge or a vertex. As mentioned earlier, the first entity that you select is taken as the datum entity from where the remaining entities will be measured. Select the first entity and place the dimension on the right or left of it. You will notice that the dimension shows a value of 0, refer to the dimension of the left vertical line in Figure 3-40.

Next, select the edge that you need to dimension using the first selected edge as datum. As soon as you select the edge, a vertical dimension will be placed between the datum and this entity. Similarly, continue placing the dimensions to create multiple horizontal ordinate dimensions, refer to Figure 3-40.

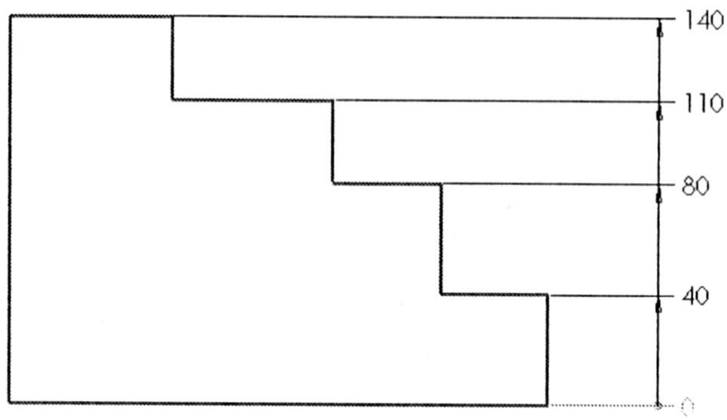

Figure 3-40 Vertical ordinate dimensions

Additional Dimensioning Options

In SolidWorks, you are also provided with some other dimensioning options other than those discussed earlier. The main additional dimensioning option is discussed next.

Dimensions Between Arcs or Circles

If you select two arcs, two circles, or an arc and a circle to add dimension, the dimension is placed from center point to center point, refer to Figure 3-41. But SolidWorks allows you to dimension the minimum and maximum distances between the selected curves.

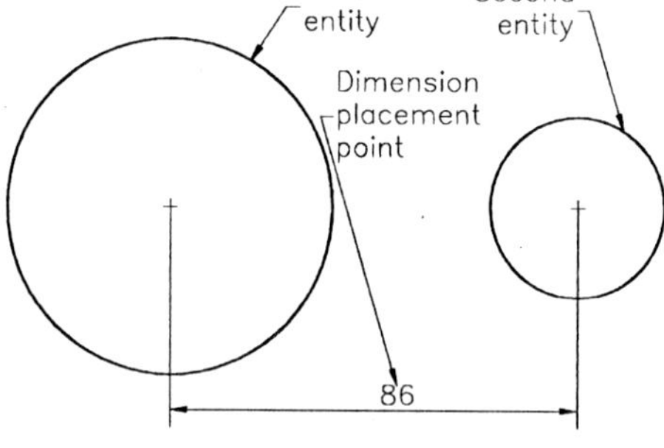

Figure 3-41 *Dimension with first arc condition as* ***Center*** *and second arc condition as* ***Center***

To add dimensions in terms of minimum or maximum distances, place the dimension between the two center points by selecting the two curves. Note that if you select the center points to add dimensions, you will not be able to change the dimensions to minimum or maximum values. After placing the dimension, right-click to display the shortcut menu. Choose the **Properties** option from the shortcut menu to display the **Dimension Properties** dialog box. You will notice that in the **Dimension Properties** dialog box, the **Center** radio buttons are selected in the **First arc condition** and **Second arc condition** areas. If you select the **Min** radio buttons from the **First arc condition** and **Second arc condition** areas, the resultant dimensioning will be displayed as the minimum distance between two circles or arcs, as shown in Figure 3-42. If you select the **Max** radio buttons, the dimension will be displayed as the maximum distance between two circles, as shown in Figure 3-43. Similarly, you can adjust the first end condition and the second end condition accordingly.

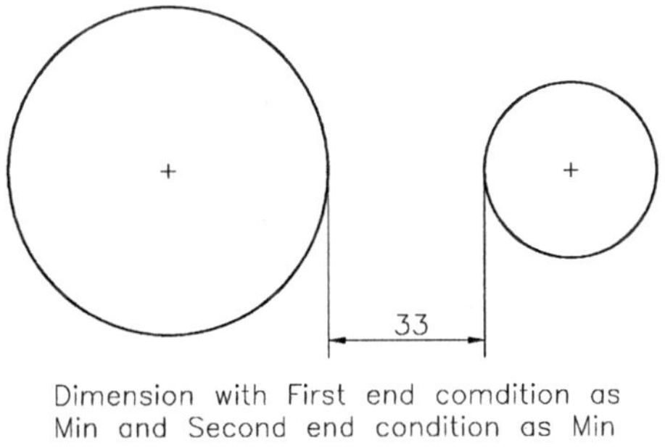

Figure 3-42** Dimension with first and second arc conditions as **Min

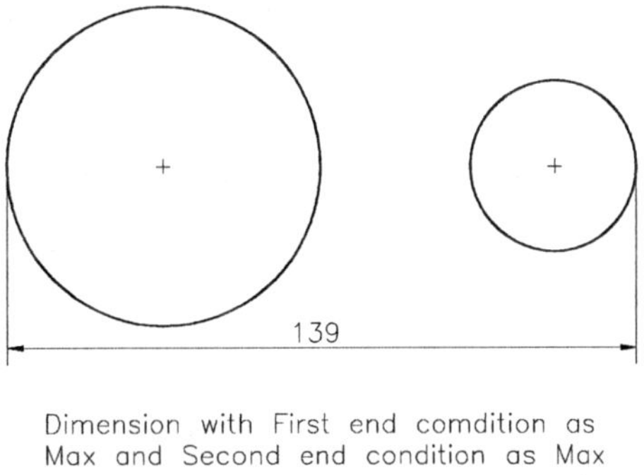

Figure 3-43** Dimension with first and second arc conditions as **Max

CONCEPT OF FULLY DEFINED SKETCH

It is very necessary for you to understand the concept of fully defined sketches. While creating a model, you first have to draw the sketch of the base feature and then proceed further for creating other features. This is the reason sketching is the first step of modeling. After creating the sketches, you have to add the required relations and dimensions to constrain the sketch with respect to the surrounding environment. After adding the required relations and dimensions, the sketch may exist in one of the six states. The six states of the sketch are discussed below.

1. Fully Defined
2. Overdefined
3. Underdefined
4. Dangling
5. No Solution Found
6. Invalid Solution Found

Fully Defined

A fully defined sketch is the one in which all the entities of the sketch and their positions are fully defined by the relations or dimensions, or both. In the fully defined sketch, all the degrees of freedom of a sketch are constrained using relations and dimensions and the sketched entities cannot move or change their size and location unexpectedly. If the sketch is not fully defined, it can change its size or position at any time during the design because all the degree of freedom are not constrained. All the entities in a fully defined sketch are displayed in black.

Overdefined

An overdefined sketch is the one in which some of the dimensions, relations, or both are conflicting or the dimensions or relations in the sketch have exceeded the required number. The overdefined sketch is displayed in red. In an overdefined sketch, you need to delete the extra and conflicting relations or dimensions. It is recommended that you do not proceed further for creating the feature with an over defined sketch. The overdefined sketch is solved to fully defined or underdefined sketch by deleting the conflicting relations or dimensions. Deleting the overdefining relations or dimension is discussed later in this chapter.

Underdefined

An underdefined sketch is the one in which some of the dimensions or relations are not defined and the degree of freedom of the sketch is not fully constrained. In this type of sketches, the entities may move or change their size unexpectedly. The sketched entities of the underdefined sketch are displayed in blue. This is the reason the sketch is displayed in blue while drawing. When you add relations and dimensions, the color of the entities in the sketch changes to black, suggesting that the sketch is fully defined. If the entire sketch is displayed in black and only some of the entities are shown in blue, this means that the entities in blue require some dimension or relation.

Tip. *In SolidWorks, it is not necessary that you fully dimension or define the sketches before you use them to create the features of the model. However, it is recommended that you fully define the sketches before you proceed further for creating the feature.*

*If you want to always use fully defined sketches before proceeding further, you can set the option by choosing **Tools > Options** from the menu bar to display the **System Option - General** dialog box. Select the **Sketch** option from the area on the left. Select the **Use fully defined sketches** check box and choose **OK** from this dialog box.*

Note

From this chapter onwards, you will work with fully defined sketches. Therefore, follow the above procedure to use the fully defined sketches in future.

Dangling

In the dangling sketch, the dimensions or relations applied to an entity lose their reference because of the deletion of the entity from which they were referenced. These entities are displayed dashed in brown color. You need to delete the dangling entities, dimensions, or relations that conflict.

No Solution Found

In the no solution found state, the sketch is not solved with the current constraints. Therefore, you need to delete the conflicting dimensions or relations and add other dimensions or relations. The sketched entity, dimension, or relation will be displayed in pink.

Tip. *The status bar of the SolidWorks window is divided into four areas while working in the sketching environment. The* ***Sketch Definition*** *area of the status bar always displays the status of the sketch, dimension, and relation. If the sketch is underdefined, the status area will display the* ***Under Defined*** *message; if the sketch is overdefined then the message displayed in the status area will be* ***Over Defined;*** *if the sketch is fully defined then the message displayed in the status area will be* ***Fully Defined.***

Invalid Solution Found

In the invalid solution found state, the sketch is solved but the sketch will result in invalid geometry such as a zero length line, zero radius arc, or self-intersecting spline. The sketch entities for this state are displayed in yellow.

Sketch Dimension or Relation Status

In SolidWorks when you are applying the dimensions and relations to the sketches, sometimes you apply the dimensions or relations that are not compatible with the geometry of the sketched entities or that make the dimensioned entity overdefined. In addition to the fully defined state, the sketch dimensions or relations may have any of the following states.

1. Dangling
2. Satisfied
3. Overdefining
4. Not Solved
5. Driven

Dangling

A dangling dimension or relation is the one that cannot be resolved because the entity to which it was referenced is deleted. The dangling dimension appears in brown color.

Satisfied

A satisfied dimension is the one that is completely defined and is displayed in black.

Overdefining

An overdefining dimension or relation overdefines one or more entities in the sketch. The overdefining dimension appears in red.

Not Solved

The not solved dimension or relation is not able to determine the position of the sketched entities. A not solved dimension appears in pink.

Driven

The driven dimension's value is driven by other dimensions in the sketch that solve the sketch. The driven dimension appears in gray.

DELETING THE OVERDEFINING DIMENSIONS

In SolidWorks, when you add a dimension that overdefines a sketch, the sketch and dimension turns red. Also, the **Make Dimension Driven?** dialog box is displayed as shown in Figure 3-44.

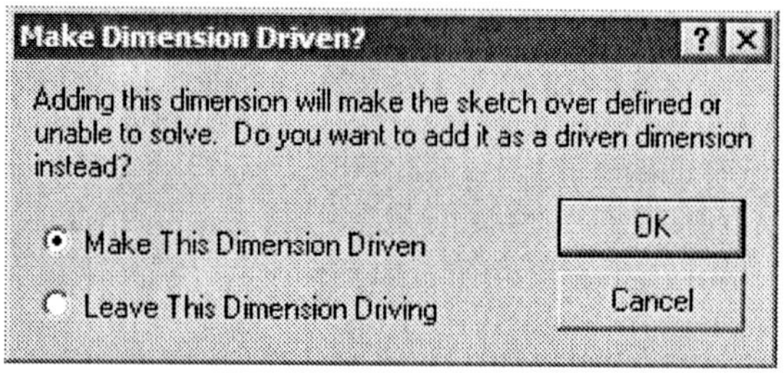

Figure 3-44 *The* ***Make Dimension Driven?*** *dialog box*

This dialog box informs you that adding this dimension will overdefine the sketch or the sketch will not be solved. You are also prompted to specify whether you want to add the dimension as driven dimension? If you select the **Make This Dimension Driven** radio button from this dialog box and choose **OK**, then that dimension will become a driven dimension. The driven dimension is displayed in gray and you cannot modify a driven dimension. Its value depends on the value of the driver dimension. If you change the value of the driver dimension, the value of the driven dimension will be automatically changed.

If you select the **Leave This Dimension Driving** radio button and choose **OK**, then some of the entities and dimensions in the sketch will be displayed in red. You need to delete either the red sketched entity or the red dimension to make sure the sketch is no more overdefined. When the sketch is not overdefined any more, the **SolidWorks** information dialog box will be displayed as shown in Figure 3-45. The information box informs you that the sketch is no more overdefined. Choose **OK** from the **SolidWorks** information dialog box. The sketch will be displayed in black or blue, depending on the current state of the sketch.

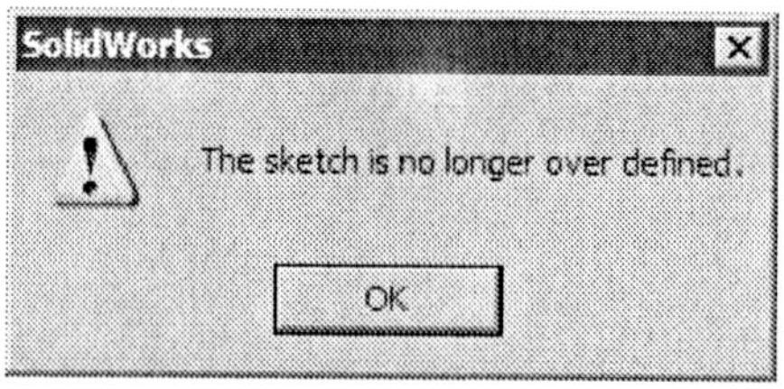

Figure 3-45 *The* ***SolidWorks*** *information dialog box*

You can also prevent the sketch from being overdefining by choosing the **Cancel** button from the **Make Dimension Driven?** dialog box. If you choose **Cancel** from the **Make Dimension Driven?** dialog box, the **SolidWorks** information dialog box will be displayed, as shown above, with a message that the sketch is no longer overdefined.

Displaying and Deleting Relations

CommandManager:	Sketch > Display/Delete Relations
Menu:	Tools > Relations > Display/Delete
Toolbar:	Sketch > Display/Delete Relations

If the sketch is overdefined after adding the dimensions and relations, then you need to delete some of the overdefining, dangling, or not solved relations or dimensions. You can view and delete the relations applied to the sketch using the **Display/Delete Relations PropertyManager**. To invoke this **PropertyManager**, choose the **Display/Delete Relations** button from the **Sketch CommandManager**. You can also right-click in the drawing area to display the shortcut menu and choose the **Display/Delete Relations** option. Figure 3-46 shows the **Display/Delete Relations PropertyManager**. The options available in the **Display/Delete Relations PropertyManager** are discussed next.

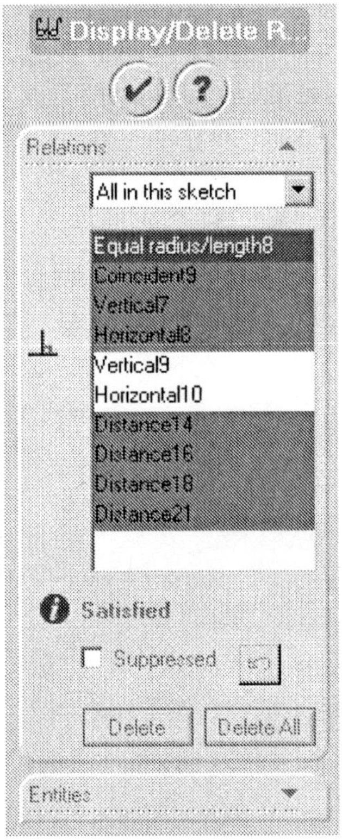

*Figure 3-46 The **Display/Delete Relations PropertyManager***

Relations

The **Relations** rollout is used to check, delete, and suppress the unwanted and conflicting relations. The status of the sketch or the selected entity is displayed in the **Information** area of this rollout. The various options of the **Relations** rollout are discussed next.

Filter

The **Filter** drop-down list is used to select the filter to show the relations in the **Sketch Relation PropertyManager**. The various options available in the **Filter** drop-down list are discussed next.

All in this sketch

The **All in this sketch** option is used to display all the relations applied to the sketch. This option is selected by default in the **Filter** drop-down list. The first relation displayed in the list will be selected by default and will appear with blue background. The status of the selected relations is displayed in the **Information** area of the **Relations** rollout. The overdefined relations are highlighted in red color. If you select a relation highlighted red in color in the **Relations** area, the status of the selected relation is displayed as **Over Defining**. The dangling relation is highlighted in brown. When you select the dangling relation, the status of the relation will be displayed as **Dangling** in the **Information** area. Similarly, the not solved relation will be highlighted in yellow and the driven relation in gray.

Dangling

The **Dangling** option is used to display only the dangling relations applied to the sketch.

Overdefining/Not Solved

The **Overdefining/Not Solved** option is used to display only the overdefining and not solved relation. The dangling relations are also not solved relations. Therefore, they will also be displayed.

External

The **External** option is used to display the relations that have a reference with an entity outside the sketch. This entity can be an edge, vertex, or origin within the same model or it can be an edge, vertex, or origin of a different model within an assembly.

Defined In Context

The **Defined In Context** option is used to display only the relations that are in the context of a design. They are the relations between the sketch entity in one part and an entity in another part. These relations are defined while working with the top-down assemblies.

Locked

The **Locked** option is used to display only the locked relations.

Broken

The **Broken** option is used to display only the broken relations.

Note

*The **Locked** and the **Broken** relations are applied while creating a part within the assembly environment. You will learn more about these relations in later chapters.*

Selected Entities

The **Selected Entities** option is used to display the relations of only the selected set of entities. When you select this option from the **Filter** drop-down list, the **Selected Entities** area is displayed in the **Relations** rollout. When you select an entity to display the relations, the name of the selected entity is displayed in the **Selected Entities** area and the relations applied to this entity are displayed in the **Relations** area. To remove the selected entity from the selection set, select the entity to be removed and right-click to display the shortcut menu. Choose the **Delete** option from the shortcut menu. If you choose the **Clear Selections** option, all the entities will be removed from the selection set.

Suppressed

The **Suppressed** check box is selected to suppress the selected relation of the current configuration. When you suppress a relation, it is displayed in gray in the **Relations** area. The status of the suppress relation is displayed as **Satisfied** or **Driven** in the information area. If you suppress over defining dimensions, the **SolidWorks** information dialog box will be displayed with a message that **The sketch is no longer over defined.** Choose the **OK** button from this dialog box.

Note

The configurations used in the part and assembly modeling will be discussed in later chapters.

Delete

The **Delete** button is used to delete the relation selected in the **Relations** area.

Delete All

The **Delete All** button is used to delete all the relations that are displayed in the **Relations** area.

Undo last relation change

The **Undo last relation change** button is used to undo the **Delete**, **Replace**, and **Suppressed** options used earlier. The replace option is discussed later in this chapter.

Entities

The **Entities** rollout is used to display the entities that are referred to in the selected relation. This rollout is also used to display the status of the selected relation and the external reference, if any. By default, the **Entities** rollout is closed. You can open this rollout by clicking on the blue arrow displayed on the right of this rollout. The **Entities** rollout is shown in Figure 3-47. The various options in the **Entities** rollout are discussed next.

Entities used in the selected relation

The **Entities used in the selected relation** area is used to display the information about entities used in the selected relation. The information about the name of the entity, the status of the entity, and the place where the entity is defined is provided. This area is divided in three columns. The three columns available in this area are discussed next.

Entity

The **Entity** column is used to display the entity or entities on which the selected relation is applied.

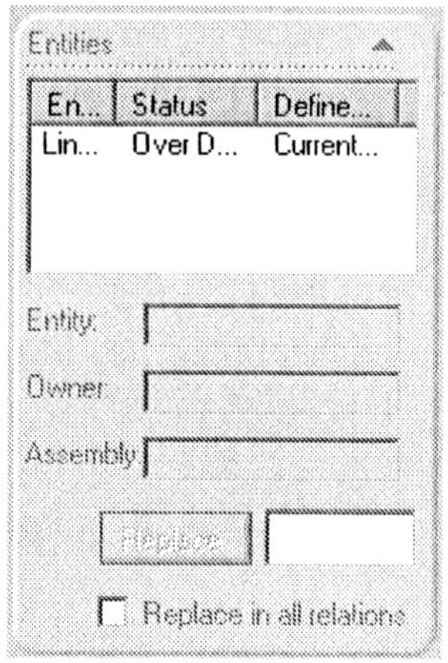

*Figure 3-47 The **Entities** rollout*

Status

The **Status** column is used to display the status of the selected relation. The status displayed in the **Status** column can be **Fully Defined**, **Dangling**, **Over Defined**, or **Not Solved**.

Defined In

The **Defined In** column is used to display the placement of the entity. The entity can be placed in any of the following places.

Current Sketch

The **Current Sketch** option is displayed in the **Defined In** column when the entity is placed in the same sketch.

Same Model

The **Same Model** option is displayed in the **Defined In** column when the entity is defined as placed in the same model. This means that the entity is placed in the same model, but outside the sketch. This entity can be an edge, vertex, or an origin of the same model.

External Model

The **External Model** option is displayed in the **Defined In** column when the entity is placed in some other model but within the assembly. This entity can be the edge, vertex, or the origin of a different model, but in the same assembly.

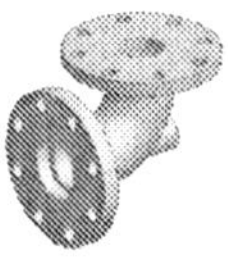

Tip. *By default, the **Override Dims on Drag** option is selected from **Tools > Sketch Settings** menu in the menu bar. As a result, if you drag a sketched entity, the value of dimension applied to the sketched entity will be changed automatically. If you clear this option, you cannot change a dimensioned sketched entity by dragging because it will not override the dimension. However, if the sketch is not fully defined, the entities that are not properly dimensioned or constrained will move.*

*By default, the **Automatic Solve** option is selected from the **Tools > Sketch Settings** menu in the menu bar. This option helps you to solve the relations and dimensions automatically when you drag or modify a sketched entity. If you clear this option, a message that **The sketch cannot be dragged because Auto Solve Mode is off. To drag the sketch, please turn the Auto Solve Mode on** appears. If you modify the dimension value using the **Modify** dialog box, the dimension will not update automatically, and you have to update the new dimension manually. To update and solve the dimension you have to choose the **Rebuild** button from the **Standard** toolbar or press CTRL+B using the keyboard.*

Entity

The **Entity** display box is used to display the name of the entity and the name of the part in which the selected entity is placed. This entity is selected in the **Entity** column of the **Entities used in the selected relation** area. The selected entity is also highlighted in the drawing area.

Owner

The **Owner** display box is used to display the name of the model in which the entity is placed when the **External Model** option is displayed in the **Defined In** column.

Assembly

The **Assembly** display box is used to display the path of the assembly in which the entity is placed when the **External Model** option is displayed in the **Defined In** column.

Replace

The **Replace** button is used to replace the selected entity from the **Entity** column with some other entity from the drawing area. When you select the entity from the drawing area, the entity will be displayed in the display box provided below the **Replace** button. Choose the **Replace** button to replace the entity. If the sketch is overdefined, you will be given a warning message. Sometimes after replacing the entity, the status of the entity is changed to not solved or overdefining. You need to undo the last operation. The **Replace in all relations** check box is selected to replace the entity in all the relations.

VIEWING AND EXAMINING RELATIONS

You can view and examine the relations applied to the sketch. To view and examine the relations applied to a particular entity, select that entity. The **PropertyManager** of that entity will be displayed. The **Relations** area in the **Existing Relations** rollout will display the relations that are already applied to the selected entity. Select the relations one by one from the **Existing Relations** area. The callout of the relations will be displayed in the drawing area, as shown in Figure 3-48.

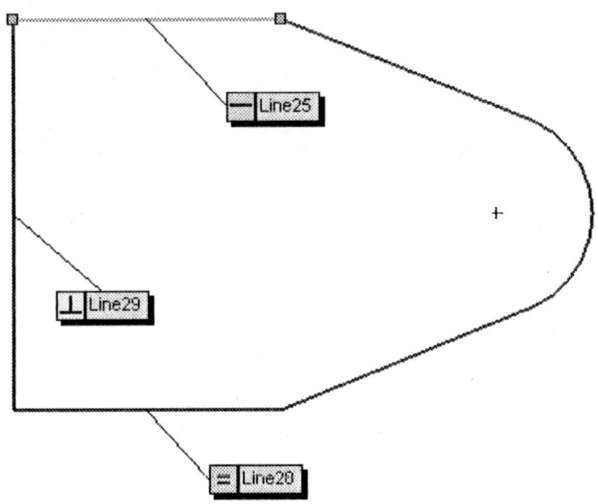

Figure 3-48 Callout of relations displayed in the drawing area

Tip. *You can also display the relation callouts in the drawing area without invoking the **Sketch Relations PropertyManager**. Double-click on the sketched entity and the relation callouts will be displayed in the drawing area.*

The callout that is displayed in the drawing area is divided in two parts. The first part displays the symbol of the relation applied to the selected entity and the second part displays the name of the entity. If you double-click the relation symbol then the **Sketch Relations PropertyManager** will be displayed. You can analyze, examine, and delete the unwanted, overdefining, dangling, and not solved relations using this option.

Note

To exit the callout, click anywhere on the screen.

OPENING AN EXISTING FILE

Toolbar:	Standard > Open
Menu:	File > Open

The **Open** dialog box is used to open an existing SolidWorks part, assembly, or drawing document. You can also use this dialog box to import files from other applications saved in some standard file formats. Choose the **Open** button from the **Standard** toolbar, or press CTRL+O keys to invoke the **Open** dialog box. The **Open** dialog box is shown in Figure 3-49. The options available in this dialog box are discussed next.

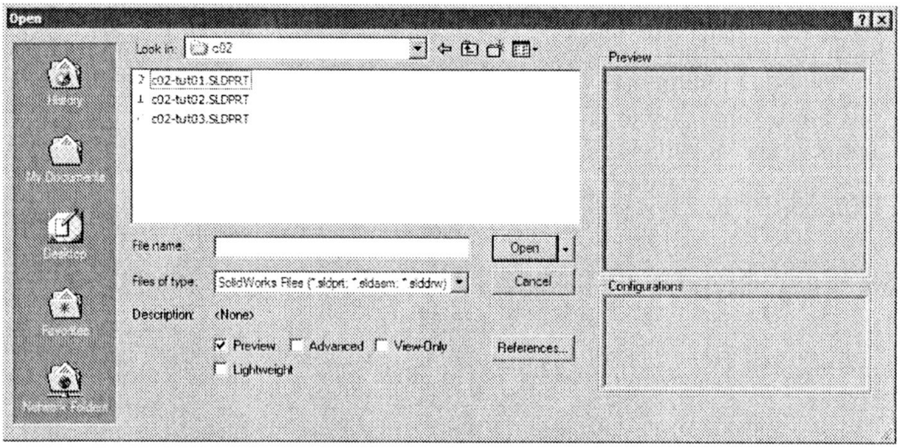

*Figure 3-49 The **Open** dialog box*

Look in drop-down list

The **Look in** drop-down list is used to specify the drive or directory in which the file is saved. The location of the file and the folder that you browse is shown in this drop-down list.

File name

The name of the file selected is shown in the **File name** edit box. You can also enter the name of the file to open in this edit box.

Files of type

The **Files of type** drop-down list is used to specify the type of file to open. Using this drop-down list, you can select a particular type of file type such as the part file, assembly file, drawing file, all SolidWorks files, and so on. You can also define the file standard format in this drop-down list to import the files saved in those file formats. You will learn more about importing the files in later chapters.

Open as read-only

The **Open as read-only** option is selected to open the document as a read only file. This option is available in the flyout that is displayed by choosing the down arrow available at the right of the **Open** button. If you modify the design in a read-only file, the changes are saved in a new file. The original file is not modified. This also allows another user to access the document while the document is open on your system.

Preview

The **Preview** check box is selected to display the **Preview** area in this dialog box. In the **Preview** area you can preview the selected part, assembly, or drawing document before opening.

Advanced

The **Advanced** check box is selected to display the configurations available in the selected file. The configurations available in the selected file are displayed in the **Configurations** area. You will learn more about configurations in the later chapters.

View-Only

The **View-Only** check box is selected to open a SolidWorks document in the view only format. When you open a view-only file, only the tools related to viewing the models are enabled. Rest of the tools are not available. This is the reason you cannot do any modification in the view-only document. You can use only the zoom, pan, or dynamically rotate tools. If you want to edit the design, right-click in the drawing area and choose the **Edit** option from the shortcut menu to edit the design.

Lightweight

The **Lightweight** check box is selected to open the assembly document using the lightweight parts. You will learn more about the lightweight parts in later chapters.

TUTORIALS

Tutorial 1

In this tutorial, you will draw the sketch of the model shown in Figure 3-50. This is the same sketch that was drawn in Tutorial 1 of Chapter 1. In this tutorial you will draw the sketch using the mirror line and then add the required relations and dimensions. The sketch is shown in Figure 3-51. The solid model is given only for reference. **(Expected time: 30 min)**

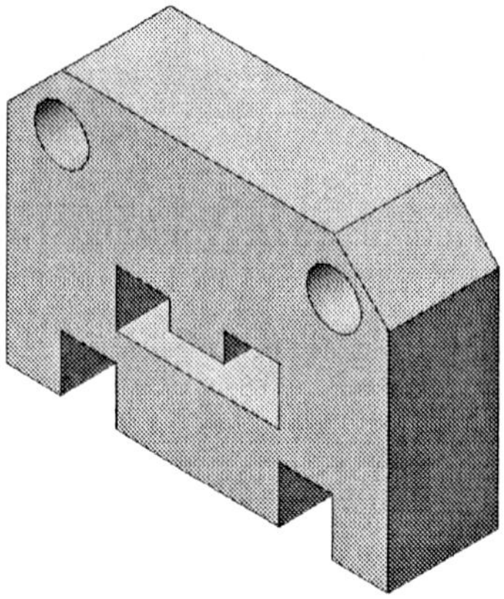

Figure 3-50 *Solid Model for Tutorial 1*

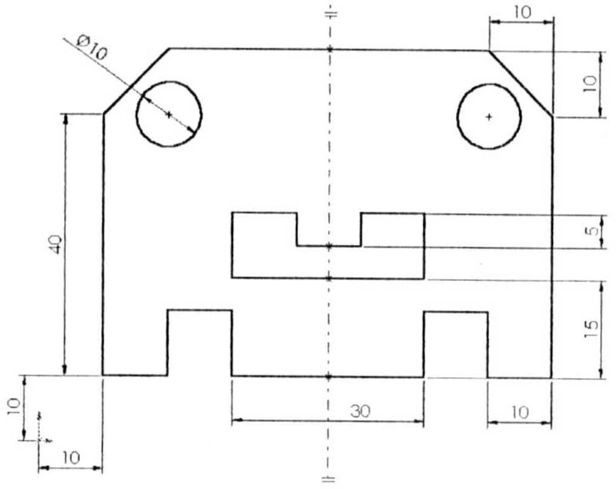

Figure 3-51 *Sketch of the model*

The steps that will be followed to complete this tutorial are listed below.

a. Start SolidWorks and then start a new part document.
b. Maximize the part document and then invoke the sketching environment.
c. Create a mirror line using the **Centerline** and the **Mirror** tool.
d. Draw the sketch of the model on one side of the mirror line so that it is automatically drawn on the other side, refer to Figures 3-52 through 3-57.
e. Add the required relations to the sketch, refer to Figures 3-58 and 3-59.
f. Add the required dimensions to the sketch and fully define the sketch, refer to Figure 3-60.
g. Save the sketch and then close the document.

Starting SolidWorks and Starting a New Part Document

1. Start SolidWorks by choosing **Start > Programs > SolidWorks 2004 > SolidWorks 2004** or by double-clicking the shortcut icon of SolidWorks available on the desktop of your computer.

   The **Welcome to SolidWorks 2004** window is displayed and also the **Tip of the Day** dialog box.

2. Close the **Tip of the Day** dialog box by choosing the **Close** button and then choose the **New Document** option from the **Welcome to SolidWorks 2004** window.

   The **New SolidWorks Document** dialog box is displayed.

3. The **Part** button is chosen by default. Choose the **OK** button from the **New SolidWorks Document** dialog box.

   A new SolidWorks part document is started. But the part document window is not maximized in the SolidWorks window.

4. Choose the **Maximize** button available on the upper right corner of the part document window to maximize the document window.

5. Choose the **Sketch** button from the **Standard** toolbar and then select the **Front** plane to invoke the sketching environment.

   In previous chapters, you used the grid and snap settings to create the sketches. However, from this chapter onwards, it is recommended that you do not use those settings. This way you can get familiar with drawing sketches at arbitrary locations and then using dimensions and relations to move them to their actual locations.

6. Invoke the **Document Properties - Grid/Snap** dialog box and then clear the **Snap to points** check box from the **Snap** area.

7. Set the units for measuring linear dimensions to millimeters and the units for angular dimensions to degree using the **Document Properties - Units** dialog box. If you selected millimeters as units while installing SolidWorks, then you can skip this point.

Creating the Mirror Line

In this tutorial, you will draw the sketch of the given model with the help of the mirror line. As mentioned earlier, the sketches that are symmetrical along any axis are recommended to be drawn using the mirror line. The mirror line is created using the **Centerline** and **Mirror** tools. When you draw an entity on one side of the mirror line, the same entity is drawn automatically on the other side of the mirror line. Also, the symmetrical relation is applied to the entities on both the sides of the mirror line. Therefore, if you modify an entity on one side of the mirror line, the same modification is reflected in the mirrored entity and vice versa. First, you need to create a mirror line.

The origin of the sketcher environment is placed in the center of the drawing area and you have to create the sketch in the first quadrant. Therefore, it is recommended that you modify the drawing area such that the area in the first quadrant in increased. This can be done using the **Pan** tool.

1. Choose the **Pan** tool button from the **View** toolbar. Press and hold the left mouse button down and drag the cursor toward the bottom left corner of the drawing area.

2. Choose the **Centerline** button from the **Sketch CommandManager**.

3. Move the line cursor to a location whose coordinates are close to 45 mm 70 mm 0 mm. You do not need to move the cursor exactly to this location. You can move it to a point close to this location.

4. Specify the start point of the centerline and move the line cursor vertically downward to draw a line of length close to 80 mm.

   As soon as you specify the endpoint of the centerline, a rubber-band line is attached to the line cursor. Double-click any where in the drawing area or right-click to display the shortcut menu and choose the **End chain** option from the shortcut menu to end line creation.

5. Choose the **Zoom to Fit** button from the **View** toolbar to fit the sketch on the screen.

6. Select the centerline and convert it into a mirror line using the **Mirror Entities** tool.

Drawing the Sketch

You will draw the sketch on the right side of the mirror line and the same sketch will be automatically drawn on the other side of the mirror line.

1. Choose the **Line** button from the **Sketch CommandManager**.

   The arrow cursor is replaced by the line cursor.

2. Move the line cursor to a location where the coordinates are close to 45 mm 10 mm 0 mm. At this point, the cursor snaps to the mirror line and turns yellow in color. Also, an orange bulb is displayed below the cursor.

3. Specify the start point of the line at this point and move the cursor horizontally toward the right. Specify the endpoint of the line when the length of the line above the line cursor shows a value close to 15. As soon as you press the left mouse button to specify the endpoint of the line, a line of the same length is drawn automatically on the other side of the mirror line. Figure 3-52 shows the mirrored entity created automatically on the other side of the mirror line.

   Note that the mirrored entity that is automatically created on the left of the mirror line is merged with the line drawn on the right. Therefore, the entire line becomes a single entity. As mentioned earlier, the mirror image of the line is merged with the line that you draw only if one of the endpoints of the line you draw is coincident with the mirror line.

4. Move the cursor vertically upward. Specify the endpoint of the line when the length of the line on the line cursor displays a value close to 10. Figure 3-53 shows the sketch after drawing the vertical lines.

5. Move the line cursor horizontally toward the right. Specify the endpoint of the line when the length of the line on the line cursor displays a value close to 10.

6. Move the line cursor vertically downward. Specify the endpoint when the length of the line on the line cursor displays a value close to 10.

7. Move the line cursor horizontally toward the right. Specify the endpoint when the length of the line on the line cursor displays a value close to 10.

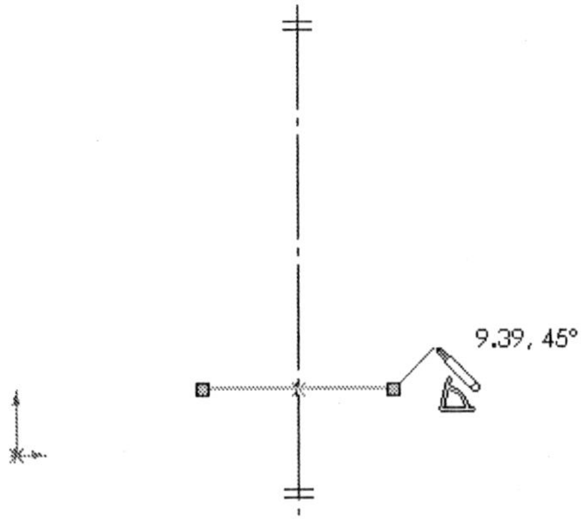

Figure 3-52 *Sketching using automatic mirroring*

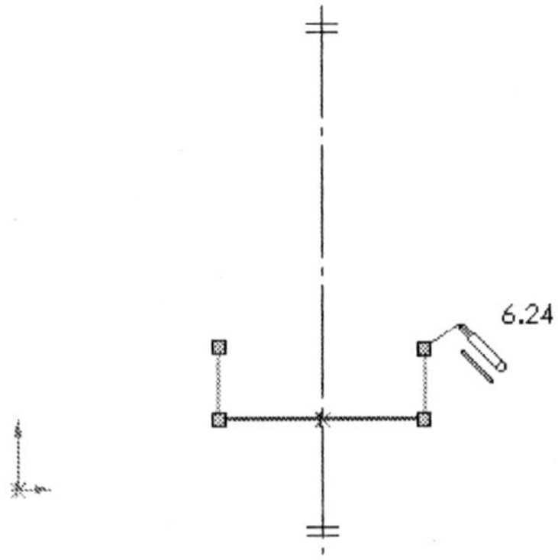

Figure 3-53 *Left line drawn by mirroring*

8. Move the line cursor vertically upward. Specify the endpoint when the length of the line on the line cursor displays a value close to 40.

9. Move the line cursor at an angle close to 135-degree and specify the endpoint of the line. Figure 3-54 shows the sketch after drawing this inclined line.

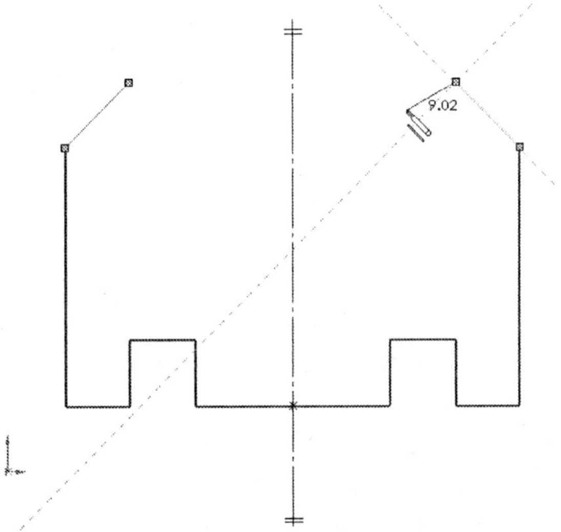

Figure 3-54 *Sketch after drawing the inclined line*

10. Move the line cursor horizontally toward the left. Specify the endpoint when the cursor snaps to the mirror line and the line cursor turns yellow in color. Double-click anywhere in the drawing area to end line creation. The sketch after completing the outer loop is shown in Figure 3-55.

    Next, you will draw the sketch of the inner cavity. To draw the sketch of the inner cavity, you will start with the lower horizontal line.

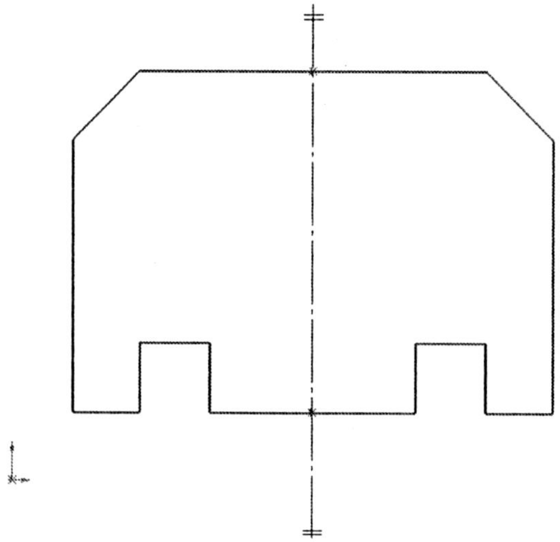

Figure 3-55 Sketch after completing the outer profile of the sketch

11. Move the line cursor to a location whose coordinates are close to 45 mm 25 mm 0 mm.

12. Specify the start point of the line at this point and move the cursor horizontally toward the right. Specify the endpoint when the length of the line above the line cursor shows a value close to 15.

13. Move the line cursor vertically upwards. Specify the endpoint when the length of the line on the line cursor displays a value close to 10.

14. Move the line cursor horizontally toward the left. Specify the endpoint when the length of the line on the line cursor displays a value close to 10.

15. Move the line cursor vertically downwards. Specify the endpoint when the length of the line on the line cursor displays a value close to 5.

16. Move the line cursor horizontally toward the left. Specify the endpoint when the line cursor snaps to the mirror line.

17. Double-click anywhere in the drawing area to end line creation. The sketch after completing the inner cavity is shown in Figure 3-56.

18. Choose the **Circle** button from the **Sketch CommandManager** to invoke the circle tool.

19. Move the circle cursor to the point where the inferencing line originating from the endpoints of the right inclined line intersect.

20. Specify the center of the circle at this point and move the circle cursor toward the left to define the radius of the circle. Press the left mouse button when the radius of the circle above the circle cursor shows a value close to 5.

The circle is automatically mirrored on the other side of the mirror line. The sketch after drawing the circle is shown in Figure 3-57.

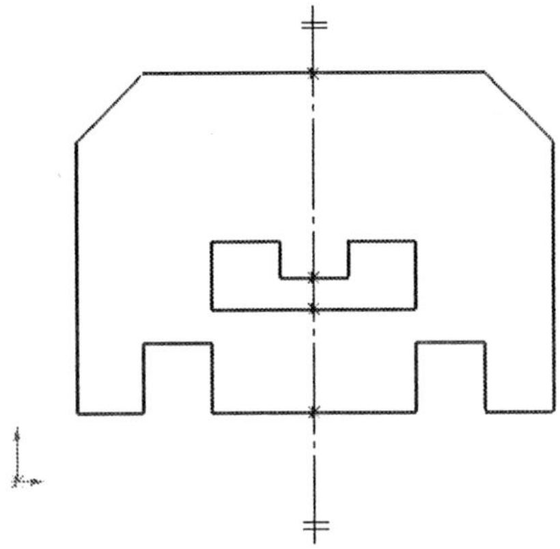

Figure 3-56 *Sketch after drawing the inner cavity*

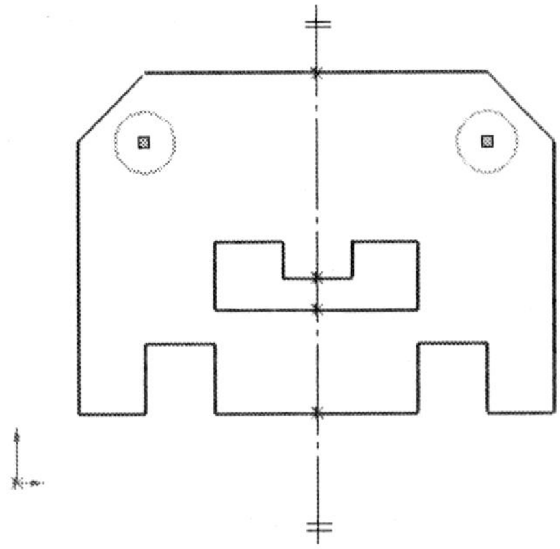

Figure 3-57 *Sketch after drawing the circle*

21. Clear the automatic mirroring option by choosing the **Mirror Entities** button again.

Adding the Required Relations

After drawing the sketch, you need to add the relations using the **Add Relations PropertyManager**. The relations are applied to a sketch to constrain its degree of freedom, to reduce the number of dimensions in the sketch, and also to capture the design intent in the sketch.

1. Press the ESC key to remove the circles created previously from the selection set.

2. Choose the **Add Relation** button from the **Sketch CommandManager** to invoke the **Add Relations PropertyManager**. The confirmation corner is also displayed at the upper right corner of the drawing area.

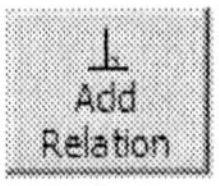

3. Select the center point of the circle on the right and then select the lower endpoint of the right inclined line. The names of the selected entities are displayed in the **Selected Entities** rollout of the **Add Relations PropertyManager**.

   The relations that can be applied to the two selected entities are displayed in the **Add Relations** rollout of the **Add Relations PropertyManager** as shown in Figure 3-58. The **Horizontal** option is highlighted, suggesting that the horizontal relation is the most appropriate relation for the selected entities.

Tip. *Sometimes unwanted inferencing lines are displayed when you are drawing a sketch. You can remove the unwanted inferencing lines by choosing **View > Redraw** from the menu bar. You can also redraw the drawing area by pressing CTRL+R key.*

Note

*The names of the entities displayed in the **Selected Entities** rollout of the **Add Relations PropertyManager** may be different from those displayed on your screen.*

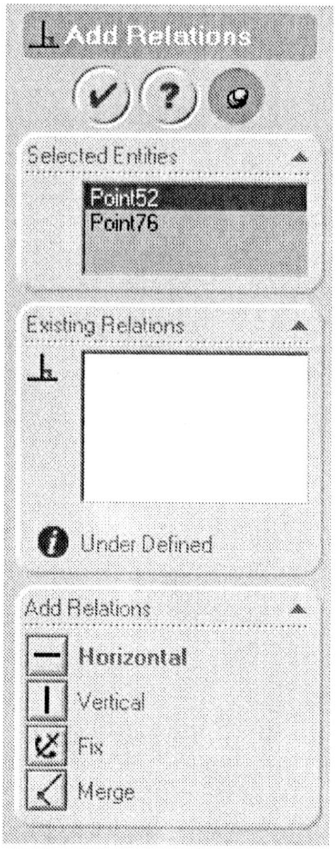

Figure 3-58 Add Relations PropertyManager

4. Choose the **Horizontal** button from the **Add Relations** rollout to apply the **Horizontal** relation to the selected entities.

5. Move the cursor to the drawing area and right-click to invoke the shortcut menu. Choose the **Clear Selections** option to remove the selected entities from the selection set.

6. Select the center point of the circle on the right and the upper endpoint of the right inclined line.

   The relations that can be applied to the selected entities are displayed and the **Vertical** option is highlighted.

7. Choose the **Vertical** button from the **Add Relations PropertyManager**. Right-click in the drawing area and choose the **Clear Selections** option.

8. Select the entities shown in Figure 3-59. Choose the **Equal** button from the **Add Relations** rollout of the **Add Relations PropertyManager**.

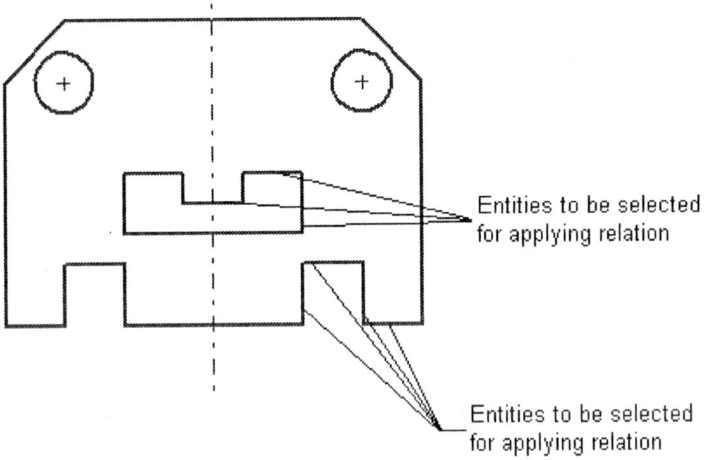

*Figure 3-59 Entities to be selected to apply the **Equal** relation*

9. Choose the **OK** button from the **Add Relations PropertyManager** or choose **OK** from the confirmation corner to close the **PropertyManager**. Specify a point in the drawing area to clear the selected entities.

Applying the Dimensions to the Sketch

Next, you will apply the dimensions to the sketch and fully define the sketch. As mentioned earlier, the sketched entities are shown in blue, suggesting that the sketch is underdefined. It will be changed to black after applying the required dimensions to the sketch, suggesting that the sketch is fully defined.

1. Choose **Tools > Options** from the menu bar to display the **System Options - General** dialog box. Select the **Input dimension value** check box, if cleared, and choose **OK** from the **System Options - General** dialog box. This check box is selected to invoke the modify dialog box to enter a new dimension value and modify the sketch as you place the dimension.

2. Choose the **Smart Dimension** button from the **Sketch CommandManager**. You can also right-click in the drawing area and choose the **Smart Dimension** option to invoke the dimension option.

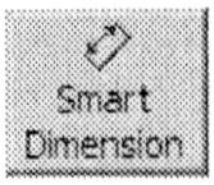

The cursor is replaced by the dimension cursor.

3. Move the dimension cursor to the lower right horizontal line. The lower right horizontal line is highlighted in red color.

4. Select the line. A linear dimension is attached to the cursor.

5. Move the cursor downward and left-click to place the dimension below the line, refer to Figure 3-60. As you place the dimension, the **Modify** dialog box is displayed.

6. Enter the dimension value of **10** in this dialog box and press ENTER. The dimension is placed and the length of line is also modified to 10.

7. Move the dimension cursor to the lower middle horizontal line, refer to Figure 3-60. Select the line when the line changes to red. A dimension is attached to the cursor.

8. Move the cursor downward and left-click to place the dimension. Enter a value of **30** in the **Modify** dialog box and press ENTER.

9. Move the cursor to the outer left vertical line and when the color of the line changes to red, select the line. A dimension is attached to the cursor.

10. Move the cursor to the left and then left-click to place the dimension. Enter a value of **40** in the **Modify** dialog box and press ENTER.

11. Select the right inclined line. A dimension is attached to the cursor. Move the cursor vertically upward to apply the horizontal dimension to the selected line. Left-click to place the dimension at an appropriate place, see Figure 3-60.

12. Enter a value of **10** in the **Modify** dialog box and press ENTER.

13. Again, select the right aligned line. A dimension is attached to the cursor. Move the cursor horizontally toward the right to apply the vertical dimension for the selected line. Left-click to place the dimension at an appropriate place, see Figure 3-60.

14. Enter a value of **10** in the **Modify** dialog box and press ENTER.

15. Move the cursor to the left circle and when the circle is highlighted in red, select it. A diameter dimension is attached to the cursor. Move the cursor outside the sketch.

16. Place the diameter dimension. Enter a value of **10** in the **Modify** dialog box and press ENTER.

17. Select the lower horizontal line of the inner cavity. A linear dimension is attached to the cursor. Select the lower horizontal line of the outer loop.

    A vertical dimension between the lower horizontal line of the inner cavity and the lower horizontal line of the outer sketch is attached to the cursor.

18. Move the cursor horizontally toward the right and place the dimension. Enter a value of **15** in the **Modify** dialog box and press ENTER.

19. Select the inner right vertical line of the cavity and place the dimension outside the sketch. Enter a value of **5** in the **Modify** dialog box and press ENTER.

20. Select the lower horizontal line of the outer sketch and the origin.

21. Move the cursor horizontally toward the left and place the dimension. Enter a value of **10** in the **Modify** dialog box.

    Notice, that some of the entities are displayed in black. This suggests that these entities are now fully defined. But you have to fully define the entire sketch. So you need to add some more dimensions.

22. Select the left vertical line of the outer sketch and the origin.

23. Move the cursor vertically downward and left-click to place the dimension. Enter a value of **10** in the **Modify** dialog box.

    Notice that all the entities are displayed in black. This suggests that the sketch is fully defined. If the sketch is not fully defined then you have to add a dimension between the outer right vertical line and the outer left vertical line. The value of the dimension should be maintained 70. The fully defined sketch, after applying all the required relations and dimensions, is shown in Figure 3-60.

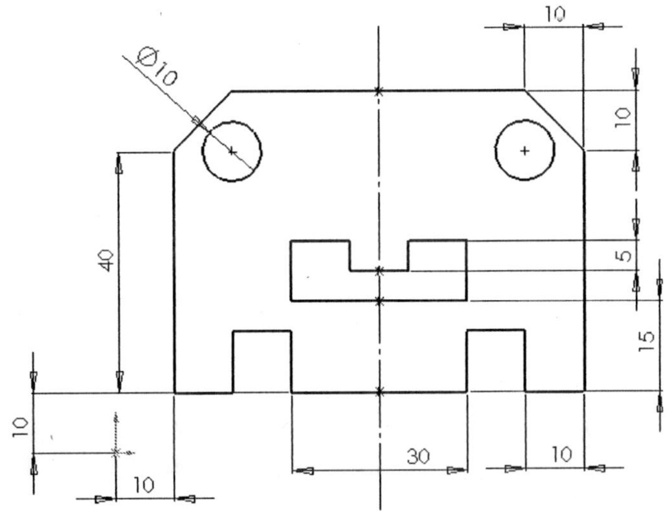

***Figure 3-60** Fully defined sketch after applying all the required relations and dimensions*

Saving the Sketch

1. Choose the **Save** button from the **Standard** toolbar to invoke the **Save As** dialog box.

2. Browse to the *My Documents/SolidWorks* folder. Choose the **Create New Folder** button from the **Save As** dialog box. Enter the name of the folder as *c03* and press ENTER.

3. Enter the name of the document as *c03-tut01.sldprt* in the **File name** edit box and choose the **Save** button. The document will be saved in the *\My Documents\SolidWorks\c03* folder.

4. Close the document by choosing **File > Close** from the menu bar.

Tutorial 2

In this tutorial you will draw the sketch of the model shown in Figure 3-61. You will draw the sketch using the mirror line and then add the required relations and dimensions. The sketch is shown in Figure 3-62. The solid model is given only for reference. **(Expected time: 30 min)**

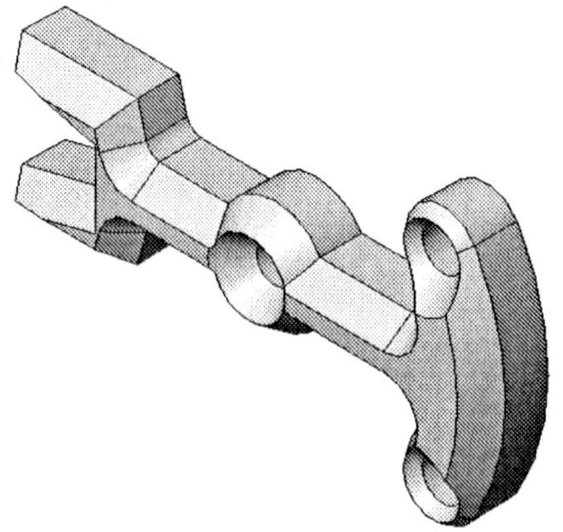

Figure 3-61 Solid Model for Tutorial 2

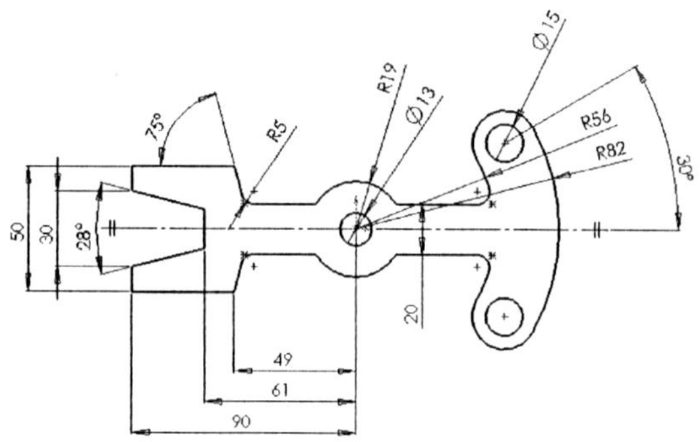

Figure 3-62 Sketch of the solid model

The steps that will be followed to complete this tutorial are listed below.

a. Start a new document file.
b. Maximize the part file document and then invoke the sketching environment.
c. Create a mirror line.
d. Draw the sketch on one side of the mirror line, refer to Figure 3-63.
e. Trim the arcs and circles and add the fillets, refer to Figures 3-64 through 3-66.
f. Add the required relations.
g. Add the required dimensions and fully define the sketch, refer to Figure 3-67.

Starting a New Document and Invoking the Sketching Environment

1. Choose the **New** button from the **Standard** toolbar to invoke the **New SolidWorks Document** dialog box.

2. The **Part** button is chosen by default, choose the **OK** button. A new SolidWorks part document is started. Maximize the document.

3. Choose the **Sketch** button from the **Standard** toolbar and select the **Front** plane to invoke the sketching environment.

Drawing the Mirror Line

Similar to the last tutorial, you will draw the sketch of the given model with the help of a mirror line.

1. Choose the **Centerline** button from the **Sketch CommandManager**.

2. Move the line cursor to a location whose coordinates are close to -102 mm 0 mm 0 mm. You do not need to move the cursor exactly to this location. You can move it to a point close to this location.

3. Specify the start point of the centerline at this point and move the line cursor horizontally toward the right. Specify the endpoint of the center line when the length of the line shows a value close to 204. Double-click anywhere in the drawing area to end line creation.

4. Choose the **Zoom to Fit** button from the **View** tollbar to fit the sketch in the drawing area.

5. Convert the centerline to mirror line and enable the automatic mirror option.

Drawing the Sketch

You will draw the sketch on the upper side of the mirror line and the same sketch will be automatically drawn on the other side of the mirror line.

1. Choose the **Centerpoint Arc** button from the **Sketch CommandManager**.

   The arrow cursor is replaced by the arc cursor.

2. Move the arc cursor close to the origin. Specify the centerpoint of the arc when the cursor turns yellow in color. Move the cursor horizontally toward the right. The cursor snaps to the mirror line. As you move the cursor, a reference circle is drawn. Specify the start point of the arc when the radius of the arc on the arc cursor shows a value close to 82.

3. Move the arc cursor in the counterclockwise direction. Specify the endpoint of the arc when the value of the angle above the arc cursor shows a value close to 30-degree. The mirror image of the sketched entity is automatically created on the other side of the mirror line.

4. Move the arc cursor to the origin. Specify the centerpoint of the arc. Move it horizontally toward the right. The cursor snaps to the mirror line and a reference circle is drawn. Specify the start point of the arc when the radius of the arc on the arc cursor shows a value close to 56.

5. Move the arc cursor in the counterclockwise direction. Specify the endpoint of the arc when the value of the angle above the arc cursor shows a value close to 30-degree. The mirror image of the sketched entity is automatically created on the other side of the mirror line.

6. Choose the **Tangent Arc** button from the **Sketch CommandManager** .

7. Move the cursor to the upper endpoint of the left arc and when the red circle is displayed, specify the first point of the tangent arc. Move the cursor close to the upper endpoint of the right arc and when the red circle is displayed, specify the second point of the tangent arc.

8. Choose the **Circle** button from the **Sketch CommandManager**.

9. Move the cursor to the centerpoint of the upper arc and when the red circle is displayed,

specify the centerpoint of the circle. Move the cursor horizontally toward the right. Press the left mouse button when the radius of the circle above the circle cursor shows a value close to 7.5.

10. Move the cursor to the origin and when the red circle is displayed, specify the center point of the circle. Move the cursor horizontally toward the right. Press the left mouse button when the radius of the circle above the circle cursor shows a value close to 6.5.

    The **SolidWorks** warning dialog box is displayed. This warning message will inform you "**Unable to create the symmetric element**". This message is displayed because the circle is created on both sides of the mirror line. Therefore, a symmetric element is not created in this case.

11. Choose **OK** from the **SolidWorks** warning dialog box.

12. Move the cursor to the origin and when the red circle is displayed, specify the center point of the circle. Move the cursor horizontally toward the right. Press the left mouse button when the radius of the circle above the circle cursor displays a value close to 13. The **SolidWorks** warning dialog box will be displayed.

13. Choose **OK** from the **SolidWorks** warning dialog box.

14. Choose the **Line** button from the **Sketch CommandManager**.

15. Move the line cursor to a location whose coordinates are close to 55 mm 10 mm 0 mm. The cursor snaps to the left arc.

16. Specify the start point of the line at this point and move the cursor horizontally toward the left. Specify the endpoint of the line when the line cursor snaps the outer circle. Double-click anywhere in the drawing area to end line creation.

17. Move the cursor to a location whose coordinates are close to -10 mm 10 mm 0 mm.

18. Specify the start point of the line on this location and move the cursor horizontally toward the left. Specify the endpoint when the length of the line above the line cursor shows a value close to 42.

19. Move the line cursor such that the line is drawn at an angle close to 105-degree. Specify the endpoint when the length of line is close to 15.

Note

*You may need to scroll down the **Line PropertyManager** to view the angle. Move the cursor to the scroll bar, press and hold down the left mouse button, and drag the cursor vertically downward to scroll down the **Line PropertyManager**.*

20. Move the line cursor horizontally toward the left. Specify the endpoint when the length of the line above the line cursor shows a value close to 34.

21. Move the line cursor vertically downward. Specify the endpoint when the length of the line above the line cursor shows a value close to 10.

22. Move the line cursor such that the line is drawn at an angle close to 346-degree. Specify the endpoint when the length of the line is around 27.

23. Move the line cursor vertically downward. Specify the endpoint when the line cursor snaps the mirror line. Double-click anywhere in the drawing area to end line creation.

The sketch after drawing the required arcs, circles, and lines is shown in Figure 3-63.

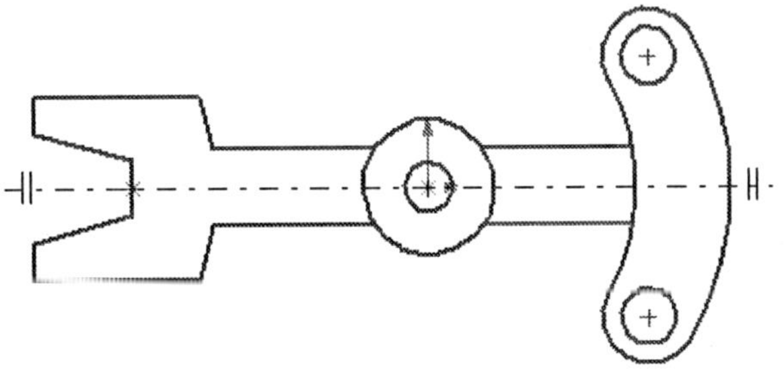

Figure 3-63 *Sketch after drawing all the entities*

Trimming the unwanted entities

After creating the sketch, you need to trim some of the unwanted sketched entities using the **Trim Entities** tool.

1. Choose the **Trim Entities** button from the **Sketch CommandManager**.

The line cursor is replaced by the trim cursor.

2. Select the entities to be trimmed as shown in Figure 3-64. The entities are dynamically trimmed.

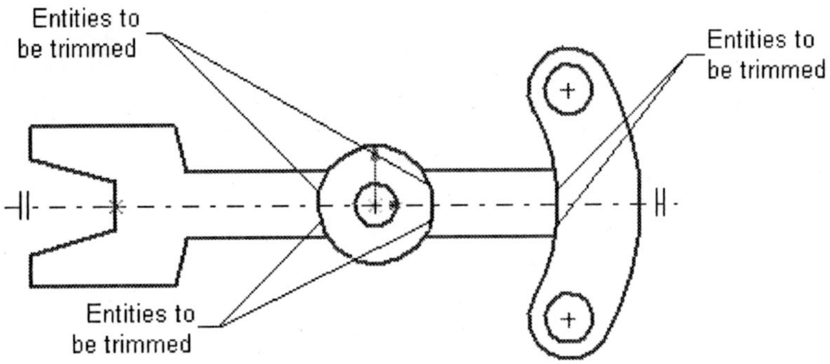

Figure 3-64 *Entities to be trimmed*

Adding the Fillets to the Sketch

Next, you need to add fillets to the sketch. The fillets are generally applied to avoid the stress concentration at sharp corners.

1. Choose the **Sketch Fillet** button from the **Sketch CommandManager**. Set the radius spinner to **5**.

2. Select the entities shown in Figure 3-65 to apply the fillet.

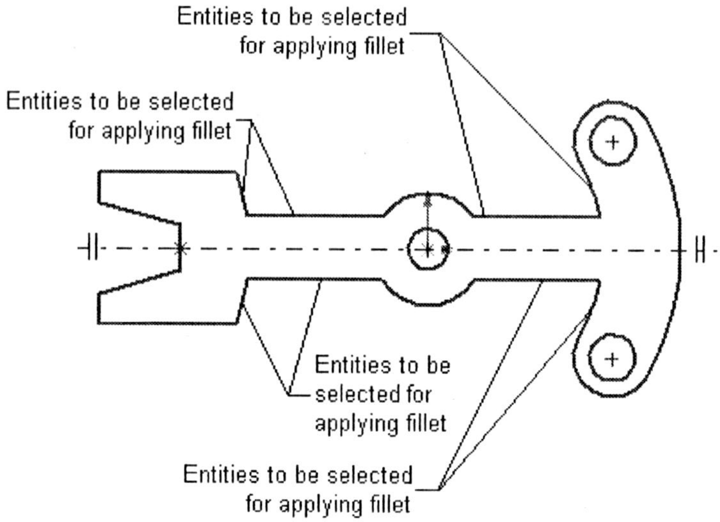

Figure 3-65 Entities to be selected to apply fillet

3. Choose the **OK** button from the **Sketch Fillet PropertyManager** to exit the **Fillet** tool. The sketch after applying the fillets to the sketch is shown in Figure 3-66.

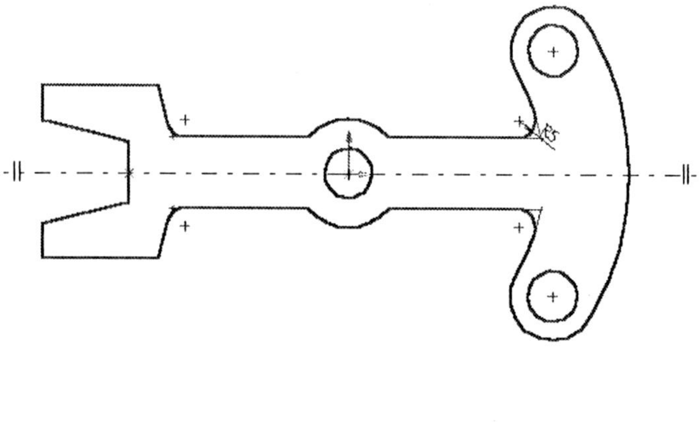

Figure 3-66 Sketch after creating the fillets

Adding Relations to the Sketch

Next, you need to add the required relations to the sketch.

1. Choose the **Add Relations** button from the **Sketch CommandManager** to invoke the **Add Relations PropertyManager**.

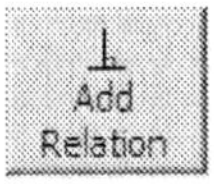

2. Select the lower left arc and the lower tangent arc. The **Tangent** button is highlighted in bold in the **Add Relations** rollout of the **Add Relations PropertyManager**. This suggests that the **Tangent** relation is the most appropriate relation for the selected entities.

3. Choose the **Tangent** button from the **Add Relations PropertyManager**.

4. Right-click in the drawing area and choose the **Clear Selections** option from the shortcut menu to clear the selections from the selection set. Select the right arc and the lower tangent arc. The **Tangent** button is highlighted.

5. Choose the **Tangent** button from the **Add Relations PropertyManager**. Clear the current selections from the selection set.

6. Select the upper right horizontal line and the upper left horizontal line that are coincident with the trimmed circle. Choose the **Collinear** button from the **Add Relations PropertyManager**.

7. Choose the **OK** button from the **Add Relations PropertyManager** or choose **OK** from the confirmation corner.

Adding Dimensions to the Sketch

Next, you will apply the required dimensions to the sketch and fully defined it.

1. Choose the **Smart Dimension** button from the **Sketch CommandManager** to invoke the dimension tool. The arrow cursor is replaced by the dimension cursor.

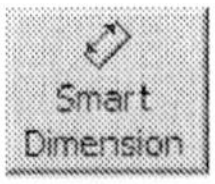

2. Select the right arc. A radius dimension is attached to the cursor. Move the cursor away from the sketch toward the right and place the dimension.

3. Enter a value of **82** in the **Modify** dialog box and press ENTER.

4. Select the left upper arc, refer to Figure 3-67. A radius dimension is attached to the cursor. Move the cursor away from the sketch toward the right and place the dimension.

5. Enter a value of **56** in the **Modify** dialog box and press ENTER.

6. Select the **Origin** and the centerpoint of the upper circle. A dimension is attached to the cursor. Next, select the right endpoint of the mirror line. An angular dimension is attached to the cursor. Place the angular dimension outside the sketch.

7. Enter the angular dimension value of **30** in the **Modify** dialog box and press ENTER.

8. Select the upper right circle; a diameter dimension is attached to the cursor. Place the dimension outside the sketch.

9. Enter the diameter dimension value as **15** in the **Modify** dialog box and press ENTER.

10. Select the upper right horizontal line that coincides the trimmed circle and the lower right horizontal line that coincides the trimmed circle, refer to Figure 3-67. A vertical dimension is attached to the cursor. Move the cursor vertically downward and left-click to place the dimension.

11. Enter a value of **20** in the **Modify** dialog box.

12. Select the smaller circle at the origin. A diameter dimension is attached to the cursor. Move the cursor upward and place the dimension outside the sketch.

13. Enter diameter dimension value as **13** in the **Modify** dialog box.

14. Select the outer trimmed circle and place the radius dimension outside the sketch.

15. Enter the radial dimension value as **19** in the **Modify** dialog box.

16. Select the upper right inclined line. A dimension is attached to the cursor. Now, select the upper left horizontal line. An angular dimension is attached to the cursor. Place the dimension above the upper left horizontal line.

17. Enter the angular dimension value as **75** in the **Modify** dialog box.

18. Select the origin and the lower endpoint of the lower right inclined line. Move the cursor vertically downward and place the dimension.

19. Enter a value of **49** in the **Modify** dialog box.

20. Select the origin and the middle left vertical line, refer to Figure 3-67. Move the cursor vertically downward and place the dimension below the previous dimension.

21. Enter a value of **61** in the **Modify** dialog box.

22. Select the origin and select the lower endpoint of the outer left vertical line. Move the cursor vertically downward and place the dimension below the last dimension.

23. Enter a value of **90** in the **Modify** dialog box.

24. Select the upper left inclined line, and the lower left inclined line, refer to Figure 3-67. An angular dimension is attached to the cursor. Move the cursor horizontally toward the left and place the dimension.

25. Enter the angular dimension value as **28** in the **Modify** dialog box.

26. Select the upper left horizontal line and the lower left horizontal line. A linear dimension is attached to the cursor. Move the cursor horizontally toward the left and place the dimension.

27. Enter a value of **50** in the **Modify** dialog box.

28. Select the lower endpoint of the upper left vertical line and the upper endpoint of the lower left vertical line. A linear dimension is attached to the cursor. Move the cursor horizontally toward the left and left-click to place the dimension.

29. Enter a value of **30** in the **Modify** dialog box.

    All the sketched entities are displayed in black, suggesting that the sketch is fully defined. The fully defined sketch is shown in Figure 3-67.

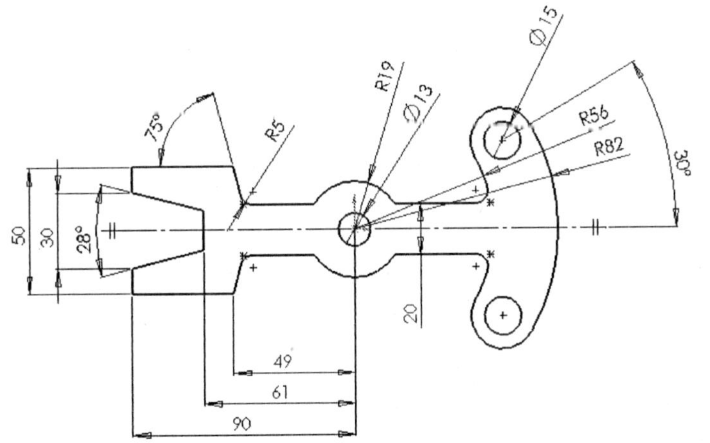

***Figure 3-67** Sketch after applying all relations and dimensions*

Saving the Sketch

1. Choose the **Save** button from the **Standard** toolbar to invoke the **Save As** dialog box.

2. Enter the name of the document as *c03-tut02.sldprt* in the **File name** edit box and choose the **Save** button.

3. Close the document by choosing **File** > **Close** from the menu bar.

Tutorial 3

In this tutorial you will draw the sketch of a revolved model shown in Figure 3-68. The sketch is shown in Figure 3-69. The solid model is given only for your reference.

(Expected time: 30 min)

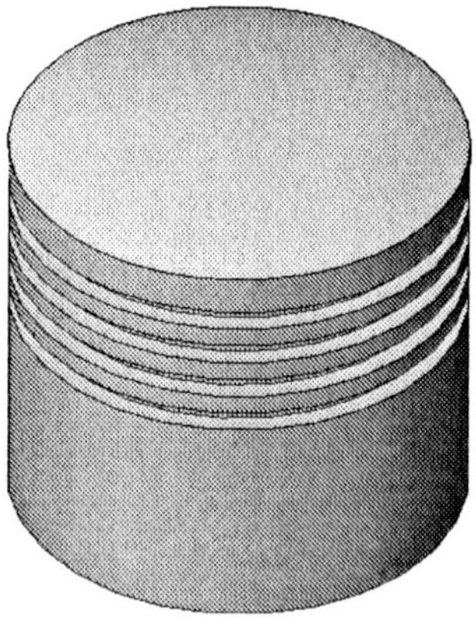

Figure 3-68 Solid model of the piston

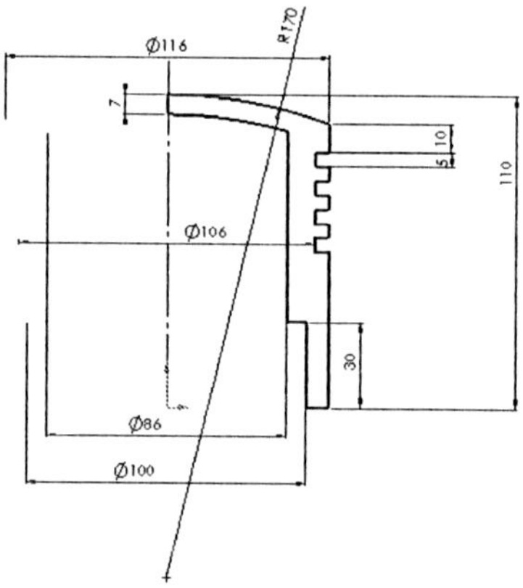

Figure 3-69 The sketch of the base feature

The steps that will be followed to complete this tutorial are listed below.

a. Start a new part document and then invoke the sketching environment.
b. Draw a centerline that will be used to add the linear diameter dimensions to the sketch of the piston.
c. Create the sketch using the various sketching tools.
d. Use the **Offset Entities** tool to offset the required lines, refer to Figure 3-70.
e. Draw the arcs and trim the unwanted entities, refer Figure 3-71.
f. Add the required relations.
g. Add the required dimensions and fully define the sketch, refer Figure 3-72.

Starting a New Part Document

1. Choose the **New** button from the **Standard** toolbar and start a new part document using the **New SolidWorks Document** dialog box.

2. Choose the **Sketch** button from the **Standard** toolbar and select the **Front** plane to invoke the sketching environment for drawing the sketch. Maximize the part document.

Creating the Sketch

To create the sketch of the revolved model, you first need to draw the centerline around which the sketch of the base feature will be revolved.

1. Choose the **Centerline** button from the **Sketch CommandManager**.

2. Draw a vertical centerline starting from the origin and with a length of 120 mm.

   Next, you will draw the sketch of the piston.

3. Right-click in the drawing area and choose the **Line** option from the shortcut menu. Move the cursor to a location whose coordinates are close to 58 mm 0 mm 0 mm. Left-click at this point to specify the start point of the line.

4. Draw a vertical line of dimension close to 100. Right-click and choose the **End chain** option from the shortcut menu.

5. Move the cursor to the lower endpoint of the line created earlier. Specify the start point of the line when the red circle is displayed. Move the cursor horizontally toward the left and create a horizontal line of dimension close to 8.

6. Move the line cursor vertically upward and create a vertical line of dimension close to 30.

7. Move the cursor horizontally toward the left and create a horizontal line of dimension close to 7.

8. Move the line cursor vertically upward and create a vertical line of dimension close to 70.

9. Right-click and choose the **3 Point Arc** option from the shortcut menu. Move the cursor near the upper endpoint of the right vertical line.

10. Specify the first point of the arc when the red circle is displayed. Move the cursor horizontally toward the left. The reference arc is attached to the cursor. Specify the second point of the arc when the value of the length is close to 116.

11. Move the cursor toward the right. Specify the third point of the arc when the value of the radius is close to 170.

12. Right-click and choose the **Line** option from the shortcut menu. Move the line cursor to a location whose coordinates are close to 58 mm 90 mm 0 mm. Specify the start point of the line at this location and move the cursor horizontally toward the left to create a line of dimension close to 5. Double-click anywhere in the drawing area to end line creation.

13. Choose the **Zoom To Fit** button from the **View** toolbar to fit the sketch to the drawing area.

Offsetting the Lines

Using the **Offset Entities** tool, you will offset the entities created earlier.

1. Select the line created previously using the select tool.

2. Choose the **Offset Entities** button from the **Sketch CommandManager** to invoke the **Offset Entities PropertyManager**. The confirmation corner is also displayed at the upper right corner of the drawing area.

3. Choose the **Keep Visible** button from the **Offset Entities PropertyManager** to pin the **PropertyManager**.

4. Set the value of **Offset Distance** spinner to **5**. Now, select the **Reverse** check box because you need to offset the entity in the reverse direction. The preview of the entity to be offset is modified in the drawing area.

5. Choose the **OK** button from the **Offset Entities PropertyManager**.

   You will notice that an entity is created at an offset distance of 5 from the original entity. Also, a dimension is applied between the newly created entity and the original entity with a value of **5**. This dimension is the offset distance between the two entities.

6. Select the newly created entity. Move the cursor vertically downwards. The preview of the entity and the direction of offset creation is also displayed. Press the left mouse button to offset the selected line.

   Repeat this procedure of offsetting the entities until you get the eight entities, including the original entity.

7. Set the value of **Offset Distance** spinner to **7** and clear the **Select chain** check box. Select the upper arc. The preview of the offset arc is shown in the drawing area.

8. Choose the **OK** button twice on the **Offset Entities PropertyManager**.

Completing the Remaining Sketch

Next, you will complete the remaining sketch using the line tool.

1. Right-click and choose the **Line** option from the shortcut menu. Move the line cursor to the left endpoint of the original line that was used to create offset lines. When the red circle is displayed, specify the start point. Specify the endpoint of the line at the endpoint of the last offset line.

2. Move the cursor to the intersection point of the upper arc and the centerline. When the cursor snaps the intersection, create a vertical line that snaps the intersection point of the lower arc and the centerline.

   The sketch after creating the entities using the various sketch tools and the offset tool is shown in Figure 3-70.

Trimming the Unwanted Entities

Next, you will trim the unwanted entities using the trim tool.

1. Choose the **Trim Entities** button from the **Sketch CommandManager**.

2. Using the left mouse button trim the unwanted entities. The sketch after trimming the unwanted entities is shown in Figure 3-71.

Adding the Required Relations

Now, you will add the required relations to the sketched entities.

1. Right-click and choose the **Select** option from the shortcut menu. Press and hold down the CTRL key from the keyboard and select one of the endpoints of the lower horizontal line

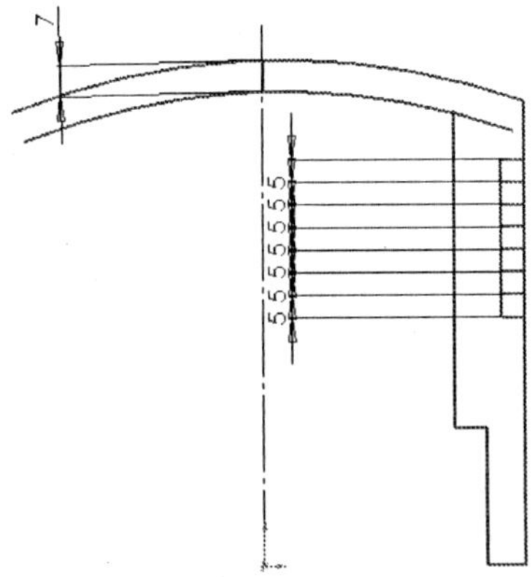

Figure 3-70 Sketch after creating various entities

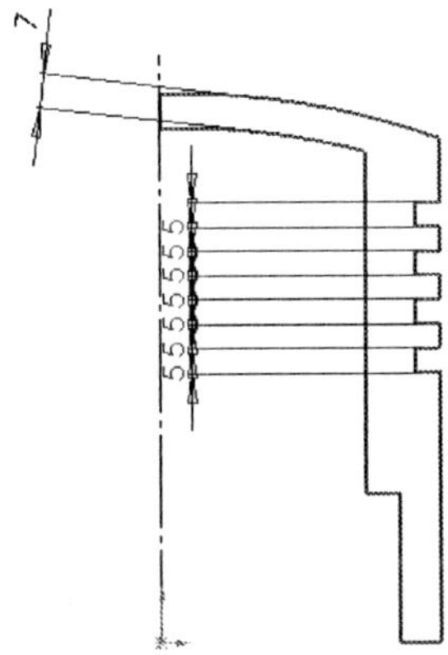

Figure 3-71 Sketch after trimming the unwanted entities

and then select the origin. Release the CTRL key after selection. Right-click and choose the **Horizontal** option from the shortcut menu to add the horizontal relation to the selected entities. Click anywhere in the drawing area to clear the selection set.

2. Press and hold down the CTRL key and select the small horizontal lines in the right of the sketch. Release the CTRL key after making the selection and right-click to display the shortcut menu. Choose the **Equal** option from the shortcut menu to apply the equal relation.

3. Similarly, apply the **Equal** relation to all the small vertical lines.

   The **SolidWorks** warning message window will be displayed. This warns you that the solution cannot be determined for the sketch. You will notice that some the entities of the sketch turn red. This indicates that the sketch is overdefined.

4. Choose **OK** from this message window.

   Another SolidWorks warning message window is displayed. This warns you that applying this relation will result in overdefining the sketch.

5. Choose **OK** from this message window also.

   As discussed earlier, the overdefined sketch is not used to create any feature; therefore, you have to delete the conflicting relations. The sketch is overdefined after applying the previous relation. Therefore, if you delete the last applied relation using the **Undo** button from the **Standard** toolbar then the sketch will not be overdefined.

6. Choose the **Undo** button from the **Standard** toolbar. Click anywhere in the drawing area to clear the selections from the selection set.

As evident from the color of the entities, which is blue, you still have to add some more relations or dimensions to fully define the sketch. The fully defined sketch is displayed in black color.

7. Zoom in using the **Zoom In/Out** tool and apply the coincident relation between the centerpoint of the upper arc and the centerline.

Adding the Dimensions to the Sketch

After drawing, editing, and applying the relations to the sketch, you will add the required dimensions to the sketch to fully define the sketch.

1. Select the dimension with a value of **7** that is placed between the upper arcs and press the DELETE key from the keyboard.

   The SolidWorks warning message window is displayed. This warns you that if you delete the offset dimension, the offset relation will also be deleted from the sketch.

2. Choose the **Yes** button from this message window.

   You have to delete this dimension because during the design and manufacturing practices the dimension between the tangents should be avoided.

3. Right-click and choose the **Smart Dimension** option from the shortcut menu to invoke this tool.

4. Select the outer upper right vertical line. A dimension is attached to the cursor. Now, select the centerline and move the cursor to the other side of the centerline. You will notice that the diameter dimension is displayed along the cursor.

5. Left-click to place the dimension above the sketch and enter a value of **116** in the **Modify** edit box and press ENTER.

6. Now, select the inner left vertical line and then select the centerline, refer to Figure 3-72. Move the cursor to the other side of the centerline and left-click to place the dimension below the sketch.

7. Enter a value of **86** in the **Modify** edit box and press ENTER.

   Add the remaining dimensions to fully defined the sketch, refer to Figure 3-72.

8. Using the left mouse button, select one of the dimensions that are created after offsetting the entities and drag the cursor toward the right and using the left mouse button place the dimension at an appropriate place. Arrange all the dimensions using the above method. The fully defined sketch is shown in Figure 3-72.

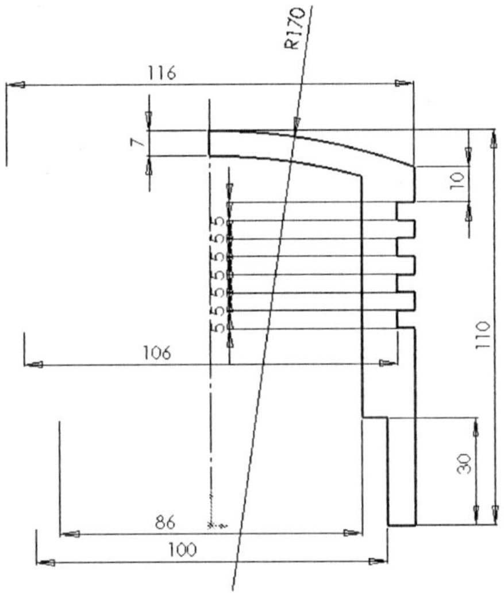

Figure 3-72 Fully defined sketch after applying all the relations and dimensions

Saving the Sketch

1. Choose the **Save** button from the **Standard** toolbar and save the sketch with the name given below.

   \My Documents\SolidWorks\c03\c03-tut03.sldprt.

2. Choose **File > Close** from the menu bar to close the document.

SELF-EVALUATION TEST

Answer the following questions and then compare your answers with the answers given at the end of this chapter.

1. Some relations are automatically applied to the sketch while it is being created. (T/F)
2. When you choose the **Add Relations** button, the **Apply Relations PropertyManager** is displayed. (T/F)
3. You can modify the arrowhead style of a selected dimension. (T/F)
4. The dimension favorite created in one document cannot be retrieved in any other document. (T/F)
5. You can do modifications in the view-only file. (T/F)

6. The __________ **PropertyManager** is displayed using the **Add Relation** button from the **Sketch CommandManager**.

7. The __________ dimension is used to dimension a line that is at an angle with respect to the X axis or the Y axis.

8. A __________ defined sketch is the one in which all the entities and their positions are described by the relations or dimensions, or both.

9. The __________ dimensions or relations are not able to determine the position of one or more sketched entities.

10. The __________ option is displayed in the **Defined In** column when the entity is defined as placed in the same sketch.

REVIEW QUESTIONS

Answer the following questions.

1. You can invoke the **Display/Delete Relations PropertyManager** using the **Display/Delete Relations** button from __________ **CommandManager**.

2. The linear diameter dimensions are applied to the sketches of __________ features.

3. You can modify the dimension to display minimum or maximum distance between two circles using the __________ dialog box.

4. The __________ sketch geometry is constrained by too many dimensions and/or relations. Therefore, you have to delete the extra and conflicting relations or dimensions.

5. The __________ relation forces two selected lines, arcs, points, or ellipses to remain equidistant from a centerline.

6. In SolidWorks, by default, the dimensioning between two arcs, two circles, or between an arc and a circle is done from

   (a) Centerpoint to centerpoint (b) Centerpoint to tangent
   (c) Tangent to tangent (d) None

7. Which relation forces the selected arc to share the same centerpoint with another arc or a point?

   (a) **Concentric** (b) **Coradial**
   (c) **Merge points** (d) **Equal**

8. Which type of dimensions force the color of the entities to red?

(a) Underdefined (b) Overdefined
(c) Dangling (d) None

9. Which dialog box is displayed when you modify a dimension?

(a) **Modify Dimensional Value** (b) **Insert a value**
(c) **Modify** (d) None

10. When you add an extra dimension to a sketch or add an extra relation that over defines the sketch, then which dialog box is displayed?

(a) **Over defining** (b) **Delete relation**
(c) **Make Dimension Driven?** (c) **Add Geometric Relations**

EXERCISES

Exercise 1

Create the sketch of the model shown in Figure 3-73. Create the sketch and apply the required relations and dimensions and fully define the sketch. The sketch is shown in Figure 3-74. The solid model is given only for reference. **(Expected time: 30 min)**

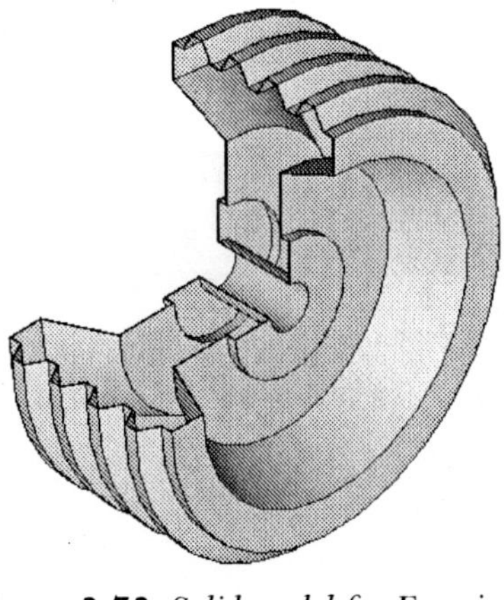

Figure 3-73 Solid model for Exercise 1

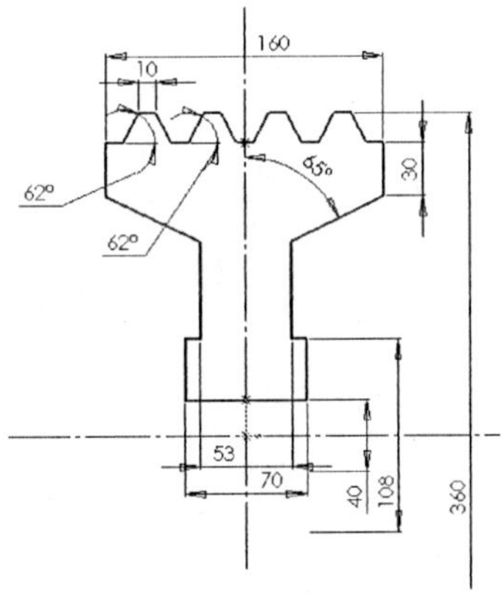

Figure 3-74 Sketch for Exercise 1

Exercise 2

Create the sketch of the model shown in Figure 3-75. Create the sketch and apply the required relations and dimensions and fully define the sketch. The sketch is shown in Figure 3-76. The solid model is given only for reference. **(Expected time: 30 min)**

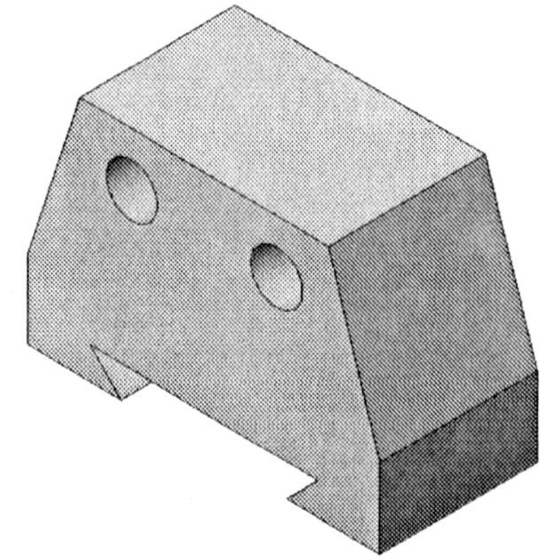

Figure 3-75 Solid model for Exercise 2

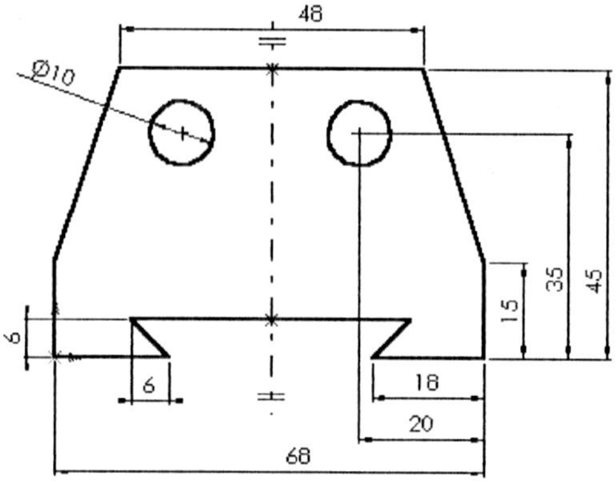

Figure 3-76 Sketch for Exercise 2

Exercise 3

Create the sketch of the model shown in Figure 3-77. Create the sketch and apply the required relations and dimensions and fully define the sketch. The sketch is shown in Figure 3-78. The solid model is given only for reference. **(Expected time: 30 min)**

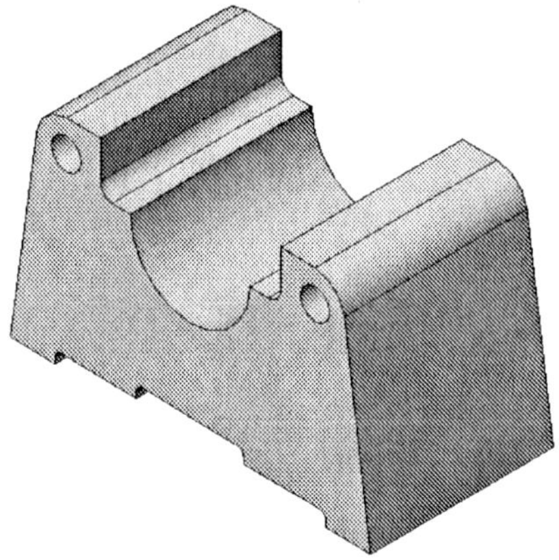

Figure 3-77 Solid model for Exercise 3

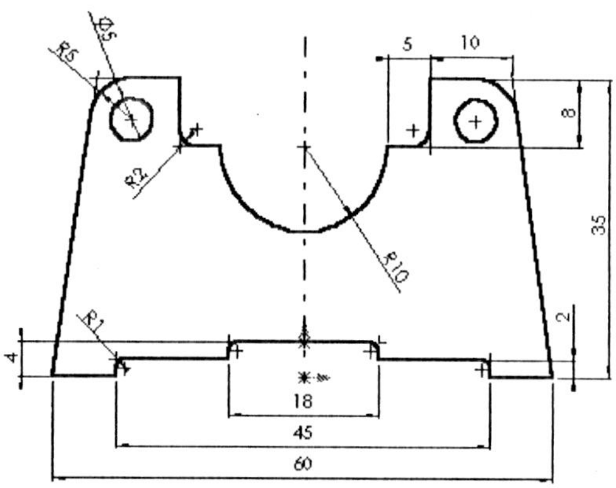

Figure 3-78 Sketch for Exercise 3

Answers to Self-Evaluation Test

1. T, **2.** F, **3.** T, **4.** F, **5.** T, **6. Add Relations**, **7.** aligned, **8.** fully defined, **9. Dangling**, **10. All in this sketch**

Chapter 4

Advance Dimensioning and Base Feature Options

Learning Objectives

After completing this chapter you will be able to:

- *Dimension the sketch using the Autodimension sketch tool.*
- *Dimension the true length of arc.*
- *Measure Distances and View Section Properties*
- *Create solid base extruded features.*
- *Create thin base extruded features.*
- *Create solid base revolved features.*
- *Create thin base revolved features.*
- *Dynamically rotate the view to display the model from all directions.*
- *Modify the orientation of the view.*
- *Change the display modes of the solid model.*
- *Apply material and textures to the models.*

ADVANCE DIMENSIONING TECHNIQUES

In this chapter, you will learn about some of the advance dimensioning techniques used in dimensioning the sketches in SolidWorks. With this release of SolidWorks, you are able to apply all the possible dimensions to a sketch using a single option. This option is known as **Autodimension Sketch**. The advanced dimensioning techniques are discussed next.

Autodimension the Sketches

CommandManager:	Dimension/Relations > Autodimension	*(Customize to Add)*
Menu:	Tools > Dimensions > Autodimension Sketch	
Toolbar:	Dimension/Relations > Autodimension	*(Customize to Add)*

The **Autodimension** tool is used to automatically apply the dimensions to the sketch. To apply autodimensions to a sketch, create the sketch using standard sketching tools and then apply the required relations to the sketch. Now, choose the **Autodimension** button from the **Dimension/Relations CommandManager**. The **Autodimension PropertyManager** is displayed as shown in Figure 4-1. The various options available in the **Autodimension Sketch PropertyManager** are discussed next.

Entities to Dimension

The **Entities to Dimension** rollout is used to specify the entities on which the dimensions have to be applied. The **All entities in sketch** radio button is selected by default. As a result, all the entities drawn in the current sketching environment are selected to apply the dimensions. The **Selected entities** radio button is selected if you have to dimension only the selected entities. When you select this radio button, the **Selected Entities to Dimension** selection area is displayed in the **Entities to Dimension** rollout. Select the entities to dimension using the select cursor. The names of the selected entities are displayed in the **Selected Entities to Dimension** selection area. If you select one or more entities before invoking the **Autodimension sketch PropertyManager**, then the **Selected entities** radio button is selected by default. Also, the names of the selected entities will be displayed in the **Selected Entities to Dimension** selection area.

Horizontal Dimensions

The **Horizontal Dimensions** rollout is used to specify the type of horizontal dimension, reference for the horizontal dimension, and the dimension placement. The various options available in the **Horizontal Dimensions** rollout are discussed next.

Scheme

The **Scheme** drop-down list is used to specify the dimensioning scheme to be applied to the sketch. The various types of dimensioning schemes available in the this drop-down list are discussed next.

Chain

The **Chain** option is used for the relative or incremental horizontal dimensioning of the sketch. When you invoke the **Autodimension PropertyManager** and select this scheme, a point or a vertical line is selected as the reference entity. This reference entity is used as a datum for the generation of dimension. The name of the selected reference

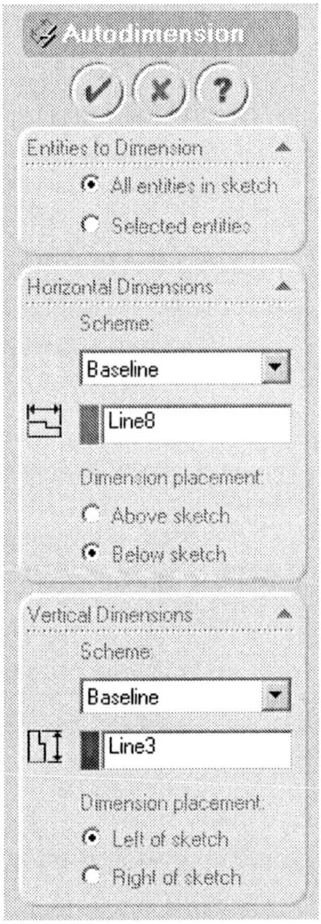

Figure 4-1** The **Autodimension PropertyManager

entity is displayed in the **Datums - Vertical Model Edge, Model Vertex, Vertical Line or Point** selection area and the reference entity is displayed in red color in the drawing area. You can also specify a user-defined reference entity.

Note
Chain dimensioning should be avoided if the tolerances relative to a common datum are required in the part.

Baseline
The **Baseline** option is used for absolute or datum vertical dimensioning of the sketch. In this dimensioning method, the dimensions are applied to the sketch with respect to the common datum. When you invoke the **Autodimension PropertyManager** and select this option, a point or a vertical line is selected as the reference entity, which is used as a datum for the generation of dimension. The name of the selected reference entity is displayed in the **Datums - Vertical Model Edge, Model Vertex, Vertical Line or Point**

selection area and the reference entity is displayed in red color in the drawing area. You can also specify a user-defined reference entity.

Ordinate

The **Ordinate** option is used for the ordinate dimensioning of the sketch. When you invoke the **Autodimension PropertyManager** and select this option, a point or a vertical line is selected as the reference entity, which is used as a datum for the generation of dimension. The name of the selected reference entity is displayed in the **Datums - Vertical Model Edge, Model Vertex, Vertical Line or Point** selection area and the reference entity is displayed in red color in the drawing area. You can also specify a user-defined reference entity.

Dimension placement

The **Dimension Placement** area is used to define the position where the generated dimensions will be placed. Two radio buttons are available in this area. The first radio button is the **Above sketch** radio button. If you use this option, the horizontal dimensions generated using the **Autodimension** tool will be placed above the sketch. The **Below sketch** radio button is selected by default and is used to place the dimensions below the sketch.

Vertical Dimensions

The **Vertical Dimensions** rollout is used to specify the type of vertical dimension, reference for the vertical dimension, and the dimension placement. The various options available in the **Scheme** area are similar to those discussed under the **Horizontal Dimensions** rollout. The remaining option is discussed next.

Dimension placement

The **Dimension Placement** area of the **Vertical Dimensions** rollout is used to define the position where the generated dimensions will be placed. The **Left of the Sketch** radio button is selected to place the dimensions on the left of the sketch. The **Right of the Sketch** radio button is selected to place the dimensions on the right of the sketch. The **Right of the Sketch** radio button is selected by default.

After specifying all the parameters in the **Autodimension Sketch PropertyManager**, choose the **OK** button or choose the **OK** icon from the **Confirmation Corner**. The dimension will be created with the selected dimension scheme. Figure 4-2 shows the autodimension created using the **Baseline** scheme taking the origin as the datum. Figure 4-3 shows the autodimension created using the **Ordinate** scheme taking the origin as the datum.

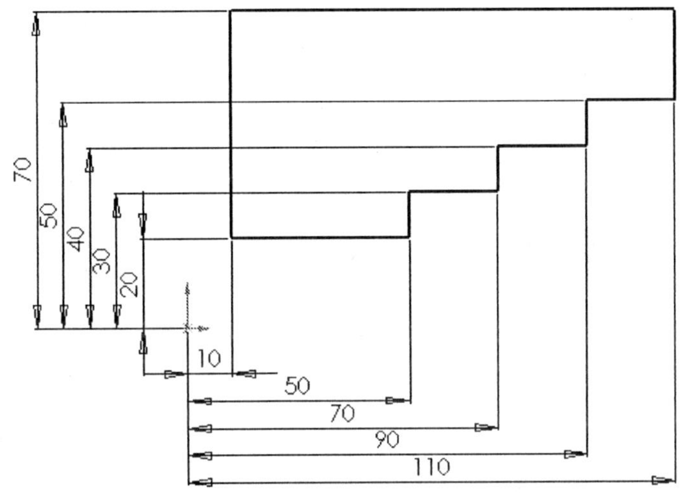

*Figure 4-2 Baseline dimension created using the **Autodimension** tool*

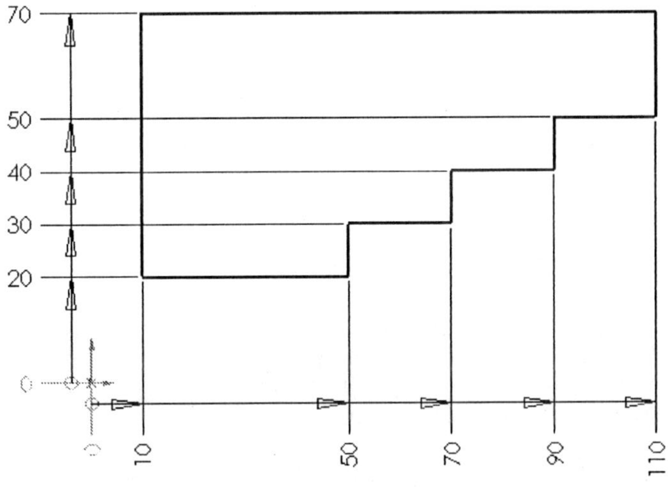

*Figure 4-3 Ordinate dimension created using the **Autodimension** tool*

Dimensioning of the True Length of an Arc

In SolidWorks you can also create the dimension of the true length of an arc, which is one of the advantages of the sketching environment of SolidWorks. To create the dimension of the true length, invoke the **Smart Dimension** tool and select the arc using the dimension cursor. A radial dimension is attached to the cursor. Move the cursor to any of the endpoints of the arc. When the cursor snaps the endpoint, use the left mouse button to specify the first endpoint of the arc. A linear dimension is attached to the cursor; move the cursor to the second endpoint of the arc and when the cursor snaps the endpoint, select it. A dimension is attached to the cursor. Move the cursor to an appropriate place to place the dimension. The dimension of the true length of the arc is shown in Figure 4-4.

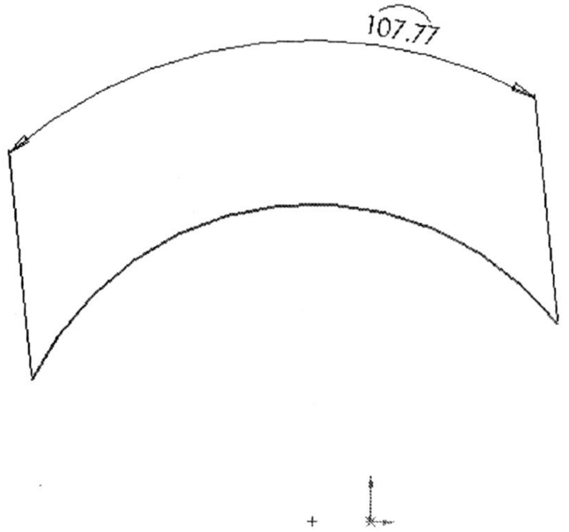

Figure 4-4 *Dimensioning the true length of an arc*

MEASURING DISTANCES AND VIEWING SECTION PROPERTIES

SolidWorks allows you to measure the distance of the entities and also view the section properties. These tools are discussed next.

Measuring Distances

CommandManager:	Tools > Measure
Menu:	Tools > Measure
Toolbar:	Tools > Measure

The **Measure** tool is used to measure perimeter, angle, radius, and distance between lines, points, surfaces, and planes in sketches, 3D models, assemblies, or drawings. To use the measure tool, you have to invoke the **Measure** dialog box by choosing the **Measure** button from the **Tools CommandManager**. The **Measure** dialog box is displayed. The name of the document in which you are working is displayed at the top of the **Measure** dialog box. The **Measure** dialog box is displayed in Figure 4-5. The current cursor is replaced by the measure cursor. Using the measure cursor, select the entity or entities to be measured. The various options in the **Measure** dialog box are discussed next.

Output coordinate system drop-down list

Using this drop-down list you can specify the coordinate system with reference to which you want to measure the selected entities. By default, the **default** option is selected in the drop-down list. If you create a coordinate system, then you can also select that coordinate system from this drop-down list. Creation of the coordinate system is discussed in later chapters.

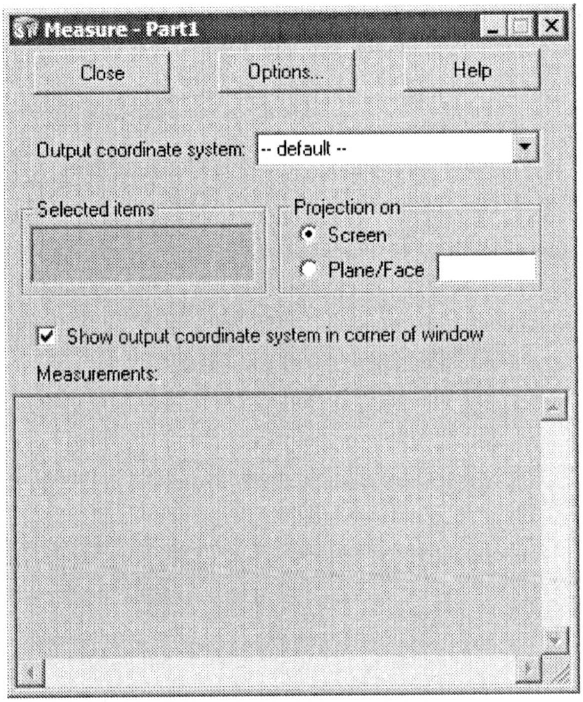

*Figure 4-7 The **Measure** dialog box*

Selected items Area

The **Selected items** area consists of a display box that displays the entities that are selected to be measured. For example, if you select two points to measure the distance between them, both the points will be displayed in this area. You can right-click in this area to remove all or the selected entity.

Projection on

The **Projection on** area is used to specify where the selected entity should be projected. You can project the selected entity on the screen or on a specific plane. The system will then calculate the measurement of the true projection.

Show output coordinate system in corner of window

The **Show output coordinate system in corner of window** check box is selected to display the coordinate system at the corner of the screen. This check box is selected by default. If you clear this check box, the coordinate system will be placed at the origin in the drawing area.

Measurements Area

When you select one or more entities to be measured, the appropriate values of the length, angle, perimeter, and so on are displayed in the **Measurements** area. New measurements update dynamically when you change the selection of entities. If the combination of selected entities is not appropriate for the measure function, then a message is displayed in the **Measurements** area. This message prompts you that the selection is invalid. If two selected entities are intersecting lines then you are provided the value of the angle between the intersecting entities and the message prompts you that the selected items intersect.

Options

The **Options** button is used to invoke the **Measurement Options** dialog box. In this dialog box, you can specify the units for linear measurement, units for angular measurement, material properties, and the accuracy level. The **Measurement Options** dialog box is shown in Figure 4-6. The options in the **Measurement Options** dialog box are discussed next.

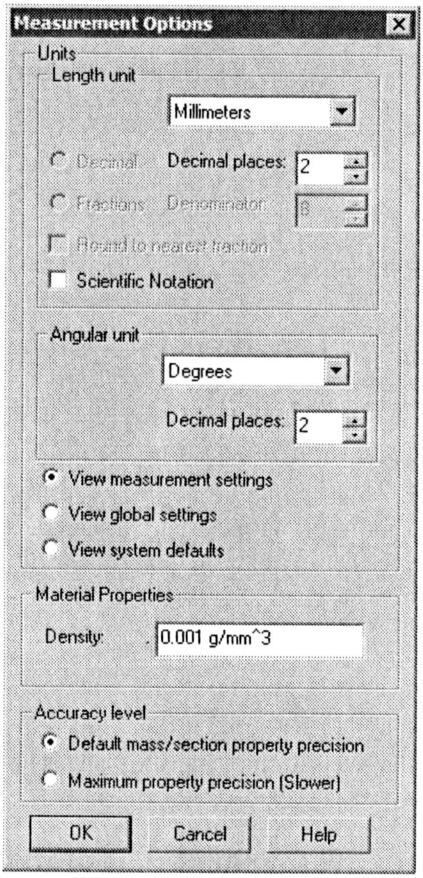

*Figure 4-6 The **Measurement Options** dialog box*

Units Area

The **Units** area of the **Measurement Options** dialog box is used to set the linear and the angular units for measurement. By default, the linear units are selected as millimeters with decimal up to two places, and the angular units are selected as degree with decimal up to two places. The various options in the **Units** area are discussed next.

Length unit Area

The **Length unit** area is used to set the units and options of linear measurements of the entities. The **Unit** drop-down list is provided at the top right corner of the **Length unit** area. In this drop-down list you can select any type of unit such as **Angstroms, Nanometers, Microns, Millimeters, Centimeters, Meters, Microinches, Miles, Inches, Feet,** and **Feet and Inches**. The **Decimal places** spinner is provided to control the decimal places. The other options in the **Length unit** area are discussed next.

Decimal. The **Decimal** radio button is available only when you select the units as **Microinches, Miles, Inches,** or **Feet and Inches** from the **Unit** drop-down list. This radio button is selected to display the dimension in the decimal form. You can also specify the decimal places using the **Decimal places** spinner provided at the right of the **Decimal** radio button.

Fractions. The **Fractions** radio button is available only when you select the units as **Microinches, Miles, Inches,** or **Feet and Inches** from the **Units** drop-down list. This radio button is selected to display the dimension in the fraction form. You can also set the value of the denominator using the **Denominator** spin box provided at the right of the **Fractions** radio button.

Round to nearest fraction. The **Round to nearest fraction** check box is selected to display the value in fractions by rounding the value to the nearest fraction.

Scientific Notation. The **Scientific Notation** check box is selected to display the value in the scientific notation units.

Angular unit Area

The **Angular unit** area is used to set the units for angular measurement. This area is provided with a drop-down list to specify the angular measurement unit such as **Degrees, Deg/Min, Deg/Min/Sec,** and **Radians**. The **Decimal places** spinner is provided to specify the decimal places.

View measurement settings

The **View measurement settings** radio button is selected to activate a temporary system of units. After selecting this radio button if you change the units, it will not affect the units specified for the document, or the default units of the system for new documents.

View global settings

The **View global settings** radio button is selected to display the units specified for the active document. If you change the unit setting in the **Length unit** and **Angular unit** areas and select the **View global settings** radio button, the units will be reset to the default units of the active document.

View system defaults

The **View system defaults** radio button is selected to display the default units of the system for new documents. If you change the unit setting in the **Length unit** and **Angular unit** areas and select the **View system defaults** radio button, then the units will be reset to the default units of the system and when you create a new document the default units will be displayed.

Material Properties Area

In the **Material Properties** area, you can change the density of the material for the current document. You can enter the density value using any unit. For example, if the units of the part are grams and millimeters, you can enter a density value using pounds and inches. The SolidWorks converts the value to the document's units when you choose **OK** from the **Measurement Options** dialog box.

Accuracy level

The **Accuracy level** area of the **Measurement Options** dialog box is used to specify the accuracy level provided by SolidWorks for the measurement. The **Default mass/section property precision** radio button is selected by default. This option performs the calculations faster, but the precision is not very high. The **Maximum property precision (Slower)** radio button is selected to increase the accuracy and precision of the measurement. This option slows the working of the computer, but gives a very precise output result.

Determining Section Properties of Closed Sketches

CommandManager:	Tools > Section Properties
Menu:	Tools > Section Properties
Toolbar:	Tools > Section Properties

The **Section Properties** tool enables you to determine the section properties of the sketch in the sketching environment or of the selected planar face in the **Part** mode and the **Assembly** mode. Remember that the section properties of only the closed sketches with nonintersecting closed loops can be determined. The section properties include the area, centroid relative to sketch origin, centroid relative to part origin, moment of inertia, polar moment of inertia, angle between the principle axes and sketch axes, and principle moment of inertia.

To calculate the section properties, create the sketch and then choose the **Section Properties** button from the **Tools CommandManager**. The **Section Properties** dialog box is displayed as shown in Figure 4-7.

When you invoke the **Section Properties** dialog box a red 3D triad is placed at the centroid of the sketch. The section properties of the sketch are displayed in the **Section Properties** dialog box. The **Selected Items** display box is used to display the name of the selected planar face whose section properties are to be calculated. When you are in the **Part** mode, select the face to calculate the section properties and choose the **Recalculate** button to display the properties. If you want to calculate the section properties of some other face, clear the previously selected face from the selection set. Now, select the new face and choose the **Recalculate** button.

The **Print** button available in the **Section Properties** dialog box is used to print the section properties. The **Copy** button is chosen to copy the section properties to the Clip board from where you can copy them to a program such as MS Word. The rest of the options in this dialog box are similar to those discussed in the previous section.

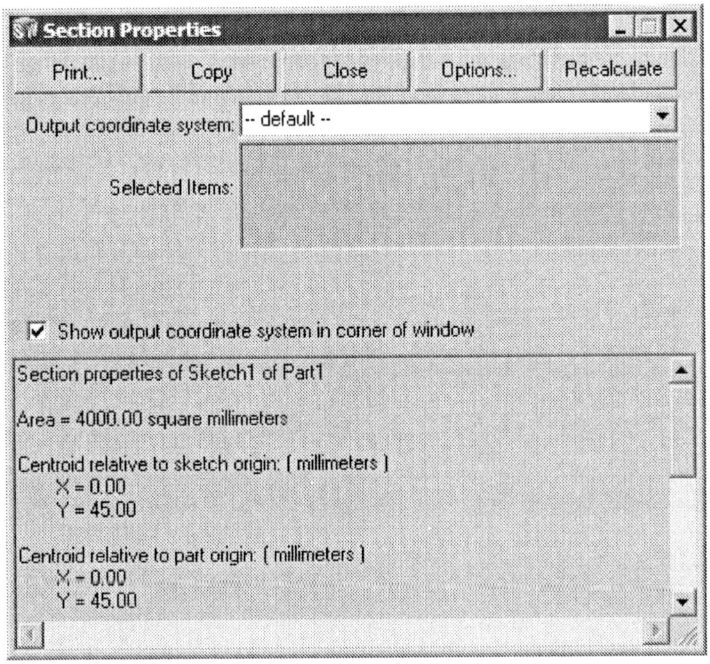

Figure 4-7 *The* ***Section Properties*** *dialog box*

CREATING BASE FEATURES BY EXTRUDING THE SKETCHES

CommandManager:	Features > Extruded Boss/Base
Menu:	Insert > Boss/Base > Extrude
Toolbar:	Features > Extruded Boss/Base

The sketches that you have drawn until now can be converted into base features by extruding using the **Extruded Boss/Base** tool. This tool is available in the **Features CommandManager**. After drawing the sketch, choose the **Features** button from the **CommandManager** to display the **Features CommandManager**. From the **Features CommandManager**, choose the **Extruded Boss/Base** tool. The sketching environment is closed and the part modeling environment is invoked. Also, the preview of the feature that will be created using the default options is displayed in trimetric view. The trimetric view gives a better display of the solid feature.

Based on the options and the sketch selected for extruding, the resultant feature can be a solid feature or a thin feature. If the sketch is closed, it can be converted into a solid feature or a thin feature. However, if the sketch is open, it can be converted only into a thin feature. The solid and thin features are discussed next.

Creating Solid Extruded Features

After you have completed drawing and dimensioning the closed sketch and converted it into a fully defined sketch, invoke the **Features CommandManager** by choosing the **Features** button and then choose the **Extruded Boss/Base** button from the **Features CommandManager**. You will notice that the view is automatically changed to trimetric view and the **Extrude PropertyManager** is displayed as shown in Figure 4-8.

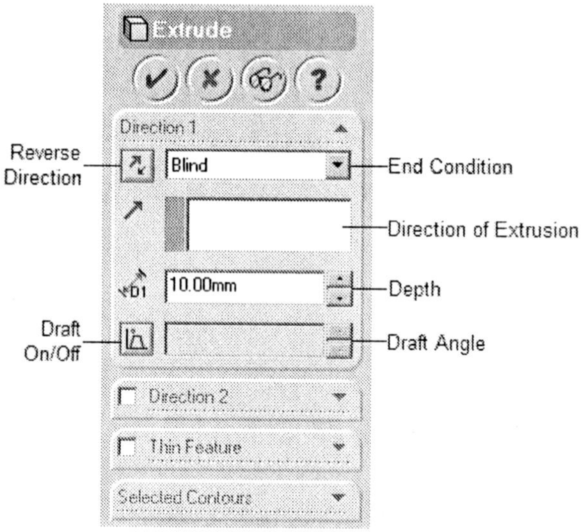

Figure 4-8 Extrude PropertyManager

You will notice that the preview of the base feature will be displayed in temporary graphics and an arrow will appear on the sketch. The arrow appears on the front of the sketch and is transparent. Figure 4-9 shows the preview of the sketch being extruded. Note that if the sketch consists of some closed loops inside the outer loop, they will be automatically subtracted from the outer loop while extruding. The various options available in the **Extrude PropertyManager** are discussed next.

Direction 1

The **Direction 1** rollout is used to specify the end condition for extruding the sketch in one direction from the sketch plane. The various options available in the **Direction 1** drop-down list are discussed next.

End Condition

The **End Condition** drop-down list provides the options to define the termination of the extruded feature. Note that because this is the first feature, some of the options available in this drop-down list will not be used at this stage. Also, some additional options will be available later in this drop-down list. The options that are available to define the termination of the base feature are discussed next.

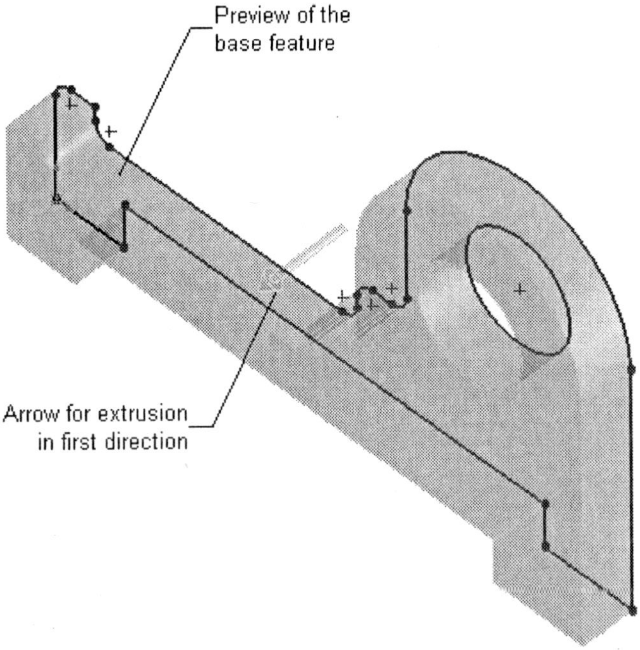

Figure 4-9 Preview of the feature being extruded

Blind

The **Blind** option is selected by default and is used to define the termination of the extruded base feature by specifying the depth of extrusion. The depth of extrusion can be specified in the **Depth** spinner that is displayed in this rollout when you select the **Blind** option. You can reverse the extrusion direction by selecting the **Reverse Direction** button provided on the left of this drop-down list. Figure 4-10 shows the preview of the feature being created by extruding the sketch using the **Blind** option.

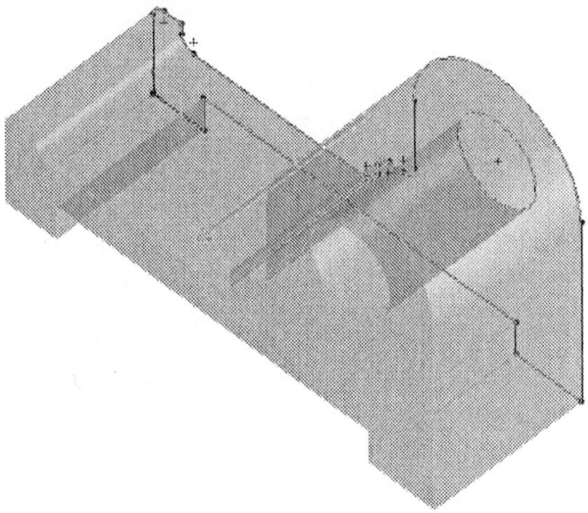

*Figure 4-10 Preview of the feature being extruded using the **Blind** option*

You can also extrude a sketch to a blind depth by dynamically dragging the feature using the mouse. Invoke the **Extrude PropertyManager** and move the mouse to the transparent arrow and when the color of the arrow changes to red, press the left mouse button. Now, move the cursor to specify the depth of extrusion and left-click to specify the termination of extruded feature. The preview of the sketch being dragged is shown in Figure 4-11. The select cursor will be replaced by the mouse cursor. Use the right mouse button to complete feature creation or choose the **OK** button from the **Extrude PropertyManager**.

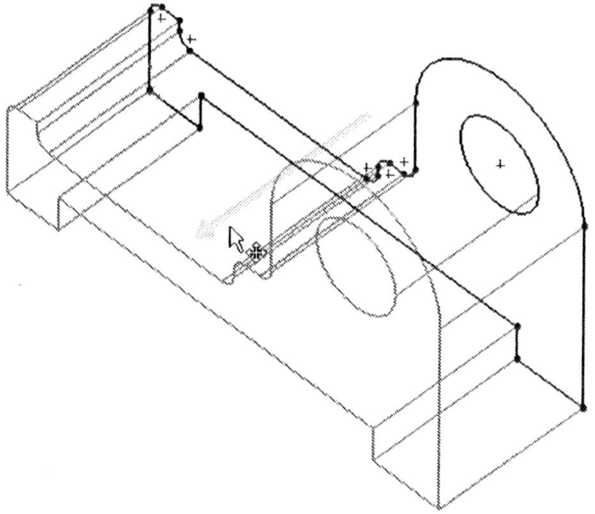

Figure 4-11 *Preview of the feature being extruded by dynamically dragging*

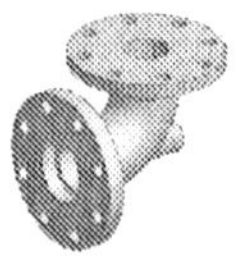

Tip. *You can also extrude an under-defined or an overdefined sketch. However, if you extrude an underdefined sketch, a - sign is displayed on the left of the sketch in the* ***Feature Manager****. Similarly, if you extrude an overdefined sketch, you will find a + sign on the left of the sketch in the* ***Feature Manager****. To check these signs, click the + sign available on the left of* ***Extrude 1*** *in the* ***Feature Manager****. The sketch will be displayed and you can see the + or the - sign.*

Mid Plane

The **Mid Plane** option is used to create the base feature by extruding the sketch equally in both the directions of the plane on which the sketch is drawn. For example, if the total depth of the extruded feature is 30 mm, it will be extruded 15 mm toward the front of the plane and 15 mm toward the back. The depth of the feature can be defined in the **Depth** spinner that is displayed below this drop-down list.

Figure 4-12 shows the preview of the feature being created by extruding the sketch using the **Mid Plane** option.

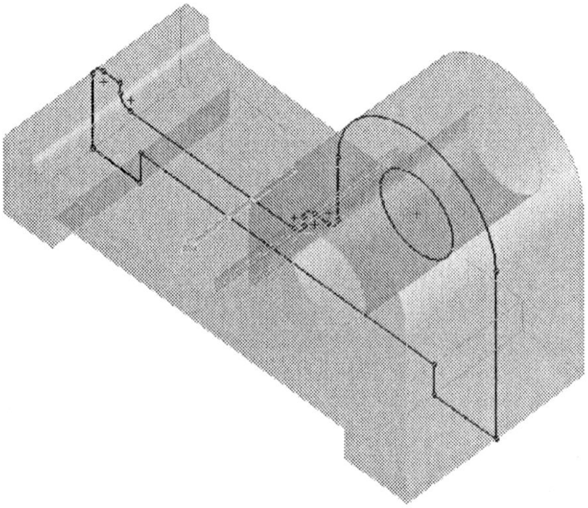

Figure 4-12 Preview of the feature being extruded using the ***Mid Plane*** *option*

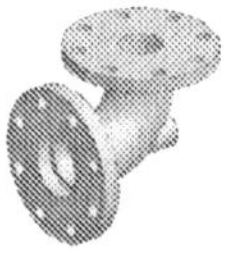

Tip. *You can also select the termination options using the shortcut menu that is displayed when you right-click in the drawing area.*

Draft On/Off

The **Draft On/Off** button is used to specify a draft angle while extruding the sketch. Applying the draft angle will taper the resultant feature. This button is not chosen by default. Therefore, the resultant base feature will not have any taper. However, if you want to add a draft angle to the feature, choose this button. The **Draft Angle** spinner and the **Draft outward** check box will be available. You can enter the draft angle for the feature in the **Draft Angle** spinner. By default, the feature will be tapered inward as shown in Figure 4-13.

If you want to taper the feature outward, select the **Draft outward** check box that is displayed below the **Draft Angle** spinner. The feature created with outward taper is shown in Figure 4-14.

Note

The ***Direction of Extrusion*** *area will be discussed in later chapters.*

Direction 2

The **Direction 2** check box is selected to invoke the **Direction 2** rollout. This rollout is used to extrude the sketch with different values in the second direction of the sketching plane. This check box will not be available if you select the **Mid Plane** termination type.

The options available in this rollout are similar to those in the **Direction 1** rollout. Note that unlike the **Mid Plane** termination option, the depth of extrusion and other parameters in both

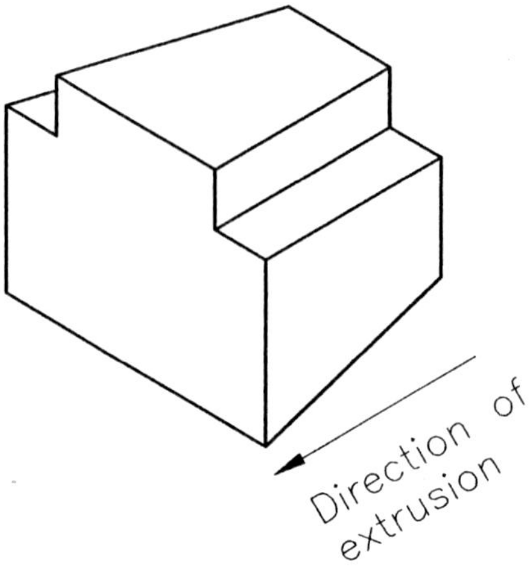

Figure 4-13 Feature created with outward draft

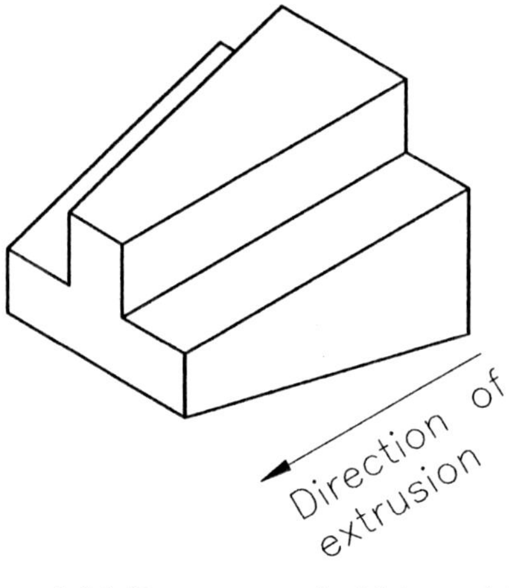

Figure 4-14 Feature created with inward draft

the directions can be different. For example, you can extrude the sketch to a blind depth of 10 mm and an inward draft of 35-degree in front of the sketching plane and to a blind depth of 15 mm and an outward draft of 0-degree behind the sketching plane as shown in Figure 4-15.

After setting the values for both the directions, choose the **OK** button or choose the **OK** icon from the confirmation corner. The feature will be created with the defined values.

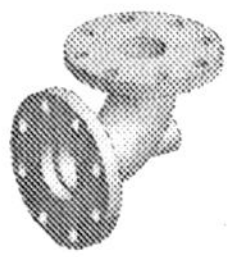

Tip. *As soon as you select the **Direction 2** check box, another arrow is displayed in the preview of the feature. You can use this arrow to define the depth of extrusion in the second direction.*

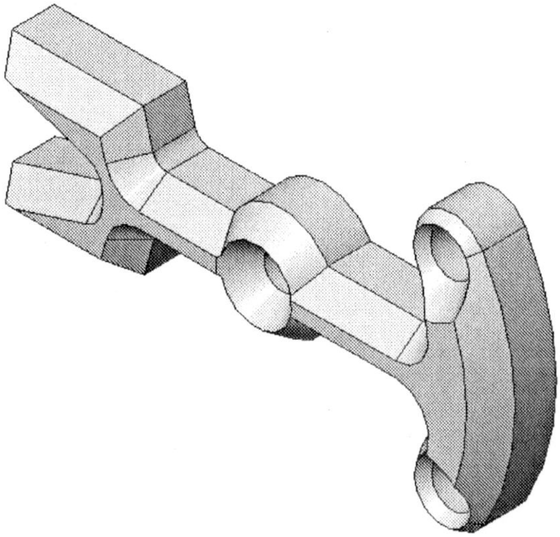

Figure 4-15 Feature created in two directions with different values

Note

*The draft will not be displayed in the preview of the feature that is displayed when you invoke the **Base-Extrude PropertyManager** unless you extrude the sketch by dynamically dragging.*

*The **Selection Contours** rollout is discussed in the later chapters of this book.*

Creating Thin Extruded Features

Thin extruded features can be created using a closed or an open sketch. If the sketch is closed, it will be offsetted inside or outside to create a cavity inside the feature as shown in Figure 4-16.

Figure 4-16 Thin feature created using a closed loop

If the sketch is open, as shown in Figure 4-17, the resultant feature will be as shown in Figure 4-18.

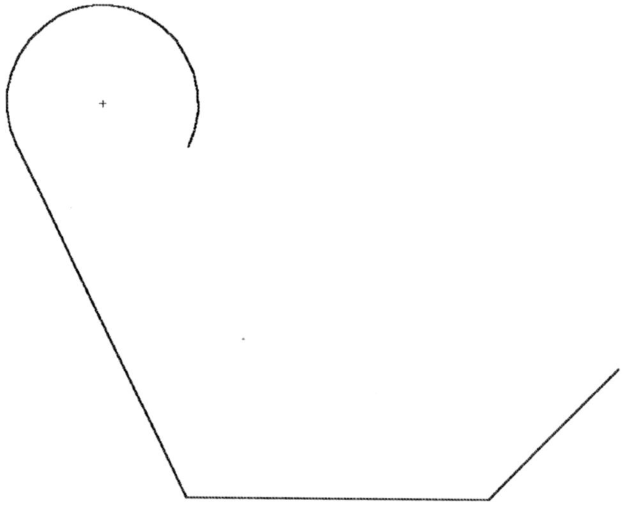

Figure 4-17 *Open loop to be converted into thin feature*

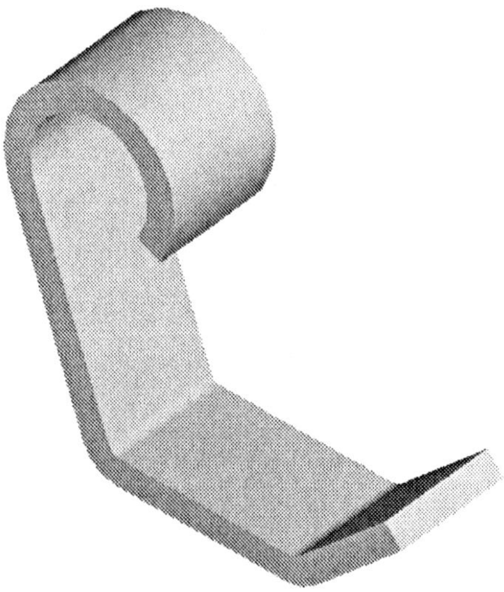

Figure 4-18 *Resultant thin feature created with fillets at sharp corners*

To convert a closed sketch into a thin feature, choose the **Thin Feature** check box to invoke the **Thin Feature** rollout. The **Thin Feature** rollout, shown in Figure 4-19, is used to create a thin feature. However, if the sketch to be extruded is open, then the **Thin Feature** rollout will be invoked automatically when you invoke the **Extrude PropertyManager**.

The options under the **Thin Feature** rollout of the **Extrude PropertyManager** are discussed next.

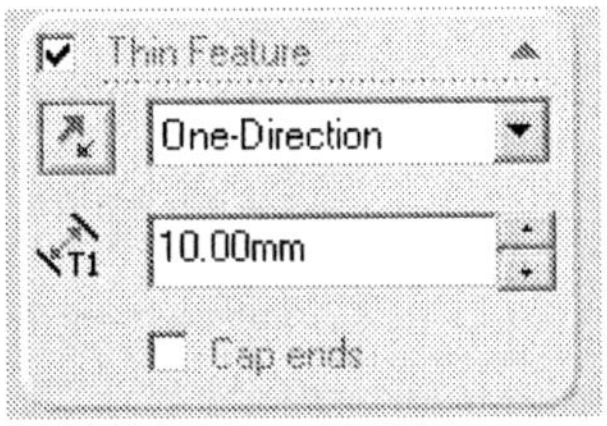

*Figure 4-19 The **Thin Feature** rollout*

Type

The options provided in the **Type** drop-down list are used to select the method to define the thickness of the thin feature. These options are discussed next.

One-Direction

The **One-Direction** option is used to add the thickness on one side of the sketch. The thickness can be specified in the **Thickness** spinner provided below this drop-down list. For the closed sketches, the direction can be inside or outside the sketch. Similarly, for open sketches, the direction can be below or above the sketch. You can reverse the direction of thickness using the **Reverse** check button available on the left of this drop-down list. This button will be available only when you select the **One-Direction** option from this drop-down list.

Mid-Plane

The **Mid-Plane** option is used to add the thickness equally on both the sides of the sketch. The value of the thickness of the thin feature can be specified in the **Thickness** spinner provided below this drop-down list.

Two-Direction

The **Two-Direction** option is used to create a thin feature by adding different thickness on both sides of the sketch. The thickness values in direction 1 and direction 2 can be specified in the **Direction 1 Thickness** spinner and the **Direction 2 Thickness** spinner respectively. These spinners are displayed below the **Type** drop-down list automatically when you select the **Two-Direction** option from this drop-down list.

Cap ends

The **Cap ends** check box is displayed only when the sketch selected to convert into a thin feature is closed. This check box is selected to cap the two ends of the thin extruded feature. Both the ends will be capped with a face of the thickness you specify. When you select this check box, the **Cap Thickness** spinner is displayed below this check box. The thickness of the end caps can be specified using this spinner.

Auto-fillet corners

The **Auto-fillet corners** check box is displayed only when you select an open sketch to convert into a thin feature. If you select this check box, all the sharp vertices in the sketch will be automatically filleted while converting into a thin feature. As a result, the thin feature will have filleted edges. The radius of the fillet can be specified in the **Fillet Radius** spinner that is displayed below the **Auto-fillet corners** check box when you select this check box.

Figure 4-20 shows the thin feature created by extruding an open sketch in both the directions. Notice that a draft angle is applied to the feature while extruding in the front direction and the **Auto-fillet corners** option is selected while creating this thin feature.

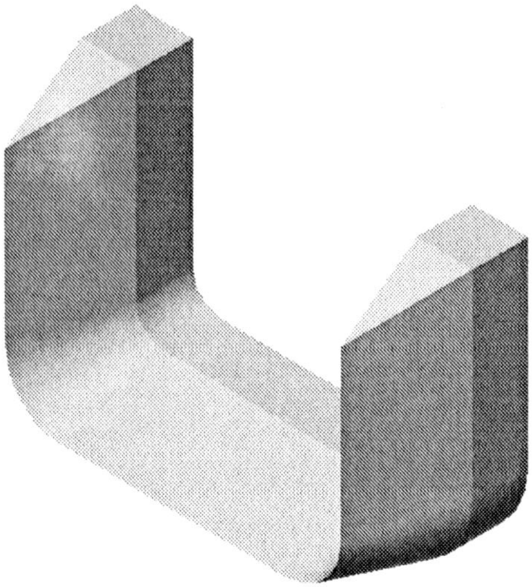

Figure 4-20 *Thin feature created in both the directions*

Note

Only the corners of the thin features that can accommodate the given radius will be filleted; other corners that cannot accommodate the given radius will not be filleted.

CREATING BASE FEATURES BY REVOLVING THE SKETCHES

CommandManager:	Features > Revolved Boss/Base
Menu:	Insert > Boss/Base > Revolve
Toolbar:	Features > Revolved Boss/Base

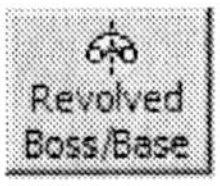

The sketches that you have drawn until now can also be converted into base features by revolving using the **Revolved Boss/Base** tool. This tool is available in the **Features CommandManager**. In the previous releases of SolidWorks, the revolved features could be created using only a centerline. However, in this release, you do not need a centerline to revolve the sketch. You can use an edge of the sketch as the revolution axis to create the revolved feature. Note that whether you use a centerline or an edge to revolve the sketch, the sketch should be drawn on one side of the centerline or the edge.

In SolidWorks, the right-hand thumb rule is referred to while determining the direction of revolution. The right-hand thumb rule states that if the thumb of your right hand points in the direction of the axis of revolution, the direction of curled fingers will determine the default direction of revolution, see Figure 4-21.

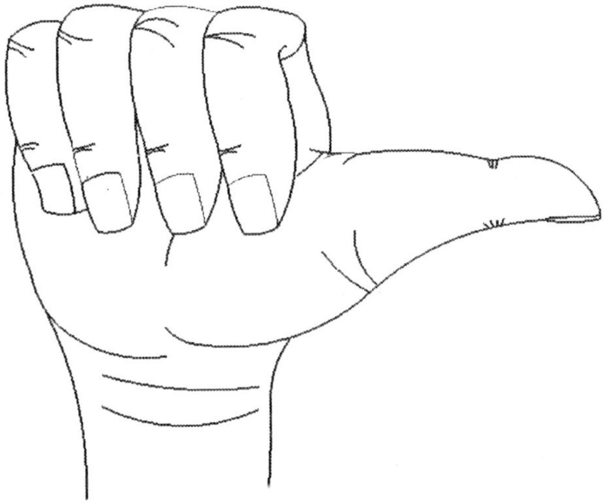

Figure 4-21 *Right-hand thumb rule*

For example, consider a case in which you draw a centerline from left to right (direction of thumb in Figure 4-21) in the drawing area. Now, if you use this centerline to create a revolved feature, the default direction of revolution will be in the direction of curled fingers.

Note

You can also reverse the direction of revolution using the button available in the ***Revolve PropertyManager****.*

After drawing the sketch, as you choose this tool, you will notice that the sketching environment is closed and the part modeling environment is invoked. Similar to extruding the sketches, the resultant feature can be a solid feature or a thin feature, based on the sketch and the options selected to revolve. If the sketch is closed, it can be converted into a solid feature or a thin feature. However, if the sketch is open, it can be converted only into a thin feature. The solid and thin features are discussed next.

Creating Solid Revolved Features

After you have completed drawing and dimensioning the closed sketch and converted it into a fully defined sketch, choose the **Revolved Boss/Base** button from the **Features** toolbar. You will notice that the view is automatically changed to a 3D view, the **Revolve PropertyManager** will be displayed as shown in Figure 4-22, and the confirmation corner will also be displayed. Also, the preview of the base feature, as it will be created using the default options, will be displayed in temporary shaded graphics. The direction arrow will also be displayed in gray color. The various options available in the **Revolve Parameters** rollout of the **Revolve PropertyManager** are discussed next.

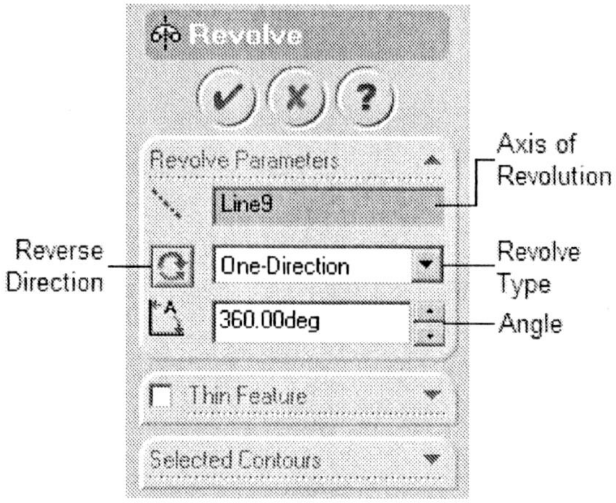

Figure 4-22** The **Revolve PropertyManager

Tip. *Even if you can revolve the sketch using an edge in the sketch, you need to draw a centerline. This is because you can create linear diameter dimensions for the revolved features only using a centerline.*

Revolve Type

The **Revolve Type** drop-down list provides the options to define the termination of the revolved feature. The options that are available in this drop-down list to terminate the revolved feature are discussed next.

One-Direction

The **One-Direction** option is used to revolve the sketch on one side of the plane on which it is sketched. The angle of revolution can be specified in the **Angle** spinner displayed below this drop-down list. The default value of the **Angle** spinner is **360deg**. Therefore, if you revolve the sketch using this value, a complete round feature will be created. You can also reverse the direction of revolution of the sketch by choosing the **Reverse Direction** button that is displayed when you select this option. Figure 4-23 shows the sketch of the piston and Figure 4-24 shows the resultant piston created by revolving the sketch through an angle of 360-degree. Note that the outer left vertical edge of the sketch that is vertically in line with the origin is used to revolve the sketch.

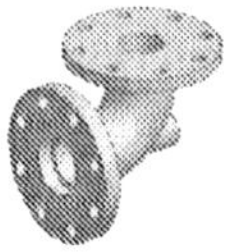

Tip. *You can dynamically specify the angle in a revolve feature by dragging the direction arrows. You can also use the right mouse button to display the shortcut menu; all the options available in the **PropertyManager** are also available in the shortcut menu.*

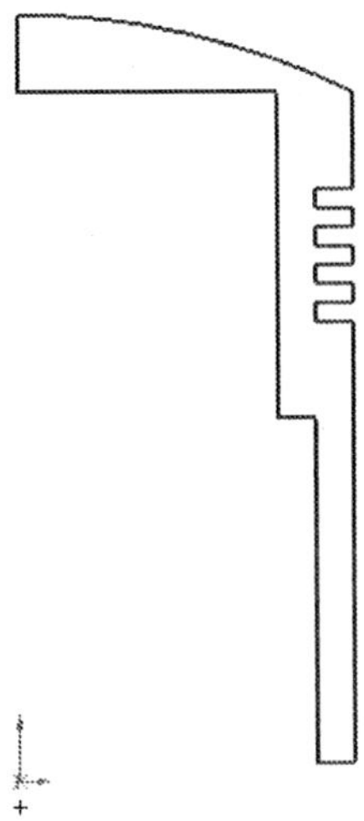

Figure 4-23 *Sketch of the piston to be revolved*

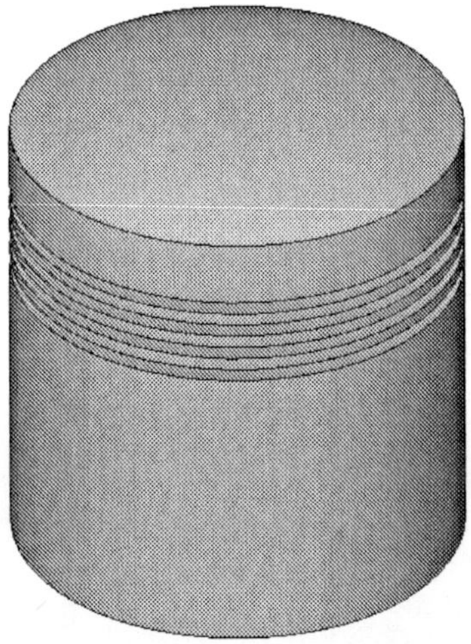

Figure 4-24 *Feature created by revolving the sketch through an angle of 360-degree*

Figure 4-25 shows a piston created by revolving the same sketch through an angle of 270-degree.

Mid-Plane

The **Mid-Plane** option is used to revolve the sketch equally on both the sides of the plane on which it is sketched. The angle of revolution can be specified in the **Angle** spinner. When you choose this option the **Reverse Direction** button will be unavailable.

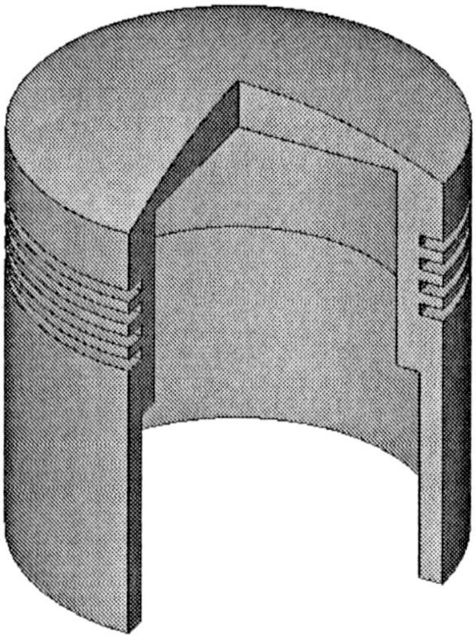

Figure 4-25 *Feature created by revolving the sketch through an angle of 270-degree*

Two-Direction

The **Two-Direction** option is used to create a revolve feature by revolving the sketch using different values on both the sides of the plane on which it is sketched. The angle values in direction 1 and direction 2 can be specified in the **Direction 1 Angle** spinner and the **Direction 2 Angle** spinner respectively. These spinners are displayed below the **Revolve Type** drop-down list automatically when you select the **Two-Direction** option from this drop-down list.

Creating Thin Revolved Features

Thin revolved features can be created using a closed or an open sketch. If the sketch is closed, it will be offsetted inside or outside to create a cavity inside the feature as shown in Figure 4-26. In this figure, the sketch is revolved through an angle of 180-degree.

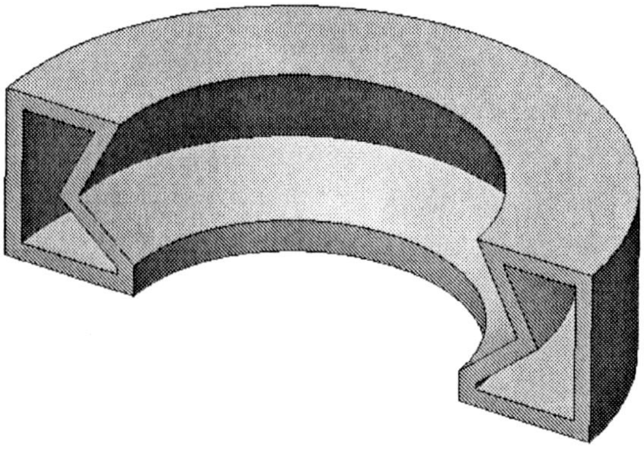

Figure 4-26 *Thin feature created by revolving the sketch through an angle of 180-degree*

If the sketch is open, as shown in Figure 4-27, the resultant feature will be as shown in Figure 4-28.

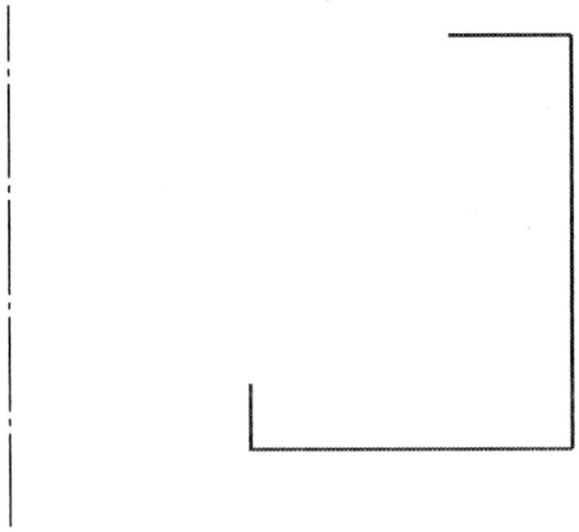

Figure 4-27 *The open sketch to be revolved and the centerline to revolve the sketch*

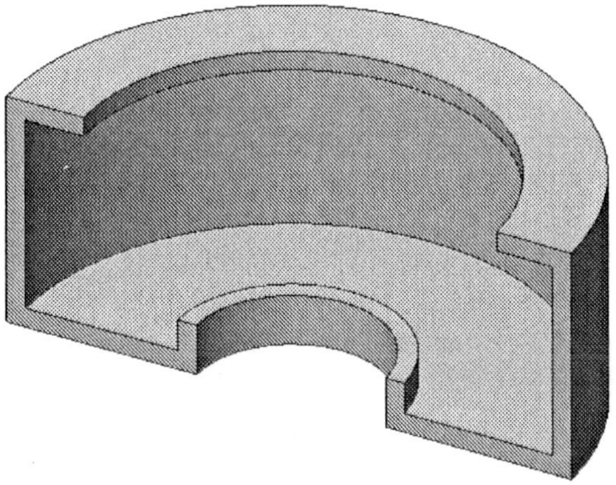

Figure 4-28 *Thin feature created by revolving the open sketch through an angle of 180-degree*

To convert a closed sketch into a thin feature, select the **Thin Feature** check box from the **Revolve PropertyManager** to invoke the **Thin Feature** rollout. However, if the sketch to be revolved is open and you invoke the **Revolved Boss/Base** tool, then the **SolidWorks** information box will be displayed. This information box will inform you that the sketch is currently open and a nonthin revolve feature requires a closed sketch. You will be given an option of automatically closing the sketch. If you choose **Yes** from this dialog box, a line segment will be automatically drawn between the first and the last segment of the sketch and the **Revolve PropertyManager** will be displayed. However, if you choose **No** from this information box, the **Revolve PropertyManager** will be displayed and the **Thin Feature** rollout will be displayed automatically. The options under the **Thin Feature** rollout of the **Revolve PropertyManager**, shown in Figure 4-29, are discussed next.

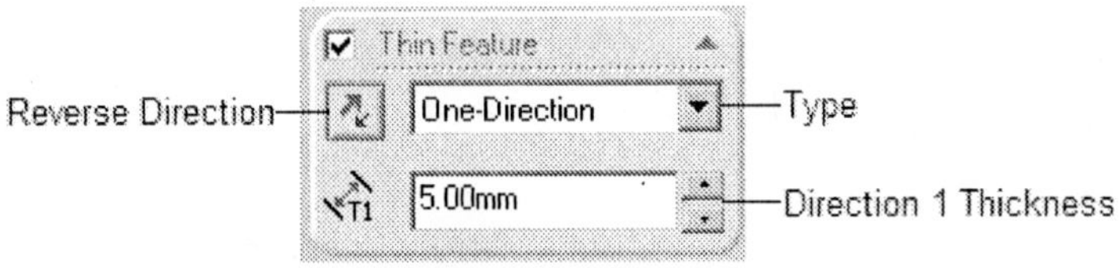

Figure 4-29 *The* ***Thin Feature*** *rollout*

Type

The options provided in the **Type** drop-down list are used to select the method to define the thickness of the thin feature. These options are discussed next.

One-Direction

The **One-Direction** option is used to add the thickness on one side of the sketch. The thickness can be specified in the **Direction 1 Thickness** spinner provided below this drop-down list. For the closed sketches, the direction can be inside or outside the sketch.

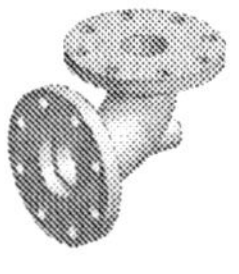

Tip. *While defining the wall thickness of a thin revolved feature, remember that the wall thickness should be added such that the centerline does not intersect with the sketch. If the centerline intersects with the sketch, the sketch will not be revolved.*

Similarly, for open sketches, the direction can be below or above the sketch. You can reverse the direction of thickness using the **Reverse Direction** button available on the right of this drop-down list. This button will be available only when you select the **One-Direction** option from this drop-down list.

Mid-Plane

The **Mid-Plane** option is used to add the thickness equally on both the sides of the sketch. The value of the thickness of the thin feature can be specified in the **Direction 1 Thickness** spinner provided below this drop-down list.

Two-Direction

The **Two-Direction** option is used to create a thin feature by adding different thickness on both the sides of the sketch. The thickness values in direction 1 and direction 2 can be specified in the **Direction 1 Thickness** spinner and the **Direction 2 Thickness** spinner respectively. These spinners are displayed below the **Type** drop-down list automatically when you select the **Two-Direction** option from this drop-down list.

DYNAMICALLY ROTATING THE VIEW OF THE MODEL

SolidWorks allows you to dynamically rotate the view in the 3D space so that the solid models in the current document can be viewed from all directions. This allows you to visually maneuver around the model so that all the features in the model can be clearly viewed. This tool can be invoked even when you are inside some other tool. For example, you can invoke this tool when the **Extrude Feature** dialog box is displayed. You can freely rotate the model in the 3D space or rotate it around a selected vertex, edge, or face. Both the methods of rotating the model are discussed next.

Rotating the View Freely in 3D Space

Toolbar:	View > Rotate View
Menu:	View > Modify > Rotate

To rotate the view freely in 3D space, choose the **Rotate View** button from the **View** toolbar. You can also invoke this tool by choosing the **Rotate View** option from the shortcut menu that is displayed when you right-click in the drawing area. When you are inside some other tool use the right mouse button in the drawing area and choose the **Zoom/Pan/Rotate** > **Rotate View** option from the shortcut menu to invoke the rotate view tool. When you invoke this tool, the cursor will be replaced by the rotate view cursor. Now, press the left mouse button and drag the cursor to rotate the view. Figure 4-30 shows a model being viewed from different directions by rotating the view.

Rotating the View Around a Selected Vertex, Edge, or Face

To rotate the view around a selected vertex, edge, or face, invoke this tool and move the rotate

Figure 4-30 *Rotating the view to display the model from different directions*

view cursor close to the vertex, edge, or the face around which you want to rotate the view. When it is highlighted, select it using the left mouse button. Next, drag the cursor to rotate the view around the selected vertex, edge, or face.

Tip. *To resume rotating the view freely after you have completed rotating it around a selected vertex, edge, or face, double-click anywhere in the drawing area. Now when you drag the cursor, you will notice that the view is rotated freely in 3D space.*

If a 3 button mouse is configured to your system, you can press and drag the middle mouse button to rotate the model freely in 3D space. Note that in this case the rotate view cursor will not be displayed.

MODIFYING THE VIEW ORIENTATION

As mentioned earlier, when you invoke the **Extruded Boss/Base** tool or the **Revolved Boss/Base** tool, the view is automatically changed to a trimetric view and the preview of the model is displayed. SolidWorks allows you to manually change the view orientation using some predefined standard views or user-defined views. To invoke these standard views, choose the **Standard Views** button from the **View** toolbar. A flyout is displayed, as shown in Figure 4-31. You can select the required standard view from this flyout.

Apart from the standard views, this flyout also provides the **Normal To** option. This option is chosen to reorient the view normal to a selected face or plane. This option will be available only in the sketching environment or when you select a planar face or a plane.

You can also invoke these standard views using the **Orientation** dialog box. This dialog box is

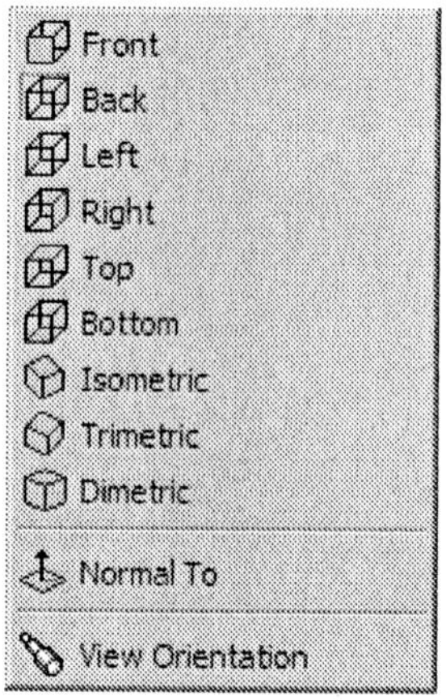

Figure 4-31 Standard View *flyout*

invoked by choosing the **View Orientation** option from the shortcut menu. This dialog box can also be invoked by pressing the SPACEBAR from the keyboard. Note that when you invoke this dialog box by pressing the SPACEBAR, the dialog box will be displayed at the location where the cursor is placed currently. The **Orientation** dialog box is shown in Figure 4-32.

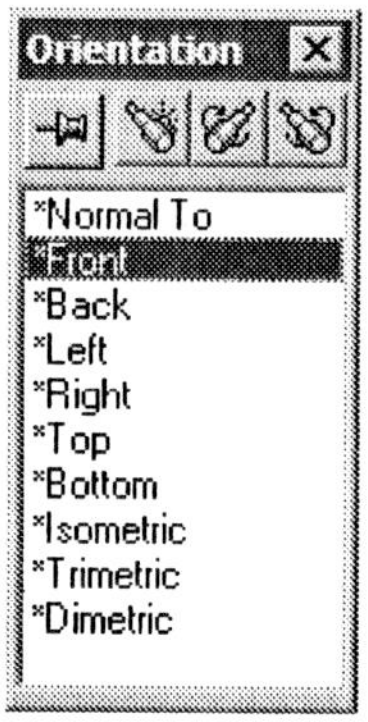

Figure 4-32 Orientation *dialog box*

You can invoke the view from this dialog box by double-clicking on it. You will notice that in addition to the standard views, two more additional views are displayed. The buttons that are available on top of this dialog box are discussed next.

Push-Pin

You will notice that the **Orientation** dialog box is automatically closed when you select a view, select a point somewhere on the screen, or invoke a tool. If you want that this dialog box should be retained on the screen, you can pin it at a location by choosing the **Push-Pin** button. This is the first button on top of this dialog box. Move the dialog box to the desired location and then choose this button. The dialog box will be pinned to that location and will not close when you perform any operation.

New View

The **New View** button is chosen to create a user-defined view and save it in the list of the views in the **Orientation** dialog box. Using the various drawing display tools and the **Rotate View** tool, modify the current view and then choose this button. When you choose this button, the **Named View** dialog box will be displayed. Enter the name of the view in the **View Name** edit box and then choose the **OK** button. You will notice that a user-defined view is created and it is saved in the list available in the **Orientation** dialog box.

Update Standard Views

The **Update Standard Views** button is chosen to modify the orientation of the standard views. For example, if you want that the view that is displayed when you invoke the **Back** option from this dialog box should be the front view, then change the current view to the back view by double-clicking it in the **Orientation** dialog box. Now, select the **Front** option from the list of views available in the **Orientation** dialog box and then choose the **Update Standard Views** button. The **SolidWorks** warning box will be displayed and you will be informed that if you change the standard view, all the other named views in the model will also be changed. If you make the change using the **Yes** button then the current view that was originally the back view will become the front view. Also, all the other views will be modified automatically.

Reset Standard Views

The **Reset Standard Views** button is chosen to reset the standard settings of all the standard views in the current drawing. When you choose this button, the **SolidWorks** warning box will be displayed and you will be prompted to confirm whether you want to reset all the standard views to their original settings or not. If you choose **Yes,** all the standard views will be reset to their default settings.

RESTORING PREVIOUS VIEW

In SolidWorks, while working on a model or while viewing it from different directions, you need to temporarily change the view of the model. Once you have finished editing or viewing the model in that view, you can restore the previous view using the **Previous View** button in the **View** toolbar. This tool saves ten previous views of the model.

DISPLAY MODES OF THE MODEL

SolidWorks provides you with various predefined modes to display the model. You can select any of these display modes from the **View** toolbar. These modes are discussed next.

Wireframe

When you choose the **Wireframe** button, all the hidden lines will be displayed along with the visible lines in the model. Sometimes it becomes difficult to recognize the visible lines and the hidden lines if you set this display mode for complex models.

Hidden Lines Visible

When you choose the **Hidden Lines Visible** button, the model is displayed in the wireframe and the hidden lines in the model will be displayed as dashed lines.

Hidden Lines Removed

When you choose the **Hidden Lines Removed** button, the hidden lines in the model will not be displayed. Only the edges of the faces visible in the current view of the model will be displayed.

Shaded With Edges

The **Shaded With Edges** mode is the default mode in which the model is displayed. In this display mode, the model is shaded and the edges of the visible faces of the model are displayed.

Shaded

This display mode is similar to the **Shaded With Edges** mode, with the only difference that the edges of the visible faces will not be displayed.

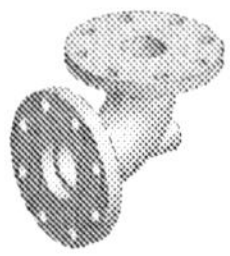

Tip. *Sometimes when you rotate the view of a large assembly, or a model with large number of features with* ***Shaded*** *or the* ***Hidden Lines Removed*** *shading modes, the regeneration of the model takes a lot of time. This can be avoided by choosing the* ***Draft Quality HLR/HLV*** *button in combination with the other shading modes. Choosing this button speeds up the regeneration time and you can easily rotate the view. This is a toggle mode and is turned on when you choose this button. This button can be chosen in combination with any of the other display modes.*

Shadows In Shaded Mode

The **Shadows In Shaded Mode** button is used to display the shadow of the model. A light appears from the top of the model to display the shadow in the current view. When you activate the shadow in the shaded mode, the performance of the system is affected during the dynamic orientation. Remember that the position of the shadow is not changed when you rotate the model in the 3D space. To change the placement of the shadow, first remove the shadow in the shaded model using the **Shadow in Shaded Mode** button and rotate the model. After rotating the model, use the same button to activate the shadow in the shaded mode. Figure 4-33 shows a T-section with shadow in the shaded mode.

Perspective

The **Perspective** button is not available in the **View** toolbar by default. You need to customize to add this button. Using this button, you can display the perspective view of a model. Figure 4-34 shows the T-section with shadow in perspective view.

You can also modify the settings of the perspective view. Choose **View > Modify > Perspective** from the menu bar to invoke the **Perspective View PropertyManager**. Using the **Object Sizes Away** spinner of this **PropertyManager**, you can modify the observer's position.

Figure 4-33 *Shadow in active shaded mode*

Figure 4-34 *T-section displayed in perspective view with shadow*

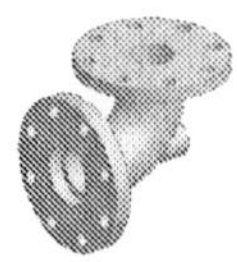

Tip. *You can also save a perspective view as a named view. To do this, invoke the perspective view and define the orientation by rotating the view. Now, press the SPACEBAR to invoke the* **Orientation** *dialog box. In this dialog box, choose the* ***New View*** *button and specify the name of the view in the* ***Named View*** *dialog box. The named view will be saved in the* **Orientation** *dialog box and you can invoke it whenever required.*

ASSIGNING MATERIAL AND TEXTURE TO THE MODELS

One of the latest enhancements of SolidWorks 2004 is its ability to let you assign material and texture to the models. The method of assigning material and textures is discussed next.

Assigning Material to a Model

Toolbar:	Standard > Edit Material *(Customize to Add)*
Menu:	Edit > Appearance > Material

New

Whenever you assign a material to a model, all the physical properties of the selected material are also assigned to the model. As a result, when you calculate the mass properties of the model, they will be based on the physical properties of the material. To apply material to a model, choose the **Edit Material** button from the **Standard** toolbar. The **Materials Editor PropertyManager** is displayed, as shown in Figure 4-35.

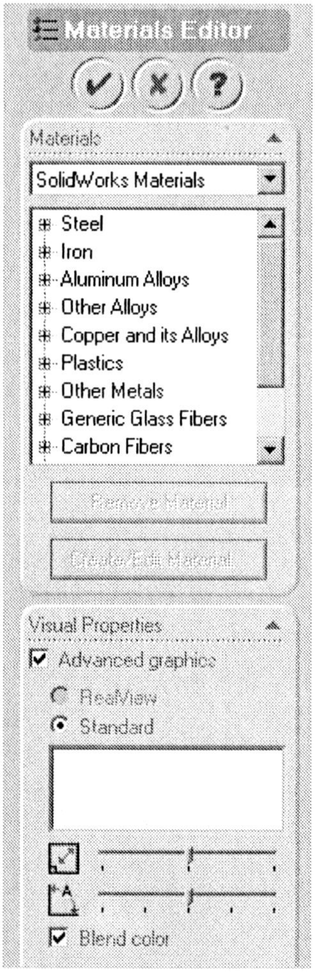

Figure 4-35** Partial display of the **Materials Editor PropertyManager

A number of material families are available in the **Materials** rollout. Click on the + sign located on the left of the material family to display all the materials under that family. Select the material from that family.

You can change the scale and angle of the textures in the selected material using the **Scale** and **Angle** spinners available in the **Visual Properties** rollout.

Assigning Textures to a Model

New

Toolbar:	Standard > Edit Texture
Menu:	Edit > Appearance > Texture

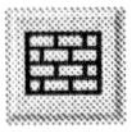

SolidWorks 2004 allows you to assign a texture to a model, feature, or a selected face. This can be done using the **Texture PropertyManager** that is displayed when you invoke the **Edit Texture** tool. Figure 4-36 shows this **PropertyManager**.

Figure 4-36 *Partial display of the* ***Texture PropertyManager***

The **Selection** rollout has some buttons available on the left. These buttons are the filters for making a selection to assign the textures. For example, if you want to assign the texture on a face of the model, choose the **Select Faces** button and clear all the remaining buttons. This allows you to select only a specified face. You can select the required face from the model. The selected face or model is displayed in the **Selected Entities** area of the **Selection** rollout.

The various textures families that you can select are available in the **Texture Tree** area of the **Texture Selection** rollout. Click on the + sign located on the left of a texture family to select a texture of that family. Figure 4-37 shows a model with Gravel type of stone applied to its top face.

Figure 4-37 *Texture applied to the top face of a model*

The preview of the selected texture is displayed on the model or the face and also in the **Texture Preview** area of the **Texture Properties** rollout. You can use the **Scale** and **Angle** spinners available in this rollout to change the scale and angle of the selected texture.

You can remove the texture applied to a face or a model by choosing the **Remove Textures** button available below the **Selected Entities** area of the **Selection** rollout.

The drop-down list available below the **Texture Tree** area of the **Texture Selection** rollout lists the last ten textures used. You can also select a texture from this drop-down list.

TUTORIALS

Tutorial 1

In this tutorial you will open the sketch drawn in Tutorial 2 of Chapter 3. You will then convert this sketch into an extruded model by extruding it in two directions as shown in Figure 4-38. The parameters for extruding the sketch are given next.

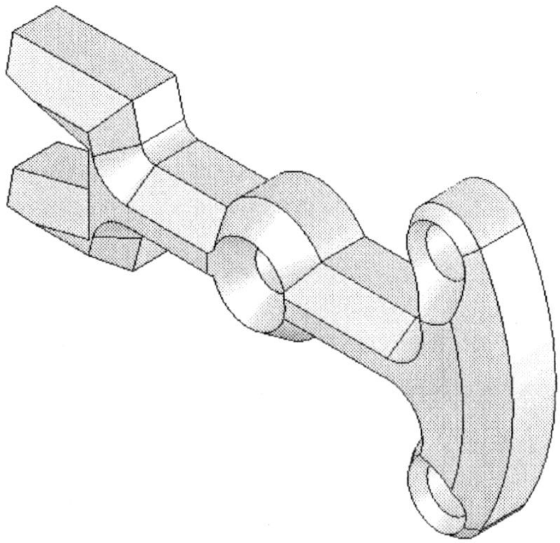

Figure 4-38 Model for Tutorial 1

<u>Direction 1</u>
Depth = 10 mm
Draft angle = 35-degree

<u>Direction 2</u>
Depth = 15 mm
Draft angle = 0-degree

After creating the model, you will rotate the view using the **Rotate View** tool and then modify the standard views such that the front view of the model becomes the top view. You will then save the model with the current settings. **(Expected time: 30 min)**

The steps that will be followed to complete this tutorial are given next.

a. Open the document of Tutorial 2 of Chapter 3, refer to Figure 4-39.
b. Save this document in the *c04* folder with a new name.
c. Invoke the **Extruded Boss/Base** tool and convert the sketch into a model, refer to Figures 4-40 and 4-41.
d. Rotate the view using the **Rotate View** tool to view the model from all the directions, refer to Figure 4-42.
e. Invoke the **Orientation** dialog box and then modify the standard view, refer to Figure 4-43.

Opening the Document of Tutorial 2 of Chapter 3

As the document that you require is saved in the *\My Documents\SolidWorks\c03* folder, you need to select this folder and then open the *c03-tut02.sldprt* document.

1. Start SolidWorks 2004 by double-clicking on its shortcut icon at the desktop of your computer. Close the **Tip of the Day** dialog box.

2. Choose the **Open Document** option from the **Welcome to SolidWorks 2004** window.

3. Browse and select the *\My Documents\SolidWorks\c03* folder.

   All the documents that were created in Chapter 3 will be displayed in this folder.

4. Select the *c03-tut02.sldprt* document and then choose the **Open** button.

   Because the sketch was saved in the sketching environment in Chapter 3, it is opened in the sketching environment.

Saving the Document in the c04 Folder

It is recommended that when you open a document of some other chapter, you should save it in the folder of the current chapter with some other name before you proceed with modifying the document. This is because if you save the document in the folder of the current chapter, the original document of the other chapter will not be modified.

1. Choose **File > Save As** from the menu bar to display the **Save As** dialog box.

   Because the *c03* folder was selected last to open the document, it will be the current folder.

2. Choose the **Up One Level** button available on the right of the **Save in** drop-down list to move to the *\SolidWorks* folder. Create a new folder with the name *c04* using the **Create New Folder** button. Make the *c04* folder current by double-clicking it.

3. Enter the new name of the document as *c04-tut01.sldprt* in the **File name** edit box and then choose the **Save** button to save the document.

   The document will be saved with the new name and the new document will now be opened in the drawing area, as shown in Figure 4-39.

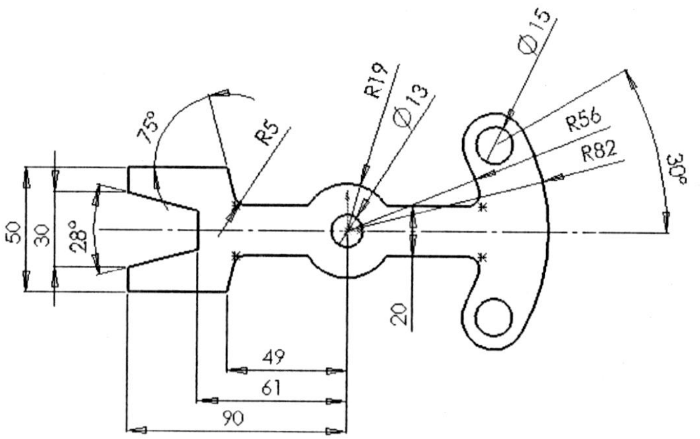

***Figure 4-39** Sketch that will be opened in the drawing area*

Extruding the Sketch

Next, you will invoke the **Extruded Boss/Base** tool and extrude the sketch using the parameters given in the tutorial description.

1. Choose the **Features** button from the **CommandManager** to display the **Features CommandManager**. From this **CommandManager**, choose the **Extruded Boss/Base** button. The sketch is automatically oriented in the trimetric view and the **Extrude PropertyManager** is displayed, as shown in Figure 4-40.

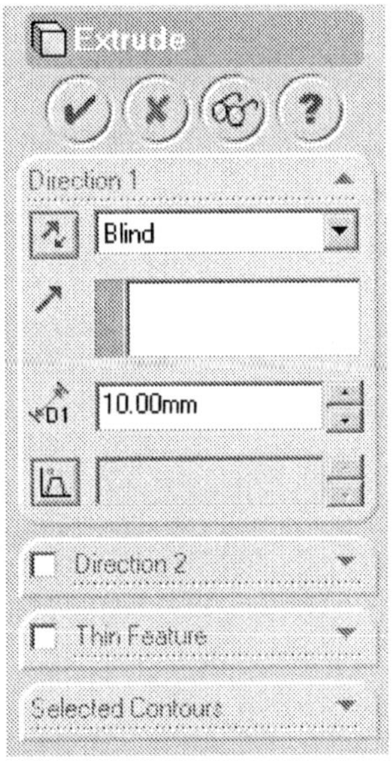

Figure 4-40 Extrude PropertyManager

Because the sketch to be converted into a feature is closed, only the **Direction 1** rollout will be displayed in the **Extrude PropertyManager**. The preview of the feature, in the temporary shaded graphics with the default values, is shown in the drawing area.

2. Choose the **Draft On/Off** button from the **Direction 1** rollout and then set the value of the **Draft Angle** spinner to **35**.

These are the settings in direction 1. Next, you need to specify the settings for direction 2.

3. Select the **Direction 2** check box to invoke the **Direction 2** rollout.

You will notice that the default values in this rollout are the same as you specified in the **Direction 1** rollout. Because the **Draft On/Off** button is chosen when you invoke the **Direction 2** rollout, you need to turn this button off. This is because you do not require the draft angle in the second direction.

4. Choose the **Draft On/Off** button from the **Direction 2** rollout to turn this option off. Set the value of the **Depth** spinner to **15** as the depth in the second direction is 15 mm. This completes all the settings for the model in both the directions.

5. Choose the **OK** button to create the feature or choose **OK** from the confirmation corner.

   It is recommended that you change the view to isometric view after creating the feature so that you can view the feature properly.

6. Choose the **Standard Views** button from the **View** toolbar to invoke the flyout. Choose **Isometric** from this flyout. Turn off the display of the origin in the model by choosing **View > Origin** from the menu bar. The isometric view of the resultant solid model is shown in Figure 4-41.

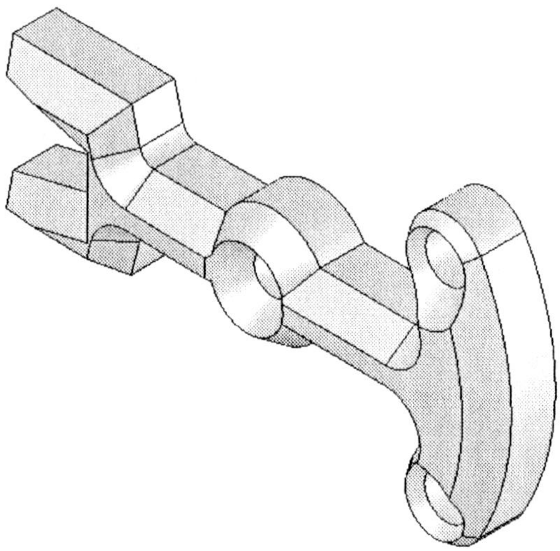

Figure 4-41 *Isometric view of the solid model*

Rotating the View

As mentioned earlier, you can rotate the view so that you can view the model from all the directions.

1. Choose the **Rotate View** button from the **View** toolbar.

   The arrow cursor will be replaced by the rotate view cursor.

2. Press and hold down the left mouse button and drag the cursor in the drawing area to rotate the view, as shown in Figure 4-42.

   You will notice that the model is being displayed from different directions. Remember that when you rotate the view, the model is not being rotated. The camera that is used to view the model is being rotated around the model.

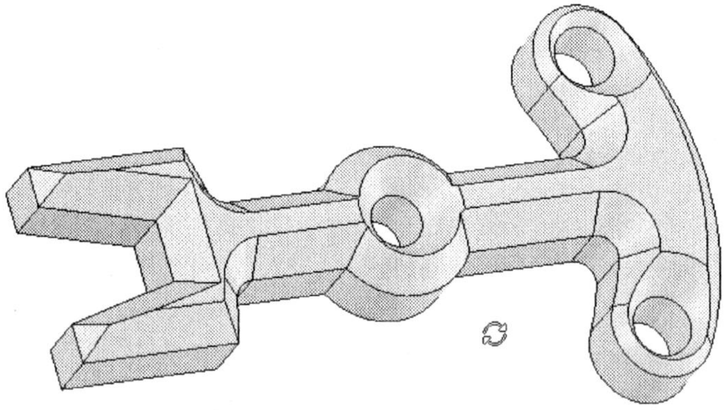

Figure 4-42 Rotating the view to display the model from different directions

3. After viewing the model from all the directions, choose **Isometric** from the flyout that is displayed when you choose the **Standard Views** button from the **View** toolbar.

Modifying the Standard Views

As mentioned in the tutorial description, you need to modify the standard views such that the front view of the model becomes the top view. This is done using the **Orientation** dialog box.

1. Press the SPACEBAR on the keyboard to invoke the **Orientation** dialog box.

   The orientation dialog box is automatically closed as soon as you perform any other operation. Therefore, you need to pin this dialog box so that it is not closed automatically.

2. Hold the **Orientation** dialog box by selecting it on the blue bar at the top of this dialog box and then drag it to the top right corner of the drawing area.

3. Choose the **Push Pin** button to pin this dialog box at the top right corner of the drawing area. Pinning the dialog box ensures that the dialog box is not automatically closed when you perform any other operation. 

4. Double-click the **Front** option in the list box of the **Orientation** dialog box.

   The current view will be automatically changed to the front view and the model will now be reoriented and displayed from the front.

5. Select the **Top** option from the list box by selecting it once.

   Make sure you do not double-click this option. This is because if you double-click this option, the model will be reoriented and displayed from the top.

6. Now, choose the **Update Standard Views** button to update the standard views.

The **SolidWorks** warning box is displayed and you will be warned that modifying the standard views will change the orientation of any named view in this document.

7. Choose **Yes** from this warning box to modify the standard views.

8. Now, double-click the **Isometric** option provided in the list box of the **Orientation** dialog box. You will notice that the isometric view is different now, see Figure 4-43.

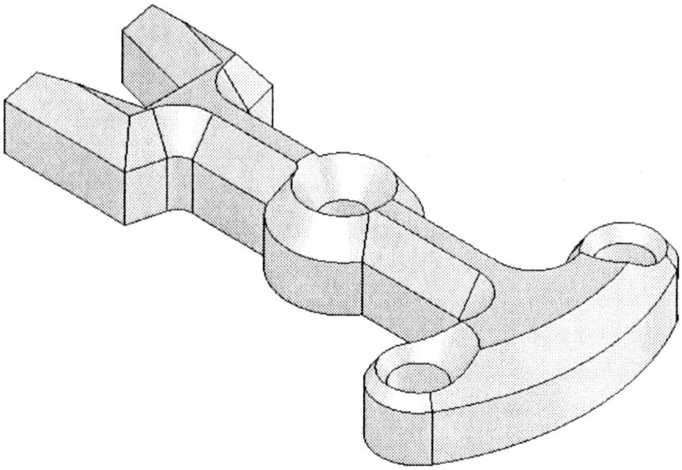

Figure 4-43 *Model displayed from modified isometric view*

9. Choose the **Push Pin** button in the **Orientation** dialog box again and pick a point anywhere in the drawing area to close the dialog box.

Saving the Model

As the name of the document was specified at the beginning, you just have to choose the save button now to save the document.

1. Choose the **Save** button from the **Standard** toolbar to save the document.

   The model will be saved with the name *\My Documents\SolidWorks\c04\c04-tut01.sldprt*.

2. Choose **File > Close** from the menu bar to close the document.

Tutorial 2

In this tutorial, you will open the sketch drawn in Exercise 1 of Chapter 3. You will then create a thin feature by revolving the sketch through an angle of 270-degree, as shown in Figure 4-44. You will offset the sketch outwards while creating the thin feature.

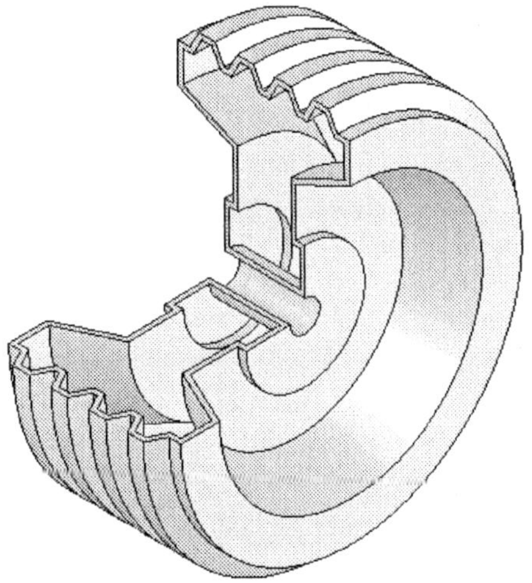

Figure 4-44 *Revolved model for Tutorial 2*

After creating the model, you will turn on the option to display the shadows and also apply Copper material to the model. **(Expected time: 30 min)**

The steps that will be followed to complete this tutorial are given next.

a. Open the sketch of Exercise 1 of Chapter 3, refer to Figure 4-45.
b. Save it in the folder of the current chapter.
c. Invoke the **Revolved Boss/Base** tool and revolve the sketch through an angle of 270-degree, refer to Figure 4-47.
d. Change the current view to isometric view and then change the display type to **Hidden Lines Removed**, refer to Figure 4-48.
e. Rotate the view using the **Rotate View** tool to display the model from different directions, refer to Figure 4-49.

Opening the Document of Exercise 1 of Chapter 3

As the document that you require is saved in the *\My Documents\SolidWorks\c03* folder, you will have to select this folder and then open the *c03-exr1.sldprt* document.

1. Choose the **Open** button from the **Standard** toolbar to display the **Open** dialog box.

   The *c04* folder will be current in this dialog box.

2. Browse and select the *\My Documents\SolidWorks\c03* folder. All the documents that were created in Chapter 3 will be displayed in this folder.

3. Select the *c03-exr01.sldprt* document and then choose the **Open** button.

   The document will be opened in the sketching environment.

Saving the Document in the c04 Folder

As mentioned earlier, it is recommended that you save the document with a new name in the folder of the current chapter so that the original document is not modified.

1. Choose **File > Save As** from the menu bar to display the **Save As** dialog box. Because the *c03* folder was selected last to open the document, it will be the current folder.

2. Browse and select the *c04* folder and double-click on it to make it current.

3. Enter the name of the document in the **File name** edit box as *c04-tut02.sldprt*. Choose the **Save** button to save the document. The sketch that will be displayed in the drawing area is shown in Figure 4-45.

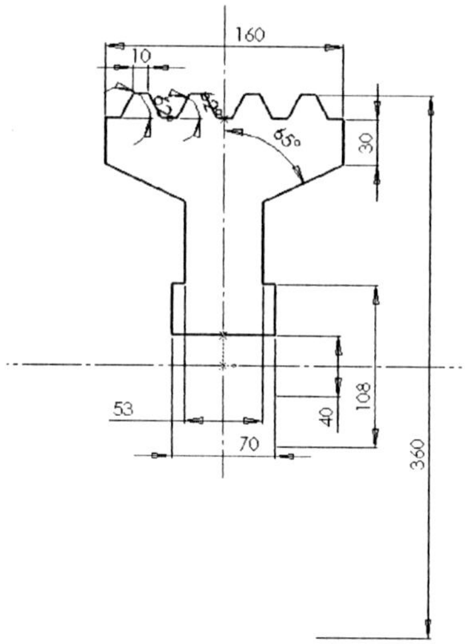

***Figure 4-45** Sketch for the revolved model*

Revolving the Sketch

The sketch consists of two centerlines. The first centerline was used to mirror the sketched entities and the second was drawn to create linear diameter dimensions. You need to revolve the sketch around the second centerline.

1. Choose the **Features** button from the **CommandManager** to display the **Features CommandManager**. From this **CommandManager**, choose the **Revolved Boss/Base** button. The sketch is automatically oriented in the trimetric view and the **Revolve PropertyManager** is displayed.

   Because the sketch has two centerlines, SolidWorks cannot determine which one to use as axis of revolution. This is the reason you are prompted to select the axis of revolution.

2. Select the horizontal centerline that was used to create linear diameter dimensions as the axis of revolution. The preview of a complete revolved feature in temporary shaded graphics will be displayed on the drawing area. Because the preview of the model is not displayed properly in the current view, you need to zoom the drawing.

3. Choose the **Zoom to Fit** button from the **View** toolbar or press the F key from the keyboard.

4. Set the value of the **Angle** spinner in the **Revolve PropertyManager** to **270**. The preview of the revolved model will also be modified accordingly. If you enter the value in the **Angle** spinner, you need to click anywhere on the screen to make sure the preview is modified.

   Note that if the horizontal centerline was drawn from left to right, then the direction of revolution has to be reversed to get the required model; refer to right-hand thumb rule. You can reverse the direction of revolution using the **Reverse Direction** button available on the left of the **Revolve Type** drop-down list.

5. Select the **Thin Feature** check box to invoke the **Thin Feature** rollout, as shown in Figure 4-46. Set the value of the **Direction 1 Thickness** spinner to **5**. You will notice that the preview of the thin feature is shown outside the original sketch.

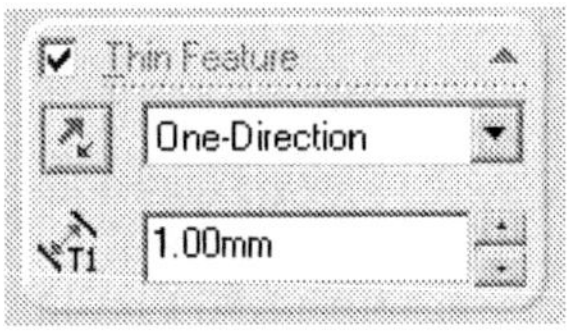

*Figure 4-46 **Thin Feature** rollout*

6. Choose the **OK** button to create the revolved feature. You will notice that the revolved feature is created.

7. Choose the **Isometric** option from the flyout that is displayed upon choosing the **Standard Views** button from the **View** toolbar.

   You will notice that the sketch that was used to create the thin feature is still displayed. You need to turn off the display of this sketch.

8. Click on the + sign located on the left of **Revolve-Thin1** in the **Feature Manager Design Tree**. The tree view expands and the sketch is displayed. Right-click on the sketch and choose **Hide** from the shortcut menu to hide the sketch. The revolved feature is shown in Figure 4-47.

Rotating the View

Next, you need to rotate the view so that you can view the model from all the directions. As mentioned earlier, the view can be rotated using the **Rotate View** tool.

1. Choose the **Rotate View** button from the **View** toolbar. The arrow cursor will be replaced by the rotate view cursor.

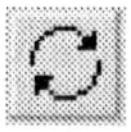

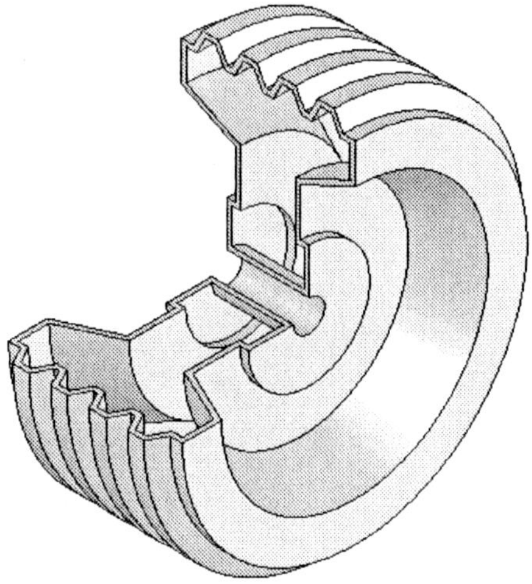

Figure 4-47 *Model created by revolving the sketch*

2. Press the left mouse button and drag the cursor in the drawing area to rotate the view.

3. Press the SPACEBAR to invoke the **Orientation** dialog box. In this dialog box, double-click on the **Isometric** option.

Displaying the Shadow

As mentioned in the tutorial description, you need to display the shadow of the model. You can turn on the display of the shadow using the **View** toolbar.

1. Choose the **Shadows In Shaded Mode** button from the **View** toolbar. The shadow of the model is displayed below the model, see Figure 4-48. 

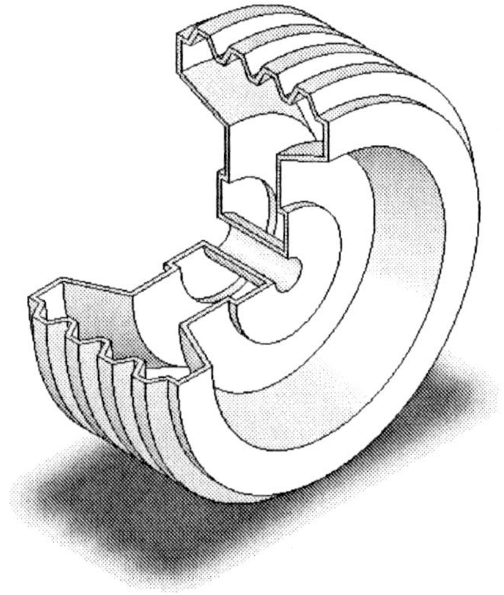

Figure 4-48 *Model with the display of shadow turned on*

Assigning Material to the Model

As mentioned earlier, you can assign material to the model using the **Materials Editor PropertyManager**, which is displayed by choosing the **Edit Material** button. Note that you can also invoke this **PropertyManager** using the **Material** option available in the **Feature Manager Design Tree**.

1. Right-click on the **Material <not specified>** option in the **Feature Manager Design Tree** and choose the **Edit Material** option. The **Materials Editor FeatureManager** is displayed.

2. Click on the + sign located on the left of the **Copper and its Alloys** option from the list box available in the **Materials** rollout. The tree view expands and the materials available under this family are displayed in the tree view.

3. Select the **Copper** option and choose **OK** from the **Materials Editor PropertyManager**. The model after assigning the material is shown in Figure 4-49.

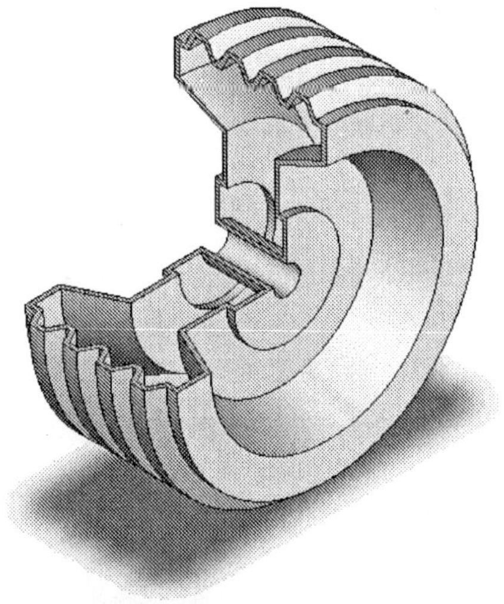

Figure 4-49 *Model after assigning Copper material*

Saving the Sketch

As the name of the document was specified at the beginning, you just need to choose the save button now to save the document.

1. Choose the **Save** button from the **Standard** toolbar to save the model. If the SolidWorks warning box is displayed, choose **Yes** from it to rebuild the model before saving.

   The model will be saved with a name *\My Documents\SolidWorks\c04\c04-tut02.sldprt.*

2. Choose **File > Close** from the menu bar to close the document.

Tutorial 3

In this tutorial, you will create the model shown in Figure 4-50. The dimensions of the model are shown in Figure 4-51. The extrusion depth of the model is 20 mm. After creating the model, rotate the view and then change the view back to isometric view before saving the model.

(Expected time: 45 min)

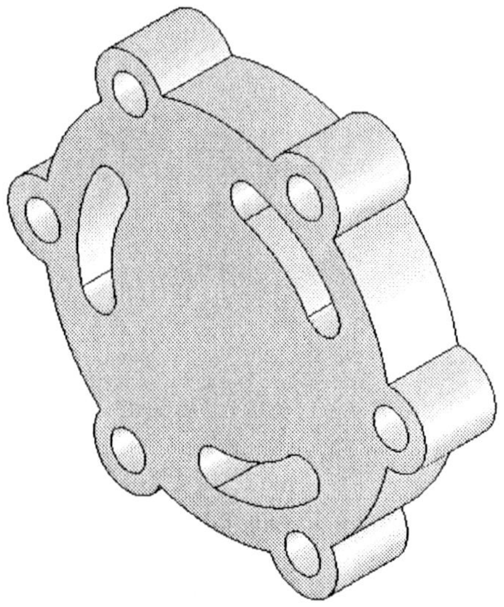

Figure 4-50 *Model for Tutorial 3*

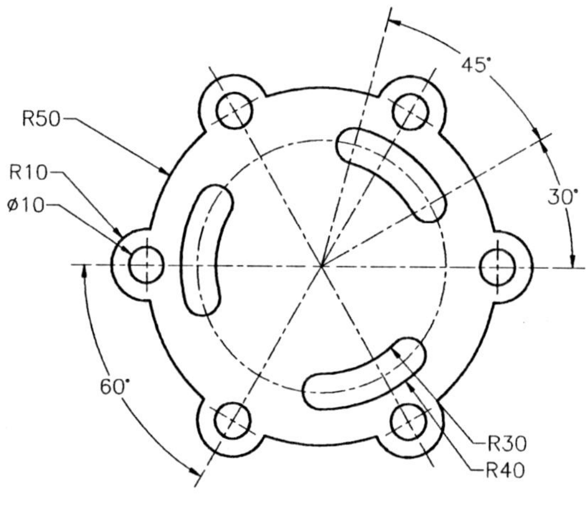

Figure 4-51 *Dimensions of the model for Tutorial 3*

The steps that will be used to complete this tutorial are given next.

a. Start a new SolidWorks part document and then invoke the sketching environment.
b. Create the outer loop and then create the sketch of three inner cavities. Finally, draw the six circles inside the outer loop for the holes, refer to Figures 4-52 through 4-56.

c. Invoke the **Extruded Boss/Base** tool and extrude the sketch through a distance of 20 mm, refer to Figure 4-57.
d. Rotate the view using the **Rotate View** tool.
e. Change the current view to isometric view and then save the document.

Starting a New Part Document

1. Choose the **New** button from the **Standard** toolbar and open a new part document using the **New SolidWorks Document** dialog box.

2. Choose the **Sketch** button from the **Standard** toolbar and select the **Front** plane to invoke the sketching environment.

Creating the Outer Loop

When the sketch consists of more than one closed loop, it is recommended that you add relations and dimensions to the outer loop first so that it is fully defined. Next, draw the inner loops one by one and add relations and dimensions to them. Therefore, you will first draw the outer loop and add relations and dimensions to it.

1. Draw a circle in the first quadrant and then dimension it so that it is forced to a diameter of 100 mm.

2. Locate the center of the circle at a distance of 70 mm along X and Y directions from the origin by adding dimensions in both the directions. Choose the **Zoom to Fit** button to fit the display on the screen.

3. Draw a horizontal centerline from the center of the circle and then draw a circle with the center point at the intersection of the centerline and the bigger circle.

4. Trim the part of the sketch so that the sketch looks similar to the one shown in Figure 4-52.

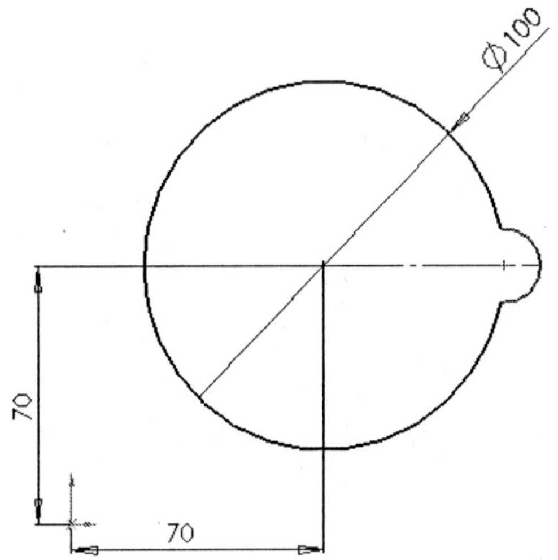

***Figure 4-52** Sketch after trimming the unwanted portion of circles*

5. Dimension the smaller arc so that it is forced to a radius of 10 mm. Add the **Coincident** relation to the center point of the smaller arc and the circumference of the bigger arc.

   You will notice that as you add the dimensions and relations to the smaller arc, it turns black in color. This suggests that the sketch drawn so far is fully defined.

   Next, you will create a circular pattern of the smaller arc. The total number of instances in the pattern is 6 and the total angle is 360-degree.

6. Select the smaller arc and then choose the **Circular Sketch Step and Repeat** button from the **Sketch CommandManager**.

   The **Circular Sketch Step and Repeat** dialog box is displayed.

7. Drag the center of the circular pattern to the center of the bigger arc.

8. Set the value of the **Number** spinner in the **Step** area to **6**. Accept all the other default values and choose the **OK** button to create the pattern.

   You will notice that all the instances of the pattern are black in color. This is because you have already applied the dimensions and relations to the original instance and so the other instances are also fully defined.

9. Trim the unwanted portion of the bigger arc. This completes the outer loop. The sketch at this stage should look similar to the one shown in Figure 4-53.

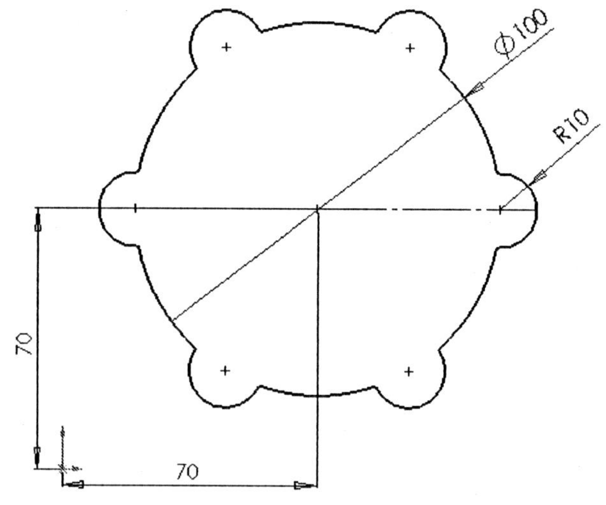

Figure 4-53 *Outer loop of the sketch*

Creating the Sketch of the Inner Cavities

Now, you need to create the sketch of the inner cavities. You will create the sketch of one of the cavities and then add the required relations and dimensions to it. Next, you will create a circular pattern of this cavity. The number of instances in the circular pattern will be 3.

1. Using the **Centerpoint Arc** tool, draw an arc with the center at the centerpoint of the bigger arc of 100 mm diameter.

2. Dimension this arc such that it is forced to a radius of 30 mm. Also, add the angular dimensions to the two endpoints of the arc, refer to Figure 4-54. The arc will turn black in color, suggesting that it is fully defined.

3. Offset the last drawn arc outward through a distance of 10 mm using the **Offset Entities** tool.

   The new arc created using the **Offset Entities** tool is also black in color. Also, a dimension with the value 10 mm will be placed between the two arcs.

4. Close the two ends of the arc using the **Tangent Arc** tool. This completes the sketch of one of the inner cavities. All the entities in the sketch at this stage should be displayed in black color as shown in Figure 4-54.

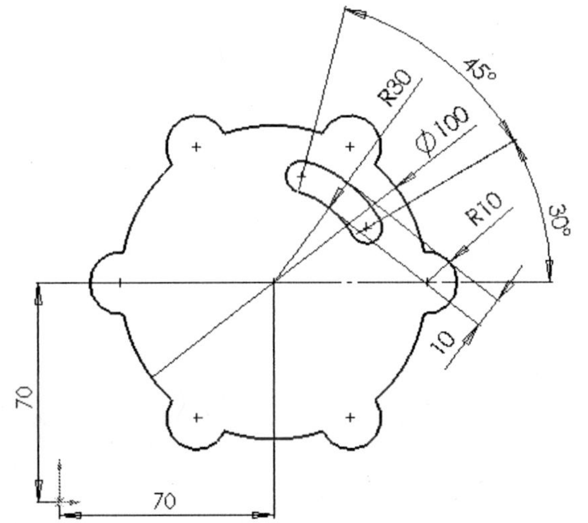

Figure 4-54 *Sketch after drawing the sketch of the inner cavity*

Next, you will create a circular pattern of the inner cavity.

5. Select all the entities in the sketch of the inner cavity and then invoke the **Circular Sketch Step and Repeat** dialog box.

6. Drag the center of the circular pattern to the center of the outer arc in the sketch.

7. Set the value of the **Number** spinner in the **Step** area to **3** and then choose the **OK** button to create the circular pattern. This completes the sketch of the inner cavities, see Figure 4-55.

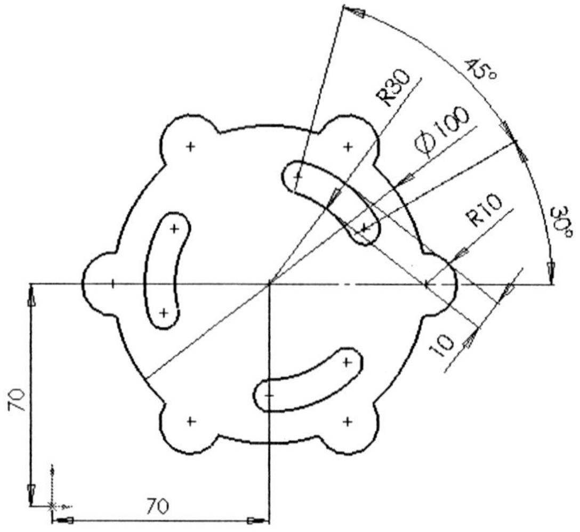

Figure 4-55 Sketch after creating the circular pattern of the inner cavity

Drawing the Sketch for the Holes

Next, you need to draw the sketch for the holes. You will draw one of the circles and then add dimension to it. Then you will create a circular pattern of the circle.

1. Taking the centerpoint of one of the smaller arcs of the outer loop, draw a circle. Add a diameter dimension to it so that the value of the diameter is forced to 10 mm.

   The circle will turn black in color when you dimension it.

2. Select the circle and invoke the **Circular Sketch Step and Repeat** dialog box.

3. Drag the center of the circular pattern to the center of the outer arc of 100 mm diameter.

4. Set the value of the **Number** spinner in the **Step** area to **6**. Choose **OK** to create the pattern. All the instances in the pattern will be displayed in black color.

   This completes the sketch of the model. The final sketch of the model is shown in Figure 4-56.

Extruding the Sketch

The next step after creating the sketch is to extrude it. The sketch will be extruded using the **Extruded Boss/Base** tool.

1. Choose the down arrow available on the right of the **Features** button in the **CommandManager**. A flyout is displayed. Choose the **Extruded Boss/Base** option from this flyout.

   The current view is changed to trimetric view and the **Extrude PropertyManager** will be displayed. Also, the preview of the model, as it will be created using the default values, will be displayed in the drawing area.

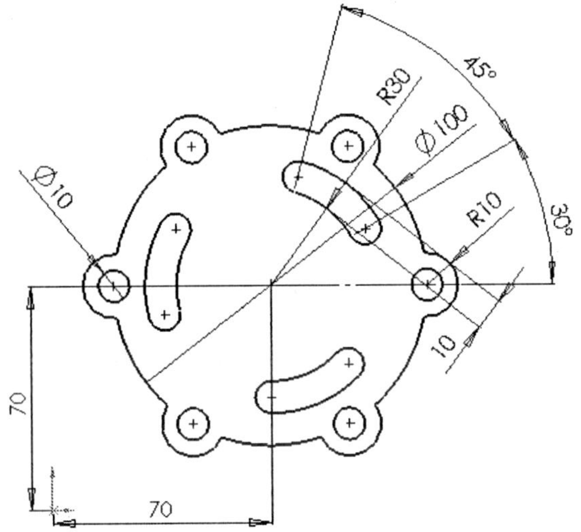

Figure 4-56 Final sketch of the model

2. Set the value of the **Depth** spinner to **20** and then choose the **OK** button to extrude the sketch.

   Because the option to display shadow was turned on in the previous tutorial, the shadow will be displayed in this tutorial also.

3. Press the SPACEBAR key and then double-click on the **Isometric** option in the **Orientation** dialog box to change the current view to the isometric view. The completed model for Tutorial 3 is shown in Figure 4-57.

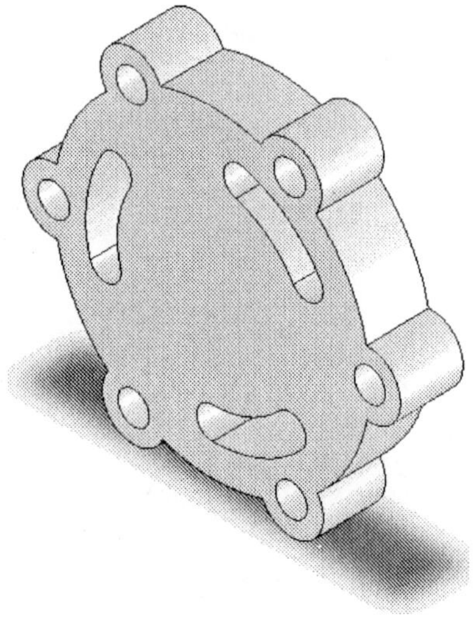

Figure 4-57 Completed model for Tutorial 3

Rotating the View

Before you start rotating the view of the model, it is recommended that you turn off the display of the shadow.

1. Choose the **Shadows In Shaded Mode** button from the **View** toolbar to turn off the display of the shadow.

2. Now, choose the **Rotate View** button from the **View** toolbar. The arrow cursor will be replaced by the rotate view cursor.

3. Press the left mouse button and drag the cursor in the drawing area to rotate the view.

4. Change the current view to isometric view using the **Orientation** dialog box.

Saving the Sketch

1. Choose the **Save** button from the **Standard** toolbar and save the model with the name given below.

   \My Documents\SolidWorks\c04\c04-tut03.sldprt.

2. Choose **File > Close** from the menu bar to close the document.

SELF-EVALUATION TEST

Answer the following questions and then compare your answers with the answers given at the end of this chapter.

1. In SolidWorks, a sketch is revolved using the **Extrude PropertyManager**. (T/F)

2. You can also specify the depth of extrusion dynamically in the preview of the extruded feature. (T/F)

3. You can invoke the drawing display tools such as **Zoom to Fit** while the preview of a model is displayed on the screen. (T/F)

4. When you rotate the view with the current display mode set to **Hidden Lines Removed**, the hidden lines in the model are automatically displayed while the view is being rotated. (T/F)

5. __________ tool is used to display the perspective view of a model.

6. The **Cap ends** check box is displayed in the **Extrude PropertyManager** only when the sketch for the thin base feature is __________.

7. The __________ check box is used to create a feature with different values in both the directions of the sketching plane.

8. The __________ check box is used to apply automatic fillets while creating a thin feature.

9. The __________ button is used to display the shadow in the shaded mode.

10. To resume rotating the view freely after you have completed rotating it around a selected vertex, edge, or face, __________ any where in the drawing area.

REVIEW QUESTIONS

Answer the following questions.

1. You can also invoke the **Rotate View** tool by choosing the **Rotate View** option from the __________ that is displayed when you right-click in the drawing area.

2. When you choose the **Wireframe** button, all the __________ lines will be displayed along with the visible lines in the model.

3. You can also modify the parallel view to perspective view by choosing __________ from the menu bar.

4. When you invoke the **Extruded Boss/Base** tool or the **Revolved Boss/Base** tool, the view is automatically changed to a __________.

5. The thin revolved features can be created using a __________ or an __________ sketch.

6. Which one of the following buttons is chosen to modify the orientation of the standard views?

   (a) **Update Standard Views** (b) **Reset Standard Views**
   (c) None (d) Both

7. Which one of the following buttons is not available in the **View** toolbar by default?

   (a) **Hidden Lines Removed** (b) **Hidden Lines Visible**
   (c) **Shaded** (d) **Perspective**

8. Which one of the following parameters will not be displayed in the preview of the model?

   (a) Depth (b) Draft angle
   (c) None (d) Both

9. If the sketch is open, it can be converted into

(a) Thin feature (b) Solid feature
(c) None (d) Both

10. In SolidWorks, which tool is used to apply automatic dimensions to the sketch?

(a) **Autodimension** (b) **Smart Dimension**
(c) None (d) Both

EXERCISES

Exercise 1

Create the model shown in Figure 4-58. The sketch of the model is shown in Figure 4-59. Create the sketch and dimension the sketch using the autodimension option. The extrusion depth of the model is 15 mm. After creating the model, rotate the view. **(Expected time: 30 min)**

Figure 4-58 *Model for Exercise 1*

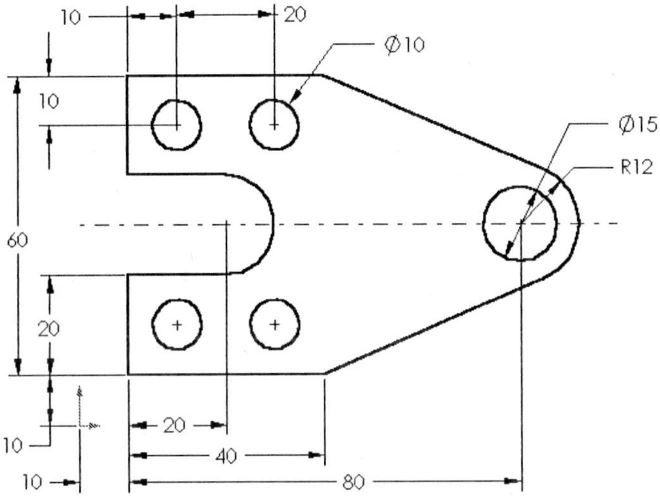

Figure 4-59 Sketch of the model for Exercise 1

Exercise 2

Create the model shown in Figure 4-60. The sketch of the model is shown in Figure 4-61. Create the sketch and dimension the sketch using the autodimension tool. The extrusion depth of the model is 25 mm. Modify the standard view such that the current front view of the model should be displayed when you invoke the top view. **(Expected time: 30 min)**

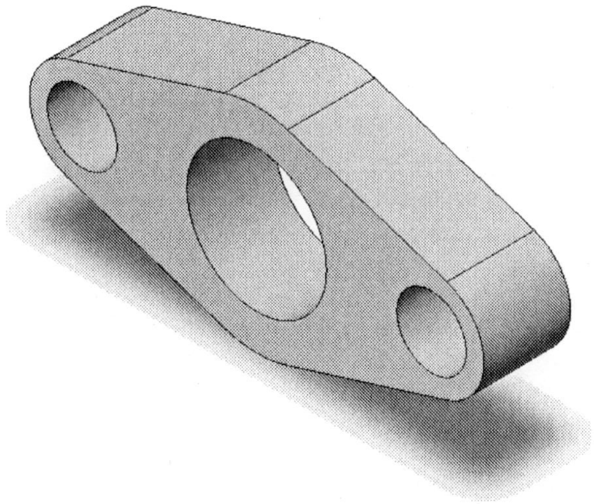

Figure 4-60 Model for Exercise 2

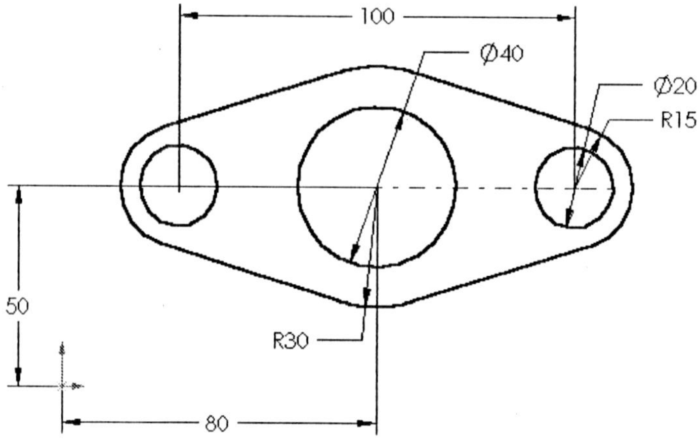

Figure 4-61 Sketch of the model for Exercise 2

Answers to Self-Evaluation Test

1. F, 2. T, 3. T, 4. F, 5. **Perspective**, 6. closed, 7. **Direction 2**, 8. **Auto-fillet corners**, 9. **Shadows In Shaded Mode**, 10. double-click

Chapter 5

Creating Reference Geometries

Learning Objectives

After completing this chapter you will be able to:

- *Create reference plane.*
- *Create reference axis.*
- *Create reference points.*
- *Create reference coordinate system.*
- *Create model using other Boss/Base options.*
- *Create model using the contour selection technique.*
- *Create a cut feature.*
- *Create multiple disjoint bodies.*

IMPORTANCE OF THE SKETCHING PLANES

In previous chapters, you created the basic models by extruding or revolving the sketches. All of these models were created on a single sketching plane, the **Front** plane. But most of the mechanical designs consist of more than one sketched features, referenced geometries, and placed features. These features are integrated together to complete a model. Most of these features lie on different planes. When you start a new SolidWorks document and try to invoke the sketching plane, you are prompted to select the plane on which you want to create the sketch. On the basis of design requirements, you can select any plane to create the base feature. To create additional sketched features, you need to select an existing plane, or a planar surface, or you have to create a plane that will be used as a sketching plane. For example, consider the model shown in Figure 5-1.

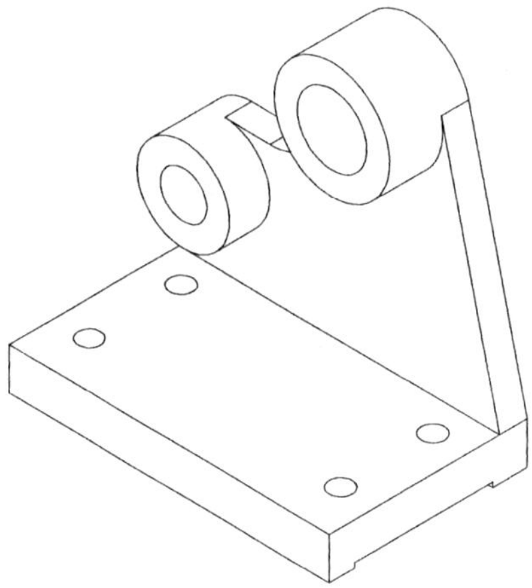

Figure 5-1 *A multifeatured model*

The base feature of this model is shown in Figure 5-2. The sketch for the base feature is drawn on the **Top** plane. After creating the base feature, you will have to create the other sketched features, placed features, and referenced features, see Figure 5-3. The boss features and cut features are the sketched features that require sketching planes where you can draw the sketch of the features.

It is evident from Figure 5-3 that the features added to the base feature are not created on the same plane on which the sketch for the base feature is created. Therefore, to draw the sketches of other sketched features, you need to define other sketching planes.

REFERENCE GEOMETRY

The reference geometry features are those that are available only to assist you in the creation of the models. The reference geometries in SolidWorks include planes, axes, points, and coordinate systems. These reference geometries act as a reference for drawing the sketches for features, defining the sketch plane, assembling the components, references for various placed features and sketched features, and so on. These features have no mass or volume.

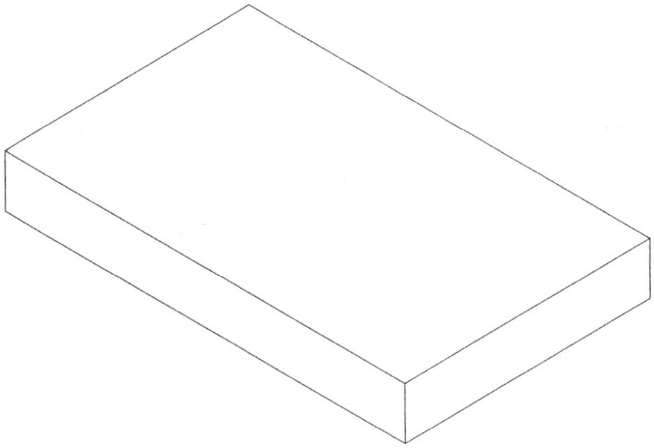

Figure 5-2 *Base feature for the model*

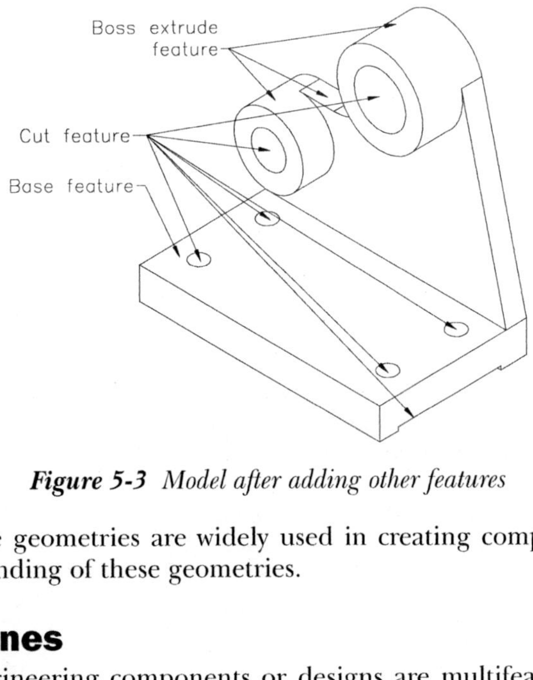

Figure 5-3 *Model after adding other features*

Because the reference geometries are widely used in creating complex models, you must have a good understanding of these geometries.

Reference Planes

Generally, all the engineering components or designs are multifeatured models. Also, as discussed earlier, all the features of a model are not created on the same plane on which the base feature is created. Therefore, you have to select one of the default planes or create a new plane that will be used as the sketching plane for the second feature. It is clear from the above discussion that you can use the default planes as the sketching plane or you can create a plane that can be used as a sketching plane. The default planes and the creation of a new plane are discussed next.

Default Planes

When you start a new SolidWorks part document, SolidWorks provides you with three default planes. These three planes are the **Front** plane, the **Top** plane, and the **Right** plane.

The orientation of the component depends on the sketch of the base feature. Therefore, it is recommended that you carefully select the sketching plane for creating the sketch for the base feature. The sketch plane for drawing the sketch of the base feature can be one of the three datum planes provided by default.

If you are creating the base feature and choose the **Sketch** tool to invoke the sketching environment, the three default planes are automatically displayed in the drawing area. However, if you are creating an additional feature, these planes are not displayed when you invoke the **Sketch** tool. You need to manually select the required plane.

To select a plane for an additional feature, choose the **Sketch** button. The **Edit Sketch PropertyManager** is displayed. In SolidWorks 2004, whenever a **PropertyManager** is displayed, the **Feature Manager Design Tree** shifts to the drawing area. Click on the + sign located on the left of the part document name in the **Feature Manager Design Tree** that is displayed in the drawing area. The tree view will expand, displaying the names of the planes. Select the required plane from the tree view.

Tip. *You can turn on the display of the default planes in the drawing area using the following procedure.*

Press and hold the CTRL key from the keyboard and one by one select the ***Front****, the* ***Top****, and the* ***Right*** *plane from the* ***FeatureManager Design Tree****. Right-click to display the shortcut menu and choose the* ***Show*** *option from this shortcut menu to display the planes in the drawing area. Choose the* ***Isometric*** *button from the* ***Standard Views*** *toolbar. The default planes are transparent and the boundary of the planes is displayed in gray color.*

To display the shaded planes, choose ***Tools > Options*** *from the menu bar to invoke the* ***System Options - General*** *dialog box. Select the* ***Display/Selection*** *option from the left of this dialog box; the name of the dialog box will be changed to* ***System Options - Display/Selection****. Select the* ***Display shaded planes*** *check box from this dialog box and choose the* ***OK*** *button.*

After displaying the planes in shaded, invoke the ***Rotate View*** *tool and drag the rotate view cursor to rotate the shaded planes. You will observe that one side of the plane is displayed in green color and the other side of the plane is displayed in red color. The green side of the plane symbolizes the positive side and the red side of the plane symbolizes the negative side. This means that when you create a extrude feature, the depth of extrusion will be assigned to the positive side of the plane by default. When you create a cut feature the depth of the cut feature is assigned to the negative direction by default.*

Note

You can also select a plane before choosing the ***Sketch*** *button to invoke the sketching environment. In this case, the* ***Edit Sketch PropertyManager*** *is not displayed and you are not prompted to select a plane for sketching. The selected plane is automatically taken as the sketching plane.*

Creating New Planes

CommandManager:	Reference Geometry > Plane
Menu:	Insert > Reference Geometry > Plane
Toolbar:	Reference Geometry > Plane

Planes or the reference planes are used to draw sketches for the sketched features. These planes are also used to create a placed feature like holes, reference an entity or a feature, and so on. You can also select a planar face of a feature that will be used as a sketching plane. Generally, it is recommended that you use the planar faces of the features as the sketching planes. However, sometimes you have to create a sketch at a plane that is at some offset distance from a plane or a planar face. In this case, you have to create a new reference plane at an offset distance from a plane or a planar face.

Consider another case where you have to define a sketching plane tangent to a cylindrical face of a shaft. You have to create a plane tangent to the cylindrical face of the shaft and this plane will be used as a sketching plane. In SolidWorks, there are six methods to create planes. To create planes, choose the **Plane** button from the **Reference Geometry CommandManager**. The **Plane PropertyManager** is displayed, as shown in Figure 5-4. The confirmation corner is also displayed at the top right corner of the drawing area.

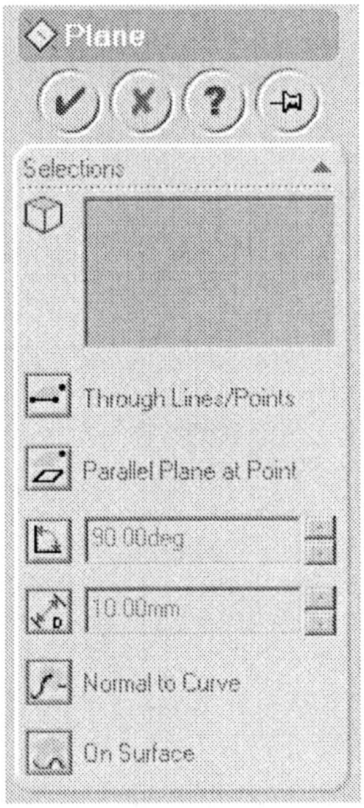

Figure 5-4 Plane PropertyManager

The options that are available in the **Plane PropertyManager** to create new planes are discussed next.

Creating a Plane Using Through Lines/Points

The **Through Lines/Points** option is used to create a plane that passes through an edge and a point, an axis and a point, or a sketch line and a point. Using this option, you can also create a plane that passes through three points. The selected point can be a sketched point or a vertex. To create a plane using this option, invoke the **Plane PropertyManager** and choose the **Through Lines/Points** button. Select the required entities from the drawing area. The name of the selected entities will be displayed in the **Reference Entities** selection area. Choose the **OK** button from the **Plane PropertyManager**. Figure 5-5 shows an edge and a vertex selected to create a plane. The resultant plane is displayed in Figure 5-6. The creation of a new plane by selecting three points is displayed in Figures 5-7 and 5-8.

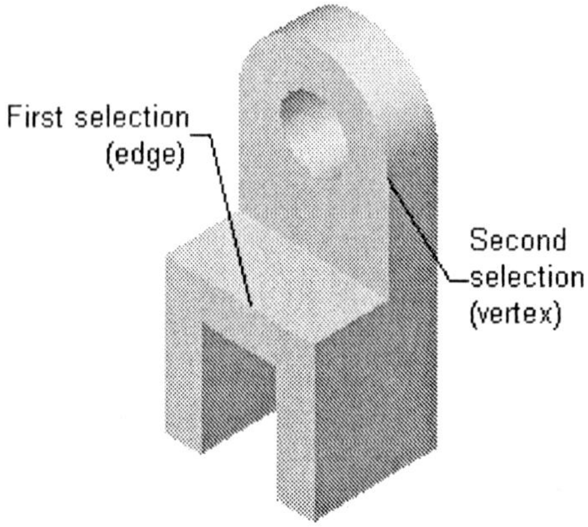

Figure 5-5 *Selecting the edge and vertex*

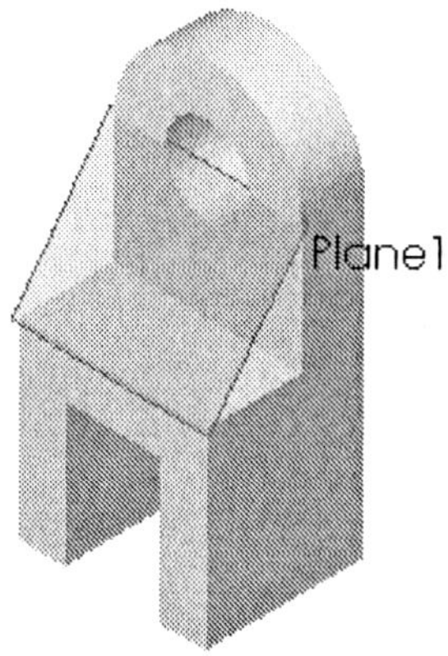

Figure 5-6 *Resultant plane*

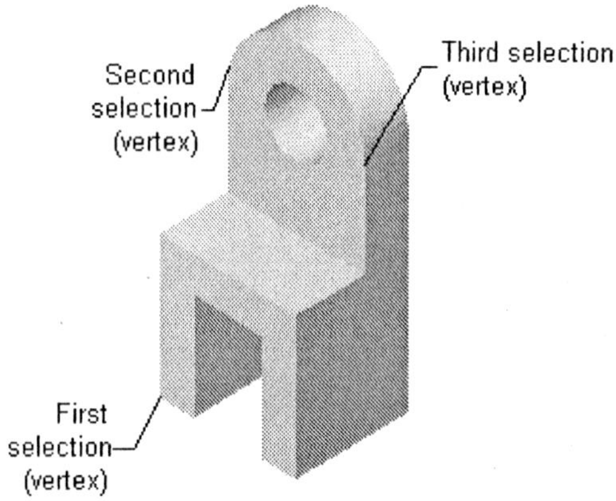

Figure 5-7 *Selecting the vertices*

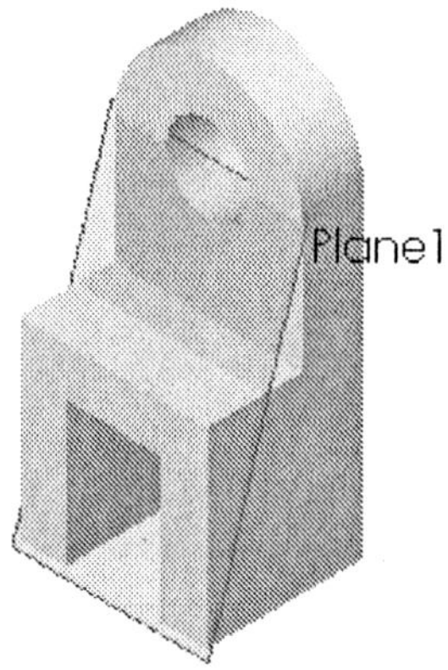

Figure 5-8 *Resultant plane*

Creating a Plane Parallel to an Existing Plane or Planar Face and Passing Through a Point

The **Parallel Plane at Point** option is used to create a plane that is parallel to another plane or a planar face and passes through a point. To create a plane using this option, invoke the **Plane PropertyManager** and then choose the **Parallel Plane at Point** button. Now, select the plane or a planar face to which the new plane will be parallel. Next, select a sketched point, an endpoint of an edge, or a midpoint of an edge. The new plane will pass through this point. Choose the **OK** option. Figure 5-9 shows a planar face and the point selected to create the parallel plane. Figure 5-10 shows the resultant plane.

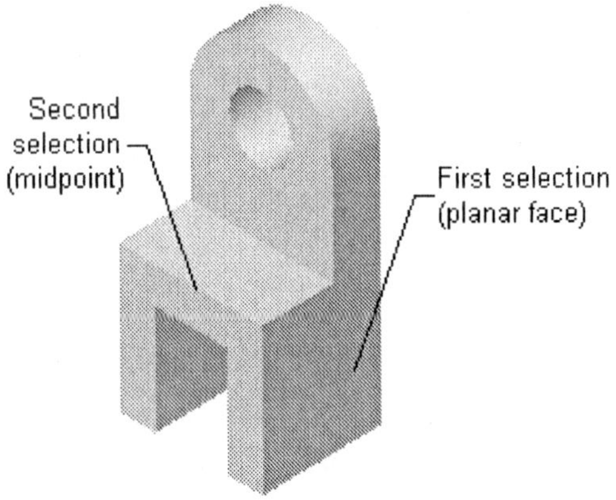

Figure 5-9 *Selecting a planar face and a vertex*

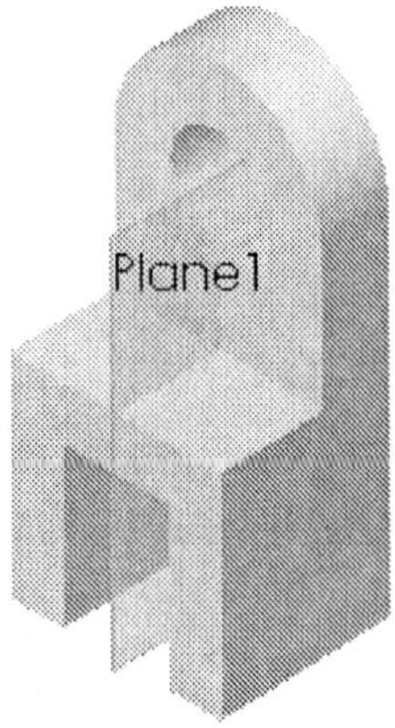

Figure 5-10 *Resultant plane*

Creating a Plane at an Angle to an Existing Plane or a Planar Face

The **At Angle** option is used to create a plane that is at an angle to the selected plane or a planar face and also passes through an edge, axis, or sketched line. To create a plane at an angle, choose the **At Angle** button from the **Plane PropertyManager**. The **Angle** spinner is invoked. The **Reverse direction** check box and the **Number of Plane to Create** spinner appears below the **Distance** spinner in the **Plane PropertyManager**, as shown in Figure 5-11.

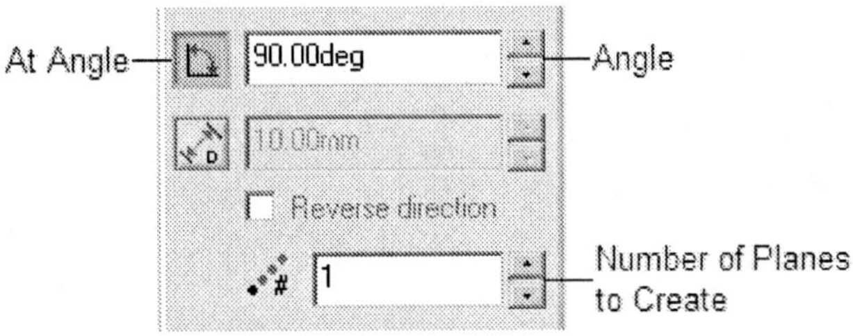

Figure 5-11 ***Plane PropertyManager*** *with the* ***At Angle*** *option selected*

Now, select an edge, an axis, or a sketched line through which the plane will pass. Next, you have to select a planar face or a plane to define the angle. After selecting the plane or the planar face, set the angle value using the **Angle** spinner. You can reverse the direction of plane creation by selecting the **Reverse direction** check box. You can also create multiple planes by increasing the value of the **Number of Planes to Create** spinner. Each plane will be incremented by the angle value defined in the **Angle** spinner. Figure 5-12 shows a planar face and edge selected. Figure 5-13 shows the resultant plane created at an angle of 45-degree to the selected plane.

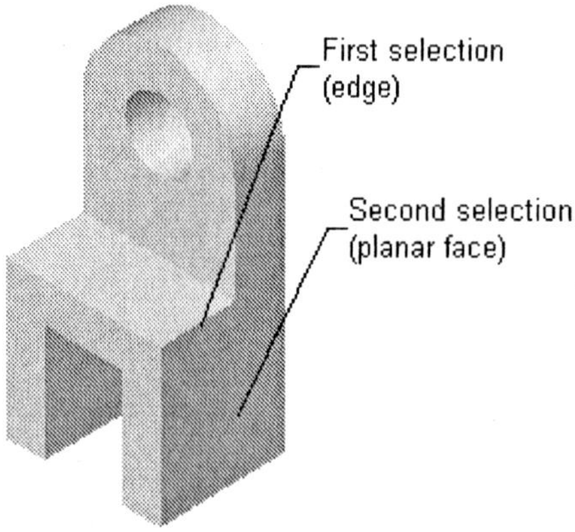

Figure 5-12 Selecting the edge and the planar face

Figure 5-13 Resultant plane

Creating a Plane at Some Offset From an Existing Plane or a Planar Face

The **Offset Distance** option is used to create a plane that is at some offset from a selected plane or planar face. To create a plane using this option, choose the **Offset Distance** button from the **Plane PropertyManager**. When you invoke this option, the **Distance** spinner is invoked. Also, the **Reverse direction** check box and the **Number of Planes to Create** spinner are displayed below the **Distance** spinner in the **Plane PropertyManager**. Select a plane or a planar face and set the value of distance in the **Distance** spinner and choose the **OK** button from the **Planar PropertyManager**. You can reverse the direction of plane creation by selecting the **Reverse direction** check box. You can also create multiple planes by increasing the value of the **Number of Planes to Create** spinner. Each plane will be placed at the offset distance specified in the **Distance** spinner. Figure 5-14 shows a plane selected to create a parallel plane and Figure 5-15 shows the resultant plane created at the required offset.

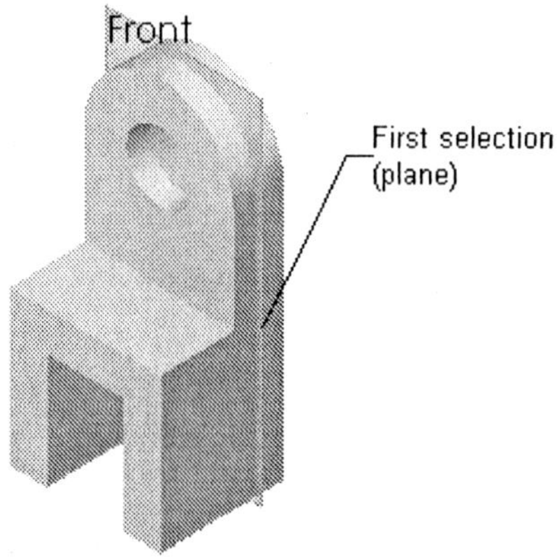

Figure 5-14 Selecting the plane

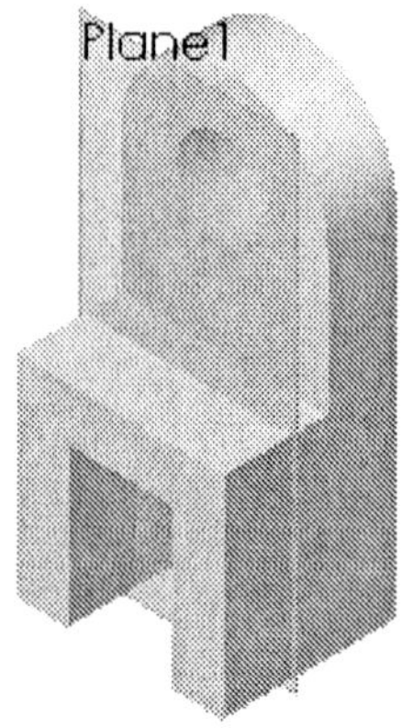

Figure 5-15 Resultant plane

Tip. *You can also create the planes by dynamically dragging an existing plane. For creating a plane by dragging, you do not need to invoke the* ***Plane PropertyManager****. Using the left mouse button select the plane from the* ***Feature Manager Design Tree*** *or the* ***drawing*** *area. Press and hold down the CTRL key and drag the cursor. You will notice that the value of distance in the* ***Distance*** *spinner of the* ***Plane PropertyManager*** *will modify and the preview of the plane is displayed in the drawing area. After dragging the plane to a required location release the left mouse button. Right-click and choose the* ***OK*** *option or choose the* ***OK*** *button from the* ***Plane PropertyManager****.*

You can also create a plane at an angle by dragging. To create a plane at an angle by dragging, select an edge or an axis and an existing plane. Now, hold the CTRL key and drag the mouse. Enter the angle value in the ***Angle*** *spinner.*

Creating a Plane Normal to Curve

This option is used to create a plane normal to a selected curve. The curve can be a sketched entity or an edge of a feature. To create a plane normal to a curve, choose the **Normal to Curve** button from the **Plane PropertyManager** and select a curve normal to which the new plane will be created. As soon as you select the curve, the preview of the plane is displayed in the drawing area. This plane is by default displayed at the endpoint closest to the point from where you selected the curve. If you want to place it at the other endpoint or at any other point on the curve, select that point. The preview of the plane will change accordingly. After selecting the plane and the point, choose the **OK** button to create the plane.

When you choose the **Normal to Curve** button from the **Plane PropertyManager**, the **Set origin on curve** check box is displayed below this button. This check box is selected to set the origin of the plane on the curve. Figure 5-16 shows a curve to create the plane and Figure 5-17 shows the resultant plane created normal to the selected curve.

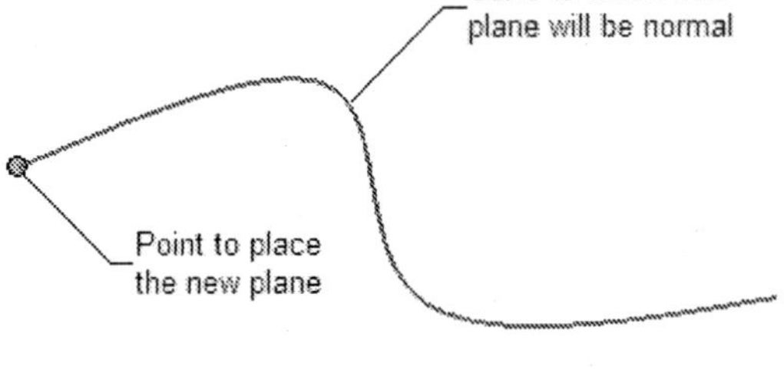

Figure 5-16 *Edge to be selected*

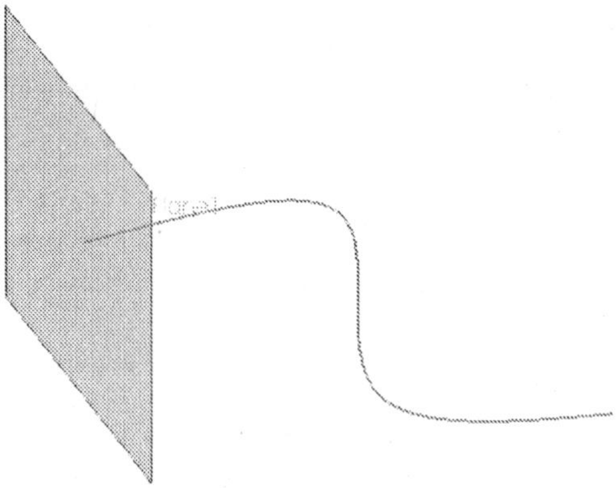

Figure 5-17 *Resultant plane*

Tip. *If you select an edge of a model or an existing curve and invoke the sketching environment, a reference plane is automatically created normal to that edge or curve and is selected as the sketching plane.*

Creating a Plane On Surface

The **On Surface** option is used to create a plane passing through a point on the selected plane or planar surface. To create a plane on surface, choose the **On Surface** button from the **Plane PropertyManager** and select the surface on which you want to create the plane. Next, select the sketched point. The preview of the plane is displayed in the drawing area and you have to right-click to choose the **OK** option. If the sketch is created on a plane at an offset distance from the selected surface, the **Project to nearest location on surface** and **Project onto surface along sketch normal** radio buttons, and the **Other Solutions** button are displayed on the **Plane PropertyManager**. Select any of the radio buttons according to requirement. You can also view the other solutions of the plane creation using the **Other Solutions** button from the **Plane PropertyManager**. Figure 5-18 shows the selection of references for the plane creation and Figure 5-19 shows the resultant plane created.

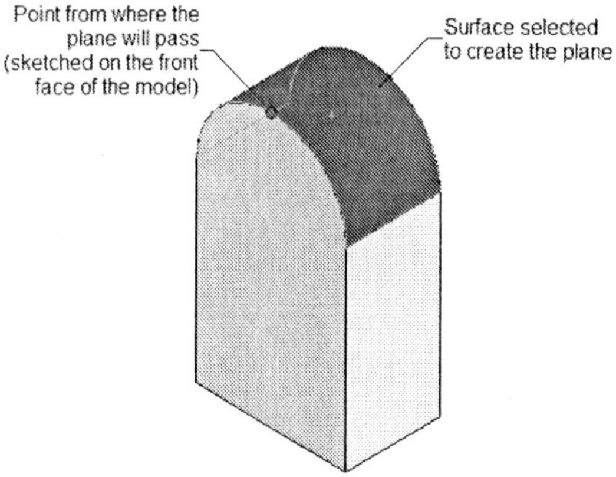

Figure 5-18 *References to be selected*

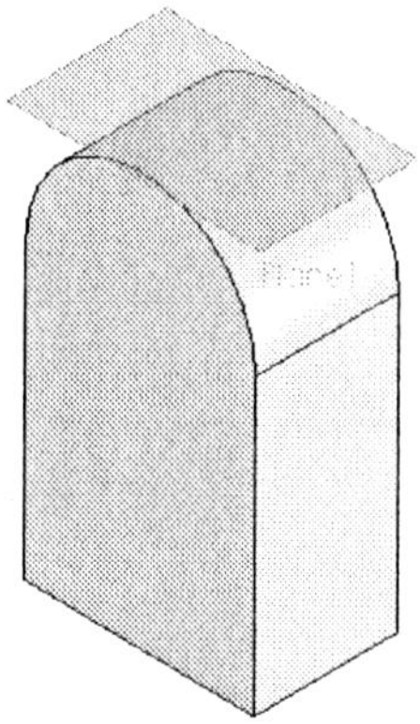

Figure 5-19 *Resultant plane*

Creating Reference Axis

CommandManager:	Reference Geometry > Axis
Menu:	Insert > Reference Geometry > Axis
Toolbar:	Reference Geometry > Axis

The **Reference Axis** option is used to create a reference axis or construction axis. These axis are the parametric lines passing through a model, feature, or reference entity. The reference axes are used to create reference planes, coordinate systems, circular patterns, and for applying mates in the assembly. These are also used as reference while sketching, or creating features. The reference axes are displayed in the model as well as in the **Feature Manager Design Tree**. When you choose the **Axis** button from the **Reference Geometry CommandManager** to create the axis, the **Axis PropertyManager** is displayed, as shown in Figure 5-20. The various options available in this dialog box are discussed next.

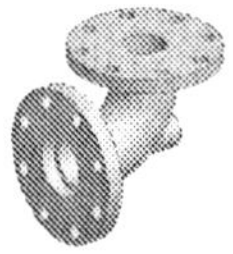

Tip. *If the axis is not displayed in the drawing area even after you have created it, choose* ***View > Axes*** *from the menu bar.*

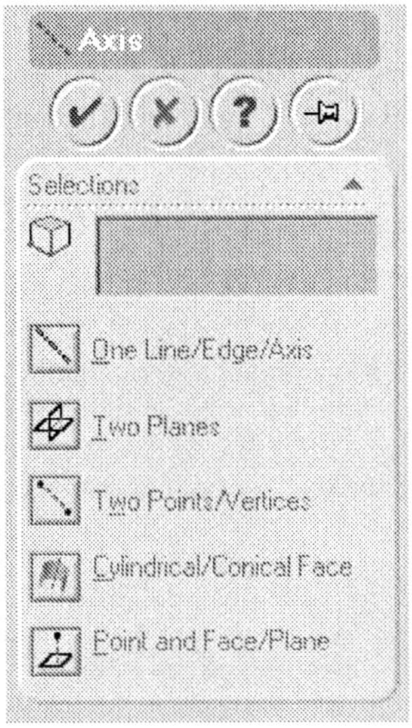

*Figure 5-20 The **Axis PropertyManager***

Creating a Reference Axis Using One Line/Edge/Axis

The **One Line/Edge/Axis** option is used to create a reference axis by selecting a sketched line or construction line, an edge, or a temporary axis. To use this option, invoke the **Axis PropertyManager** and choose the **One Line/Edge/Axis** button. Select a sketched line, edge, or a temporary axis. The name of the selected entity is displayed in the **Reference Entities** selection area and the preview of the reference axis is displayed in the drawing area. Choose the **OK** button to create the reference axis. Figure 5-21 shows a construction line selected as a reference for creating the axis. Figure 5-22 shows an axis created using this option.

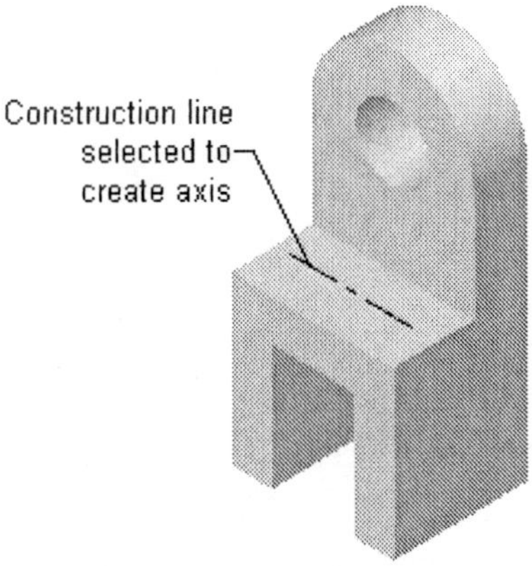

Figure 5-21 Line to be selected

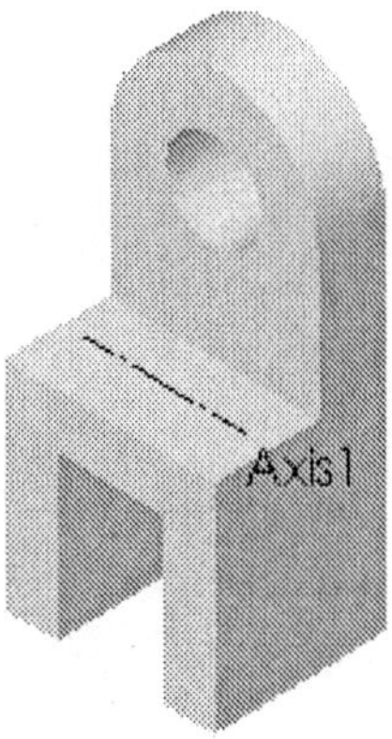

Figure 5-22 Resultant reference axis

Creating a Reference Axis Using Two Planes

Using the **Two Planes** option, you can create a reference axis at the intersection of two planes. To create a reference axis using this option, invoke the **Axis PropertyManager** and choose the **Two Planes** button. Now, select two planes, two planar faces, or a plane and a planar face that you want to use to create the axis. The preview of the axis is displayed in the drawing area. Choose the **OK** button from the **Axis PropertyManager**. Figure 5-23 shows two planes selected and Figure 5-24 shows the resultant reference axis created using this option.

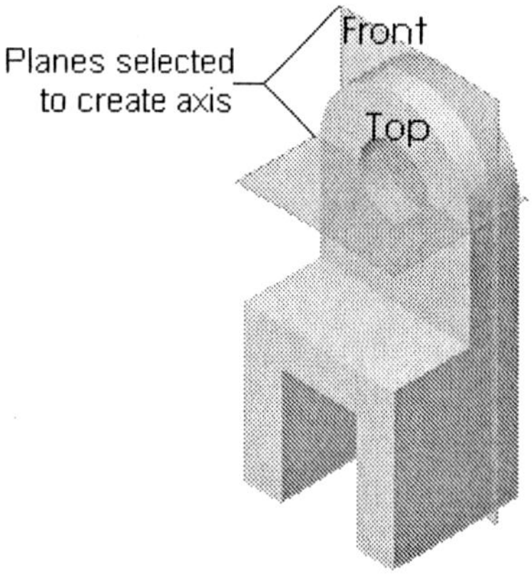

Figure 5-23 Planes to be selected

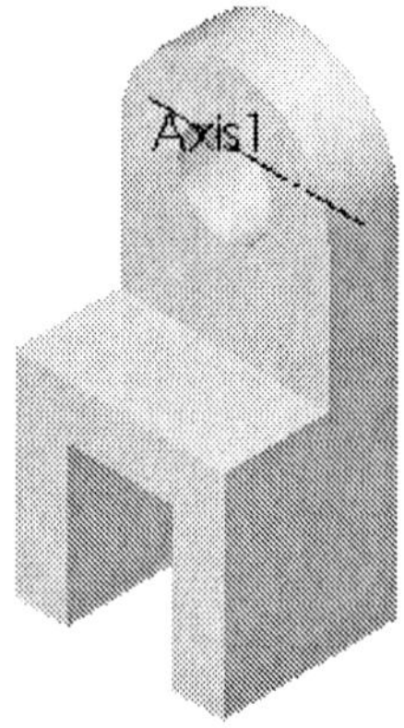

Figure 5-24 Resultant reference axis

Creating a Reference Axis Using Two Points or Vertices

Using the **Two Points/Vertex** option, you can create a reference axis that passes through two points or two vertices. To create a reference axis using this option, invoke the **Axis PropertyManager** and choose the **Two Points/Vertex** button. Now, select two points or two vertices through which you want the reference axis to pass. The preview of the reference axis is displayed in the drawing area. Choose the **OK** button from the **Axis PropertyManager**. Figure 5-25 shows two vertices to be selected and Figure 5-26 shows the resultant reference axis created using this option.

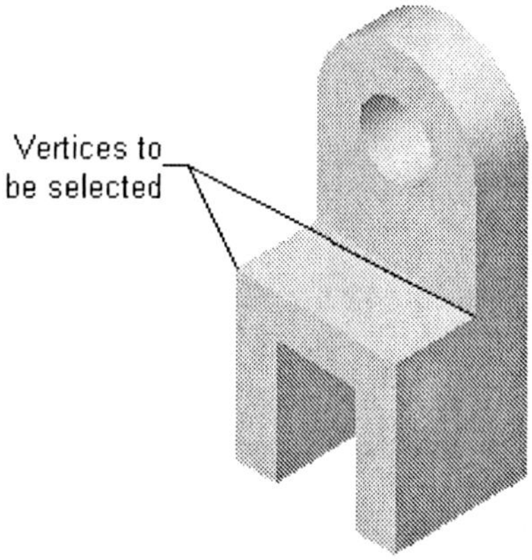

Figure 5-25 Vertices to be selected

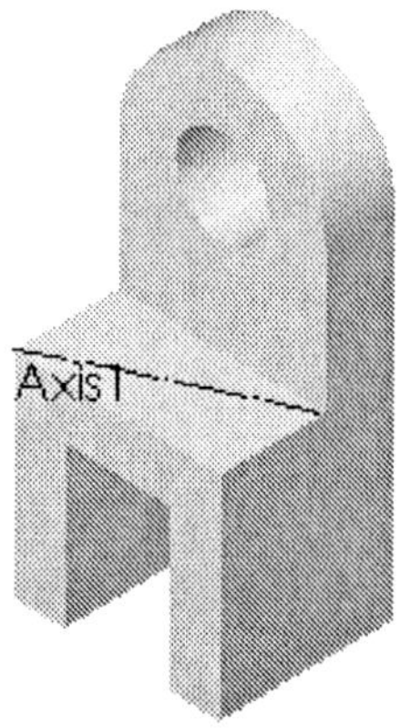

Figure 5-26 Resultant reference axis

Creating a Reference Axis Using a Cylindrical or a Conical Surface

Using the **Cylindrical/Conical Surface** option, you can create a reference axis that passes through the center of a cylindrical or a conical surface. To create a reference axis using this option, invoke the **Axis PropertyManager** and choose the **Cylindrical/Conical Surface** button. Now, select the cylindrical or the conical surface through the center of which the axis needs to pass. The preview of the reference axis is displayed in the drawing area. Choose the **OK** button from the **Axis** dialog box. Figure 5-27 shows a cylindrical surface selected and Figure 5-28 shows the resultant reference axis created using this option.

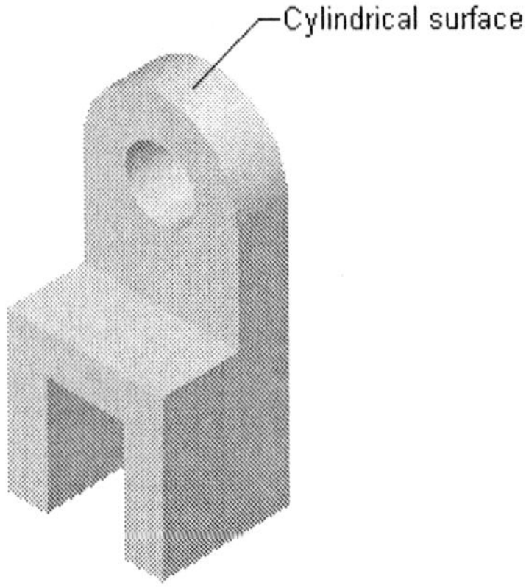

Figure 5-27 *Cylindrical surface to be selected*

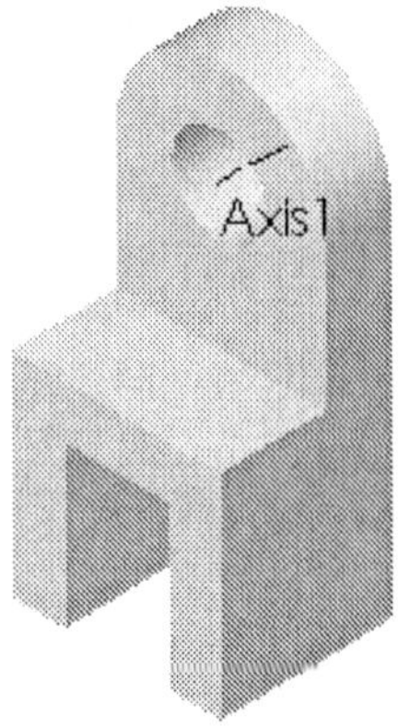

Figure 5-28 *Resultant reference axis*

Creating a Reference Axis on a Surface Passing Through a Point

Using the **Point and Surface** option you can create a reference axis that passes through a point and is normal to the selected surface. If the selected surface is a nonplanar surface, the selected point should be created on the surface. To create a reference axis using this option, invoke the **Reference Axis** dialog box. Select the **Point and Surface** radio button from this dialog box. Now, select a point, vertex, or midpoint and then select a surface. The preview of the axis will be displayed in the drawing area. Choose the **OK** button from the **Reference Axis** dialog box. The newly created axis will be normal to the selected surface. Figure 5-29 shows the point and surface selected and Figure 5-30 shows the resultant axis created using this option.

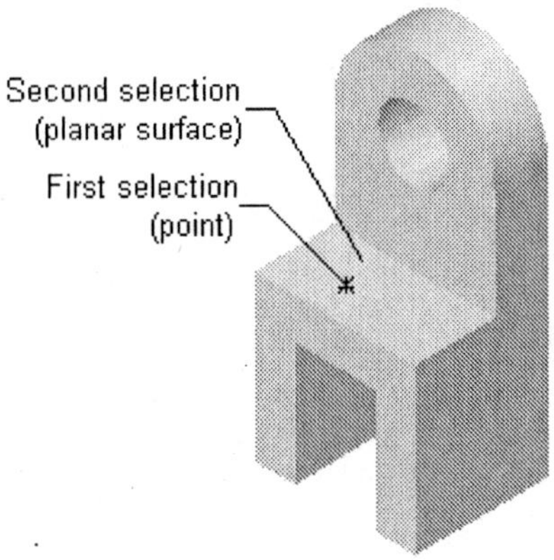

Figure 5-29 *Point and surface to be selected*

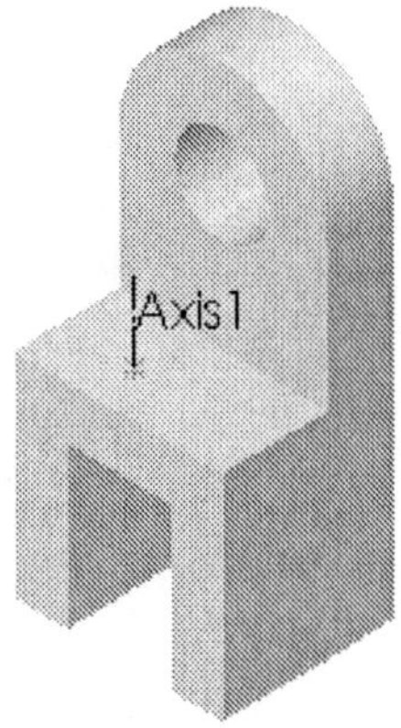

Figure 5-30 *Resultant reference axis*

Note

*When you create a circular feature, a temporary axis is automatically created. You can display the temporary axis by choosing **View** > **Temporary Axis** from the menu bar.*

Creating Reference Points

CommandManager:	Reference Geometry > Point
Menu:	Insert > Reference Geometry > Point
Toolbar:	Reference Geometry > Point

Reference points are also created to assist you in designing. They work as an aid for creating another reference geometry or feature. To create a reference point, choose the **Point** button from the **Reference Geometry CommandManager**. The **Point PropertyManager** is displayed, as shown in Figure 5-31.

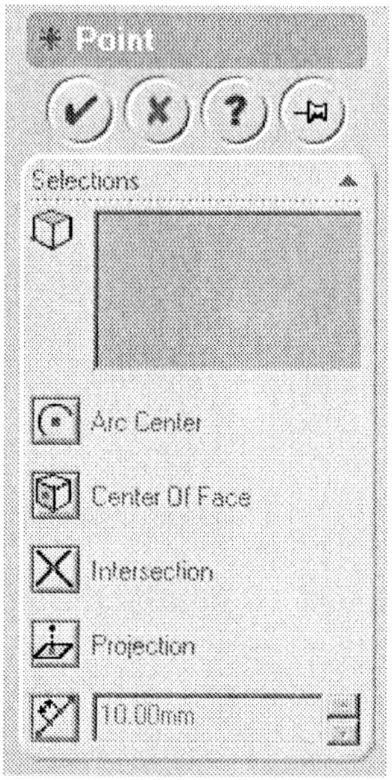

*Figure 5-31 The **Point PropertyManager***

This **PropertyManager** allows you to use five methods of creating reference points. These methods are discussed next.

Creating a Reference Point at the Center of an Arc or a Curved Edge

The **Arc Center** option is used to create a reference point at the center of a sketched arc or a curved edge. When you invoke the **Point PropertyManager** and choose the **Arc Center** button, you will be prompted to select an arc or a circular edge to define the reference point. As soon as you select a sketched arc or circle, or a curved edge, the preview of the reference point is displayed at its center. Choose **OK** to confirm the creation of the reference point. Figure 5-32 shows a curved edge selected to create the reference point and the preview of the resultant reference point.

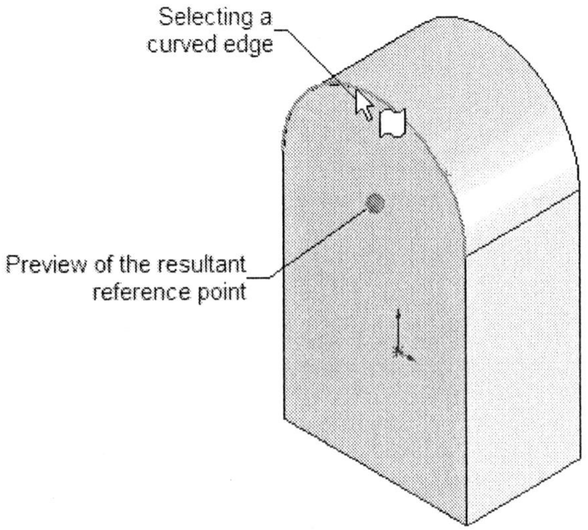

Figure 5-32 Reference point at the center of a curved edge

Creating a Reference Point at the Center of a Face

The **Center Of Face** option allows you to create a reference point at the center of a face. When you choose this button, you are prompted to select a face to define the reference point. You can select a planar or a curved face. The resultant point is automatically placed at the point of the center of gravity of the selected face.

Creating a Reference Point at the Intersection of Two Edges, Sketched Segments, or Reference Axes

The **Intersection** option allows you to create a reference point at the intersection of two edges, sketched segments, or reference axes.

Creating a Reference Point by Projecting an Existing Point

The **Projection** option allows you to create a reference point by projecting a point from some other plane to a specified plane. The point that you can project can be a sketched point, endpoints or center points of sketched entities, or a reference point.

Creating Single or Multiple Reference Points Along the Distance of a Sketched Curve or an Edge

The **Along curve distance or multiple reference point** option allows you to create single or multiple reference points along the distance of a selected curve. When you choose this button, the **Selections** rollout of the **Point PropertyManager** expands and provides additional options. These options are discussed next.

Enter the distance/percentage value according to distance

This spinner is used to specify the distance or the percentage value between the individual reference points along the selected curve.

Distance

This radio button is selected to define the gap between the individual points in terms of distance value. If this radio button is selected, the value entered in the **Enter the distance/percentage value according to distance** spinner will be in terms of linear units of the current document.

Percentage

This radio button is selected to define the gap between the individual points in terms of the percentage of the length of the selected curve. If this radio button is selected, the total length of the selected curve will be taken as 100%. Now, the value entered in the **Enter the distance/percentage value according to distance** spinner will be in terms of the percentage of selected curve.

Evenly Distribute

This radio button is selected to evenly distribute the specified number of reference points through the length of the selected curve. If this radio button is selected, the **Enter the distance/percentage value according to distance** spinner will not be available.

Enter number of reference points to be created along the selected entity

This spinner is used to specify the number of reference points to be created. The specified number of reference points will be placed with the gap defined in the **Enter the distance/percentage value according to distance** spinner along the selected curve.

Creating Reference Coordinate System

CommandManager:	Reference Geometry > Coordinate System
Menu:	Insert > Reference Geometry > Coordinate System
Toolbar:	Reference Geometry > Coordinate System

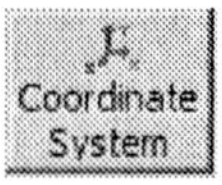

In SolidWorks you may need to define some reference coordinate systems other than the default coordinate system for creating features, analyzing the geometry, analyzing the assemblies, and so on. To create a user-defined coordinate system, choose the **Coordinate System** button from the **Reference Geometry CommandManager**. The **Coordinate System PropertyManager** is displayed, as shown in Figure 5-33. Also, a coordinate system is displayed at the origin of the current document.

To create a new coordinate system, you need to select a point that will be selected as the origin for the new coordinate and then define the directions of the X and Y, Y and Z, or Z and X axes. This is the reason when you invoke the **Coordinate System PropertyManager**, the **Origin** selection area is highlighted. After selecting the origin, you can define the direction of any two axes. The direction of the third axis will be automatically determined. You can select edges, points, or reference axes to define these directions.

You can reverse the directions of the axes by choosing the buttons available on the right of their respective selection areas.

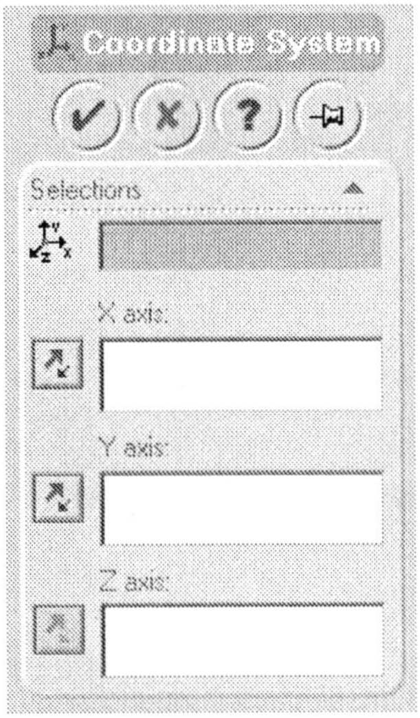

Figure 5-33 Coordinate System PropertyManager

OTHER BOSS/BASE OPTIONS

Some of the boss/base extrusion options were discussed in the previous chapter. In this chapter, the remaining boss/base extrusion options are discussed.

End Condition

The various options available in the **End Condition** drop-down list are discussed next.

Through All

The **Through All** option is available in the **End condition** drop-down list only after you create a base feature. After creating a base feature, define a new sketching plane and create the sketch using the standard sketching tools. Now, choose the **Extruded Boss/Base** button from the **Features CommandManager** to invoke the **Extrude PropertyManager**. The preview of the extruded feature that will be created using the default settings is displayed in temporary graphics in the drawing area. Select the **Through All** option from the **End Condition** drop-down list and the preview of the extruded feature extends from the sketching plane through all existing geometric entities. You can also reverse the direction of extrusion using the **Reverse Direction** button available on the left of the **End Condition** drop-down list.

While creating an extruded feature using the **Through All** option, the sketch extrudes through all the existing geometries. You will observe that the **Merge result** check box is displayed in the **Extrude PropertyManager**. This check box is selected by default. Therefore, the newly created extruded feature will merge with the base feature. If you clear this check box, this extruded feature will not merge with the existing base feature, resulting in the creation of another body. The creation of a new body can be confirmed by observing the **Solid Bodies** folder in the **Feature Manager Design Tree**. The value of the number of disjoint bodies in the model is displayed in parentheses on the right of the **Solid Bodies** folder. You can click on the + sign on the left of the **Solid Bodies** folder to expand the folder. To collapse the folder back, click the - sign.

Figure 5-34 displays a sketch created on the sketching plane at an offset distance from the right planar face of the model. Figure 5-35 displays the feature created by extruding the sketch using the **Through All** option.

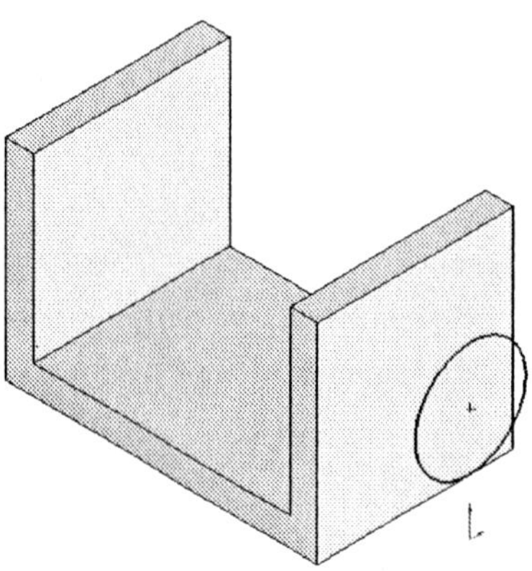

Figure 5-34 *A sketch created at an offset distance from the right planar surface*

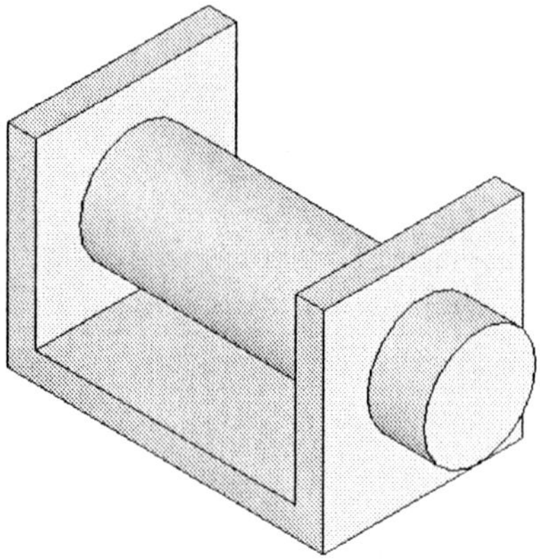

Figure 5-35 *Sketch extruded using the* ***Through All*** *option*

Tip. *The feature created from multiple disjoint closed contours results in creation of disjoint bodies.*

Note

It is recommended that while creating additional features after the base feature, you should always select the ***Merge results*** *check box in the* ***Feature PropertyManager****.*

Up To Next

The **Up To Next** option is used to extrude the sketch from the sketching plane to the next surface that intersects the feature. After creating a base feature, create a sketch by selecting or creating a sketching plane. Invoke the **Extrude PropertyManager**; the preview of the base feature is displayed with the default options. Select the **Up To Next** option from the **End Condition** drop-down list. You can also reverse the direction of feature creation using the **Reverse Direction** button. The preview of the feature will be modified and the sketch will be displayed as extruded from the sketching plane to the next surface that intersects the feature geometry. Figure 5-36 shows the sketch that will be extruded using the **Up To Next** option and Figure 5-37 shows the resultant feature.

Up To Vertex

The **Up To Vertex** option is used to define the termination of the extruded feature at a virtual plane parallel to the sketching plane and passing through the selected vertex. You can also select a point on an edge, a sketched point, or a reference point. Figure 5-38 shows a sketch created on a plane at an offset distance and Figure 5-39 shows the model in which the sketch is extruded up to the selected vertex.

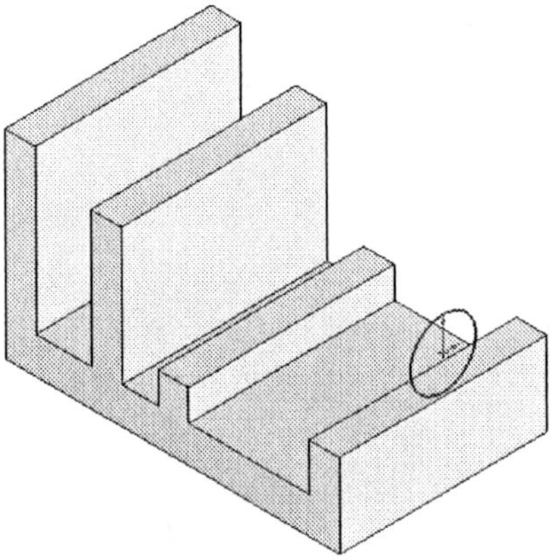

Figure 5-36 A sketch created on the right plane as the sketching plane

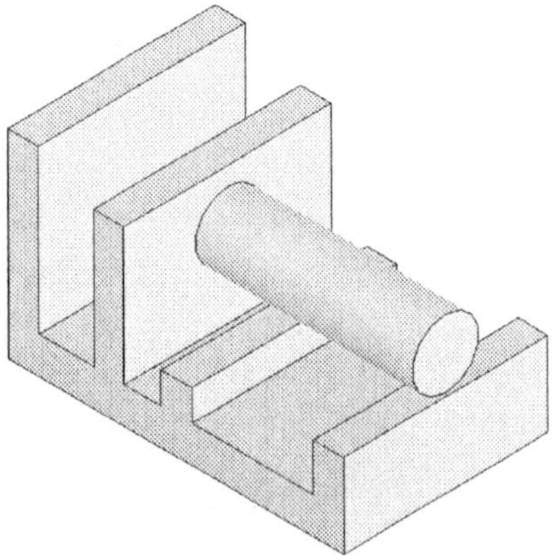

*Figure 5-37 Sketch extruded using the **Up To Next** option*

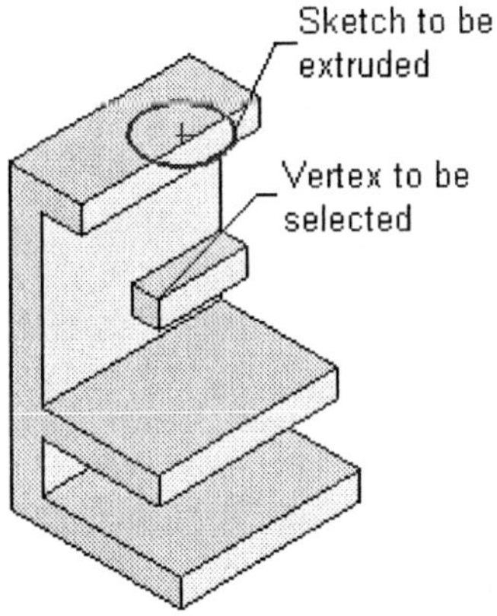

Figure 5-38 Sketch created on a plane created at an offset distance and vertex to be selected

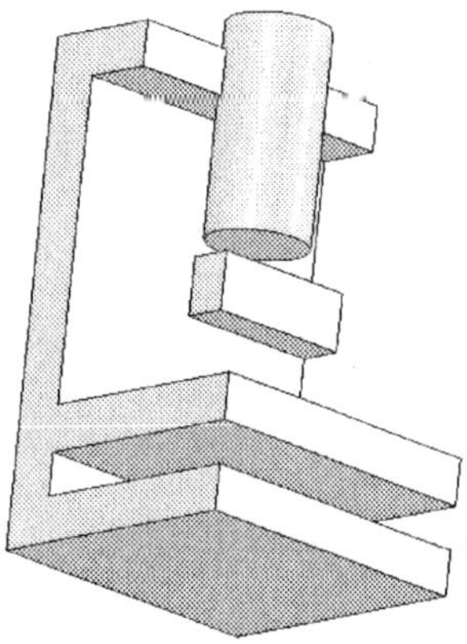

*Figure 5-39 Sketch extruded using the **Up To Vertex** option.*

Up To Surface

The **Up To Surface** option is used to define the termination of the extruded feature using a selected surface or a face. To create an extruded feature using this option, create a sketch using the normal sketching options and then invoke the **Extrude PropertyManager**. Select the **Up To Surface** option from the **End Condition** drop-down list. The **Face/Plane** selection area is displayed and you are prompted to select a face or a surface. Select a surface up to which you want to extrude the feature. Figure 5-40 shows the sketch created at an offset distance and the surface to be selected. Figure 5-41 shows the resultant feature extruded up to the selected surface.

Offset From Surface

The **Offset From Surface** option is used to define the termination of the extruded feature on a virtual surface created at an offset distance from the selected surface. To create an extruded feature using the **Offset From Surface** option, create a sketch and invoke the **Extrude PropertyManager**. Select the **Offset From Surface** option from the **End Condition** drop-down

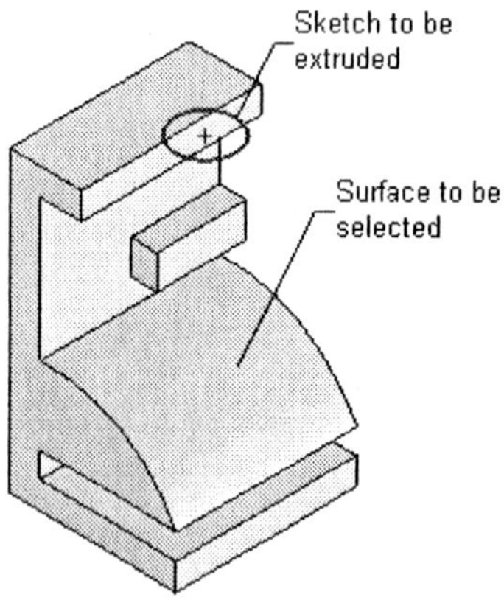

Figure 5-40 Sketch created on a plane created at an offset distance and surface to be selected

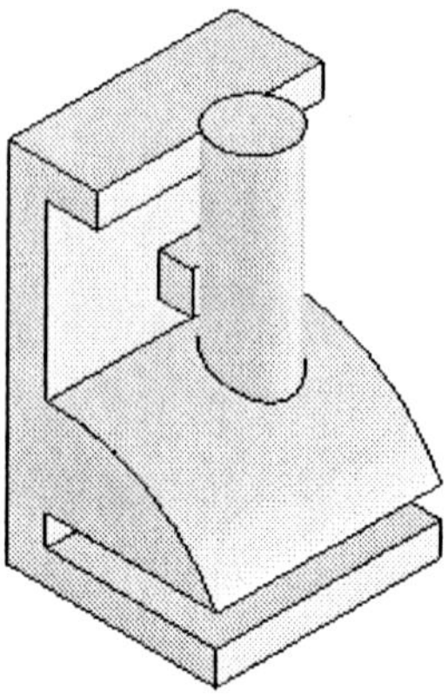

*Figure 5-41 Sketch extruded using the **Up To Surface** option.*

list. The **Face/Plane** selection area is displayed along with the **Offset Distance** spinner. You are prompted to select a face or a surface. Select the surface and set the offset distance in the **Offset Distance** spinner. You can reverse the direction of offset by selecting the **Reverse offset** check box from the **Direction 1** rollout. If the **Translate surface** check box is cleared, the virtual surface created for the termination of the extruded feature will have a concentric relation with the selected surface. Therefore, it reflects the true offset of the selected surface. If the **Translate surface** check box is selected, the center of the virtual surface is at the offset distance from the selected surface. Therefore, a reference surface is created to define the termination of the extruded feature and it does not reflect the true offset of the selected surface. Figure 5-42 shows the front view of the sketch extruded with termination at an offset distance from the selected cylindrical surface with the **Translate surface** check box cleared. Figure 5-43 shows the front view of the extruded feature with the **Translate surface** check box selected.

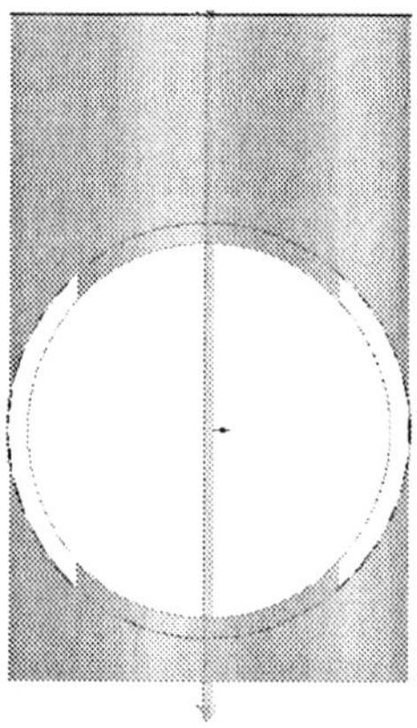

*Figure 5-42 Sketch extruded using the **Offset From Surface** option with the **Translate surface** check box cleared*

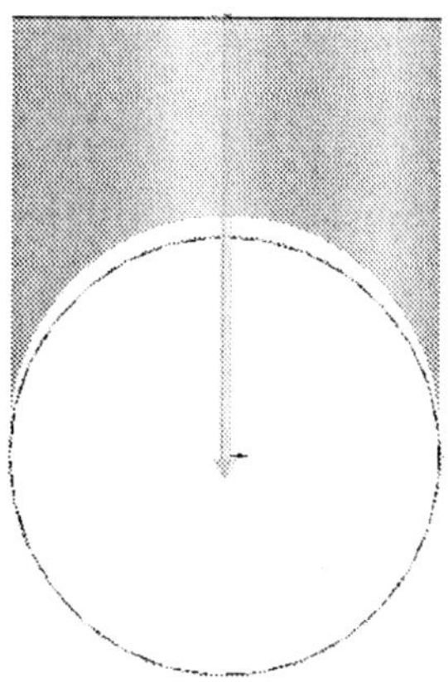

*Figure 5-43 Sketch extruded using the **Offset From Surface** option with the **Translate surface** check box selected*

Up To Body

The **Up To Body** option is used to define the termination of the extruded feature to another body. To create an extruded feature using the **Up To Body** option, invoke the **Extrude PropertyManager** and select the **Up To Body** option from the **End Condition** drop-down list. The **Solid/Surface Body** selection area is displayed. Select the body to terminate the feature and choose the **OK** button. Figure 5-44 shows the sketch for the extruded feature and a body up to which the sketch will be extruded. Figure 5-45 shows the resultant feature.

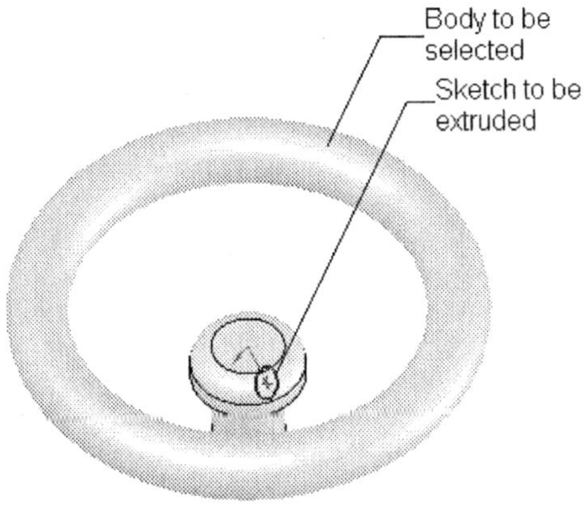

Figure 5-44 Sketch to be extruded and the body to be selected for the extrude feature

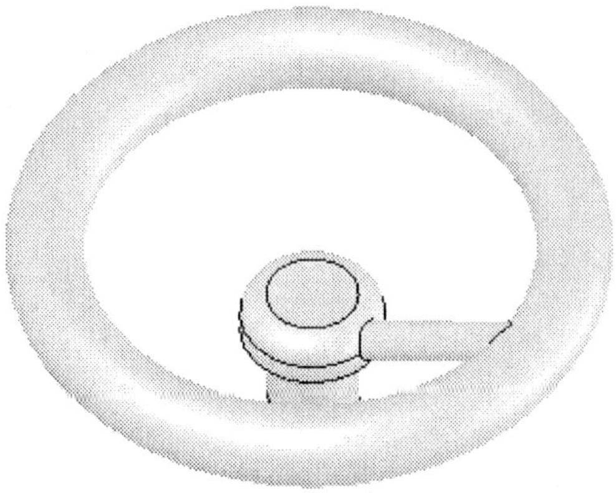

*Figure 5-45 Sketch extruded using the **Up To Body** option*

Direction of Extrusion

SolidWorks 2004 allows you to define a direction for extruding the sketches. As mentioned in the previous chapter, the direction of extrusion is generally normal to the sketching plane. However, from this release onwards, you can also define the direction of extrusion using a sketched line, an edge, or a reference axis. Note that the entity that you want to use for defining the direction of extrusion should not be drawn on the sketch plane parallel to the plane on which the sketch to be extruded is drawn.

To define the direction of extrusion, click in the **Direction of Extrusion** selection box in the **Direction 1** area of the **Extrude PropertyManager**. Next, select an edge, a sketched line segment, or an axis. Figure 5-46 shows a sketch drawn on the top face of a rectangular block and a line sketched on the left face of the block to define the direction of extrusion. Figure 5-47 shows the resultant extruded feature.

MODELING USING THE CONTOUR SELECTION METHOD

Modeling using the contour selection method allows you to use partial sketches for creating the features. Using this method, you can create sketches of the entire model in a single sketching environment and then manipulate the sketches by sharing them between various features. To understand this concept, consider the multifeatured solid model shown in Figure 5-48.

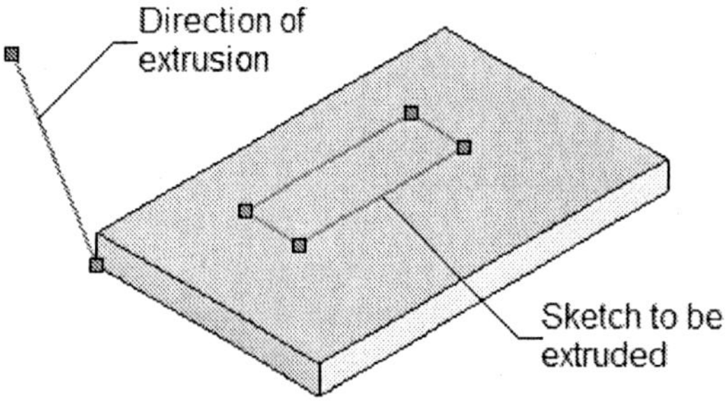

Figure 5-46 *Sketch to be extruded and the direction of extrusion*

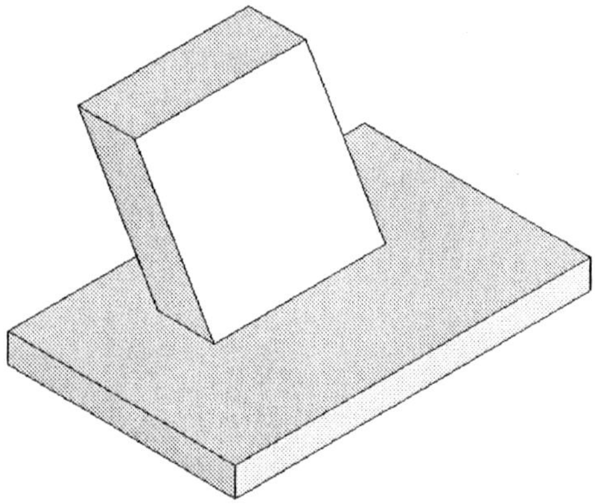

Figure 5-47 *Resultant extruded feature*

For a multifeatured model as shown above, ideally you first need to create the sketch for the base feature and convert that sketch into the base feature. Next, you need to create the sketch for the second sketched feature and so on. In other words, you have to create the sketch for each sketched feature. But using the contour selection method, you can share the contour created using the sketch for creating features. Figure 5-49 shows the sketch to be created for modeling using the contour selection method.

Figure 5-48 *Multifeatured solid model*

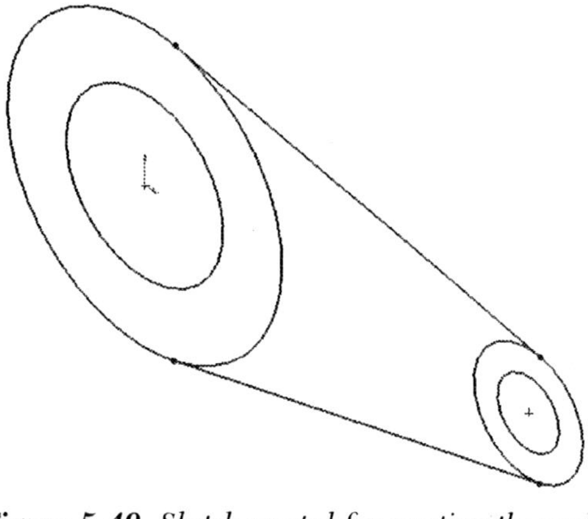

Figure 5-49 *Sketch created for creating the model*

After creating the entire sketch right-click in the drawing area to invoke the shortcut menu. Choose the **Contour Selection Tool** option from the shortcut menu. If this option is not displayed by default, choose the down arrows in the shortcut menu to expand it. This option will be displayed now. When you choose this option, the select cursor is replaced by the contour selection cursor and contour selection confirmation corner is displayed. Using the

Tip. *When you move the contour selection cursor in the sketch, the areas where the contour selection is possible are dynamically highlighted.*

contour selection cursor, specify a point between the two circles on the left. The area between the two circles is selected, as shown in Figure 5-50. Invoke the **Extrude PropertyManager** and extrude the selected contour using the **Mid Plane** option as shown in Figure 5-51.

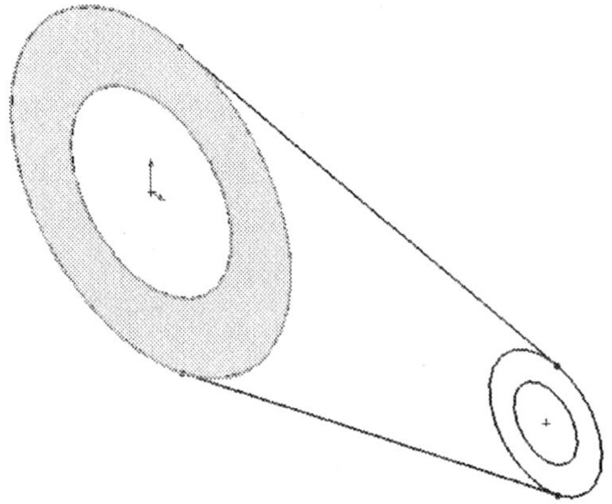

Figure 5-50 *Contour selected for creating the extruded feature*

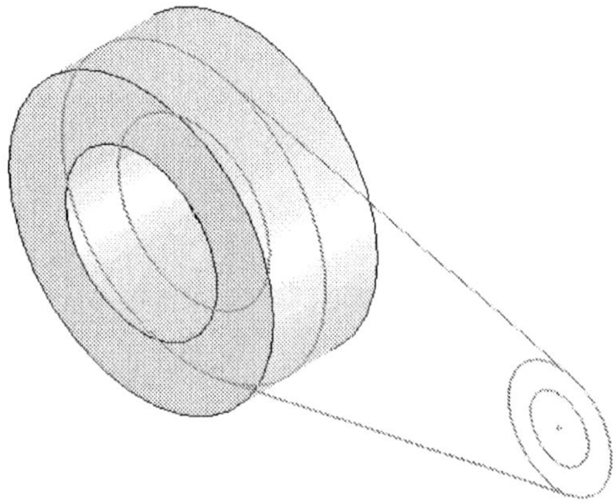

Figure 5-51 *Isometric view of the feature created by extruding the selected contour*

Now, right-click in the drawing area and again choose the **Contour Selection Tool** option from the shortcut menu. Using the contour selection cursor, select any entity in the sketch and then select the middle contour of the sketch as shown in Figure 5-52. Invoke the **Extrude PropertyManager** and extrude the selected contour using the **Mid Plane** option. Again, invoke the contour selection tool and select an entity in the sketch. Similarly, again invoke the contour selection tool and select any entity of the sketch. Next, specify a point between the two circles on the right and extrude the same using the **Mid Plane** option.

After creating the model using this option, you will notice that the sketches are displayed in the model. This is the reason you need to hide them. Click the **+** sign located on the left of any of the extruded features to expand the tree view. Select the sketch icon and right-click to invoke the shortcut menu. Choose the **Hide** option. Figure 5-53 shows the model after creating all the features using contour selection and after hiding the sketch.

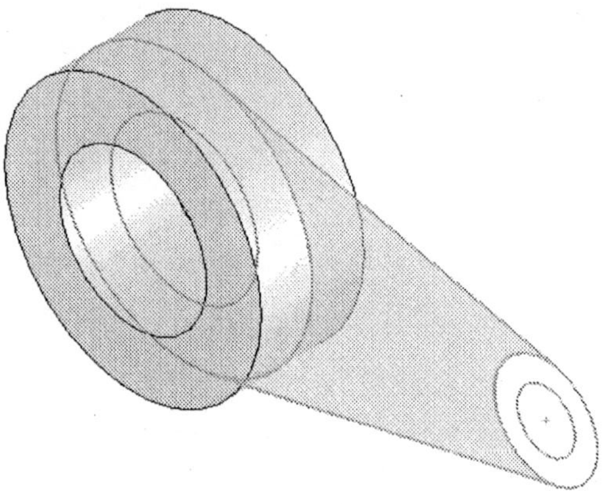

Figure 5-52 *Contour selected for second feature*

Figure 5-53 *Final model*

One of the latest enhancements in SolidWorks 2004 is its ability to allow you to select the model edges as a part of the contour. For example, consider Figure 5-54. This figure shows a line drawn on the top face of a rectangular block. You can use the edges of the top face of the rectangular block that form a contour with the line as a sketch to be extruded, see Figure 5-55.

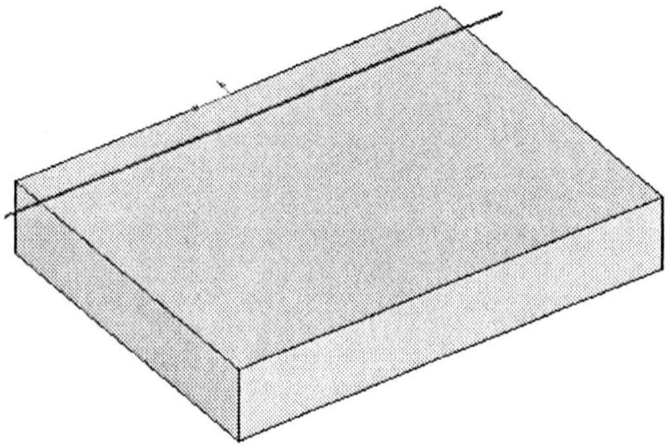

Figure 5-54 *Line drawn on the top face of a rectangular model*

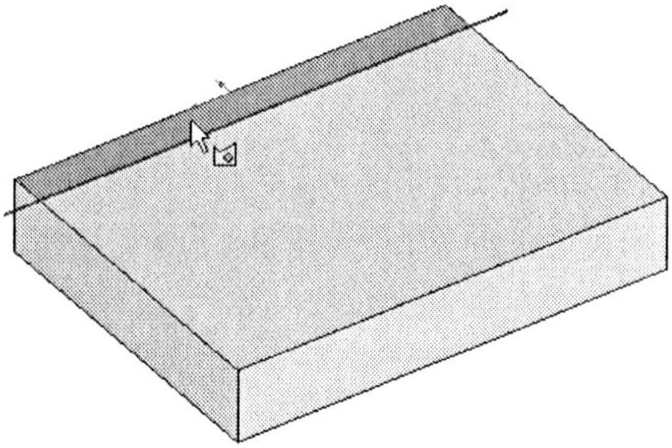

Figure 5-55 *Selecting the contour formed by the line and the model edges*

Figure 5-56 shows the resultant extruded feature.

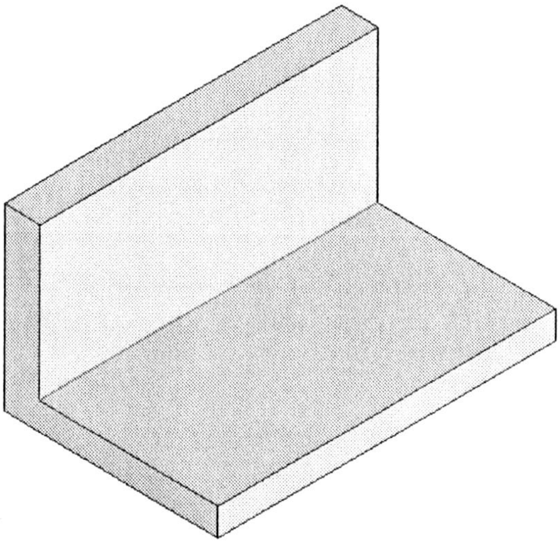

Figure 5-56 *Extruded feature created using the model edges as a part of the contour*

Note

*If you click the +sign to expand the extruded feature in the **FeatureManager Design Tree** you will notice that instead of showing the icon of a simple sketch, it will show the icon of contour selected sketch.*

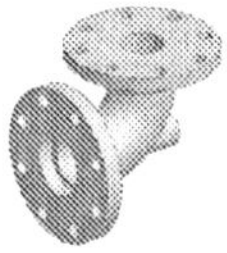

Tip. *When you select the contour using the contour selection tool and invoke the* ***Extrude PropertyManager****, you can observe the name of the selected contour in the selection area of the* ***Contour Selection*** *rollout.*

You can select the contours for all the sketched features such as revolve, cut, sweep, loft, and so on.

You can also select a single sketched entity from a sketch using the contour selection tool instead of selecting the contour for creating the sketched features.

CREATING CUT FEATURES

The cut is a material removal process. You can define a cut feature by extruding a sketch, revolving a sketch, sweeping a section along a path, lofting sections, or using a surface. You will learn more about sweep, loft, and surface in later chapters. The cut feature can be created only if a base feature exists. The extruded and revolved cut features are discussed next.

Creating Extruded Cuts

CommandManager:	Features > Extruded Cut
Menu:	Insert > Cut > Extrude
Toolbar:	Features > Extruded Cut

To create an extruded cut feature, create a sketch for the cut feature and then choose the **Extruded Cut** button from the **Features CommandManager**. The **Cut-Extrude PropertyManager** is displayed, as shown in Figure 5-57. Also, the preview of the cut feature with default options is displayed in the drawing area.

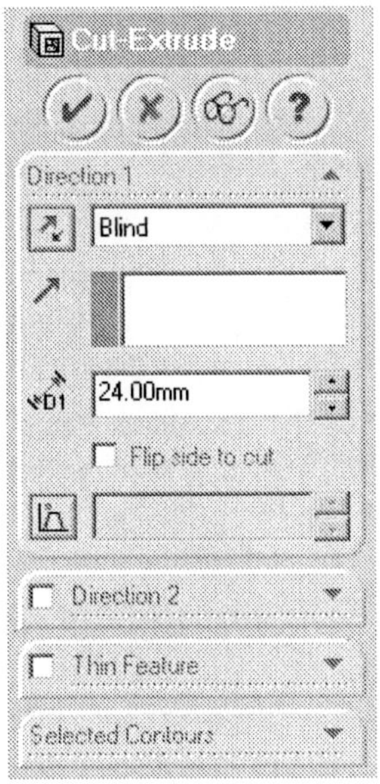

Figure 5-57 The ***Cut-Extrude PropertyManager***

Figure 5-58 shows the preview of the cut feature when you invoke the **Cut-Extrude PropertyManager** after creating a sketch. Remember that when you are creating a cut feature, the current view is not automatically changed to a 3D view. You need to change it manually. The material to be removed is displayed in temporary graphics. Figure 5-59 shows the model after creating the cut feature.

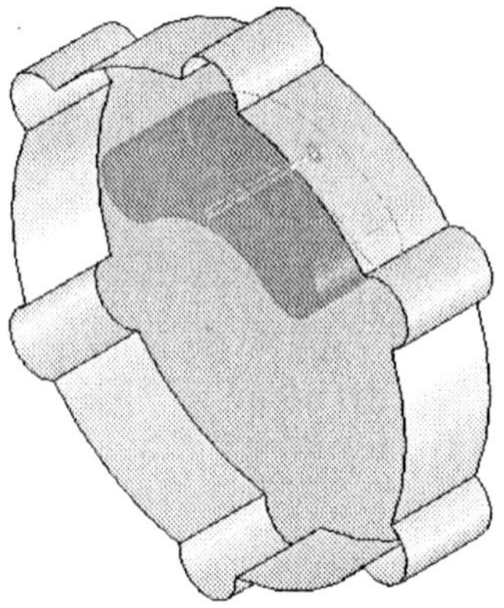

Figure 5-58 *The preview of the cut feature*

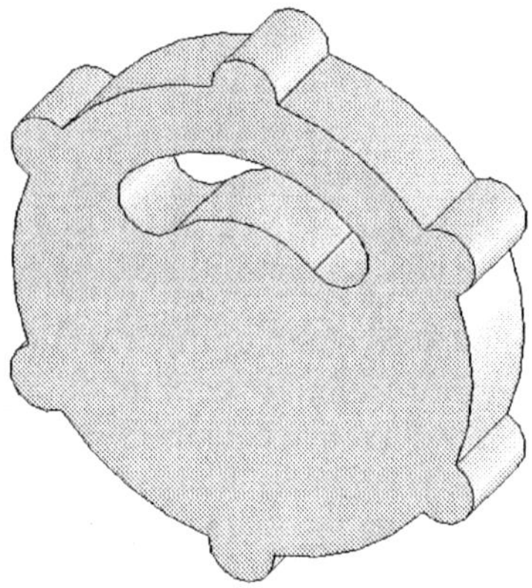

Figure 5-59 *Cut feature added to the model*

The various options available in the **Cut-Extrude PropertyManager** are discussed next.

Direction 1

The **Direction 1** rollout is used to define the termination of the extrude in the first direction. The various options available in the **Direction 1** rollout are discussed next.

End Condition

The **End Condition** drop-down list available in the **Direction 1** rollout is used to specify the type of termination option available. The feature termination options available in this drop-down list are **Blind, Through All, Up To Next, Up To Vertex, Up To Surface**, and **Mid Plane**. These option are the same as those discussed for boss/base options. By default, the **Blind** option is selected in the **End Condition** drop-down list. Therefore, the **Distance** spinner is displayed to specify the depth. If you choose the **Through All** or the **Through Next** options, the spinner will not be displayed. The type of spinner or the selection area that is displayed depends on the option selected from the **End Condition** drop-down list. The **Reverse Direction** button is used to reverse the direction of feature creation. If you choose the **Mid Plane** option from the **End Condition** drop-down list, the **Reverse Direction** button is not available.

Flip side to cut

The **Flip side to cut** check box is selected to define the side of material removal. By default, the **Flip side to cut** check box is cleared. Therefore, the material will be removed from inside the profile of the sketch drawn for the cut feature. If you select this check box, the material will be removed from outside the profile of the sketch. Figure 5-60 shows a cut feature with the **Flip side to cut** check box cleared and Figure 5-61 shows a cut feature with the **Flip side to cut** check box selected.

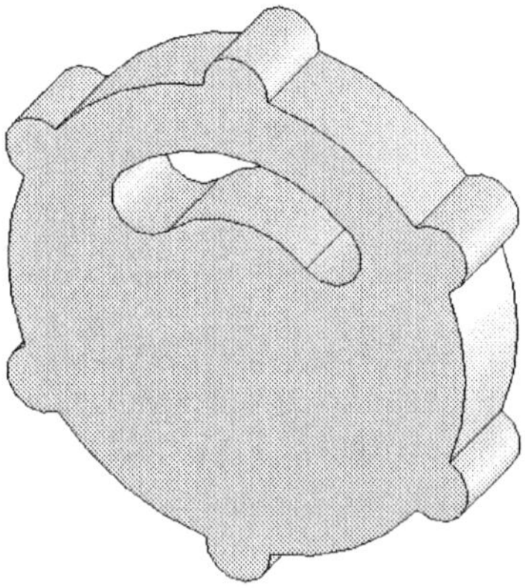

Figure 5-60 *Cut feature with* ***Flip side to cut*** *check box cleared*

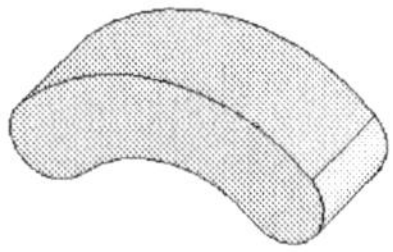

Figure 5-61 *Cut feature with* ***Flip side to cut*** *check box selected*

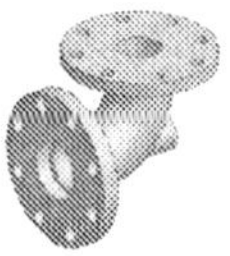

Tip. *You can also reverse the direction of material removal by clicking the arrow available on the sketch while creating the cut feature.*

Draft On/Off

The **Draft On/Off** button is used to apply the draft angle to the extruded cut feature. The **Draft Angle** spinner available on the right of the **Draft On/Off** button is used to set the value of the draft angle. By default, the **Draft outward** check box is cleared. Therefore, the draft is created inwards with respect to the direction of feature creation. If you select this check box, the draft added to the cut feature will be created outwards with respect to the direction of feature creation. Figure 5-62 shows the draft added to the cut feature with the **Draft outward** check box cleared and Figure 5-63 shows the draft added to the cut feature with the **Draft outward** check box selected.

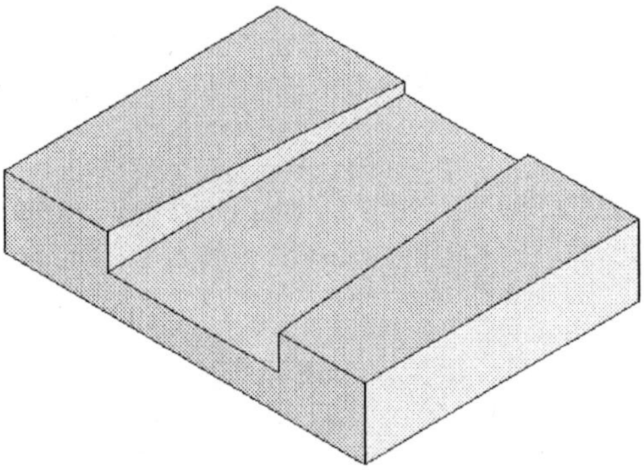

Figure 5-62 *Cut feature with the* ***Draft outward*** *check box cleared*

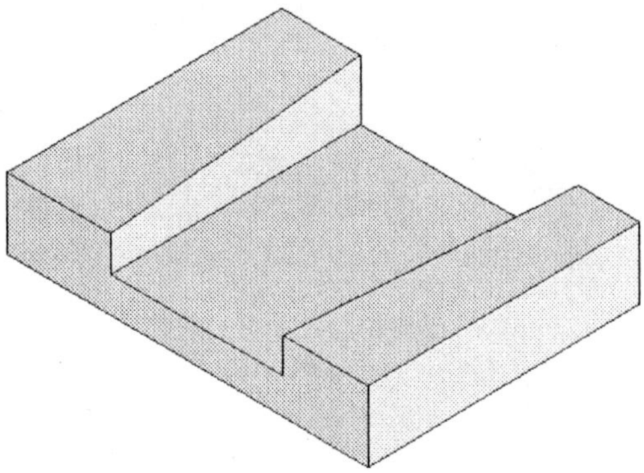

Figure 5-63 *Cut feature with the* ***Draft outward*** *check box selected*

The **Direction 2** rollout is used to specify the termination of feature creation in the second direction. The options available in the **Direction 2** rollout are the same as those discussed for the **Direction 1** rollout.

The **Selected Contour** rollout is used to add the feature to the selected contours.

Thin Feature

The **Thin Feature** rollout is used to create a thin cut feature. When you create a cut feature, you have to apply the thickness to the sketch in addition to the end condition. This rollout is used to specify the parameters to create the thin feature. To create a thin cut invoke the cut tool after creating the sketch and specify the end conditions in the **Direction 1** and **Direction 2** rollouts. Now, select the check box available in the **Thin Feature** rollout to activate the **Thin Feature** rollout. The options available in this rollout are the same as those discussed for the thin boss feature.

Tip. *The sketch used for the cut feature can be a closed loop or an open sketch. Note that if the sketch is an open sketch, the sketch should completely divide the model in two or more than two parts.*

Handling Multiple Bodies in Cut Feature

While creating the cut feature, sometimes because of geometric conditions, feature termination, or end conditions, the cut feature results in the creation of multiple bodies. Figure 5-64 shows a sketch created on the top planar surface of the base feature to create a cut feature. Figure 5-65 shows the cut feature created with the end condition as **Through All**. Using this type of sketch and end condition, if you choose **OK** from the **Extrude-Cut PropertyManager** the **Bodies to Keep** dialog box is displayed. As multiple bodies are created while applying the cut feature, this dialog box is used to define which body do you want to keep.

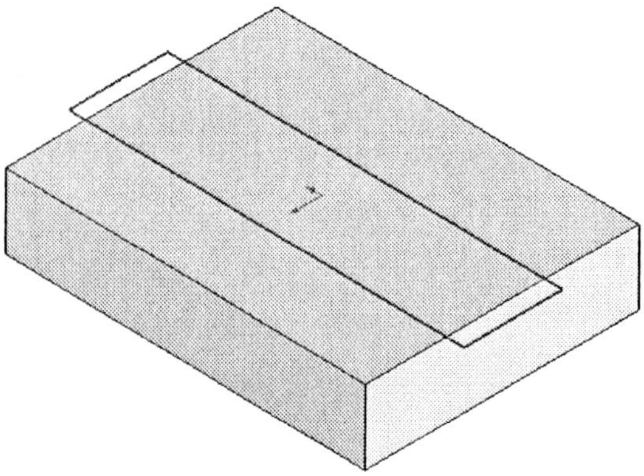

Figure 5-64 Sketch created for the cut feature.

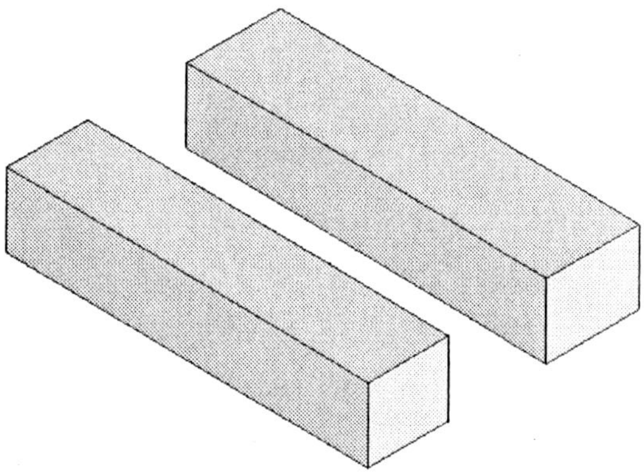

Figure 5-65 Multiple bodies created using the cut feature

The **Bodies to Keep** dialog box is displayed in Figure 5-66. By default, the **All Bodies** radio button is selected. Therefore, if you choose **OK** from this dialog box, all the bodies created

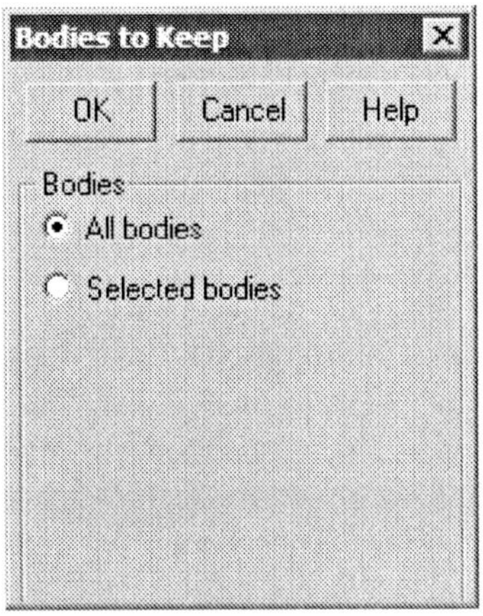

Figure 5-66 *The* ***Bodies to Keep*** *dialog box*

after the cut feature will remain in the model. If you want the cut feature to consume any of the body, select the **Selected bodies** radio button. When you select this radio button, the dialog box expands, as shown in Figure 5-67.

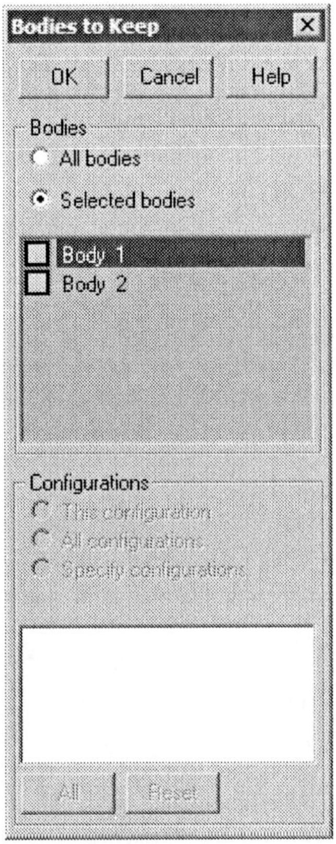

Figure 5-67 *The* ***Bodies to Keep*** *dialog box with the* ***Selected bodies*** *option selected*

You can select the check box provided on the left of the name of the body to keep that body. The selected body is displayed in green in temporary graphics. Select the bodies to keep and choose the **OK** button from the **Bodies to Keep** dialog box. Figure 5-68 shows a sketch created for the cut feature. Figure 5-69 shows the cut feature created using the thin option and the **All bodies** option selected from the **Bodies to Keep** dialog box.

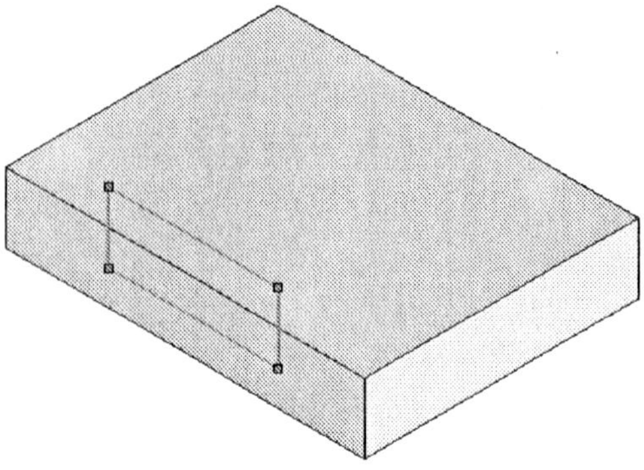

Figure 5-68 *Sketch to create a cut feature using the thin option*

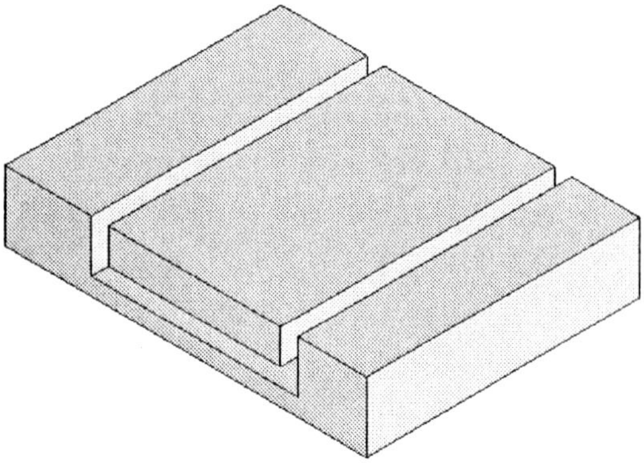

Figure 5-69 *A thin cut feature created with all the resultant bodies retained*

Note

You will learn more about configurations in the advanced chapters.

Creating Revolved Cuts

CommandManager:	Features > Revolved Cut
Menu:	Insert > Cut > Revolve
Toolbar:	Features > Revolved Cut

Revolved cuts are used to remove the material by revolving a sketch around a selected axis. Similar to revolved boss/base features, you can define the revolution axis using a centerline or using an edge in the sketch. When you invoke the **Revolved Cut** tool, the **Cut-Revolve PropertyManager** is displayed, as shown in Figure 5-70. The various options available in this **PropertyManager** are similar to those discussed earlier. Figure 5-71 shows a sketch for a revolved cut feature and Figure 5-72 shows the resultant cut feature. Note that in Figure 5-72, a texture is applied to the cut feature.

Note

*You can also select the tool to create the feature first and then select the plane to create the sketch. This way, after you exit the sketching environment, the feature tool is activated automatically and you can define the parameters in its respective rollout. For example, you can choose the **Revolved Cut** tool without creating any sketch. In this case, the **Revolve PropertyManager** is displayed and you are prompted to select a plane or a planar face to create a sketch or to select a sketch. Select a plane or a planar face and create the sketch. As soon as you exit the sketching environment after creating the sketch, the **Cut-Revolve PropertyManager** is automatically displayed.*

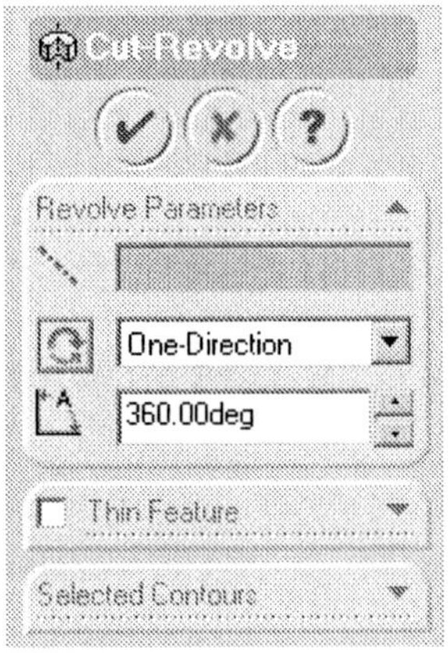

Figure 5-70 Cut-Revolve PropertyManager

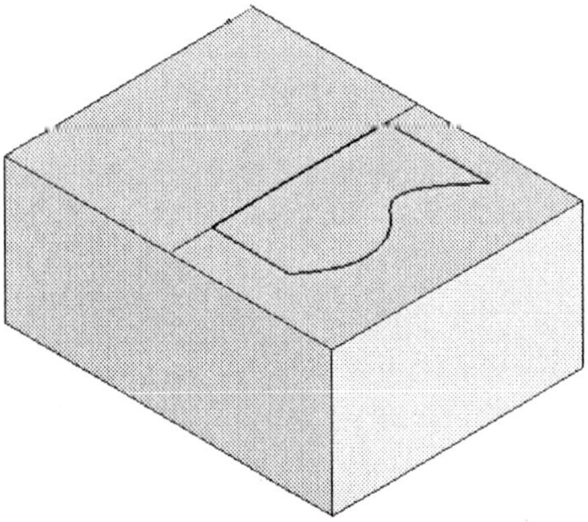

Figure 5-71 Sketch for the revolved cut feature

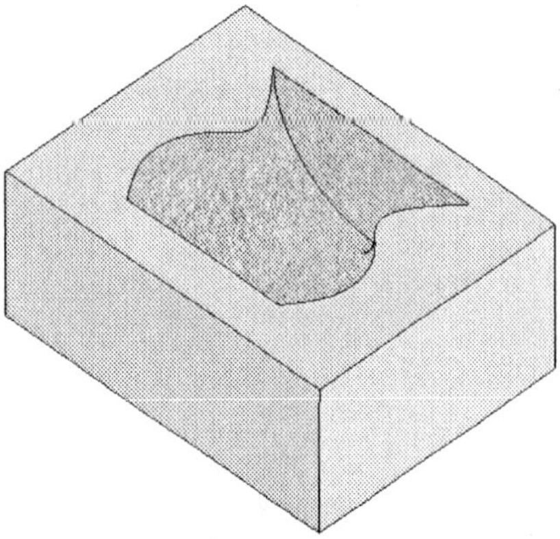

Figure 5-72 Resultant cut feature with a texture

CONCEPT OF THE FEATURE SCOPE

As discussed earlier, you can create different disjoint bodies in a single part file in SolidWorks. After creating two or more than two disjoint bodies, when you create another feature, the **Feature Scope** rollout is displayed in the **PropertyManager**. This rollout is used to define the bodies that will be affected by the creation of the feature. The feature scope option is used with the Extrude boss and cut, Revolve boss and cut, Sweep boss and cut, Loft boss and cut, Boss cut and thicken, Surface cut, and Cavity features.

In the **Feature Scope** rollout, the **Selected bodies** radio button and the **Auto-select** check box is selected by default. With the **Auto-select** check box selected, all the disjoint bodies are selected and are affected by the feature creation. If you clear the **Auto-select** check box, a selection area is invoked. You can select the bodies that you want to be affected. The name of the selected body is displayed in the selection area. If you select the **All bodies** radio button then all the bodies available in the part file are selected and will be affected by the creation of the feature.

TUTORIALS

Tutorial 1

In this tutorial you will create the model shown in Figure 5-73. You will create the model by drawing the sketch of the front view of the model and then select the contours to extrude them. As a result, in this tutorial you will learn the procedure of modeling using the contours selection method. The dimensions of the model are shown in Figure 5-74.

(Expected time: 30 min)

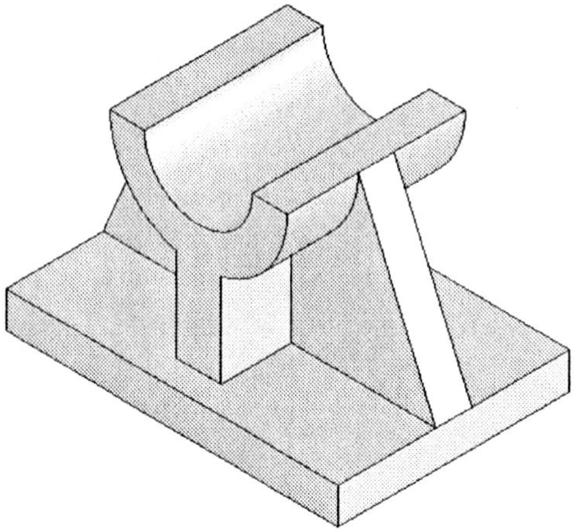

Figure 5-73 *Solid model for Tutorial 1*

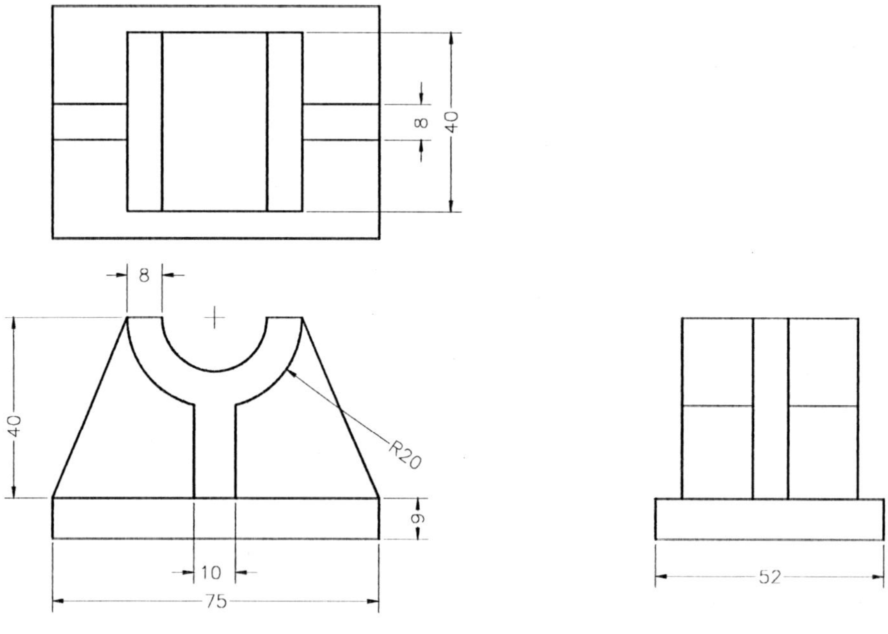

Figure 5-74 *Dimensions and views for Tutorial 1*

It is clear from the figures that the given model is a multifeatured model. It consists of various extrude features. You first need to draw the sketch for each feature and then convert that sketch into a feature. In conventional methods, you have to create a separate sketch for each sketched feature. But in this tutorial, you will use the contour selection method. Using this method, you have to create only one sketch and you will select the contours and share the same sketch for creating all the features.

The steps to be followed to complete this tutorial are listed next.

a. Create the sketch on the default plane and apply the required relations and dimensions, refer to Figure 5-75.
b. Invoke the extrude option and extrude the selected contour, refer to Figures 5-76 and 5-77.
c. Select the second set of contour and extrude it to the required distance, refer to Figures 5-78 and 5-79.
d. Select the third set of contour and extrude it to the required distance, refer to Figures 5-80 and 5-81.
e. Save the document and then close the document.

Creating the Sketch of the Model

1. Start SolidWorks and then start a new part document using the **New SolidWorks Document** dialog box.

2. Create the sketch of the front view of the model on the **Front** plane. Add the required relations and dimensions to the sketch as shown in Figure 5-75. Make sure that you do not exit the sketching environment.

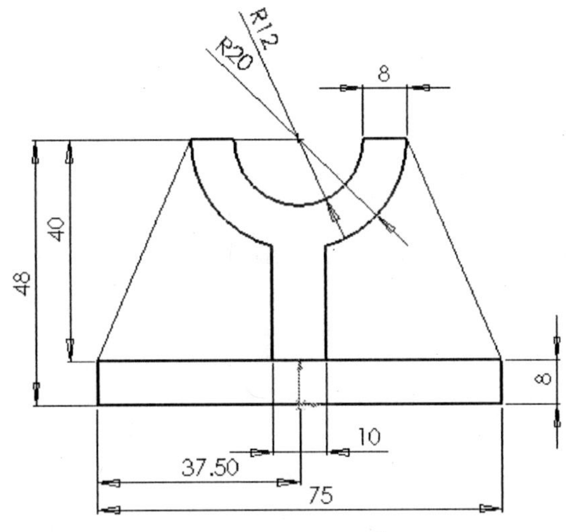

***Figure 5-75** Fully defined sketch for creating the model*

Selecting and Extruding the Contours of the Sketch

In this tutorial, you will use the contour selection method to create the model. Therefore, you first need to select one of the contours from the given sketch and extrude it. For a better representation of the sketch, you will also orient the sketch to isometric view.

1. Choose the **Isometric** option from the flyout that is displayed when you choose the **Standard Views** button from the **View** toolbar.

2. Right-click in the drawing area to invoke the shortcut menu and choose the **Contour Select Tool** option. The select cursor will be replaced by the contour selection cursor and selection confirmation corner is displayed.

3. Move the cursor to the lower rectangle of the sketch. The area of the rectangle will be highlighted. This indicates that this rectangle is a closed profile.

4. Specify a point when the rectangular area is highlighted. Figure 5-76 shows the lower rectangle selected using the contour selection tool.

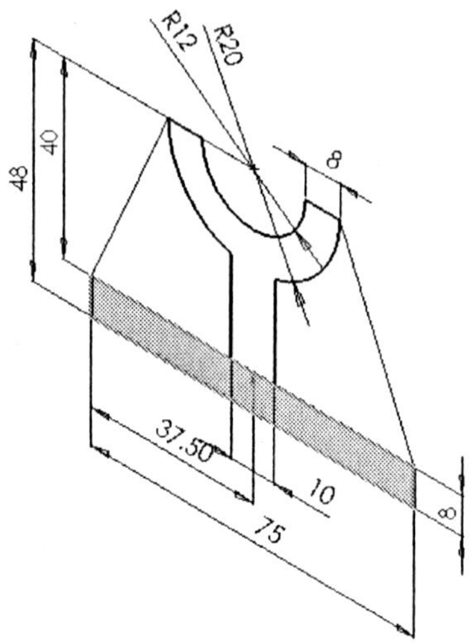

Figure 5-76 *Lower rectangle selected as a contour*

5. Choose the **Extruded Boss/Base** button from the **Features CommandManager**. The **Extrude PropertyManager** is invoked and the preview of the base feature is displayed in the drawing area in temporary graphics.

   The name of the selected contour is displayed in the selection area of the **Selected Contours** rollout.

6. Right-click in the drawing area and choose the **Mid Plane** option from the shortcut menu. The preview of the feature is modified dynamically when you choose the **Mid Plane** option.

7. Set the value of the **Depth** spinner to **52** and choose the **OK** button from the **Extrude PropertyManager**. The base feature of the model by extruding the selected contour is shown in Figure 5-77.

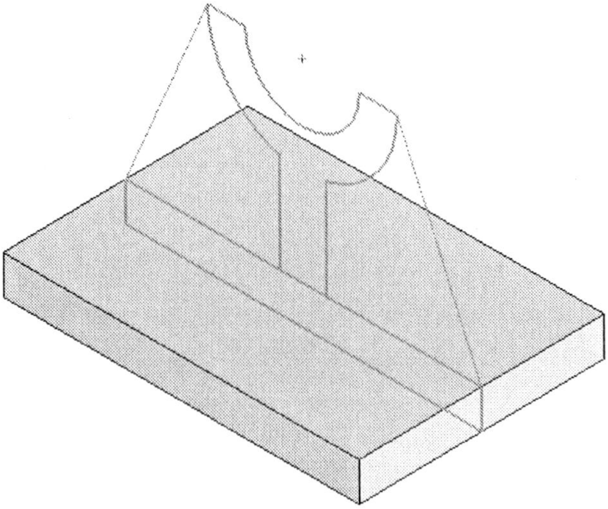

Figure 5-77 *Base feature of the model*

8. Right-click in the drawing area and choose the **Contour Select Tool** option from the shortcut menu. The select cursor is replaced by the contour selection cursor.

9. Using the contour selection cursor, select any entity of the sketch to invoke the selection mode of the sketch.

10. Using the left mouse button select the middle contour of the sketch, as shown in Figure 5-78. The selected region will be highlighted.

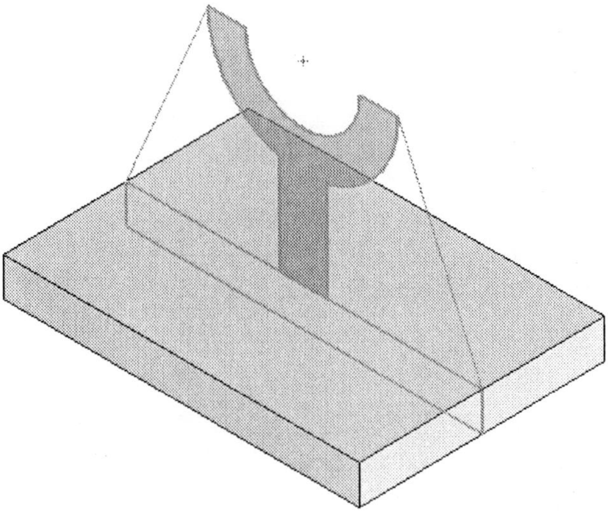

Figure 5-78 *Middle contour is selected using the contour selection tool*

11. Invoke the **Extrude PropertyManager**. Right-click in the drawing area and choose the **Mid Plane** option from the shortcut menu.

12. Set the value of the **Depth** spinner to **40** and choose the **OK** button from the **Extrude PropertyManager**. The feature created by selecting the middle contour is shown in Figure 5-79.

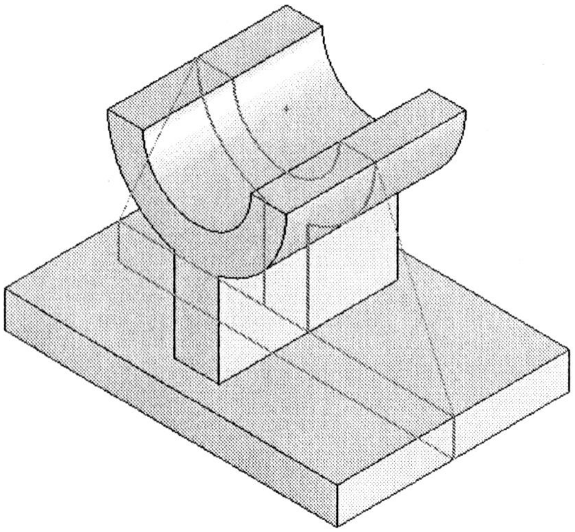

Figure 5-79 *Second feature created by extruding the middle contour*

13. Again invoke the **Contour Select Tool** and select a sketch entity. Next, select the right contour of the sketch. Press and hold down the CTRL key and select the left contour of the sketch, see Figure 5-80.

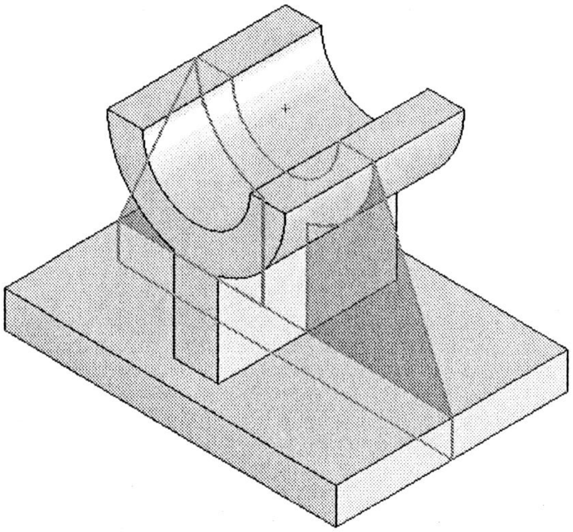

Figure 5-80 *The right and the left contours selected*

14. Invoke the **Extrude PropertyManager**. Right-click and choose the **Mid Plane** option from the shortcut menu.

15. Set the value of the **Depth** spinner to **8** and choose the **OK** button from the **Extrude PropertyManager**.

The model is completed, but the sketch is displayed in the model. Therefore, you need to hide the sketch.

16. Move the cursor to any of the sketched entity and when the entity turns red in color, select the entity. The selected sketched entity will be displayed in green. Now, right-click and choose the **Hide** option from the shortcut menu.

The isometric view of the final model with hidden sketch is shown in Figure 5-81. The **FeatureManager Design Tree** of the model is shown in Figure 5-82.

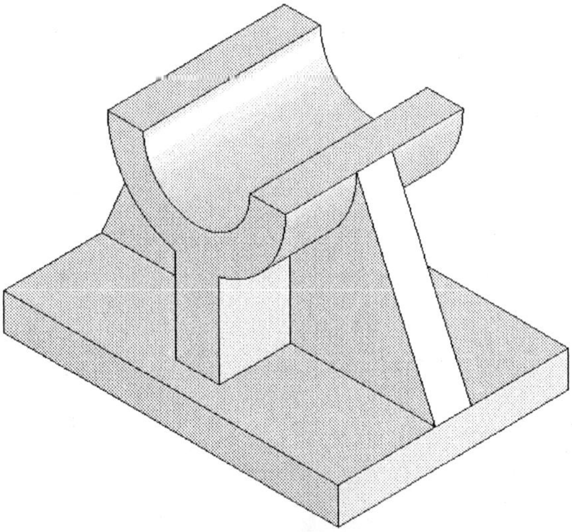

Figure 5-81 *Final solid model*

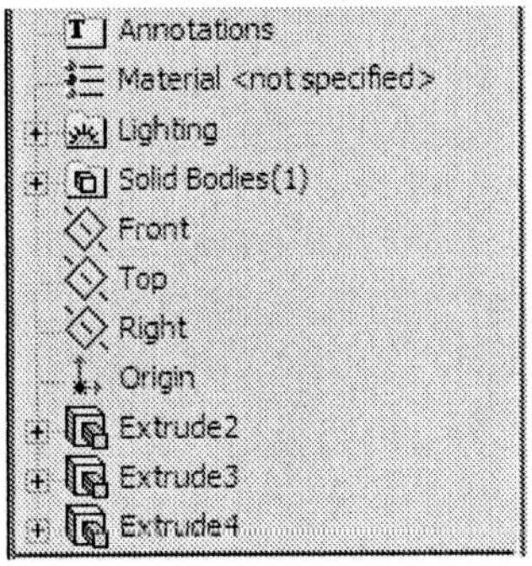

Figure 5-82 *The **FeatureManager Design Tree***

Saving the Model

1. Choose the **Save** button from the **Standard** toolbar and save the model with the name given below.

   \My Documents\SolidWorks\c05\c05-tut01.sldprt

2. Choose **File > Close** from the menu bar to close the document.

Tutorial 2

In this tutorial you will create a model for which the dimensions are shown in Figure 5-83. You will use a combination of the conventional modeling method and the contour selection method to create this model. The solid model is shown in Figure 5-84.

(Expected time: 30 min)

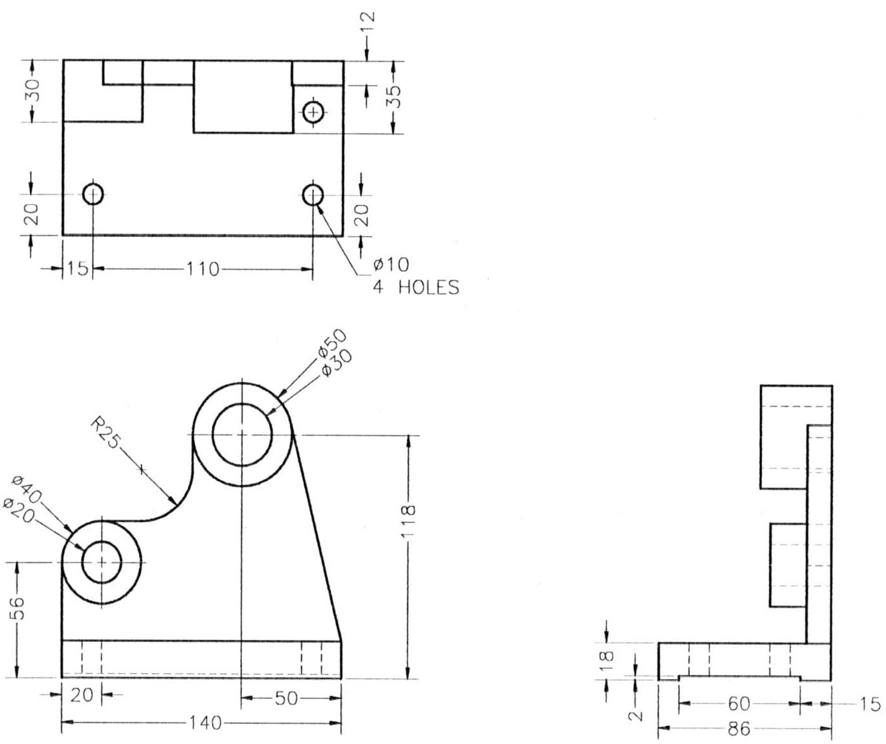

Figure 5-83 *Drawing views of the solid model*

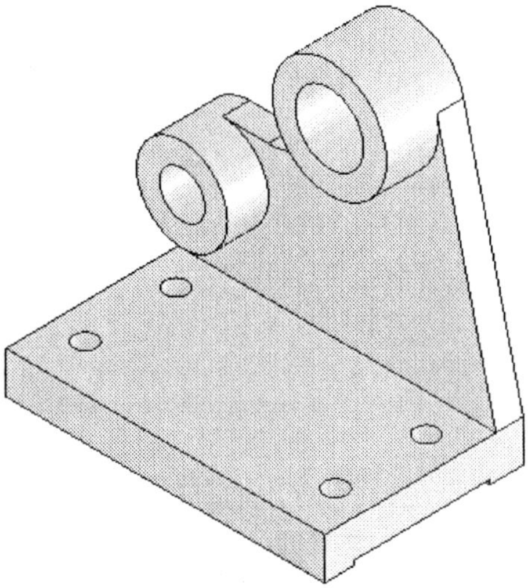

Figure 5-84 *Solid model for Tutorial 2*

The steps that will be used to complete the model are listed next.

a. Create the sketch of the front view of the model, refer to Figure 5-85.
b. Extrude the selected contours, refer to Figures 5-86 through 5-88.
c. Add recess feature to the model by creating the sketch on the right planar face, refer to Figures 5-89 and 5-90.
d. Create four holes using the cut feature on the top face of the base feature, refer to Figures 5-91 and 5-92.
e. Save and close the document.

Creating the Sketch for Contour Selection Modeling

1. Start a new SolidWorks part document. Draw the sketch of the front view of the model on the **Front** plane using the standard sketching tools.

2. Apply the required relations and dimensions to fully define the sketch, see Figure 5-85. Orient the view to isometric view because it will help you in the selection of contours.

3. Change the current view to isometric view.

4. Right-click and choose the **Contour Select Tool** option from the shortcut menu. The select cursor is replaced by the contour selection cursor.

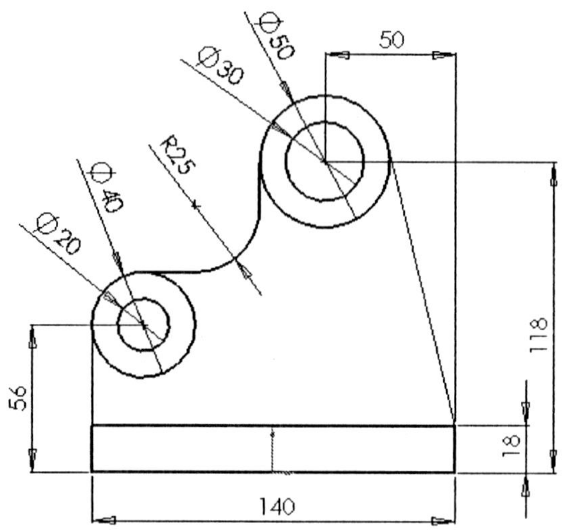

Figure 5-85 *Fully defined sketch*

5. Using the contour selection cursor select the area enclosed by the lower rectangle as shown in Figure 5-86.

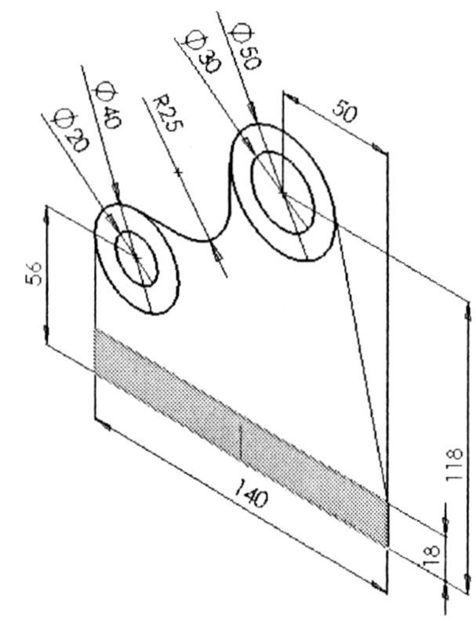

Figure 5-86 *Lower rectangle selected as contour*

6. Choose the **Extruded Boss/Base** button from the **Features CommandManager** to invoke the **Extrude PropertyManager**.

7. Set the value of the **Depth** spinner to **86** and choose the **OK** button from the **Extrude PropertyManager**. The base feature created after extruding the selected contour is shown in Figure 5-87.

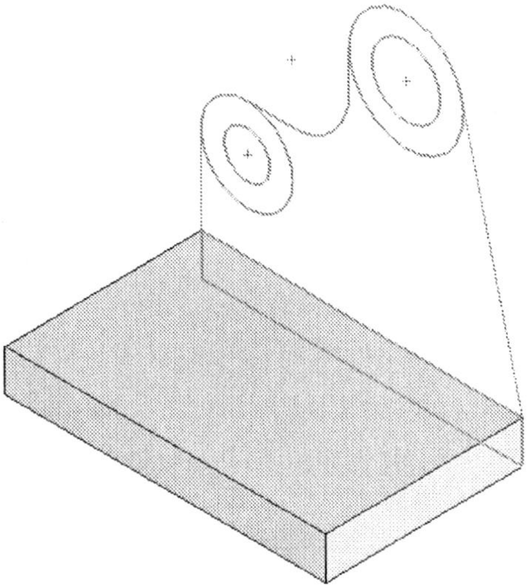

Figure 5-87 *Base feature created after extruding the selected contour*

8. Using the contour selection tool and the extrude tool create the other features and then hide the sketch. The model created after extruding all the contours and hiding the sketch is shown in Figure 5-88.

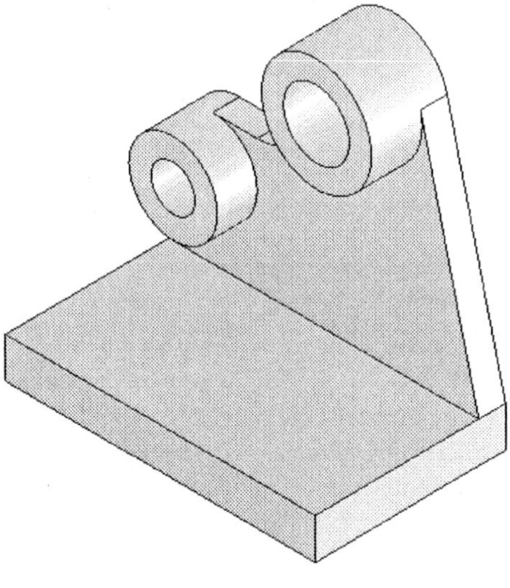

Figure 5-88 *Model created after extruding all the contours*

Creating the Recess at the Base of the Model

After creating the extruded features of the model, you need to create the recess provided at the base of the model. The recess will be created by a cut extrude feature created using a sketch created at the right planar face of the model.

1. Select the right planar face of the base feature as the sketching plane. The selected face will be displayed in green color.

2. Right-click in the drawing area and choose the **Insert Sketch** option from the shortcut menu to invoke the sketching environment.

   Now you need to orient the view such that the selected face is normal to your eye view.

3. Choose the **Normal To** option from the flyout that is displayed when you choose the **Standard Views** button to orient the selected plane normal to the view.

4. Using the standard sketching tools, create the sketch for the recess and apply the required relations and dimensions to the sketch. The fully defined sketch for the cut feature is shown in Figure 5-89.

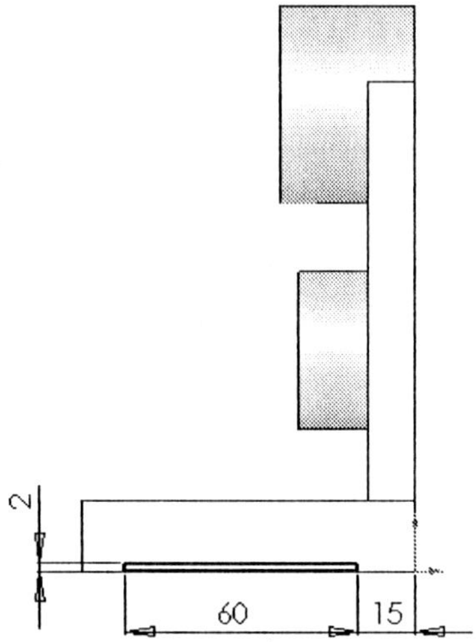

Figure 5-89 *Sketch for the cut feature*

5. Choose the **Extruded Cut** button from the **Features CommandManager** to invoke the **Cut-Extrude PropertyManager**. The preview of the cut feature will be displayed in the drawing area in temporary graphics.

6. Right-click in the drawing area and choose **Through All** from the shortcut menu.

7. Choose **OK** from the **Cut-Extrude PropertyManager** to complete feature creation. The isometric view of the model after creating the cut feature is shown in Figure 5-90.

Creating the Holes

Next, you need to create the holes on the base of the model. These holes will be created as the extruded cut features. You will create the sketch of the hole feature on the top planar face of the base feature of the model. To create the sketch of the holes, you will

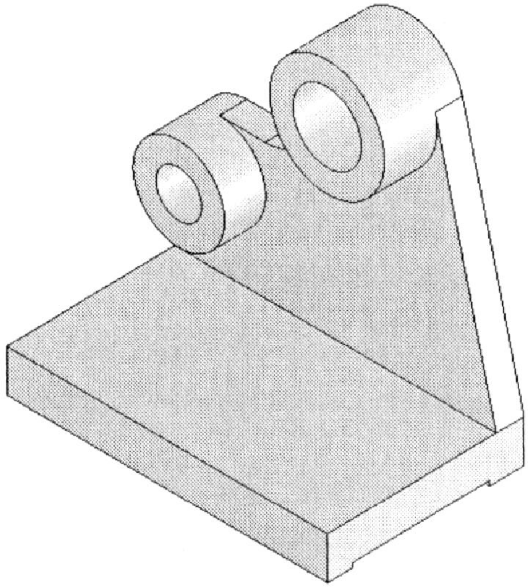

Figure 5-90 Cut feature added to the model

first draw a circle and then using the **Linear Sketch Step and Repeat** option create the remaining circles.

1. Select the top planar face of the base feature as the sketching plane. Right-click and choose the **Insert Sketch** option from the shortcut menu.
2. Orient the current view normal to the viewing direction. Using the standard sketching tools, create the sketch for the holes and apply the required relations and dimensions to fully define the sketch. You may need to apply the horizontal relations between the center points of the top circles to fully define the sketch. The fully defined sketch is shown in Figure 5-91.

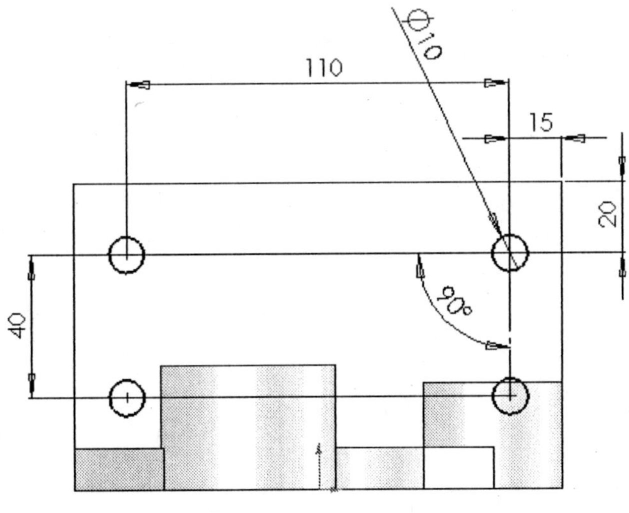

Figure 5-91 Holes sketched for the cut feature

3. Change the current view to isometric view and then choose the **Extruded Cut** button from the **Features CommandManager** to invoke the **Cut-Extrude PropertyManager**.

4. Right-click in the drawing area and choose the **Through All** option from the shortcut menu. Choose the **OK** button from the **Cut-Extrude PropertyManager**. The isometric view of the final model after hiding the sketch is shown in Figure 5-92. The **FeatureManager Design Tree** of the model is shown in Figure 5-93.

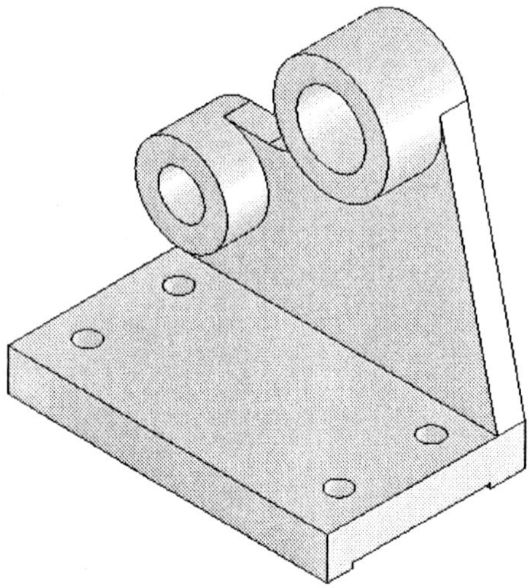

Figure 5-92 *Final solid model*

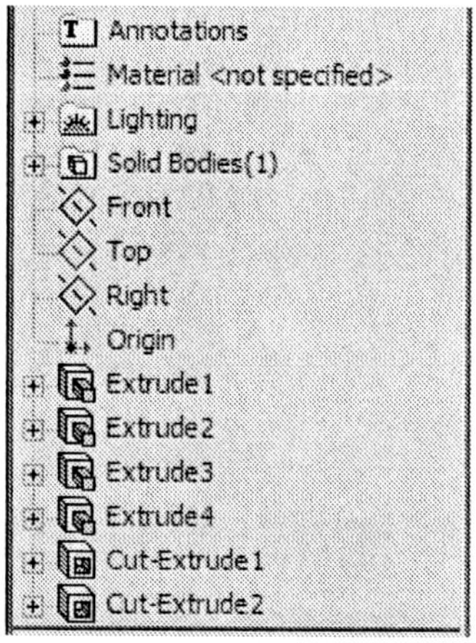

Figure 5-93 *The* ***FeatureManager Design Tree***

Saving the Model

1. Choose the **Save** button from the **Standard** toolbar and save the model with the name given below.

   \My Documents\SolidWorks\c05\c05-tut02.sldprt

2. Choose **File > Close** from the menu bar to close the document.

Tutorial 3

In this tutorial, you will create the model whose dimensions are shown in Figure 5-94. The solid model is shown in Figure 5-95. **(Expected Time: 30 min)**

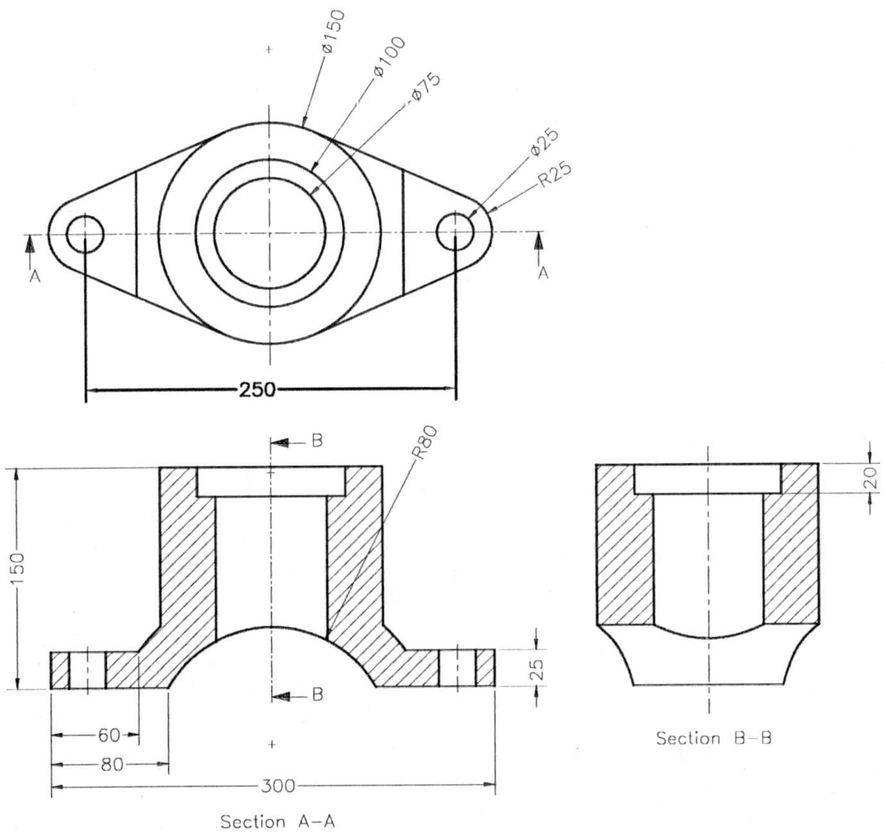

Figure 5-94 *Top view, sectioned front view, and sectioned right view of the model*

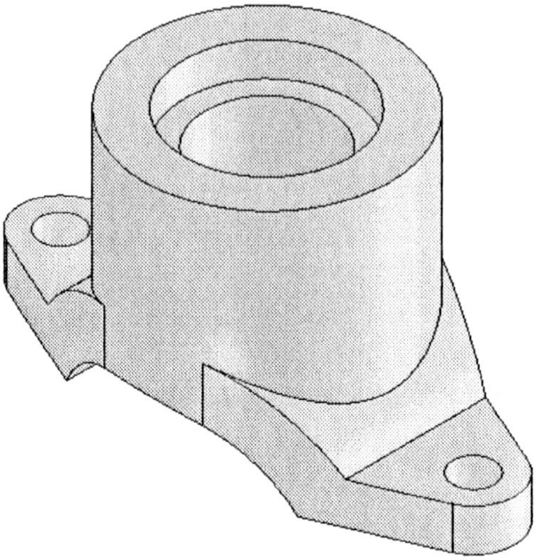

Figure 5-95 *Solid model for Tutorial 3*

The steps that will be followed to complete this tutorial are listed next.

a. Create the base feature by extruding the sketch created on the front plane, refer to Figures 5-96 and 5-97.
b. Extrude the sketch created on the top plane to create a cut feature, refer to Figures 5-98 and 5-100.
c. Create a plane at an offset distance of 150 from the top plane.
d. Create a sketch on the newly created plane and extrude it to the selected surface, refer to Figures 5-101 and 5-102.
e. Create a contour bore using the cut revolve option, refer to Figures 5-103 and 5-104.
f. Create the holes using the cut feature, refer to Figures 5-105 and 5-106.
g. Save the document and close it.

Creating the Base Feature

It is evident from the model that the base of the model comprises of a complex geometry. You need to create the base feature and then apply the cut feature to the base of the model to get the desired shape. You will create the base feature on the **Front** plane as the sketching plane. After creating the sketch, you will extrude the sketch using the mid plane option to complete feature creation.

1. Start a new SolidWorks part document and invoke the sketching environment to draw the sketch for the base feature.

2. Using the standard sketching tools, create the sketch of the base feature and then apply the required relations and dimensions to the sketch as shown in Figure 5-96.

3. Invoke the **Extrude PropertyManager**. Right-click in the drawing area and choose the **Mid Plane** option from the shortcut menu.

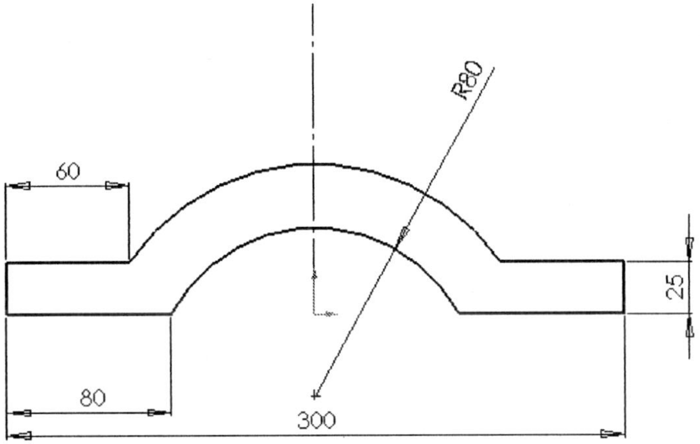

Figure 5-96 Sketch of the base feature

4. Set the value of the **Depth** spinner to **150** and choose the **OK** button from the **Extrude PropertyManager**. The isometric view of the base feature of the model is shown in Figure 5-97.

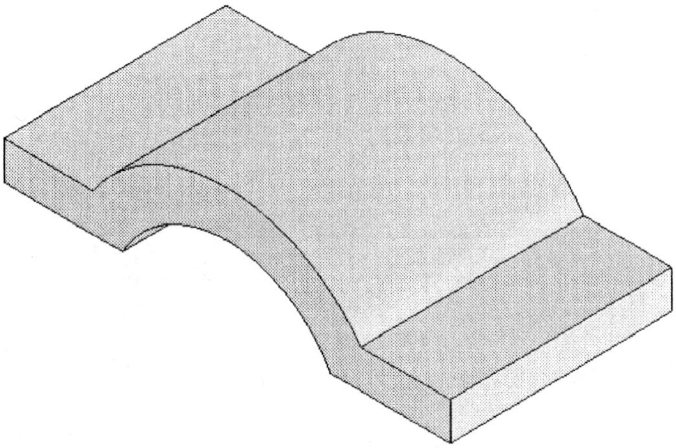

Figure 5-97 Base feature of the solid model

Creating the Cut Feature

Now, you need to create a cut feature to get the required shape of the base feature. The sketch for this cut feature will be created using a reference plane defined tangent to the curved face of the previous feature.

1. Choose the **Plane** button from the **Reference Geometry CommandManager**. The **Plane PropertyManager** is displayed.

2. Select the upper curved face of the existing feature. The **On Surface** button is automatically chosen. Now, move the cursor close to the midpoint of the curved edge of the upper curved face. The point is highlighted in red, see Figure 5-98. Select this point. The preview of a plane tangent to the curved face and passing through the midpoint of the curved edge is displayed. Choose **OK** to create the reference plane.

3. Create the sketch for the cut feature using the standard sketching tools and then apply the required relations and dimensions to the sketch as shown in Figure 5-99.

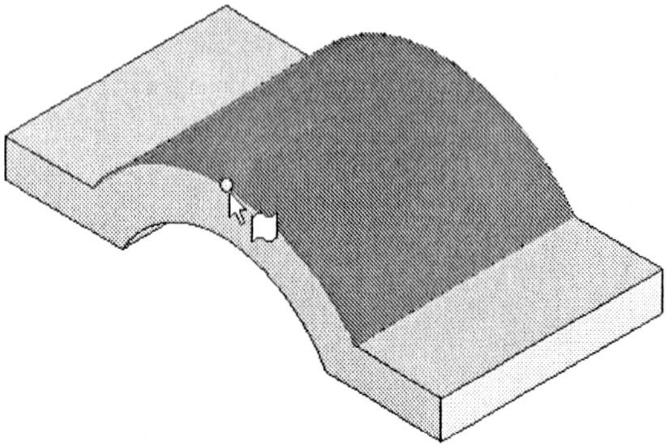

Figure 5-98 *Selecting the point to define the tangent plane*

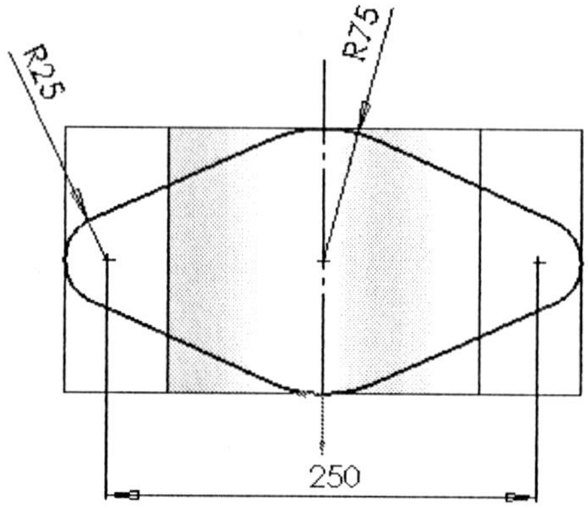

Figure 5-99 *Fully dimensioned sketch for the cut feature*

4. Choose the **Extruded Cut** button from the **Features CommandManager** to invoke the **Cut-Extrude PropertyManager**. Change the current view to isometric view.

   You will also observe that direction of material removal is not in the required direction. Therefore, you need to flip the direction of material removal.

5. Select the **Flip side to cut** check box. You will observe that the direction of material removal is also changed in the preview.

6. Right-click in the drawing area and choose the **Through All** option from the shortcut menu and choose the **OK** button from the **Cut-Extrude PropertyManager**.

   Because the reference plane is displayed in the drawing area, you need to hide it.

7. Right-click on **Plane1** in the drawing area and choose **Hide** from the shortcut menu. The display of the reference plane is turned off. The model after creating the base feature is shown in Figure 5-100.

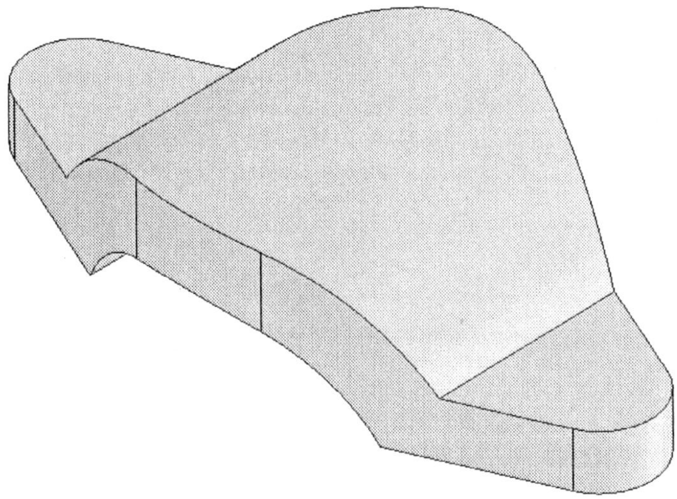

Figure 5-100 Cut feature added to the base feature

Creating a Plane at an Offset Distance for the Extruded Feature

After creating the base of the model you will create a plane at an offset of 150 mm from the **Top** plane. This newly created plane will be used as a sketching plane for the next feature.

1. Choose the **Plane** button from the **Reference Geometry CommandManager** to invoke the **Plane PropertyManager**.

2. Choose the **Offset Distance** button from the **Plane PropertyManager**. The **Distance** spinner, the **Reverse direction** check box, and the **Number of Planes to Create** spinner is displayed in the **Plane PropertyManager**.

3. Click on the + sign located on the left of the **Feature Manager Design Tree** that is now displayed in the drawing area. The tree view expands and the three default planes are now visible in the tree view.

4. Select the **Top** plane from the **FeatureManager Design Tree**. The preview of the newly created plane at a default offset is displayed in the drawing area.

5. Set the value of the **Distance** spinner to **150** and choose the **OK** button from the **Plane PropertyManager**. The required plane is created.

Creating the Extruded Feature

After creating the plane at an offset distance from the **Top** plane, you will create the sketch for the next feature.

1. Select the reference plane you just created if it is not selected and invoke the sketching environment.

2. Set the current view normal to the eye view.

3. Draw the sketch of the circle and apply the required relations to the sketch as shown in Figure 5-101.

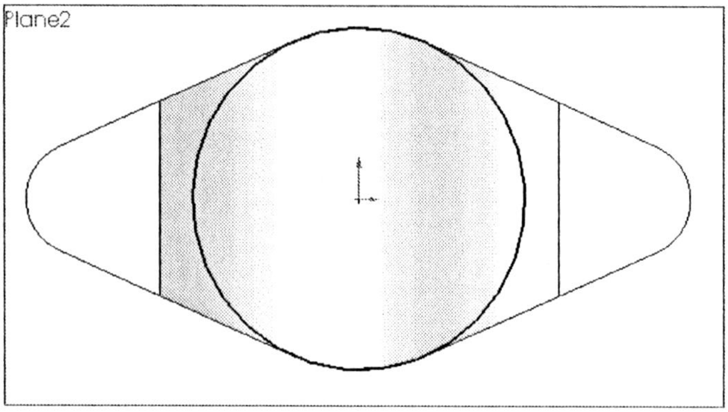

Figure 5-101 *Sketch created on the newly created plane*

4. Change the current view to the isometric view and invoke the **Extrude PropertyManager**. You will observe from the preview that the direction of feature creation is opposite to the required direction. Therefore, you need to change the direction of feature creation.

5. Choose the **Reverse Direction** button provided on the left of the **End Condition** drop-down list to reverse the direction of feature creation. The preview of the feature is changed when you reverse the direction of feature creation.

6. Right-click in the drawing area and choose the **Up To Surface** option from the shortcut menu. You are prompted to select a face or a surface to complete the specification of Direction 1 and the **Face/Plane** selection area is displayed under the **End Condition** drop-down list in the **Direction 1** rollout.

7. Select the upper curved surface of the model using the left mouse button. You will observe that the preview shows the feature extruded up to the selected surface.

8. Choose the **OK** button from the **Extrude PropertyManager**.

   Because the plane is displayed in the drawing area, you need to turn off its display.

9. Select **Plane2** from the **FeatureManager Design Tree** or from the drawing area and invoke the shortcut menu. Choose the **Hide** option from the shortcut menu. The model created after creating the extruded feature is shown in Figure 5-102.

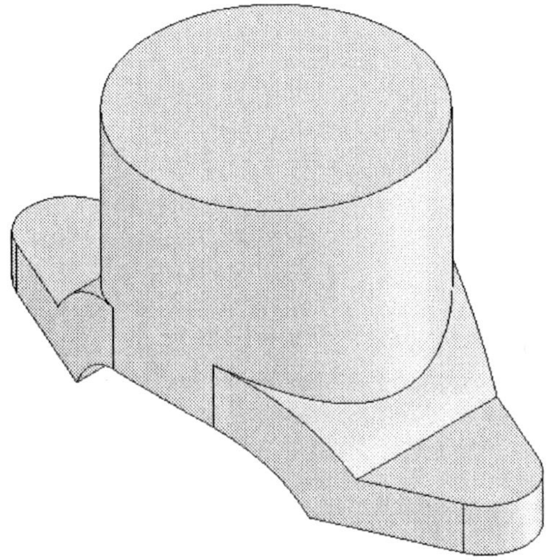

Figure 5-102 Sketch extruded up to a selected surface

Creating the Counterbore Hole

Next, you need to create the counterbore hole. It will be created as a revolved cut feature using a sketch created on the front plane.

1. Invoke the sketching environment by selecting the **Front** plane as the sketching plane and orient the plane normal to the eye view.

2. Create the sketch of the counterbore hole using the standard sketching tools. Add the required relations and then add linear diameter dimensions, as shown in Figure 5-103.

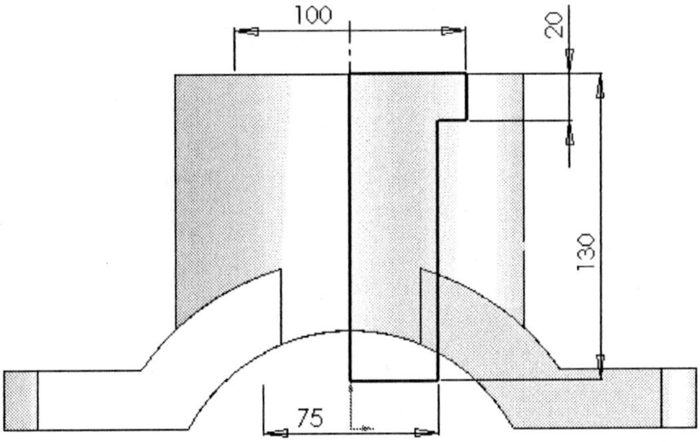

Figure 5-103 Fully defined sketch for the counterbore

3. Set the current view to the isometric view and then choose the **Revolved Cut** button from the **Features CommandManager**. The **Cut-Revolve PropertyManager** is displayed.

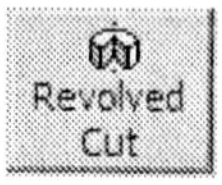

The preview of the cut feature is displayed in the drawing area in temporary graphics. The value of the angle in the **Angle** spinner is set to **360** by default. Therefore, you do not need to set the value of the **Angle** spinner.

4. Choose the **OK** button from the **Cut-Revolve PropertyManager**. Figure 5-104 shows the model after creating the revolve cut feature.

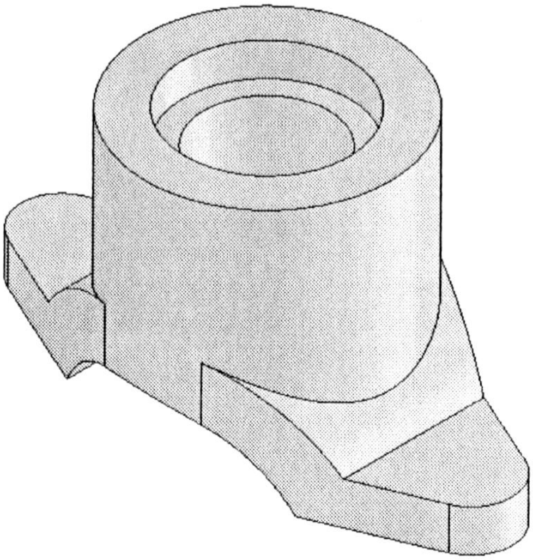

Figure 5-104 *Cut feature added to the model*

Creating the Holes

After creating all the features, you will create the holes using the extruded cut feature to complete the model. The sketch for the cut feature will be created using the top planar surface of the base feature as the sketching plane.

1. Select the top planar surface of the base feature and invoke the sketching environment. Orient the model so that the selected face of the model is oriented normal to the eye view.

2. Create the sketch using the standard sketching tools and apply the required relations and dimensions as shown in Figure 5-105.

3. Change the current view to the isometric view. Choose the **Extruded Cut** button from the **Features CommandManager** to invoke the **Cut-Extrude PropertyManager**.

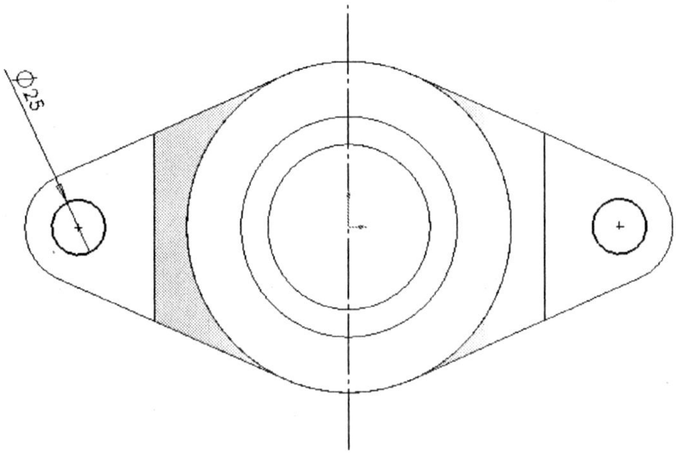

Figure 5-105 Fully defined sketch for cut feature

4. Right-click and choose the **Through All** option from the shortcut menu and choose the **OK** button from the **Cut-Extrude PropertyManager**. The final model is shown in Figure 5-106.

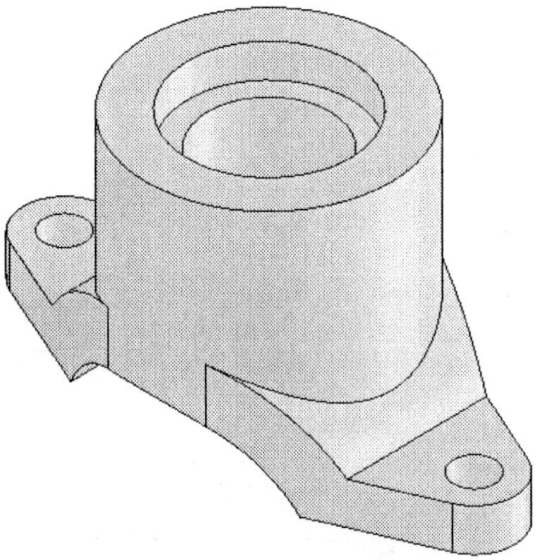

Figure 5-106 Final model

The **FeatureManager Design Tree** of the model is shown in Figure 5-107.

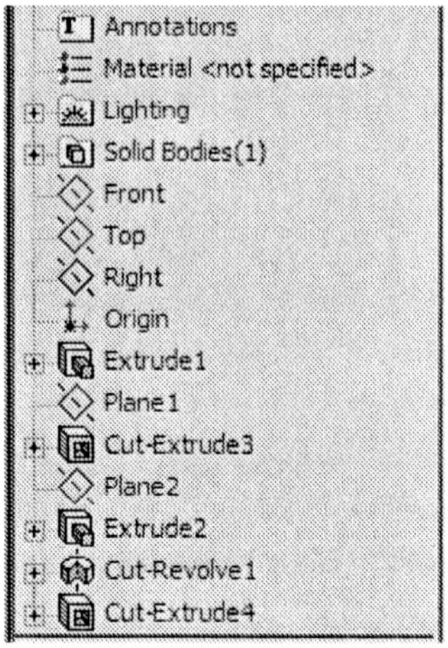

Figure 5-107 FeatureManager Design Tree

Saving the Model

1. Choose the **Save** button from the **Standard** toolbar and save the model with the name given below.

   \My Documents\SolidWorks\c05\c05-tut03.sldprt

2. Choose **File > Close** from the menu bar to close the document.

SELF-EVALUATION TEST

Answer the following questions and then compare your answers with the answers given at the end of this chapter.

1. When you create a sketch for the first time in the sketching environment, the sketch is created on the default plane, which is the **Front** plane. (T/F)

2. When you start a new SolidWorks part document, SolidWorks provides you with two default planes. (T/F)

3. You can choose the **Plane** button from the **Reference Geometry CommandManager** to invoke the **Plane PropertyManager**. (T/F)

4. You cannot create a plane at an offset distance by dynamically dragging. (T/F)

5. When you create a circular feature, a temporary axis is automatically created. (T/F)

6. The __________ option is used to extrude a sketch from the sketching plane to the next surface that intersects the feature.

7. The __________ option available in the **End Condition** drop-down list is used to define the termination of the extruded feature up to another body.

8. The __________ check box is used to merge the newly created body with the parent body.

9. Using the __________ option, you can create a reference axis that passes through the center point of a cylindrical or a conical surface.

10. Sometimes multiple bodies are created while applying the cut feature. Therefore, the __________ dialog box is displayed that allows you to define which body to keep.

REVIEW QUESTIONS

Answer the following questions.

1. If the __________ check box is cleared, the virtual surface created for the termination of the extruded feature will have a concentric relation with the selected surface.

2. The __________ option is selected from the shortcut menu to select the contours.

3. The __________ option is available in the **End condition** drop-down list only after you create the base feature.

4. The __________ check box is used to define the side of material removal.

5. The __________ check box is used to create an outward draft in a cut feature.

6. Which check box is selected while creating any other feature in single body modeling?

   (a) **Combine results** (b) **Fix bodies**
   (c) **Merge results** (d) **Union results**

7. Which button is used to add a draft angle to a cut feature?

   (a) **Add Draft** (b) **Create Draft**
   (c) **Draft On/Off** (c) None of these

8. Which **PropertyManager** is invoked to create a cut feature by extruding a sketch?

   (a) **Extruded Cut** (b) **Cut-Extrude**
   (c) **Extrude-Cut** (d) **Cut**

9. Which one of the following options is used to define the termination of feature creation at an offset distance to a selected surface?

   (a) **Distance To Surface** (b) **Normal From Surface**
   (c) **Distance From Surface** (d) **Offset From Surface**

10. Which one of the following options is used to define the termination of feature creation to the selected surface?

(a) **To Surface** (b) **Selected Surface**
(c) **Up To Surface** (d) None of these

EXERCISES

Exercise 1

Create the model whose dimensions are shown in Figure 5-108. The solid model is given in Figure 5-109. **(Expected time: 30 min)**

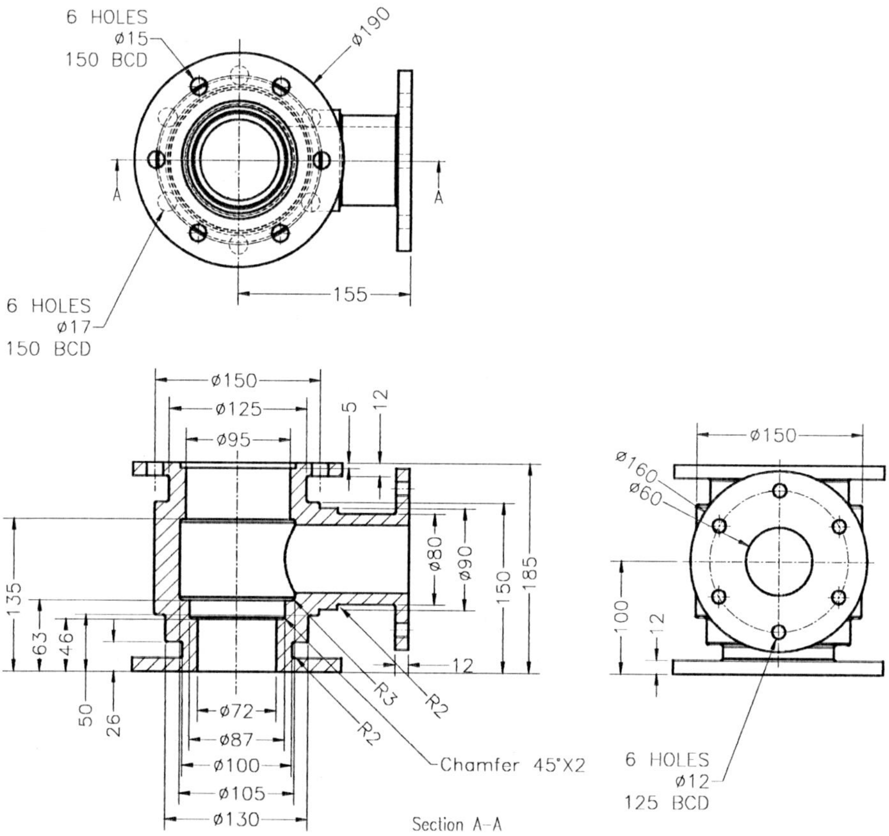

Figure 5-108 Dimensions of the model for Exercise 1

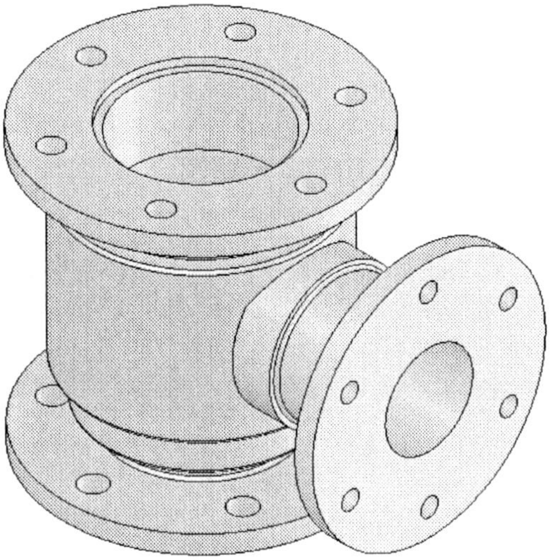

Figure 5-109 Model for Exercise 1

Exercise 2

Create the model shown in Figure 5-110 . The dimensions of the model are given in Figure 5-111. **(Expected time: 30 min)**

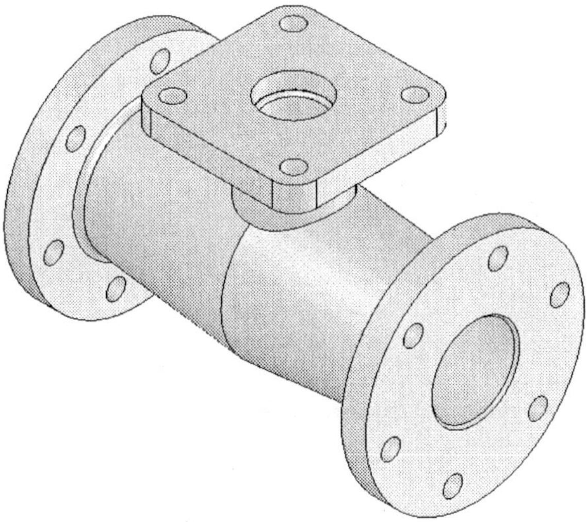

Figure 5-110 Model for Exercise 2

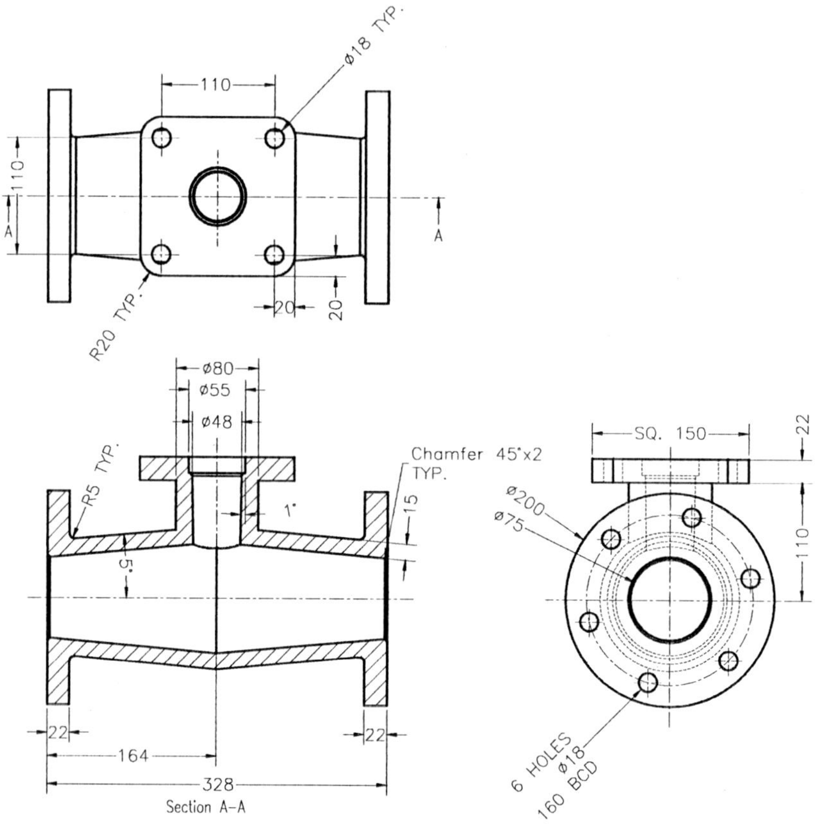

Figure 5-111 Dimensions for Exercise 2

Answers to Self-Evaluation Test

1. F, **2.** F, **3.** T, **4.** F, **5.** T, **6. Up To Next**, **7. Up To Body**, **8. Merge results**, **9. Cylindrical/Conical Surface**, **10. Bodies to Keep**

Chapter 6

Advanced Modeling Tools-I

Learning Objectives

After completing this chapter you will be able to:

- *Create holes using the Simple Hole option.*
- *Create standard holes using Hole Wizard.*
- *Apply simple and advanced fillets.*
- *Understand various selection methods.*
- *Chamfer the edges and vertices of the model.*
- *Create the Shell feature.*
- *Create the Wrap feature.*

ADVANCED MODELING TOOLS

This chapter discusses various advanced modeling tools available in SolidWorks that assist you in creating a better and accurate design by capturing the design intent in the model. For example, in the previous chapters you learned how to create a hole using the cut option. But in this chapter you will create holes using the **Simple Hole** option and the **Hole Wizard** option. Using the hole wizard, you can create standard holes classified on the basis of the industrial standard, screw type, and size. Hole wizard of SolidWorks is one of the largest standard industrial virtual hole generation machine available in any CAD package. You will also learn about some other advanced modeling tools such as fillet, chamfer, and shell in this chapter.

Creating Simple Hole

CommandManager:	Features > Simple Hole	*(Customize to Add)*
Menu:	Insert > Features > Hole > Simple	
Toolbar:	Features > Simple Hole	*(Customize to Add)*

The **Simple Hole** option is used to create a simple hole feature. In previous chapter you learned to create holes by sketching a circle and then using the cut option to complete hole creation. But using the simple hole option you do not need to create a sketch of the hole. This is the reason the holes created with this option act as a placed feature. To create a hole using this option, first you need to select the plane on which you want to place the hole feature. If you invoke this tool before selecting a plane, then the **Hole PropertyManager** is displayed that prompts you to select a placement plane. Now, choose the **Simple Hole** button from the **Features CommandManager** after customizing it. The **Hole PropertyManager** is displayed as shown in Figure 6-1.

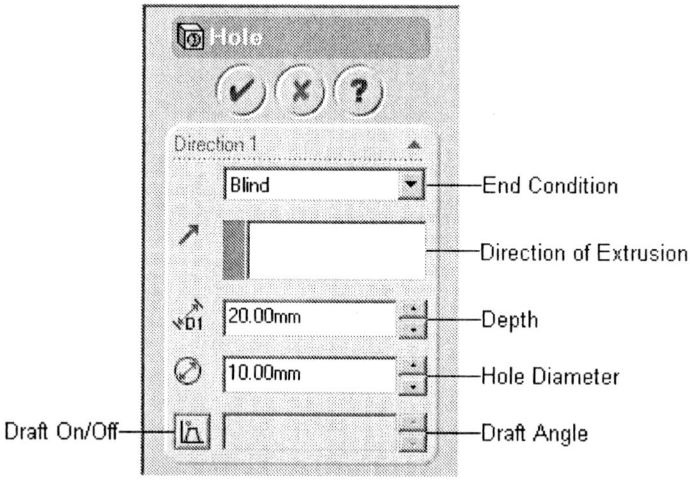

Figure 6-1** The **Hole PropertyManager

When you invoke the **Hole PropertyManager**, the preview of the hole feature is displayed in the drawing area in temporary graphics with the default values. The preview of the hole feature with default values is shown in Figure 6-2. Using the **End Condition** drop-down list,

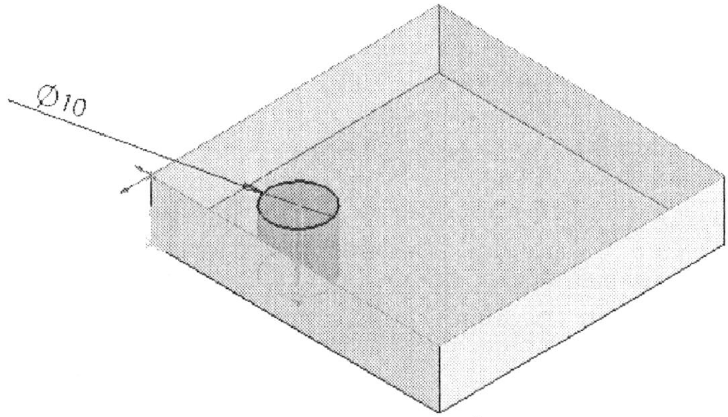

***Figure 6-2** Preview of the hole being created using the **Simple Hole** option*

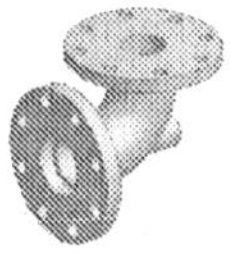

Tip. *It is recommended that before creating the hole feature, you create a point that will define the placement of the hole feature. While creating the hole feature, you can dynamically move the hole feature by selecting the center point of the sketch and dragging the cursor to the point sketched earlier. The cursor will snap to that point and coincident relation will be applied between the sketched point and the center point of the hole.*

specify the termination type and set the value of the hole diameter in the **Hole Diameter** spinner. You can also specify the direction of extrusion using the **Direction of Extrusion** selection area. You can also specify a draft angle in the hole feature using the **Draft On/Off** button and you can set the value of the draft angle using the **Draft Angle** spinner. The preview of the draft angle is also displayed in the drawing area in temporary graphics. After setting all the parameters choose the **OK** button from the **Hole PropertyManager**.

The hole feature created using this option is placed on the selected plane but the placement of the hole is not yet defined. Therefore, select the hole feature from the **FeatureManager Design Tree** and right-click to display the shortcut menu. Choose the **Edit Sketch** option from the shortcut menu. The sketching environment is invoked and you can apply the relations and dimensions to define the placement of the hole feature on the selected face and exit the sketching environment.

Creating Standard Holes Using Hole Wizard

Enhanced

CommandManager:	Features > Hole Wizard
Menu:	Insert > Features > Hole > Wizard
Toolbar:	Features > Hole Wizard

The **Hole Wizard** tool is used to add standard holes to the model. The holes applied using the hole wizard include the standard counterbore, countersink, drilled, tapped, and pipe tap holes. You can also create a user-defined counterbored drilled hole, counter-drilled drilled hole, counterbored hole, counterdrilled hole, countersunk hole, countersunk drilled hole, simple hole, simple drilled hole, tapered hole, and tapered drilled hole. You can control all the parameters of the holes including the termination options and you can also modify the holes according to your requirement after placing them. Thus, it results in the placement of standard parametric holes using this option. There are two methods to place a hole feature using the **Hole Wizard** tool. The first method is the preselection method. In this method you have to select a placement face or a placement plane before invoking this tool. The placement face can be a planar face or a curved face. Then choose the **Hole Wizard** button from the **Features CommandManager** to invoke the **Hole Definition** dialog box. The preview of the hole feature is displayed in the graphics area. If the preview is not visible properly then move the **Hole Definition** dialog box to see the preview. If you change the settings and parameters or the type of hole, the preview of the hole will also get modified dynamically. The various options available in the **Hole Definition** dialog box are discussed next.

Counterbore Tab

The **Counterbore** tab, as shown in Figure 6-3, available in the **Hole Definition** dialog box is used to define the parameters of a counterbore hole. You can also add the frequently used counterbore holes in the favorites list option in this tab. The various options available in the **Counterbore** tab are discussed next.

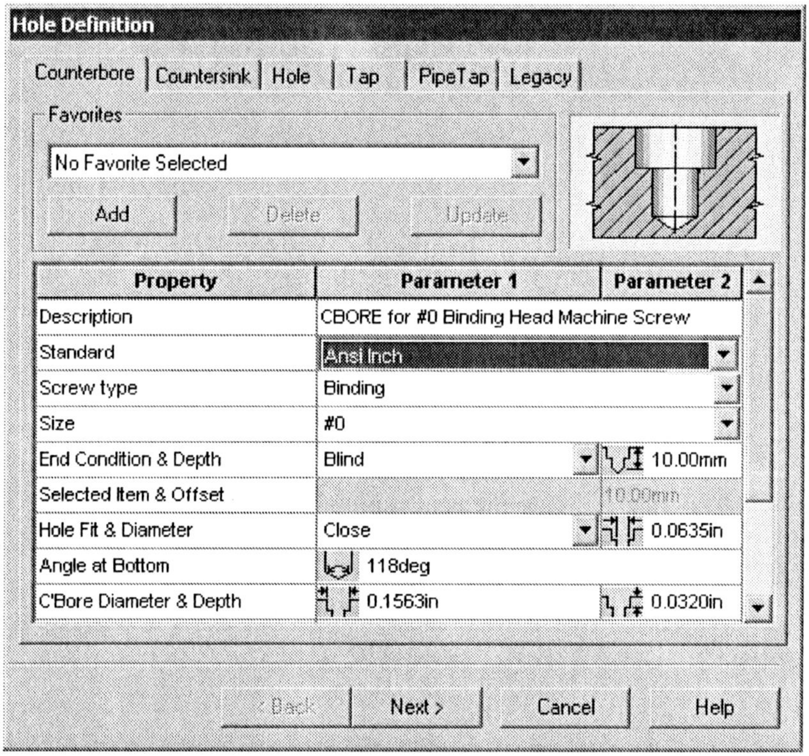

Figure 6-3 *The* ***Counterbore*** *tab of the* ***Hole Definition*** *dialog box*

Property

The **Property** column of the **Hole Definition** dialog box is used to define the various properties of the standard hole. The properties include the description, standard, screw type, size, and so on. The parameters available for all the properties are divided in two columns. The names of the columns are **Parameter 1** and **Parameter 2**. These two columns are used to define the values and other related information to create the hole. The various options available in the **Property** column are discussed next.

Description

The **Description** option is used to display the name and the description of the standard hole.

Standard

The **Standard** option is used to specify the industrial dimensioning and hole standard. You can select the standard from the **Standard** drop-down list available in the **Parameter 1** column. By default, the **Ansi Inch** standard is selected. Various other dimensioning standards are available in this drop-down list such as **Ansi Metric**, **BSI**, **DIN**, **ISO**, **JIS**, **DME Mould Bases**, **Hasco Metric Mould Bases**, **PCS Mould Bases**, **Progressive Mould Bases**, and **Superior Mould Bases**.

Screw type

The **Screw type** option is used to define the type of fastener to be inserted in the hole. The standard holes created using the **Hole Wizard** tool depend on the fastener to be inserted in that hole and the size of the fastener. Therefore, creation of a hole based on the fastener to be used in that hole is a good practice and is a unique feature of SolidWorks. You can select the screw type from the **Screw Type** drop-down list available in the **Parameter 1** column. The type of screws available in the drop-down list depends on the standard selected from the **Standard** drop-down list.

Size

The **Size** option is used to define the size of the fastener that will be inserted in the hole that is created using the **Hole Wizard** tool. The size of the fastener is selected from the **Size** drop-down list available in the **Parameter 1** column. The sizes of the fasteners available in the **Size** drop-down list depend on the standard selected from the **Standard** drop-down list.

End Condition and Depth

The **End Condition and Depth** option is used to define the end condition of feature termination. To define the end condition for feature termination, select the end condition option from the **End Condition** drop-down list available in the **Parameter 1** area. By default, the **Blind** option is selected in this drop-down list. Therefore, if the end condition is **Blind** you have to specify the depth of feature termination. This is the reason the **Depth** area available on the right of the **End Condition** drop-down list in the **Parameter 2** column is replaced by the edit box when clicked once. You can specify the depth of feature termination in this edit box.

Note

A preview area is provided on the top right of the ***Hole Definition*** *area. You can observe the preview of the hole being created. The preview of the hole feature is changed dynamically as you change the parameters of the hole feature.*

Selected Item and Offset

The **Selected Item and Offset** option is used to define the surface for feature termination. The **Selected Item** selection area in the **Parameter 1** column is active only when the **Up To Vertex**, **Up To Surface**, or **Offset From Surface** option is selected from the **End Condition** drop-down list. The **Offset** area available on the right of the **Select Item** selection area is replaced by the **Offset** edit box and you can enter the value of offset in this edit box. This option is available only when you select the **Offset From Surface** option from the **End Condition** drop-down list.

Hole Fit and Diameter

The **Hole Fit and Diameter** option is used to specify the type of **Fit** to be applied in the hole to be created. The fit types are available in the **Hole Fit** drop-down list in the **Parameter 1** column. The type of fits available in the **Hole Fit** drop-down list are **Close**, **Normal**, and **Loose**. The **Diameter** area on the right of the **Hole Fit** drop-down list in the **Parameter 2** area is replaced by the **Diameter** edit box by clicking once in this area. The value of the diameter in this area is the default value according to the standard. The value changes depending upon the type of hole fit selected from the **Hole Fit** drop-down list. You can also modify this value according to the requirement.

Angle at Bottom

The **Angle at Bottom** option is used to specify the angle at the bottom of the hole feature. The **Angle at Bottom** area available in the **Parameter 1** column is replaced by the **Angle at Bottom** edit box by clicking once. You can modify the default value for the angle at bottom by entering a new value in this edit box.

C'Bore Diameter and Depth

The **C'Bore Diameter and Depth** option is used to specify the diameter of the counterbore and the depth of the counterbore. The **C'Bore Diameter** area is available in the **Parameter 1** column and is replaced by the **C'Bore Diameter** edit box by clicking once to modify the default value. The **Depth** area available on the right of the **C'Bore Diameter** area is replaced by the **Depth** edit box by clicking once and you can modify the default depth of the counterbore.

Head Clearance

The **Head Clearance** option available in the **Property** column is used to specify the clearance distance between the head of the fastener and the placement plane of the hole feature. The **Head Clearance** area available in the **Parameter 1** column is replaced by the **Head Clearance** edit box by clicking once and you can modify the default head clearance value.

Near Side C'Sink Dia. and Angle

The **Near Side C'Sink Dia. and Angle** option is used to specify the diameter and the angle for the countersink on the upper face, which is the placement plane of the hole feature. The diameter is specified in the **Near Side C'Sink Dia.** area of the **Parameter 1** column and the angle is specified in the **Angle** area of the **Parameter 2** column.

Under Hd C'Sink Dia. and Angle

The **Under Hd C'Sink Dia. and Angle** option is used to specify the diameter and the angle for the countersink to be applied at the end of the counterbore head. The diameter is specified in the **Under Hd C'Sink Dia.** area in the **Parameter 1** column and the angle is specified in the **Angle** area in the **Parameter 2** column.

Far Side C'Sink Dia. and Angle

The **Far Side C'Sink Dia. and Angle** option is used to specify the diameter and the angle for the countersink to be applied at the end of the hole feature. This option is available only if the **Through All** or the **Up To Next** option is selected from the **End Condition** drop-down list. The diameter is specified in the **Far Side C'Sink Dia.** area in the **Parameter 1** column and the angle is specified in the **Angle** area in the **Parameter 2** column.

Favorites

The **Favorites** area available in the **Hole Definition** dialog box is used to add the frequently used holes to the favorite list. If you add a hole to the favorite list, you will not have to perform the same settings to add similar types of holes every time. To add a particular hole to the favorite list, choose the **Add** button available in the **Favorite** area of the **Hole Definition** dialog box. The **New Favorite Name** dialog box is displayed with the default name of the hole in the **Favorite Name** edit box as shown in Figure 6-4.

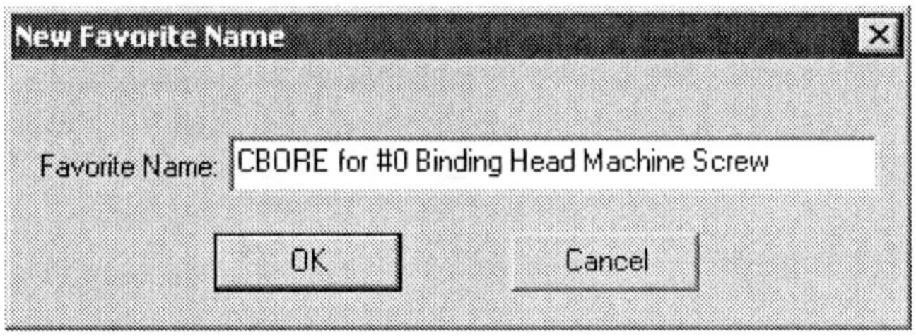

*Figure 6-4 The **New Favorite Name** dialog box*

You can enter a new name in the **Favorite Name** edit box and choose the **OK** button to add the hole to the favorite list. If you have to apply a hole feature from the favorite list, you just need to select the name of the hole feature from the **Favorites** drop-down list. The current parameters of the hole will be changed to the parameters of the hole feature that is selected from the **Favorites** drop-down list. Figure 6-5 shows the counterbore holes created using the **Hole Wizard** tool.

Countersink Tab

The **Countersink** tab, as shown in Figure 6-6, available in the **Hole Definition** dialog box is used to define the parameters of the countersink hole. Most of the options available in the **Property** column are the same as those in the **Property** column of the **Counterbore** tab. However, there are some additional options to define the angle and diameter of the countersink. These options of the **Property** column of the **Countersink** tab are discussed next.

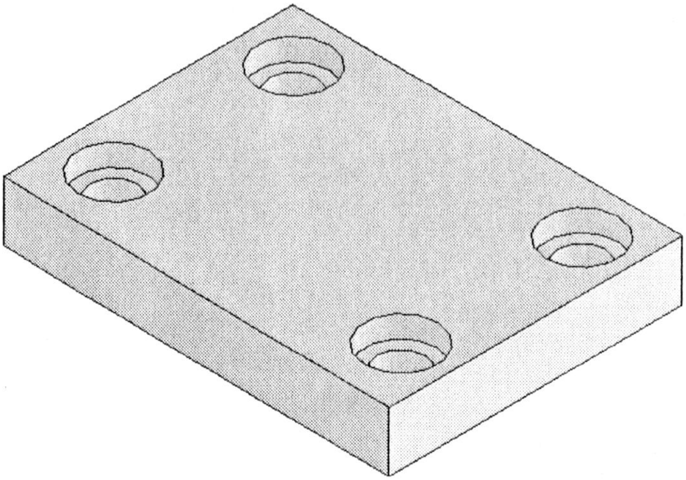

Figure 6-5 A model with counterbore holes

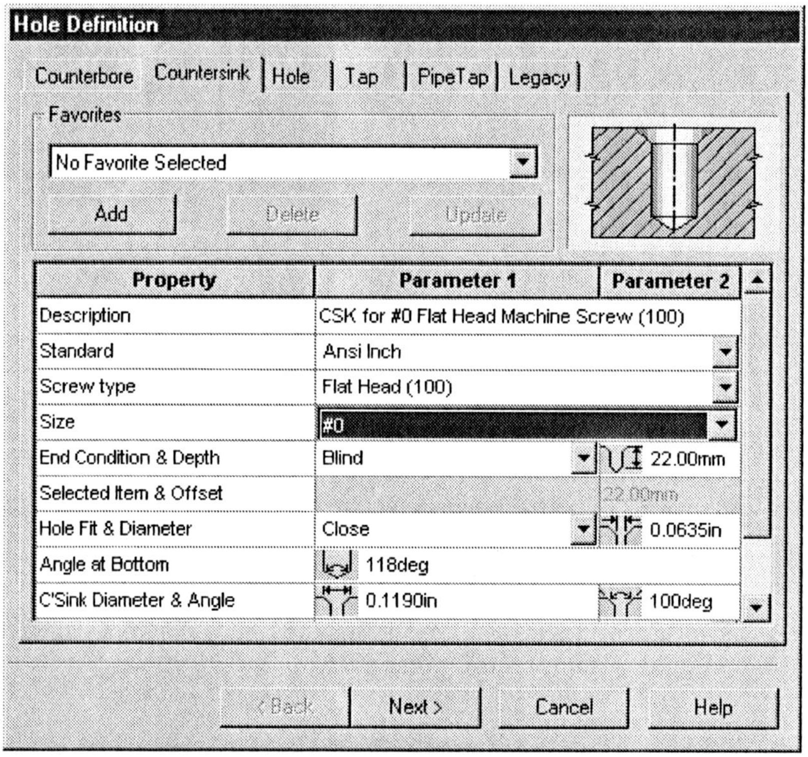

*Figure 6-6 The **Countersink** tab of the **Hole Definition** dialog box*

Property

The **Property** column of the **Hole Definition** dialog box is used to define the parameters of the countersink hole. As mentioned earlier, most of the options are the same as those discussed in the **Property** column of the **Counterbore** tab. The options that have been discussed earlier are not included here. Some options that are used to create only a counterbore and are not used in countersink are not available in this tab. Only the additional options available in the **Property** column of the **Countersink** tab are discussed next.

C'Sink Diameter and Angle

The **C'Sink Diameter and Angle** option is used to define the diameter and the angle of the countersink. The **C'Sink Diameter** area in the **Parameter 1** column is changed to the **C'Sink Diameter** edit box when clicked once. The value of the countersink diameter is specified in this edit box. The **Angle** area available in the **Parameter 2** area is replaced by the **Angle** edit box when clicked once. The value of the angle of countersink is specified in this edit box.

Head Clearance and Type

The **Head Clearance and Type** option is used to define the head clearance distance between the placement surface of the hole feature and the top face of the fastener and to define the type of the counter to be added. The **Head Clearance** area available in the **Parameter 1** column is replaced by the **Head Clearance** edit box when clicked once. The value of the clearance distance is specified in this edit box. The **Type** drop-down list available on the right of the **Head Clearance** area is used to specify if the clearance distance is compensated by extending the countersink or by adding a counterbore.

Figure 6-7 shows the countersink holes created using the **Hole Wizard**.

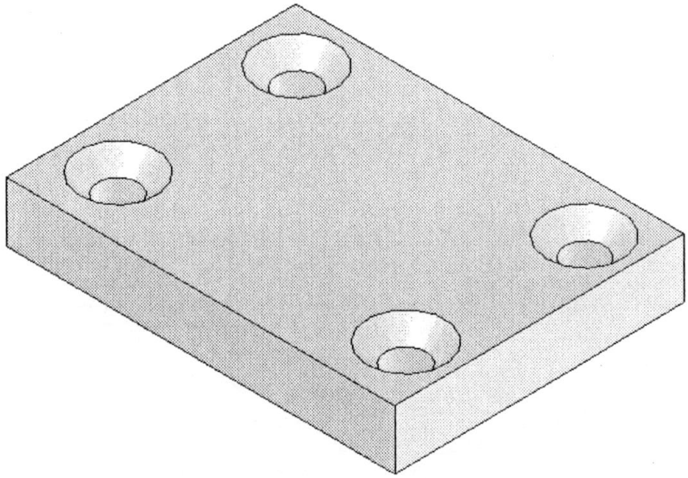

Figure 6-7 *A model with countersink holes*

Hole Tab

The **Hole** tab, shown in Figure 6-8, available in the **Hole Definition** dialog box is used to define the parameters to add a standard drilled hole to the model. Using this tab you can add simple holes that can be drilled using the standard drills available in the market. The options available in the **Property** column of this tab are used to specify the parameters for creating a standard drilled hole.

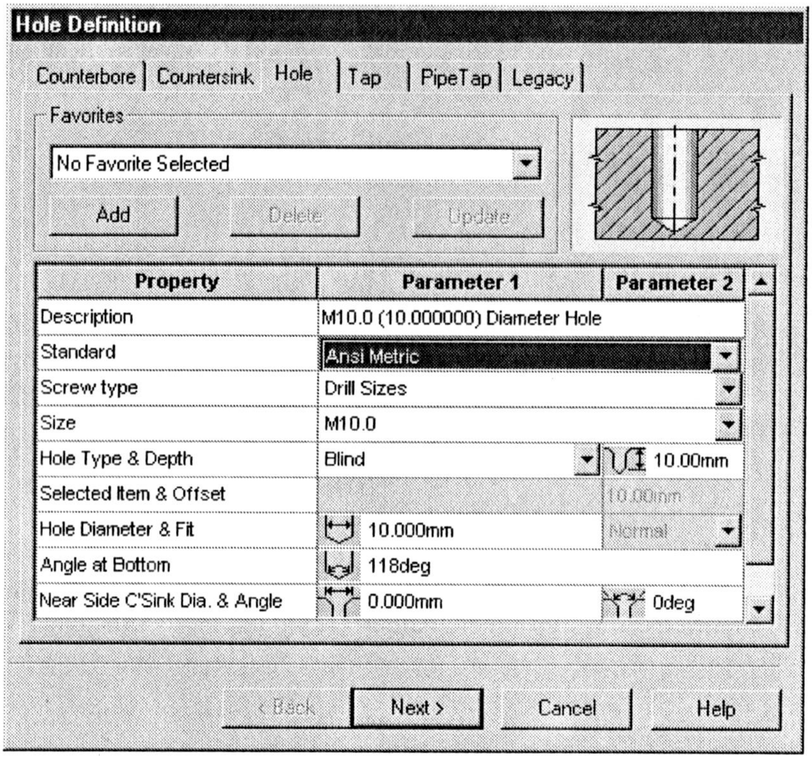

*Figure 6-8 The **Hole** tab of the **Hole Definition** dialog box*

Figure 6-9 shows the drilled holes created using the **Hole Wizard** tool.

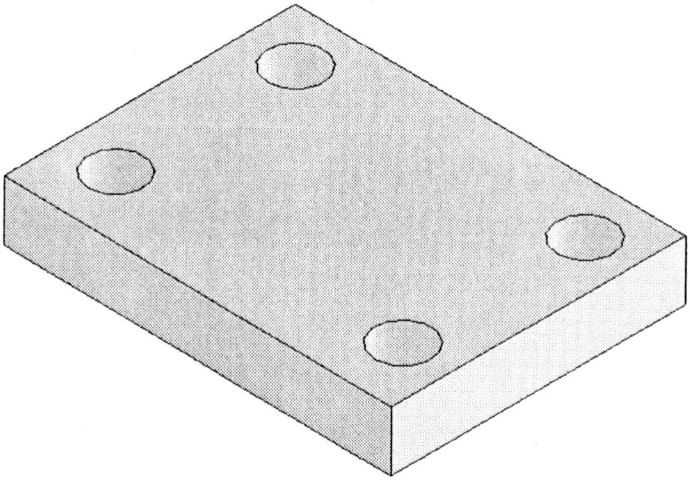

Figure 6-9 A model with drilled holes

Tap Tab

The **Tap** tab, shown in Figure 6-10, available in the **Hole Definition** dialog box is used to define the parameters to add a tapped hole to the model. Tap hole is basically a hole in which

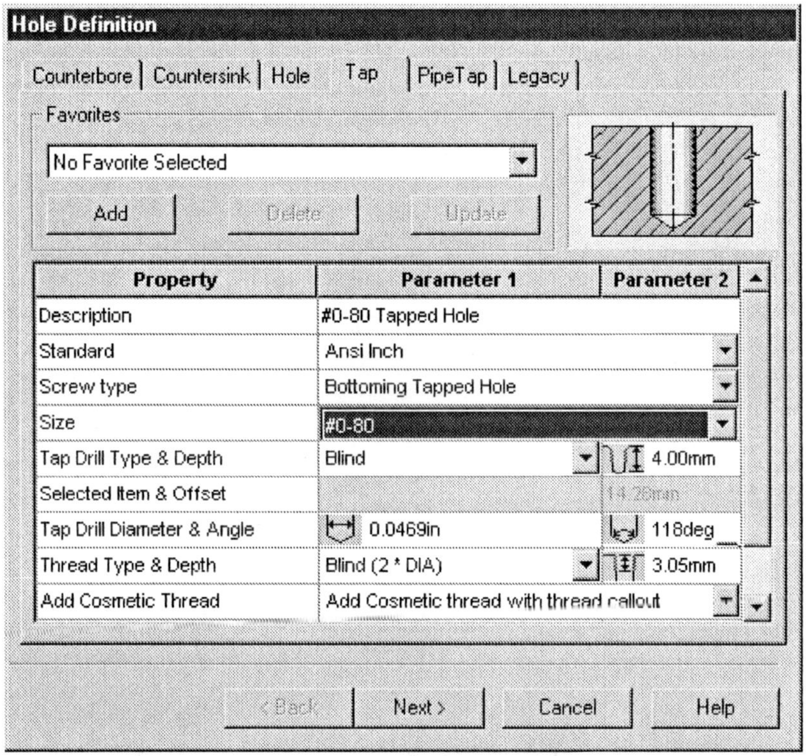

***Figure 6-10** The **Tap** tab of the **Hole Definition** dialog box*

internal threading (also known as tapping) is shown. Using this tab, you can add the standard tapped holes in your design. Almost all the options available in the **Property** column of this tab are the same as those discussed earlier. The remaining options are discussed next.

Thread Type and Depth

The **Thread Type and Depth** option is used to define the type of depth and the value of the depth of the thread. The **Thread Type** drop-down list available in the **Parameter 2** column is used to specify the type of thread depth. The options available in the **Thread Type** drop-down list depend on the feature termination option selected from the **Tap Drill Type** drop-down list. The depth of the thread is specified in the **Depth** area in the **Parameter 2** column.

Add Cosmetic Thread

The enhancement in the **Hole Wizard** tool is the introduction of graphics threads. The **Add Cosmetic Thread** option is used to add the cosmetic threads while creating a tapped hole. The **Add Cosmetic Thread** drop-down list is available in the **Parameter 1** column. If you select the **No Cosmetic thread** option from this drop-down list, the cosmetic thread will not be added to the tapped hole. If you select the **Add Cosmetic thread with thread callout** option from this drop-down list, the graphic cosmetic thread and the thread callout will be added to the tapped hole feature. The **Add Cosmetic thread without thread callout** option is used to add only the graphic cosmetic thread to the hole feature. Figure 6-11 shows the tapped holes created using the **Hole Wizard**.

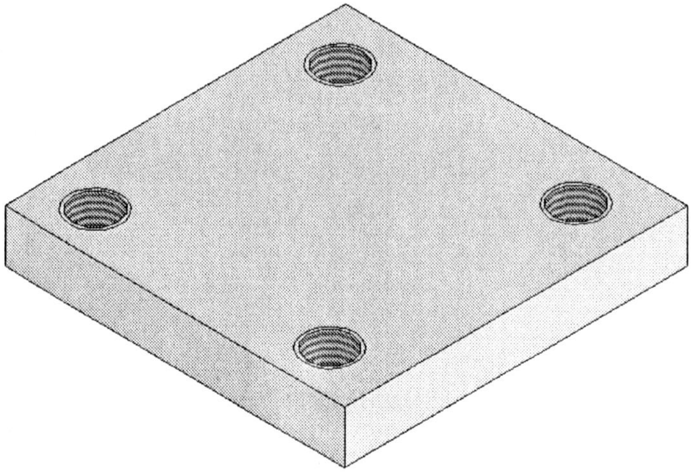

Figure 6-11 Model with tapped holes

Pipe Tap Tab

The **Pipe Tap** tab, shown in Figure 6-12, is used to specify the parameters for the tapered pipe tap hole. The options available in this tab are the same as those discussed in earlier tabs.

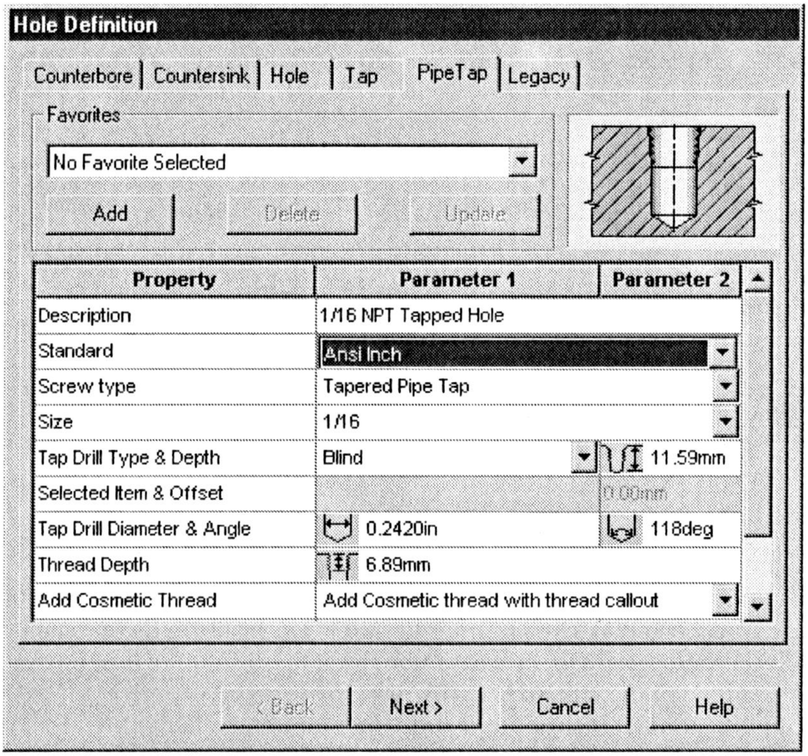

*Figure 6-12 The **Pipe Tap** tab of the **Hole Definition** dialog box*

Legacy Tab

The **Legacy** tab of the **Hole Definition** dialog box, shown in Figure 6-13, is used to add a user-defined hole feature. The options available in this tab are discussed next.

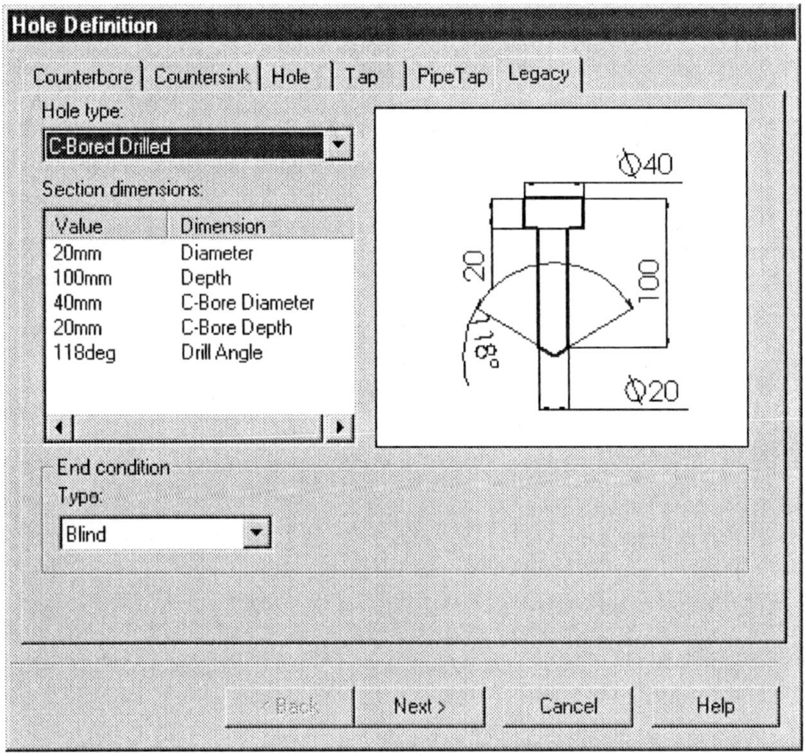

*Figure 6-13 The **Legacy** tab of the **Hole Definition** dialog box*

Hole type

The **Hole type** drop-down list is used to specify the type of hole you want to create. The type of holes available in this drop-down list are **C-Bored Drilled**, **C-Drilled Drilled**, **Counterbored**, **Counterdrilled**, **Countersunk**, **C-Sunk Drilled**, **Simple, Drilled**, **Tapered**, and **Tapered Drilled**. The preview of the selected hole type is displayed in the preview area provided at the right of the **Hole Definition** dialog box.

Section dimensions

The **Section dimensions** area is used to display the various dimensions of the selected hole and their default values. You can also modify the values by replacing the **Value** area by the **Value** edit box.

End condition

The **Type** drop-down list available in the **End condition** area is used to specify the end condition of feature termination.

As discussed earlier, the preview of the hole feature is dynamically updated in the drawing area while you are setting the parameters of the hole feature. This is because you had already selected the placement plane for the placement of the hole feature. If you do not select a placement plane for the placement of the hole feature, the preview of the hole feature is not displayed in the drawing area. After setting all the parameters of the hole feature, choose the **Next** button. The **Hole Placement** dialog box is displayed as shown in Figure 6-14.

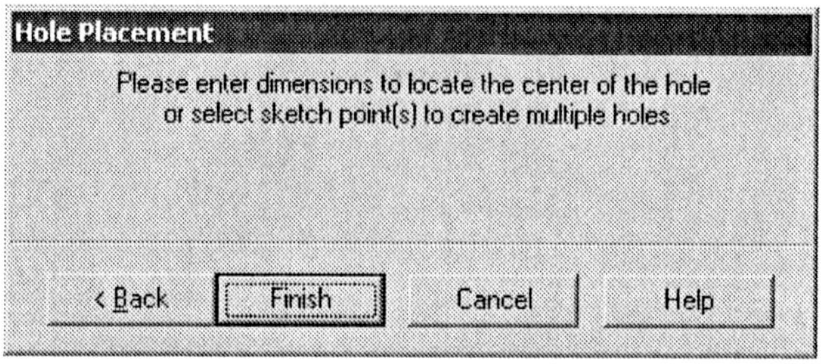

***Figure 6-14** The **Hole Placement** dialog box*

The select cursor is replaced by the placement cursor. Using the placement cursor you can place more holes in the current hole features. As discussed earlier, if the placement plane is selected earlier, the hole is already placed on the selected placement plane. If the placement plane is not selected earlier then you can specify a point to place the hole feature. You can also constrain the placement of the placement point of the hole feature using the relations and dimensions. Choose the **Finish** button to complete feature creation. Figure 6-15 shows a base plate with holes created using the **Hole Wizard** tool.

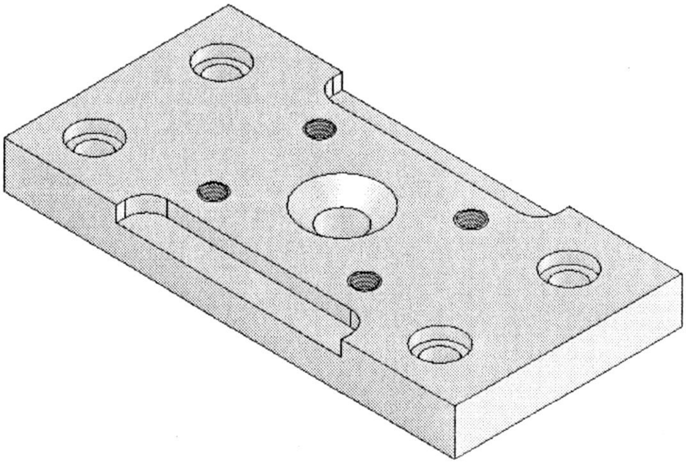

***Figure 6-15** Base Plate with holes created using the **Hole Wizard** option*

Note

*If you create a pattern feature of a tapped hole feature, the graphic threads are not displayed in the other instances of the pattern, except the parent instance. Therefore, to add graphic in other pattern instances use the **Texture PropertyManager**. You will learn more about patterns in the later chapters.*

> **Tip**. *The hole feature created using the **Hole Wizard** tool consists of two sketches. The first sketch is the sketch of the placement point and the second sketch is the sketch of the profile of hole feature. If you preselect the placement plane before invoking the **Hole Definition** dialog box, the resulting placement sketch will be a 2D sketch. Instead of preselecting the placement plane if you select the placement point after invoking the **Placement Point** dialog box, the resulting placement sketch is a 3D sketch. You will learn more about 3D sketches in the later chapters.*
>
> *In the modern modeling practice, the creation of threads is avoided in models because it results in the creation of complex geometry. When you generate the views from that model, the views generated contain complex geometry, which is difficult to understand. Therefore, it is a better practice to avoid the creation of threads in the model and add the cosmetic threads. Using the cosmetic threads you will get the thread convention in the drawing views, which is recommended, instead of creating complete thread.*
>
> *If a cosmetic thread is added in a tapped hole, the cosmetic thread is also displayed along with the placement and hole profile sketches. You can edit the cosmetic threads by selecting it from the **FeatureManager Design Tree** and right-clicking to display the shortcut menu. Choose the **Edit Definition** option to display the **Cosmetic Thread** dialog box. The **Cosmetic Thread** dialog box and the cosmetic threads are discussed in later chapters.*
>
> *You can also view the convention of the thread if the cosmetic thread is added to a tapped hole feature. Orient the model to the top view to observe the thread convention from the top view. Orient the model in the front view, back view, or any side view to observe the thread convention from the side views.*

Creating Fillets

Enhanced

CommandManager:	Features > Fillet
Menu:	Insert > Features > Fillet/Round
Toolbar:	Features > Fillet

Fillet

In SolidWorks, you can add fillets as a feature in the model using the **Fillet** tool. As discussed earlier you can also add fillets within the sketch. But adding the fillets in a sketch is not a good practice according to the design point of view. This is because you have to keep the sketch as simple as possible. Using the fillet tool you can round an internal or external face or edge of a model. You can also use the advanced fillet options to add advanced fillets to the model. You can preselect the face, edge, or feature on which the fillet has to be applied. You can also select the entity to be filleted after invoking the fillet tool. Choose the **Fillet** button from the **Features CommandManager** to invoke the **Fillet PropertyManager**. The **Fillet PropertyManager** is shown in Figure 6-16. The preview of the fillet feature is also displayed in the drawing area if the entities to be filleted are selected and the **Full Preview** radio button is selected by default. If preselection is not done, you are prompted to select the edges, faces, features, or loops to add the fillet feature. Using

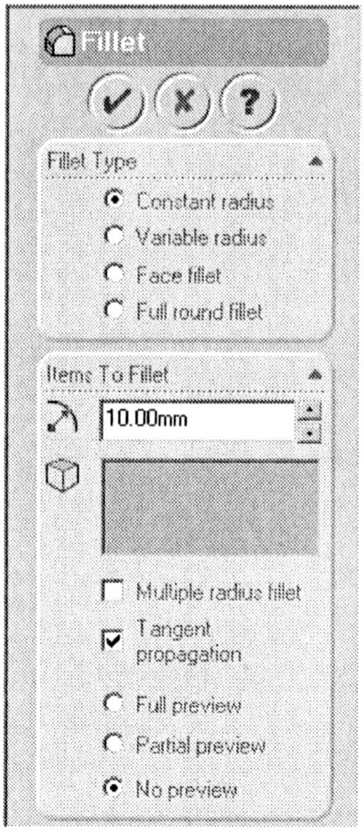

Figure 6-16 *The* ***Fillet PropertyManager***

the select cursor, select the entity to be filleted. A fillet callout is also displayed along the preview of the fillet. Figure 6-17 shows the preview of the fillet feature with fillet callout.

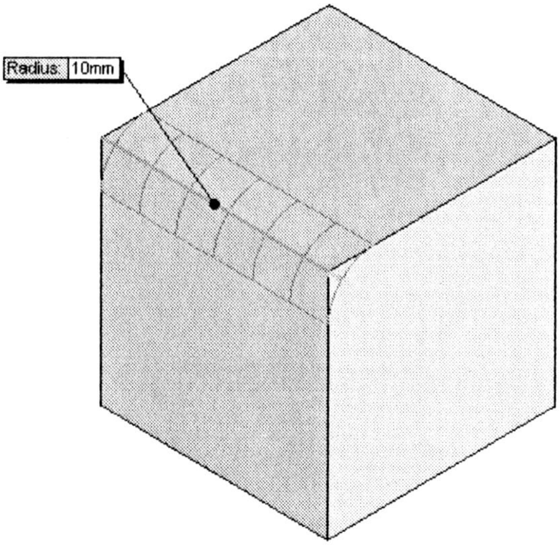

Figure 6-17 *Preview of the fillet feature*

Using the fillet tool you can create various types of fillets. Types of fillets that can be created using the fillet tool are given below.

1. Constant radius fillet
2. Variable radius fillet
3. Face Fillet
4. Full round fillet

All the above-mentioned fillets are discussed next.

Constant Radius Fillet

The constant radius fillet option creates a fillet of constant radius along the selected entity. This is the default option selected in the **Fillet Type** rollout. You can set the value of the fillet radius using the **Fillet** spinner provided in the **Items To Fillet** rollout or click the value area of the fillet callout. The value area of the callout will be replaced by the radius edit box. Enter the value of the radius and press the ENTER key from the keyboard. The preview of the fillet will be changed dynamically when the value of the radius of the fillet is changed. The names of the selected entities are displayed in the **Edges**, **Faces**, **Features**, and **Loops** selection area. The entities that you can select to add the fillet feature are faces, edges, features, and loops. Now, choose the **OK** button from the **Fillet PropertyManager**. Figures 6-18 through 6-23 show the selection of different entities and the resultant fillet creation from the selected entities.

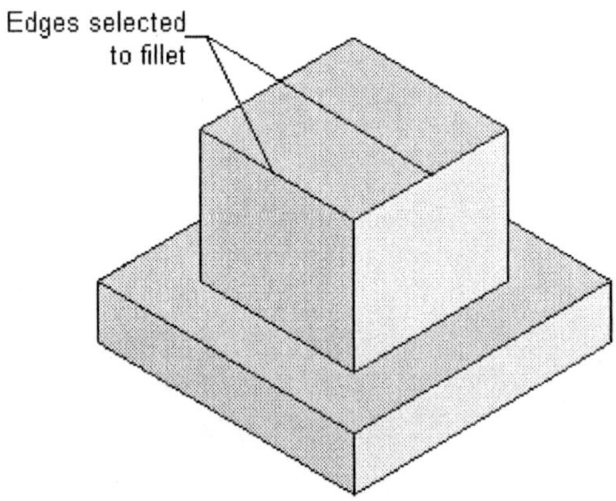

Figure 6-18 Selecting the edges

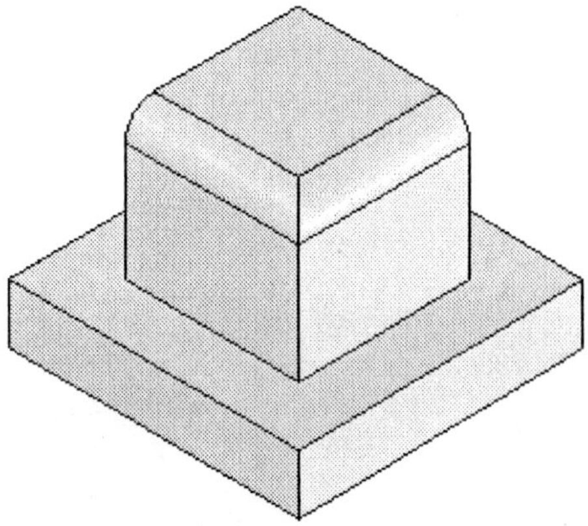

Figure 6-19 Resultant fillet

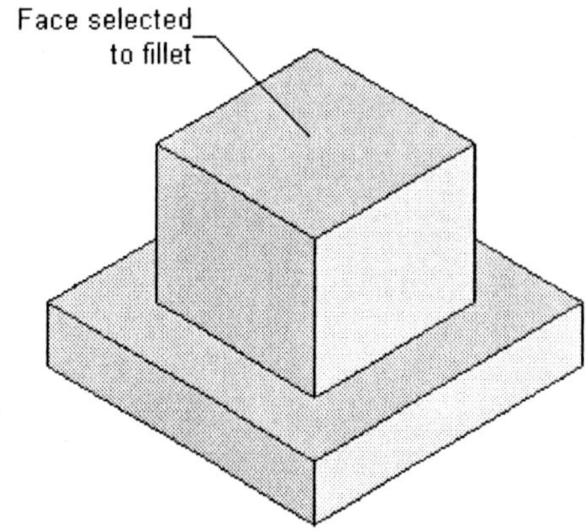

Figure 6-20 *Selecting the face*

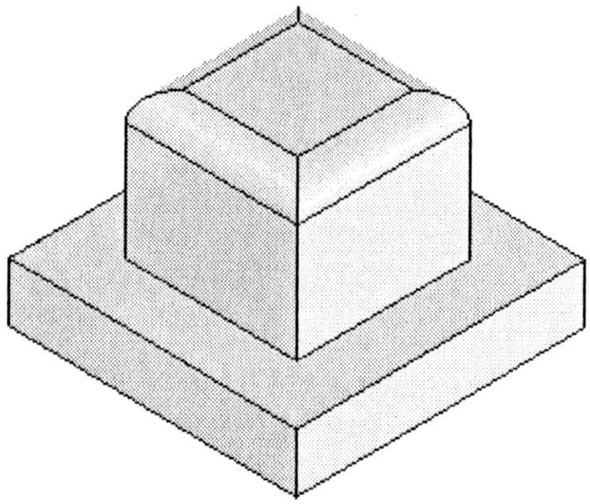

Figure 6-21 *Resultant fillet*

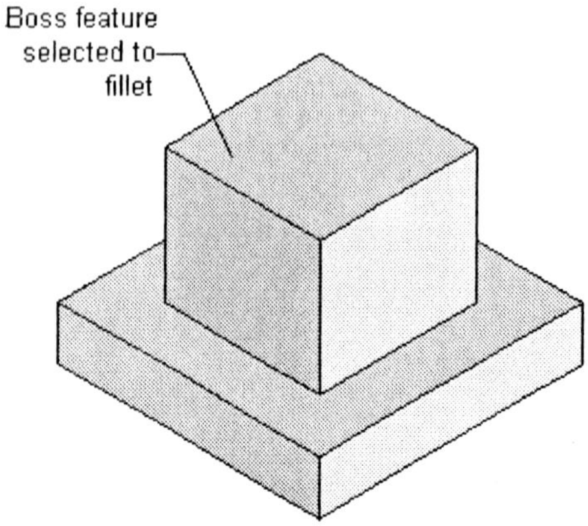

Figure 6-22 *Selecting the feature*

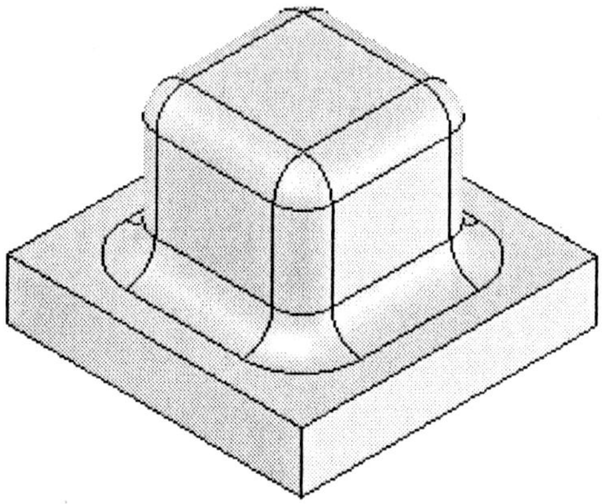

Figure 6-23 *Resultant fillet*

Tip. *You can preselect a feature for creating a fillet feature or select the feature after invoking the* ***Fillet*** *tool. For postselection of the feature to be filleted, expand the* ***FeatureManager Design Tree****, which is displayed in the drawing area, and select the feature. You can also select the feature from the drawing area. The preview of the filleted feature is displayed in the drawing area.*

Multiple Radius Fillet

Using the **Multiple radius fillet** option provided in the **Fillet PropertyManager** you can specify a fillet of different radii to all the selected edges. For creating a fillet feature using the multiple radius option, preselect the edges, faces, or features or select them after invoking the **Fillet PropertyManager**. After invoking the **Fillet** tool, select the **Multiple radius fillet** check box. The preview of the fillet feature with the default value is displayed in the drawing area. You will notice that you are provided with different callouts for each selected entity. Figure 6-24 shows the preview of the fillet feature with **Multiple radius fillet** check box selected. The names of the selected entities are displayed in the **Edges, Faces, Features and Loops** display box. The boundaries of the currently selected entity in the selection box are highlighted in green with a thick line. You can set the value of each selected entity by using

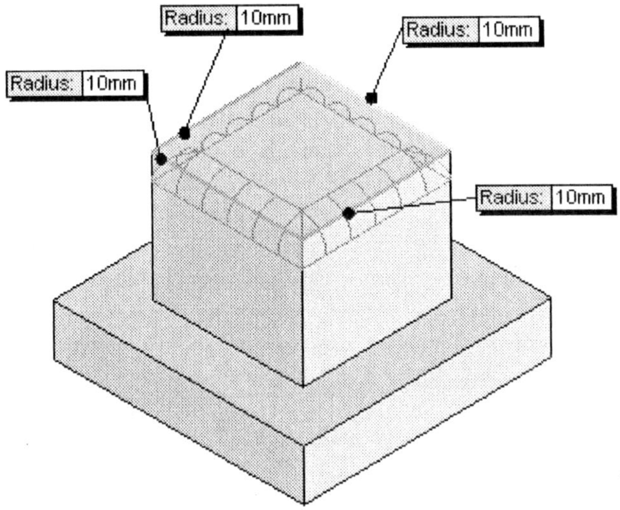

Figure 6-24 *Preview of the fillet feature with* ***Multiple radius fillet*** *check box selected*

the **Radius** spinner or specify the value of fillet radius in the radius callout as shown in Figure 6-25. As you modify the value of the radius, the preview of the fillet feature modifies dynamically in the drawing area. Figure 6-26 shows the fillet created using the multiple radius fillet.

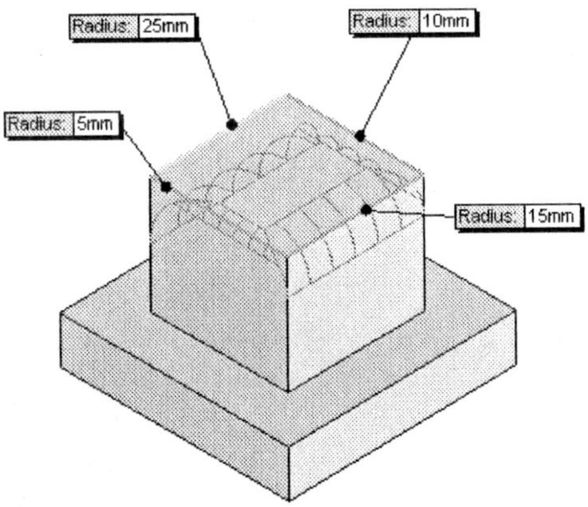

Figure 6-25 *Different radii specified in each radius callout*

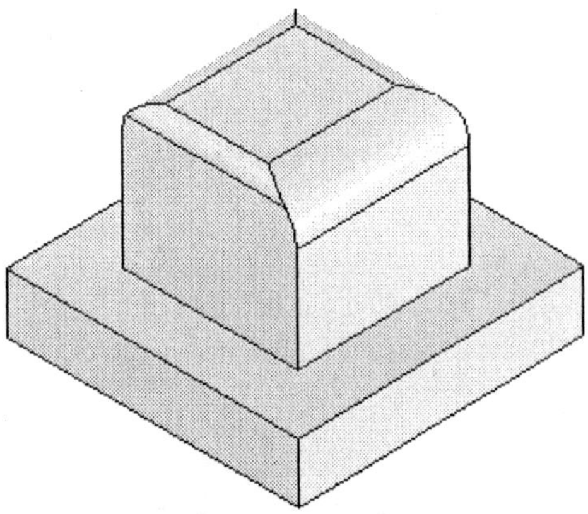

Figure 6-26 *Resultant fillet*

Fillet With and Without Tangent Propagation

In SolidWorks you can add a fillet feature to a model with or without the tangent propagation. When you invoke the **Fillet PropertyManager**, you can observe that by default the **Tangent propagation** check box is selected. Therefore, if you select an edge, face, feature, or loop to fillet, then it will automatically select other entities that are tangential to the selected entity. Thus, it will apply the fillet feature to all the entities that are tangential to the selected one. If you clear the **Tangential propagation** check box, the fillet is applied only to the selected

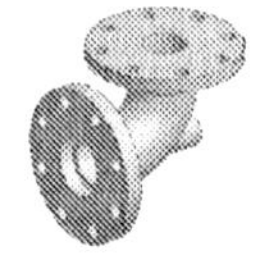

> **Tip**. *The* ***Full preview*** *radio button available in the* ***Fillet PropertyManager*** *is used to preview the fillet feature before actually creating the feature. If you select the* ***Partial preview*** *radio button, then you can view only the partial preview of the fillet feature. If you select a face to add a fillet feature and select the* ***Partial preview*** *radio button then you cannot preview the fillet feature created to all the edges adjacent to the selected face. You can preview only the fillet on the single edge of the selected face. Use the A key from the keyboard to cycle the preview of the fillet feature on other edges of the selected face. If you select the* ***No preview*** *radio button, you cannot see the preview of the fillet feature.*

entity. Figure 6-27 shows the entity to be selected to add a fillet feature. Figures 6-28 and 6-29 show the fillet feature created with the **Tangent propagation** check box cleared and selected respectively.

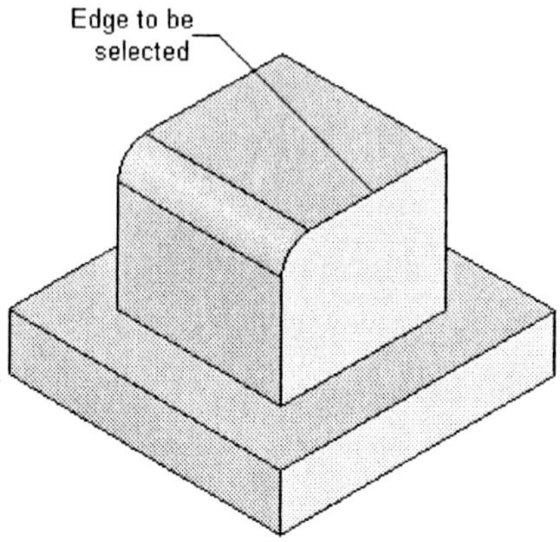

Figure 6-27 *Edge to be selected to apply the fillet feature*

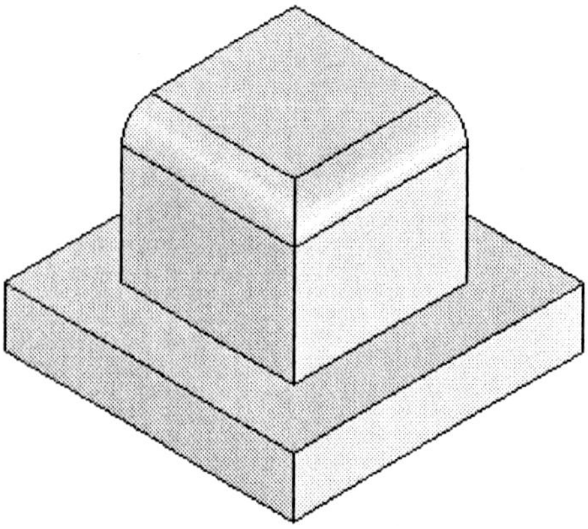

Figure 6-28 *Fillet feature created with the* ***Tangent propagation*** *check box cleared*

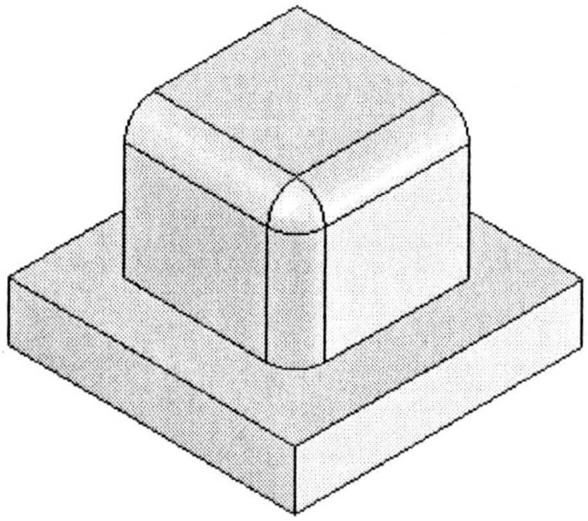

Figure 6-29 *Fillet feature created with the* ***Tangent propagation*** *check box selected*

Setback Fillets

The setback fillet is created where three or more than three edges are merged into a vertex. This type of fillet is used to smoothly blend transition surfaces generated from the edges to

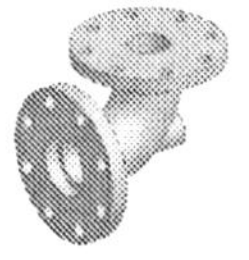

Tip. *You can also drag and drop the fillet features created on one edge to the other edge. Using the left mouse button select the fillet feature from the* ***FeatureManager Design Tree*** *or from the drawing area and hold down the left mouse button and drag the cursor and release the left mouse button to drop the feature on the required edge or face. You can also copy the fillet feature and paste it on the selected entity.*

the fillet vertex. This smooth transition is created between all the selected edges and the vertex selected for the setback type of fillet. To create a setback fillet, invoke the **Fillet PropertyManager** and select three or more than three edges to apply the fillet. Note that the edges should share the same vertex. The preview of the fillet will be displayed in the drawing area. Now, use the left mouse button at the black arrow on the **Setback Parameters** rollout to expand the rollout. The **Setback Parameters** rollout is displayed as shown in Figure 6-30. This rollout will be used to specify the setback parameters.

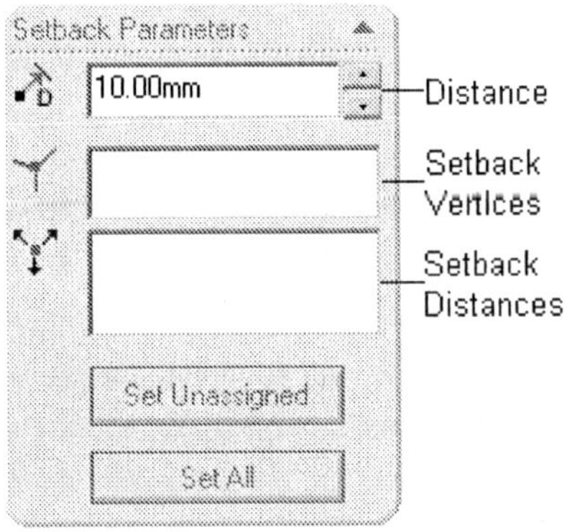

Figure 6-30 *The* ***Setback Parameters*** *rollout*

Using the left mouse button click once in the **Setback Vertices** selection area to invoke the setback vertex selection command. Now, select the vertex where the edges meet. Figure 6-31 shows the selected edges and the vertex assigned the setback parameters.

When you select the vertex for the setback fillet, you will observe that callouts with unassigned setback distances are displayed in the drawing area. The name of the selected vertex is displayed in the **Setback Vertices** selection area. The names of the selected edges are displayed in the **Setback Distances** edit box. Select the name of the edge in the **Setback Distances** edit box to assign a setback distance to that edge. A red color arrow is displayed along that edge. Using the **Distance** spinner provided in the **Setback Parameters** rollout, assign a setback distance. Similarly, assign the setback distance to all the edges. You can also assign the setback distance directly by specifying the value in the setback callouts displayed in the drawing area. As discussed earlier, the preview of the fillet is updated automatically when you assign any value. The **Set Unassigned** button available in the **Setback Parameters** rollout is used to assign the setback distance displayed in the **Distance** spinner to all the unassigned edges. The **Set All** button is used to assign the setback distance displayed in the **Distance** spinner to all the edges. Figure 6-32 shows the preview of the setback fillet and Figure 6-33 shows a setback fillet on the right of the model and a normal fillet on the other side of the model.

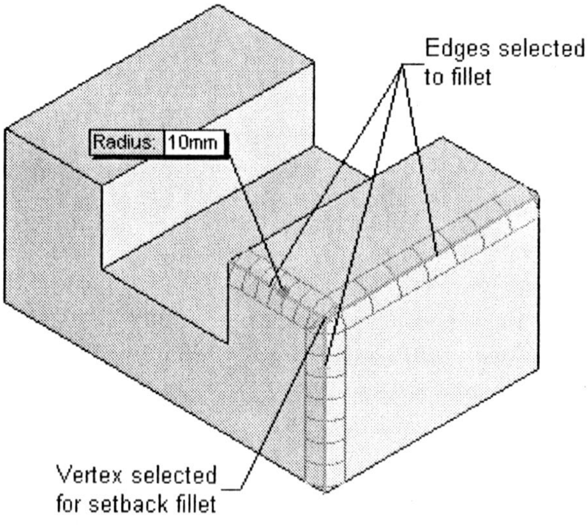

Figure 6-31 Edges and vertex to be selected to apply the setback fillet feature

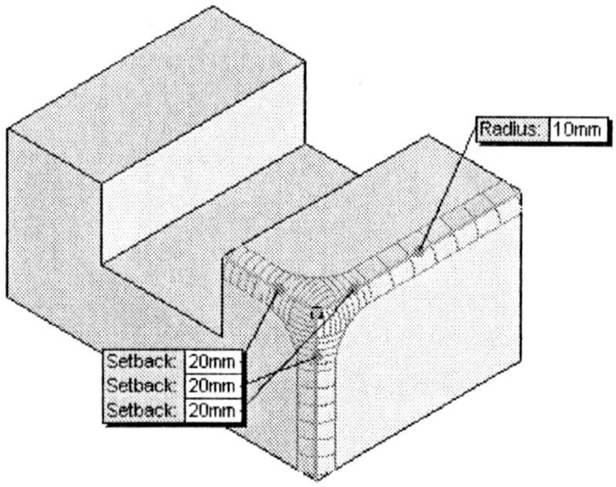

Figure 6-32 Preview of the setback fillet

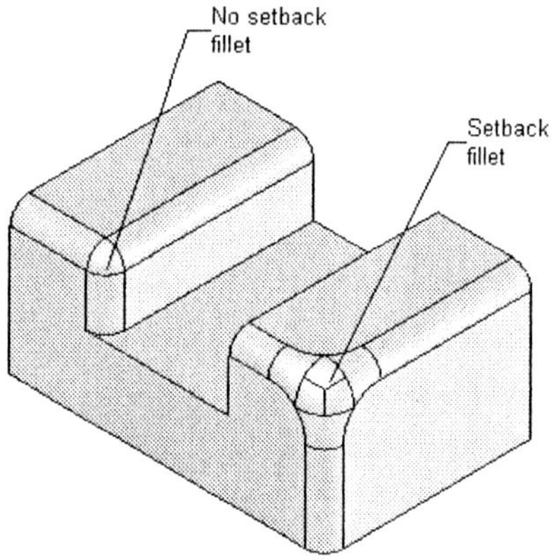

Figure 6-33 Simple and setback fillet features

Other Fillet Options

You are also provided with various other fillet options using which you can create an accurate and aesthetic design. The other fillet options available in the **Fillet PropertyManager** are **Keep features**, **Round corners**, **Controlling the Overflow type**, and so on. These options are discussed next.

Keep feature

If you have boss or cut features in a model and the fillet created is large enough to consume those features, it is recommended that you select the **Keep features** check box available in the **Fillet Options** rollout. This check box is selected by default, but you should confirm before creating any fillet feature. If you clear this check box, the fillet feature will consume the features that will obstruct its path. Note that the features that are consumed by the fillet feature are not deleted from the model. They disappear from the model because of some geometric inconsistency. When you rollback, suppress, or delete the fillet, the consumed features will reappear. You will learn more about rollback and suppress in later chapters. Figure 6-34 shows the model and the edge to be selected for fillet.

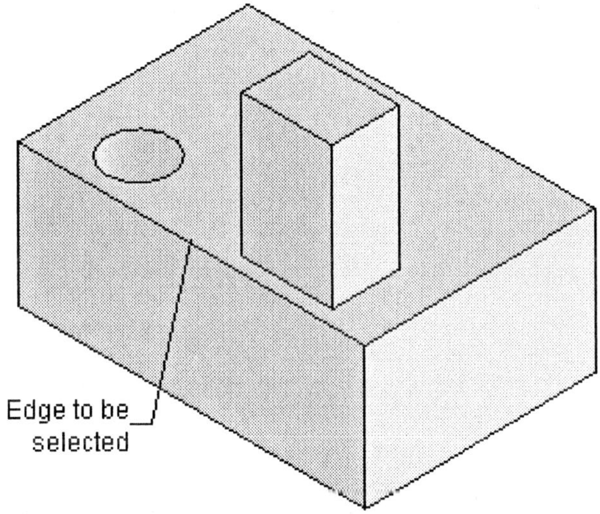

Figure 6-34 *Edge to be selected to apply the fillet feature*

Figure 6-35 shows the fillet feature with the **Keep features** check box cleared. Figure 6-36 shows the fillet feature with the **Keep features** check box selected.

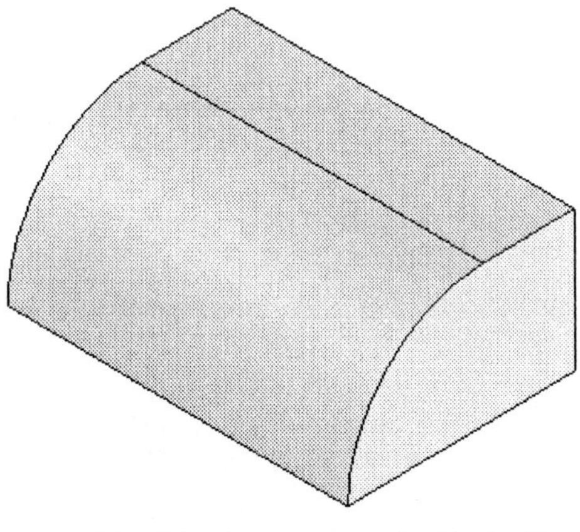

Figure 6-35 *Fillet feature created with the* ***Keep features*** *check box cleared*

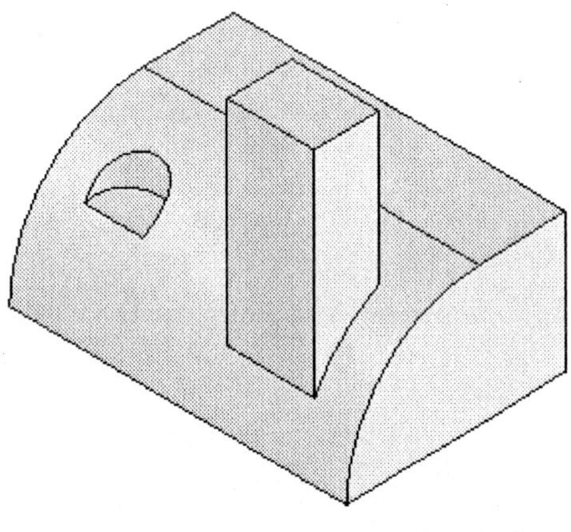

Figure 6-36 *Fillet feature created with the* ***Keep features*** *check box selected*

Round corners

The **Round corners** option is used to round the edges at the corners of the fillet feature. To create a fillet feature with round corners, select the **Round corners** check box from the **Fillet Options** rollout after specifying all the parameters of the fillet feature. Figure 6-37 shows a fillet feature created with the **Round corners** check box cleared and Figure 6-38 shows a fillet feature created with the **Round corners** check box selected.

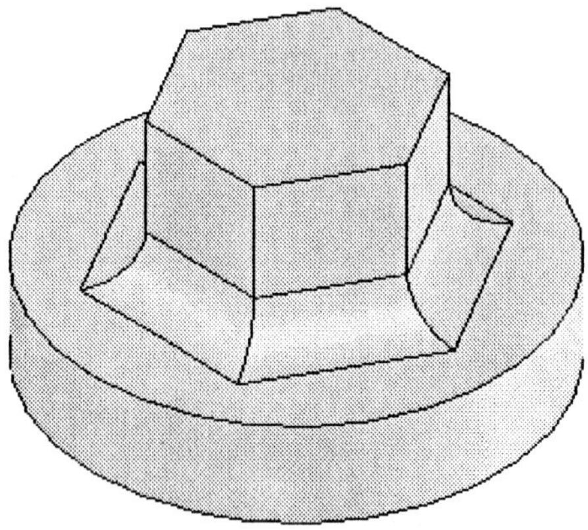

Figure 6-37 Fillet feature created with the ***Round corners*** *check box cleared*

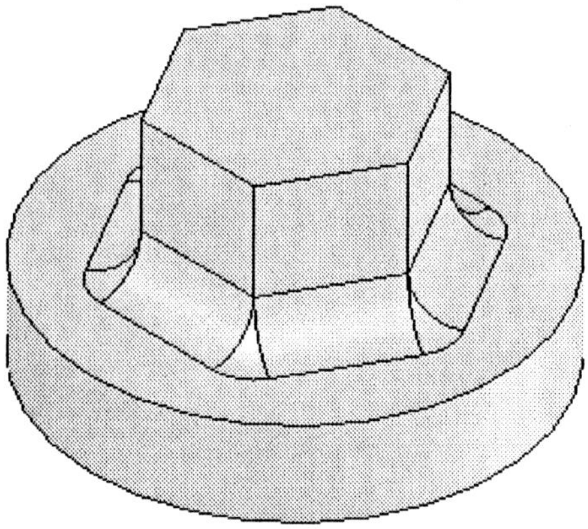

Figure 6-38 Fillet feature created with the ***Round corners*** *check box selected*

Overflow type

The **Overflow type** area is used to specify the physical condition that the fillet feature should adopt when it extends beyond an area. By default, SolidWorks automatically adopts the best possible flow type to accommodate the fillet, depending on the geometric conditions. This is because the **Default** option is selected by default in the **Overflow type** area. The other two options available in this area are discussed next.

Keep edge

The **Keep edge** radio button is selected when the fillet feature extends beyond a specified area. Therefore, to accommodate the fillet feature this option will divide the fillet into multiple surfaces and the adjacent edges are not disturbed. A dip is created at the top of the fillet feature. Figure 6-39 shows a fillet feature created with the **Keep edge** radio button selected from the **Overflow type** area.

Keep surface

The **Keep surface** radio button available in this area is selected to accommodate the fillet feature by trimming the fillet feature. This will maintain the smooth rounded fillet surface but it disturbs the adjacent edges. As this option maintains the smooth fillet surface, it extends the adjacent surface. Figure 6-40 shows a fillet feature created with the **Keep surface** radio button selected from the **Overflow type** area.

Variable Radius Fillet

The variable radius fillet is created by specifying different radii along the length of selected edge at specified intervals. Depending on the options, you can create a smooth transition or a straight transition between the vertices to which the radii are applied. To create a variable radius fillet invoke the **Fillet PropertyManager**. Select the **Variable radius** radio button from the **Fillet Type** rollout. The **Variable Radius Parameters** rollout is automatically displayed in the **Fillet PropertyManager** as shown in Figure 6-41.

You are prompted to select the edge to fillet. Using the left mouse button select the edge or

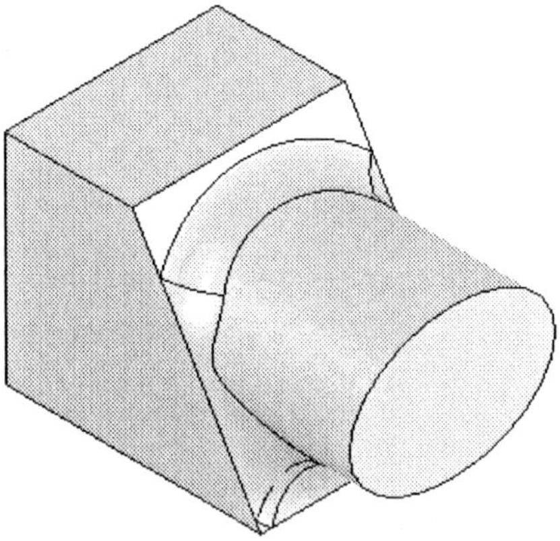

Figure 6-39 *Fillet feature created with the overflow type as* ***Keep edge***

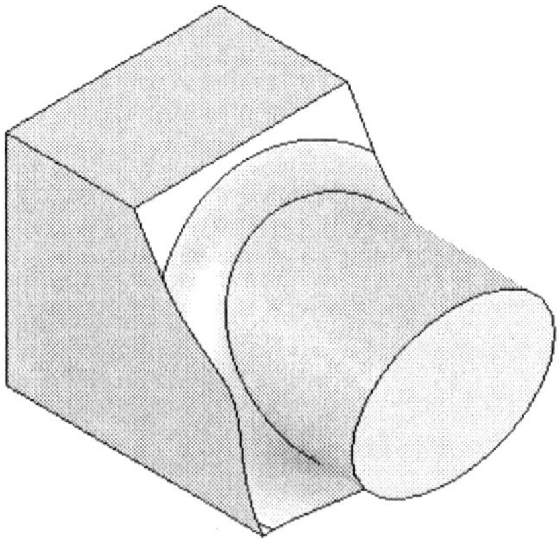

Figure 6-40 *Fillet feature created with the overflow type as* ***Keep surface***

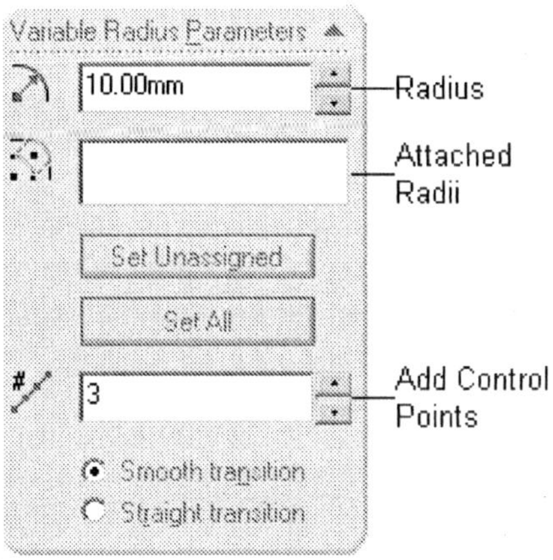

Figure 6-41 *The* ***Variable Radius Parameters*** *rollout*

edges that you want to fillet. The name of the selected edge is displayed in the **Edges, Faces, Features and Loops** selection area. By default, the radius is applied at the start point and the endpoint. The variable radius callouts are displayed at the ends of the selected edge as shown in Figure 6-42.

The names of the vertices on which the callouts are added are displayed in the **Attached Radii** display box. You will find three red points on the selected edge because the value of the control point in the **Add Control Points** spinner is set to **3**. You can create a number of control points using the **Add Control Points** spinner. These control points are also called movable points because you can change the position of these control points. The additional radii are specified on these points on the selected edge.

Using the left mouse button select the three control points available on the selected edge. As you select the control point the **Radius and Position** callouts are displayed for each control point as shown in Figure 6-43. The name of the selected point is also displayed in the **Attached Radii** display box.

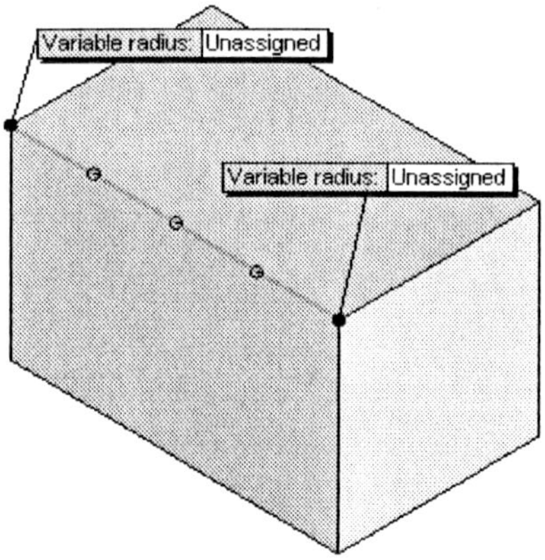

Figure 6-42 Variable radius callouts displayed at the vertices of the selected edge

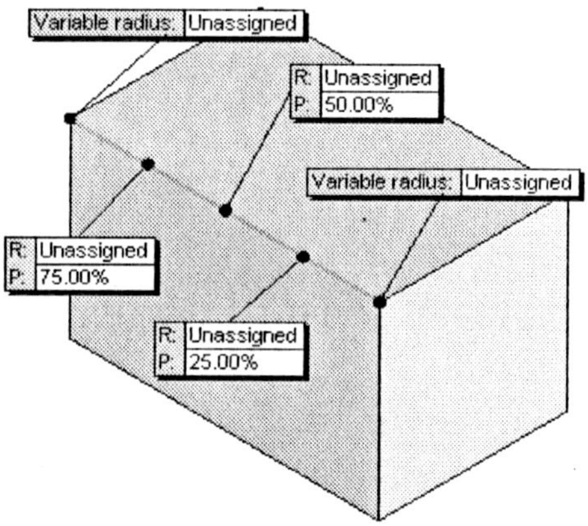

***Figure 6-43 Radius and Position** callouts are displayed after selecting the control points*

You will observe that the position of three points is described in terms of percentage. You can modify the position of the points by modifying the value of percentage in the **Position** area of the **Radius and Position** callout. By following this procedure you can also modify the placement of other points. You will observe that radius value is not assigned to any of the callouts. Therefore, you need to specify the value of the radius in the callouts. Using the left mouse button select the name of the vertex or point; the name of the selected item is highlighted in yellow in its respective callout. Using the **Radius** spinner set the value of the radius for the selected item. You can also specify the value of the radius in the radius area of the callout. Set the value of each unassigned radius. You can also use the **Set Unassigned** button to assign the value displayed in the **Radius** spinner to all the unassigned radii. The **Set All** button is used to assign the same value that is displayed in the **Radius** spinner to all the radii. Figure 6-44 shows the preview of the fillet feature with modified positions of the control points with radius values specified to all the points and vertices. Figure 6-45 shows the resultant fillet feature.

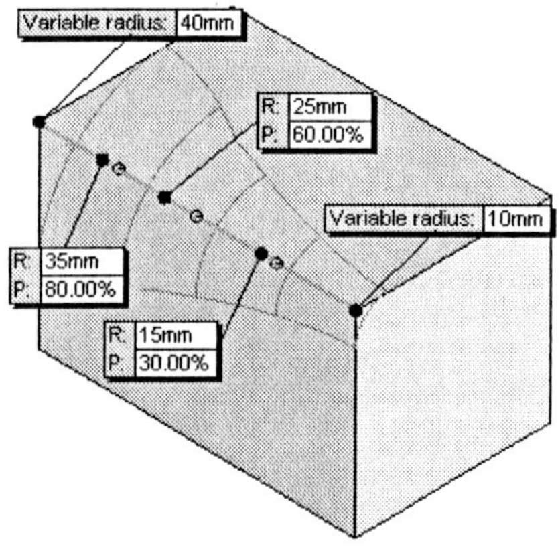

Figure 6-44 Preview of the variable radius fillet

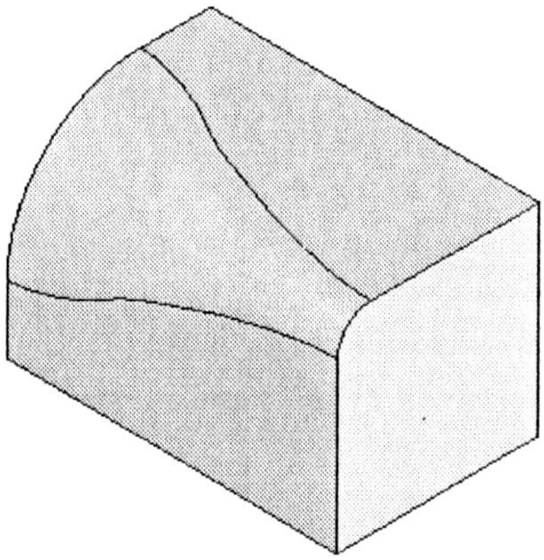

Figure 6-45 Resultant fillet

Smooth transition

The **Smooth transition** radio button when selected creates a smooth transition by smoothly blending the points and vertices on which you have defined the radius on the selected edge.

Straight transition

The **Straight transition** radio button when selected creates a linear transition by blending the points and vertices on which you have defined the radius on the selected edge. In this case the edge tangency is not maintained between one fillet radius and the adjacent face.

Figures 6-46 and 6-47 show the fillets created with smooth transition and straight transition options respectively.

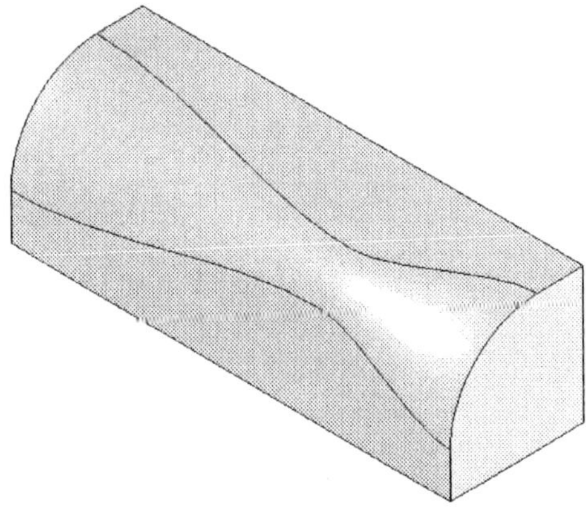

Figure 6-46 *Variable radius fillet with smooth transition*

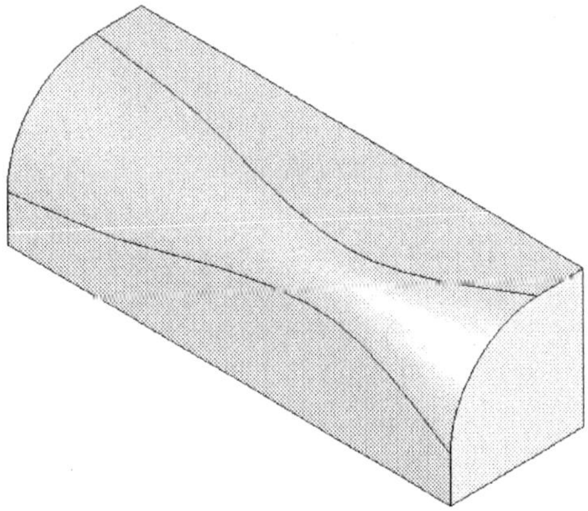

Figure 6-47 *Variable radius fillet with straight transition*

Face Fillet

Using the **Face Fillet** you can add a fillet between two sets of faces. It blends the first set of face with the second set of face. It adds or removes the material according to the geometric conditions. It can also completely or partially remove the faces to accommodate the fillet feature. To create a face fillet feature, invoke the **Fillet PropertyManager** and select the **Face fillet** radio button from the **Fillet Type** rollout. The **Item To Fillet** rollout is modified and provides the **Face Set 1** and **Face Set 2** selection areas. The **Fillet PropertyManager** is shown in Figure 6-48 with the **Face fillet** radio button selected.

You are prompted to select the faces to fillet for face set 1 and face set 2. Using the left mouse button select the first set of faces. You can select even more than one face in a set. The name of the selected faces is displayed in the **Face Set 1** display box and the selected faces are displayed in green color. The **Face Set 1** callout with radius is displayed in the drawing area. Click in the **Face Set 2** selection area to invoke the selection tool and select the second set of faces. The second set of selected faces is displayed in magenta color and the **Face Set 2** callout is displayed in the drawing area. With this release of SolidWorks the preview of the face fillet is displayed in the drawing area. Now, set the value of the radius in the **Radius** spinner. The **Tangent propagation** check box is used to create the face fillet tangent to the adjacent faces. This check box is selected by default. If you clear this check box, the fillet is

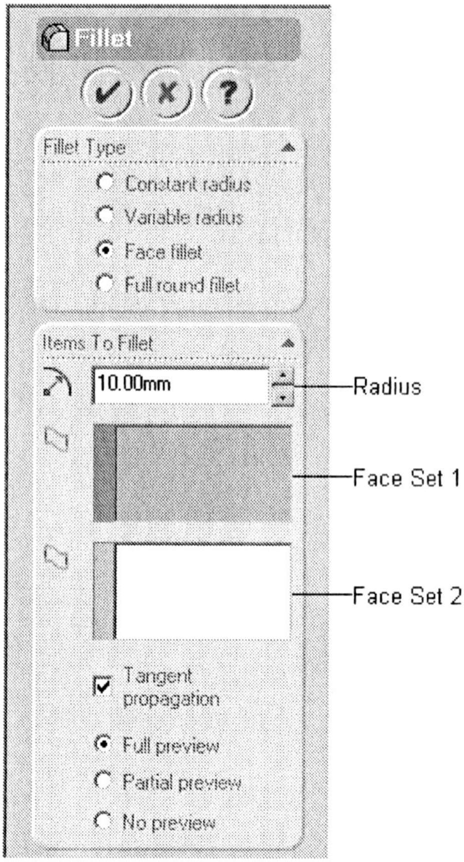

Figure 6-48 *The* ***Fillet PropertyManager*** *with* ***Face fillet*** *radio button selected*

not forced to be tangent to the adjacent faces. Figure 6-49 shows the faces to be selected to apply the face fillet. Figure 6-50 displays the resultant fillet with three faces of the slot completely eliminated after applying the fillet.

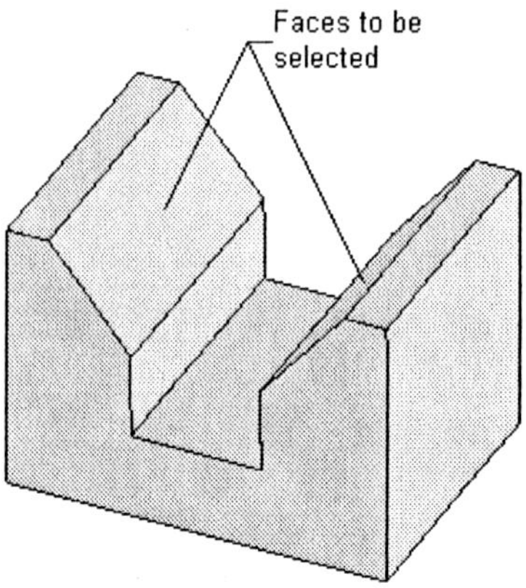

Figure 6-49 *Faces to be selected*

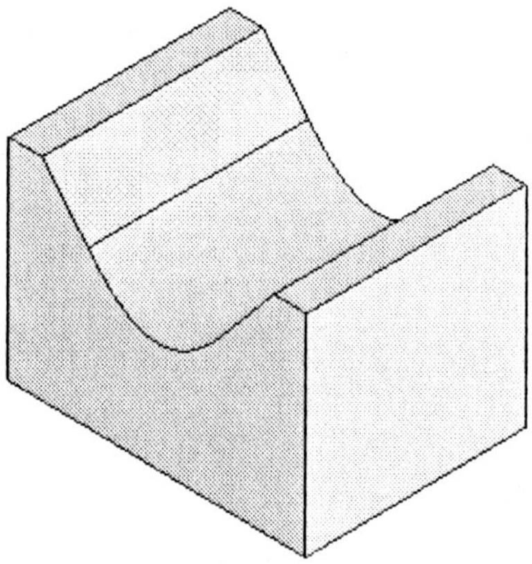

Figure 6-50 *Resultant fillet*

Face Fillet Using the Hold Line

Using the face fillet created with the hold line you can specify the radius and the shape of the fillet by determining a hold line. A hold line can be a set of edges, or a split line projected on a face. You will learn more about split lines in the later chapters. The radius of the fillet is determined by the distance between the hold line and the edges or faces selected to be filleted. To create a face fillet using the hold line invoke the **Fillet PropertyManager**. Using the left mouse button click on the black arrow available at the top right of the **Fillet Options** rollout to display the rollout as shown in Figure 6-51.

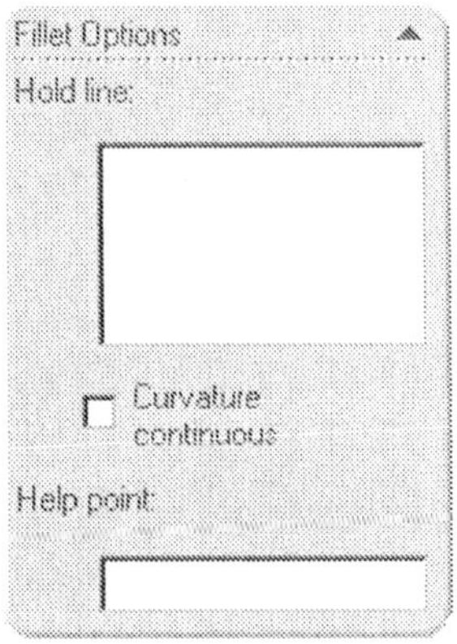

Figure 6-51 *The* ***Fillet Options*** *rollout*

The selection mode is active in the **Face Set 1** selection area and you are prompted to select the faces to fillet for face set 1 and face set 2. Using the left mouse button select the faces for the face set 1. The names of the selected faces are displayed in the **Face Set 1** selection area and the selected faces are displayed in green. Now, click in the **Face Set 2** selection area to activate the selection mode and select the faces to add in the face set 2. The selected faces are displayed in magenta. The preview of the face fillet with the default settings is displayed in the drawing area. By default, the **Tangent propagation** check box is selected. Therefore, you do not need to select the tangent faces in both the face sets. This is because it will automatically select the faces tangent to the selected face. Now, using the left mouse button click in the **Hold line** display box and select the hold line or lines. The preview of the face fillet modifies automatically. Now, choose the **OK** button from the **Fillet PropertyManager**. Figure 6-52 shows an example in which the faces and the hold line are selected. Figure 6-53 shows the resultant face fillet using the hold line.

Curvature Continuous in the Face Fillet With Hold Line

The **Curvature continuous** check box is selected to apply the face fillet feature with continuous curvature throughout the fillet feature. Note that fillet with continuous curvature is possible only by creating a face fillet feature with hold line. You have to specify the hold lines on both set of faces. Figure 6-54 shows a model in which face fillet using the hold line is created on both the pillars. On the right pillar the face fillet is created with the **Curvature continuous** check box cleared and on the left pillar the face fillet is created with the **Curvature continuous** check box selected.

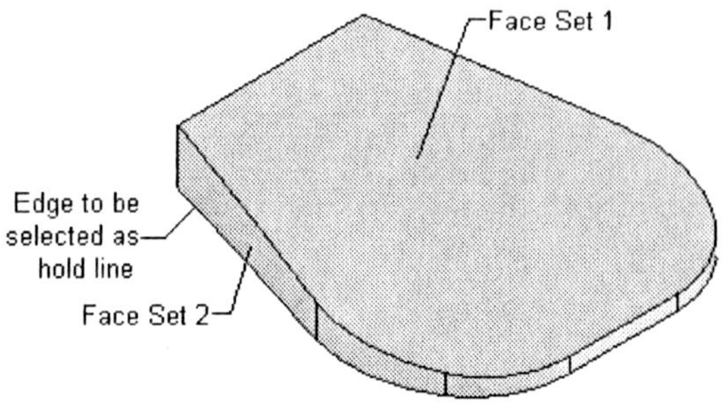

Figure 6-52 Faces and hold line to be selected

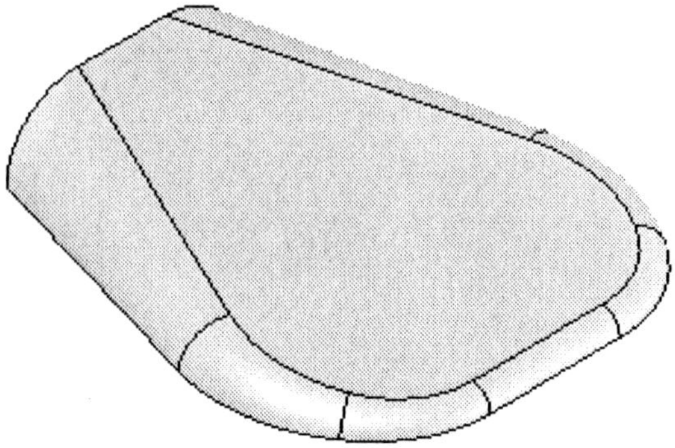

Figure 6-53 Resultant fillet

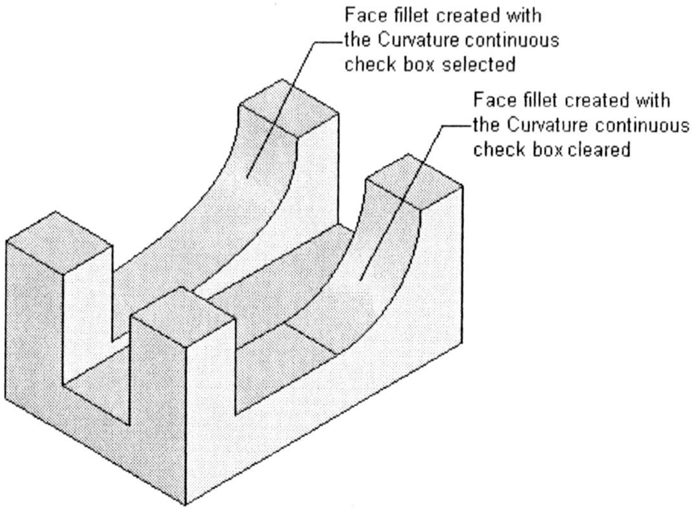

*Figure 6-54 Face fillet created with the **Curvature continuous** check box selected and cleared*

Full Round Fillet

The full round fillet is used to add a semicircular fillet feature. To create a full round fillet invoke the **Fillet PropertyManager** and select the **Full round fillet** radio button. The **Item To Fillet** rollout is modified as shown in Figure 6-55. The selection mode is active in the **Side Face Set 1** selection area and you are prompted to select faces for the center and side face sets. Select the first face for the side face set 1. The selected face is displayed in green. Now, click in the **Center Face** selection area and select the center face. The selected face is displayed in blue. Next, click in the **Side Face Set 2** selection area and select the face for the side face set 2. The preview of the full round fillet is displayed in the drawing area. Choose the **OK** button from the **Fillet PropertyManager**. Figure 6-56 shows the two faces being selected; the third face to be selected is the left face parallel to the first selected face. Figure 6-57 shows the resultant full round fillet.

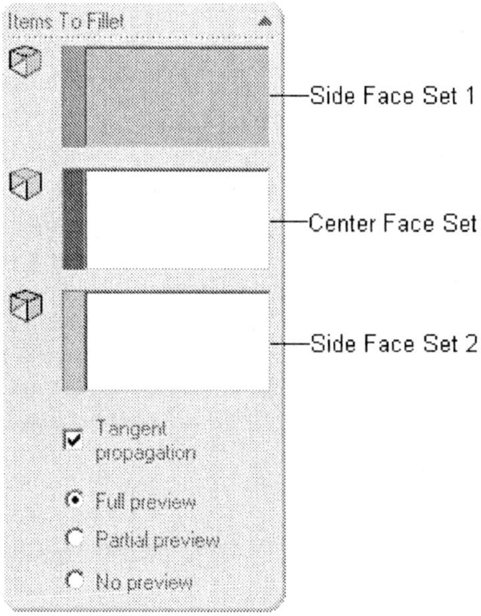

Figure 6-55 *The **Item to Fillet** rollout when the **Full round fillet** radio button is selected from the **Fillet Type** rollout*

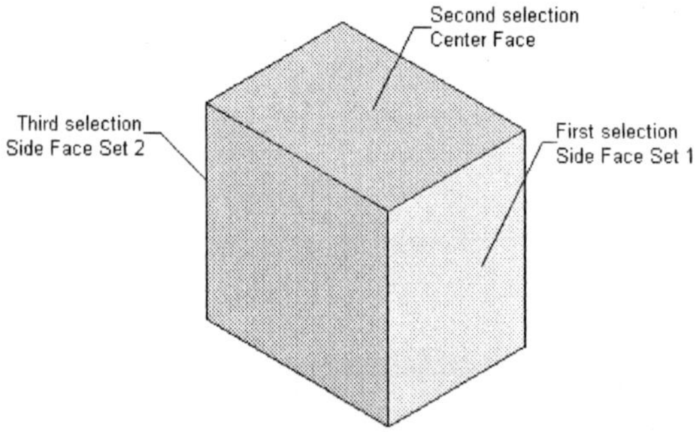

Figure 6-56 *Faces selected to create full round fillet.*

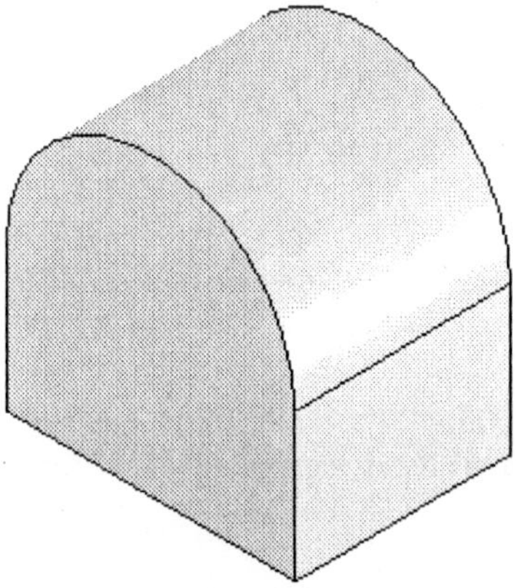

Figure 6-57 *Resultant full round fillet*

Selection Methods

As you have learned about basic and advanced modeling tools, it is necessary for you to learn about some selection methods using which you can increase your productivity and speed of modeling. The selection methods that can increase your speed of creating fillets and chamfers are discussed next.

Select Other

The **Select Other** option is the most common tool to cycle through the entities for selection. This option is used when the selection is difficult in a multifeatured complex model. Before invoking any other tool, select any entity and right-click to invoke the shortcut menu. Choose the **Select Other** option from the shortcut menu. The select cursor is replaced by the cycle through cursor. Also, any one face of the model is selected and the boundary of the face is highlighted in green. You can cycle through the other faces using the right-click and when the required face is selected, use the left mouse button to confirm the selection.

Select Loop

The **Select Loop** option is used to select the loops and using this tool you can also cycle through various loops before confirming the selection. This option is very useful when you are working with a complex model and you have to select a loop from that model. Using the select cursor, select any of the edge of the loop and right-click to invoke the shortcut menu. Choose the **Select Loop** option from the short cut menu. The loop that is possible by selecting that edge will be highlighted in green and an arrow will be displayed in yellow color. Move the cursor on that arrow and when the arrow is displayed in red, use the left mouse button to cycle through the loops. Repeat this until you select the required loop. Figure 6-58 shows a loop selected using the **Select Loop** option. Figure 6-59 shows the second loop selected when the left mouse button is used on the arrow to cycle through the loops.

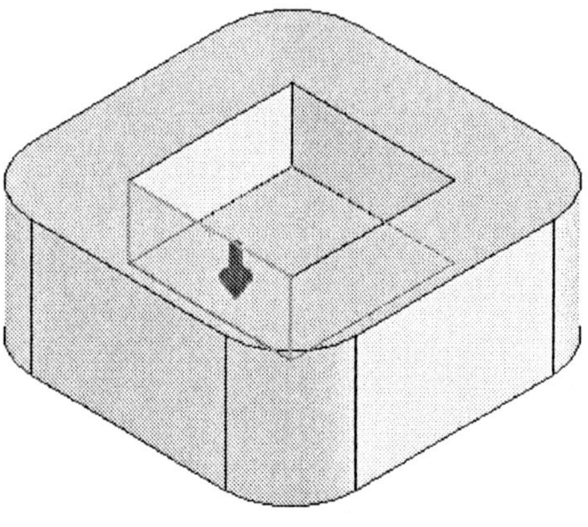

Figure 6-58 *Loop selected using the* ***Select Loop*** *tool*

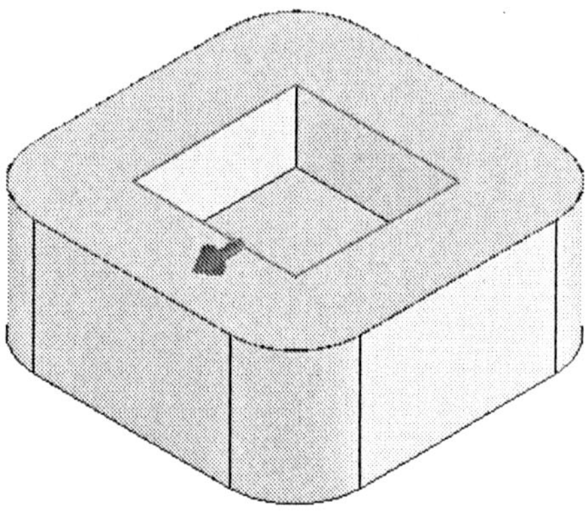

Figure 6-59 *Second loop is selected while cycling through the loops*

Select Partial Loop

Selecting the partial loop is one of the enhancements of SolidWorks 2004. Using this technique you can select the partial loop created by joining two edges. To select a partial loop, select two edges as shown in Figure 6-60. Invoke the shortcut menu and select the **Select Partial Loop** option from the shortcut menu. Generally, the selection point on second edge defines the major or minor partial loop selection. Figure 6-61 shows the resultant partial loop selected.

Note

For better visualization of the selected partial loop the display of the model edges is turned off in Figures 6-60 and 6-61.

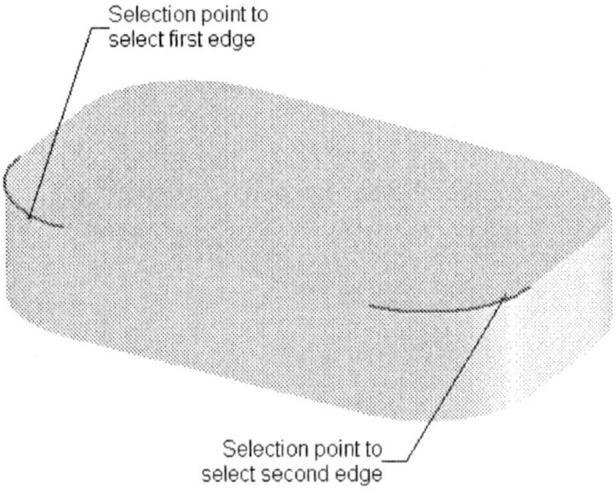

Figure 6-60 Selecting edges to define the partial loop

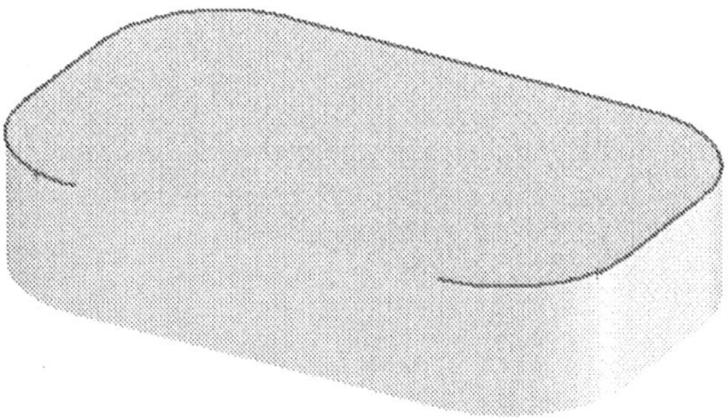

Figure 6-61 Resultant partial loop that is selected

Select Tangency

The **Select Tangency** option is used to automatically select the edges or faces that are tangent to the selected face. This option is available in the shortcut menu only when a face or edge is tangent to the selected face or edge. For using this option select any face or edge using the select tool and right-click to invoke the shortcut menu and choose the **Select Tangency** option from the shortcut menu.

Tip. *The* ***Select Midpoint*** *option available in the shortcut menu is generally used in the sketching environment or while creating 3D sketches. You will learn more about 3D sketches in the later chapters.*

Creating Chamfer

Enhanced

CommandManager:	Features > Chamfer
Menu:	Insert > Features > Chamfer
Toolbar:	Features > Chamfer

Chamfering is defined as a process in which the sharp edges are bevelled in order to reduce the area of stress concentration. This process also eliminates the sharp edges and corners that are not desirable. In SolidWorks chamfer is created using the **Chamfer** tool. This tool is invoked by choosing the **Chamfer** button available in the **Features CommandManager**. When you invoke the **Chamfer** tool, the **Chamfer PropertyManager** is displayed as shown in Figure 6-62. Various types of chamfers created using the **Chamfer PropertyManager** are discussed next.

Creating Edge Chamfer

The chamfers that are applied to the edges are known as edge chamfer. To create an edge chamfer invoke the **Chamfer PropertyManager** and then select the edges to chamfer. When you select an edge to chamfer, the preview of the chamfer feature with a distance and angle callout is displayed in the drawing area. The name of the selected edge is displayed in the **Edges and Faces or Vertex** selection area. Also, the selected entity is highlighted in green

Tip. *For creating a fillet or a chamfer feature if you select a face, the fillet or chamfer is applied to all the edges of that face. Consider a case in which you have a blind cut feature on the top face of a block. You want to fillet only the upper edges of the cut feature. You can use the* ***Select Loop*** *option to execute feature creation. If you select the top face of the model, then all the edges of the top of the block and the edges of the cut will be filleted. Therefore, after selecting the top face of the block press and hold down the CTRL key and select any one upper edge of the cut. Now, apply the fillet feature to this combination of selection. You will observe that only the upper loop of edges of the cut will be filleted instead of the whole upper face. In the same way you can also fillet only the edges of the block when you are also having a slot on the top face of the model.*

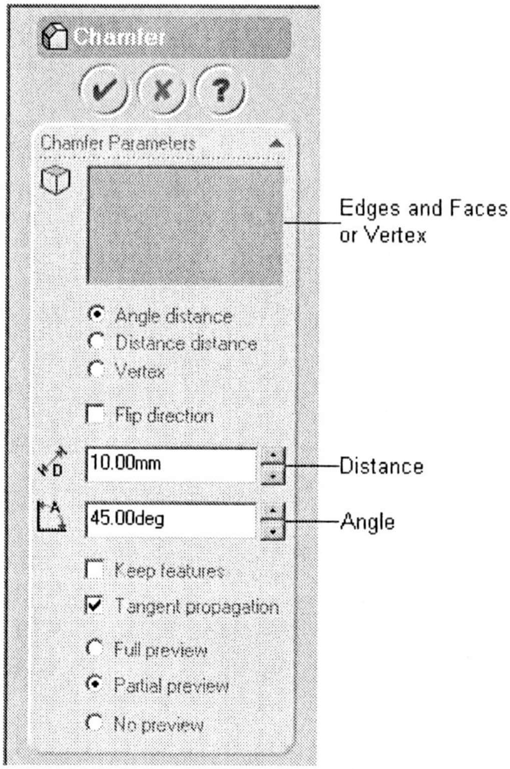

Figure 6-62 *The* ***Chamfer PropertyManager***

color and a yellow arrow is also displayed in the preview. With this release of SolidWorks the **Tangent propagation** check box and the preview of the feature is included in the **Chamfer PropertyManager**. The **Tangent propagation** check box is selected by default. Therefore, the edges tangent to the selected edge are selected automatically. By default the **Partial preview** button is selected by default; you can select the **Full preview** button to display the full preview of chamfer feature. Figure 6-63 shows the edge to be selected for chamfering and Figure 6-64 shows the full preview of the chamfer feature.

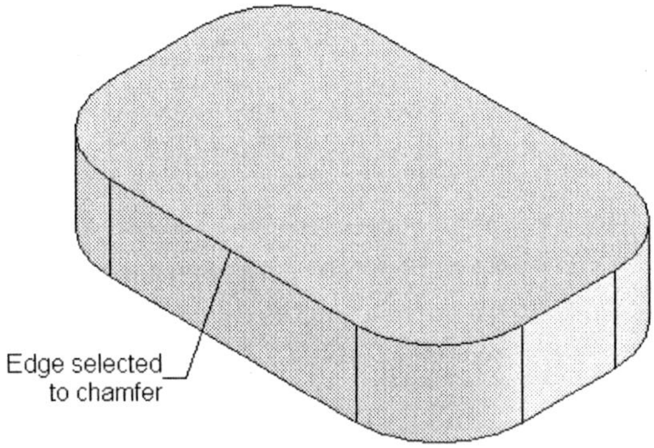

Figure 6-63 *Edge selected to chamfer*

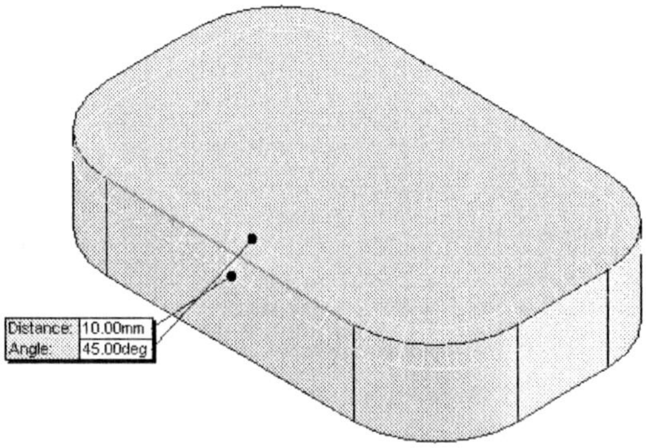

Figure 6-64 *Preview of the chamfer feature*

By default, the **Angle distance** radio button is selected. Therefore, the distance and angle callout is displayed in the drawing area. You can set the value of the distance and angle using the **Distance** spinner and the **Angle** spinner or you can insert the value of distance and angle directly in the **Distance** and **Angle** callouts. The **Flip direction** check box is used to specify the direction of distance measurement. If you select the **Flip direction** check box, the arrow is also flipped in the preview in the drawing area. You can also flip the direction by clicking the arrow in the drawing area.

If you select the **Distance distance** radio button from the **Chamfer Parameters** rollout, the **Flip direction** check box is replaced by the **Equal distance** check box. The **Angle** and **Distance** callouts are replaced by **Distance 1** and **Distance 2** callouts. By default, the **Equal distance** check box is cleared. Set the value of the chamfer distance in the **Distance 1** spinner or specify the value in the callout. Now, set the value of distance 2 in the respective spinner or callout If you need to specify the name distance for creating the chamfer, select the **Equal distance** check box. The **Distance 2** spinner disappears from the **Chamfer Parameters** rollout.

After specifying all the parameters choose the **OK** button from the **Chamfer PropertyManager**. Figure 6-65 shows the chamfer created on a model.

Creating Vertex Chamfer

Using the chamfer tool you can also add chamfer to a selected vertex. It will chop the selected vertex to the specified distance. To create the vertex chamfer invoke the **Chamfer PropertyManager** and select the **Vertex** radio button from the **Chamfer Parameters** rollout. Select the vertex; the preview of the chamfer is displayed in the drawing area with **Distance** callouts. Figure 6-66 shows the vertex to be selected and Figure 6-67 shows the preview of the vertex chamfer.

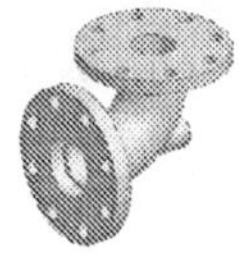

Tip. *You can also select the face for applying the chamfer feature. If you select the face for applying the chamfer, then the chamfer is applied to all the edges of the selected face.*

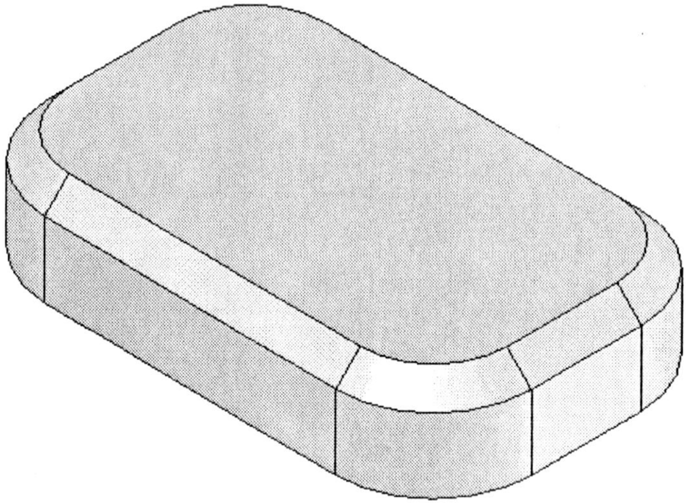

Figure 6-65 *Chamfer created on a base plate*

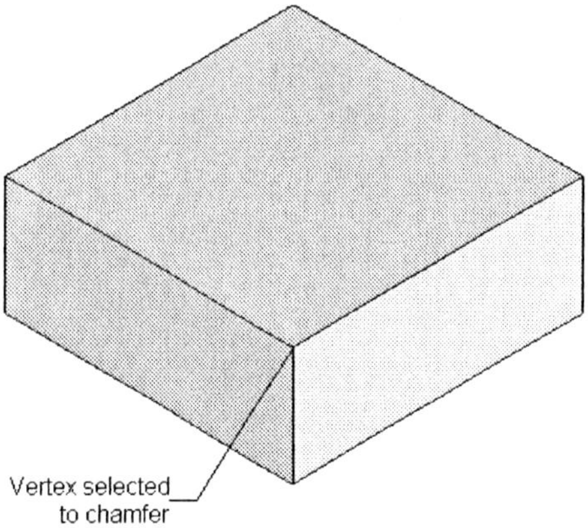

Figure 6-56 *Vertex to be selected*

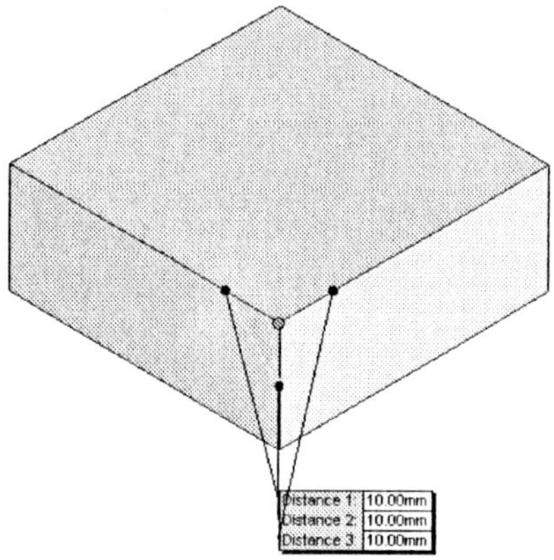

Figure 6-57 *Preview of the vertex chamfer*

Set the value of the chamfer distance along each edge in the **Distance 1**, **Distance 2**, **Distance 3** spinners. You can also specify the value of chamfer distance in the distance callouts. If you want to specify equal distance to all the edges then select the **Equal distance** check box. After specifying all the parameters choose the **OK** button from the **Chamfer PropertyManager**. Figure 6-68 shows the vertex chamfer feature applied to the base feature.

Chamfer With and Without Keep Feature Option

If you have boss or cut features in a model and the chamfer created is large enough to consume those features, it is recommended that you select the **Keep features** check box. If this check box is cleared, the chamfer feature will consume the features that will obstruct its path. Note that the features that are consumed by the chamfer feature are not deleted from the model. They are removed from the model because of some geometric inconsistency. When you rollback or delete that chamfer, the consumed features will

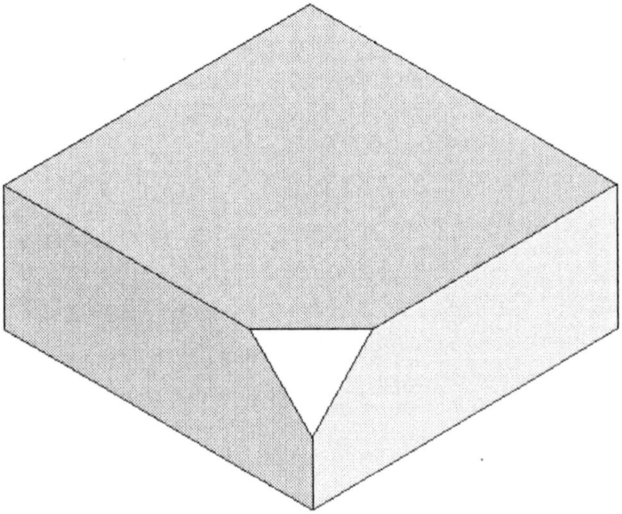

Figure 6-68 *Vertex chamfer created on a base feature*

reappear. Figure 6-69 shows the chamfer feature with the **Keep features** check box cleared. Figure 6-70 shows the chamfer feature with the **Keep features** check box selected.

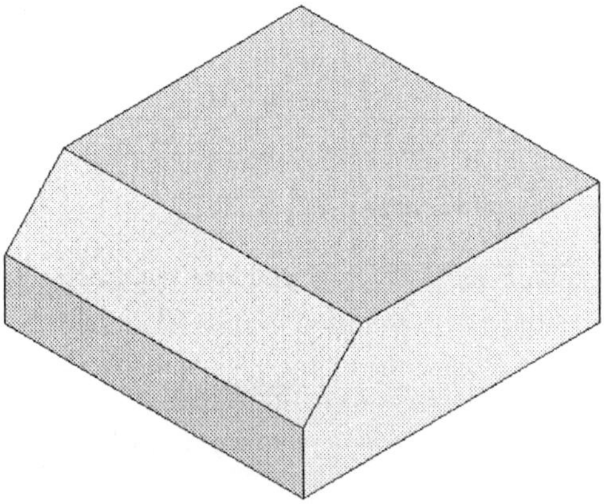

Figure 6-69 *Chamfer feature with the* ***Keep features*** *check box cleared*

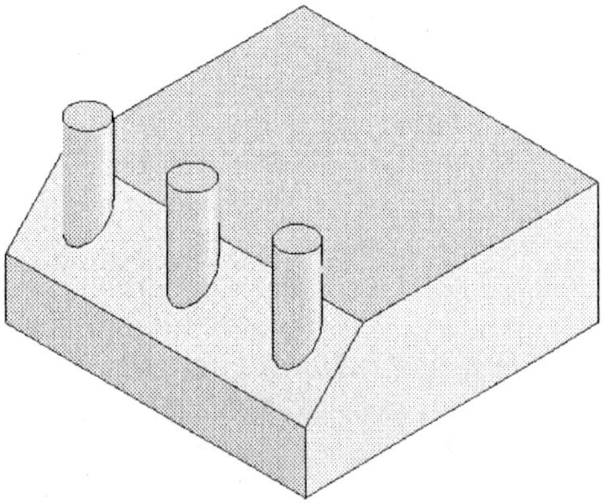

Figure 6-70 *Chamfer feature with the* ***Keep features*** *check box selected*

Creating the Shell Feature

Enhanced

CommandManager:	Features > Shell
Menu:	Insert > Features > Shell
Toolbar:	Features > Shell

Shelling is defined as the process in which the material is scooped out from a model and the resultant model is hollowed from inside. The resultant model will be a hollow model with walls of specified thickness and cavity inside. The selected face or faces of the model are also removed in this operation. If you do not select any face to remove, it will create a closed hollow model. You can also specify multiple thickness to the walls. You can create shell feature using the **Shell** tool. Use the **Shell** button from the **Features CommandManager** The **Shell1 PropertyManager** is invoked as shown in Figure 6-71.

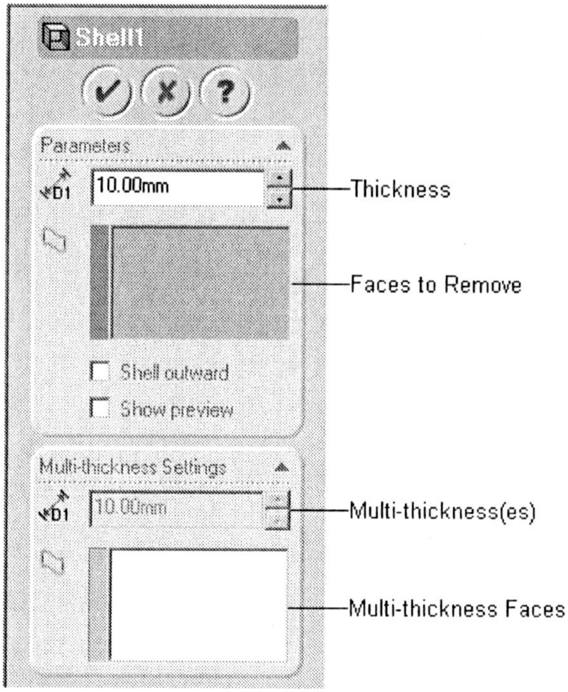

Figure 6-71 *The* ***Shell1 PropertyManager***

You are prompted to select the faces to remove. Select the face or faces from the model that you want to remove. The selected faces will be highlighted in green color and their names are displayed in the **Faces to Remove** selection area. Set the value of the wall thickness in the **Thickness** spinner and choose the **OK** button from the **Shell1 PropertyManager**. Figure 6-72 shows the face selected to remove and Figure 6-73 shows the resultant shell feature created.

If none of the faces are selected to be removed, then the resultant model will be hollowed from inside with no face removed. Figure 6-74 shows a model in the **Hidden Line Visible** mode with a shell feature in which faces are not selected to be removed.

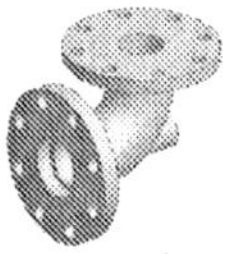

Tip. *If the thickness of the shell feature is more than the radius of the fillet feature then the fillet will not be included in the shell feature. Therefore, it results in sharp edges after adding the fillet. Same is with the chamfer feature. The faces selected to be removed in the shell feature can be a planar face or a curved face. But creating a shell by removing a curved face depends upon the geometry of the curved face to adopt the specified shell thickness and other geometric conditions.*

The ***Shell outward*** *check box is selected to create the shell feature on the outer side of the model.*

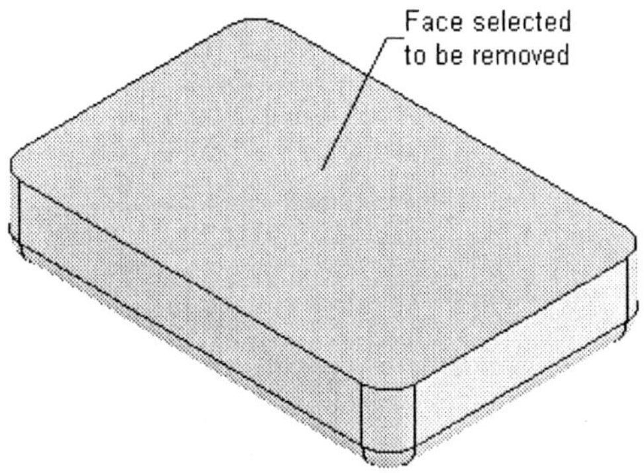

Figure 6-72 Face selected to be removed

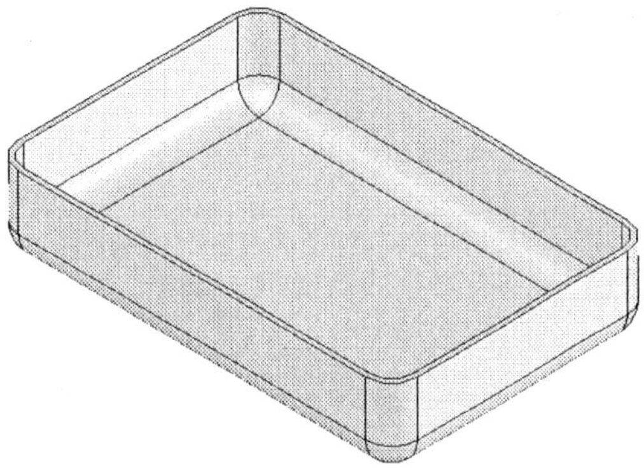

Figure 6-73 The resultant shell feature

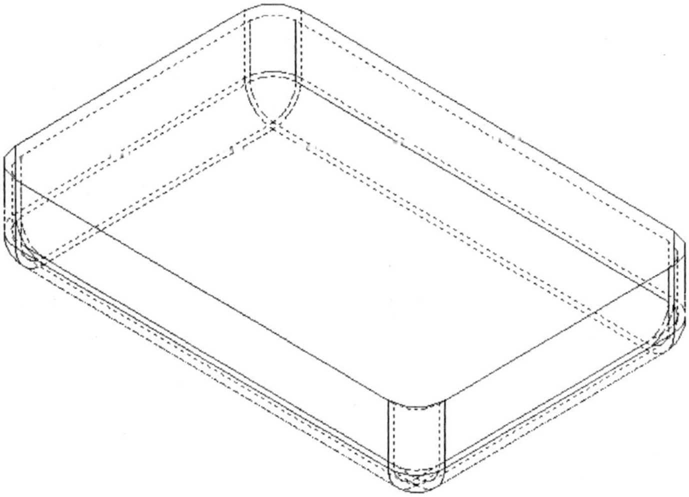

Figure 6-74 Shell feature with faces not selected to remove

The in-built artificial intelligence of shelling command in SolidWorks enables the shell feature to decide how much quantity of material to be removed depending upon the geometric conditions. Refer to Figure 6-75 in which the wall thickness of the shell feature is less for uniform shelling of the entire model. Refer to Figure 6-76 in which the wall thickness is more because of which it cannot accommodate the uniform shelling of the entire model. Therefore, the shell feature will not remove the material from those area where the material removal is not possible.

With this release of SolidWorks, you can also display the preview of the shell feature. To display the preview of the shell feature, select the **Show preview** check box from the **Parameters** rollout of the **Shell1 PropertyManager**.

Creating Multi-thickness Shell

Using this option available in the shell tool you can specify different thickness values to the

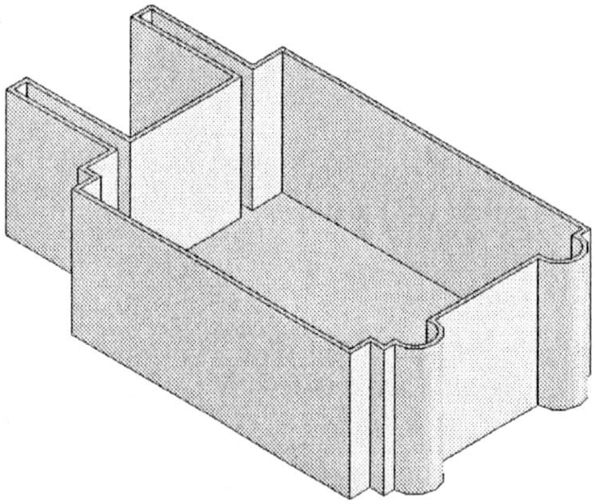

Figure 6-75 *Shell feature with smaller shell thickness*

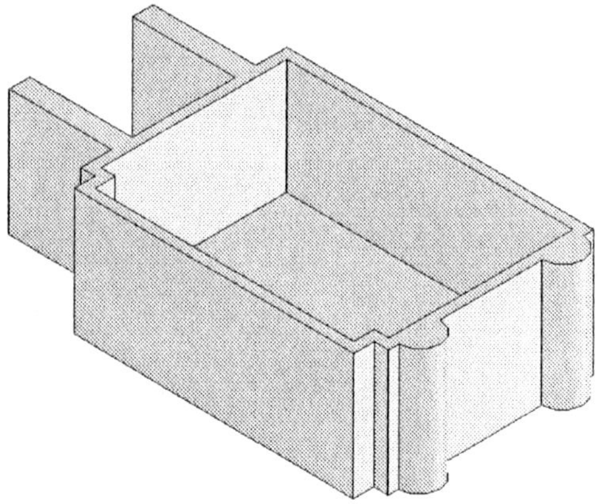

Figure 6-76 *Shell feature with larger shell thickness*

selected faces. To use this option, invoke the **Shell1 PropertyManager** and select the faces to be removed and then specify the uniform thickness in the **Thickness** spinner of the **Parameters** rollout. Click once in the **Multi-thickness(es) Faces** selection area to activate the selection mode.

Select the faces on which you want to specify the special thickness. Set the value of the thickness using the **Multi-thickness(es)** spinner and choose the **OK** button. Figure 6-78 shows the faces to be selected to specify the multi-thickness shell. Figure 6-79 shows the resultant shell feature.

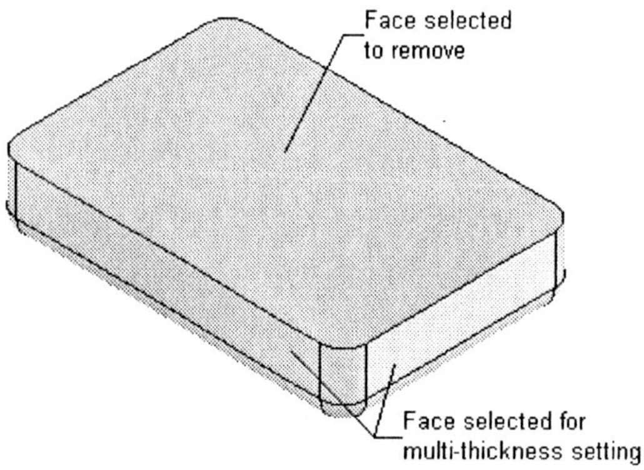

Figure 6-77 *Faces to be selected to apply multithickness shell*

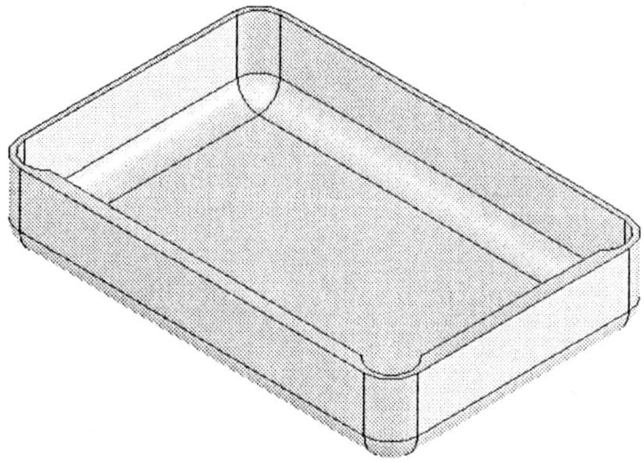

Figure 6-78 *Resultant multithickness shell*

Error Diagnostics

With this release of SolidWorks, the **Error Diagnostics** option is available while creating the shell feature. This option activates automatically if the shell feature creation fails. With the help of this option you could figure out the possible reasons for failure of shell feature.

While creating the shell feature, you specify all the parameters in the **Shell1 PropertyManager** and choose the **OK** button. If the shell feature creation fails because of geometric inconsistency, then the **Shell1 - Rebuild Errors** dialog box is displayed that informs you about the possible errors because of which the feature creation failed. You will learn more about the **Rebuild Errors** dialog box later in this chapter. The **Error Diagnostics** rollout shown in Figure 6-79 is displayed in the **Shell 1 PropertyManager**.

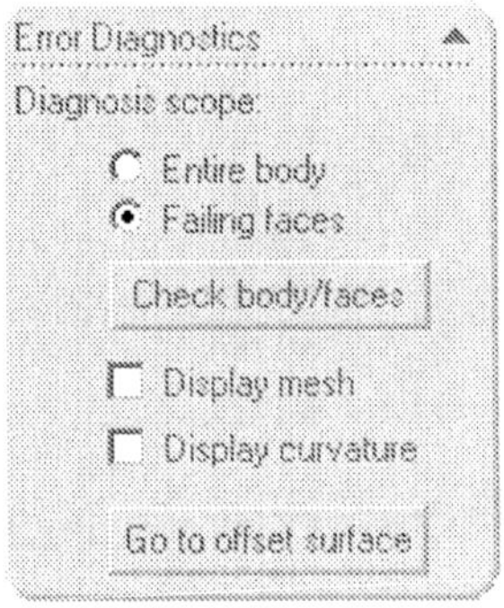

*Figure 6-79 The **Error Diagnostics** rollout*

The radio buttons available in the **Diagnosis scope** area of the **Error Diagnostics** rollout are used to define the diagnosis to be done on the entire body or only on the faces that failed to be shelled. The **Check body/faces** button is used to run the diagnostic tool. When you choose this button, then areas of the model that are responsible for feature failure are highlighted using callouts. The **Display mesh** check box is used to display the surface curvature mesh. The **Display curvature** check box is selected to display the surface curvature. The **Go to offset surface** button is used to display the **Offset Surface PropertyManager** and the offsetted surfaces.

Creating the Wrap Feature

New

CommandManager:	Features > Wrap *(Customize to Add)*
Menu:	Insert > Features > Wrap
Toolbar:	Features > Wrap *(Customize to Add)*

One of the major enhancements in the **Part** mode of SolidWorks 2004 is the introduction of the **Wrap** tool. Using this tool you can emboss, deboss, or scribe a closed multiloop sketch on a selected planar or curved face that is tangent to the plane on which the selected sketch is created. To create a wrap feature, create a closed multiloop or a single loop sketch and choose the **Wrap** button from the **Features CommandManager**. The **Message PropertyManager** is displayed and you are prompted to select a plane or face on which you need to create a closed counter or select an existing sketch. Select an existing sketch; the **Wrap1 PropertyManager** is displayed, as shown in Figure 6-80.

You will notice that the **Emboss** radio button is selected by default. This radio button is selected to create an embossed wrap feature. Next, select the face on which you need to wrap the sketch. As soon as you select the face, the preview of the wrap feature that will be created

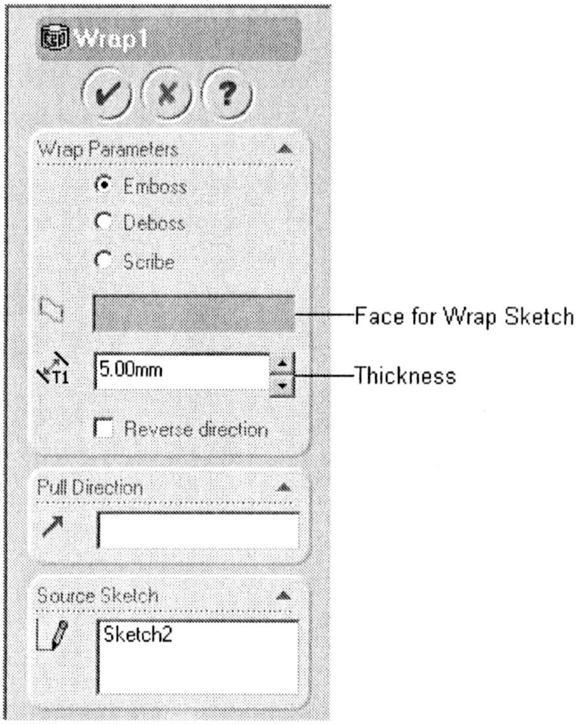

*Figure 6-80 The **Wrap1 PropertyManager***

with the default settings is displayed in the drawing area. Set the value of thickness using the **Thickness** spinner. Choose the **OK** button from the **Wrap1 PropertyManager**. Figure 6-81 shows the sketch and the face to be selected to create the wrap feature. Figure 6-82 shows the resultant embossed wrap feature.

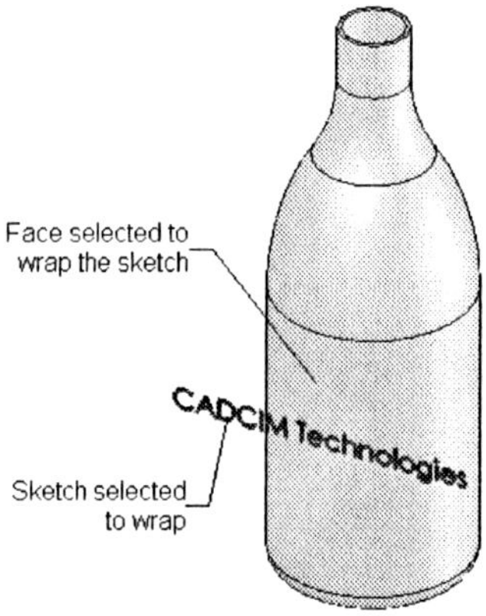

***Figure 6-81** Sketch and the face selected to create the wrap feature*

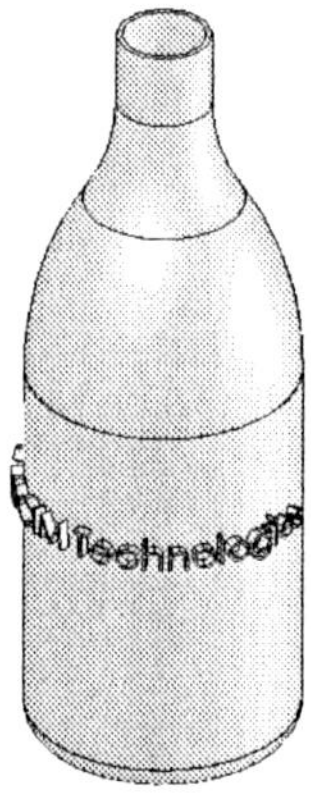

***Figure 6-82** Resultant embossed wrap feature*

Using the **Deboss** radio button available in the **Wrap Parameters** rollout you can engrave the sketch on a selected planar or curved face. Figure 6-82 shows a wrap feature created using the **Deboss** option. The **Scribe** radio button is selected to project the selected sketch on a planar or a curved face. The projected sketch will split the face on which it is projected. Using the **Pull Direction** rollout you can specify the normal direction in which you need to create the wrap feature. Figure 6-84 shows the wrap feature created using the **Scribe** option.

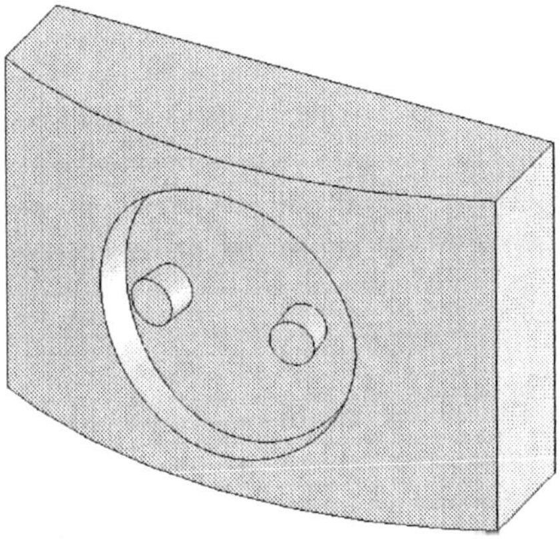

Figure 6-83 Wrap feature created using the ***Deboss*** *option*

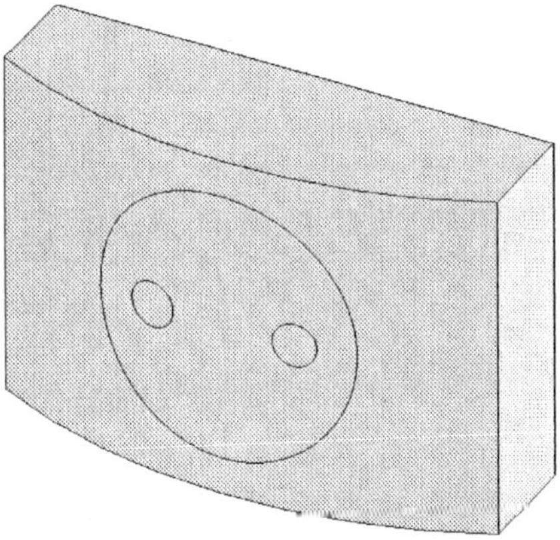

Figure 6-84 Wrap feature created using the ***Scribe*** *option*

TUTORIALS

Tutorial 1

In this tutorial, you will create the model of the Plummer Block Casting shown in Figure 6-85. The dimensions of the model are shown in Figure 6-86. **(Expected time: 30 min)**

The steps to be followed to complete this tutorial are listed next.

a. Create the base feature of the model on the **Front** plane, refer to Figures 6-87 and 6-88.
b. Create the second feature, which is a cut feature, on the top planar face of the base feature, refer to Figures 6-89 through 6-91.
c. Create the rectangular recess at the bottom of the base as a cut feature, refer to Figure 6-92.
d. Create the square cuts that will act as the recess for the head of the square head bolts, refer to Figure 6-92.
e. Create a hole feature using the **Hole PropertyManager** and modify the placement of the hole feature, refer to Figure 6-93.
f. Create the cut feature on the second top planar face of the base feature, refer to Figure 6-94.
g. Add the fillet feature to the model, refer to Figures 6-95 and 6-96.
h. Add chamfers to the model, refer to Figures 6-97 and 6-98.

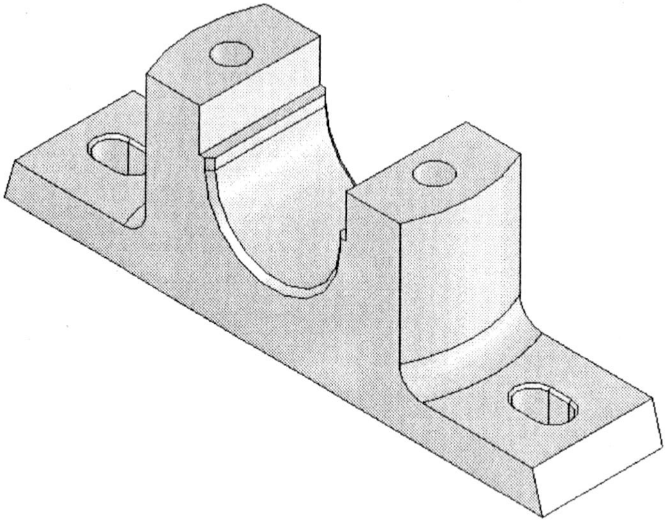

Figure 6-85 Solid model of the Plummer Block Casting

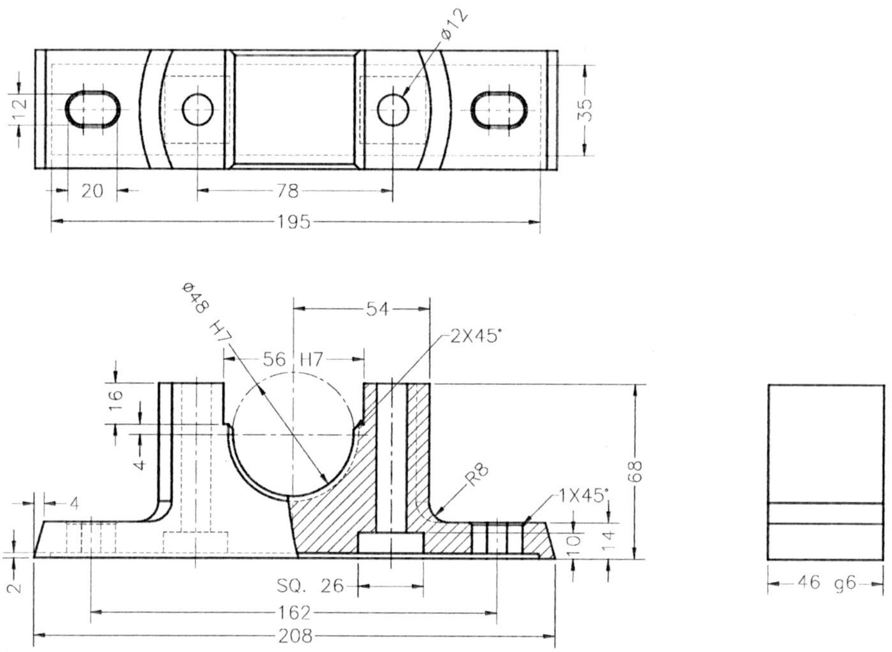

Figure 6-86 Dimensions of the Plummer Block Casting

Creating the Base Feature

1. Start SolidWorks and then start a new SolidWorks part document using the **New SolidWorks Document** dialog box.

2. Invoke the **Extruded Boss/Base** tool and select the **Front** plane as the sketching plane.

3. Draw the sketch of the front view of the model using the automatic mirroring option to capture the design intent of the model.

4. Add the required relations and dimensions to the sketch as shown in Figure 6-87.

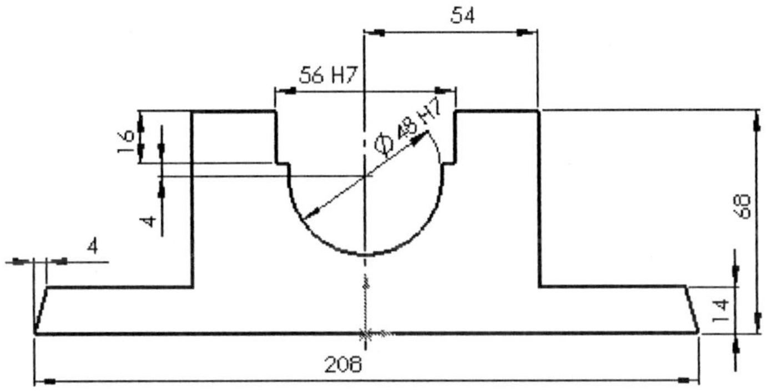

Figure 6-87 *Fully defined sketch of the base feature*

Note

As shown in the above figure, tolerances are applied to some dimensions. After adding the dimensions to the sketch, select the dimension and add the tolerance to the dimension using the ***Dimension PropertyManager*** *as discussed in the earlier chapters.*

Using the ***Dimension Properties*** *dialog box, change the radial dimension of the arc to the diameter dimension.*

You need to extrude the sketch to a distance of 46 mm using the **Mid Plane** option. You will extrude the sketch using the **Mid Plane** option because it is recommended that parts that are to be assembled should have default planes in the center of the model.

5. Exit the sketching environment. The **Extrude PropertyManager** is displayed.

6. Right-click in the drawing area and select the **Mid Plane** option from the shortcut menu to extrude the sketch symmetrically to both the sides of the sketching plane.

7. Set the value of the **Depth** spinner to **46** and choose the **OK** button from the **Extrude PropertyManager**.

8. Change the view orientation to isometric. The resultant base feature created after extruding the sketch to a given depth is shown in Figure 6-88.

Creating the Second Feature

The second feature of the model is a cut feature. The sketch of the cut-extrude feature is created on the top planar face of the base feature. This sketch will be extruded up to the specified plane to create the resultant cut feature.

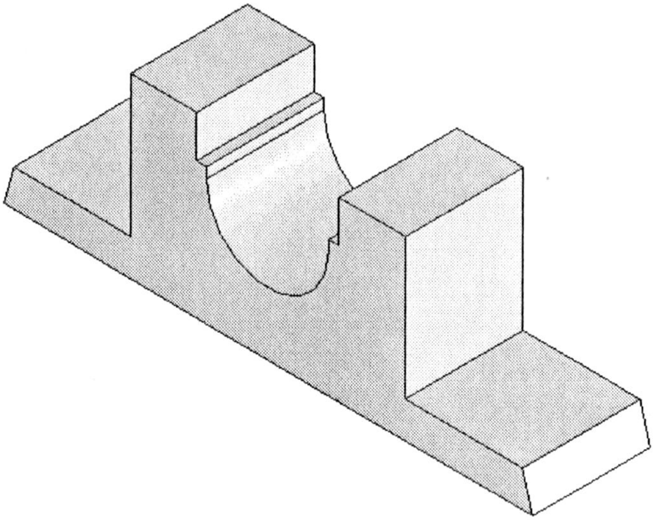

Figure 6-88 *Isometric view of the base feature of the model*

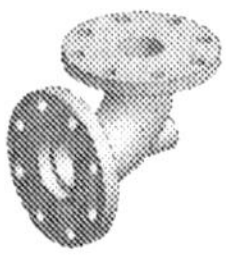

Tip. *Refer to Figure 6-86; you will observe that the dimension that reflects the depth of the extruded feature has a tolerance applied to it. Therefore, double-click the extruded feature in the* ***FeatureManager Design Tree*** *or in the drawing area. All the dimensions applied to the model are displayed. Select the dimension that reflects the depth of the extruded feature and apply the tolerance using the* ***Dimension PropertyManager****.*

1. Invoke the **Extruded Cut** tool and selected the top planar face of the base feature as the sketching plane.

2. Orient the model normal to the sketching plane and using the standard sketching tools create the sketch for the cut feature and apply the required relations to the sketch.

   The sketch of the cut-extrude feature is shown in Figure 6-89.

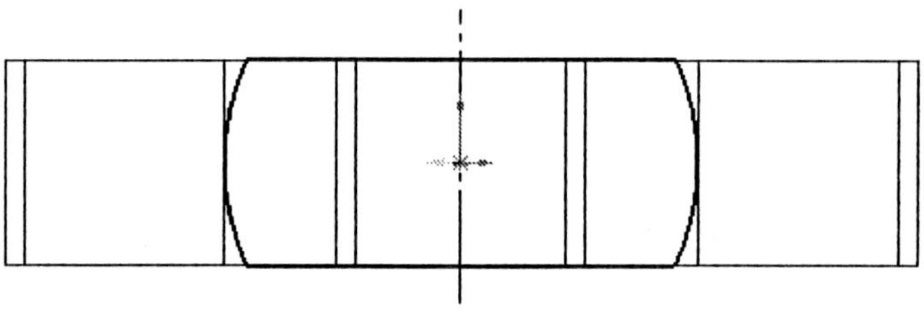

Figure 6-89 *Sketch of the second feature*

3. Exit the sketching environment. The **Cut-Extrude PropertyManager** is displayed.

   The preview of the cut feature is displayed in the drawing area with the default values of the blind option. You need to extrude the cut feature up to the selected surface. Therefore, orient the model in the isometric view because the selection of feature termination surface can be easily selected in the isometric view.

4. Change the view orientation to isometric.

   You will observe that the preview of the cut feature is inside the model, but you need to remove the outer part of the sketch profile. Therefore, you need to change the side of the cut feature.

5. Select the **Flip side to cut** check box. The preview of the cut feature is modified dynamically.

6. Right-click in the drawing area and select the **Up To Surface** option from the shortcut menu.

   You are prompted to select a face or a surface to complete the specification of Direction 1.

7. Select the surface for feature termination shown in Figure 6-90. Right-click and choose **OK** from the shortcut menu to exit the tool.

   The model after creating the cut feature is shown in Figure 6-91.

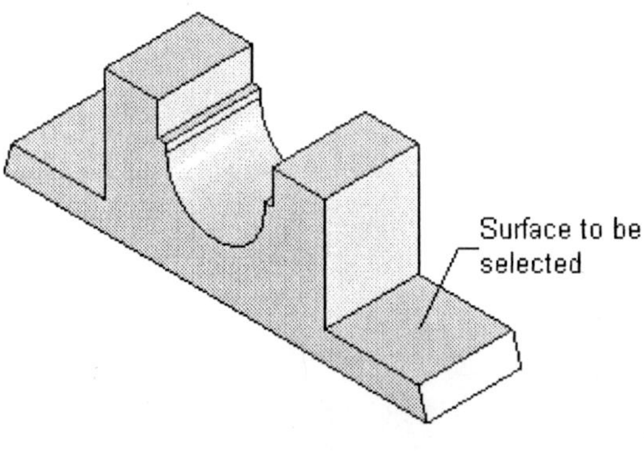

Figure 6-90 Surface to be selected for the cut feature

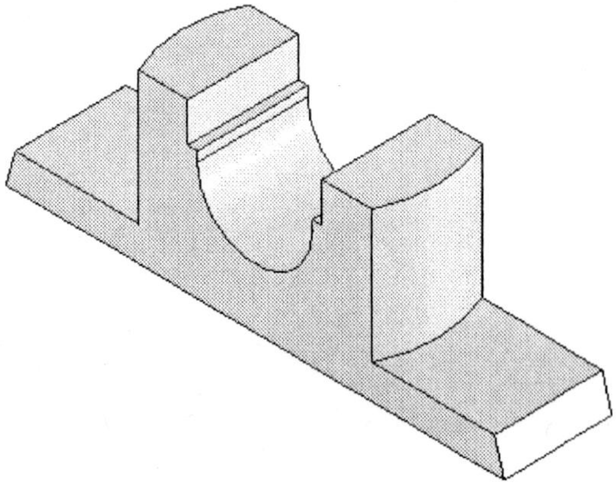

Figure 6-91 Model after creating the cut feature

Creating the Rectangular Recess

The third feature is the rectangular recess. This feature is created using a rectangular cut feature that will be created on the bottom face of the model.

1. Orient the model using the **Rotate View** tool and select the bottom face of the base feature as the sketching plane.

2. Invoke the **Extruded Cut** tool. The sketching environment in invoked.

3. Using the standard sketching tools, create a rectangle of 195 mm x 35 mm as the sketch of the rectangular recess. Apply the required relations and dimensions to the sketch.

4. Exit the sketching environment.

5. Set the value of the **Depth** spinner to **2** and choose **OK** from the **Cut-Extrude PropertyManager**.

Creating the Recess for the Head of Square-headed Bolt

It is evident from Figure 6-86 that the bolt to be inserted in the part will be a square-headed bolt. Therefore, you need to create the recess for the head of the square-headed bolt. A square of 26 mm length is used to create this recess.

1. Rotate the model and select the upper face of the recess created in the previous section as the sketching plane. Invoke the **Extruded Cut** tool and orient the model normal to the sketching plane.

2. Using the standard sketching tools create the sketch. The sketch includes two squares of length 26 mm and the distance between the centers of the squares is 78 mm, refer to Figure 6-86. Apply the required relations and dimensions.

3. Exit the sketching environment and extrude the sketch to a depth of 10 mm. Exit the tool. The rotated model after adding this cut feature is shown in Figure 6-92.

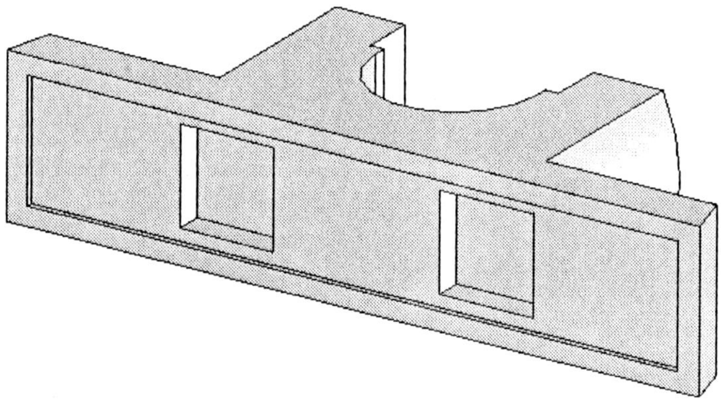

Figure 6-92 *Model after creating recess for square-headed bolts*

Creating the Hole Feature

After creating the recess for the head of the square-headed bolt, you need to create the hole using the **Hole** tool. To create a hole using the hole tool, you first need to draw sketch points on the placement plane of the hole feature. After drawing the sketched point, select the hole placement plane and invoke the **Hole** tool.

1. Invoke the sketching environment using the top right planar surface of the base feature as the sketching plane.

2. Create two sketched points on the right and the left top planar faces of the base feature.

   Next, you need to define the placement of these sketched point. It is evident from Figure 6-86 that the center point of the hole is coincident with the center point of the square recess feature. Therefore, you will coincide the sketched points with the center of the squares. First you will have to change the model view display from **Shaded** to **Hidden Lines Visible** so that the square recess feature is visible.

3. Orient the view normal to the viewing direction and choose the **Hidden Lines Visible** button from the **View** toolbar.

4. Invoke the **Centerline** tool from the **Sketch CommandManager** and move the cursor to the lower left corner of the left square. When the cursor is displayed in yellow color, specify the start point of the line. Move the cursor to the upper right corner of the square and when the cursor is displayed in yellow color, specify the endpoint of the line.

5. Right-click in the drawing area and choose the **Select** option from the shortcut menu. This will end line creation and also exit the **Line** tool. A centerline is created as a diagonal of the square.

6. Press and hold down the CTRL key from the keyboard and select the centerline created earlier. Now, select the left sketched point.

   The **Properties PropertyManager** is displayed. Using this you will add the relation between the left sketched point and the centerline. The centerline is created as the diagonal of the square and it is a general geometric fundamental that the midpoint of the diagonal of a square or a rectangle lies at the center. Therefore, you will add a relation such that the sketched point will be placed at the midpoint of the diagonal.

7. Choose the **Midpoint** button from the **Add Relations** rollout of the **Properties PropertyManager**. Choose the **OK** button from the **Properties PropertyManager**.

8. Similarly, create the diagonal centerline for the right square. Add the midpoint relation between this centerline and the right sketched point.

9. Exit the sketcher environment and change the model display mode to shaded. Select the right top planar surface of the base feature as the placement plane for the hole feature.

10. Choose **Insert > Features >Hole > Simple** from the menu bar to invoke the **Hole PropertyManager**. The preview of the hole feature with default settings is displayed in the drawing area.

11. Right-click in the drawing area and choose the **Through All** option from the shortcut menu to define feature termination.

12. Set the value of the **Hole Diameter** spinner to **12**.

13. Select the center point of the hole and drag the cursor to the right sketched point. Release the left mouse button when the cursor turns yellow in color. Choose the **OK** button from the **Hole PropertyManager**

14. Using the same procedure create the second hole feature on the left side of the model. The model after creating both the hole features is displayed in Figure 6-93.

15. Similarly, create the cut feature on the second top planar face of the base feature, refer to Figure 6-86 for dimensions of the sketch.

    Note that the dimension of 162 mm between the cut on the left and the cut on the right is given from the center of the sketches. To get this dimension, you need to create vertical centerlines in both the sketches. Using the dimensions and relations, you need to make sure that the centerlines pass through the center of the sketches. Now, add dimension between these two vertical centerlines. The model after creating the cut feature is shown in Figure 6-94.

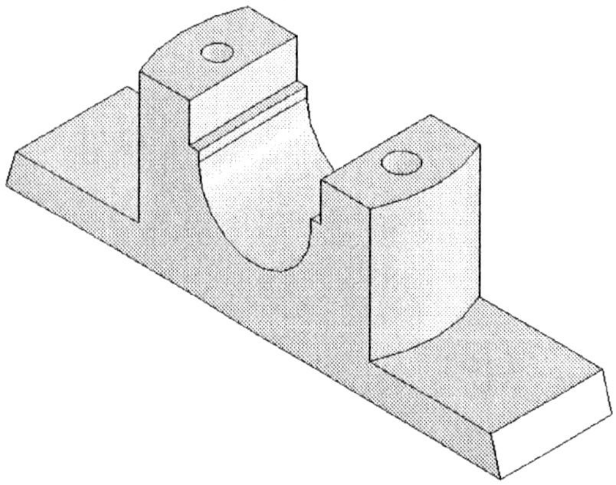

Figure 6-93 *Model after adding the hole features*

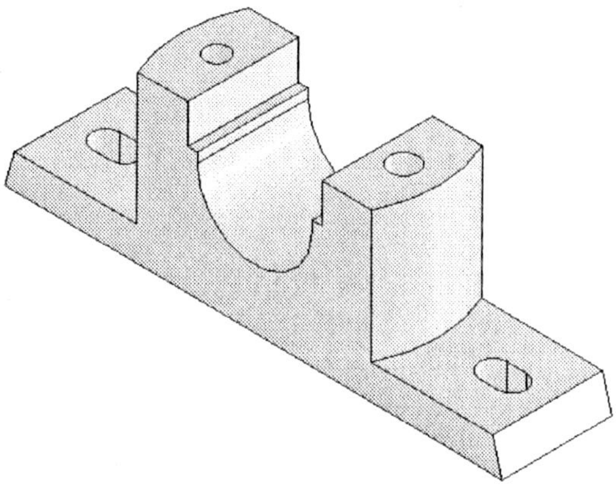

Figure 6-94 *Model after adding the cut feature*

Adding Fillet to the Model

After creating all the other features, you will now add the fillet feature to the model.

1. Choose the **Fillet** button from the **Features CommandManager** to invoke the **Fillet PropertyManager**.

After invoking the **Fillet PropertyManager** you are prompted to selected edges, faces, features, or loops to fillet. As evident from the model you will select only the edges to apply the fillet feature.

2. Select the edges shown in Figure 6-95.

   As soon as you select the edges the preview of the fillet feature with the default values is shown in the drawing area. A radius callout is also displayed along the selected edge. Now, you need to modify the default radius value of the fillet feature.

Note

If the preview of the fillet feature is not displayed in the drawing area then select the ***Full Preview*** *radio button available in the* ***Items To Fillet*** *rollout of the* ***Fillet PropertyManager****.*

3. Set the value of the **Radius** spinner to **8** and choose the **OK** button from the **Fillet PropertyManager**.

   The model after adding fillet is shown in Figure 6-96.

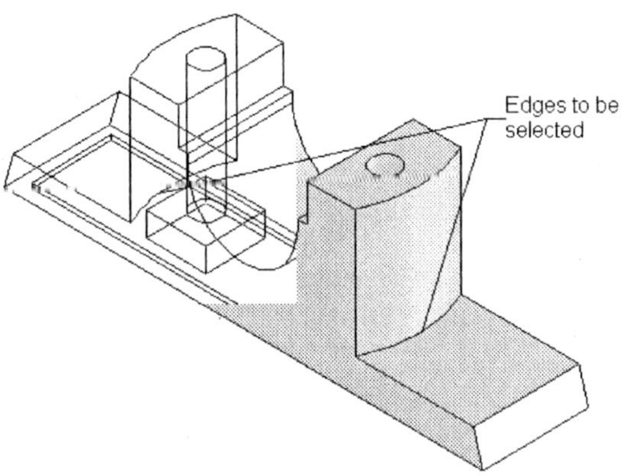

Figure 6-95 *Edges to be selected for the fillet feature*

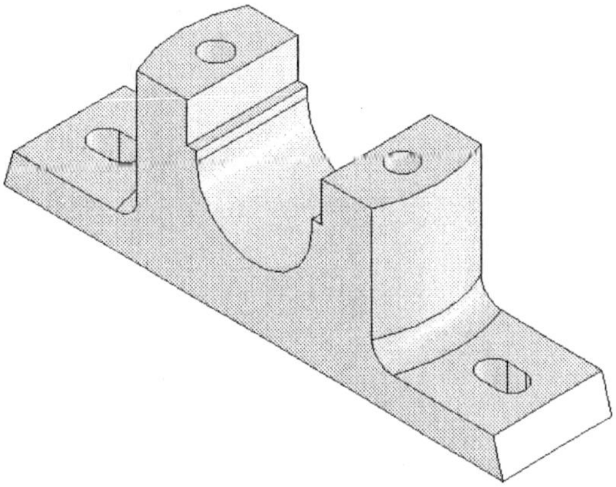

Figure 6-96 *Model after adding fillet*

Adding Chamfer to the Model

The next feature that you need to add to this model is chamfer.

1. Choose the **Chamfer** button from the **Features CommandManager** to invoke the **Chamfer PropertyManager**.

2. Select the edges of the cut features as shown in Figure 6-97. The edges that are tangent to the selected edges are selected automatically because the selection mode of the chamfer feature uses tangent propagation.

   As soon as you select the edges the preview of the chamfer feature is displayed in the drawing area with the default values. The angle and distance callouts are also displayed. Now, you will set the required value of the chamfer.

   The required chamfer parameters are 1 mm and 45-degree. The **Angle distance** radio button is selected by default in the **Chamfer Parameters** rollout. The value of angle in

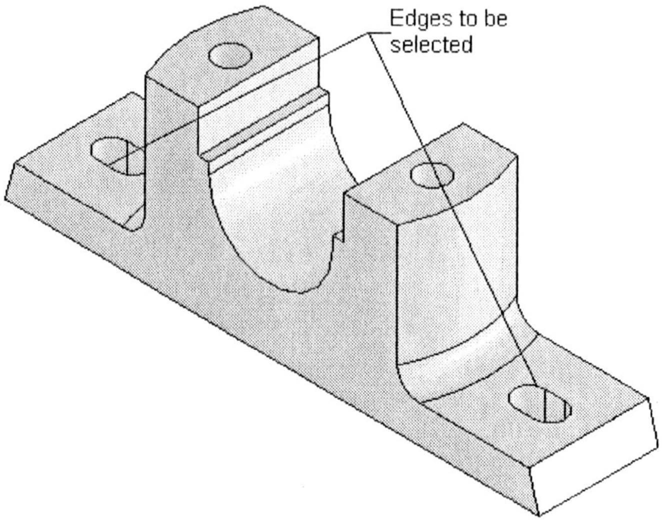

Figure 6-97 Edges to be selected

the **Angle** spinner is set as 45-degree. Therefore, you do not need to modify this value. You will set only the value of the distance in the **Distance** spinner.

3. Set the value of the distance in the **Distance** spinner to **1** and choose the **OK** button from the **Chamfer PropertyManager**.

4. Following the same procedure add chamfer to the semicircular edges on the front and back faces of the base feature. Refer to Figure 6-86 for the parameters of the chamfer feature.

The final solid model is shown in Figure 6-98. The **FeatureManager Design Tree** of the model is shown in Figure 6-99.

Saving the Model

Next, you need to save the document.

1. Choose the **Save** button from the **Standard** toolbar and save the document with the name given below.

   \My Documents\SolidWorks\c06\c06-tut01.sldprt.

2. Choose **File > Close** from the menu bar to close the document.

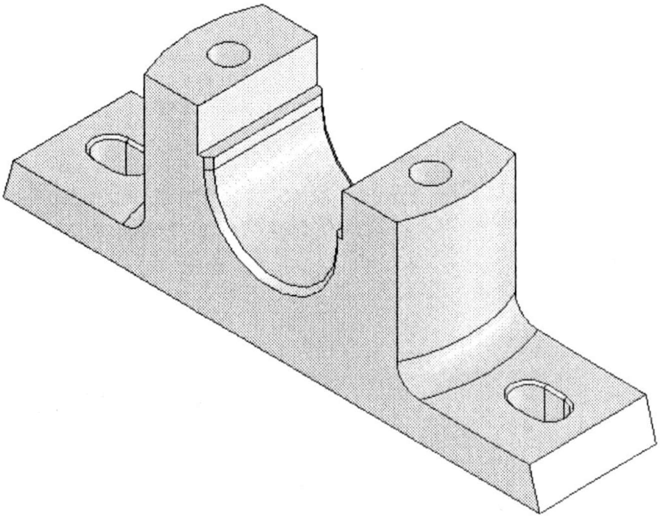

Figure 6-98 Final solid model

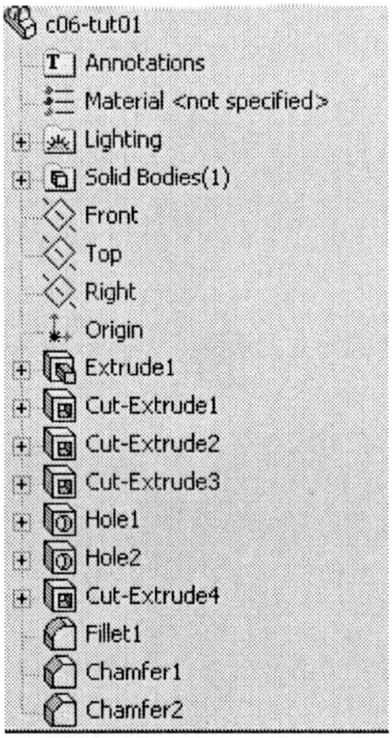

*Figure 6-99 The **FeatureManager Design Tree** of the model*

Tutorial 2

In this tutorial you will create the model shown in Figure 6-100. For better understanding of the model the section view of the model is shown in Figure 6-101. The dimensions of the model are shown in Figure 6-102. **(Expected time: 30 min)**

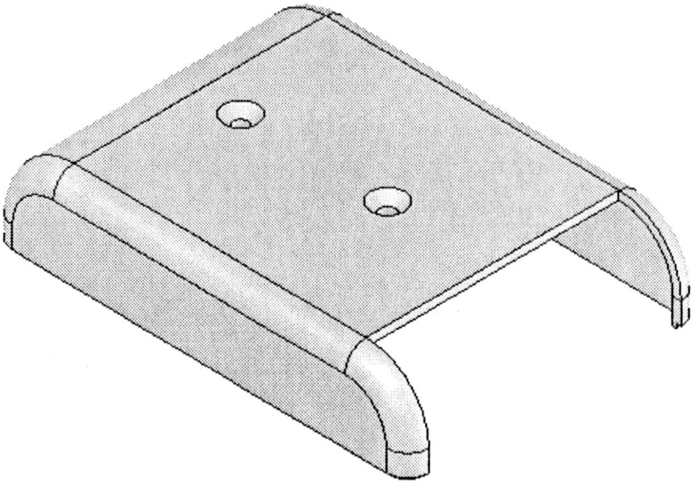

Figure 6-100 *Solid model for Tutorial 2*

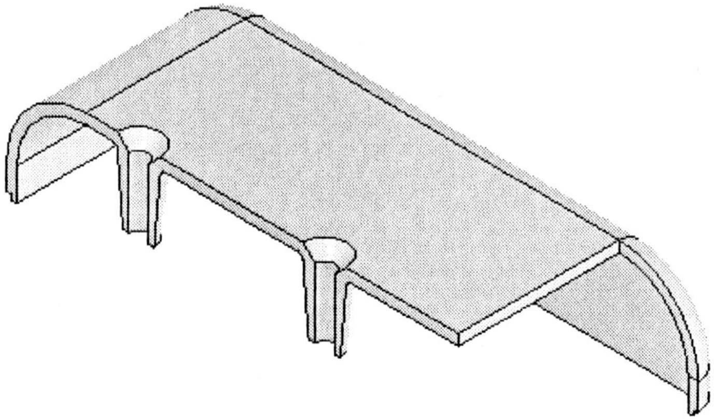

Figure 6-101 *Section view of the model*

The steps to be followed to complete this tutorial are listed next.

a. Create the base feature of the model by extruding a rectangle of 100 mm x 70 mm to a distance of 20 mm.
b. Add fillet to the base feature, refer to Figures 6-103 through 6-106.
c. Create the shell feature to create a thin walled part and remove some of the faces, refer to Figures 6-107 and 6-108.
d. Create a reference plane at an offset distance from the top planar face of the base feature and extrude the sketch created on the new plane, refer to Figure 6-109.
e. Using the **Hole Wizard** tool add the countersink hole to the model, refer to Figure 6-110.
f. Add fillet to the extruded feature, refer to Figures 6-111 and 6-112.
g. Create the lip of the component by extruding the sketch, refer to Figure 6-113.

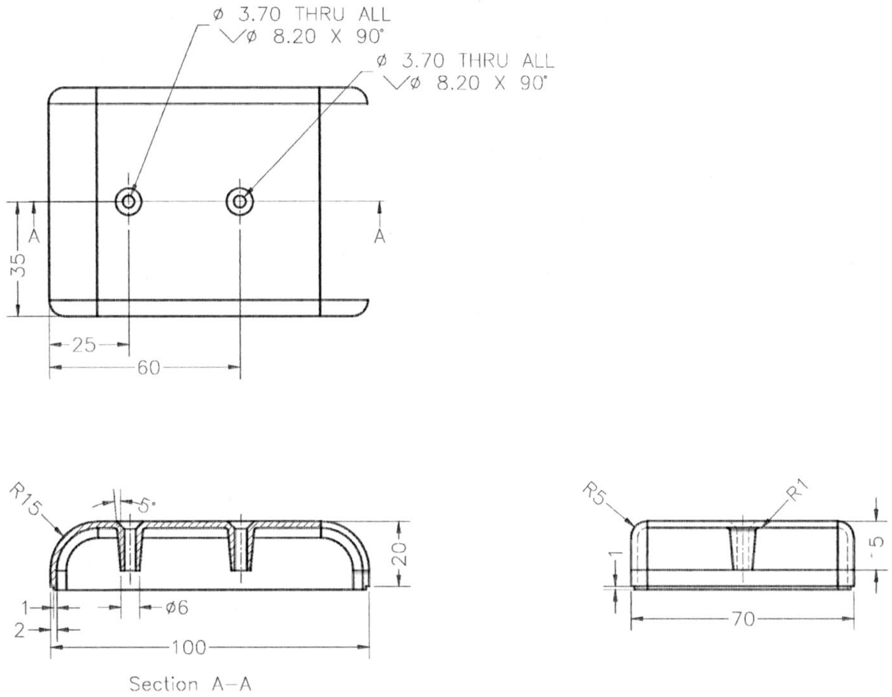

Figure 6-102 *Top view, Front section view, and right-side view with dimensions*

Creating the Base Feature

The base feature of the model will be created by extruding a rectangle of 100 mm x 70 mm to a distance of 20 mm. It is evident from the model that the sketch of the base feature will be created on the top plane. Therefore, you have to select the top plane as the sketching plane.

1. Start a new SolidWorks part document and invoke the **Extruded Boss/Base** tool. Select the **Top** plane from the **FeatureManager Design Tree**.

2. Orient the sketch plane normal to the viewing direction.

3. Using the **Rectangle** tool, draw a rectangle and force it to a dimension of 100 mm x 70 mm. Add the other required dimensions to fully define the sketch.

4. Exit the sketching environment and extrude the rectangle to a depth of 20 mm.

Creating the Fillet Features

After creating the base feature, you need to add fillet to the model. In this model, you need to add three fillet features. Two fillet features will be added at this stage of the design process and the remaining one fillet feature will be added at a later stage of the design process.

1. Choose the **Fillet** button from the **Features CommandManager** to invoke the **Fillet PropertyManager**.

2. Select the edges of the model as shown in Figure 6-103.

   As soon as you select the edges of the model, the preview of the fillet with the default values and the radius callout are displayed in the drawing area.

3. Set the value of the **Radius** spinner to **15** and choose the **OK** button from the **Fillet PropertyManager**.

   Figure 6-104 shows the model after adding the first fillet feature.

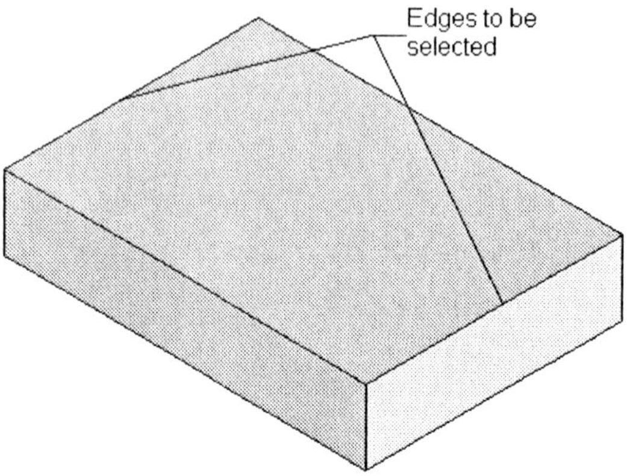

Figure 6-103 *Edges to be selected*

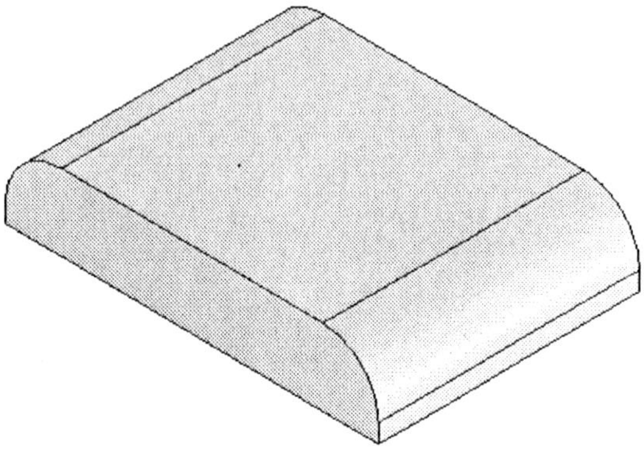

Figure 6-104 *Fillet added to the base feature*

Now, you will add the second fillet feature to the model.

4. Invoke the **Fillet PropertyManager** and set the value of the **Radius** spinner to **5**.

5. Select the edges of the model as shown in Figure 6-105. Right-click and choose **OK** from the shortcut menu to complete feature creation. The model after adding the second fillet feature is shown in Figure 6-106.

Tip. *The edges tangent to the selected edges are filleted automatically because by default, the* ***Tangent propagation*** *check box is selected in the* ***Fillet PropertyManager****.*

Creating the Shell Feature

It is evident from Figures 6-100 and 6-101 that a shell feature is required to create a thin walled structure. As discussed earlier, the shell feature is used to scoop out the material from the model, leaving behind a thin-walled hollow part.

1. Choose the **Shell** button from the **Features CommandManager** to invoke the **Shell1 PropertyManager**. You are prompted to select the faces to remove.

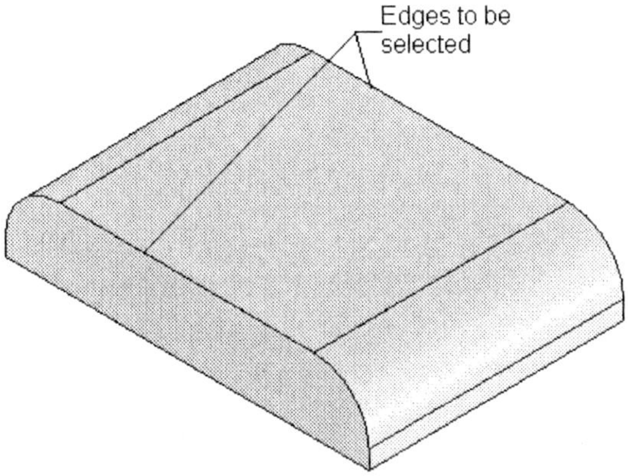

Figure 6-105 Edges to be selected

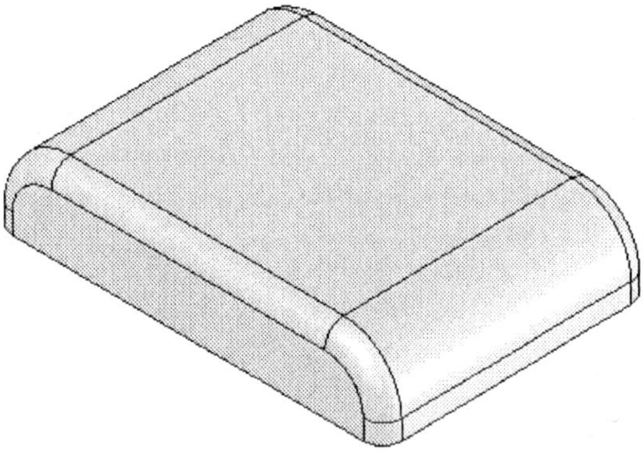

Figure 6-106 Second fillet added to the model

2. Rotate the model and select the faces to remove as shown in Figure 6-107. The names of the selected faces are displayed in the **Faces to Remove** selection area.

3. Set the value of the **Thickness** spinner to **2** and choose the **OK** button from the **Shell1 PropertyManager**.

   The model after creating the shell feature is shown in Figure 6-108.

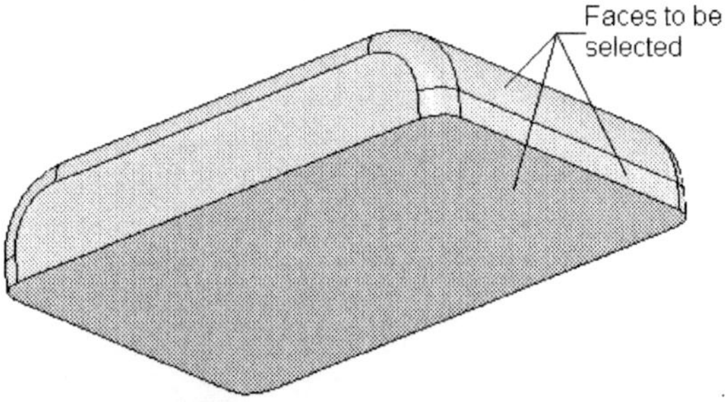

Figure 6-107 Faces to be selected to remove

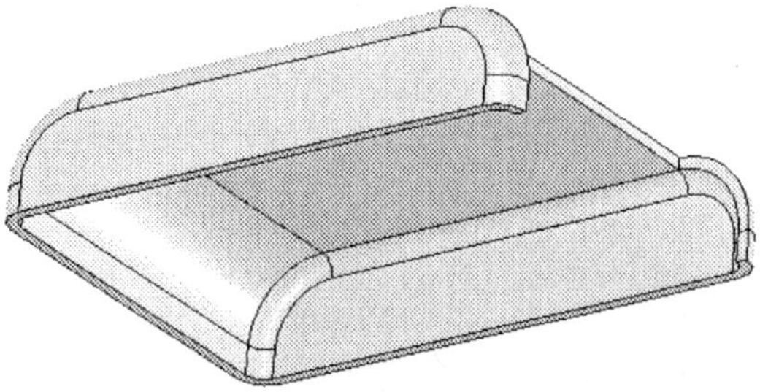

Figure 6-108 Model after creating the shell feature

Creating the Extruded Feature

The next feature that you are going to create is an extruded feature. But, before creating this extruded feature, you need to create a reference plane at an offset distance from the top planar face of the base feature.

1. Invoke the **Plane PropertyManager** and create a plane at an offset distance of 15 mm from the top planar face of the base feature. You have to select the **Reverse direction** check box from the **Plane PropertyManager** to create the plane.

2. Invoke the **Extruded Boss/Base** tool and select the newly created plane as the sketching plane and create the sketch using the standard sketching tools. The sketch consists of two circles of 6 mm diameter. For other dimensions refer to Figure 6-102.

3. Exit the sketching environment and extrude the sketch using the **Up To Next** option and add an outward draft of 5-degree. Hide the reference plane.

   Figure 6-109 shows the rotated model after creating the extruded feature with draft.

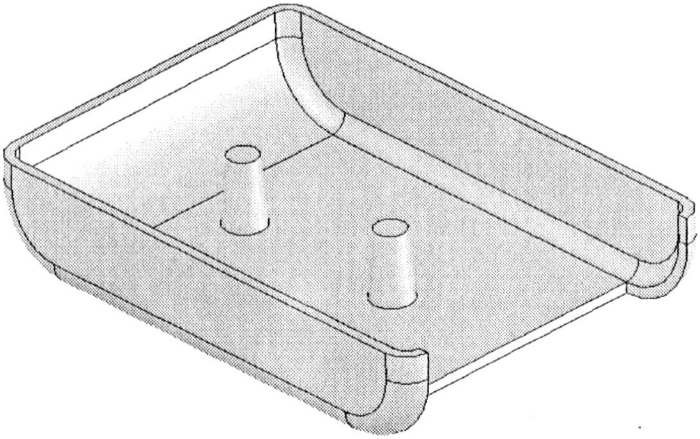

Figure 6-109 *Model after creating the extruded feature*

Adding the Countersink Hole Using the Hole Wizard Tool

The next feature that you are going to create is a **M3.5 Flat Head Machine Screw** countersink hole, refer to Figure 6-102. In SolidWorks you are provided with one of the largest standard hole generating technique known as **Hole Wizard**. Using the **Hole Wizard** tool, you can create the standard holes to the model that can accommodate standard fasteners.

1. Rotate the model and select the top planar face of the base feature and then choose the **Hole Wizard** button from the **Features CommandManager**.

   The **Hole Definition** dialog box is displayed and the preview of the hole feature on the selected face with the default values is shown in the drawing area.

Tip. *If you select the placement plane for the placement of the hole feature before invoking the* ***Hole Definition*** *dialog box then the placement sketch will be a 2D sketch. If you select the placement plane after invoking the* ***Hole Definition*** *dialog box then the hole placement sketch will be a 3D sketch. You will learn more about 3D sketches in the later chapters.*

2. Choose the **Countersink** tab of the **Hole Definition** dialog box.

Now, you will set the parameters to define the standard hole. The preview of the standard hole will be modified dynamically as you set the parameters for the hole feature.

3. Select the **Ansi Metric** option from the **Standard** drop-down list.

4. Select the **Flat Head Screw - ANSI B18.6.7M** option from the **Screw type** drop-down list

5. Select the **M3.5** option from the **Size** drop-down list and choose the **Through All** option from the **End Condition & Depth** drop-down list.

6. Choose the **Next** button from the **Hole Definition** dialog box. The **Hole Placement** dialog box is displayed and the sketcher environment is invoked. The select cursor is replaced by the point cursor.

7. Using the point cursor, specify another point anywhere on the top planar face of the base feature to create another hole.

   Because both the placement points are not properly placed, you need to add the required relations and dimensions to define the proper location of these placement points. Before doing that, you need to change the model display from **Shaded With Edges** to **Hidden Lines Visible** for a better display.

8. Choose the **Hidden Lines Visible** button from the **View** toolbar to display the model with hidden lines visible.

9. Press the ESC key to invoke the select cursor.

10. Right-click in the drawing area and choose the **Add Relation** option from the shortcut menu. The **Add Relations PropertyManager** is displayed.

11. Select the left placement point and then select the upper left hidden circle. Choose the **Concentric** button from the **Add Relations** rollout.

12. Right-click in the drawing area and choose **Clear Selection** from the shortcut menu. Select the right placement point and the upper right hidden circle. Choose the **OK** option from the confirmation corner.

13. Choose the **Finish** button from the **Hole Placement** dialog box and choose the **Shaded With Edges** button from the **View** toolbar.

    The isometric view of the model after adding the hole feature is shown in Figure 6-110.

Adding Fillet to the Model

Now, you need to add fillet to the edges of the extruded feature with draft that was created earlier.

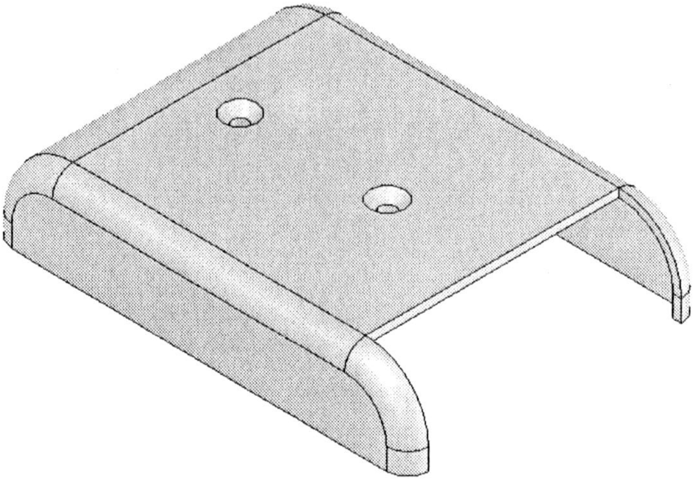

Figure 6-110 Model after adding the hole feature using the ***Hole Wizard*** *tool*

1. Rotate the model and invoke the **Fillet PropertyManager**. Select the edges of the model, as shown in Figure 6-111.

2. Set the value of the **Radius** spinner to **1** and choose the **OK** button to end feature creation.

   The model after adding fillet is shown in Figure 6-112.

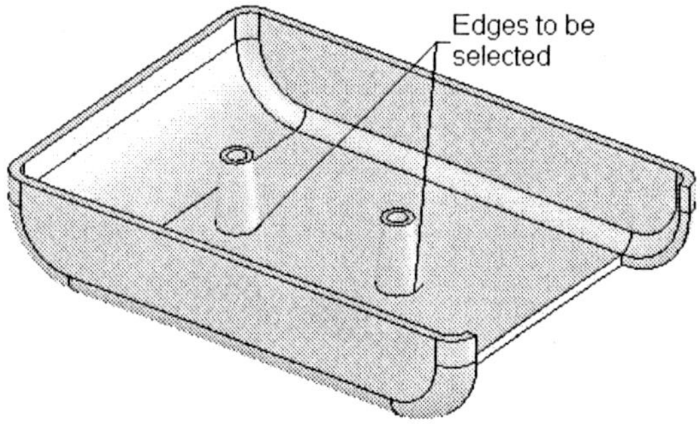

Figure 6-111 Edges to be selected

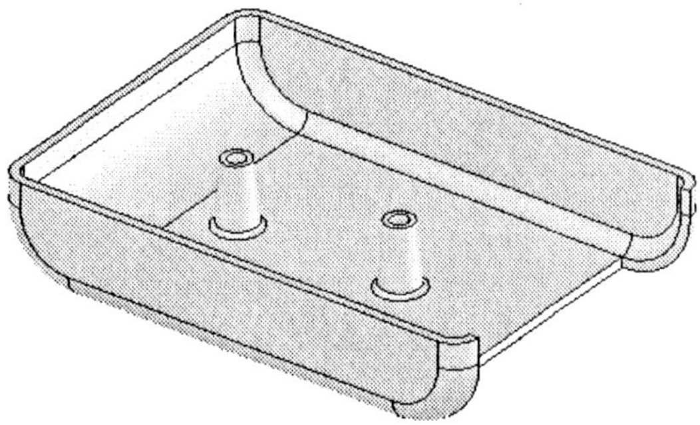

Figure 6-112 Model after adding fillet

Adding a Lip to the Model

The last feature that you need to add to the model is a lip. It will be created by extruding an open sketch using the thin option.

1. Invoke the **Extrude Boss/Base** tool and select the bottom face of the base feature as the sketching plane and invoke the sketching environment.

2. Using the select tool, select any one of the inner edge of the model on the current sketching plane and right-click to display the shortcut menu. Now, choose the **Select Tangency** option from the shortcut menu.

3. Choose the **Convert Entities** button from the **Sketch CommandManager**.

   The selected edges are converted into sketched entities.

4. Exit the sketching environment. The **Extrude PropertyManager** is displayed and the **Thin Feature** rollout is invoked automatically because you are extruding an open sketch.

5. Set the value of **Distance** spinner in the **Direction 1** rollout to **1** and set the value of the **Thickness** spinner in the **Thin Feature** rollout to **1**.

6. Choose the **OK** button from the **Extrude PropertyManager**.

The rotated view of the final model is shown in Figure 6-113. The **FeatureManager Design Tree** of the model is shown in Figure 6-114.

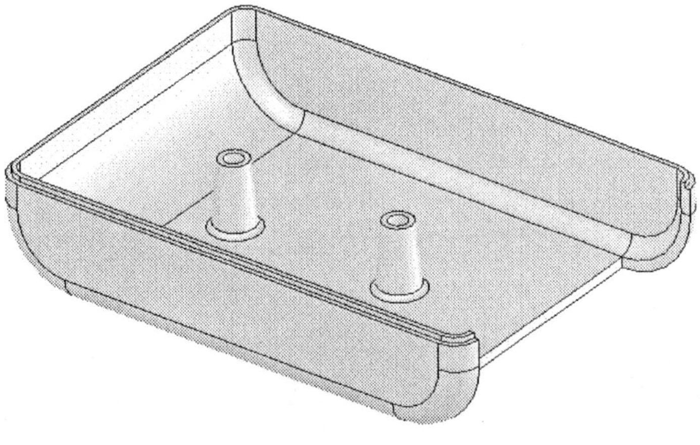

Figure 6-113 *Final model rotated to display the maximum features*

Saving the Model

Next, you need to save the document.

1. Choose the **Save** button from the **Standard** toolbar and save the part document with the name given below.

   \My Documents\SolidWorks\c06\c06-tut02.sldprt

2. Choose **File > Close** from the menu bar to close the document.

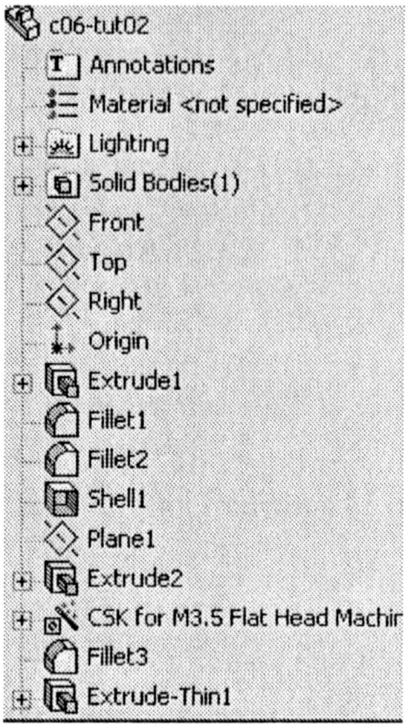

***Figure 6-114** The **FeatureManager Design Tree** of the model*

SELF-EVALUATION TEST

Answer the following questions and then compare your answers with the answers given at the end of this chapter.

1. Using the **Hole PropertyManager** you can create counterbore, countersink, and tapped holes. (T/F)

2. The hole features created using the **Hole Wizard** tool and the **Hole PropertyManager** are not parametric. (T/F)

3. You cannot define a user-defined hole using the **Hole Wizard** tool. (T/F)

4. You cannot preselect the edges or faces for creating a fillet feature. (T/F)

5. In SolidWorks you can create a multithickness shell feature. (T/F)

6. The __________ check box is selected to create the shell feature on the outer side of the model.

7. The __________ is created by specifying different radii along the length of selected edge at specified intervals.

8. The names of the faces to be removed in the shell features are displayed in the __________ selection area.

9. If you want to specify different distances while creating the chamfer then clear the __________ check box.

10. The __________ check box is selected to apply the face fillet feature with continuous curvature throughout the fillet feature.

REVIEW QUESTIONS

Answer the following questions.

1. The __________ option is used to add standard holes to the model.

2. After specifying all the parameters to a hole feature using the **Hole Definition** dialog box, the __________ dialog box is displayed to specify the placement of the hole feature.

3. The __________ radio button is selected to create a smooth transition by smoothly blending the points and vertices on which you have defined the radius on the selected edge.

4. The __________ tab available in the **Hole Definition** dialog box is used to define the parameters to add a standard drilled hole.

5. By default, the __________ radio button is selected in the **Chamfer PropertyManager**.

6. If you preselect the placement surface for creating a hole feature using the **Hole Wizard** tool then the resultant placement sketch will be a

   (a) 2D sketch (b) Planar sketch
   (c) Bezier spline (d) 3D sketch

7. Using which options you do not use radius while creating a fillet feature?

   (a) Face fillet with hold line (b) Constant radius fillet
   (c) Variable radius fillet (d) Full round fillet

8. Which radio button from the **Variable Radius Parameters** rollout is used to create smooth transition while creating a variable radius fillet?

   (a) **Straight transition** (b) **Parametric transition**
   (c) **Smooth transition** (d) **Surface transition**

9. If you do not select any face to be removed while creating a shell feature, then what will be the resultant model?

   (a) Remains complete solid model (b) Thin walled hollow model
   (c) Automatically removed one face (d) None of these

10. Which dialog box is displayed when you choose the **Hole Wizard** button from the **Features CommandManager**?

    (a) **Hole** (b) **Hole Definition**
    (c) **Hole Wizard** (d) **Hole Parameters**

EXERCISES

Exercise 1

Create the model shown in Figure 6-115. The dimensions of the model are shown in Figure 6-116. **(Expected time: 30 min)**

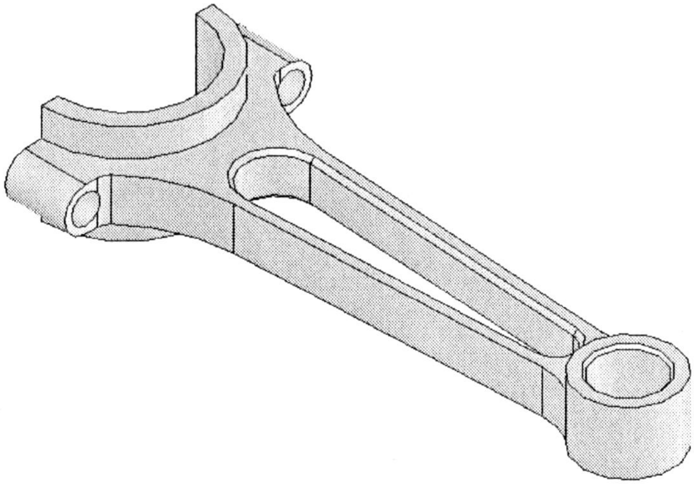

Figure 6-115 *Solid model for Exercise 1*

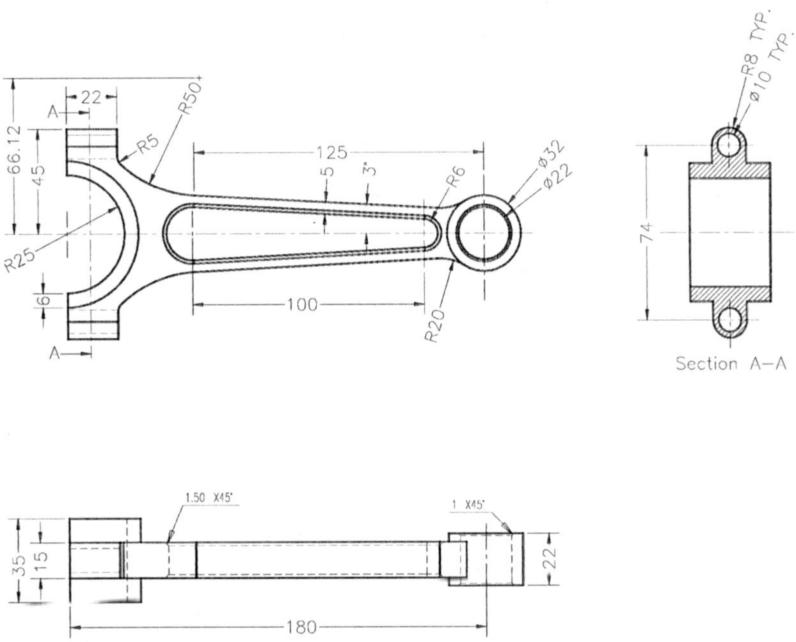

Figure 6-116 *Views and dimensions of the model for Exercise 1*

Exercise 2

Create the model shown in Figure 6-117. The dimensions of the model are shown in Figure 6-118. **(Expected time: 30 min)**

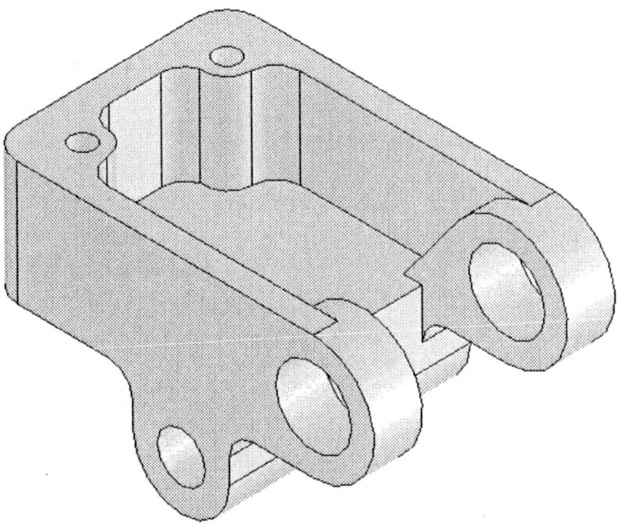

Figure 6-110 *Solid model for Exercise 2*

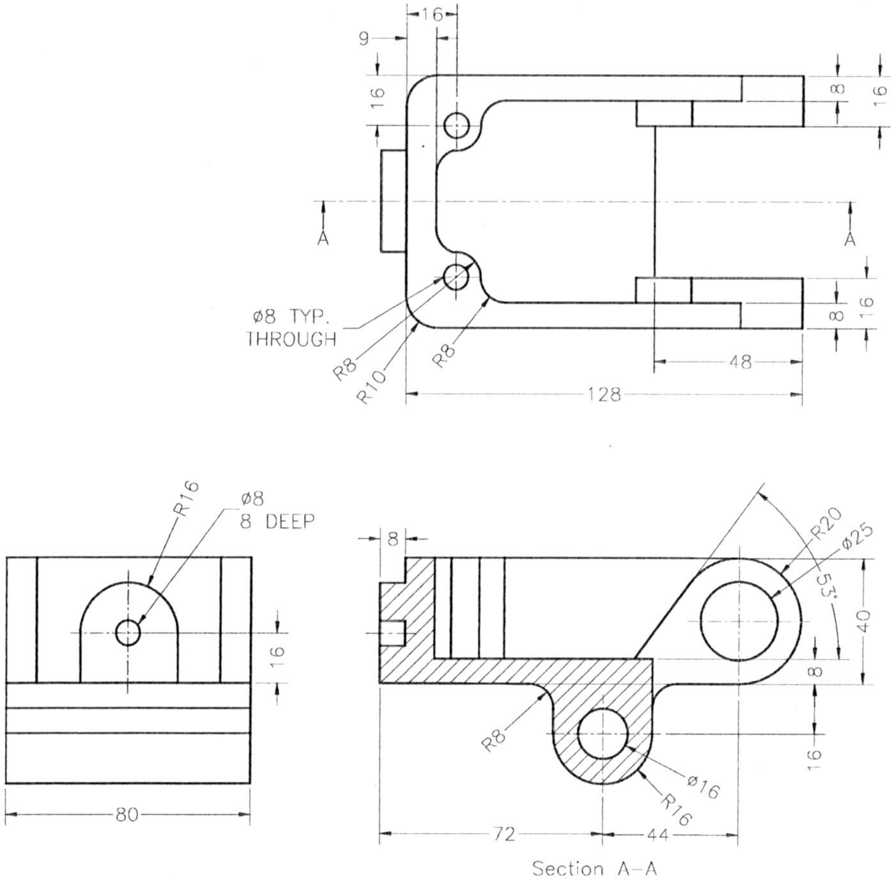

Figure 6-118 Views and dimension of the model for Exercise 2

Answers to Self-Evaluation Test

1. T, **2.** F, **3.** F, **4.** F, **5.** T, **6. Shell outward, 7.** Variable radius fillet, **8. Faces to Remove, 9. Equal distance, 10. Curvature continuous**

Chapter 7

Advanced Modeling Tools-II

Learning Objectives

After completing this chapter you will be able to:

- *Mirror features, faces, and bodies.*
- *Create Linear Pattern.*
- *Create Circular Pattern.*
- *Create Sketch-driven Pattern.*
- *Create Curve-driven Pattern.*
- *Create Table-driven Pattern.*
- *Create Rib Feature.*
- *Create Dome Feature.*
- *Create Shape Feature.*
- *Display the section view of the model.*

ADVANCED MODELING TOOLS

Some of the advanced modeling options were discussed in Chapter 6, Advanced Modeling Tools-I. In this chapter you will learn about some more advanced modeling tools using which you can capture the design intent of the model. The rest of the advanced modeling tools are discussed in later chapters.

Creating Mirror Feature

CommandManager:	Features > Mirror
Menu:	Insert > Pattern/Mirror > Mirror
Toolbar:	Features > Mirror

The **Mirror** tool is used to copy the selected feature, face, or body about a specified mirror plane. The mirror plane that is used to mirror a feature, face, or body can be a reference plane or a planar face. To use this tool, choose the **Mirror** button from the **Features CommandManager** or choose **Insert > Pattern/Mirror > Mirror** to invoke the **Mirror PropertyManager** as displayed in Figure 7-1. The confirmation corner is also displayed in the drawing area.

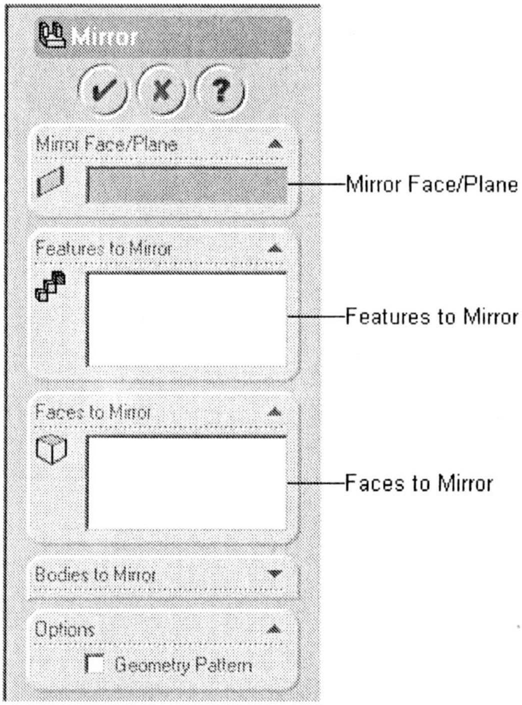

*Figure 7-1 The **Mirror PropertyManager***

Using this tool you can mirror features, faces, and bodies. The options that are used to mirror features, faces, and bodies are discussed next.

Mirroring Features

Using this option you can mirror the selected feature along the specified mirror plane or face. To mirror the features, invoke the **Mirror PropertyManager**. You are prompted to select a plane or a planar face about which to mirror, followed by the features to mirror. Select a plane or a planar face that will act as mirror plane or mirror face. After selecting the mirror plane or face, the selection mode of the **Features to Mirror** selection area is invoked and you are prompted to select features to mirror. Select the feature or features from the drawing area or from the **FeatureManager Design Tree** which is displayed in the drawing area. When you select the features to be mirrored, the preview of the mirrored image is displayed in the drawing area. After selecting all the required features, choose the **OK** button from the **Mirror PropertyManager**. Figure 7-2 shows the mirror plane and features to be mirrored and Figure 7-3 shows the resultant mirrored features.

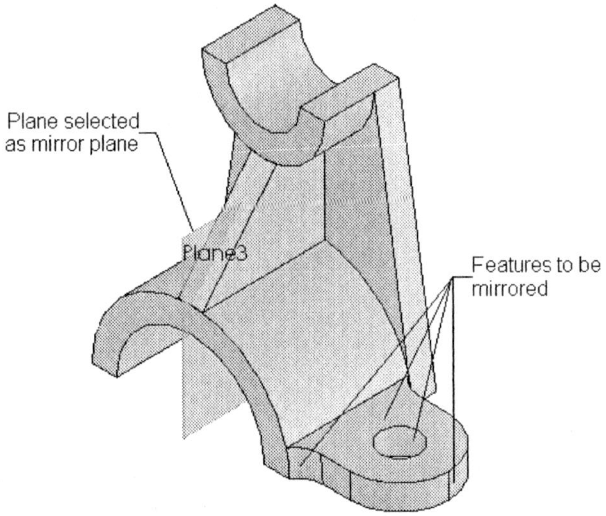

Figure 7-2 *The features and the plane to be selected*

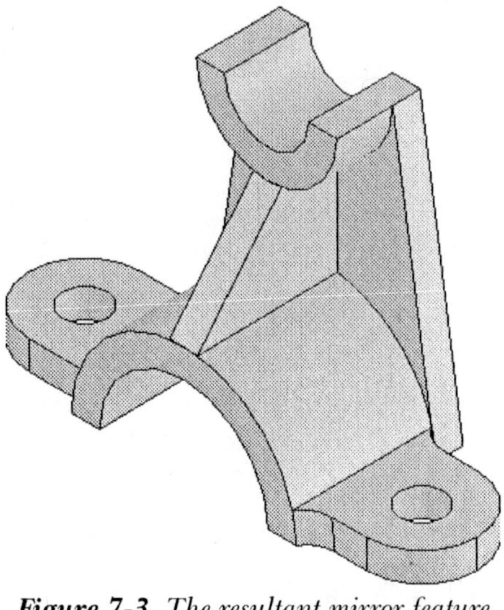

Figure 7-3 *The resultant mirror feature*

Tip. *You can also preselect the mirror plane or mirror face and the features to be mirrored before invoking the **Mirror PropertyManager**.*

Mirroring with and without Geometric Pattern

When you are creating a mirror feature you are provided with an option known as geometric pattern. This option is available in the **Options** rollout as shown in Figure 7-4.

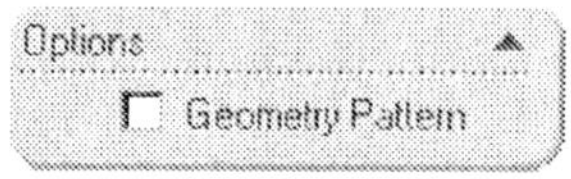

Figure 7-4 *The **Options** rollout*

By default, the **Geometry Pattern** check box is cleared. Therefore, if you mirror a feature that is related to some other entity, the same relationship will be applied to the mirrored feature. Consider a case in which an extruded cut is created using the **Offset From Surface** option. If you mirror the cut feature along a plane, the same relationship will be applied to the mirrored cut feature. The mirrored cut feature will be created with the same end condition of feature termination. Figure 7-5 shows a hole feature created on the right and mirrored along **Plane 1** with the **Geometry Pattern** check box cleared.

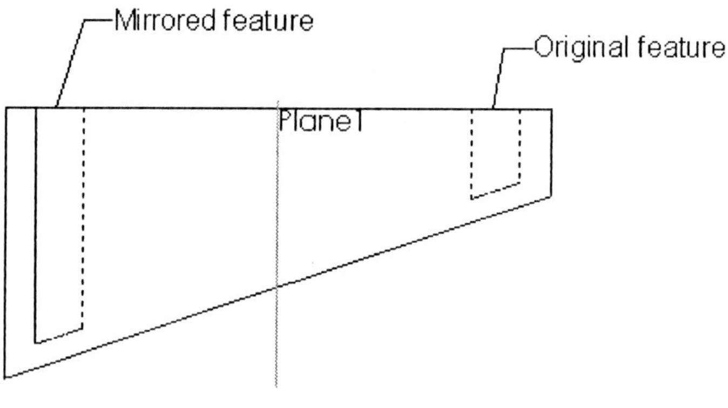

Figure 7-5 *Mirror feature created with the **Geometry Pattern** check box cleared*

If you select the **Geometry Pattern** check box, the mirror feature created will not depend on the relational references. It will just create a replica of the selected geometry. Figure 7-6 shows the mirror feature created with the **Geometry Pattern** check box selected.

Mirroring Faces

Using this option you can mirror the faces along a mirror plane or mirror face. To use this

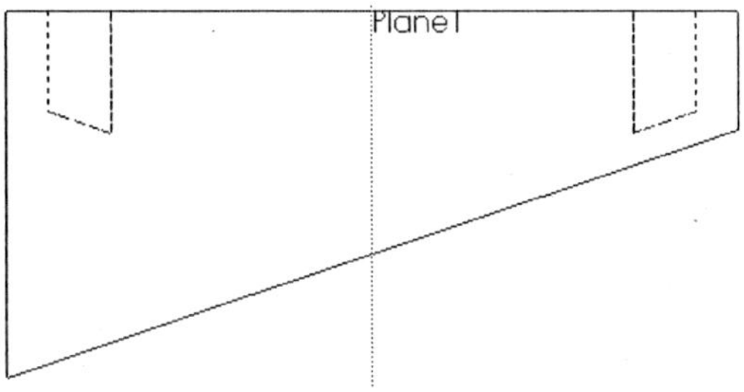

*Figure 7-6 Mirror feature created with the **Geometry Pattern** check box selected*

option, invoke the **Mirror PropertyManager**. You are prompted to select a plane or a planar face about which to mirror. Select the planar face or a plane about which the selected faces will be mirrored. Now, click once in the **Faces to Mirror** selection area to invoke the selection mode and select the faces to be mirrored. The selected faces must form a closed body. If the selected faces do not form a closed body, feature creation is not possible. Use the **OK** button from the **Mirror PropertyManager** to end feature creation. Figure 7-7 shows the faces and the mirror plane to be selected. Figure 7-8 shows the resultant mirror feature creation.

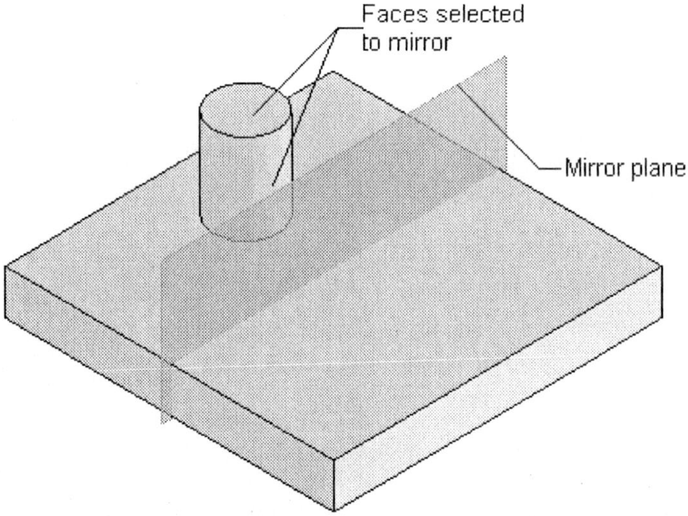

Figure 7-7 Plane and faces selected to mirror

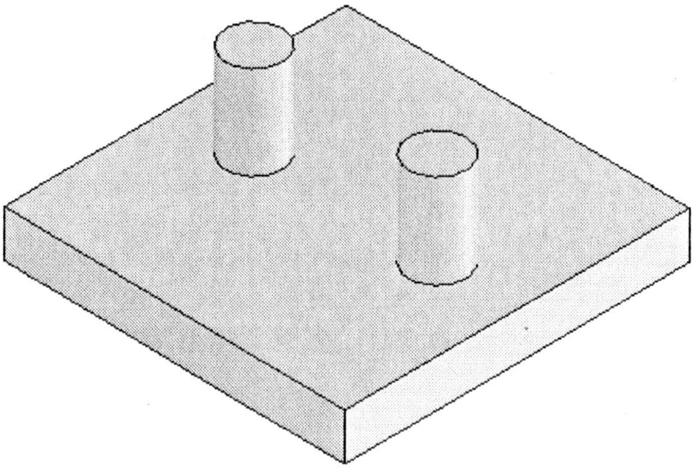

Figure 7-8 *The resultant mirror creation*

Note

Following are some of the factors that should be considered while creating a mirror feature by mirroring the faces along the selected plane or planar face:

1. If the replica of the faces is not coincident to the parent part body, SolidWorks will give an error while creating the mirror feature.

2. If the replica of the faces exists on faces other than the original face, SolidWorks will give an error while creating the mirror feature.

3. If the selected faces form a complex geometry, SolidWorks will give an error while creating the mirror feature.

4. If the mirrored faces exist on more than one face, SolidWorks will give an error while creating the mirror feature.

5. The selected faces should form a closed body. If the selected faces do not form a closed body, SolidWorks will give an error while creating the mirror feature.

Mirroring Bodies

As discussed in the earlier chapters, the multibodies environment is supported in SolidWorks. Therefore, using the **Mirror** tool you can also mirror the disjoint bodies. To mirror a body along a plane, invoke the **Mirror PropertyManager** and select a plane or a planar face that will act as mirror plane. Invoke the **Bodies to Mirror** rollout and select the body from the drawing area or expand the **FeatureManager Design Tree** flyout and select the body to mirror from the **Solid Bodies** folder. The name of the selected body is displayed in the **Solid/Surface Bodies to Mirror** selection area. The preview of the mirrored body is displayed in the drawing area. Choose the **OK** button from the **Mirror PropertyManager**. Figure 7-9 shows the plane and the body selected to mirror. Figure 7-10 shows the resultant mirror feature.

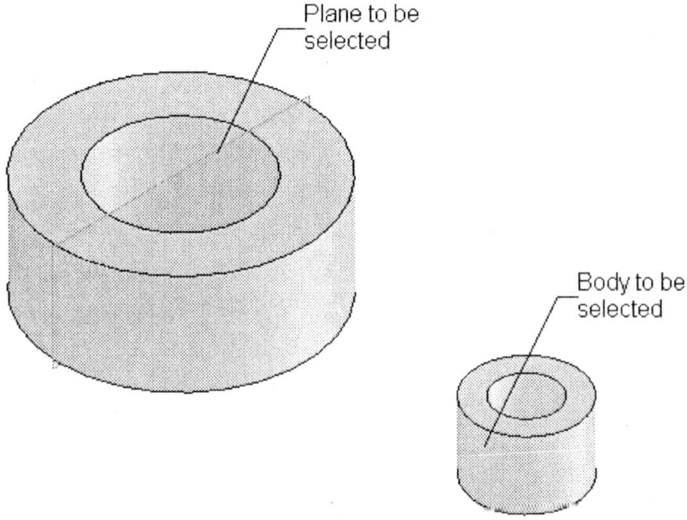

Figure 7-9 *Selecting the mirror plane and body to be mirrored*

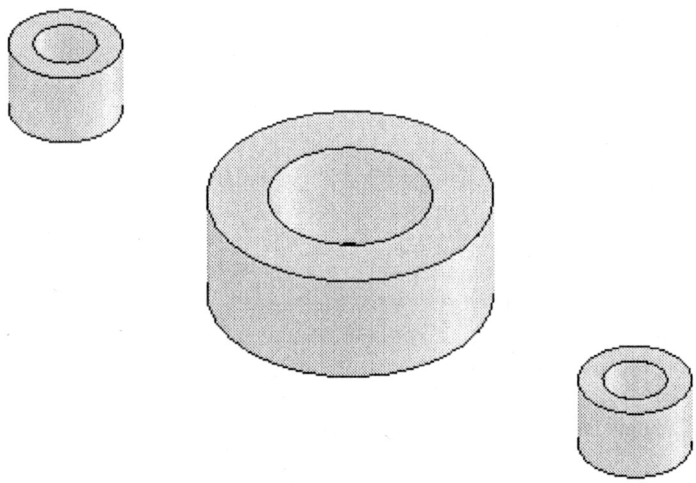

Figure 7-10 *Resultant mirror feature*

The options available in the **Options** rollout while mirroring the bodies are discussed next. The **Options** rollout with these options is displayed in Figure 7-11.

Merge solids

The **Merge solids** option is used to merge the mirrored body with the parent body. For example, consider a case in which you mirror a body along a selected plane or a planar face of the same body and the resultant mirrored body is joined to the parent body. In this

*Figure 7-11 The **Options** rollout*

case if you select the **Merge solids** check box, the resultant mirrored body will merge with the parent body to become a single body. If the **Merge solids** check box is cleared, the resultant body will be joined with the parent body but it will not merge with the parent body and results in two separate bodies.

Knit surfaces

If you mirror a surface body, then the **Knit Surface** check box is selected to knit the mirrored and the parent body together. You will learn more about surfaces and surface bodies in the later chapters.

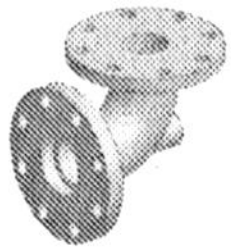

Tip. *As discussed earlier, the design intent is captured in the model using the mirror option. Therefore, if you modify the parent feature, face, or body the same will be reflected on the mirrored feature, face, or body.*

*If you want to mirror all the features of the model using the **Features to Mirror** option, you have to select all the features. But if you use the **Bodies to Mirror** option, you have to select the body from the **Solid Bodies** folder. By selecting the body, all the features are added to the mirror image.*

Creating Linear Pattern Feature

CommandManager:	Features > Linear Pattern
Menu:	Insert > Pattern/Mirror > Linear Pattern
Toolbar:	Features > Linear Pattern

As discussed in earlier chapters, you can arrange the sketched entities in a particular arrangement or pattern. In the same manner you can also arrange the features, faces, and bodies in a particular pattern. In SolidWorks, you are provided with various types of patterns such as linear patterns, circular patterns, sketch-driven patterns, curve-driven patterns, and table-driven patterns. In this section you will learn how to create linear patterns. The other type of patterns are discussed later in this chapter.

To create a linear pattern, choose the **Linear Pattern** button from the **Features CommandManager** or choose **Insert > Pattern/Mirror > Linear Pattern** from the menu bar. The **Linear Pattern PropertyManager** is invoked and the confirmation corner is also displayed. A partial view of the **Linear Pattern PropertyManager** is displayed in Figure 7-12.

The various options available in the **Linear Pattern PropertyManager** are discussed next.

Linear Pattern in One Direction

When you invoke the **Linear Pattern PropertyManager**, the **Direction 1** rollout and the **Features** rollout are invoked by default. Also, you are prompted to select an edge or axis

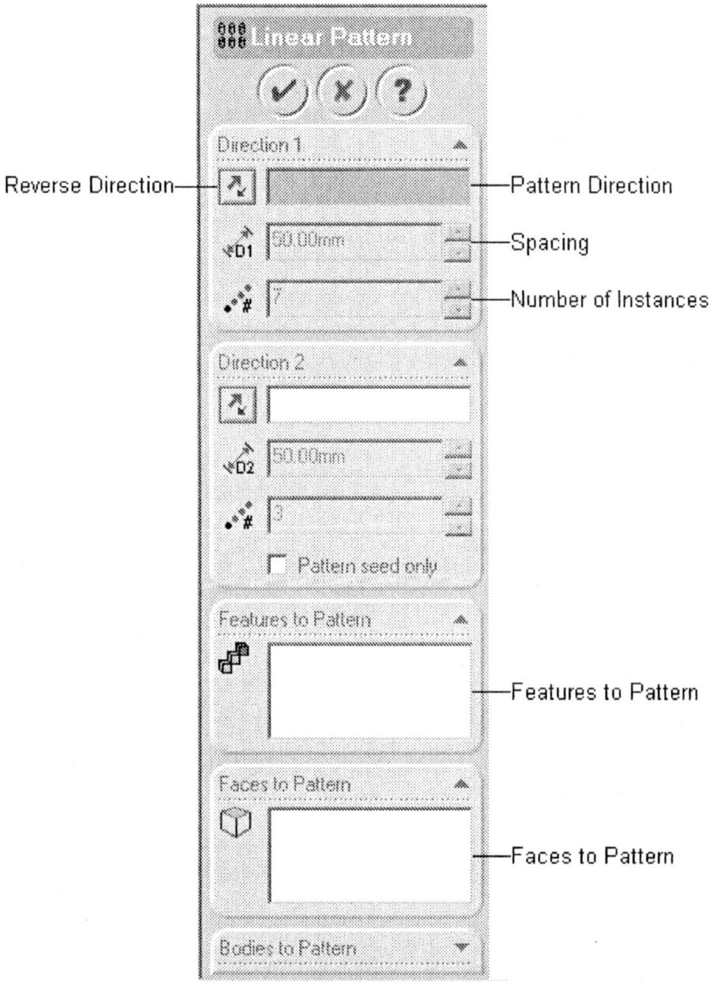

Figure 7-12** Partial view of the **Linear Pattern PropertyManager

for direction reference and face of feature for pattern features. Therefore, first you need to select an edge or an axis as a direction reference. Select an edge or an axis as the direction reference. The name of the selected reference will be displayed in the **Pattern Direction** selection area of the **Direction 1** rollout. The selected reference is displayed in green and the **Direction 1** callout is attached to the selected reference. The **Direction 1** callout has two edit boxes to define the number of instances and the spacing. You are also provided with the **Reverse Direction** arrow along with the selected reference. Now, select a face of the feature to be patterned. The name of the selected feature will be displayed in the **Features to pattern** area of the **Features to Pattern** rollout. The preview of the pattern is displayed in the drawing area with the default values. Set the value of the center to center spacing between the pattern instances in the **Spacing** spinner. Set the value of the number of instances to be patterned in the **Number of Instances** spinner. You can also set these values in the **Direction 1** callout. Using the **Reverse Direction** button from the **PropertyManager** or using the **Reverse Direction** arrow from the drawing area you can reverse the direction of pattern feature creation.

Figure 7-13 shows the model before pattern creation and Figure 7-14 shows the model after pattern creation.

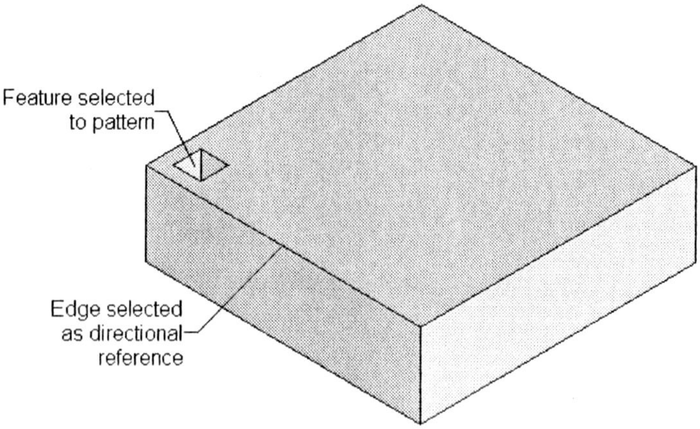

***Figure 7-13** Edge and the feature to be selected*

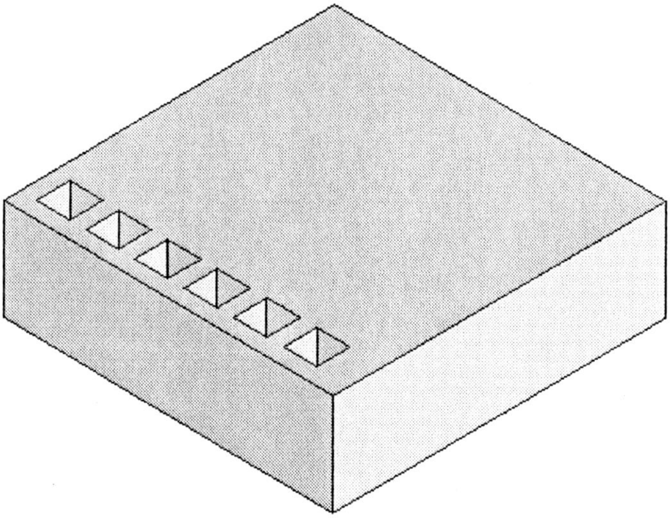

***Figure 7-14** Linear pattern created using the **Direction 1** option*

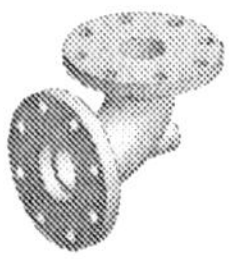

Tip. *When you select the feature to be patterned, the dimensions of the feature are also displayed in the drawing area. You can also select the dimensions as the directional reference.*

As discussed earlier, you can mirror the faces and bodies, you can also pattern the faces and bodies. It should be noted that the selected faces must form a closed body, otherwise the patterning of faces will give an error.

Linear Pattern in Two Directions

As discussed earlier you can create a linear pattern of features, faces, and bodies by defining a single direction using the **Direction 1** rollout. You can also define the parameters in the **Direction 2** rollout to define the pattern in the second direction. The **Direction 2** rollout is shown in Figure 7-15. When you define the pattern in the second direction the entire row created by specifying the parameters in the first direction is patterned in the second direction.

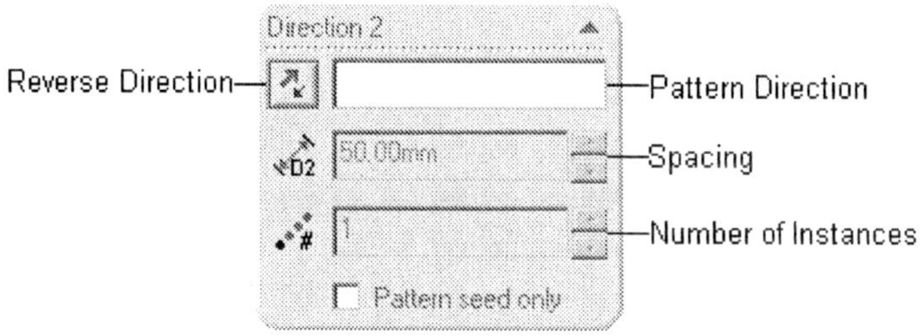

*Figure 7-15 The **Options** rollout*

To create a pattern by specifying the parameters in both the directions, invoke the **Linear Pattern PropertyManager**. Select the feature to be patterned and select the first directional reference. Next, specify the parameters in the **Direction 1** rollout. Now, select the second directional reference. If the **Direction 2** rollout is not invoked by default in the **Linear Pattern PropertyManager**, use the black arrow in the **Direction 2** callout to invoke the rollout. The options available in the **Direction 2** rollout are the same as those discussed in the **Direction 1** rollout. Figure 7-16 shows the directional references and feature to be selected. Figure 7-17 shows the linear pattern created using the **Direction 1** and **Direction 2** rollouts.

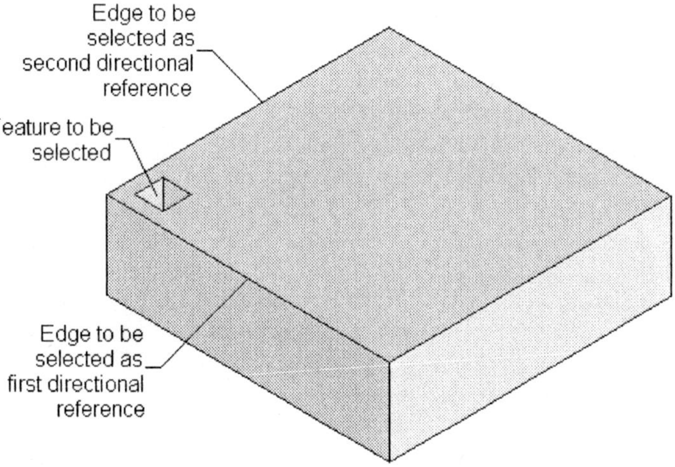

Figure 7-16 References and feature to be selected

By default, all the rows of the instances created in the first direction are patterned in the second direction also. This is because the **Pattern seed only** check box in the **Direction 2**

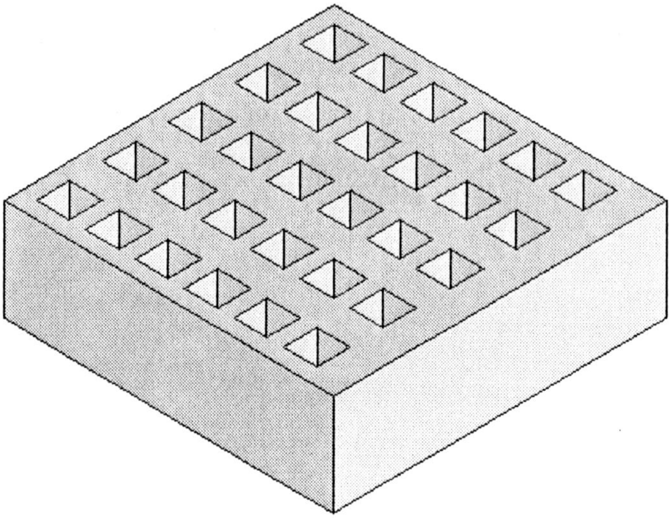

Figure 7-17 *Linear pattern created using the* ***Direction 1*** *and* ***Direction 2*** *rollouts*

rollout is cleared. You can select this check box to pattern only the original selected feature (also called seed feature) in the second direction. Figure 7-18 shows the pattern created with the **Pattern seed only** check box selected.

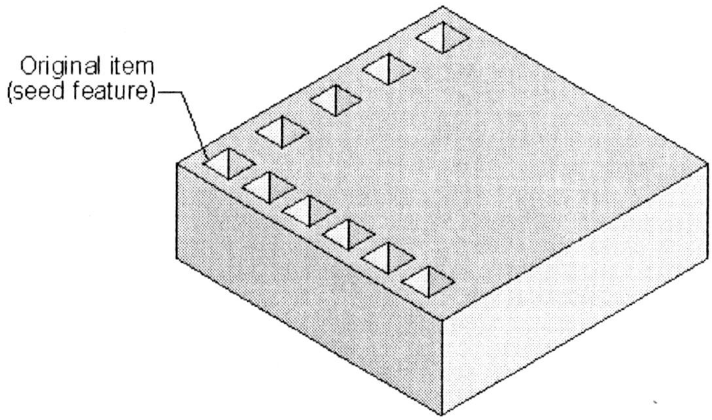

Figure 7-18 *Linear pattern created with the* ***Pattern to seed*** *check box selected*

Instances to Skip

Using the **Instances to Skip** option you can skip some of the instances from the pattern. The instances that are skipped are not actually deleted. These only disappear from the pattern feature and you can resume these instances at any time of your design cycle. To skip the pattern instances, invoke the **Instances to Skip** rollout from the **Linear Pattern PropertyManager**. The **Instances to Skip** rollout is displayed in Figure 7-19.

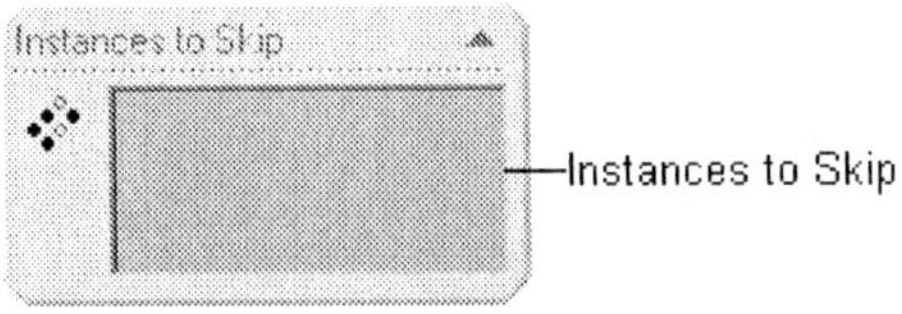

Figure 7-19 *The* ***Instances to Skip*** *rollout*

As soon as you invoke this rollout, pink dots are displayed at the center of all pattern instances except the parent instance. Therefore, you cannot skip the parent instance. Now, move the cursor to the pink dot of the instance to be skipped. The cursor will be replaced by the instance to skip cursor and the position of that instance, in the form of a matrix, is displayed in tooltip below this cursor. Use the left mouse button to skip that instance. The pink dot will be replaced by a red dot and the preview of that instance will disappear from the pattern. The position of the skipped instance is displayed in the **Instances to Skip** selection area of the **Instances to Skip** rollout. Figure 7-20 shows a pattern created with some instances skipped.

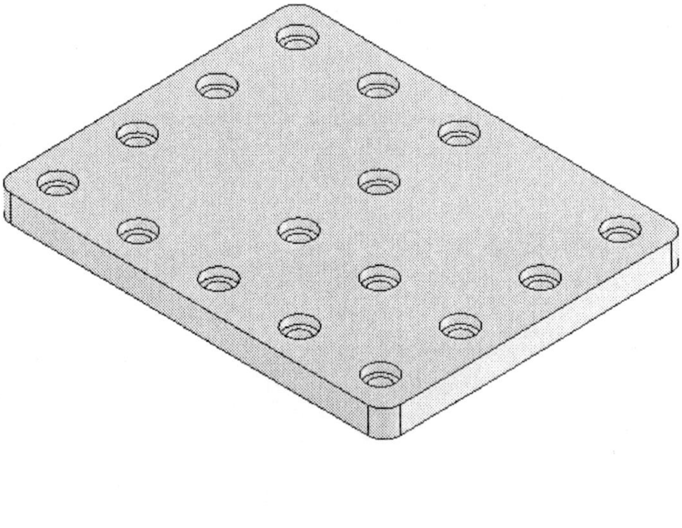

Figure 7-20 *Linear pattern created with some instances skipped*

You can resume the skipped instances by deleting the position of the instance from the **Instances to Skip** selection area or selecting the red dot from the drawing area.

Pattern Using Varying Sketch

The **Vary Sketch** option is used in a pattern where the shape and size of each pattern instance is controlled by the relations and dimension of the sketch of that feature. In this type of pattern, the dimension of the sketch of the feature to be patterned is selected as the directional reference. The dimension selected as the dimensional reference drives the shape and size of the sketch of the feature to be patterned. In Figure 7-21 a cut feature is created on

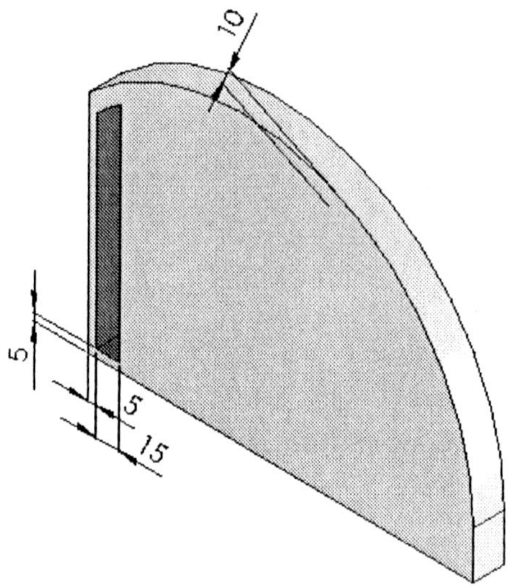

Figure 7-21 *Cut feature created on the base feature*

the base feature. Figure 7-22 shows the linear pattern created using the varying sketch option.

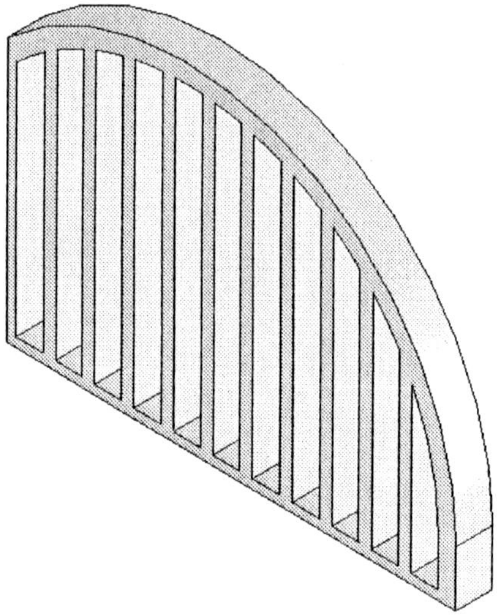

Figure 7-22 *Linear pattern created with the* ***Vary sketch*** *check box selected*

For creating this type of pattern, the sketch of the feature to be patterned should be in relation with the geometry along which it will vary. The dimensions of the sketch should allow the sketch to change the shape and size easily. You should also provide a linear dimension that will drive the entire sketch and will be the directional reference. From the **FeatureManager Design Tree** select the feature to be patterned and then invoke the **Linear Pattern PropertyManager**. Now, select the dimension to specify the directional reference and set the value of spacing and the number of instances. Invoke the **Options** rollout and select the **Vary sketch** check box. As you select the **Vary sketch** check box, the preview of the pattern disappears from the drawing area. Choose the **OK** button from the **Linear Pattern PropertyManager** to end feature creation.

The **Geometry pattern** option available in the **Options** rollout of the **Linear Pattern PropertyManager** is the same as that discussed earlier in the **Mirror PropertyManager**.

Creating Circular Pattern Feature

CommandManager:	Features > Circular Pattern
Menu:	Insert > Pattern/Mirror > Circular Pattern
Toolbar:	Features > Circular Pattern

As discussed in earlier chapters, you can arrange the sketched entities in a circular pattern using the **Circular Sketch Step and Repeat** option. In this section, you will learn how to create the circular pattern of a feature, face, or body. You can create the circular pattern of a feature, face, or body using the **Circular Pattern** tool. To invoke this tool, choose the **Circular Pattern** button from the **Features CommandManager** or choose **Insert > Pattern/Mirror > Circular Pattern** from the menu bar. The **Circular Pattern PropertyManager** is invoked and the confirmation corner is displayed as shown in Figure 7-23.

After invoking the **Circular Pattern PropertyManager**, you are prompted to select an edge or an axis for direction reference, and a face of the feature to be patterned. If the base feature of the model is a circular feature, the temporary axis is created automatically at the center of the model. Choose **View > Temporary Axes** from the menu bar to display the temporary axis. Otherwise, before invoking the **Circular Pattern** tool, you have to create an axis. Now, select the axis that will act as the pattern axis and the feature to pattern. The preview of the pattern feature with the default values is displayed in the drawing area. The **Direction 1** callout is displayed with the **Reverse Direction** arrow in the drawing area. By default, the **Equal Spacing** check box is cleared. Therefore, you need to set the value of the incremental angle between the instances in the **Total Angle** spinner. Set the value of number of instances to pattern in the **Number of Instances** spinner. If you select the **Equal Spacing** check box then you need to enter the value of total angle along which all the instances of the pattern will be accommodated. The angular spacing between the instances will be automatically calculated.

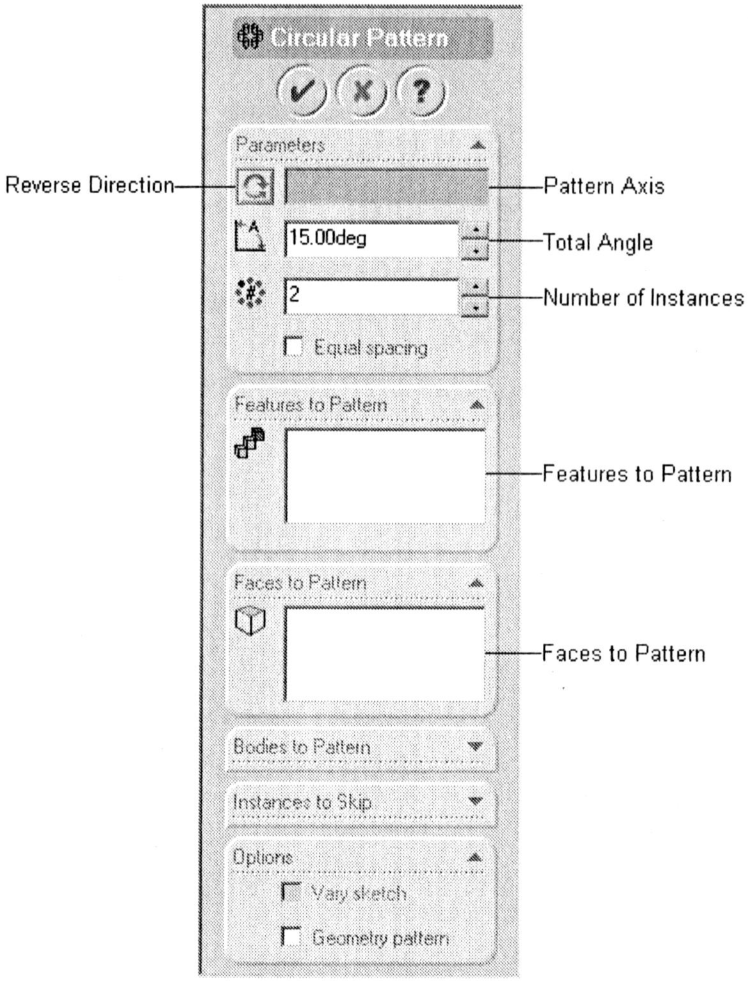

Figure 7-23** The **Circular Pattern PropertyManager

The **Reverse Direction** button available on the left of the **Pattern Axis** selection area is used to change the direction of rotation while creating the pattern feature. By default, the direction of pattern creation is clockwise. If you choose this button then the resultant pattern will be created in the counterclockwise direction. You can also change the direction of pattern creation using the **Reverse Direction** arrow from the drawing area. Figure 7-24 shows the feature and the temporary axis being selected. Figure 7-25 shows the resultant pattern feature.

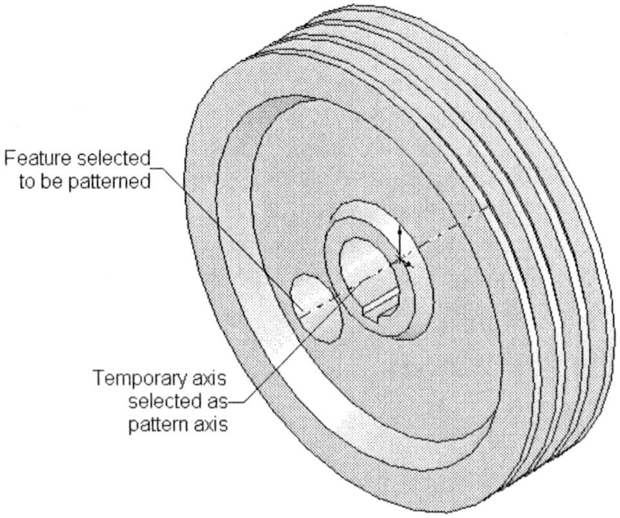

Figure 7-24 *The reference to be selected for creating a circular pattern*

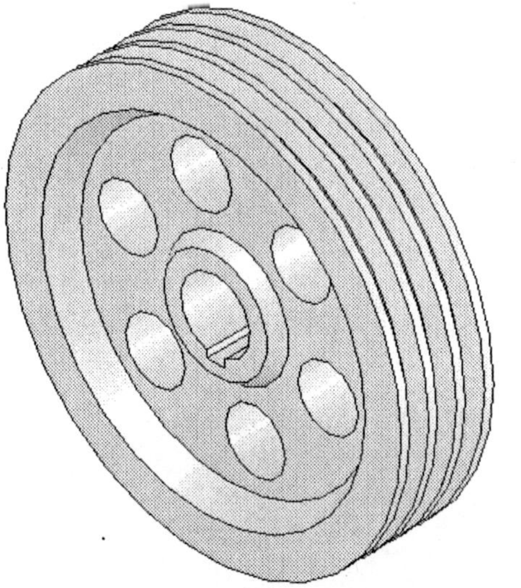

Figure 7-25 *Resultant circular pattern*

Tip. *Instead of setting the value of angle and number of instances in the respective spinners, you can also the values in the callout. To enter the values in the callout is a better and timesaving practice.*

It is always a good practice to create patterns of features instead of creating complex sketches or repeatedly creating the same feature again and again. It also helps in capturing the design intent of the model. The patterns created in ***Part*** *mode are very useful in assembly modeling. You will learn more about assembly modeling in later chapters.*

Circular Pattern Using Dimensional Reference

You can also create circular pattern by selecting an angular dimension. To create a pattern using this option you have to create an angular dimension in the sketch of the feature that is to be patterned. Invoke the **Circular Pattern PropertyManager** and select the feature to be patterned. The dimensions of the feature to be patterned will be displayed in the drawing area as shown in Figure 7-26. Now, select the angular dimension and set the value of total angle and spacing in the **Circular Pattern PropertyManager**. Figure 7-27 shows the pattern created by selecting the angular dimension as angular reference. The other options available in the **Circular Pattern PropertyManager** are the same as those discussed earlier for the **Linear Pattern PropertyManager**.

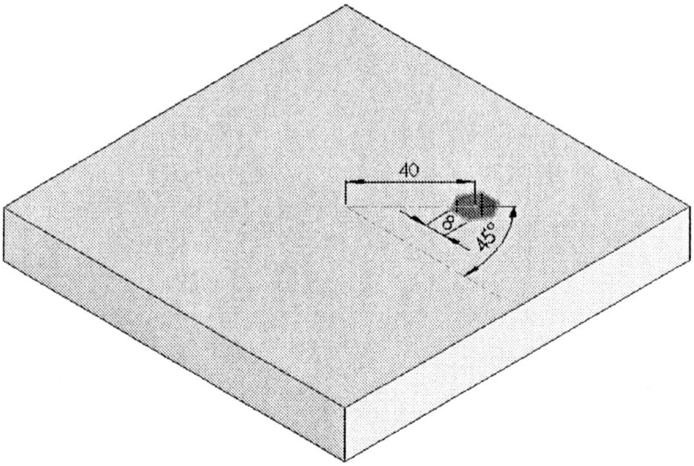

Figure 7-26 *Dimensions displayed after selecting the feature to be patterned*

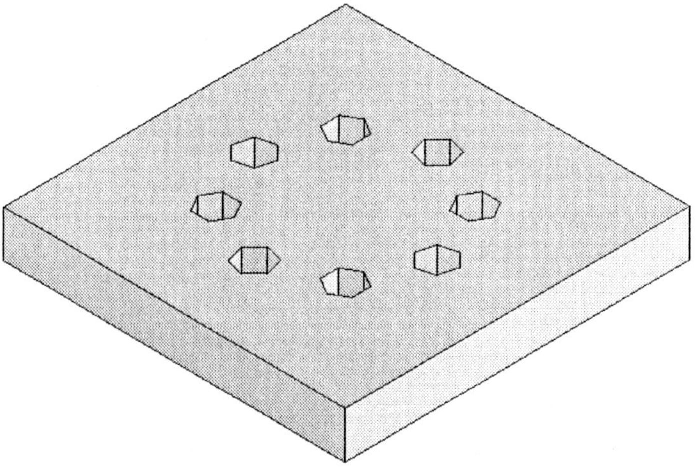

Figure 7-27 *Circular pattern created by selecting the angular dimension as the angular reference*

Creating Sketch-driven Pattern

CommandManager:	Features > Sketch Driven Pattern	*(Customize to Add)*
Menu:	Insert > Pattern/Mirror > Sketch Driven Pattern	
Toolbar:	Features > Sketch Driven Pattern	*(Customize to Add)*

A sketch-driven pattern is created when the features, faces, or bodies are to be arranged in a nonuniform manner, which is neither rectangular nor circular. For creating a sketch-driven pattern, first you have to create an arrangement of the sketch points in a single sketch. This arrangement of sketch points will drive the instances in the pattern feature. After creating the sketch and the feature to pattern, invoke the **Sketch Driven Pattern PropertyManager** as shown in Figure 7-28. You are prompted to select a sketch for pattern layout, and face of feature to pattern.

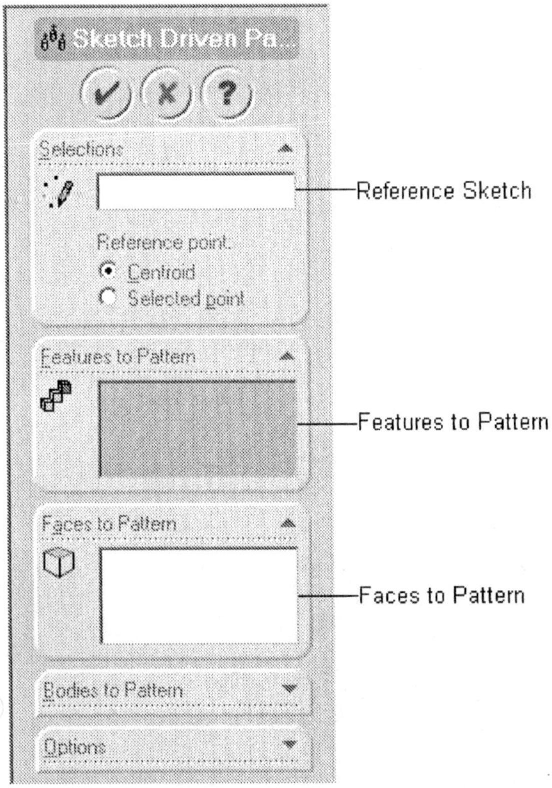

Figure 7-28** The **Sketch Driven Pattern PropertyManager

Select the feature or features to be patterned. Now, click in the **Reference Sketch** display box in the **Selections** rollout and select any one of the sketched point from the drawing area. You can also select the sketch from the **FeatureManager Design Tree** flyout. Choose the **OK** button from the **Sketch Driven Pattern PropertyManager**.

Figure 7-29 shows the feature and the sketch point selected and Figure 7-30 shows the resultant pattern.

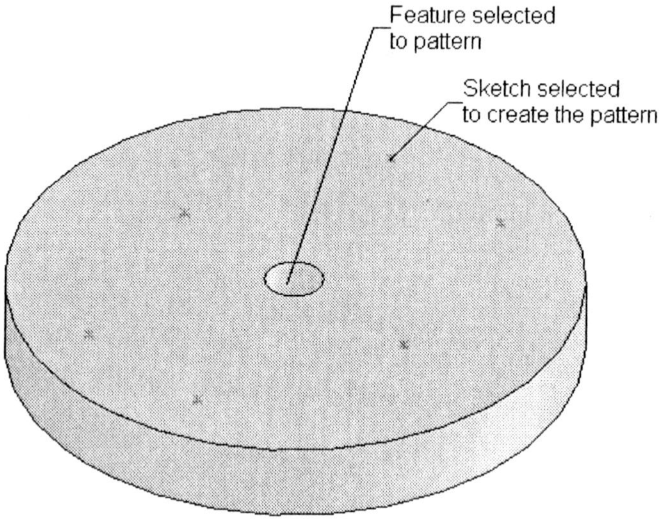

Figure 7-29 *The feature and the sketch point to be selected*

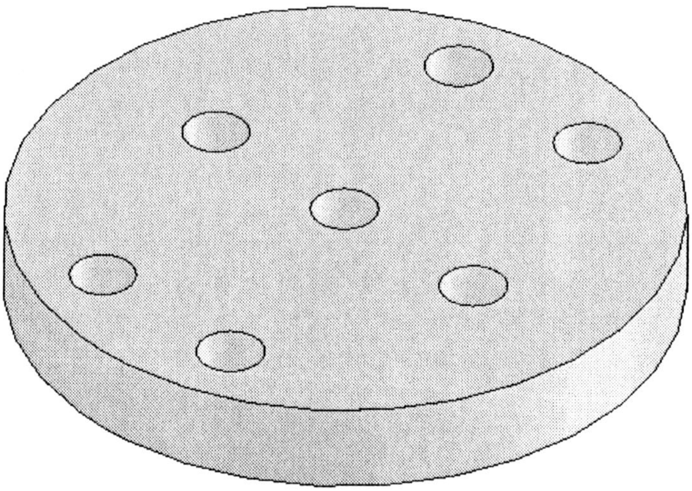

Figure 7-30 *The resultant pattern feature*

The options available in the **Sketch Driven Pattern PropertyManager** are discussed next.

Sketch Driven Pattern Using Centroid

When you invoke the **Sketch Driven Pattern PropertyManager**, the **Centroid** radio button is selected by default in the **Reference Point** area of the **Selections** rollout. Therefore, the pattern created is with reference to the centroid.

Sketch Driven Pattern Using Selected point

If you select the **Selected point** radio button from the **Reference Point** area of the **Selections** rollout, the pattern created is with reference to the selected point. When you select this radio button, a **Reference Vertex** selection area is displayed. Select a vertex; the pattern will be created in reference to that vertex.

Creating Curve-driven Pattern

CommandManager:	Features > Curve Driven Pattern	*(Customize to Add)*
Menu:	Insert > Pattern/Mirror > Curve Driven Pattern	
Toolbar:	Features > Curve Driven Pattern	*(Customize to Add)*

The **Curve Driven Pattern** option is used to pattern the features, faces, or bodies along a selected reference curve. The reference curve can be a sketched entity or an edge and can be an open profile or a closed loop. To create a pattern using this option, choose the **Curve Driven Pattern** button from the **Features CommandManager** or choose **Insert > Pattern/Mirror > Curve Driven Pattern** from the menu bar. When you choose this button, the **Curve Driven Pattern PropertyManager** is displayed as shown in Figure 7-31. Figure 7-32 shows the feature and the curve that will be used to create the pattern. Figure 7-33 shows the curve-driven pattern created.

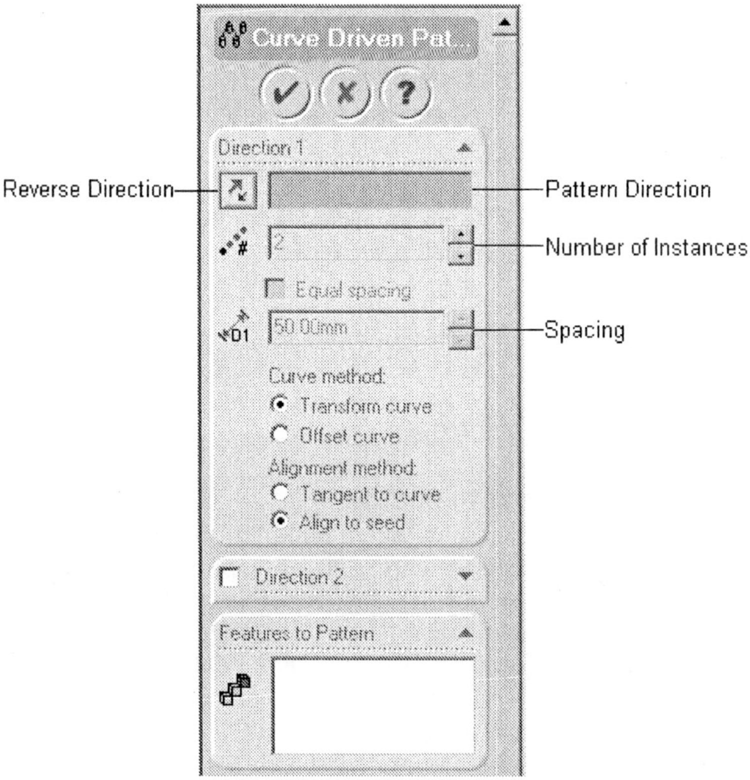

*Figure 7-31 Partial view of the **Curve Driven Pattern PropertyManager***

After invoking the **Curve Driven Pattern PropertyManager**, you are prompted to select an edge, curve, or sketch segment for pattern layout and select a face of feature to be patterned. Select the reference curve along which the feature, face, or body will be patterned. When you

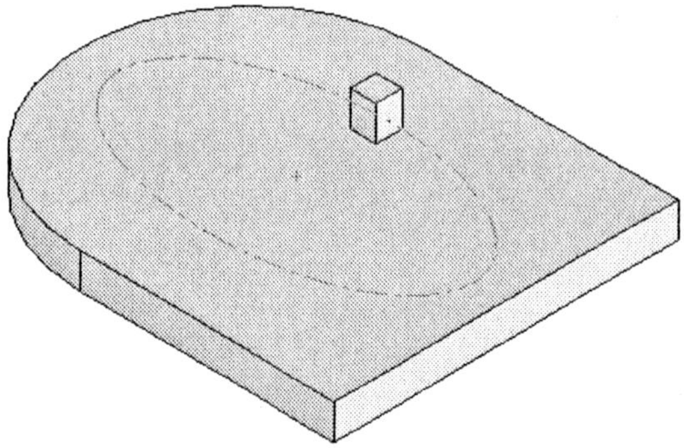

Figure 7-32 *The feature and curve to be used to create the curve-driven pattern*

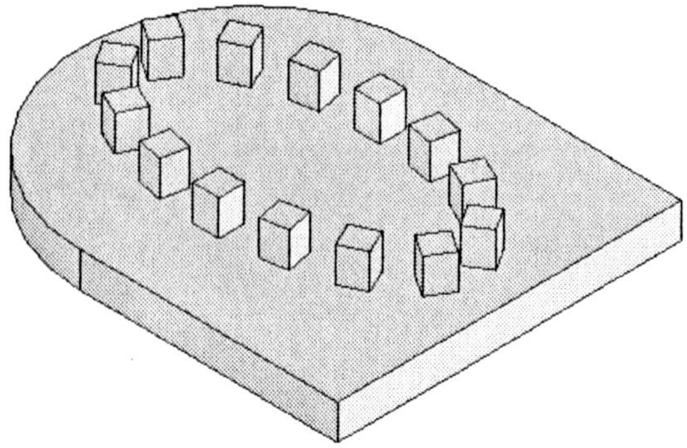

Figure 7-33 *The resultant pattern feature*

select the reference curve, the name of curve is displayed in the **Pattern Direction** selection area and the **Direction 1** callout is displayed. As discussed earlier, the **Direction 1** callout is divided in two areas. Select the feature to pattern; the preview of the pattern is displayed in the drawing area. Set the various parameters available in the **Direction 1** rollout and choose the **OK** button from the **Curve Driven Pattern PropertyManager**.

The various options available in the **Direction 1** rollout are discussed next.

Equal Spacing

The **Equal Spacing** check box is used to accommodate all the instances of the pattern along the selected curve. By default, this check box is cleared. Therefore, you have to specify the distance between the instances and the total number of instances to be created along the selected curve. When you select this check box, the **Spacing** spinner is not available and you have to specify only the total number of instances. The distance between the instances is calculated automatically.

Curve method and Alignment method

The **Curve method** area of the **Direction 1** rollout is used to specify the type of curve method to be followed while patterning along the curve. The two options available in this area are in the form of radio buttons. The options are the **Transform curve** method and the **Offset curve** method. The **Alignment method** area of the **Direction 1** rollout is used to specify the type of alignment method to be applied. The two alignment methods are the **Tangent to curve** method and the **Align to seed** method. Figure 7-34 shows the curve-driven pattern created with the **Transform curve** and the **Tangent to curve** radio buttons selected. Figure 7-35 shows the curve-driven pattern created with the **Transform curve** and **Align to seed** radio buttons selected.

Figure 7-36 shows the curve-driven pattern created with the **Offset curve** and the **Tangent to curve** radio buttons selected. Figure 7-37 shows the curve-driven pattern created with the **Offset curve** and **Align to seed** radio buttons selected.

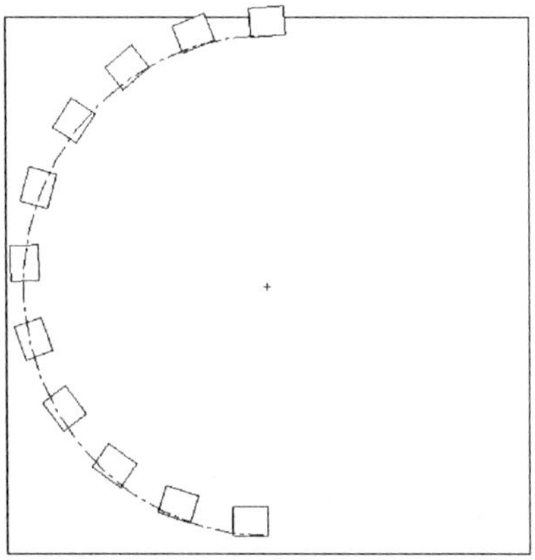

Figure 7-34 *Pattern created with* ***Transform curve*** *and* ***Tangent to curve*** *radio buttons selected*

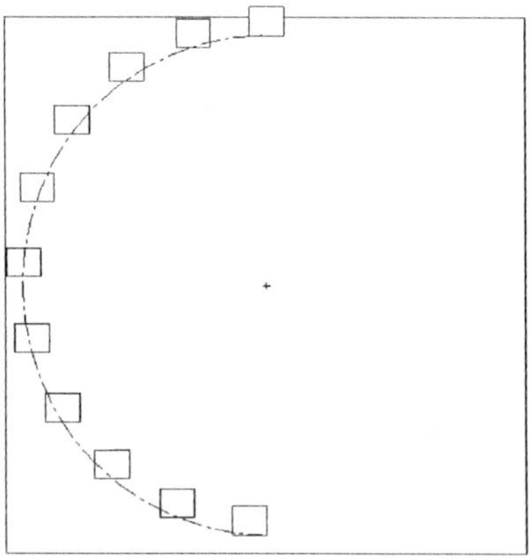

Figure 7-35 *Pattern created with* ***Transform curve*** *and* ***Align to seed*** *radio buttons selected*

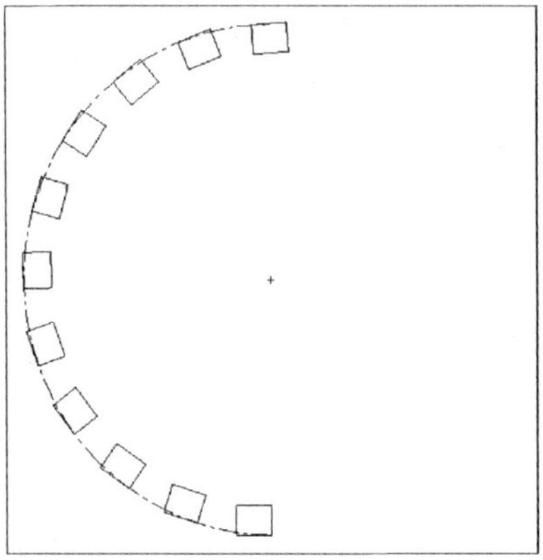

Figure 7-36 *Pattern created with* ***Offset curve*** *and* ***Tangent to curve*** *radio buttons selected*

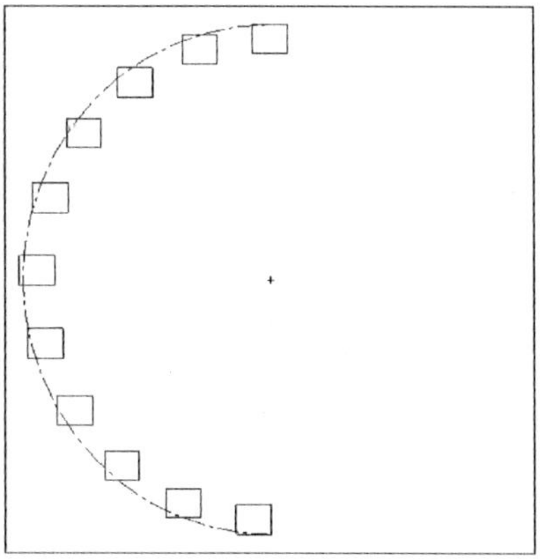

Figure 7-37 *Pattern created with* ***Offset curve*** *and* ***Align to seed*** *radio buttons selected*

Other options available in the **Curve Driven PropertyManager** are the same as those discussed earlier for the mirror and other pattern features. By selecting the check box available in the **Direction 2** rollout you can also specify the parameters in the second direction. Figure 7-38 shows the curve-driven pattern feature created with pattern defined in direction 1 and in direction 2.

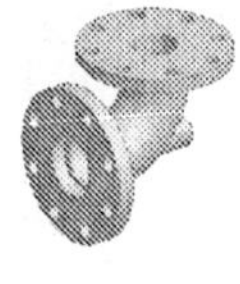

Tip. *In SolidWorks you can pattern a pattern feature. You can also pattern a mirror feature. The mirror of the pattern feature is also possible in SolidWorks.*

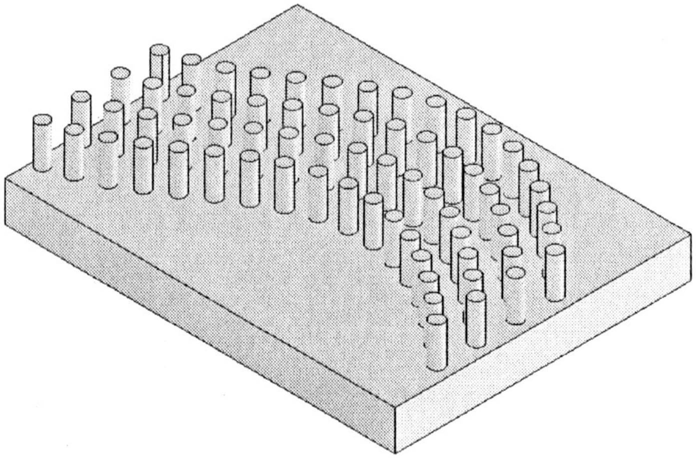

Figure 7-38 A curve-driven pattern created by specifying parameters in both the directions

Creating Table-driven Pattern

CommandManager:	Features > Table Driven Pattern	*(Customize to Add)*
Menu:	Insert > Pattern/Mirror > Table Driven Pattern	
Toolbar:	Features > Table Driven Pattern	*(Customize to Add)*

The **Table Driven Pattern** is created by specifying the X and Y coordinates with reference to a coordinate system. The instances of the selected features, faces, or bodies are created at the points specified using the X and Y coordinates. For creating this type of pattern, you first need to create a coordinate system using the **Coordinate System** button from the **Reference Geometry** toolbar. The coordinate system defines the direction along which the selected feature will be patterned. Choose the **Table Driven Pattern** button from the **Features CommandManager** or choose **Insert > Pattern/ Mirror > Table Driven Pattern** from the menu bar. The **Table Driven Pattern** dialog box is displayed as shown in Figure 7-39.

Select the feature to be patterned and the coordinate system from the drawing area or from the **FeatureManager Design Tree**. Enter the coordinates for creating the instances in the **Coordinate points** area of the **Table Driven Pattern** dialog box. As you enter the coordinates for the instances, the preview of the pattern is displayed in the drawing area. After entering all the coordinate points, choose the **OK** button from the **Table Driven Pattern** dialog box. Figure 7-40 shows the feature and the coordinate system to be selected. Figure 7-41 shows the table-driven pattern created after entering the coordinate values in the **Table Driven Pattern** dialog box.

You can also save the table-driven pattern file and retrieve the same coordinates by simply browsing the saved file using the **Browse** button from the **Table Driven Pattern** dialog box. You can also simply write the coordinates in a text file and browse the same file while creating

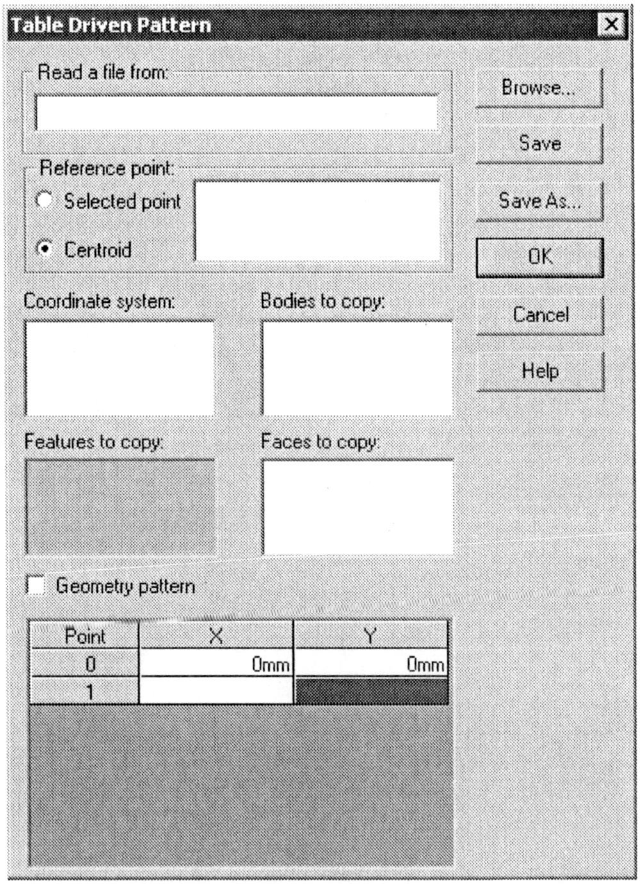

Figure 7-39 *The* ***Table Driven Pattern*** *dialog box*

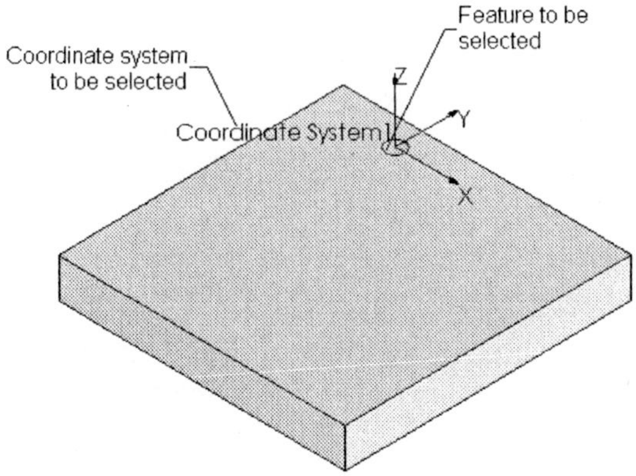

Figure 7-40 *Feature and the coordinate system to be selected for creating a table-driven pattern*

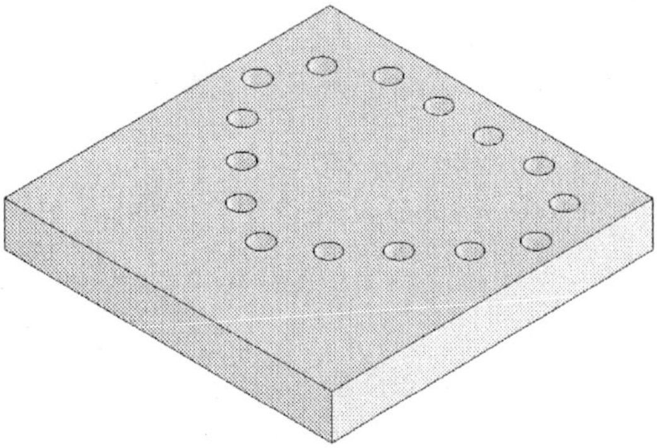

Figure 7-41 *Resultant pattern after specifying the coordinate points*

a table driven pattern. The other options available in this dialog box are the same as discussed earlier.

Creating the Rib Feature

CommandManager: Features > Rib
Menu: Insert > Features > Rib
Toolbar: Features > Rib

Ribs are defined as thin walled structures that are used to increase the strength of the entire structure of the component so that it does not fail under an increased load. In SolidWorks, the ribs are created using an open sketch as well as a closed sketch. To create a rib feature, invoke the **Rib PropertyManager** and select the plane on which you need to draw the sketch for creating the rib feature. Create the sketch and exit the sketching environment. Specify the rib parameters in the **Rib PropertyManager** and view the detailed preview using the **Detailed Preview** button. The **Rib** tool is invoked by choosing the **Rib** button from the **Features CommandManager** or by choosing **Insert > Features > Rib** from the menu bar. After invoking the **Rib** tool create the sketch and exit the sketching environment; the **Rib PropertyManager** is displayed as shown in Figure 7-42.

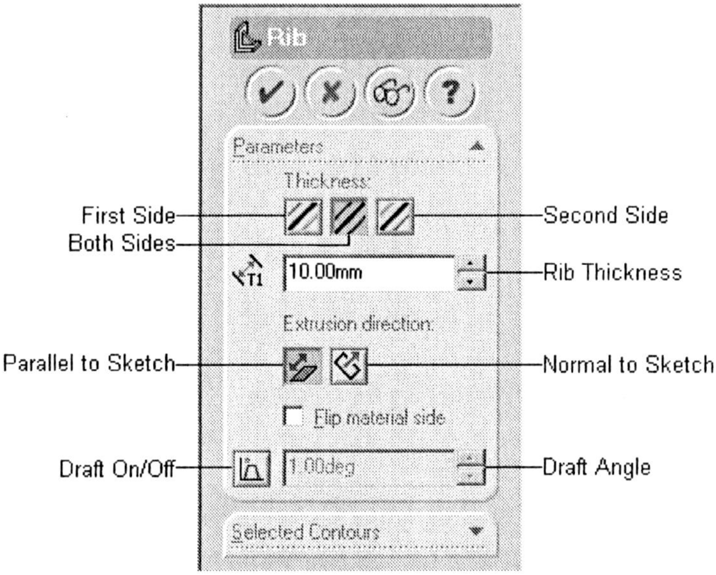

*Figure 7-42 The **Rib PropertyManager***

The preview of THE rib feature with direction arrow and the confirmation corner is displayed in the drawing area. Figure 7-43 shows the sketch created for the rib feature and Figure 7-44 shows the resultant rib feature. The various options available in the **Rib PropertyManager** are discussed next.

Tip. *You can also create the rib feature by first creating the sketch and then invoking the **Rib** tool within the sketching environment. If you exit the sketching environment after creating the sketch for a rib feature, invoke the **Rib** tool and select the sketch from the drawing area.*

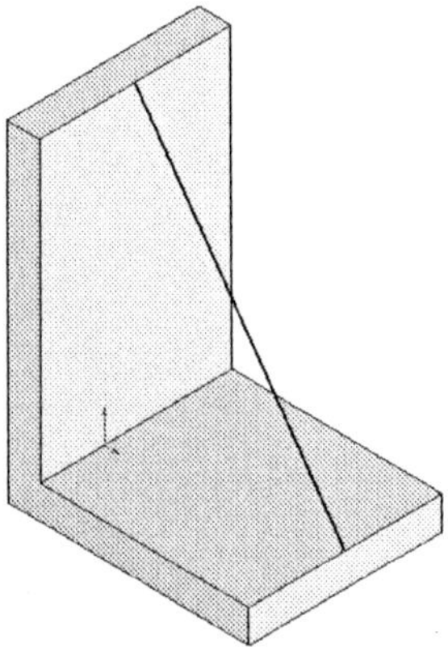

Figure 7-43 Sketch for the rib feature

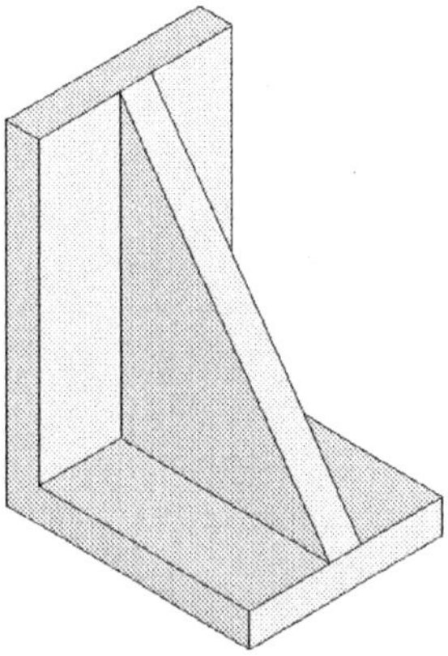

Figure 7-44 Resultant rib feature

Thickness

The **Thickness** area of the **Parameters** rollout is used to specify the side of the rib thickness and the thickness of the rib feature. The buttons available in this area are used to control the side on which you want to add the rib thickness. By default, the **Both Sides** button is chosen. Therefore, the rib is created on both the sides of the sketch. The **Rib Thickness** spinner in this area of the **Parameters** rollout is used to specify the rib thickness. You can choose the **First Side** or **Second Side** button to create ribs on either side of the sketch. The thickness of the rib is specified using the **Rib Thickness** spinner.

Extrusion direction

The **Extrusion direction** area of the **Parameters** rollout is used to specify the method of extruding the closed or open sketch. When you invoke the **Rib PropertyManager**, by default the option that is suitable for creating the rib feature will be active, depending on the geometric conditions. The options available in this area are discussed next.

Parallel to Sketch

The **Parallel to Sketch** option is used to extrude the sketch in a direction that is parallel to the sketch as well as parallel to the sketching plane. When you invoke the **Rib PropertyManager** and the sketch created for the rib feature is a continuous single entity open sketch, then this option is selected by default. Figure 7-45 shows an open sketch suitable for creating a rib using the **Parallel to Sketch** option. Figure 7-46 shows the rib feature created using that sketch.

Normal to Sketch

The **Normal to Sketch** option is generally used to create a rib feature when the sketch of the rib feature is a closed loop sketch or it consists of multiple sketched entities. The sketch with multiple entities can be closed loops or open profiles. If you create a sketch with a closed loop or a sketch with multiple sketched entities and invoke the **Rib** tool, the **Normal to Sketch** button will be selected by default. You can also choose the **Normal to Sketch** button from the **Extrusion Direction** area to use this option. Figure 7-47 shows a multiple entity sketch for the rib feature. Figure 7-48 shows the resultant rib feature.

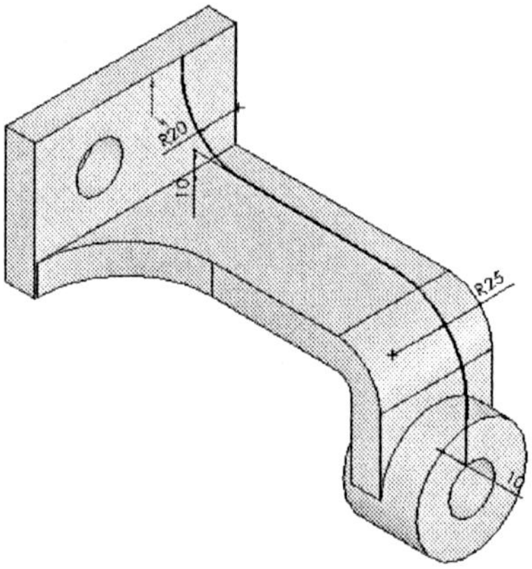

Figure 7-45 *An open sketch for the rib feature*

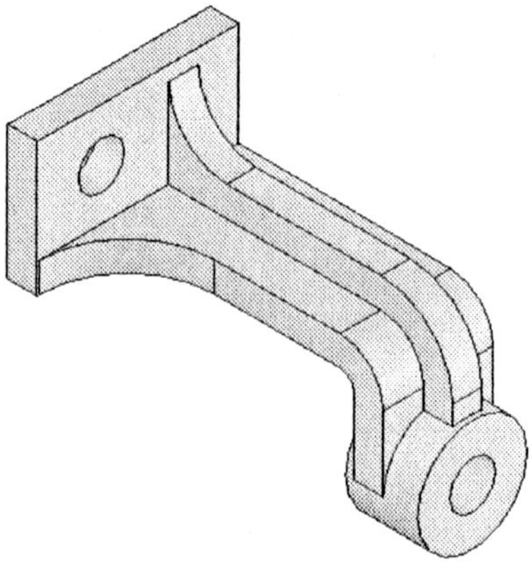

Figure 7-46 *Resultant rib feature*

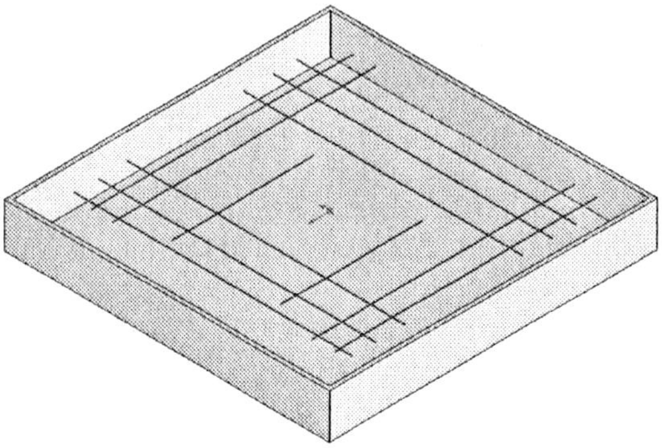

Figure 7-47 *Multiple entities for the rib feature*

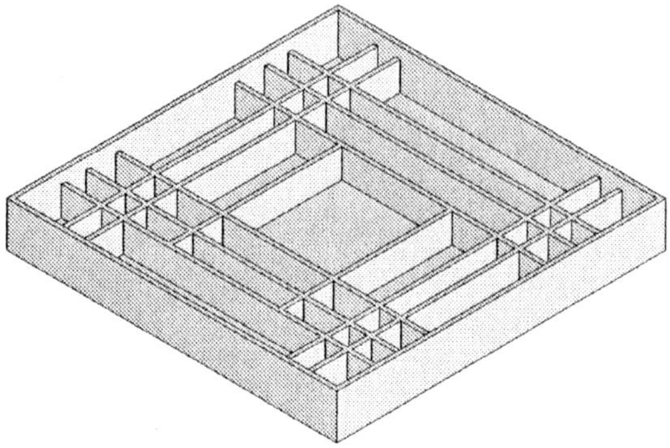

Figure 7-48 *Resultant rib feature*

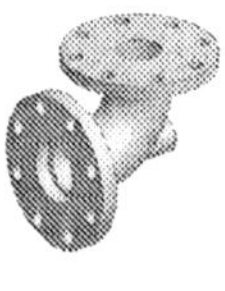

Tip. *You will observe that the endpoints of the sketched lines created in Figure 7-47 do not merge with the model edges. But the rib created using this sketch merges with the model edges. This is because while creating the sketch for the rib feature you do not have to create a complete sketch. The ends of the rib feature automatically extend to the next surface.*

When this option is selected, the **Type** area is displayed under the **Draft Angle** spinner. The options available in the **Type** area are discussed next.

Type

The **Type** area is available only when you choose the **Normal to Sketch** button from the **Extrusion direction** area of the **Parameters** rollout. The **Type** area is provided with two radio buttons: **Linear** and **Natural**. These options are used if the endpoints of the open sketch for the rib are not coincident with the faces of the existing feature. If the **Linear Radio** button is selected, the rib is created by extending the sketch normal to the sketched entity direction. The sketch will be extended up to a point where it meets a boundary.

On the other hand, if the **Natural** radio button is selected, the rib feature is created by extending the sketch along the direction of sketched entities. For example, consider the sketch shown in Figure 7-49. This figure shows a multiple entity sketch created for the rib feature.

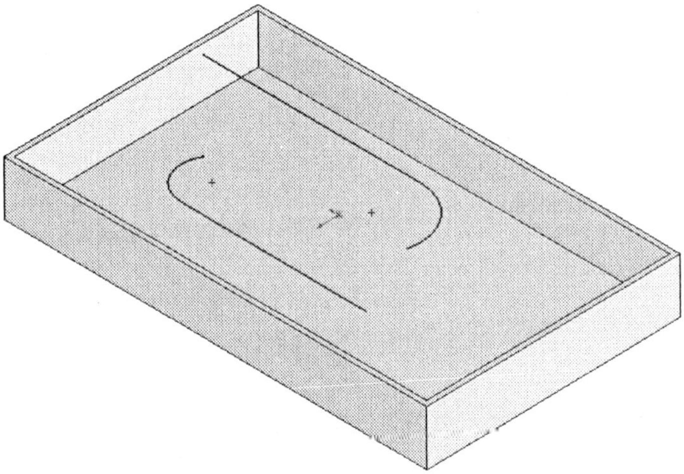

Figure 7-49 *Sketch for the rib feature*

Figure 7-50 shows a rib feature created by extending the sketch normal to the arc and the line. This is because the **Linear** radio button is selected. Similarly, in Figure 7-51, the feature is created by extending the sketch along the line and arc using the **Natural** radio button. This is the reason a circular feature is created at the end where the sketch has the arc.

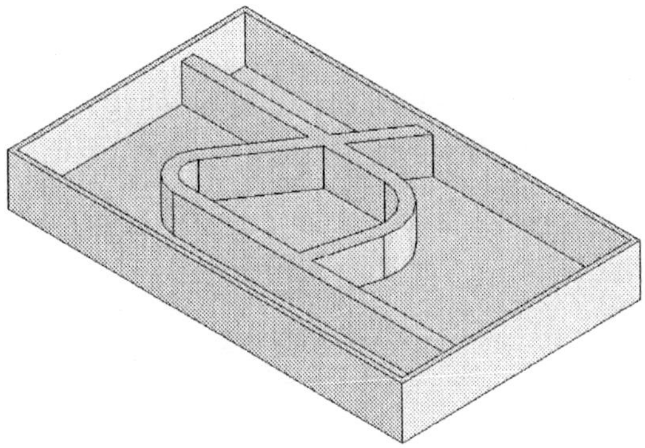

Figure 7-50 *Rib feature created with* ***Linear*** *radio button selected from the* ***Type*** *area of the* ***Rib PropertyManager***

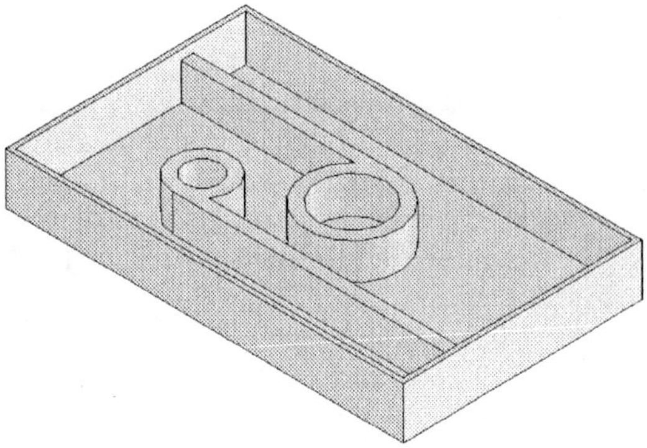

Figure 7-51 *Rib feature created with* ***Natural*** *radio button selected from the* ***Type*** *area of the* ***Rib PropertyManager***

Flip material side

The **Flip material side** check box is selected to reverse the direction of material addition while creating the rib feature. You can also reverse the direction of material addition using the **Flip material side** arrow available in the drawing area.

Draft On/Off

The **Draft On/Off** button is used to add the taper to the faces of the rib feature. When you choose the **Draft On/Off** button, the **Draft Angle** spinner is invoked. If you are creating a rib feature using multiple sketch entities, you can add only a simple draft to the rib feature. Figure 7-52 shows the draft angle added to the rib feature. By default, the draft is added inwards to the rib feature.

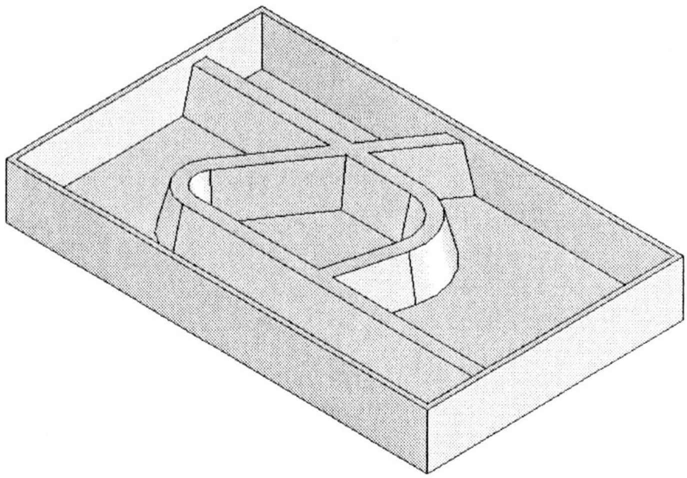

Figure 7-52 *Draft angle added to the rib feature*

Using the **Draft outward** check box you can add the draft outwards. If the rib feature to be created consists of a single continuous sketch and if you choose the **Draft On/Off** button, the **Next Reference** button is displayed below the **Draft Angle** spinner. A reference arrow is also displayed in the drawing area. Using the **Next Reference** button you can cycle through the reference along which you want to add the draft angle.

Figure 7-53 shows the preview of the rib feature and Figure 7-54 shows the resultant rib feature with **Draft outward** check box selected.

Figure 7-55 shows the preview of the rib feature and Figure 7-56 shows the resultant rib feature with **Draft outward** check box cleared.

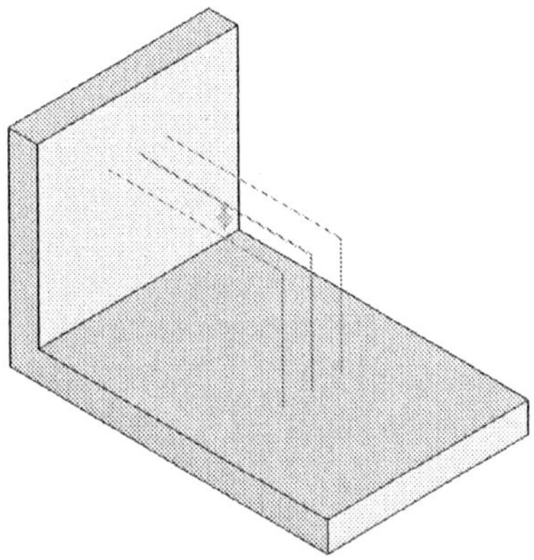

Figure 7-53 Sketch selected to create a rib feature with outward draft

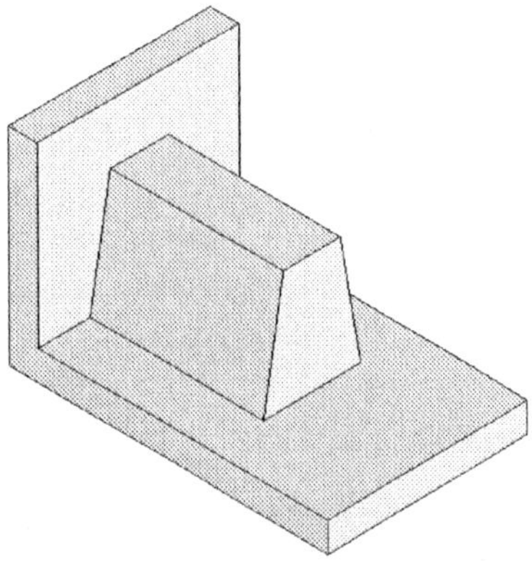

Figure 7-54 Resultant rib feature

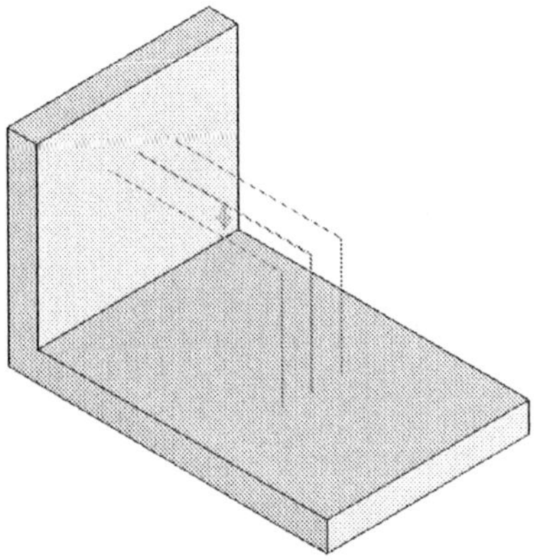

Figure 7-55 Sketch selected to create a rib feature with inward draft

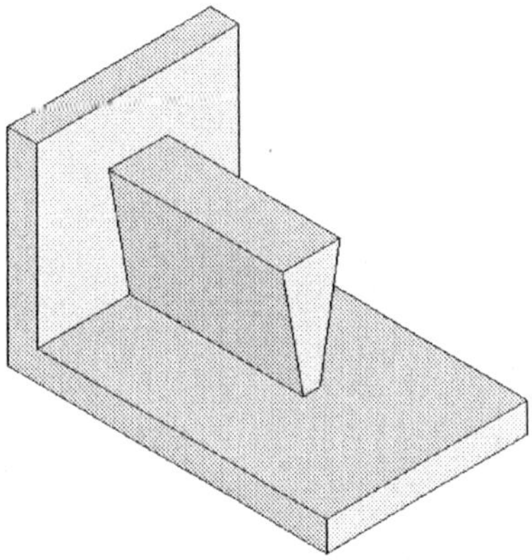

Figure 7-56 Resultant rib feature

Creating the Dome Feature

CommandManager:	Features > Dome	*(Customize to Add)*
Menu:	Insert > Features > Dome	
Toolbar:	Features > Dome	(Customize to Add)

Using the **Dome** tool you can create a dome feature on the selected face. Depending upon the feature creation direction a dome can be of a convex or concave shape. To create a dome feature choose the **Dome** button from the **Features CommandManager** or choose **Insert >Features > Dome** from the menu bar. When you choose this button the **Dome PropertyManager** is displayed as shown in Figure 7-57.

After invoking the **Dome PropertyManager** you are prompted to select a face or faces on which you want to add the dome feature. Select the face on which you need to create the dome feature; the name of the selected face will be displayed in the **Faces to Dome** selection

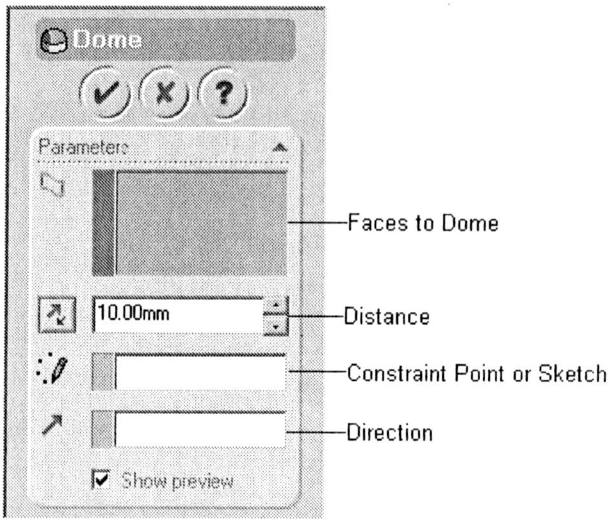

Figure 7-57 *The **Dome PropertyManager***

area. The selected face can be a planar face or a curved face. The preview of the dome feature with the default values is displayed in the drawing area. Using the **Distance** spinner you can set the height of the dome feature. The preview of the dome feature modifies dynamically when you modify the height using the **Distance** spinner. The height of the dome feature is calculated from the centroid of selected face to the top of the dome feature. After specifying the height of the dome feature choose the **OK** button from the **Dome PropertyManager**. Figure 7-58 shows the planar face to be selected for the dome feature. Figure 7-59 shows the resultant dome feature created.

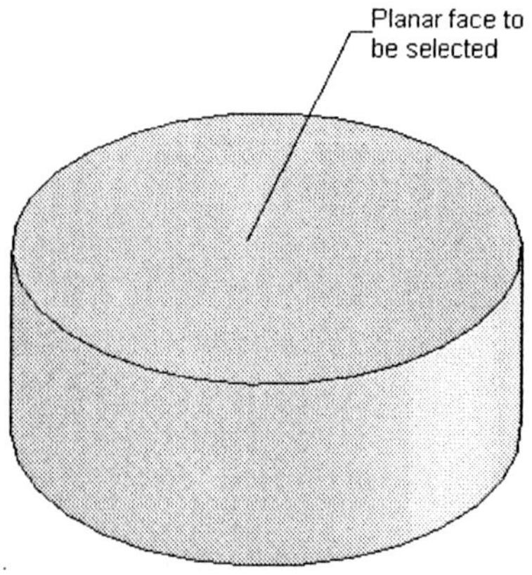

Figure 7-58 *Planar face to be selected*

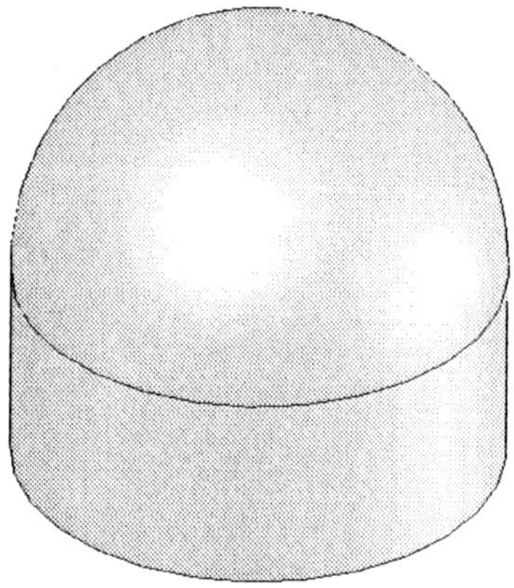

Figure 7-59 *Resultant dome feature*

If the selected planar face belongs to a circular or an elliptical feature, the **Elliptical dome** check box is displayed in the **Dome PropertyManager**. If you select this check box then you can create an elliptical dome. Figure 7-60 shows a circular dome created by selecting a circular planar face. Figure 7-61 shows an elliptical dome created by selecting the **Elliptical dome** check box and selecting a circular planar face.

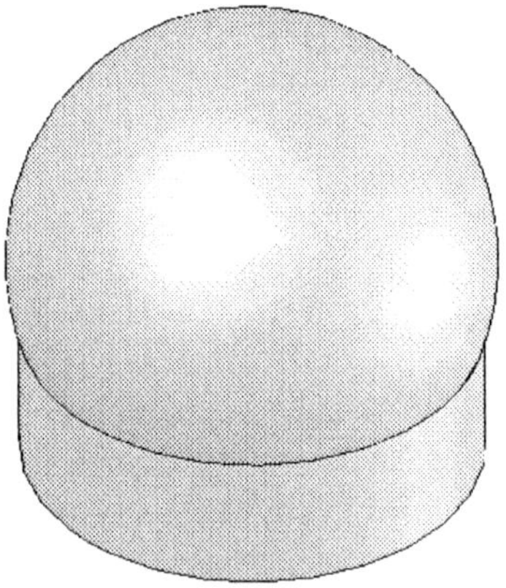

Figure 7-60 *Circular dome*

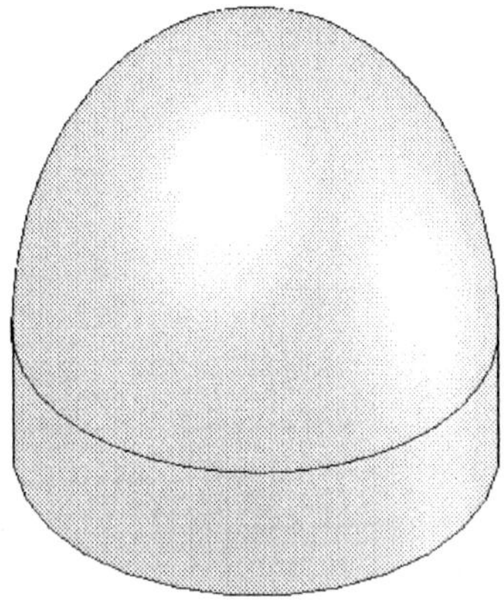

Figure 7-61 *Elliptical dome*

The **Reverse direction** button available in the **Dome PropertyManager** is used to create a concave dome. A concave dome is the one in which the material is removed by creating a cavity in the form of a dome. For removing the material using the dome feature, select the **Reverse direction** button. Figure 7-62 shows the dome feature created with the **Reverse direction** button selected.

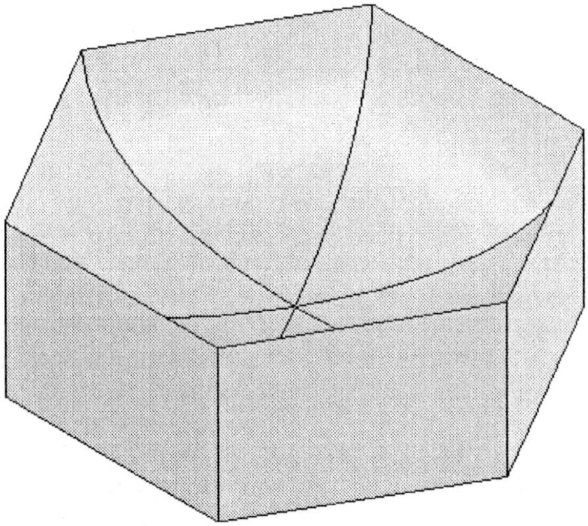

Figure 7-62 *A dome feature created with the* ***Reverse direction*** *button selected*

Other options available in the **Dome PropertyManager** are discussed next.

Constraint Point or Sketch

The **Constraint Point or Sketch** option is provided in the **Dome PropertyManager** to constraint the dome creation to the selected reference. The selected reference can be a point or a sketched point, or the endpoint of an entity. To create a dome feature using the **Constraint Point or Sketch** option invoke the **Dome PropertyManager** and select the face on which you need to add the dome feature. Now, click once in the **Constraint Point or Sketch** selection area and select the constraint point from the drawing area and choose the **OK** button from the **Dome PropertyManager**. Figure 7-63 shows the face on which you need to add the dome feature and the point to be selected as the constraint point. Figure 7-64 shows the resultant dome feature.

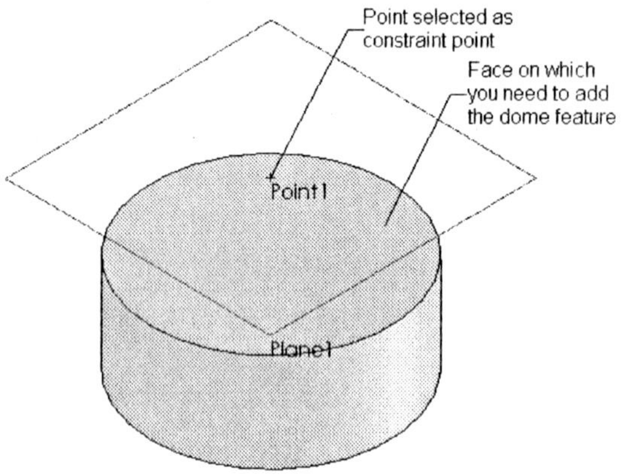

Figure 7-63 *Face and the point to be selected*

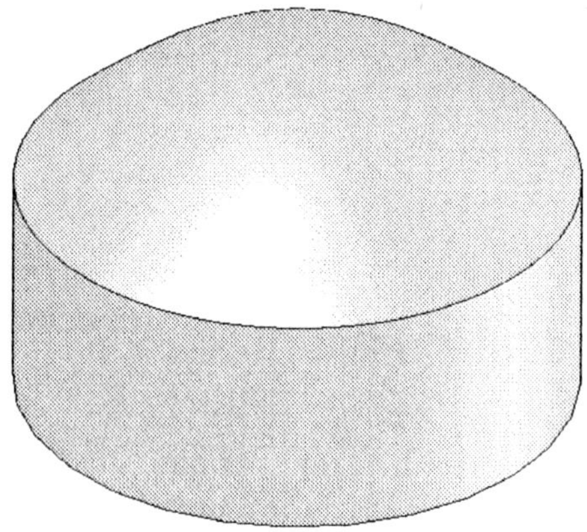

Figure 7-64 *Resultant dome feature*

Note

The dome feature will be created depending upon the height of the dome and the position of the constraint point.

Direction

The **Direction** option available in the **Dome PropertyManager** is used to specify the direction vector in which you need to create the dome feature. For creating the dome feature by defining the direction vector invoke the **Dome PropertyManager**. Select the face on which you need to create the dome feature. Now, click once in the **Direction** selection area and select an edge as the directional reference from the drawing area. The selected edge will be displayed in cyan. Set the value of the height of the dome and choose the **OK** button from the **Dome PropertyManager**. Figure 7-65 shows the faces on which you need to add the dome and the edge to be selected as direction vector. Figure 7-66 shows the resultant dome.

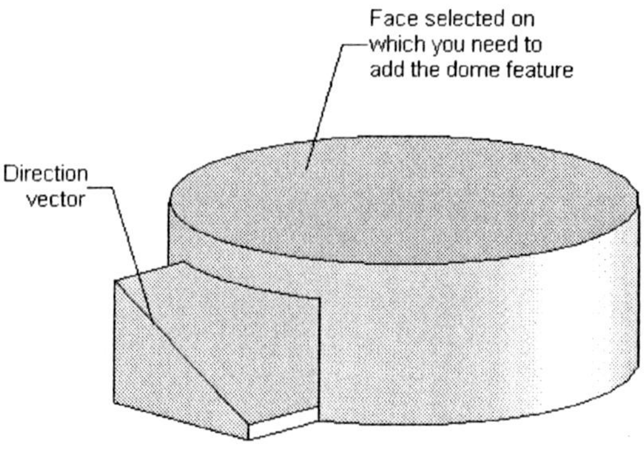

Figure 7-65 *Face and direction to be selected*

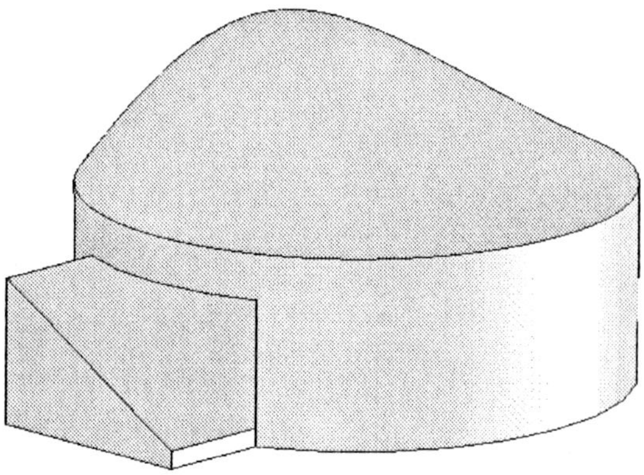

Figure 7-66 *Resultant dome feature*

Creating the Shape Feature

CommandManager:	Features > Shape	*(Customize to Add)*
Menu:	Insert > Features > Shape	
Toolbar:	Features > Shape	(Customize to Add)

SolidWorks provides you with a tool using which you can create free form shapes by manipulating the faces of the models. This free form styling tool is named as the **Shape** tool. By using the **Shape** tool, the selected face acts as a rubber membrane in which the pressure is applied, because of which the face inflates or deflates like a balloon. You can also adjust the shape of the membrane by changing the stretch and bend characteristics of the membrane. To create a shape feature choose **Insert > Features > Shape** from the menu bar or select the **Shape** button from the **Features CommandManager** after customizing it. When you invoke this tool the **Shape Feature** dialog box is displayed as shown in Figure 7-67.

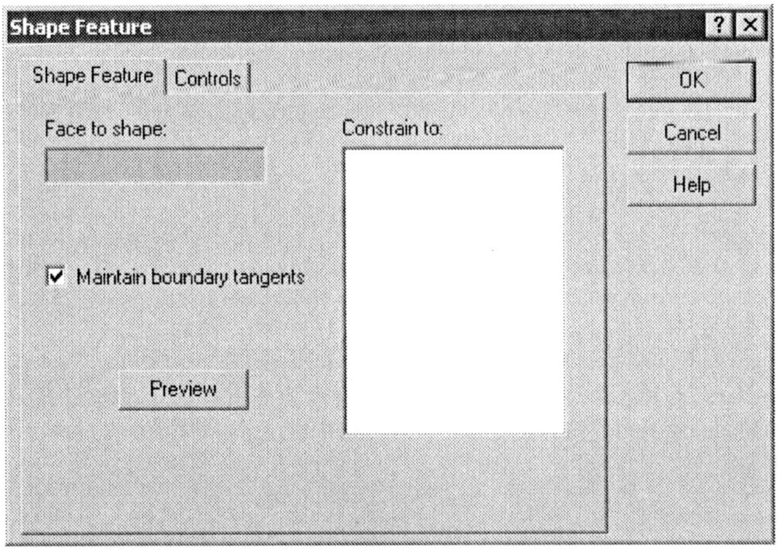

Figure 7-67 *The **Shape Feature** dialog box*

After invoking the **Shape Feature** dialog box you need to select the face that you need to manipulate. The selected face will be highlighted in green and the name of the selected face will be displayed in the **Face to shape** selection area. You can also constrain the selected face along a curve, or a sketch or an edge. To select the constraining curve click once in the **Constrain to** selection area and select the constraining reference from the drawing area. The **Maintain boundary tangents** check box is selected by default and is used to maintain the tangency between the shape feature and the boundary of the selected face. Choose the **Preview** button to display the preview of the shape feature. The preview of the shape feature in the form of surface mesh will be displayed in the drawing area. Figure 7-68 shows the face to be manipulated and the sketch that will constrain the shape feature. Figure 7-69 shows the preview in the form of mesh.

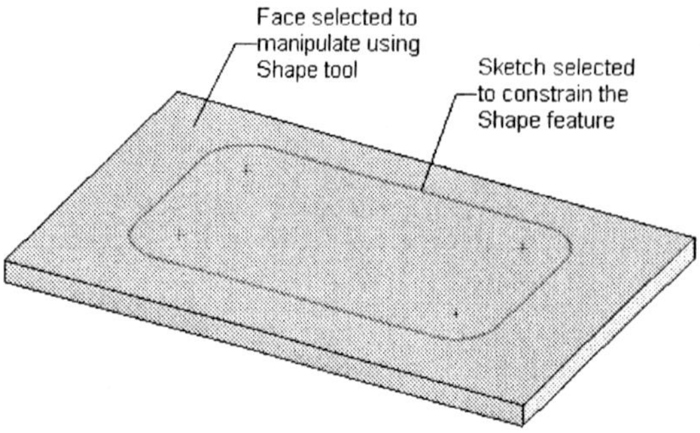

Figure 7-68 Face and the sketch to be selected

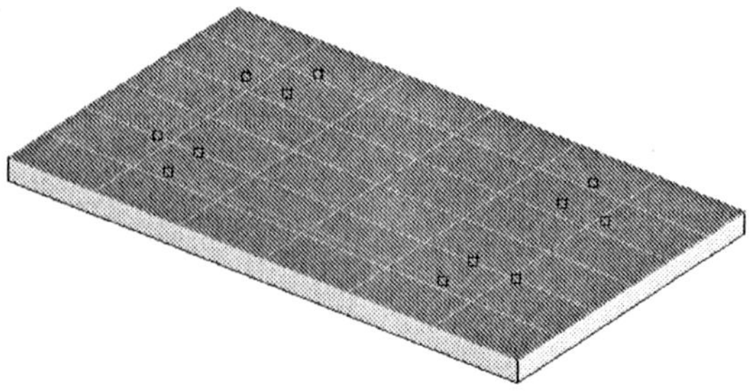

Figure 7-69 Preview in the form of mesh

Tip. *The constraint that will constrain the shape feature can be a point, sketch point, endpoint, vertex, sketch, edge, or a curve. The constrain can be created on the same face or any other face at an offset or at an angle.*

Next, you need to set the controls to manipulate the shape feature. Choose the **Control** tab to set the parameters of the shape feature. The options available in the **Control** tab are discussed next.

Gains

The **Gains** area of the **Control** tab is used to set the pressure and the value of the curve influence while creating the shape feature. These options are discussed next.

Pressure

The **Pressure** slider available in the **Gains** area is used to specify the pressure on the selected face. By default this slider is set at **0**. The positive pressure inflates and the negative pressure deflates.

Curve Influence

The **Curve Influence** slider is used to specify the percentage of the influence of the constraining curve on the shape feature creation.

Characteristics

The **Characteristics** area is provided with the options such as stretch and bend that are used to adjust the smoothness of the shape feature.

Advanced controls

The **Advanced controls** area is used to set the resolution of the mesh. Higher the resolution better will be the shape of the feature and it will take more time to rebuild the feature. Using the **Resolution** slider available in the **Advanced controls** area you can set the value of the resolution.

Figure 7-70 shows the model with top face manipulated using the shape feature.

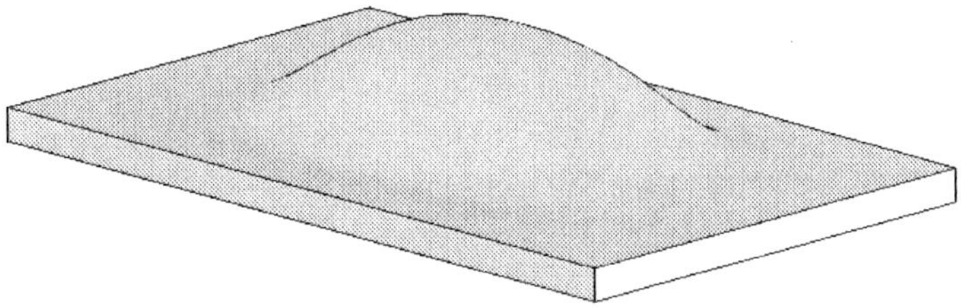

Figure 7-70 Shape feature added on the top face of the model

Displaying the Section View of the Model

CommandManager:	View > Section View	*(Customize to Add)*
Menu:	View > Display > Section View	
Toolbar:	View > Section View	

The **Section View** tool is used to display the section view of the model by cutting the model using a plane or face. To display the section view of a model choose the **Section View** button from the **View CommandManager** or choose **View** > **Display** > **Section View** from the menu bar. If the **View CommandManager** is not available by default then you need to invoke it. Move the cursor on any one of the buttons available in **CommandManager** and right-click to invoke the shortcut menu. Choose the **Customize CommandManager** option from the shortcut menu. The **Customizing CommandManager** menu is displayed. Select the **View** check box to display the **View CommandManager** and click once anywhere in the drawing area. When you invoke this tool the **Section View PropertyManager** is displayed as shown in Figure 7-71. By default, the **Front** plane is automatically selected in the **Section View PropertyManager** and the section view of the model created by using the front plane as the section plane is displayed in the drawing area. Drag handle is provided in the drawing area to dynamically adjust the offset distance of the section plane as shown in Figure 7-72.

If you need to select the **Right** plane or **Top** plane as the section plane then choose the respective button available in the **Section 1** rollout. You can also select a face or a user-defined plane as the section plane. You can also specify the offset distance using the **Offset Distance** spinner. You will observe that as you modify the offset distance, the preview of the section view is automatically modified. You can also rotate the section plane along the X axis and Y axis using the **X Rotation** and **Y Rotation** spinners.

The **Edit Color** button is used to specify the color of the preview of the section cap. This color is displayed only when the **Section View PropertyManager** is active. When you exit the **PropertyManager** the color is not displayed in the section view.

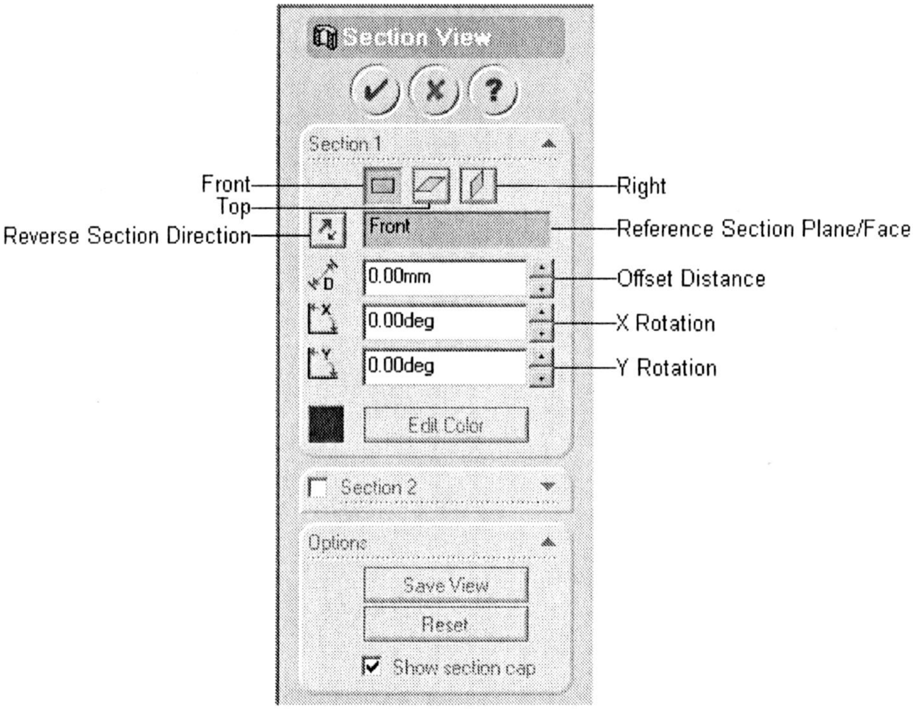

*Figure 7-71 The **Section View PropertyManager***

If you need to create a half section view then you need to invoke the **Section 2** rollout and specify the section plane in this rollout. You can also specify the offset distance and rotation of the plane, as discussed earlier for **Section 1**. After setting all the parameters choose the **OK** button from the **Section View PropertyManager**. Figure 7-73 shows the preview of the model after defining the section plane.

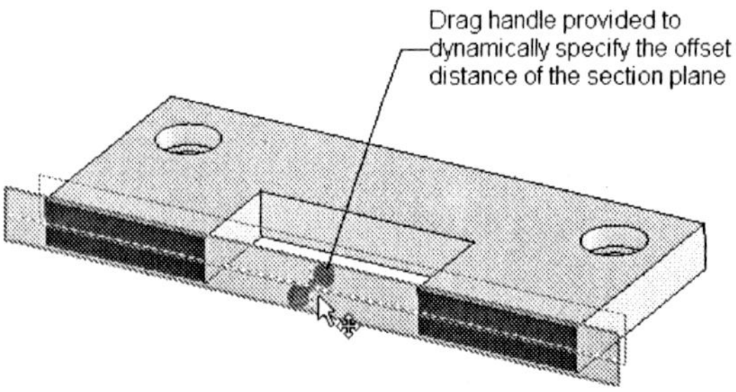

***Figure 7-72** Drag handle for dynamic offset*

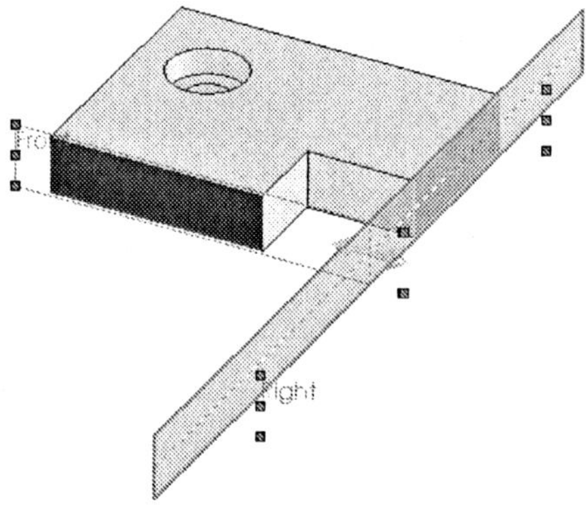

***Figure 7-73** Section preview after selecting second section plane*

You can also define the third section plane using the **Section 3** rollout. This rollout is displayed only after you invoke the **Section 2** rollout.

The options available in the **Options** rollout are used to save the section view as a named view. At any stage of your design cycle you can retrieve the saved section view from the **Orientation** dialog box. To save a named view choose the **Save View** button from the **Options** rollout. The **Named View** dialog box is displayed; specify the name of the view in this dialog box and choose the **OK** button from this dialog box. Using the **Reset** button you can reset the section view setting to the default settings. The **Show section cap** check box is selected by default, if you clear this check box then the preview of the section view is not capped. Figure 7- 74 shows the section view of a model.

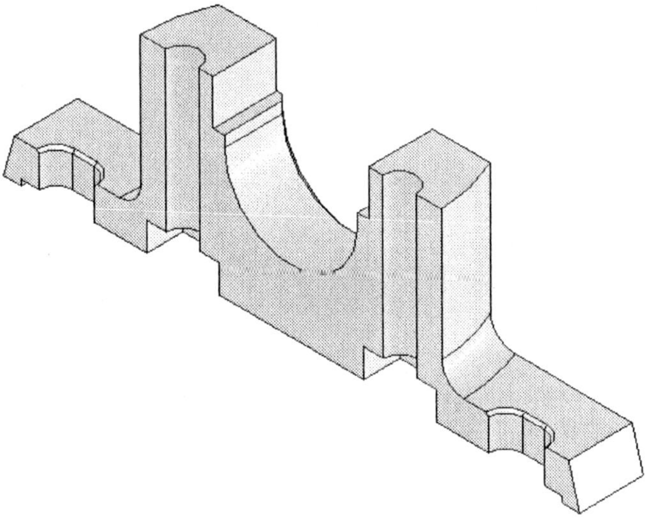

Figure 7-74 *Section view of a model*

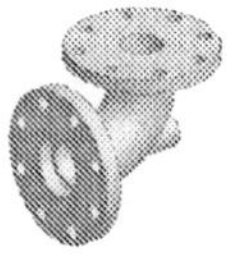

Tip. *You can also rotate the section plane dynamically. To rotate the section plane dynamically move the cursor on the edge of the section plane. The cursor will be replaced by the rotate cursor, drag the cursor to dynamically rotate the section plane.*

*To switch back to the full view mode, you need to choose the **Section View** button from the **View CommandManager** or select any face of the sectioned model and right-click to invoke the shortcut menu. Choose the **Section View** option from the shortcut menu.*

*If you need to modify the section view, select any face of the sectioned model and invoke the shortcut menu. Choose the **Section View Properties** option from the shortcut menu. The **Section View PropertyManager** is displayed and you can modify the section view.*

TUTORIALS

Tutorial 1

In this tutorial you will create the model shown in Figure 7-75. The dimensions of the model are shown in Figure 7-76. **(Expected time: 30 min)**

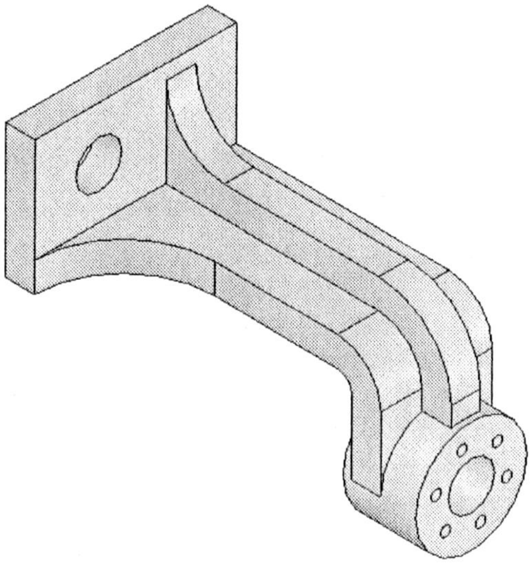

Figure 7-75 Solid model for Tutorial 1

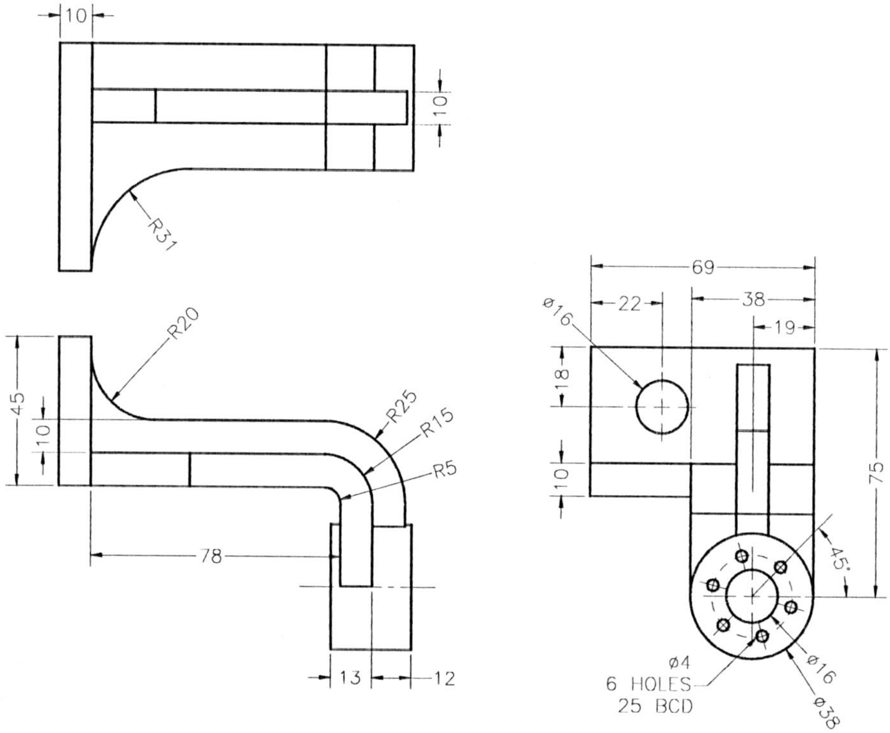

Figure 7-76 Views and dimensions of the model for Tutorial 1

The steps to be followed to complete this tutorial are discussed next:

a. Create the base feature of the model by extruding a rectangle of 69 mm x 45 mm created on **Right** plane to a depth of 10 mm, refer to Figure 7-77.

b. Create the second feature, which is created by extruding the sketch created on the back face of the base feature, refer to Figure 7-78(a).
c. The third feature of the model is a circular feature, refer to Figure 7-78(b).
d. Create a hole feature placed concentric to the circular feature.
e. Create the hole on the specified BCD and then pattern the hole feature using the circular pattern option.
f. Create the hole feature on the base feature, refer to Figure 7-79.
g. Create a fillet feature to add the required fillets, refer to Figures 7-80 and 7-81.
h. Create the rib feature, refer to Figures 7-82 and 7-83.

Creating the Base Feature

1. Start SolidWorks and then start a new part document from the **New SolidWorks Document** dialog box.

   It is evident from the model that the sketch of the base feature of the model is created on the right plane. Therefore, you need to select the **Right** plane from the **FeatureManager Design Tree** to create the base feature.

2. Select the **Right** plane from the **FeatureManager Design Tree** and choose the **Extruded Boss/Base** button from the **Features CommandManager**. The sketcher environment is invoked and the **Right** plane is oriented normal to the view.

3. Draw the sketch of the base feature of the model. The sketch of the base feature consists of a rectangle of dimensions 69 mm x 45 mm.

4. Add the required relations and dimensions to the sketch. Exit the sketching environment.

5. Set the value of the **Depth** spinner to 10 mm and exit the **Extrude PropertyManager**. The base feature of the model is shown in Figure 7-77.

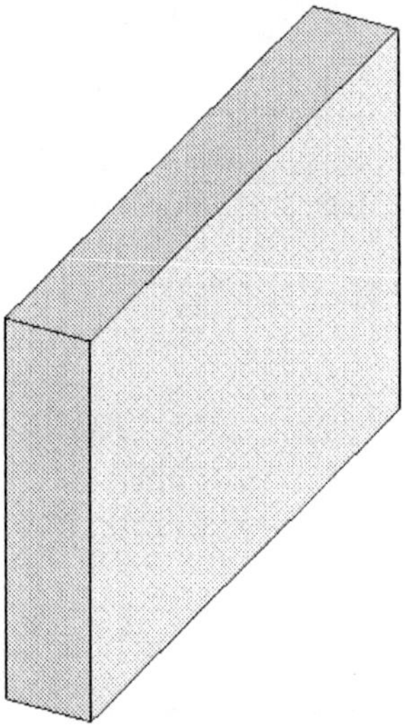

Figure 7-77 *Base feature of the model*

Creating the Second Feature of the Model

The second feature of the model is also an extruded feature. Create the sketch for the second feature on the back face of the base feature and extrude this sketch to the given depth.

1. Select the back face of the base feature as the sketching plane and invoke the **Extruded Boss/Base** tool.

2. Create the sketch of the second feature and add the required relations and dimensions.

3. Exit the sketching environment and select the **Reverse Direction** button from the **PropertyManager**.

4. Set the value of the **Depth** spinner to **38** and end the feature creation.

The model created after adding the second feature is shown in Figure 7-78(a).

Note

You can also create the first and second features using the contour selection method.

Creating the Third Feature

The third feature of this model is a circular extruded feature. This feature is created by extruding a circular sketch on both the sides of the sketching plane. The sketch for this feature is created on the right planar face of the second feature.

1. Select the right planar face of the second feature as the sketching plane and invoke the **Extruded Boss/Base** tool.

2. Create the sketch using the **Circle** tool. Add the required relations and dimensions to the sketch.

3. Exit the sketching environment and set the value of the **Depth** spinner available in the **Direction 1** rollout to **12**. Since you have to extrude the sketch in both the directions with variable values, therefore, you have to invoke the **Direction 2** rollout. Set the value of the **Depth** spinner available in the **Direction 2** rollout to **13** and end the feature creation.

Figure 7-78(b) shows the model after adding the third feature.

Creating the Fourth Feature

The fourth feature of this model is a hole feature. You will create a hole using the **Simple Hole** option on the right face of the third feature. To create a hole using this option, you first need to create a point on which the hole will be placed.

1. Select the right face of the third feature as the sketch plane for creating the sketch point and invoke the sketching environment. The center point of the hole to be created using the **Hole PropertyManager** will be placed coincident to this sketched point.

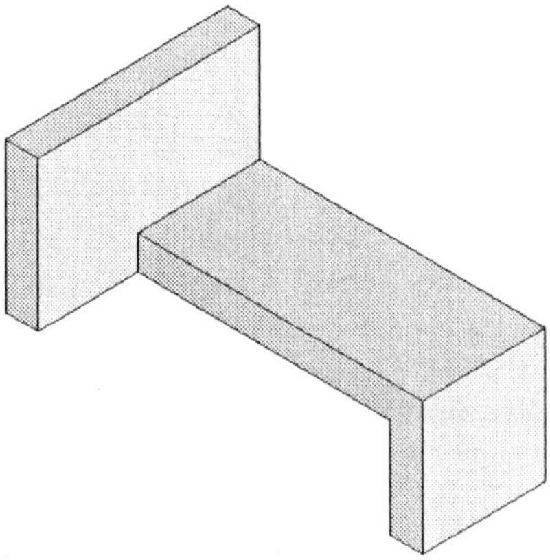

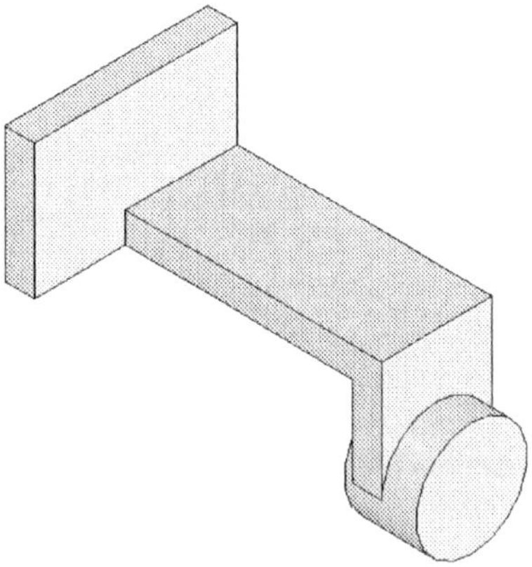

Figure 7-78 (a) *Second feature added to the model* ***Figure 7-78 (b)*** *Third feature added to the model*

2. Create a point on the right face of the circular feature.

3. Add the **Concentric** relation between the sketched point and the circular edge of the third feature and exit the sketching environment.

4. Now, select the face on which the point is placed. Choose the **Simple Hole** button from the **Features CommandManager** or choose **Insert** > **Features** > **Hole** > **Simple** from the menu bar to invoke the **Hole PropertyManager**.

5. Choose the **Through All** option from the **End Condition** drop-down list and set the value of the **Hole Diameter** spinner to **16**.

6. Select the centerpoint of the hole feature and drag the cursor to the sketched point drawn earlier. Release the left mouse button when the cursor is displayed in yellow color.

7. Choose the **OK** button from the **Hole PropertyManager**.

Creating the Fifth Feature

1. Using the procedure given in the previous section, create the fifth feature, which is also a hole feature placed on the same placement plane. This hole feature is created using the **Through All** option with the diameter of the hole as **4**. After creating the hole feature define the placement of the feature by adding the required relations and dimensions.

Patterning the Hole Feature

After creating the fifth feature, which is a hole feature, you will pattern it using the **Circular Pattern** tool.

1. In case all the buttons of the **CommandManager** are not visible. Select the black arrow; a flyout appears. You can choose the required button from the flyout that appears. Choose the **Circular Pattern** button from the **CommandManager** flyout.

The **Circular Pattern PropertyManager** is invoked and you are prompted to select an edge or an axis for direction reference; select a face of the feature for pattern features.

To create a circular pattern you need an edge or an axis that will be used as the central axis. As discussed in the earlier chapters, when you create a circular feature a temporary axis is automatically created passing through the center of the circular feature. In this tutorial, the feature to be patterned is created on a circular feature. Therefore, the temporary axis of the circular feature will be used as central axis. Therefore, first you will display the temporary axis.

2. Choose **View** > **Temporary Axes** from the menu bar to display the temporary axis.

   Temporary axes are displayed in the model.

3. Select the temporary axis that passes through the center of the first hole feature.

   The **Direction 1** callout is displayed and is attached to the selected axis.

4. Click in the **Features to Pattern** selection area to invoke the selection mode.

5. Select the smaller hole from the drawing area or expand the **FeatureManager Design Tree**, which is displayed in the drawing area, and select the **Hole2** feature.

   The preview of the pattern feature is displayed with the default settings.

6. Select the **Equal spacing** check box if cleared and set the value of the **Number of Instances** spinner to **6**. Choose the **OK** button from the **PropertyManager**.

7. Choose **View** > **Temporary Axes** to hide the temporary axes.

Creating the Seventh Feature

1. The seventh feature of this model is a hole feature. Using the procedure given to create the fourth feature you will create this hole feature. This hole feature will be placed on the right planar face of the base feature. Therefore, after selecting the right planar face of the base feature, place the hole feature. After placing the hole feature define the placement of the hole feature by adding the required relations and dimensions. Figure 7-79 shows the model after adding the hole features.

Creating the Fillet Feature

Next, you need to create the fillet feature. It is evident from the model that the fillets to be added to the model are of different radii. In SolidWorks you can specify different radii to individual selected edges, faces, or loops in a single fillet feature.

1. Choose the **Fillet** button from the **Features CommandManager** to invoke the **Fillet PropertyManager**.

   You are prompted to select edges, faces, features, or loops to fillet.

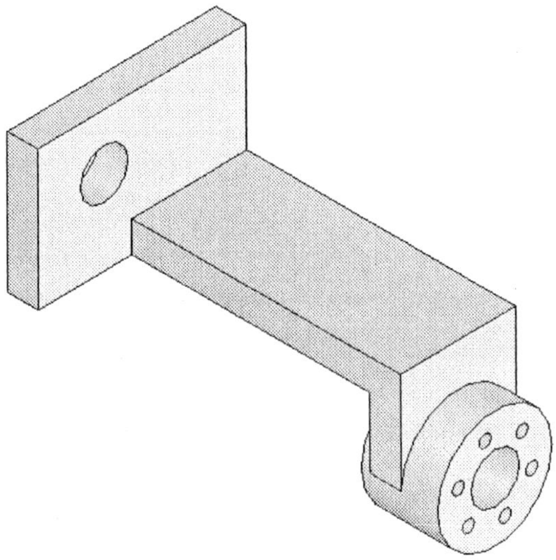

Figure 7-79 *Model after adding all the hole features*

2. Select the **Multiple radius fillet** check box from the **Items To Fillet** rollout of the **Fillet PropertyManager**.

3. Select the edges to fillet as shown in Figure 7-80. As the **Multiple radius fillet** check box is selected, therefore, each selected edge is having a separate **Radius** callout.

4. Modify the value of the radii as required in the respective **Radius** callouts.

5. Choose the **OK** button from the **Fillet PropertyManager**.

The isometric view of the model after adding the fillet feature is shown in Figure 7-81.

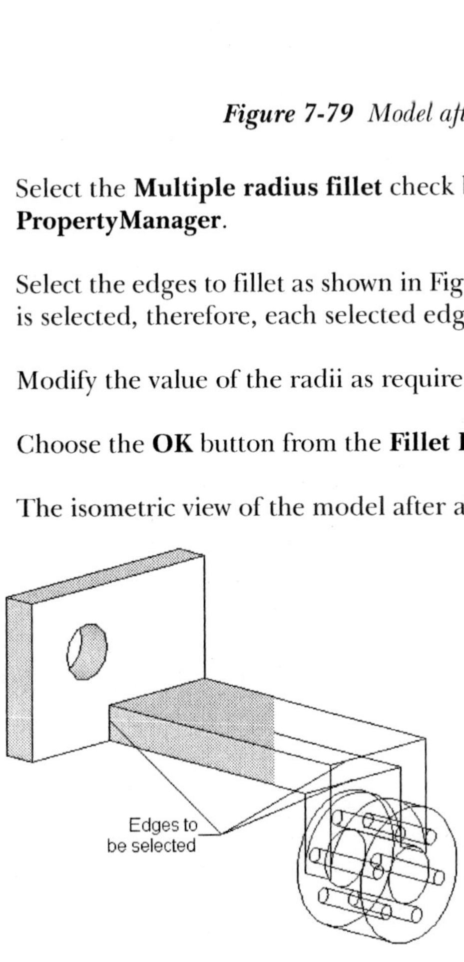

Figure 7-80 *Edges to be selected*

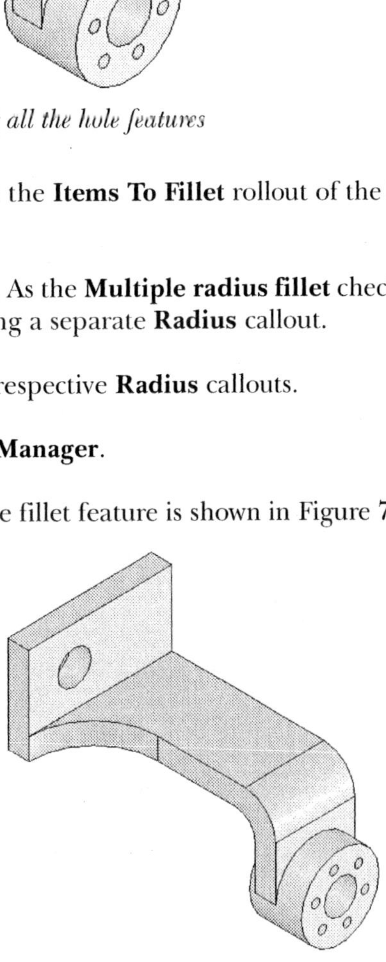

Figure 7-81 *Model after adding fillet feature*

Creating the Rib Feature

The next feature that you are going to create is a rib feature. The sketch of the rib feature will be created on a sketching plane created at an offset distance from the back planar face of the model. Therefore, first you need to create a reference plane at an offset distance from the back planar face of the model.

1. Invoke the **Plane PropertyManager**.

2. Rotate the model and select the back planar face of the model. Choose the **Reverse direction** check box under the **Distance** spinner and set the value of the **Distance** spinner to **19**.

3. Choose the **OK** button from the **Plane PropertyManager** to end feature creation.

   A new plane is created at an offset distance from the back planar face of the model.

4. Choose the **Rib** button from the **Features CommandManager**.

5. Create the sketch for the rib feature on the newly created plane and add the required relations and dimensions to the sketch as shown in Figure 7-82.

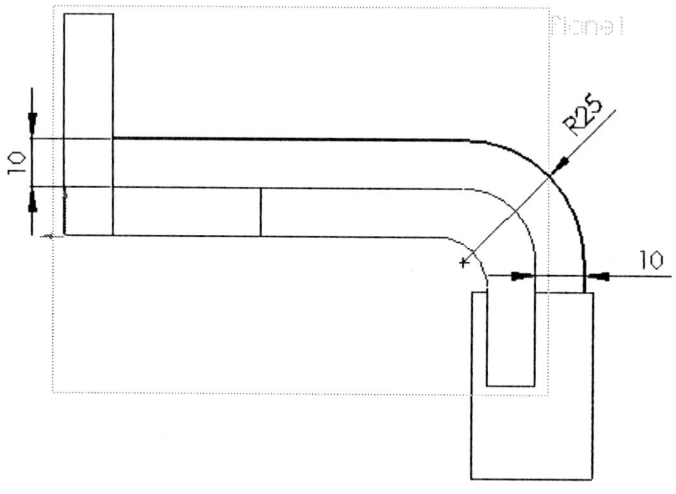

Figure 7-82 *Sketch for the rib feature*

6. Exit the sketching environment. The **Rib PropertyManager** is displayed.

   The preview of the rib feature is displayed in the drawing area and you will observe that the direction of material addition is displayed by an arrow in the drawing area. This direction of material addition is opposite to the required direction. Therefore, you need to flip the direction of material addition.

7. Select the **Flip material side** check box to flip the direction of material addition.

   Since the default value of the rib thickness is 10, which is the required value, you do not need to change the value of the rib thickness.

8. Choose **OK** from the **Rib PropertyManager**. Also, hide the newly created plane.

   The last feature of the model is the fillet feature. Add the fillet feature on the left edge of the rib using the **Fillet** tool. Figure 7-83 shows the isometric view of the final model. The **FeatureManager Design Tree** of the model is shown in Figure 7-84.

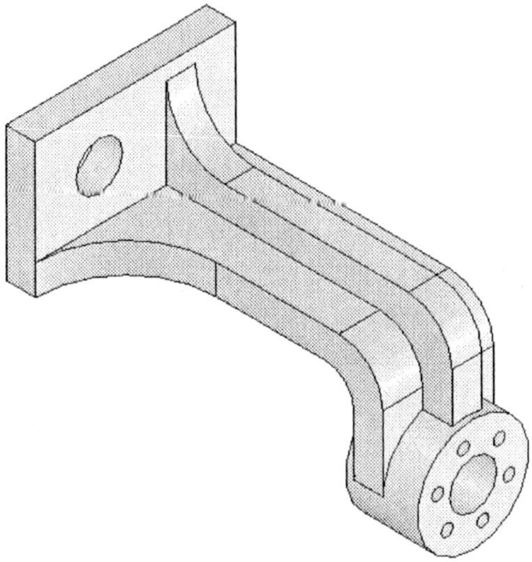

Figure 7-83 *The final solid model*

Saving the Model

1. Create a *c07* folder in the *SolidWorks* folder and choose the **Save** button from the **Standard** toolbar and save the model with the name given below.

   \My Documents\SolidWorks\c07\c07-tut01.sldprt

2. Choose **File > Close** from the menu bar to close the document.

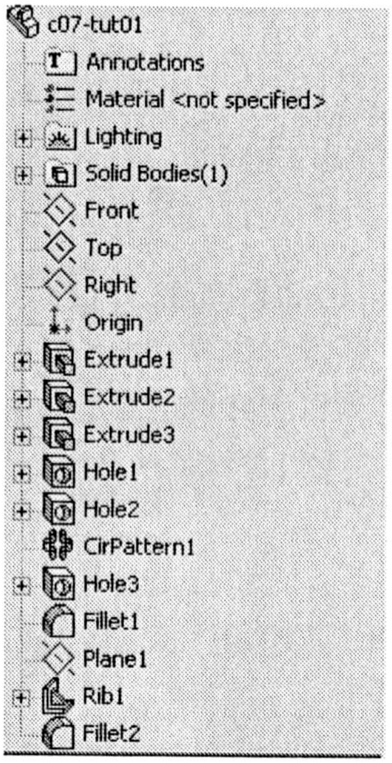

*Figure 7-84 The **FeatureManager Design Tree***

Tutorial 2

In this tutorial you will create the model shown in Figure 7-85. The dimensions of the model are shown in Figure 7-86. **(Expected time: 30 min)**

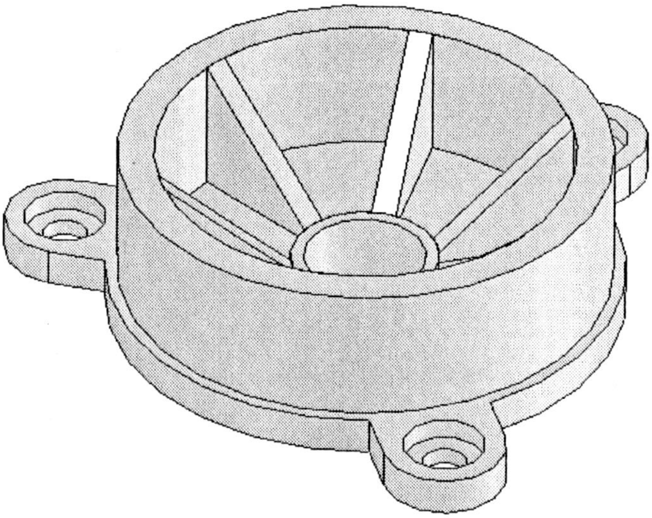

Figure 7-85 Model for Tutorial 2

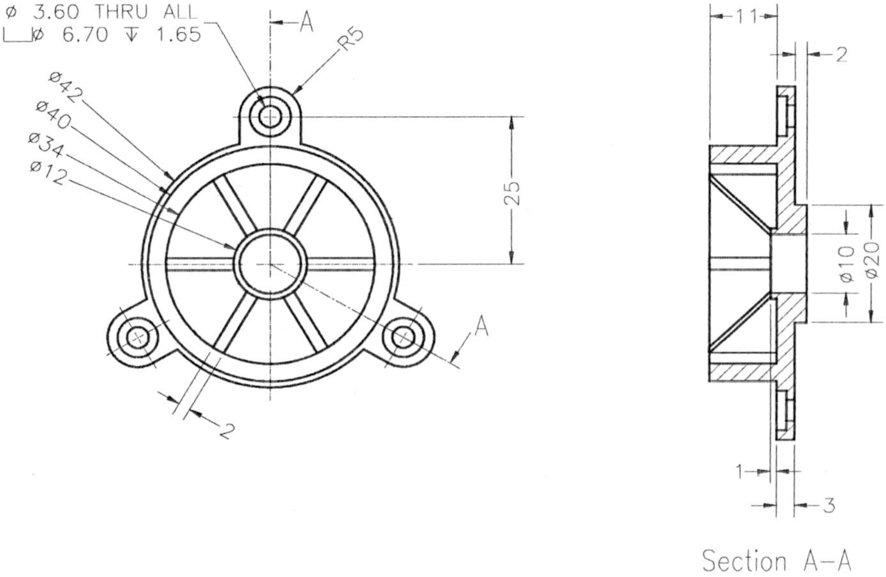

Figure 7-86 *Front view and aligned section view with dimensions*

The steps to be followed to complete this tutorial are discussed next:

a. Create the base feature of the model. The base feature will be created by revolving the sketch along a centerline, refer to Figures 7-87 and 7-88.
b. Create the second feature, which is on the outer periphery of the base feature, by extruding the sketch from the sketch plane to a selected surface, refer to Figures 7-89 and 7-90.
c. Place a counterbore hole feature on the top face of the second feature using the hole wizard.
d. Pattern the second and third features along the temporary axis using the **Circular Pattern** tool, refer to Figure 7-91.
e. Create the rib feature, refer to Figure 7-92.
f. Pattern the rib feature along a temporary axis using the **Circular Pattern** tool, refer to Figure 7-93.

Creating the Base Feature

Start a new SolidWorks part document. First you need to create the base feature of the model. The base feature of the model will be created by revolving the sketch along the axis of revolution. The axis of revolution will be a centerline and the sketch for the base feature will be drawn on the **Right** plane.

1. Invoke the **Revolved Boss/Base** tool and select the **Right** plane as the sketching plane.

2. Create the sketch for the base feature and add the required relations and dimensions to the sketch as shown in Figure 7-87.

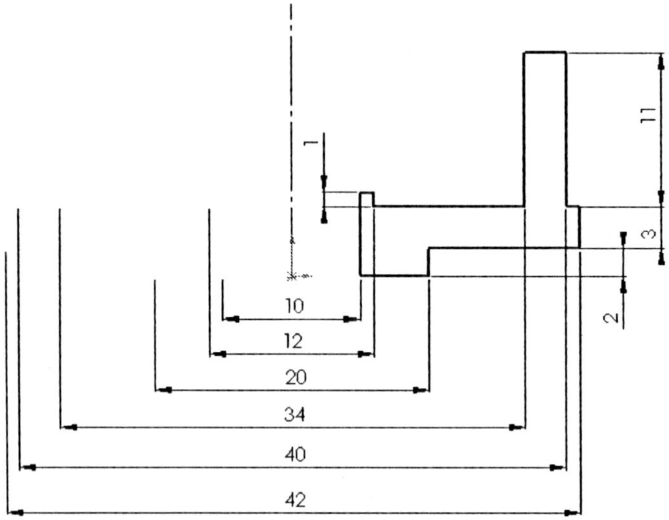

Figure 7-87 Sketch for the base feature

3. Exit the sketching environment and set the value of the **Angle** spinner to **360**.

4. Choose the **OK** button from the **Revolve PropertyManager**. The base feature created after revolving the sketch is shown in Figure 7-88.

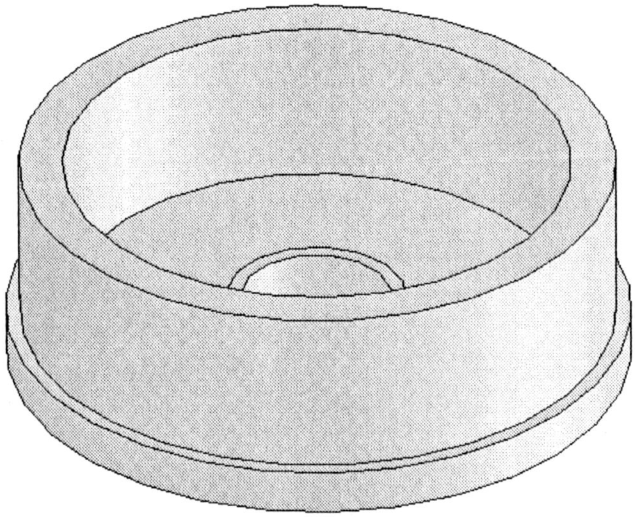

Figure 7-88 Base feature of the model

Creating the Second Feature

The second feature is created by extruding a sketch up to the selected surface.

1. Invoke the **Extruded Boss/Base** tool and select the face shown in Figure 7-89 as the sketching plane.

2. Create the sketch of the second feature and add the required relations and dimensions to the sketch.

3. Exit the sketching environment and using the **Up To Surface** option extrude the sketch. The surface to be selected is shown in Figure 7-90.

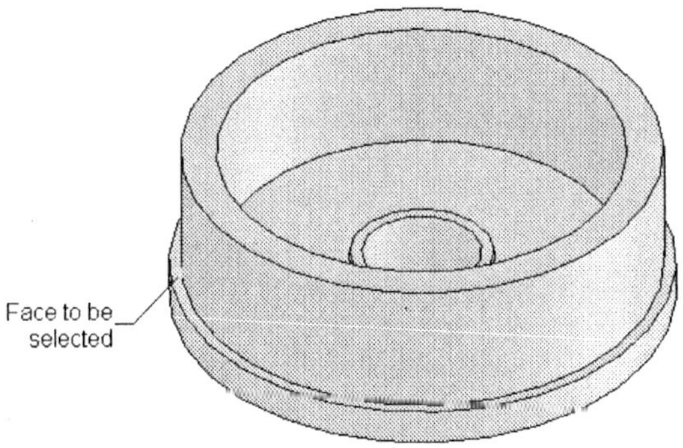

Figure 7-89 *Face to be selected*

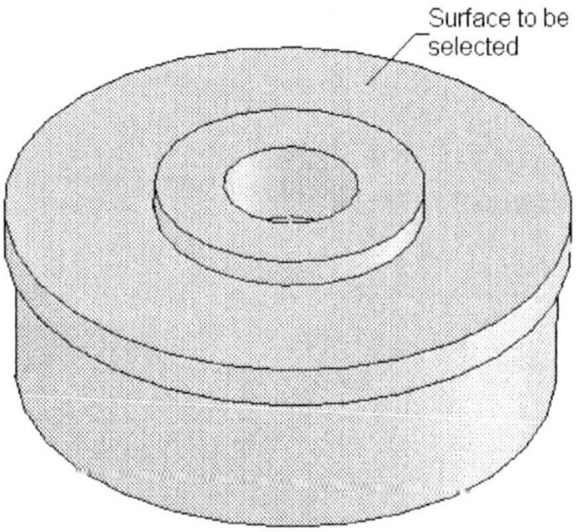

Figure 7-90 *Surface to be selected*

4. Choose the **OK** button from the **Extrude PropertyManager**.

Creating the Hole Feature

It is evident from Figure 7-86 that a counterbore hole needs to be added to the model. The counterbore hole will be added using the **Hole Wizard**. Before invoking this tool, first you need to select the placement plane for the hole feature.

1. Select the top face of the second feature as the placement plane for the hole feature.

2. Choose the **Hole Wizard** button from the **Features CommandManager** to invoke the **Hole Definition** dialog box.

The preview of the hole feature with the default settings in the **Hole Definition** dialog box is displayed in the drawing area.

3. Invoke the **Counterbore** tab if it is not invoked by default and select the **Ansi Metric** option from the **Standard** drop-down list to specify the standard to be used.

4. Choose the **Socket Button Head Cap Screw - ANSI B18.3.4M** option from the **Screw Type** drop-down list to specify the type of screw.

5. Select the **M3** option from the **Size** drop-down list to specify the size of the fastener to be used in the hole. Leave all the default options as they are and choose the **Next** button from the **Hole Definition** dialog box.

The **Hole Placement** dialog box is displayed.

6. Invoke the **Add Relations PropertyManager** and apply a concentric relation between the centerpoint of the hole feature and the circular edge of the second feature.

7. Choose the **Finish** button from the **Hole Placement** dialog box to end feature creation.

Patterning the Features

After creating the second feature and the third feature, you need to pattern both the features around the temporary axis using the **Circular Pattern** tool.

1. Choose the **Circular Pattern** button from the **Features CommandManager** to invoke the **Circular Pattern PropertyManager**.

Since you are creating a circular pattern, you need to define an axis as the direction reference. Therefore, display the temporary axes by choosing **View > Temporary Axes** from the menu bar.

2. Select the temporary axis that is passing through the center of the model and invoke the selection mode in the **Features to Pattern** area. The **Direction 1** callout is displayed.

3. Select the second and third features from the drawing area or from the **FeatureManager Design Tree** that is displayed in the drawing area.

4. Set the value of the **Number of Instances** spinner to **3** and make sure the **Equal spacing** check box is selected. Choose the **OK** button from the **Circular Pattern PropertyManager**.

5. Hide the temporary axes. The model after patterning the features is shown in Figure 7-91.

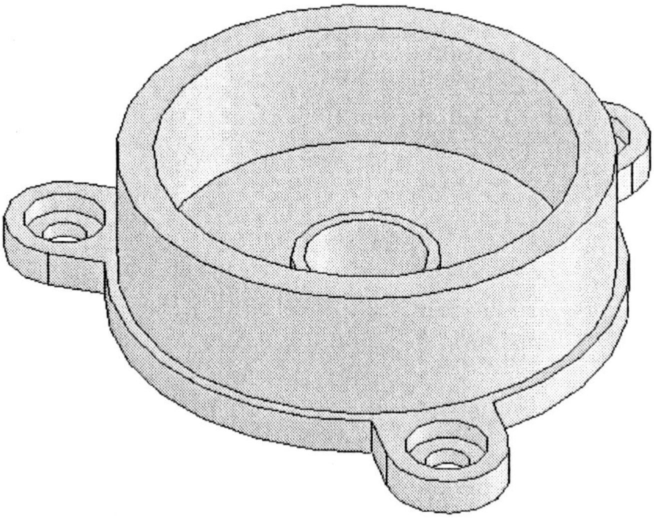

Figure 7-91 *Model after patterning the features*

Creating the Rib Feature

The next feature is a rib. The sketch for the rib feature will be created on the **Front** plane.

1. Choose the **Rib** button from the **Features CommandManager** and select the **Front** plane from the **FeatureManager Design Tree**.

2. Set the display model to wireframe and create the sketch for the rib feature and add the required relations as shown in Figure 7-92.

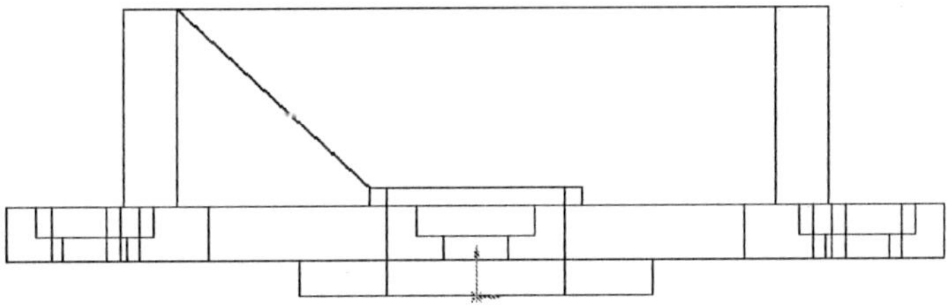

Figure 7-92 Sketch for the rib feature

3. Exit the sketching environment and set the value of the **Rib Thickness** spinner to **2**. Leave all the other default options as they are and choose the **OK** button from the **Rib PropertyManager**.

4. Change the model display mode to shaded.

5. Using the **Circular Pattern** tool, create 6 instances of the rib feature. The final model after creating all the features is shown in Figure 7-93. The **FeatureManager Design Tree** of the model is shown in Figure 7-94.

Saving the Model

1. Choose the **Save** button from the **Standard** toolbar and save the model in the *c07* folder with the name given below.

   \My Documents\SolidWorks\c07\c07-tut02.sldprt

2. Choose **File > Close** from the menu bar to close the document.

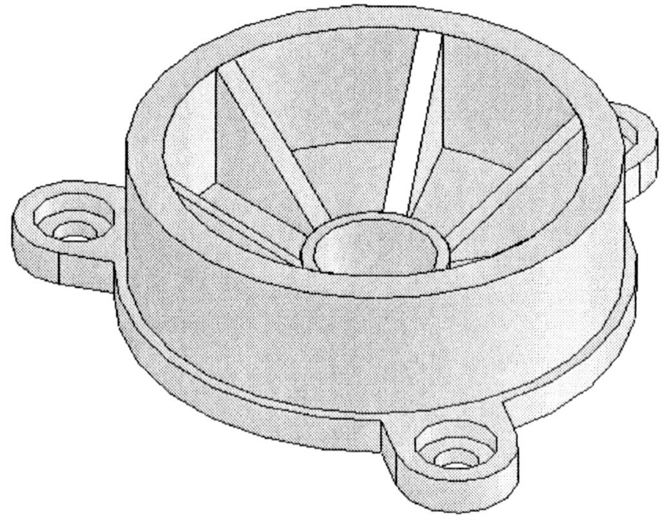

Figure 7-93 *Final solid model*

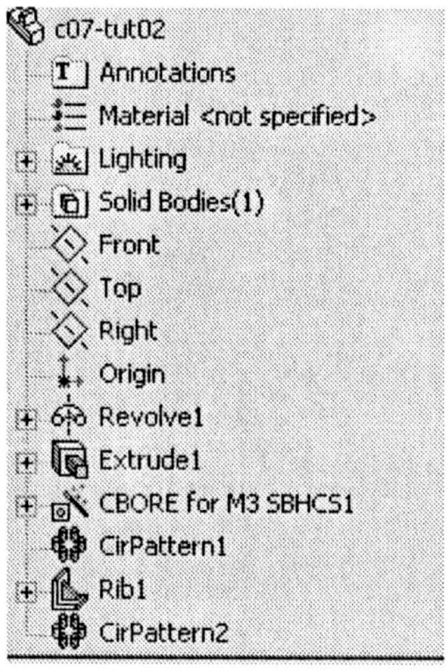

Figure 7-94 *The* ***FeatureManager Design Tree***

Tutorial 3

In this tutorial you will create the cylinder head of a two-stroke automobile engine shown in Figure 7-95. The dimensions of the model are shown in Figure 7-96. After creating the model you will also create a section view of the model using the **Section View** tool.

(Expected time: 1 hr)

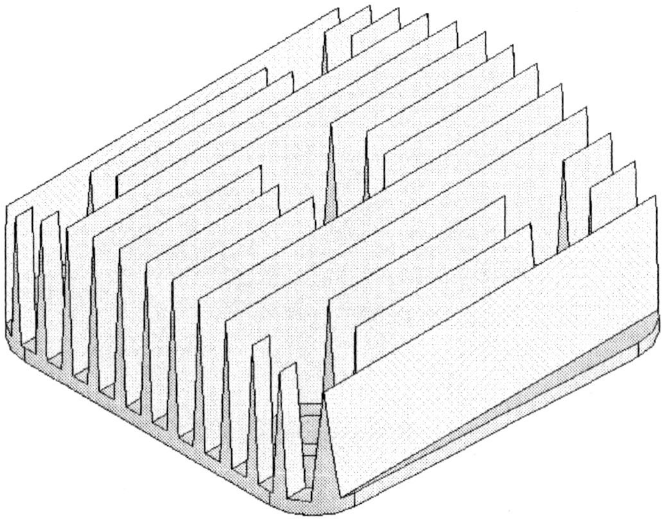

Figure 7-95 *Model for Tutorial 3*

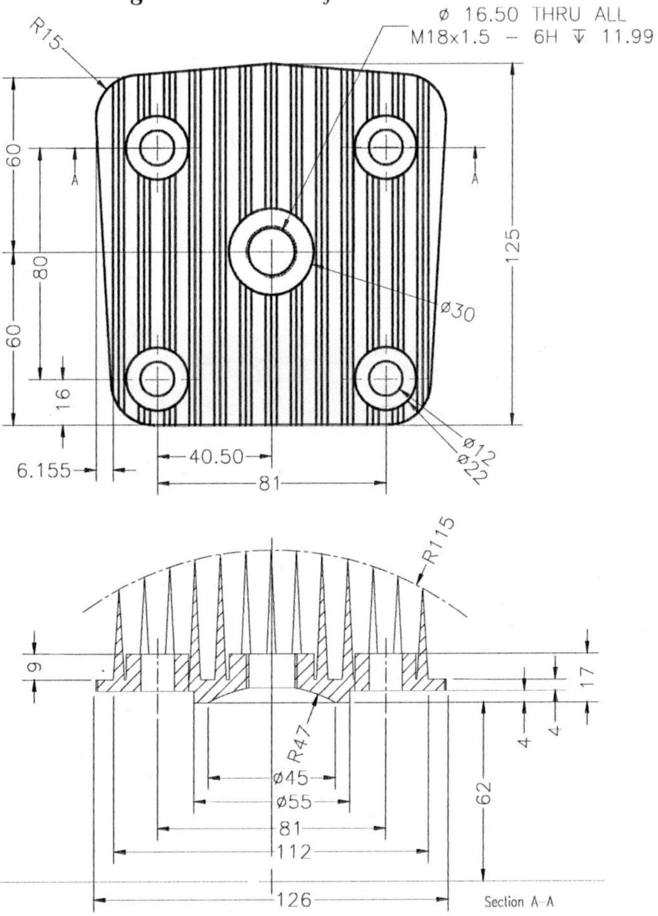

Figure 7-96 *Top view and the section front view with dimensions*

The steps to be followed to complete this tutorial are discussed next.

a. Create the base feature of the model by extruding a polygon to the given depth, refer to Figure 7-97.
b. Add the fillet to the base feature.
c. Create a circular feature at the bottom face of the base feature.
d. Create the revolve cut feature to create the dome of the cylinder head, refer to Figures 7-98 and 7-99.
e. Create the left fin of the cylinder head by extruding the sketch. The sketch for this feature should be carefully dimensioned and defined, refer to Figure 7-100.
f. Use the **Vary sketch** option to pattern the fins, refer to Figure 7-101.
g. Create other cut and extrude features to complete the model, refer to Figure 7-102.
h. Create a tap hole using the hole wizard, refer to Figure 7-103.
i. Create the section view of the model, refer to Figure 7-104.

Creating the Base Feature

1. Create a new SolidWorks part document.

   The base feature of the model will be created by extruding the sketch created on the **Top** plane.

2. Invoke the **Extruded Boss/Base** tool and select the **Top** plane as the sketching plane.

3. Create the sketch for the base feature and add the required relations and dimensions to the sketch as shown in Figure 7-97.

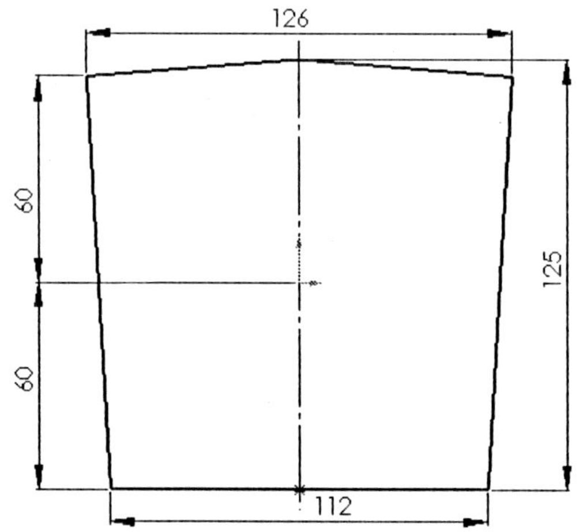

***Figure 7-97** Sketch for the base feature*

4. Exit the sketching environment and extrude the sketch to a depth of 4 mm.

Creating the Second Feature

The second feature of the model is the fillet feature. You have to fillet all the vertical edges of the base feature using the given radius.

1. Invoke the **Fillet** tool and set the value of the **Radius** spinner to **15** and select all the vertical edges of the base feature to add the fillet feature.

2. Choose the **OK** button from the **Fillet PropertyManager**.

Creating the Third Feature

After creating the base and adding fillet at the corners of the base, now you will create the third feature of the model, which is a circular extruded feature. This sketch of the feature will be created on the bottom face of the base feature. The sketch will be extruded to the given depth.

1. Invoke the **Extruded Boss/Base** tool and select the bottom face of the base feature as the sketching plane.

2. Create a circle of 55 mm diameter with its centerpoint at origin.

3. Exit the sketching environment and extrude the sketch to a depth of 4 mm.

Creating the Fourth Feature

The fourth feature is a revolved cut feature. The sketch for the revolved cut feature will be created on the **Front** plane. After creating the sketch, apply the required relations and dimensions to the sketch.

1. Invoke the **Revolved Cut** tool and select the **Front** plane from the **FeatureManager Design Tree**.

2. Create the sketch for the revolved cut feature and add the required relations and dimensions as shown in Figure 7-98. You need to apply the vertical relation between the center point of the arc and the origin to fully define the sketch.

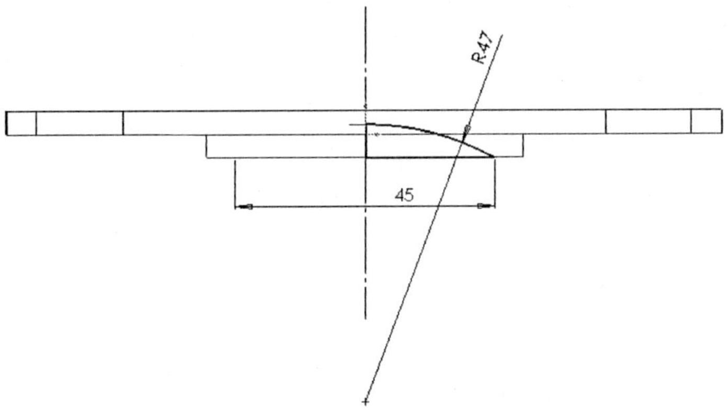

Figure 7-98 *Sketch for the revolve cut feature*

Note

When you create a sketch for the revolved cut feature for this tutorial, create a horizontal centerline such that the start point of the centerline will be merged with the upper endpoint of the arc. After creating the centerline make sure a tangent relation between the arc and the centerline exists. This is done to maintain the tangency of the arc.

3. Select the vertical centerline and exit the sketching environment. Make sure that the **Angle** spinner value is **360**.

4. Choose the **OK** button from the **Cut-Revolve PropertyManager**.

   The rotated model after creating the fourth feature is shown in Figure 7-99.

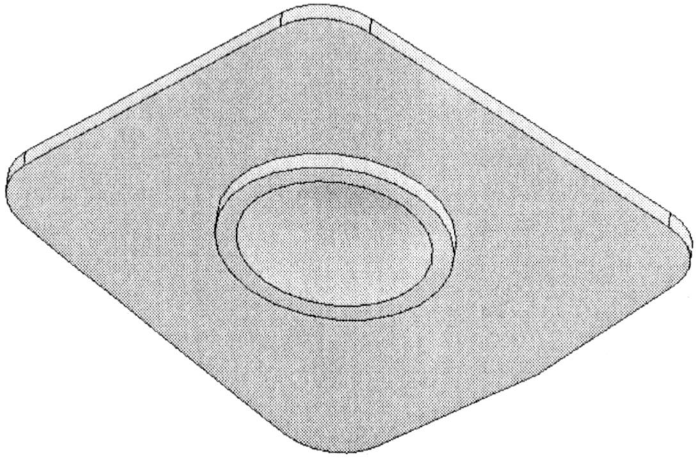

Figure 7-99 *Cut revolve feature added to the model*

Creating the Fifth Feature

Next, you will create the left fin of the cylinder head. It will be created by extruding a sketch to both the directions using the **Through All** option. The sketch of this feature is created on the **Front** plane and the sketch will be dimensioned and defined such that the length of the fin is driven by a construction arc and a horizontal dimension. The detailed step by step procedure of creating, dimensioning, and defining the sketch is discussed next.

1. Invoke the **Extruded Boss/Base** tool and select the **Front** plane from the **FeatureManager Design Tree**.

2. Using the **Line** tool create the triangle and then create a vertical centerline that passes through the upper vertex to the triangle, refer to Figure 7-100.

3. Invoke the **3 Pt Arc** tool and create the arc as shown in Figure 7-100. Select the arc and select the **For Construction** check box from the **Options** rollout of the **Arc PropertyManager**.

4. Invoke the **Add Relations PropertyManager** and add the coincident relation between the upper vertex of the triangle and the centerline.

5. Add the midpoint relation between the lower endpoint of the centerline and the horizontal line of the triangle. Make sure that the **Coincident** relation between the upper vertex of the triangle and the centerline exists. Also, add the vertical relation to the centerline, if it is missing.

6. Add the coincident relation between the upper vertex of the triangle and the arc.

7. Add the required dimensions and relations to fully define the sketch as shown in Figure 7-100.

Tip. *It is evident from Figure 7-86 that one of the horizontal dimension is having a value of 6.155. By default the primary unit precision is set to two decimal places. Therefore, for defining a dimension value with more number of decimal places, you have to select the dimension and set the precision value to the required decimal places from the **Primary Unit Precision** drop-down list from the **Dimension PropertyManager**.*

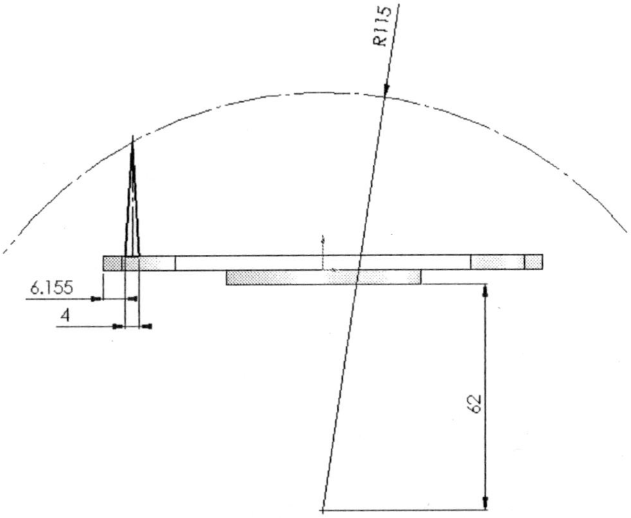

***Figure 7-100** Sketch for the fin of the cylinder head*

8. Exit the sketching environment and extrude the sketch in both the directions using the **Through All** option. You will notice that the fin extends out of the base feature at both the ends. This unwanted material of fin will be removed later in this tutorial.

Patterning the Fifth Feature

Now, you will pattern the fin using the **Linear Pattern** tool. You will use the **Vary sketch** option to pattern the fin. Using the **Vary sketch** option while patterning, the geometry of each instance of the pattern vary according to the driven dimension and the relation added to the sketch of the feature to be patterned.

1. Choose the **Linear Pattern** button from the **Features CommandManager** to invoke the **Linear Pattern PropertyManager**.

   You are prompted to select the directional reference.

2. The fifth feature is selected by default in the **Features to Pattern** selection area. If the fifth feature is not selected then you need to select it from the drawing area.

3. Select the horizontal dimension with the value of **6.155** as the directional reference in the **Pattern Direction** selection area of the **Direction 1** rollout.

4. Set the value of the **Spacing** spinner to **9**. Set the value of the **Number of Instances** spinner to **13**. Choose the **Reverse Direction** button.

5. Invoke the **Options** rollout and select the **Vary Sketch** check box from this rollout.

   The preview of the pattern is not displayed in the drawing area.

6. Choose the **OK** button from the **Linear Pattern PropertyManager**.

   The model after adding the pattern feature is shown in Figure 7-101.

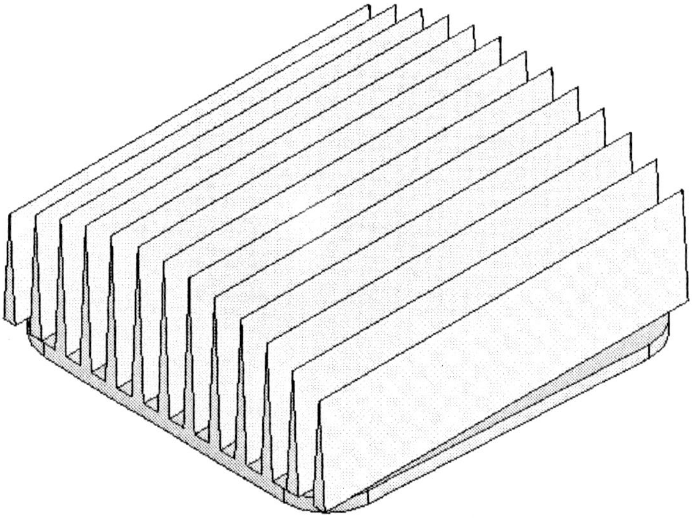

Figure 7-101 *Model after patterning the fin of the cylinder head*

Creating the Cut Feature

The next feature that you are going to create is a cut feature. Using the **Rotate View** tool rotate the solid model and you will observe that the fins of the cylinder head that you patterned in the last feature are extending beyond the boundary of the base feature. Therefore, to trim the extended portion of the fins you will create a cut feature.

1. Invoke the **Extruded Cut** tool and select the top planar face of the base feature as the sketching plane.

2. Create the sketch using the standard sketch tools. The sketch for this feature will be the outer profile of the base feature.

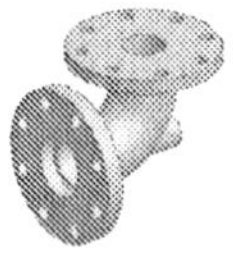

> **Tip**. *You can create the sketch using the outer profile of the base feature using the **Convert Entities** tool. Select the lower flat face of the base feature and choose the **Convert Entities** button from the **Sketch** toolbar. You will notice that the sketch similar to the outer boundary of the base feature will be placed on the sketching plane.*

3. Exit the sketching environment and choose the **Reverse Direction** button from the **Direction 1** rollout and select the **Through All** option from the **End Condition** drop-down list.

   Since the direction of the side from which material has to be removed is the opposite to the required direction, therefore, you need to flip the direction of material removal.

4. Select the **Flip side to cut** check box from the **Direction 1** rollout and choose the **OK** button from the **Cut-Extrude PropertyManager**.

5. Using the standard modeling tools shape the model as shown in Figure 7-102.

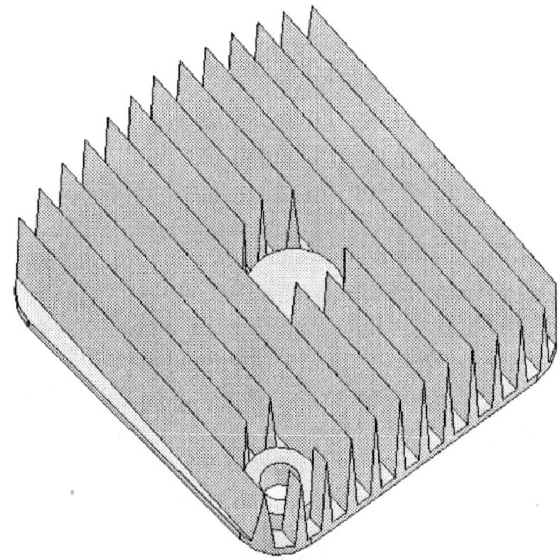

Figure 7-102 *Model after adding other extrude and cut features*

Patterning the Remaining Features

After creating all the features now you have to pattern the cut, extrude, and hole features created at the lower left corner of the model.

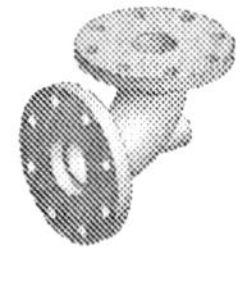

Tip. *The sketch sharing option is also available in SolidWorks. This allows you to use a sketch used earlier in creating a sketch feature to be used again while creating any other sketch feature. For sharing the sketch select the sketch from the* ***FeatureManager Design Tree*** *after you have invoked the feature tool. You need to expand the created sketched feature to select the sketch for sharing from the* ***FeatureManager Design Tree****.*

1. Invoke the **Linear Pattern PropertyManager** and select the cut, extrude, and hole features, created on the lower left corner of the model.

2. Select the two directional references to pattern the features in both the directions and set the values of the distances between the instances and the number of instances, refer to Figure 7-96.

3. Choose the **OK** button from the **Linear PropertyManager**.

Creating a Tapped Hole

The last feature of the model is a hole feature. You will create a tapped hole using the **Hole Wizard** and then define the placement of the hole.

1. Select the top face of the middle circular extrude feature as the placement plane for the hole feature.

2. Invoke the **Hole Definition** dialog box by choosing the **Hole Wizard** button from the **Features CommandManager** and select the **Tap** tab. Select **ANSI Metric** from the **Standard** drop-down list.

3. Select the **M18x1.5** option from the **Size** drop-down list to define the size of the tap hole.

4. Select the **Through All** option from the **Tap Drill Type & Depth** drop-down list. Also, select the **Through All** option from the **Thread Type & Depth** drop-down list.

5. Select the **Add Cosmetic thread with thread callout** option from the **Add Cosmetic Thread** drop-down list. Choose the **Next** button from the **Hole Definition** dialog box. The **Hole Placement** dialog box is displayed.

   The tapped hole will be placed by default on the placement plane. This default location is not the required position to place the hole. You need to define the placement of the tapped hole concentric with the center circular feature.

6. Invoke the **Add Relations PropertyManager** and add the **Concentric** relation between the centerpoint of the tapped hole and the circular extruded feature of diameter 55 mm.

7. Choose the **Finish** button from the **Hole Placement** dialog box to end the tapped hole feature creation.

   The rotated final model is shown in Figure 7-103.

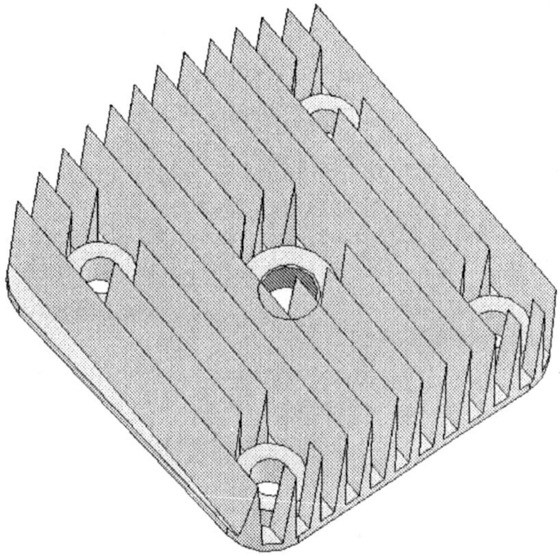

Figure 7-103 *Final solid model*

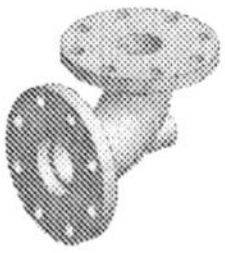

Tip. *You will observe that a graphic thread is displayed in the tapped hole. You can zoom the area for the better visualization of the graphic thread.*

You will also observe that the cosmetic thread is displayed with the tapped hole feature. If you orient the model in the top view you will observe the thread convention in the top view. If you orient the model in the front, back, right, or left views then you will view the side convention of the thread.

You can also hide the cosmetic thread by selecting the cosmetic thread from the drawing area and right-clicking to invoke the shortcut menu. Choose the ***Hide*** *option from the shortcut menu to hide the cosmetic thread.*

Displaying the Section View of the Model

Next, you will display the section view of the model. The section view of the model is created using the **Section View PropertyManager**.

1. Orient the model in the isometric view.

2. Choose the **Section View** button from the **View CommandManager**.

   The **Front** view is selected as the section plane in the **Section View PropertyManager** by default. The preview of the section view using the **Front** plane as the section plane is displayed in the drawing area.

3. Choose the **OK** button from the **Section View PropertyManager** to view the section of the model.

   The section view of the model is shown in Figure 7-104.

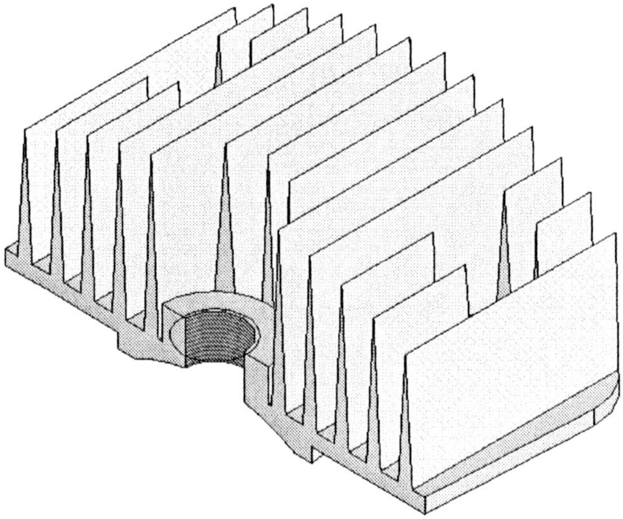

Figure 7-104 Section view of the model

7. Choose the **Section View** button again from the **View CommandManager** to return to the full view mode. The **FeatureManager Design Tree** is shown in Figure 7-105.

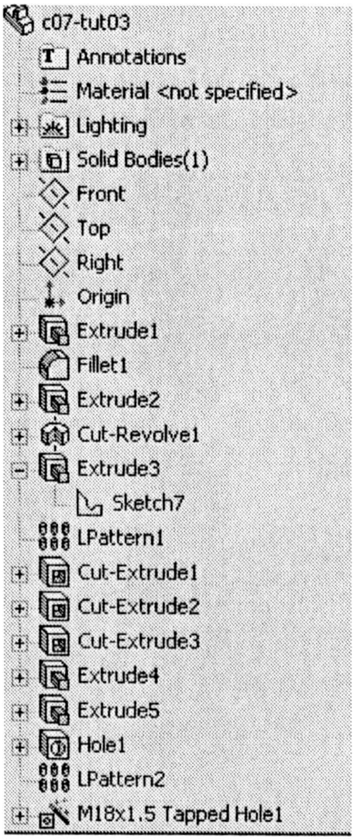

Figure 7-105 The **FeatureManager Design Tree**

Saving the Model

Since the document has not been saved until now, therefore, when you choose the **Save** button from the **Standard** toolbar, the **Save As** dialog box will be displayed. You can enter the name of the document in this dialog box.

1. Choose the **Save** button from the **Standard** toolbar and save the model in the *c07* folder with the name given below.

   \My Documents\SolidWorks\c07\c07-tut03.sldprt

2. Choose **File > Close** from the menu bar to close the document.

SELF-EVALUATION TEST

Answer the following questions and then compare your answers with the answers given at the end of this chapter.

1. To invoke the **Mirror PropertyManager** choose **View > Pattern/Mirror > Mirror** from the menu bar. (T/F)
2. If you modify the parent feature then the same change will not reflect on the mirrored feature. (T/F)
3. You cannot preselect the mirror plane and the feature to pattern before invoking the **Mirror** tool. (T/F)
4. You can mirror a single face using the **Mirror** tool. (T/F)
5. You can also pattern a pattern feature. (T/F)
6. The __________ **PropertyManager** is used to view the section view.
7. Using __________ check box you can remove the material using the **Dome** tool.
8. __________ option is used to create a pattern by specifying the coordinates.
9. __________ option is used to create a pattern with respect to the sketched points.
10. Using the __________ rollout you can delete the pattern instances.

REVIEW QUESTIONS

Answer the following questions.

1. The __________ check box is used to accommodate all the instances of pattern along the selected curve.

2. Enter the coordinates for creating the instances in the __________ area of the **Table Driven Pattern** dialog box.

3. You have to invoke __________ to create a rib feature.

4. __________ check box is selected to create an elliptical dome feature.

5. Using __________ check box from the **Section View PropertyManager** you can create a section using an invisible plane normal to the eye view as the section plane.

6. When you choose the **Mirror** button from the **Features** tool, which **PropertyManager** is displayed?

   (a) **Mirror Feature PropertyManager** (b) **Mirror All PropertyManager**
   (c) **Mirror PropertyManager** (d) **Copy/Mirror PropertyManager**

7. Which option is used to mirror the exact geometry of the feature independent of the relationships between the geometries?

   (a) **Same Mirror** (b) **Geometry Pattern**
   (c) **Geometry Copy** (d) **Copy Geometry**

8. Which pattern is created along the sketched lines, arcs, or splines?

   (a) Curve-driven pattern (b) Sketch-driven pattern
   (c) Geometry-driven pattern (d) Linear pattern

9. Which dialog box is invoked to create a pattern by specifying the coordinate points?

   (a) **Sketch Driven Pattern** (b) **Table Driven Pattern**
   (c) **Mirror** (d) None of these

10. Which plane is selected by default when you invoke the **Section View PropertyManager** to view a section of the model?

   (a) **Right** (b) **Top**
   (c) **Front** (d) **Plane 1**

EXERCISES

Exercise 1

Create the model shown in Figure 7-106. The dimensions of the model are shown in Figure 7-107. **(Expected time: 1 hr)**

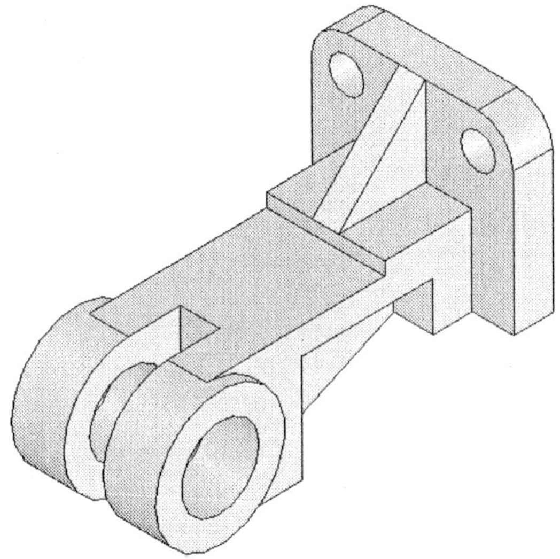

Figure 7-106 *Solid model for Exercise 1*

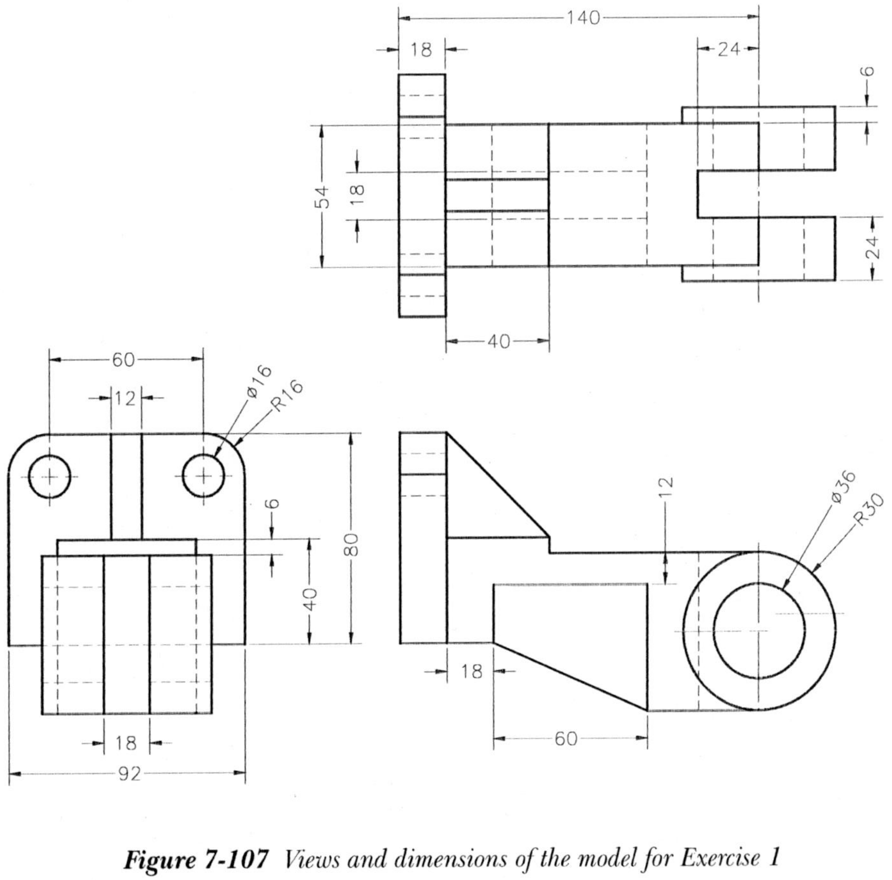

Figure 7-107 *Views and dimensions of the model for Exercise 1*

Exercise 2

Create the model shown in Figure 7-108. The dimensions of the model are shown in Figure 7-109.

(Expected time: 1 hr)

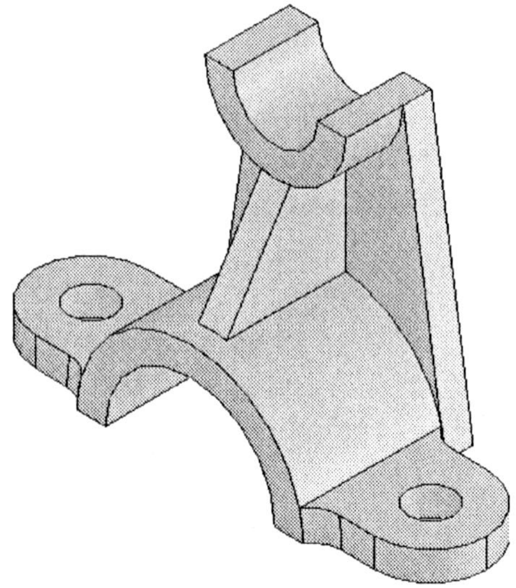

Figure 7-108 Solid model for Exercise 2

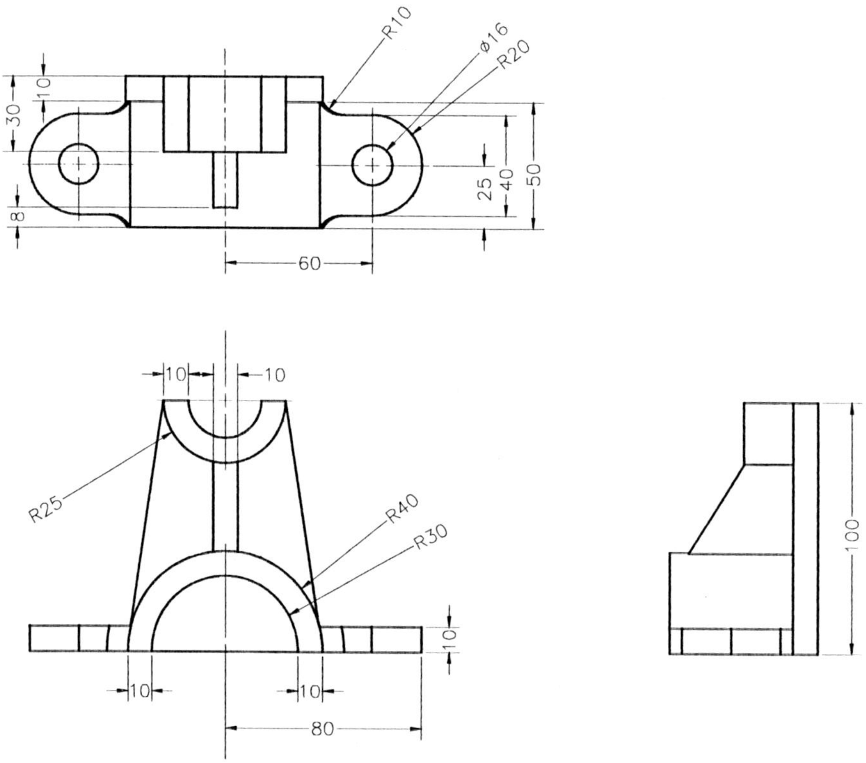

Figure 7-109 Views and dimensions of the model for Exercise 2

Exercise 3

Create the model shown in Figure 7-110. After creating the model create the section view of the model using the **Right** plane. Figure 7-111 shows the section view of the model. The dimensions of the model are shown in Figure 7-112. **(Expected time: 45 min)**

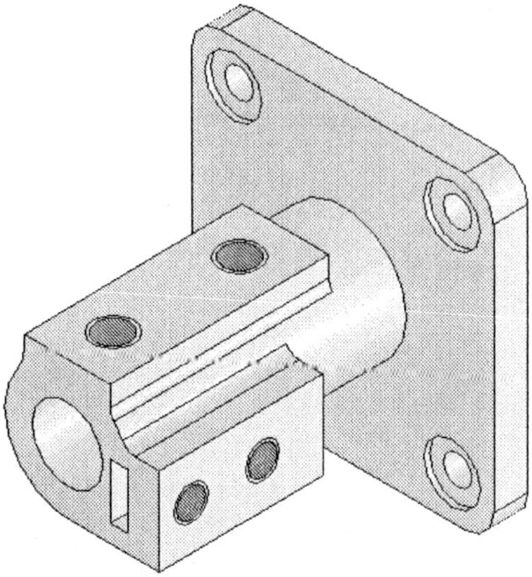

Figure 7-110 *Solid model for Exercise 3*

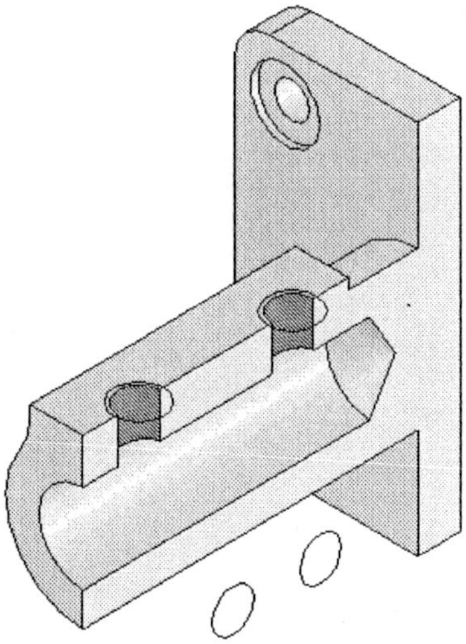

Figure 7-111 *Section view of the model*

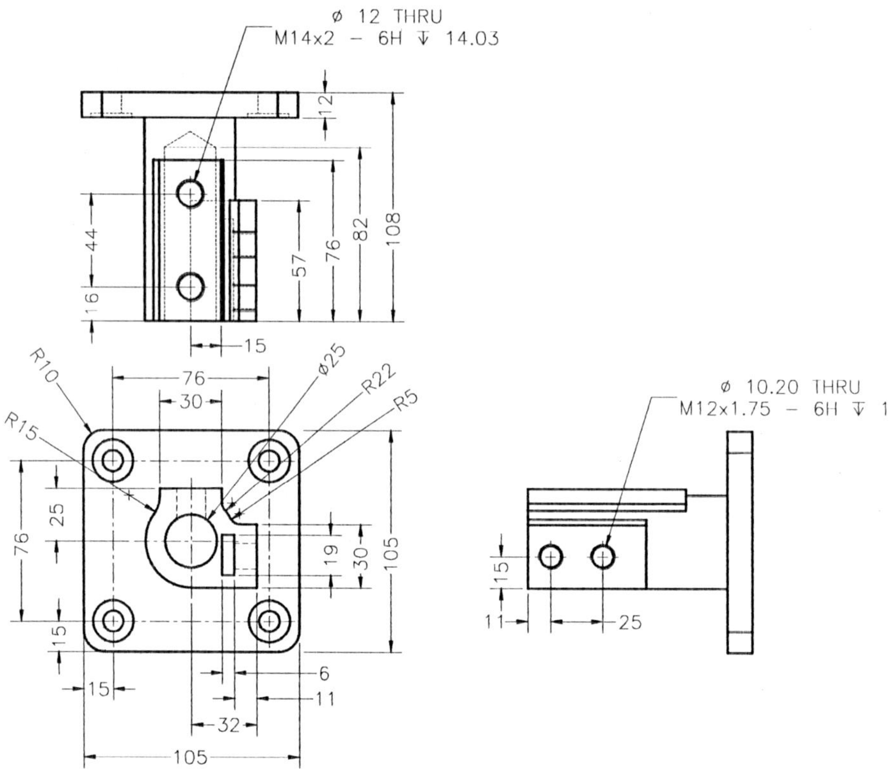

Figure 7-112 Views and dimensions of the model for Exercise 3

Answers to Self-Evaluation Test

1. F, **2.** F, **3.** F, **4.** F, **5.** T, **6. Section View**, **7. Reverse Direction**, **8. Table Driven Pattern**, **9. Sketch Driven Pattern**, **10. Instances to Skip**

Chapter 8

Editing Features

Learning Objectives

After completing this chapter you will be able to:

- *Edit the Features.*
- *Edit the Sketches of the Sketched Feature.*
- *Edit the Sketch Plane.*
- *Edit using the Move/Size Features option.*
- *Cut, Copy, and Paste the Features and Sketches.*
- *Copy Features using the Drag and Drop Method.*
- *Delete the Features.*
- *Delete the Bodies.*
- *Suppress and Unsuppress the Features.*
- *Move/Copy Bodies.*
- *Reorder the Features.*
- *Roll Back the Model.*
- *Rename the Features.*
- *Create the Folders.*
- *Use What's Wrong? Functionality.*

EDITING THE FEATURES OF THE MODEL

Editing is one of the most important aspect of the product design cycle. Almost all the designs require editing either during their creation or after they are created. As discussed earlier, SolidWorks is a feature-based parametric software. Therefore, the design created in SolidWorks is a combination of individual features integrated together to form a solid model. All these features can be edited individually. For example, Figure 8-1 shows a base plate with some holes.

Now, if you have to replace the four inner holes with four counterbore holes, all you need to do is to use one editing operation. Using the editing operation, you will change the drilled holes to the counterbore holes. For editing the holes, you have to select the hole feature and right-click to invoke the shortcut menu. Then you will have to choose the **Edit Feature** option from the shortcut menu to invoke the **Hole Definition** dialog box. Finally you have to set the new parameters and end feature creation. The drilled holes will be automatically replaced by the counterbore holes. Figure 8-2 shows the base plate with drilled holes modified to the counterbore holes.

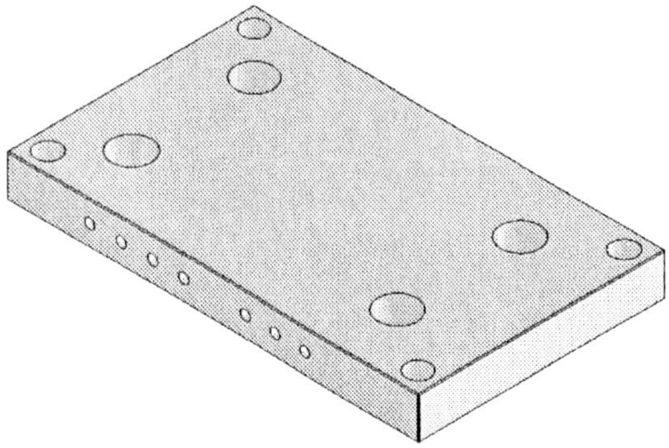

Figure 8-1 *Base plate with drilled holes*

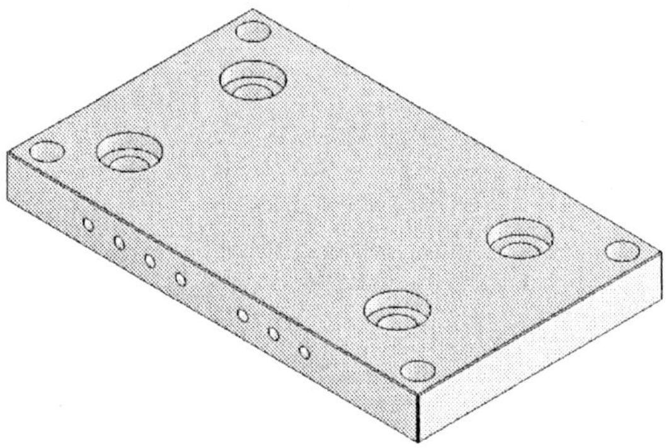

Figure 8-2 *Modified base plate*

Similarly, you can also edit reference geometry and the sketches of the sketched features. The feature created using the reference geometry also modify automatically when you modify the reference geometry. For example, if you create a feature on a plane at some angle and then edit the angle of the plane, the resultant feature is automatically modified. In SolidWorks, you can perform editing tasks using various methods. All the methods of editing are discussed as under.

Editing Using Edit Feature Option

The editing using the **Edit Feature** option is the most commonly used method of editing in SolidWorks. To edit any feature of the model using this option, select that feature either from the **FeatureManager Design Tree** or from the drawing area. After selecting the feature, right-click to invoke the shortcut menu and choose the **Edit Feature** option from the shortcut menu as shown in Figure 8-3. Depending on the feature selected, the **PropertyManager** or dialog box will be invoked and you can modify the parameters of that feature. The

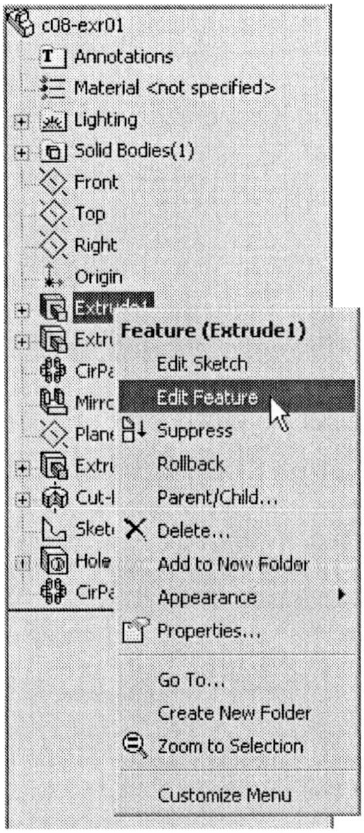

***Figure 8-3** Selecting the **Edit Feature** option from the shortcut menu*

PropertyManager will also have the sequence number of the feature. The **Extrude PropertyManager** is displayed in Figure 8-4 with sequence number 1. After editing the

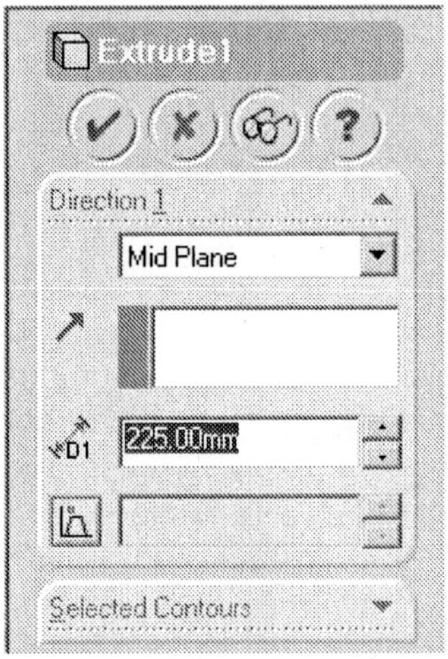

Figure 8-4** The **Extrude PropertyManager

parameters, choose the **OK** button to complete feature creation. The feature will be modified automatically.

Editing the Sketches of the Sketched Features

In SolidWorks, you can also edit the sketches of the sketched features using the **Edit Sketch** option. To edit the sketch of a sketched feature select the feature either from the **FeatureManager Design Tree** or from the drawing area and right-click to invoke the shortcut menu. Choose the **Edit Sketch** option from the shortcut menu. When you choose this option, you will enter the sketching environment. Using the standard sketching tools, edit the sketch of the sketched feature. After editing the sketch, exit the sketching environment. To exit the sketching environment you can also select the **Rebuild** button from the **Standard** toolbar. Using this you will exit the sketching environment and the model will also be rebuilt. You can also choose CTRL+B to rebuild the model.

Tip. *You can also use the + sign available on the left of a sketched feature to expand that sketched feature in the **FeatureManager Design Tree**. The sketch icon will be displayed when you expand a sketched feature. Select the sketch icon and invoke the shortcut menu. Select the **Edit Sketch** option from the shortcut menu to enter the sketching environment to edit the sketch.*

Changing the Sketch Plane of the Sketches

You can also change the sketch plane of the sketches of the sketched features. To edit the sketch plane, expand the sketched feature by clicking on the + sign on the left of that feature in the **FeatureManager Design Tree**. Select the sketch icon in the **FeatureManager Design Tree**. Right-click to invoke the shortcut menu and choose the **Edit Sketch Plane** option from the shortcut menu. Figure 8-5 shows the **Edit Sketch Plane** option being chosen from the shortcut menu. The **Sketch Plane PropertyManager** will be displayed as shown in Figure 8-6. The name of the current sketch plane is displayed in the **Sketch Plane/Face** selection area. Now, select any other plane or face as the sketching plane. After selecting the new sketching plane, choose the **OK** button from the **Sketch Plane PropertyManager**.

Tip. *If you select a sketch plane on which the relations and dimensions do not find any reference to be placed, the **Rebuild Errors** dialog box will be displayed. You, will have to undo the last step using the **Undo** button from the **Standard** toolbar, invoke the **Sketch Plane PropertyManager** again, and then select the appropriate plane. You will learn more about the **Rebuild Errors** dialog box later in this chapter.*

Editing by Double-Clicking the Entities and Features

You can also edit a feature, a reference geometry, or a sketch by double-clicking the feature either from the **FeatureManager Design Tree** or from the drawing area. For example, if you double-click an extrude feature, all the dimensions of the feature and the sketch used to create that feature are displayed in the drawing area. Remember that all the dimensions of the sketch will be displayed in black color and all the dimensions of the feature will be displayed in blue color. Double-click the dimension that you need to modify, the **Modify** dialog box is invoked. Set the new value in the **Modify** dialog box and press the ENTER key from the

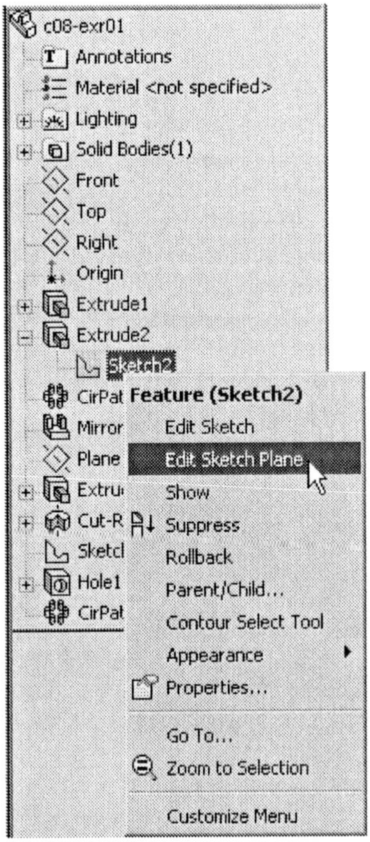

Figure 8-5 The ***Edit Sketch Plane*** *option being selected*

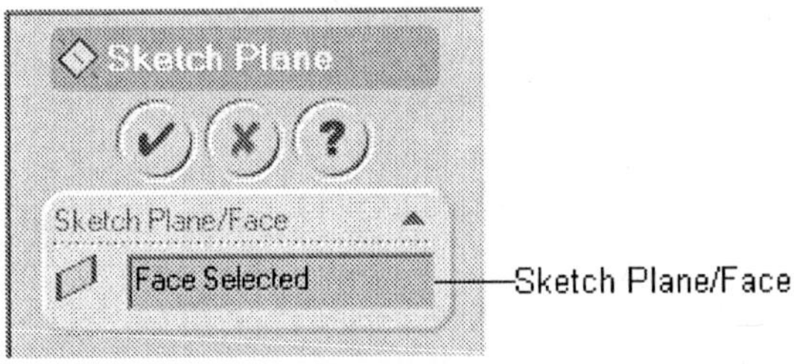

Figure 8-6 The ***Sketch Plane PropertyManager***

keyboard or choose the **Save the current value and exit the dialog** button from the dialog box. You will notice that the value of the dimension is modified but the model is not modified relative to the modified value. Therefore, you have to rebuild the model using the **Rebuild** option. To rebuild the model, choose the **Rebuild** button from the **Standard** toolbar or choose CTRL+B from the keyboard.

Editing Using the Move/Size Features

CommandManager:	Features > Move/Size Features	*(Customize to Add)*
Toolbar:	Features > Move/Size Features	*(Customize to Add)*

Using this option you can dynamically modify the feature and the sketch of the sketched feature without invoking the sketching environment. To edit the feature or sketch using this option, choose the **Move/Size Features** button from the **Features CommandManager**. You will notice that the **Move/Size Features** button is chosen in the **Features CommandManager**. Select any face of the feature to modify. The selected face will be highlighted in green and if the selected feature is a sketched feature then the sketch of that feature will also be highlighted in green. You will be provided with the resize handle, rotate handle, and the move handle. Figure 8-7 shows the move handle, resize handle, and the rotate handle provided for the selected feature.

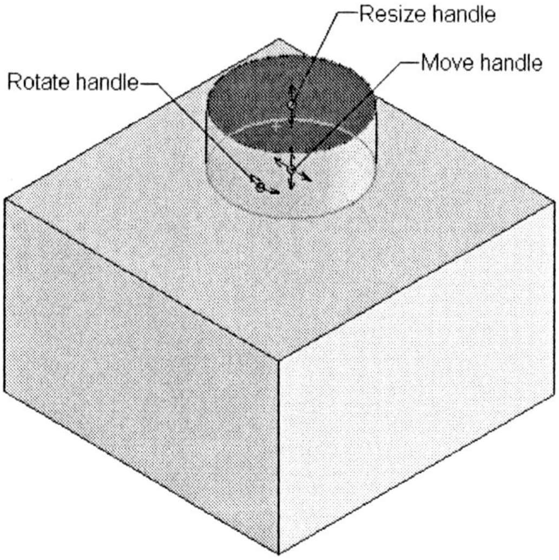

***Figure 8-7** Feature selected with **Move/Size Features** button selected*

To resize the feature, move the cursor to the resize handle; the select cursor will be replaced by the resize cursor and the name of the resize cursor will be displayed as the tooltip. Press and hold down the left mouse button at this location and drag the cursor to resize the feature. The feature will be resized with an increment of 10. Release the left mouse button when you have resized the feature. You will notice that the feature will be dynamically resized. Figure 8-8 shows the cursor being dragged to resize the feature. Figure 8-9 shows the resultant modified feature.

The rotate handle is used to rotate the selected feature. To rotate the feature using the rotate handle, move the cursor to the rotate handle and press and hold down the left mouse button. Drag the cursor to rotate the feature. You can drag the feature clockwise or counterclockwise. The preview of the feature will be displayed in the drawing area. The feature will be rotated with an increment of 10-degree. Release the left mouse button after rotating the feature to a specified location.

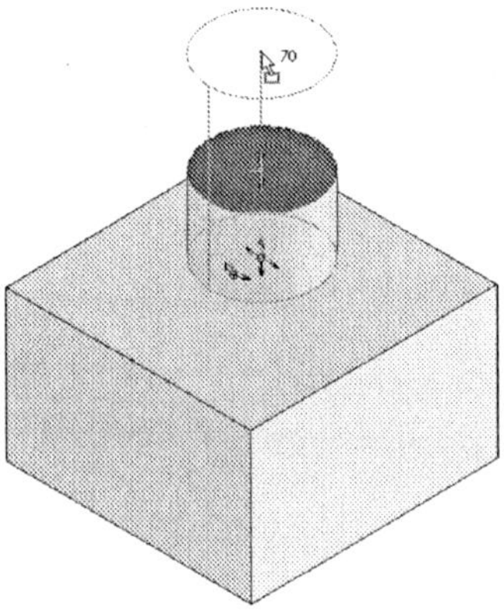

Figure 8-8 *Dragging the resize handle to resize the feature*

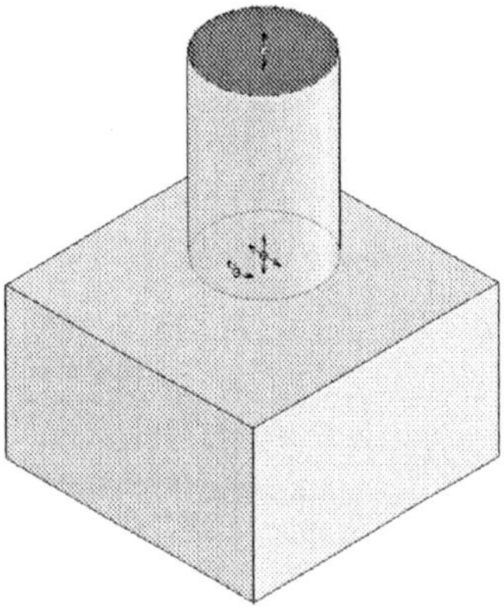

Figure 8-9 *Resultant modified feature*

Note

If you rotate a sketched feature in which the sketch of the feature is fully defined or partially defined using relations and dimensions, the ***Move Confirmation*** *dialog box will be displayed. The* ***Move Confirmation*** *dialog box is displayed in Figure 8-10. This dialog box informs you that the external constraints in the feature are being moved, and asks whether want to delete those constraints or keep the constraints by recalculating or making them dangling. The relations or dimensions that do not find the external reference after the placement are made dangling.*

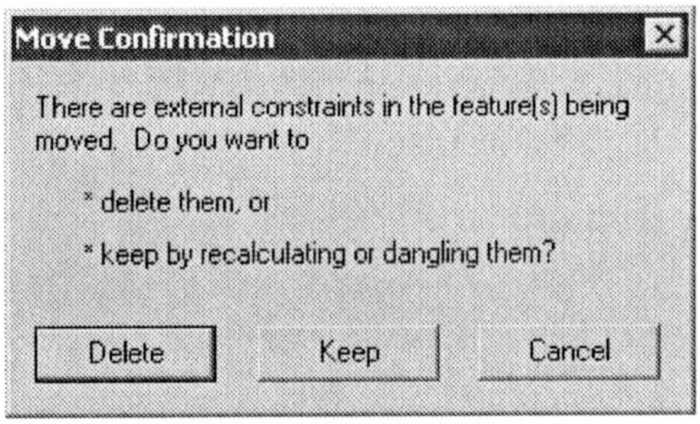

Figure 8-10 *The* ***Move Confirmation*** *dialog box*

After rotating the feature, if the **Move Confirmation** dialog box is invoked then you have to choose either the **Delete** button or the **Keep** button, based on the geometric and dimensional conditions. Figure 8-11 shows the preview of rotating the feature. Figure 8-12 shows the resultant rotated feature.

To move a feature using the move handle, select a face of the feature and move the cursor to the move handle. Press and hold down the left mouse button and drag the cursor to move the feature. Release the left mouse button after moving the feature to the required position. If the dimensions or relations of the feature are not able to be defined after moving the feature, the **Move Confirmation** dialog box will be displayed. Choose the options available in this dialog box. Using this option, you can also change the placement plane of the feature or the sketch

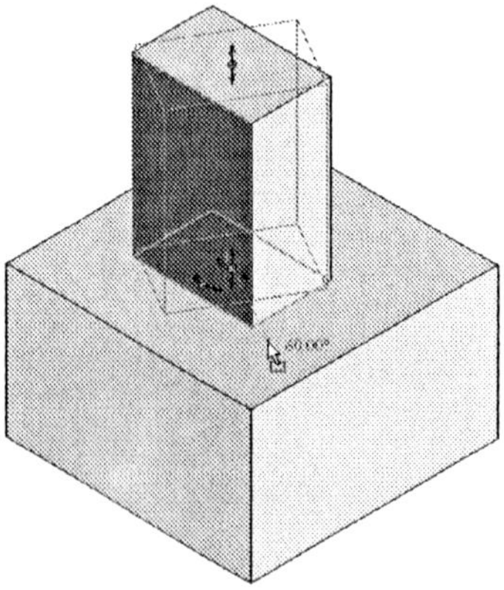

Figure 8-11 *Preview of feature being rotated*

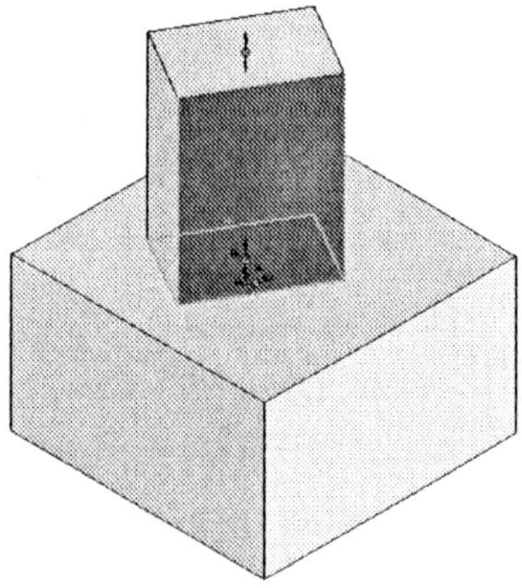

Figure 8-12 *Resultant rotated feature*

plane of the feature. Figure 8-13 shows the feature being moved to another face. Figure 8-14 shows the resultant moved feature.

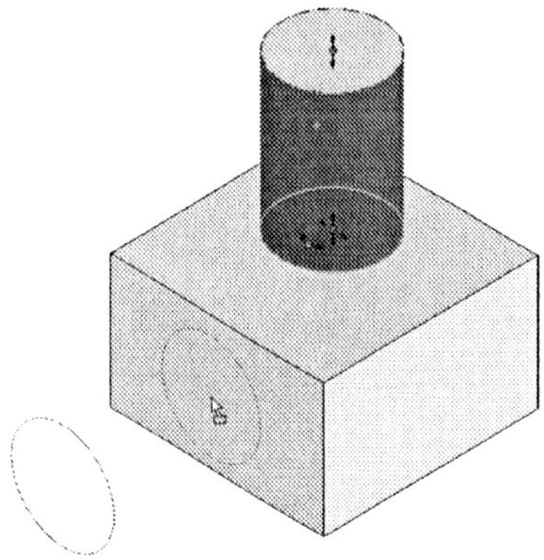

Figure 8-13 *The feature being moved*

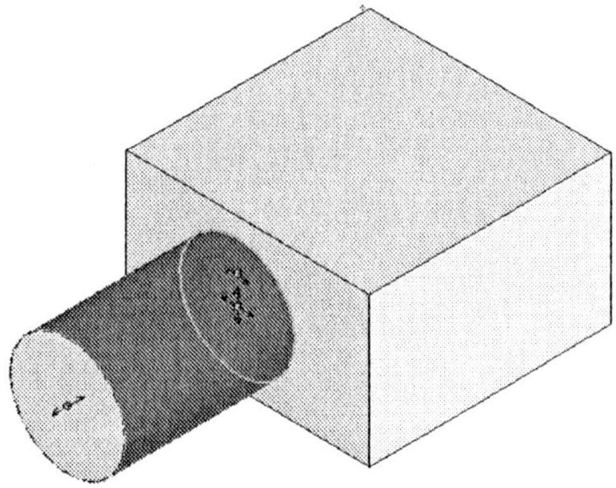

Figure 8-14 *Resultant moved feature*

You can also modify the sketches of the sketched feature using this option. To modify the sketch, move the cursor to the sketch highlighted in light green color. A vertical line appears on the right of the cursor. Press and hold down the left mouse button and drag the cursor. After modifying the sketch to the required size, release the left mouse button. Figure 8-15 shows the preview of the sketch being modified. Figure 8-16 shows the resultant modified model.

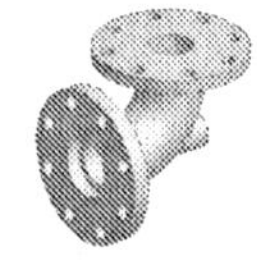

Tip. *When you resize, move, or rotate a feature using the **Move/Size Features** tool, the feature(s) that has a relationship with that feature will also be modified. In other words, if a child-parent relationship is established between the features, the child features will also be modified when you modify the parent feature.*

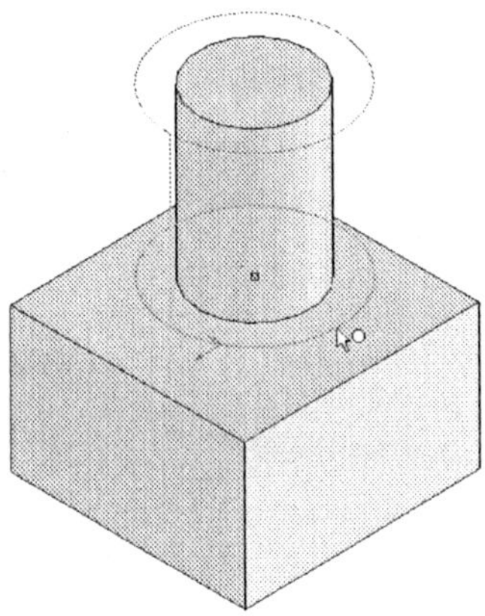

Figure 8-15 *Preview of the sketch being modified*

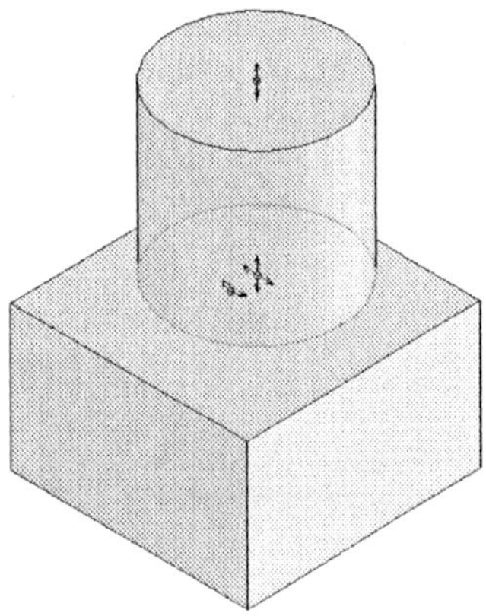

Figure 8-16 *Resultant modified model*

Tip. *To clear the selected feature, click once anywhere in the drawing area or choose the ESC key from the keyboard. You will notice that the* ***Move/Size Features*** *button is still chosen in the* ***Features*** *toolbar, which indicates that the dynamic move/size features option is still active. To exit this option you have to choose the* ***Move/Size Features*** *button again.*

Editing the Sketches With the Move/Size Features Tool Active

When the **Move/Size Features** tool is active and you choose the **Edit Sketch** option to edit the sketch of the sketched feature you will enter the sketching environment. The feature whose sketch you have to modify will be displayed in transparent yellow temporary graphics as shown in Figure 8-17.

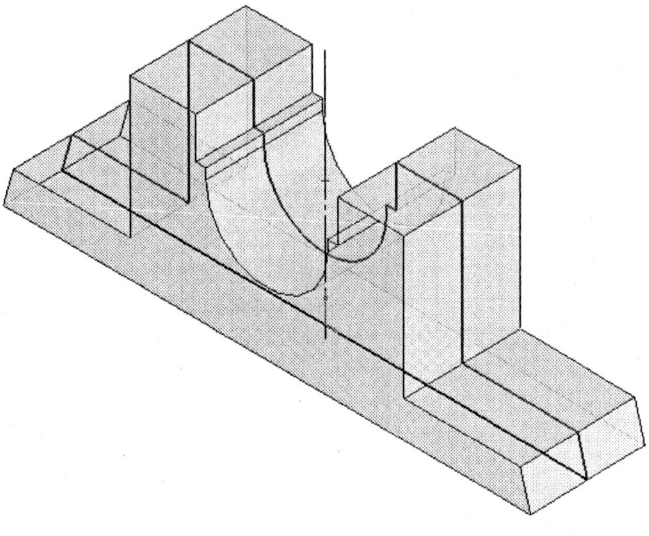

Figure 8-17 *Preview of the sketch being modified*

Tip. *If you want to modify the sketch by dragging and the sketch is fully or partially defined using dimensions, the* ***Override Dims on Drag/Move*** *option should be selected. To select this option choose* ***Tools > Sketch Settings > Override Dims on Drag/Move*** *from the menu bar. If this option is not selected, you cannot move or drag a dimensioned sketched entity.*

Now, modify the sketch according to the requirement. Consider a case in which two circles are added to the model and the height of the inclined line is increased. The model shown in temporary graphics will be modified dynamically in the sketcher environment. Therefore, you can have a better understanding of the model, that how it will be displayed after modifying the sketch. Note that the temporary graphic will show the preview of the feature created by the section you add only if you drag one of the entity in the sketch. Otherwise, the new addition will not be displayed as a preview in the temporary graphics. It is recommended that you invoke the **Modify/Size Features** button before modifying the sketch. After modifying the sketch exit the sketcher environment. If the sketch of the model is dimensioned then you can modify the sketch by modifying the dimension. Figure 8-18 shows two circles added to the model and the length of the inclined line being increased. Figure 8-19 shows the resultant modified model in the sketching environment.

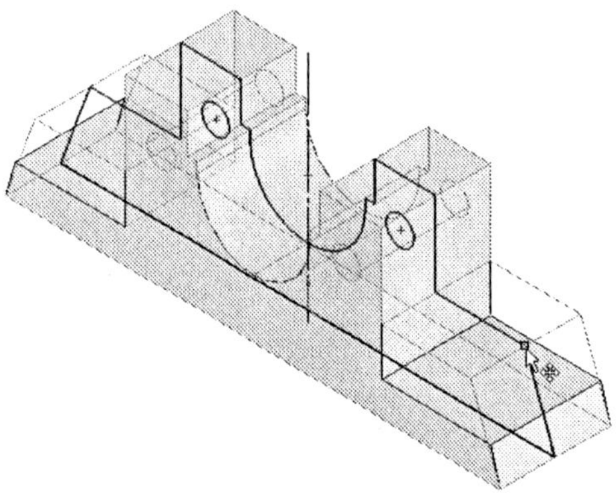

Figure 8-18 *Sketch being modified*

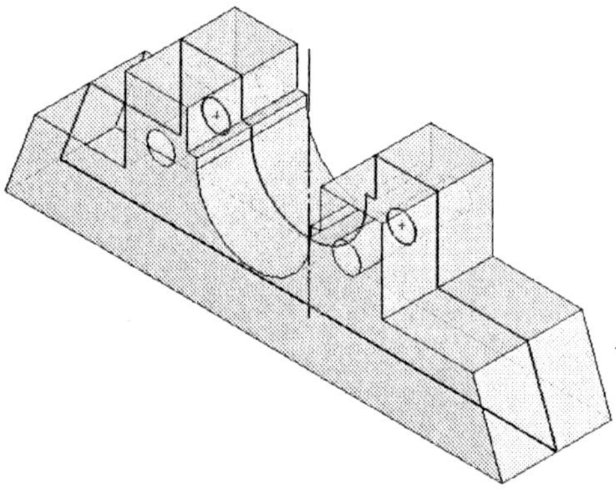

Figure 8-19 *Resultant modified model in the sketching environment*

Editing the Features and Sketches by Cut, Copy, and Paste

SolidWorks allows you to adapt the windows functionality of cut, copy, and paste to copy and paste the features and sketches. The method of using this functionality is the same as used in other windows-based applications. Select the feature or sketch to cut or copy. To cut the selected item, choose **Edit > Cut** from the menu bar or use the shortcut key, CTRL+X. The **Delete Confirmation** dialog box is displayed because when you cut the selected item, it is deleted from the document. Choose the **Yes** button from this dialog box. You will learn more about deleting later in this chapter. After you cut an item, select the placement plane or placement reference where you want to place the feature. Choose **Edit > Paste** from the menu bar or use the shortcut key CTRL+V from the keyboard. Sometimes the **Copy Confirmation** dialog box is displayed as shown in Figure 8-20. In that dialog box, you are prompted either to delete the external constraints or leave them dangling. This dialog box is

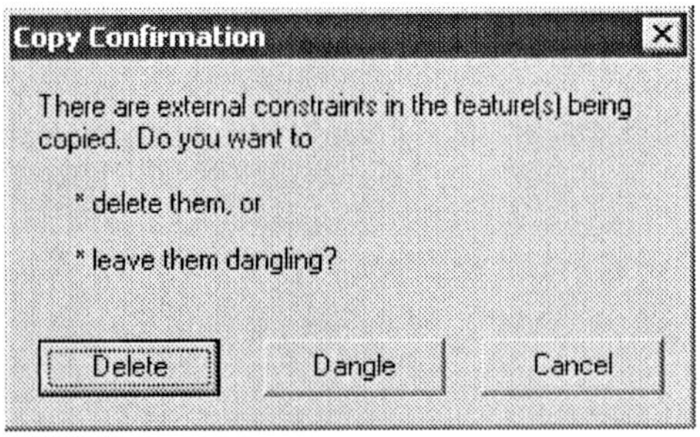

***Figure 8-20** The **Copy Confirmation** dialog box*

Tip. *For pasting a selected sketch, you have to select a plane or a planar face as the reference. For pasting a sketched feature, a simple hole, or a hole created using the hole wizard, you have to select a plane or a planar face as the reference. For pasting chamfers and fillets, you have to select an edge, edges, or face as the references.*

displayed only when the item to paste has some external references in the form of relations and dimensions. The feature will be pasted on the selected reference.

If you copy and paste an item, the selected item will remain at its position and a copy of that item will be pasted on the selected reference. To copy an item, select the feature or sketch. Choose **Edit** > **Copy** from the menu bar or press CTRL+C from the keyboard. Select the reference where you want to paste the selected item and choose **Edit** > **Paste** from the menu bar or press CTRL+V from the keyboard to paste. You can paste the selected item any number of times. If you select another item and copy on the clipboard, the last copied item will be deleted from the memory of the clipboard.

Cut, Copy, and Paste Features and Sketches from One Document to the Other

You can also cut or copy the features and sketches from one document and paste them in another document. For example, you have to copy a sketch created in the current document and paste it in a new document. You just need to select the sketch and use CTRL+C from the keyboard to copy the item to the clipboard. Create a new document in the **Part** mode and select the plane on which you want to paste the sketch. Press CTRL+V from the keyboard to paste the sketch on the selected plane. Using the same procedure, you can also copy features from one document to the other.

Copying Features Using Drag and Drop

SolidWorks also provides you with drag drop functionality of windows to copy and paste the item within the document. Press and hold down the CTRL key from the keyboard and select and drag that item from the drawing area or from the **FeatureManager Design Tree**. Drag the cursor to a location where you want to paste the item and release the left mouse button on that location. If the item to be pasted is defined using the dimensions or relations, the **Copy Confirmation** dialog box will be displayed to delete or make those constraints dangle. Figure 8-21 shows the feature being dragged. Figure 8-22 shows the resultant pasted feature.

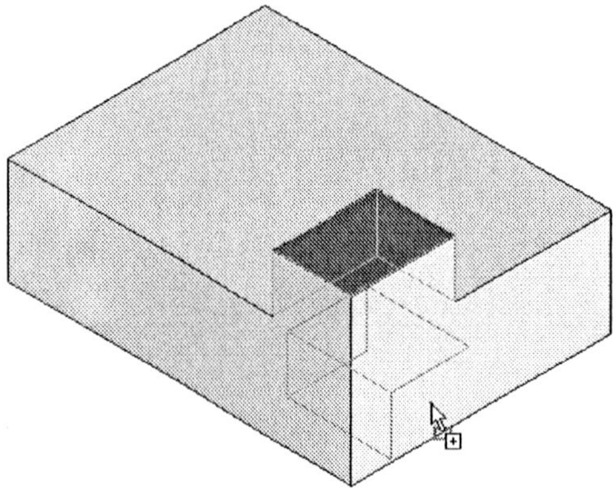

Figure 8-21 Feature being dragged

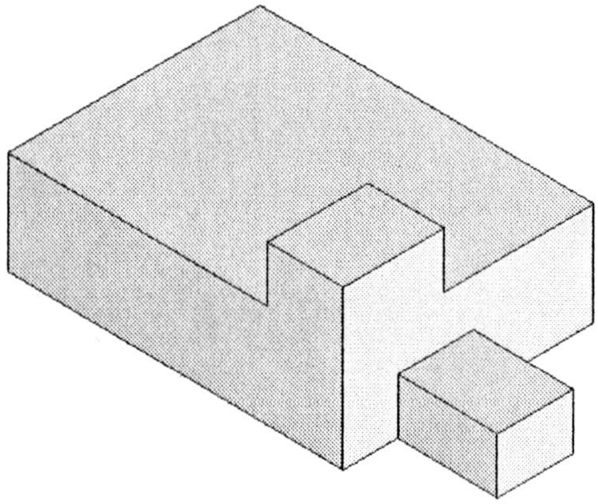

Figure 8-22 Resultant pasted feature

Dragging and Dropping Features from One Document to Other

You can also drag and drop features and sketches from one document to the other. For pasting the items from one document to the other, you should open both the documents in the SolidWorks session. Choose **Windows > Tile Vertical/Tile Horizontal** from the menu bar. Display both the documents at the same time in the SolidWorks window. Press and hold down the CTRL key from the keyboard, select and drag the feature or sketch to the other document, and place it on the required reference. Figure 8-23 shows the fillet feature being dragged to be applied on the edge of the model in the second document.

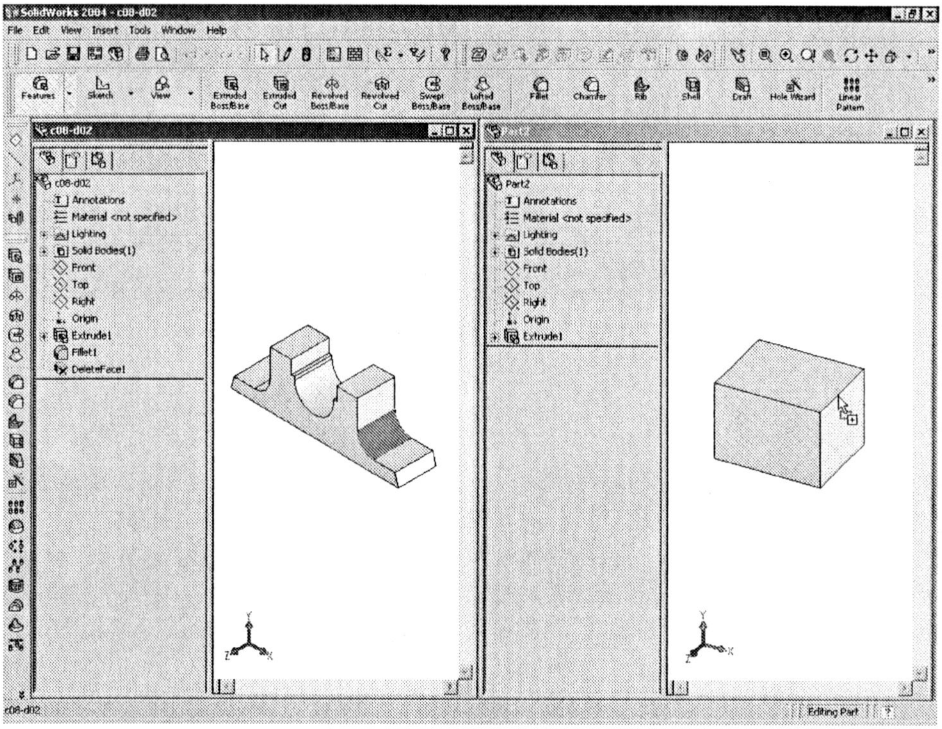

Figure 8-23 Fillet feature being dragged to be pasted in the second document

Deleting the Features

You can delete the unwanted features from the model by selecting the feature either from the **FeatureManager Design Tree** or from the drawing area. After selecting the feature to be deleted, choose the DELETE key from the keyboard or right-click to invoke the shortcut menu and choose the **Delete Feature** option. When you delete a feature the **Confirm Delete** dialog box is displayed as shown in Figure 8-24. The features that are dependent on the feature to be deleted are also displayed in the **Confirmation Delete** dialog box and it informs you that all the dependent features to this parent feature will also be deleted. You are provided with the **Also delete all child features** check box. By default this check box is cleared. If you select this check box, all the child features related to the parent feature are also deleted. When you delete a sketched feature, the sketches that are related to that feature will not be deleted. These sketches are known as absorbed features. If you need to delete the absorbed features along with the parent feature, select the **Also delete absorbed features** check box from the **Confirm Delete** dialog box. Choose the **Yes** button to delete the selected features or else choose the **No** button to cancel the delete operation. You can also delete a selected feature by choosing **Edit > Delete** from the menu bar.

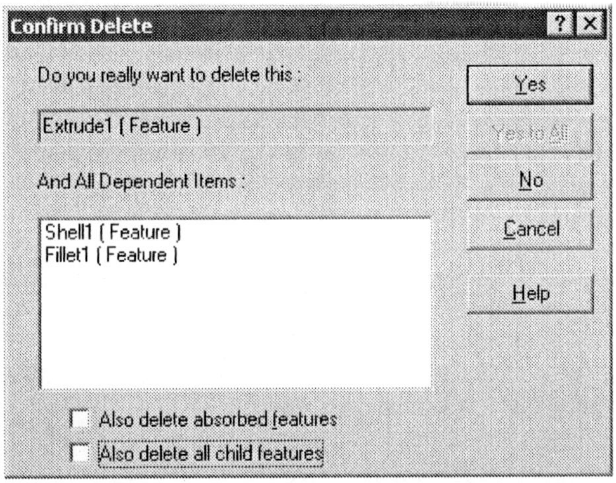

Figure 8-24 *The **Confirm Delete** dialog box*

Deleting the Bodies

CommandManager:	Features > Delete Solid/Surface	*(Customize to Add)*
Menu:	Insert > Features > Delete Body	
Toolbar:	Features > Delete Solid/Surface	*(Customize to Add)*

As discussed earlier, the multibody environment is supported in SolidWorks. Therefore, you can create multiple disjoint bodies in SolidWorks. You can also delete the unwanted bodies. The bodies to be deleted can be solid bodies or surface bodies. You will learn more about surface bodies in later chapters. To delete a body, choose **Insert > Features > Delete Body** from the menu bar. You can invoke this tool by selecting the **Delete Solid/Surface** button from the **Features CommandManager**

but first you need to customize the **CommandManager**. The **Delete Body PropertyManager** is invoked as shown in Figure 8-25. You are prompted to select the solid and/or surface bodies to be deleted.

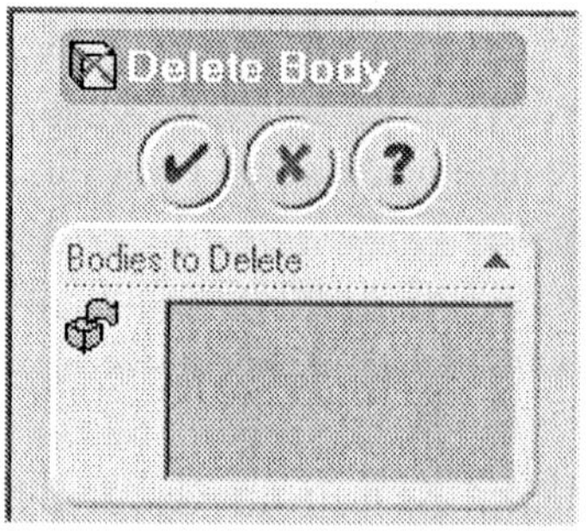

*Figure 8-25 The **Delete Body PropertyManager***

Select the body or bodies to be deleted from the drawing area or from the **Solid Bodies** folder available in the **FeatureManager Design Tree**, which is displayed in the drawing area. The selected body is displayed in green. The name of the selected body is displayed in the **Bodies to Delete** selection area. Choose the **OK** button from the **Delete Body PropertyManager**. A new item with the name **Body-Delete1** appears in the **FeatureManager Design Tree**. This item stores the deleted bodies. Therefore, at any point of your design cycle you can delete or suppress this item to resume the deleted body back in your design. You will learn more about suppressing the feature later in this chapter.

Tip. *You can also choose the **Delete Body** option from the shortcut menu. To delete a body using the shortcut menu select the body and right-click to invoke the shortcut menu. Choose the **Delete** option from **Body** area of the shortcut menu. The **Delete Body PropertyManager** will be displayed. Choose the **OK** button from the **Delete Body PropertyManager** to delete the body.*

Suppressing the Features

CommandManager:	Features > Suppress	*(Customize to Add)*
Menu:	Edit > Suppress > This Configuration	
Toolbar:	Features > Suppress	*(Customize to Add)*

Sometimes, you do not want a feature to be displayed in the model or in the drawing views of that model. Instead of deleting those features, they can be suppressed. When you suppress a feature, it is neither visible in the model nor in the drawing views. If you create an assembly using that model, the suppressed feature will not be displayed even in the assembly. You can anytime resume the feature by unsuppressing it. When you suppress a feature, the features that are dependent on the suppressed feature are also suppressed. To suppress a feature, select it from the **FeatureManager Design Tree** or from the drawing area. Choose the **Suppress** button from the **Features CommandManager** after customizing it or right-click and choose the **Suppress** option from the shortcut menu. The suppressed feature will be removed from the display of the model and the icon of the feature will be displayed in gray in the **FeatureManager Design Tree**.

When you suppress a feature, then all the features dependent on that feature are also suppressed.

Note

You can work with different configurations in a single document of SolidWorks. Therefore, you have to specify in which configuration you are suppressing a feature. This is specified when you choose the suppress option from the menu bar. You will learn more about configurations later in this book.

Unsuppressing the Features

CommandManager:	Features > Suppress	*(Customize to Add)*
Menu:	Edit > Unsuppress > This Configuration	
Toolbar:	Features > Unsuppress	*(Customize to Add)*

The suppressed features can be unsuppressed using this option. To resume the suppressed feature, select the suppressed feature from the **FeatureManager Design Tree** and choose the **Unsuppress** button from the **Features CommandManager** after customizing it or choose **Edit > Unsuppress > This Configuration** from the menu bar. You can also choose this option from the shortcut menu after selecting the suppressed feature. As discussed earlier, when you suppress a feature, the dependent features are also suppressed. But when you resume a suppressed feature, the dependent features remain suppressed. Therefore, you have to unsuppress all the features independently. The method of resuming the parent feature and the dependent features using a single option is discussed next.

Unsuppressing Features With Dependents

CommandManager:	Features > Unsuppress with Dependents *(Customize to Add)*
Menu:	Edit > Unsuppress with Dependents > This Configuration
Toolbar:	Features > Unsuppress with Dependents *(Customize to Add)*

Using this option you can resume the suppressed feature along with the dependents of the suppressed parent feature. To resume the suppressed feature using this option, select the suppressed feature from the **FeatureManager Design Tree**. Choose **Unsuppress with Dependents > This Configuration** from the **Features** toolbar. You will observe that the dependent suppressed features are also unsuppressed.

Hiding the Bodies

While working in the multibody environment, you can also hide the bodies. The hidden body is not displayed in the model and assembly, and drawing views. The bodies that are dependent on the hidden body do not hide. To hide a body, expand the **Solid Bodies** folder available in the **FeatureManager Design Tree** and select the body to hide. Right-click to invoke the shortcut menu and choose the **Hide Solid Body** option from the shortcut menu. The selected body will disappear from the drawing area. The icon of the hidden body is displayed in wireframe in the **Solid Bodies** folder. Select the hidden body from that folder and choose the **Show Solid Body** option from the shortcut menu to show the hidden body.

Move/Copy Bodies

CommandManager:	Features > Move/Copy Bodies	*(Customize to Add)*
Menu:	Insert > Features > Move/Copy Bodies	
Toolbar:	Features > Move/Copy Bodies	*(Customize to Add)*

In SolidWorks you can also move or copy the bodies. To move or copy the bodies choose the **Move/Copy Bodies** button from the **Features CommandManager** after customizing it or choose **Insert > Features > Move/Copy Bodies** from the menu bar. The **Move/Copy Body PropertyManager** will be displayed as shown in Figure 8-26. The confirmation corner is also displayed in the drawing area. You will observe a filled square at the document origin. The various options available in the **Move/Copy Body PropertyManager** are discussed next.

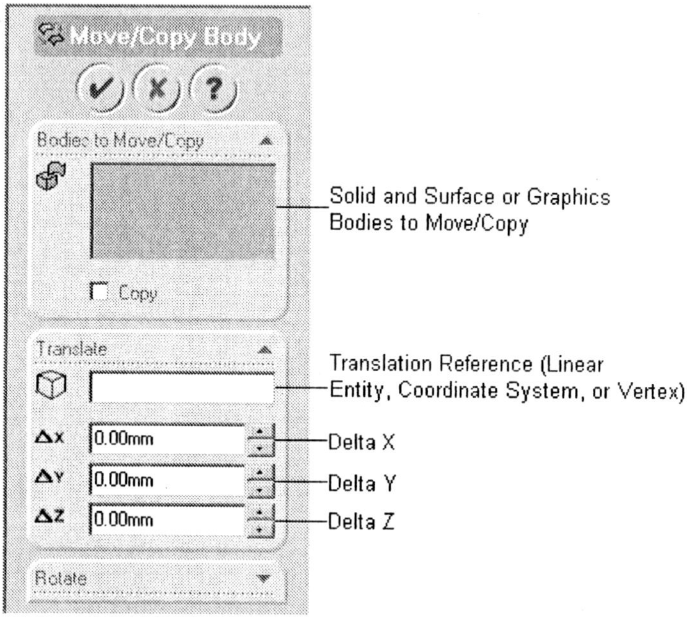

***Figure 8-26** The Move/Copy Body PropertyManager*

Bodies to Move/Copy

The **Bodies to Move/Copy** rollout is used to define the body to copy or move. When you invoke the **Move/Copy Body PropertyManager**, you are prompted to select the bodies to move/copy and set the options. Move the cursor to the body to be selected. The cursor will be replaced by the body selection cursor and the edges of the body are highlighted in red. The name of the body is also displayed in the tooltip. Use the left mouse button to select the body. The selected body is highlighted in green. The name of body is displayed in the **Solid and Surface or Graphics Bodies to Move/Copy** selection area. You can also select the body from the **Solid Bodies** folder after invoking the **FeatureManager Design Tree** flyout.

Copy

Since the **Copy** check box is cleared by default, if you select the **Copy** check box, you can create the copies of the selected body. When you select this check box the **Number of Copies** spinner is displayed below the **Copy** check box. Set the number of copies using this spinner.

Translate

The **Translate** rollout available in the **Move/Copy Body PropertyManager** is used to define the translational parameters to move the selected body. Set the value of the destination in the **Delta X**, **Delta Y**, and **Delta Z** spinners. When you set the value, the preview of moved body is displayed in temporary graphics in the drawing area. You can also move or copy the selected body between points. To move or copy a body by specifying two points, select the **Translation Reference (Linear Entity, Coordinate System, Vertex)** selection area. The selection mode in this area is active. Select the vertex from which you want the translation to start. When you select the first vertex as translation reference, the **Delta X**, **Delta Y**, and **Delta Z** spinners are replaced by the **To Vertex** selection area. The selection mode in the **To Vertex** selection area is active. Select the second translation reference. You will observe the preview of the body move between two selected vertices. The placement of the body also depends on the sequence of selection of vertices. Therefore, you have to be very careful while selecting the two vertices. Figure 8-27 shows the sequence for the selection of references. Figure 8-28 shows the resultant copied body.

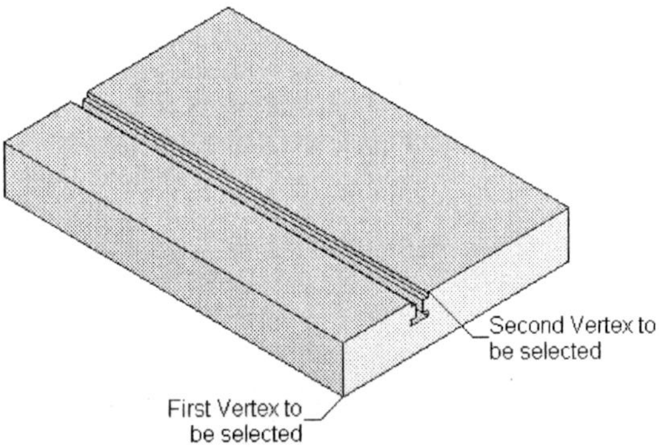

Figure 8-27 *Sequence of selection*

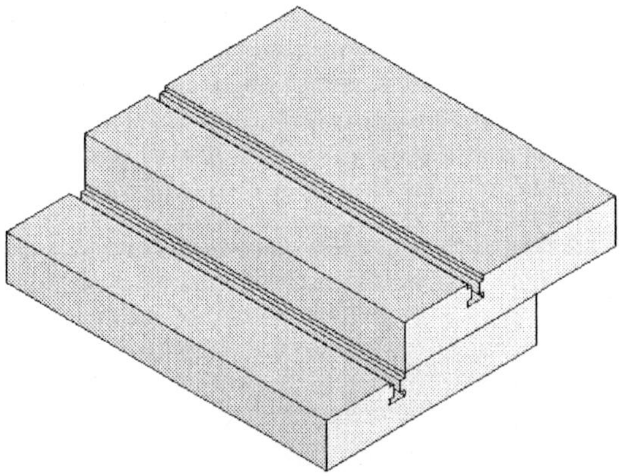

Figure 8-28 *Resultant copied body*

Rotate

The **Rotate** spinner available in the **Move/Copy Body PropertyManager** is used to define the parameters to rotate the body. To open this rollout, click once on the black arrow provided at the right of this rollout. The **Rotate** rollout is shown in Figure 8-29.

As discussed earlier, a filled square is placed at the origin when you invoke the **Move/Copy PropertyManager**. This filled square is clearly visible if you hide the origin by choosing **View > Origin** from the menu bar. It indicates the origin along which the selected body will be rotated. You can adjust the position of this temporary moveable origin using the **X Rotation Origin** spinner, **Y Rotation Origin** spinner, and **Z Rotation Origin** spinner. The **X Rotation Angle** spinner is used to set the value of the angular increment to rotate or copy the body

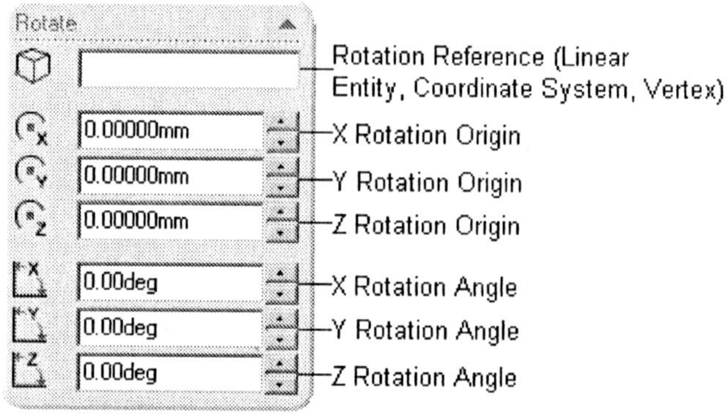

Figure 8-29 *The **Rotate** rollout*

along the X axis. The **Y Rotation Angle** spinner is used to rotate or copy the body along the Y axis. The **Z Rotation Angle** spinner is used to rotate or copy the body along the Z axis.

To rotate or copy the selected body along an edge, click once in the **Rotation Reference (Linear Entity, Coordinate System, Vertex)** selection area to invoke the selection. Select the edge along which you want to rotate the selected body. When you select an edge, all the other spinners disappear from the rollout and the **Angle** spinner is invoked in the **Rotate** rollout. Set the value of the angular increment in this spinner.

Instead of selecting an edge, you can also select a vertex along which the body will rotate or copy. Then you have to specify the axis along which you want to rotate the body.

Reordering the Features

Reordering the features is defined as a process of changing the position of the features created in the model. Sometimes, after creating a model it may be required to change the order in which the features of the model were created. For reordering the features, the features are dragged and placed before or after another feature in the **FeatureManager Design Tree**.

For reordering the features, select the feature in the **FeatureManager Design Tree** and drag the feature to the required position. When you drag the feature to reorder, the bend arrow pointer is displayed, which suggests that feature dragging is possible. If you drag the child feature above the parent feature, the reorder error pointer will be displayed. If you drag a child feature above a parent feature, the **SolidWorks** warning box will be displayed as shown in Figure 8-30. Choose **OK** from this dialog box.

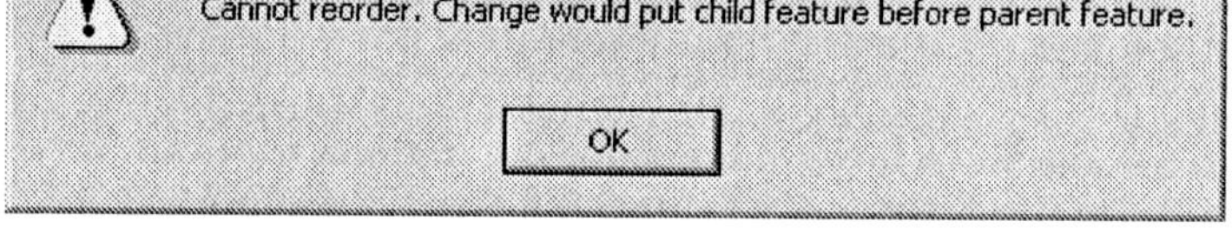

Figure 8-30 *The **SolidWorks** warning box*

Consider a case in which you have created a rectangular block and a pattern of through holes created on the base feature, as shown in Figure 8-31. Now, if you create a shell feature and remove the top face, the front face, and the right face of the model, the model will appear as shown in Figure 8-32.

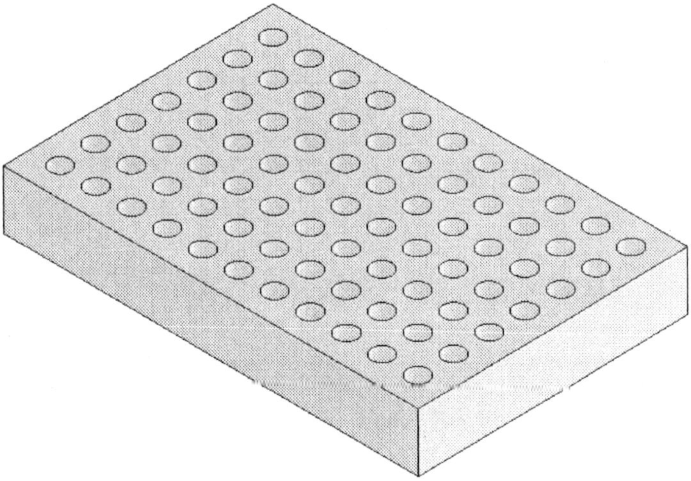

Figure 8-31 *Model created creating a pattern of the holes on the base feature*

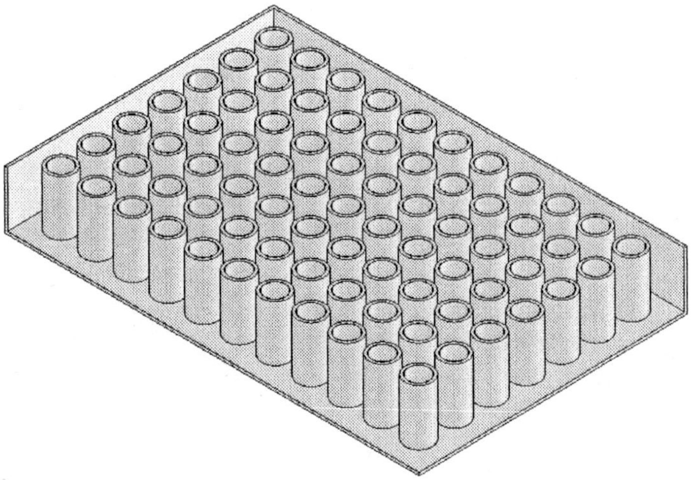

Figure 8-32 *Shell feature added to the model*

But this was not the desired result. This is the reason you need to reorder the shell feature before the holes. Select the shell feature in the **FeatureManager Design Tree** and drag it above the holes. All the features will be automatically adjusted in the new order as shown in Figure 8-33.

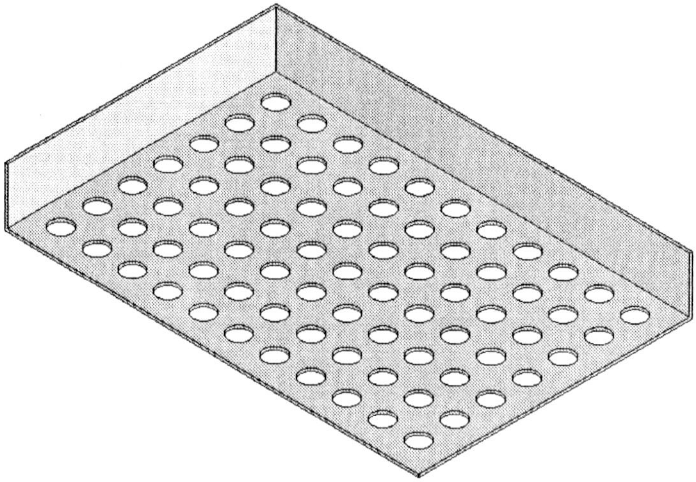

Figure 8-33 Model after reordering the features.

Rolling Back the Model

Rolling back the model is defined as a process in which you rollback the model to an earlier stage. When you rollback a feature or features, those features are suppressed and you can add new features when the model is in the rollback state. The newly added features are added before the features that are rolled back. While working with a multifeatured model you want to edit a feature that was created at the starting of the design cycle of the model. It is recommended that you should rollback the model up to that feature. This is because after each editing operation the time of regeneration will be minimized. Rolling back is done using the **Rollback Bar** available in the **FeatureManager Design Tree** as shown in Figure 8-34.

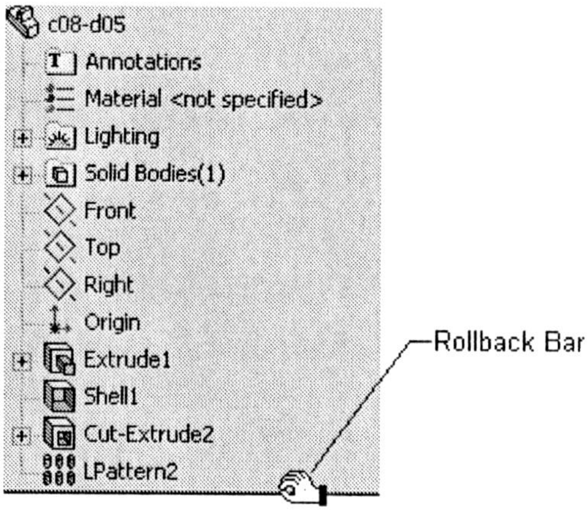

Figure 8-34** The rollback bar of the* ***FeatureManager Design Tree

Using the select tool select the **Rollback Bar**; after selection it will be changed to blue color and the select cursor will be replaced by the hand pointer. Drag the hand pointer to the feature up to which you want to rollback the model and release the pointer. To resume or roll the model, drag the **Rollback Bar** to the last feature of the model. You can also rollback using the menu bar. Select the feature up to which you want to rollback the model and choose **Edit > Rollback** from the menu bar.

Tip. *If you want to rollback the model to the previous step choose* ***Edit > Roll to Previous*** *from the menu bar. If you want to rollback the entire model to its original position, choose* ***Edit > Roll to End*** *from the menu bar.*

You can also choose the ***Roll Forward***, ***Roll Previous***, ***Roll End*** *options from the shortcut menus invoked by selecting the features placed below the* ***Rollback Bar***. *These options are used to control the roll and rollback of the model.*

You can also rollback the model using the keyboard, to rollback using the keyboard select the ***Rollback Bar*** *and use the CTRL+ALT+Up Arrow key to roll forward. Use CTRL+ALT+Down Arrow key to roll backward.*

Renaming the Features

The name of the features is displayed in the **FeatureManager Design Tree**. By default the naming of the features is done according to the sequence in which the features are created. You can also rename the features according to your convenience. To rename a feature, select the feature from the **FeatureManager Design Tree** and click once on the selected feature. An edit box will be displayed in the **FeatureManager Design Tree**. Enter the name of the feature and press the ENTER key or click anywhere on the screen.

Creating Folders in the FeatureManager Design Tree

You can also add the folders in the **FeatureManager Design Tree** and the features displayed in the **FeatureManager Design Tree** are added in the folder. This is done to reduce the length of the **FeatureManager**. Consider a case in which the base of the model consists of more than one feature. You can add a folder named Base Feature and add all the features used to create the base in that folder. To add a folder in the **FeatureManager Design Tree** select any feature and right-click to invoke the shortcut menu. Choose the **Create New Folder** option from the shortcut menu. A new folder is created above the selected feature. Specify the name of the folder and click anywhere on the screen. Now, you can drag and drop the features to the newly created folder. You can also rename the folder by selecting it and then clicking it once. Now, enter the name in the edit box and press the ENTER key.

To add the selected feature in a new folder, choose **Add to New Folder** from the shortcut menu. A new folder will be created in the **FeatureManager Design Tree** and the selected feature will be added to the newly created folder. You can also delete the folder by selecting the folder and invoking the shortcut menu. Select the **Delete** option from the shortcut menu. Using the options in this shortcut menu you can also rollback and suppress the features available in the selected folder.

What's Wrong? Functionality

When you modify a sketch or a feature and rebuild the model, sometimes the model does not rebuild properly because of the geometric errors resulting from the modification. Therefore, you are provided with a **Rebuild Errors** dialog box as shown in Figure 8-35. The possible errors in the feature are displayed in this dialog box.

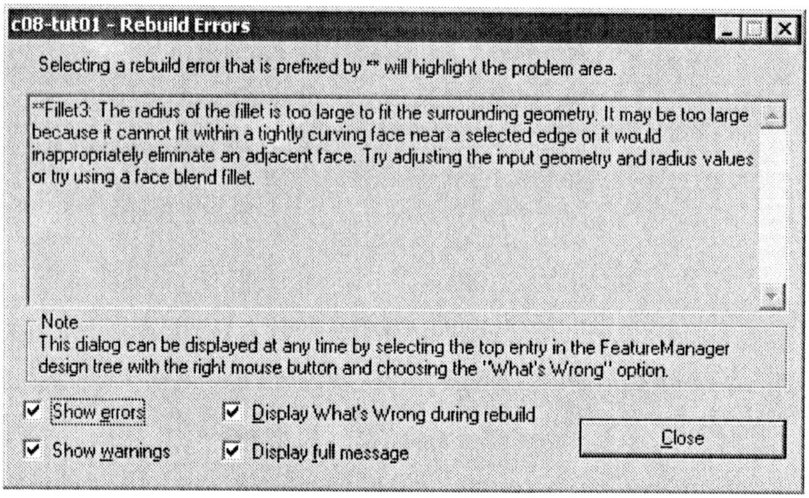

***Figure 8-35** The **Rebuild Errors** dialog box*

The **Show errors** check box is selected by default to display the errors in the selection area provided in the **Rebuild Errors** dialog box. The **Show warnings** check box is selected by default to display the warnings messages. The **Display What's Wrong during rebuild** check box is selected by default and is used to display the errors at every rebuild of the model unless the error is fixed. The **Display full message** check box is used to display the complete error message in the **Rebuild Errors** dialog box. After reading the error from this dialog box, choose the **Close** button to exit the dialog box. The errors are also displayed in the **FeatureManager design Tree**. The **FeatureManager Design Tree** with an error in a feature is displayed in Figure 8-36.

If their is an error in the model or in an assembly then symbol appears on the left of the name of the model or assembly in the **FeatureManager Design Tree**. If a feature is having an error then symbol appears along on the left of the feature in the **FeatureManager Design Tree**. If there is an error in the child feature then symbol appears at the left of the parent feature and also on the name of the document in the **FeatureManager Design Tree**. If some warning message appears for a feature then symbol appears on the left of that feature in the **FeatureManager Design Tree**.

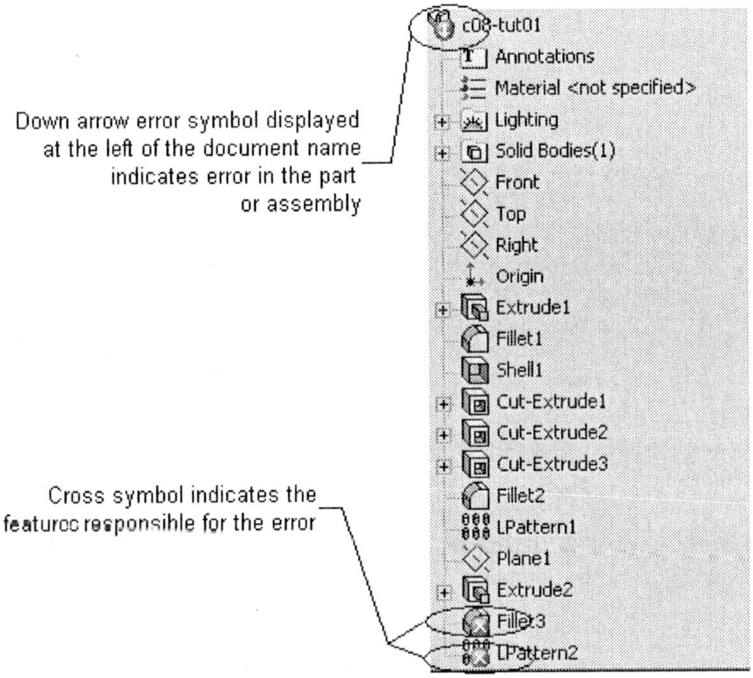

***Figure 8-36** The **FeatureManager Design Tree** with a feature having errors*

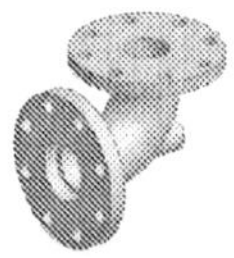

Tip. *You can also invoke the **Rebuild Errors** dialog box by selecting the feature having errors from the **FeatureManager Design Tree** and choose the **What's Wrong?** option from the shortcut menu. To view all the errors in a model, select the name of the document from the top of the **FeatureManager Design Tree** and invoke the **Rebuild Errors** dialog box. All the errors in the model are displayed in the **Rebuild Errors** dialog box.*

*While creating the features, if there are some geometrical constraints in creating the feature, the **Rebuild Errors** dialog box is displayed.*

TUTORIALS

Tutorial 1

In this tutorial you will create the model shown in Figure 8-37. After creating some of the features of the model you will dynamically modify the model and then undo the modification. The dimensions of the model are shown in Figure 8-38. **(Expected time: 30 min)**

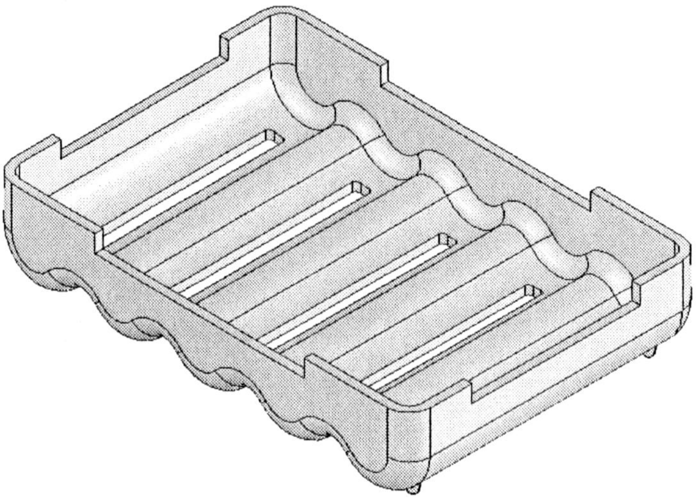

Figure 8-37 *Model for Tutorial 1*

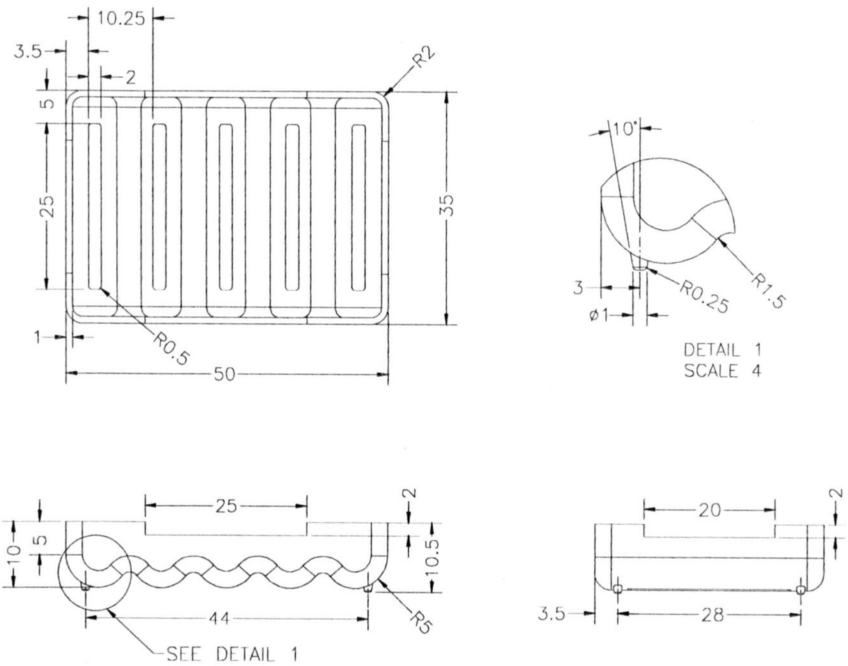

Figure 8-38 *Views and dimensions of the model for Tutorial 1*

The steps to be followed to create the model are listed next.

a. Create the base feature of the model by extruding the profile to a given distance, refer to Figures 8-39 and 8-40.

b. Add the fillets to the base feature, refer to Figures 8-41 and 8-42.

c. Add the shell feature to the model and remove the top face of the base feature, refer to Figures 8-43 and 8-44.
d. Dynamically modify the model, refer to Figures 8-45 through 8-47.
e. Create the cuts on the sides of the model, refer to Figure 8-48.
f. Create the slots on the lower part of the base and add the fillet to the slots feature, refer to Figure 8-48.
g. Pattern the slots and the fillet feature, refer to Figure 8-48.
h. Create a plane at an offset distance from the **Top** plane.
i. Create the standoffs using the extrude and **Fillet** tool and pattern the standoffs, refer to Figure 8-49.

Creating the Base Feature

You will create the sketch of the base feature on the **Front** plane and extrude the sketch using the **Mid Plane** option.

1. Start SolidWorks and start a new SolidWorks Part document using the **New SolidWorks Document** dialog box.

2. Invoke the **Extruded Boss/Base** tool and draw the sketch of the base feature on the **Front** plane. Add the required relations and dimensions to the sketch as shown in Figure 8-39.

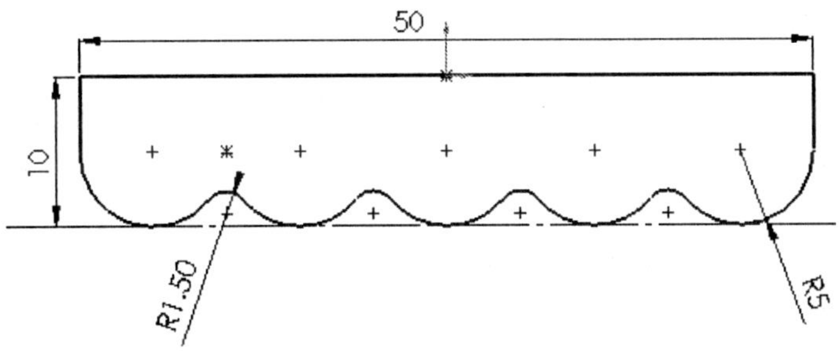

Figure 8-39 *Sketch for the base feature*

3. Extrude the sketch to a distance of 35 mm using the **Mid Plane** option.

   The base feature of the model is shown in Figure 8-40.

Add fillet to the Base Feature

After creating the base feature you will fillet the lower edges of the base feature.

1. Invoke the **Fillet PropertyManager**, rotate the model, and select the edges of the base feature as shown in Figure 8-41.

2. Set the value of the **Radius** spinner to **2.5** and choose the **OK** button from the **Fillet PropertyManager**.

   The model after adding fillet to the edges of the model is shown in Figure 8-42.

Figure 8-40 Base feature of the model

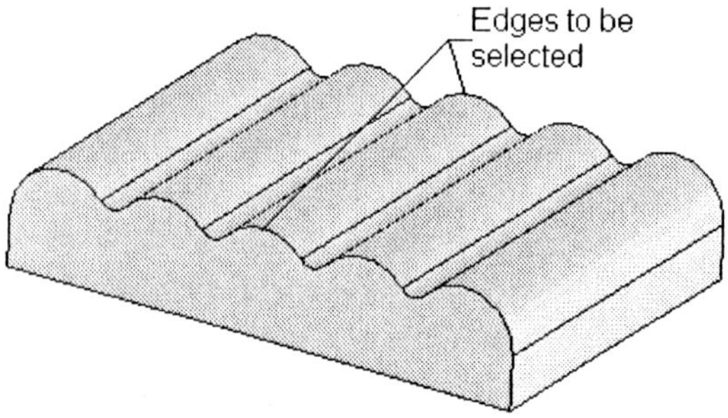

Figure 8-41 Edges to be selected

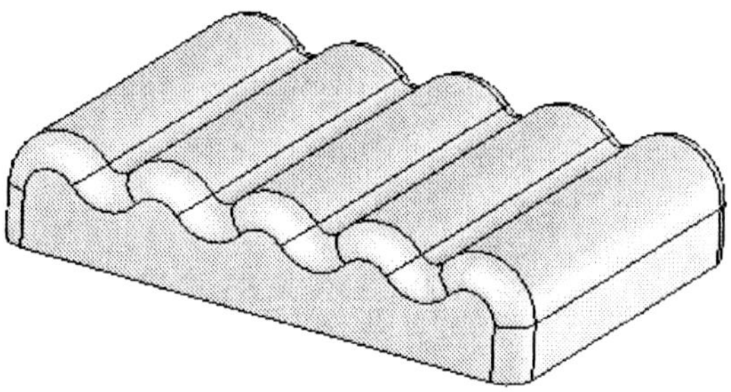

Figure 8-42 Fillet added to the model

Adding Shell to the Model

After creating the fillet feature, you need to shell the model using the **Shell** tool.

1. Orient the model in the isometric view and invoke the **Shell1 PropertyManager**.
2. Select the top planar face of the model as shown in Figure 8-43.
3. Set the value of the **Thickness** spinner to **1** and choose the **OK** button from the **Shell PropertyManager**.

The model after adding the shell feature is shown in Figure 8-44.

Dynamically Editing the Features

After creating and adding shell to the base of the model you will learn how to edit the features dynamically using the **Move/Size Features** tool.

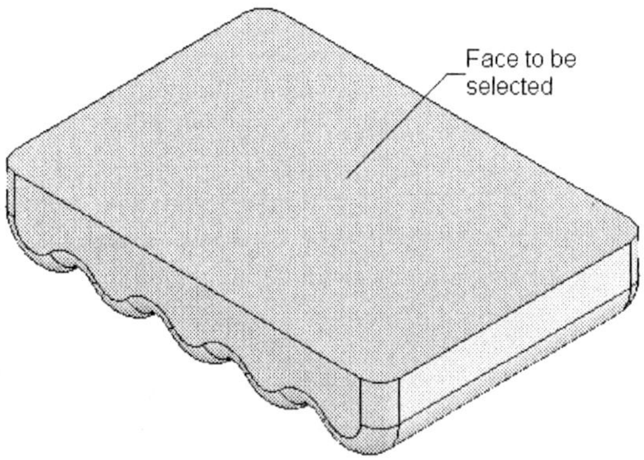

Figure 8-43 Face to be selected

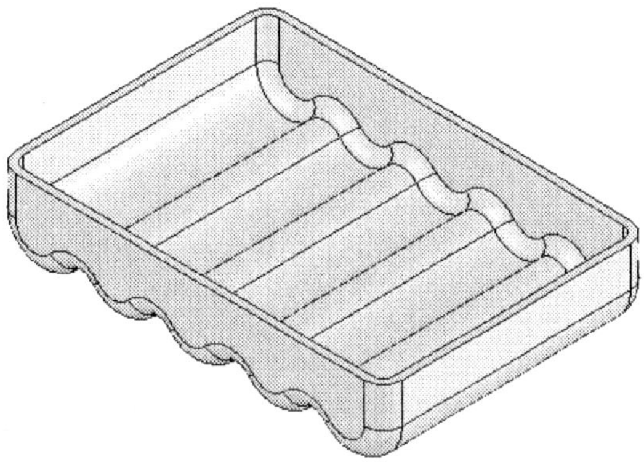

Figure 8-44 Shell feature added to the model

1. Choose the **Move/Size Features** button from the **Features CommandManager** after customizing it to invoke the dynamic dragging tool.

2. Select the right planar face of the base feature from the drawing area. The selected face will be highlighted in green. The sketch of the selected feature is also displayed in the drawing area, along with various editing handles, as shown in Figure 8-45.

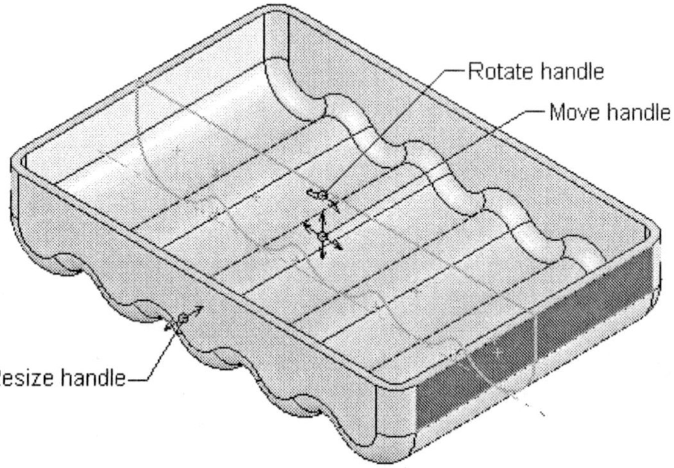

Figure 8-45 Editing handles for editing the base feature

Tip. *You can also select the feature to be edited from the* ***FeatureManager Design Tree****.*

If you want to edit the sketches using the ***Move/Size Features*** *tool, then you need to select the* ***Override Dimension*** *option by choosing* ***Tools > Sketch Settings > Override Dims on Drag/Move*** *from the menu bar. If a check mark is displayed at the left of this option in the menu bar, then the option is already selected.*

3. Select the **Resize** handle from the drawing area and drag the cursor to resize the feature.

   The preview of the resized feature and the dimension of the feature are displayed in the drawing area. As you drag the cursor the preview and the dimension updates automatically.

4. Release the left mouse button after dragging the feature to some distance. Figure 8-46 shows the preview of the dragged feature. Figure 8-47 shows the edited feature.

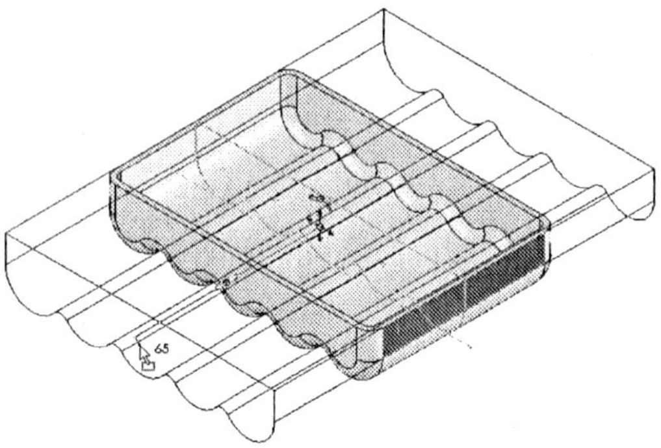

Figure 8-46 *Dragging the* ***Resize*** *handle*

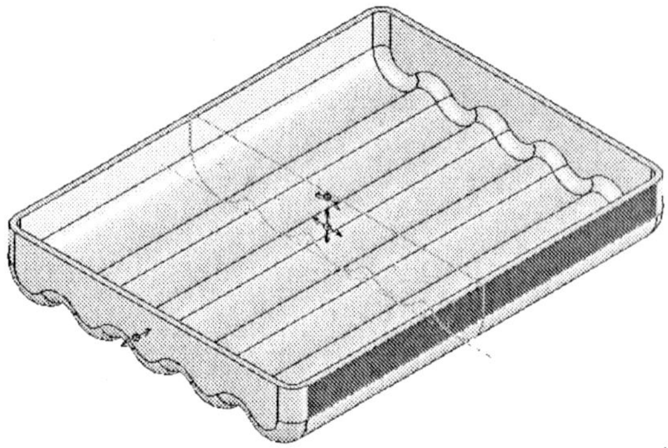

Figure 8-47 *Resultant edited feature*

5. Select the **Move/Size Features** button again to disable the dynamic editing tool and click anywhere in the drawing area to clear all the selections from the selection set.

   You have edited the model by dynamically dragging, but the original depth value of the feature was 35 mm. To bring the base feature back to its original size you need to edit the feature.

6. Double-click the base feature in the **FeatureManager Design Tree** or from the drawing area. All the dimensions of the feature are displayed in the drawing area.

7. Double-click the dimension that reflects the depth of the base feature which appears in blue.

8. The **Modify** dialog box will be displayed. Set the value of the **Dimension** spinner to **35** and press the ENTER key from the keyboard.

9. Choose the **Rebuild** button from the **Standard** toolbar or choose CTRL+B from the keyboard to rebuild the model.

   Using the **Cut-Extrude, Fillet,** and **Linear Pattern** tools create the remaining features of the model. The model after creating the features using these tools is displayed in Figure 8-48.

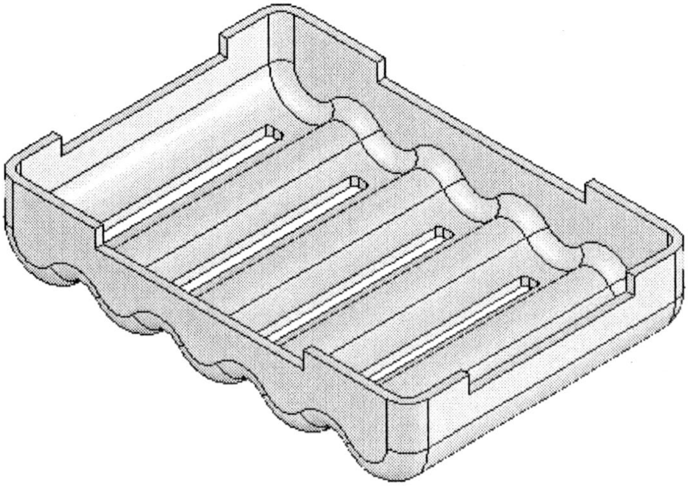

Figure 8-48 Model after creating other features

Creating the Standoffs

Now, you need to create the standoffs for the model. It is created by extruding a sketch. The sketch is created on a sketch plane at an offset distance from the **Top** plane. You also need to specify a draft angle while creating this feature.

1. Create a reference plane at an offset distance of 10.5 mm from the **Top** plane. You need to select the **Reverse direction** check box from the **Plane PropertyManager**.

2. Select the newly created sketching plane and create the sketch of the standoff and apply the required relations and dimensions. The sketch consists of a circle of diameter 1 mm. For more dimensions refer to Figure 8-38.

3. Extrude the sketch using the **Up To Next** option with an outward draft angle of 10-degree. Hide the newly created plane.

   This creates the standoffs of the model.

4. Rotate the model and add a fillet of radius 0.25 to the base of the standoff.

   The rotated and zoom view of complete standoff is displayed in Figure 8-49.

5. Pattern the filleted standoff feature using the **Linear Pattern** tool. The isometric view of the final model created is shown in Figure 8-50.

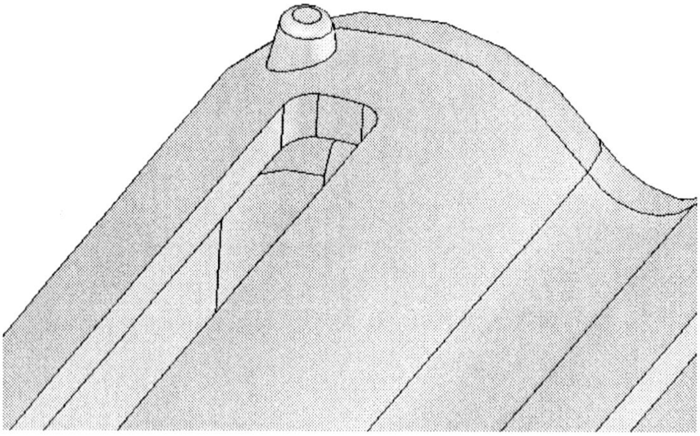

Figure 8-49 *Rotated and zoom view of the model to show the standoff*

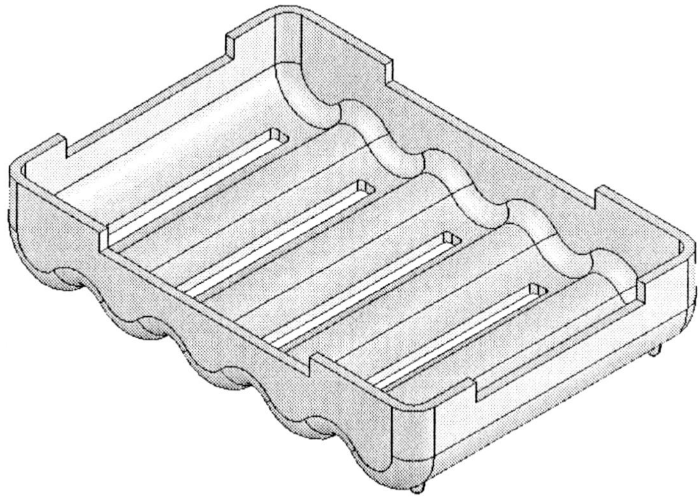

Figure 8-50 *Final model*

The **FeatureManager Design Tree** of the model is shown in Figure 8-51.

Saving the Model

1. Choose the **Save** button from the **Standard** toolbar and save the model with the name given below.

   \My Documents\SolidWorks\c08\c08-tut01.sldprt

2. Choose **File > Close** from the menu bar to close the document.

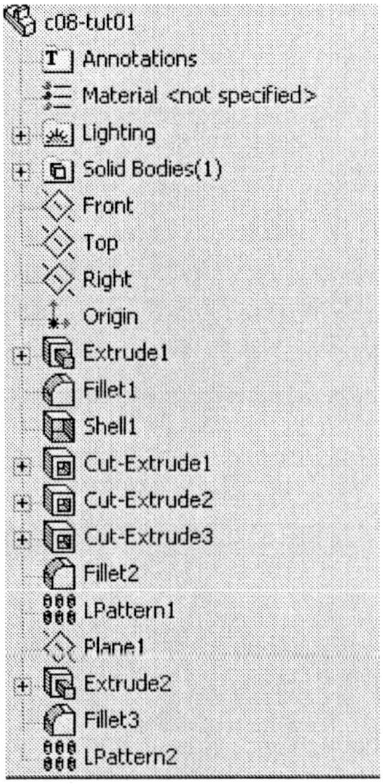

Figure 8-51 *The* ***FeatureManager Design Tree*** *of the model*

Tutorial 2

In this tutorial you will create the model shown in Figure 8-52. After creating the model you will edit the model with the **Move/Size Features** option selected. The views and dimensions of the model are shown in Figure 8-53. **(Expected time: 45 min)**

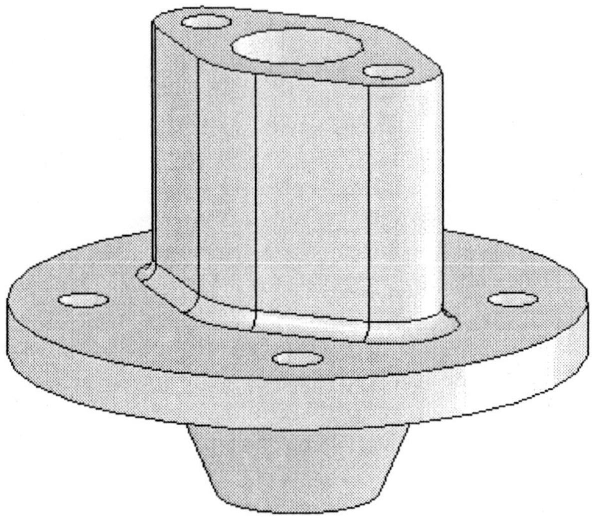

Figure 8-52 *The dimetric view of the model for Tutorial 2*

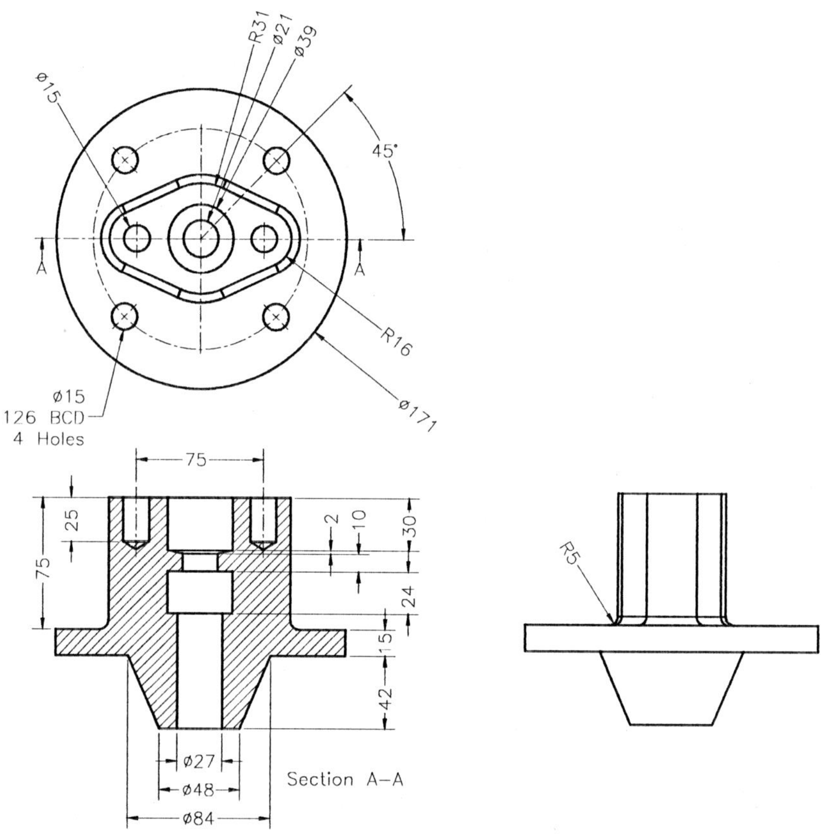

Figure 8-53 *Views and dimensions of the model for Tutorial 2*

The steps to be followed to complete this tutorial are listed next.

a. Create the base feature of the model by revolving the sketch along the central axis of the sketch, refer to Figures 8-54 and 8-55.
b. Create the sketch of the second feature on the top face of the base feature and extrude it to a given dimension, refer to Figures 8-56 and 8-57.
c. Create the revolve cut feature, refer to Figures 8-58 and 8-59.
d. Create the holes using the **Simple Hole** tool and pattern the hole using the **Circular Pattern** tool, refer to Figure 8-59.
e. Create a drilled hole feature using the **Hole Wizard** tool, refer to Figure 8-59.
f. Mirror the hole feature along the **Right** plane, refer to Figure 8-59.
g. Create the fillet feature, refer to Figure 8-59.
h. Edit the model using the **Move/Size Features** tool, refer to Figures 8-60 through 8-62.

Creating the Base Feature

First, you will create the base feature of the model by revolving the sketch created on the **Front** plane.

1. Start a new SolidWorks Part document using the **New SolidWorks Document** dialog box.

2. Invoke the **Revolved Boss/Base** tool and draw the sketch of the base feature on the **Front** plane, Add the required relations and dimensions to the sketch as shown in Figure 8-54.

3. Exit the sketching environment.

   Because the default value in the **Angle** spinner is 360-degree, as required, you do not need to set any parameters in the **Revolve PropertyManager**.

4. Choose the **OK** button from the **Revolve PropertyManager**.

   The base feature created after revolving the sketch is shown in Figure 8-55.

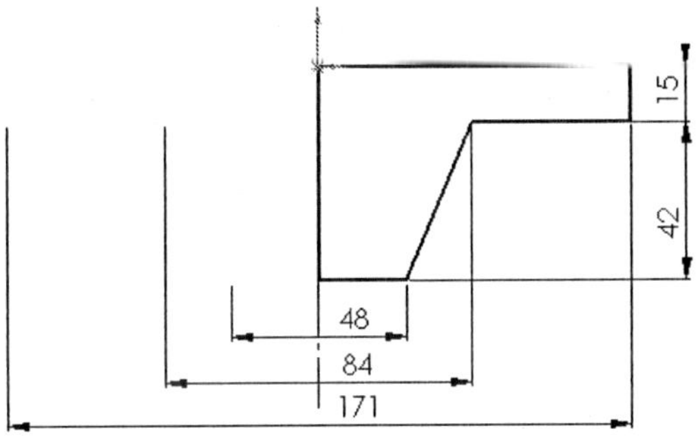

Figure 8-54 *Sketch for the base feature*

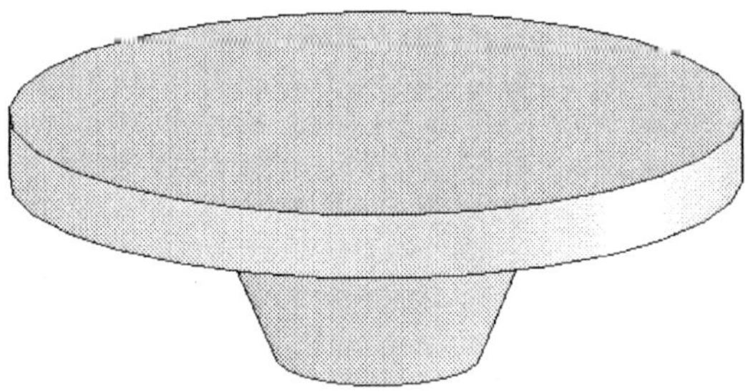

Figure 8-55 *Dimetric view of the base feature*

Creating the Second Feature

The second feature of this model is an extruded feature. This feature is created by extruding a sketch created on the top planar face of the base feature.

1. Invoke the **Extruded Boss/Base** tool and select the top planar face of the base feature as the sketching plane.

2. Create the sketch of the second feature and apply the required relations and dimensions to the sketch as shown in Figure 8-56. Make sure the sketch is symmetrical about the centerline. If it is not, the sketch will not be properly mirrored.

3. Extrude the sketch to a distance of 75 mm.

   The isometric view of the model after creating the second feature is displayed in Figure 8-57.

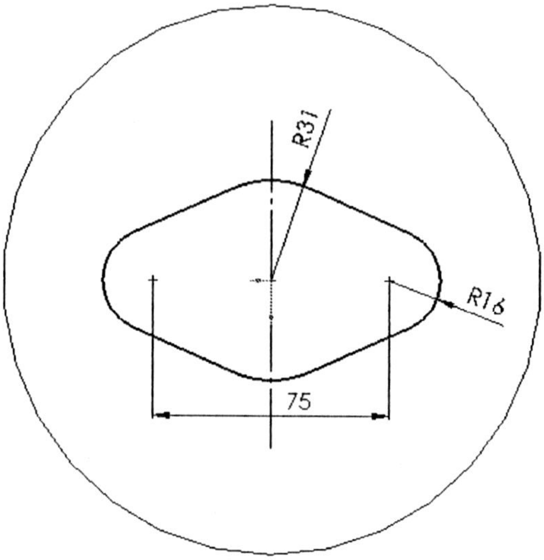

***Figure 8-56** Sketch for the second feature*

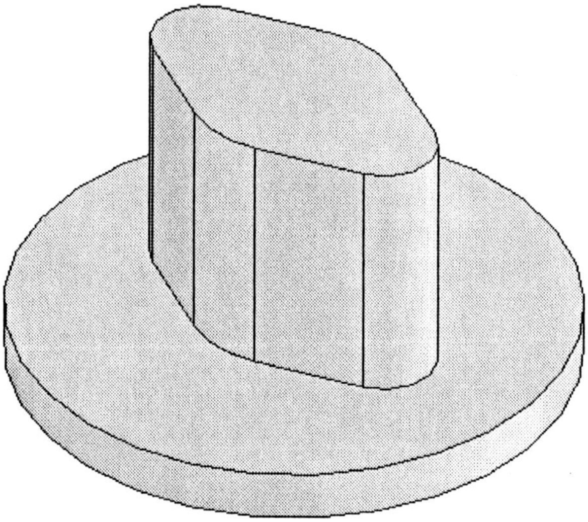

***Figure 8-57** Second feature added to the model*

Creating the Third Feature

The third feature of the model is created by revolving a sketch using the cut option. The sketch for this feature will be created on the **Front** plane.

1. Invoke the **Revolved Cut** tool and select the **Front** plane as the sketching plane.

2. Create the sketch for the revolved cut feature and apply the required relations and dimensions to the sketch as shown in Figure 8-58.

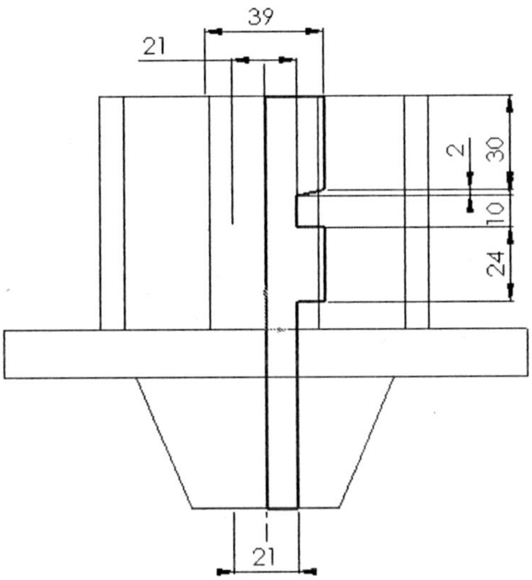

Figure 8-58 Sketch for the revolve cut feature

3. Exit the sketching environment and create a revolved cut feature with a default angle value of 360-degree.

Creating the Remaining Features

1. Create the other features of the model by referring to Figure 8-53 and using the **Simple Hole**, **Hole Wizard**, and **Mirror** tools.

   The isometric view of the model after creating all the other features is displayed in Figure 8-59.

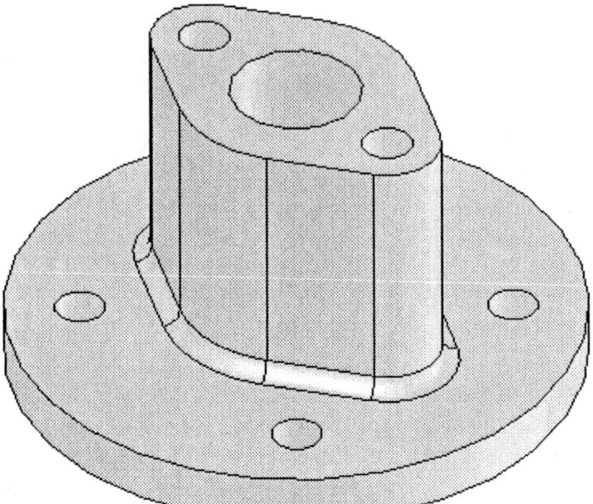

Figure 8-59 Isometric view of the final model

Editing the Sketch of the Model With the Move/Size Features tool

After creating the model, you will edit the sketch of the second feature with the **Move/Size Features** tool. Therefore, before proceeding further you have to invoke this option.

1. Choose the **Move/Size Features** button from the **Features CommandManager** to invoke the **Move/Size Features** tool.

2. Select the second feature from the drawing area and right-click to invoke the shortcut menu.

3. Choose the **Edit Sketch** option from the shortcut menu.

   The sketch of the second feature with the preview of the feature in temporary graphics is displayed in the drawing area as shown in Figure 8-60.

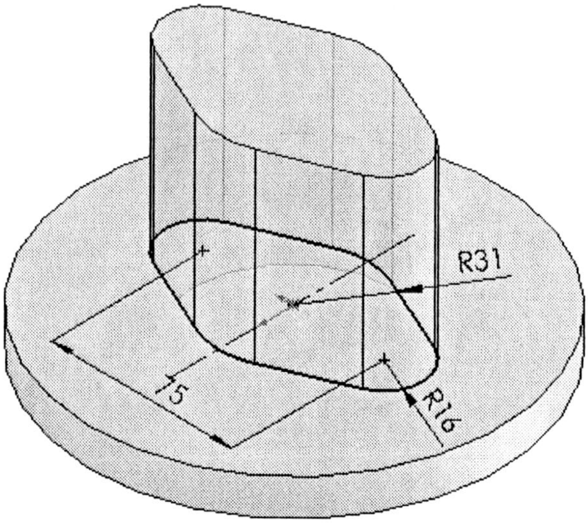

Figure 8-60 *Sketch and the preview of the second feature*

Now, you will modify the sketch by dragging the sketch entities. You will observe that the preview of the second feature that is displayed in temporary graphics is also modified.

4. Choose **Tools** > **Sketch Settings** > **Override Dims on Drag/Move** option from the menu bar, if it is not selected.

5. Select the center point of the right arc of sketch and drag the cursor toward right. As you drag the cursor, the preview of the feature is also modified dynamically. Release the left mouse button. The new dimensions appears on the sketch. Continue this process until the value of the distance between the center of the right and left arcs shows a value close to 135.

   Figure 8-61 shows the sketch being dragged. Figure 8-62 shows the final dragged sketch.

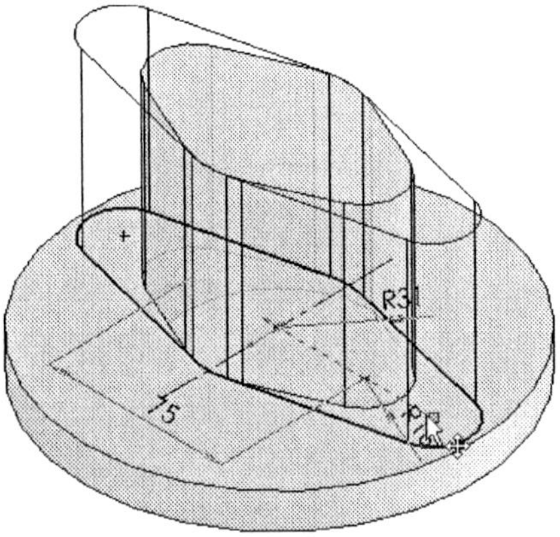

Figure 8-61 Dragging the centerpoint of right arc

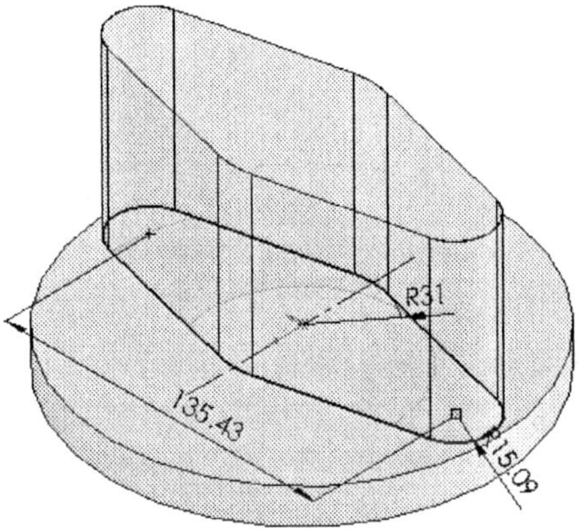

Figure 8-62 Sketch after dragging

After modifying the sketch by dragging, you need to change the modified dimensions back to original dimensions.

5. Double-click the linear dimension value between the center points of the right and left arcs and enter **75** in the **Modify** edit box and press the ENTER key.

   Similarly, edit the other two dimensions using the **Modify** edit box if necessary.

6. Use CTRL+B from the keyboard to rebuild the model.

   The **FeatureManager Design Tree** of the model is shown in Figure 8-63.

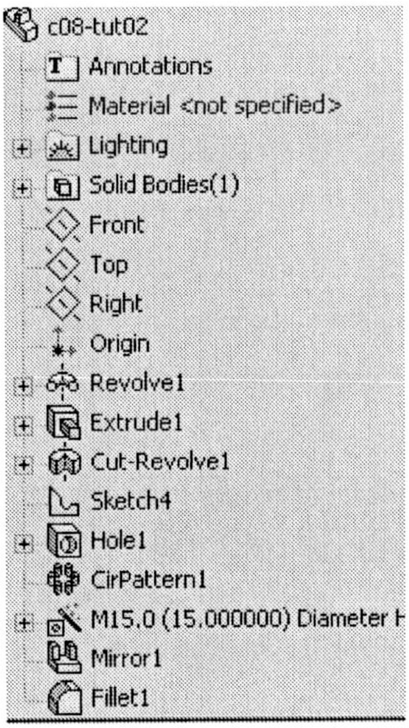

Figure 8-60 The ***FeatureManager Design Tree***

Saving the Model

Now, you need to save the model.

1. Choose the **Save** button from the **Standard** toolbar and save the model with the name given below.

\My Documents\SolidWorks\c08\c08-tut02.SLDPRT.

2. Choose **File > Close** from the menu bar to close the document.

Tutorial 3

In this tutorial you will create the model shown in Figure 8-64. While creating the model you will also perform some editing operations to the model. The views and dimensions of the model are displayed in Figure 8-65. **(Expected time: 45min)**

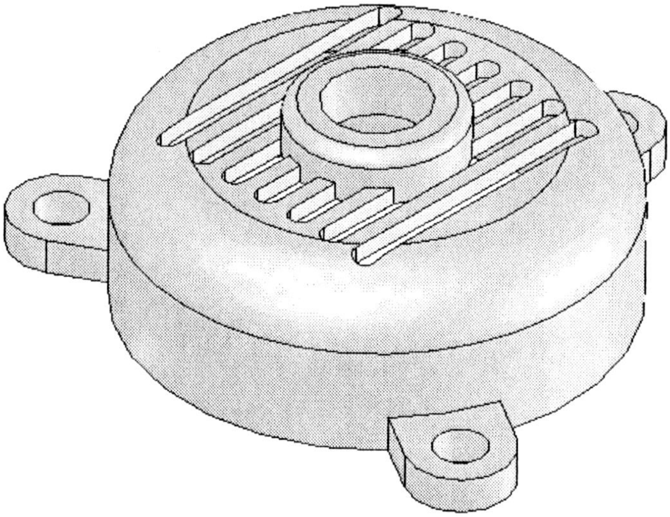

Figure 8-64 *Model for Tutorial 3*

The steps to be followed to complete this tutorial are listed next.

a. Create a base feature of the model by revolving the sketch created on the **Front** plane, refer to Figures 8-66 and 8-67.
b. Shell the model using the **Shell** tool, refer to Figure 8-68
c. Create the sketch on the **Top** plane and extrude it to the given distance, refer to Figure 8-69.
d. Pattern the extrude feature using the **Circular Pattern** tool, refer to Figure 8-70.
e. Edit the circular pattern, refer to Figure 8-71.
f. Create a cut feature on the top planar face of the base feature, refer to Figure 8-72.
g. Pattern the cut feature using the **Linear Pattern** tool, refer to Figure 8-73.
h. Unsuppress the suppressed features and create the remaining features of the model, refer to Figures 8-74 and 8-75.

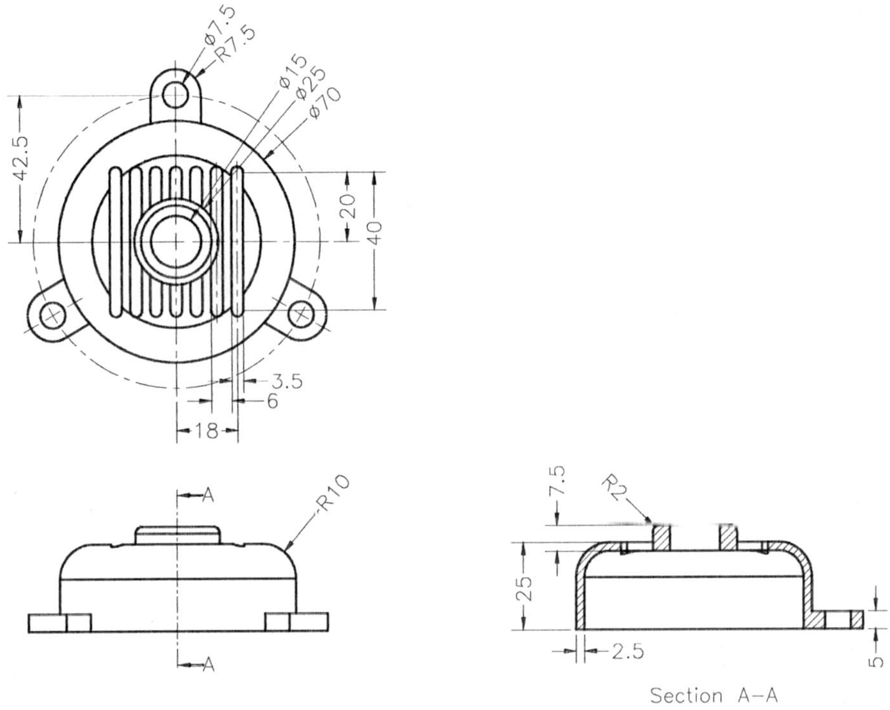

Figure 8-65 Views and dimensions of the model for Tutorial 3

Creating the Base Feature

First, you need to create the base feature of the model by revolving the sketch created on the **Front** plane.

1. Start a new SolidWorks Part document using the **New SolidWorks Document** dialog box.

2. Invoke the **Revolved Boss/Base** tool and draw the sketch of the base feature on the **Front** plane. Add the required relations and dimensions to the sketch as shown in Figure 8-66.

3. Exit the sketching environment and create the base feature of the model. The base feature of the model is shown in Figure 8-67.

Shelling the Base Feature

After creating the base feature, you need to shell the model using the **Shell** tool. You will also remove the bottom face of the base feature, leaving behind a thin walled model.

1. Invoke the **Shell1 PropertyManager** and set the value of the **Thickness** spinner to **2.5**.

2. Rotate the model and select the bottom face of the model to remove.

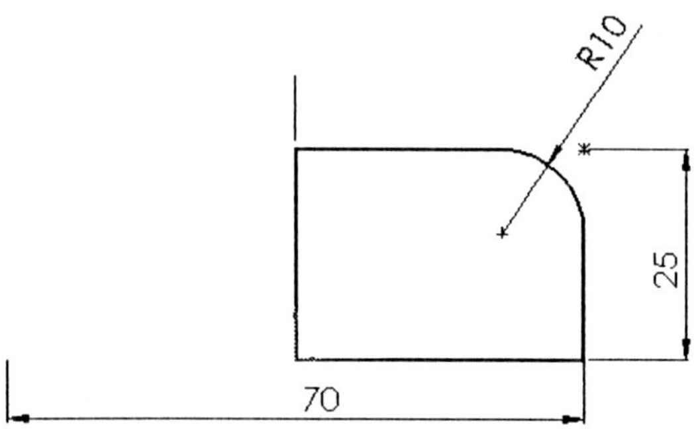

Figure 8-66 ***Sketch for the base feature***

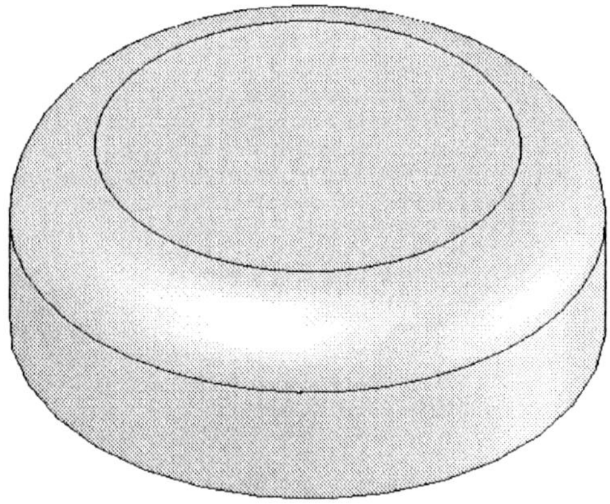

Figure 8-67 *Base feature of the model*

3. Choose the **OK** button from the **Shell1 PropertyManager**. The model after adding the shell feature is displayed in Figure 8-68.

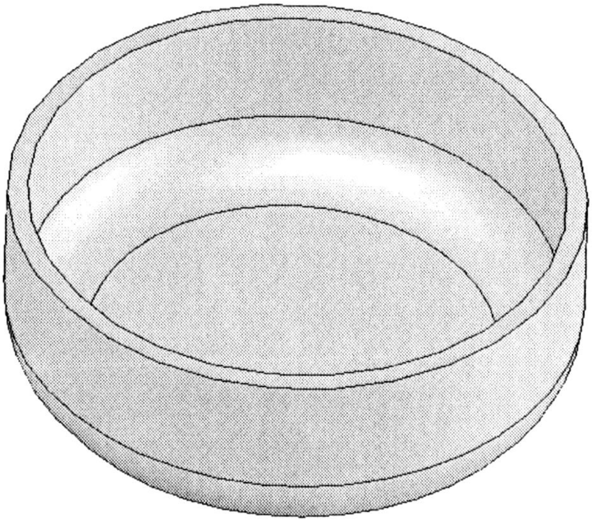

Figure 8-68 *Shell feature added to the model*

Creating the Third Feature

After adding the shell feature to the model, you need to create the third feature of the model, which is an extruded feature. The sketch for this feature will be created on the **Top** plane.

1. Invoke the **Extruded Boss/Base** tool and select the **Top** plane as the sketching plane.

2. Orient the model in the top view.

3. Create the sketch of the third feature and add the required relations and dimensions to the sketch as shown in Figure 8-69.

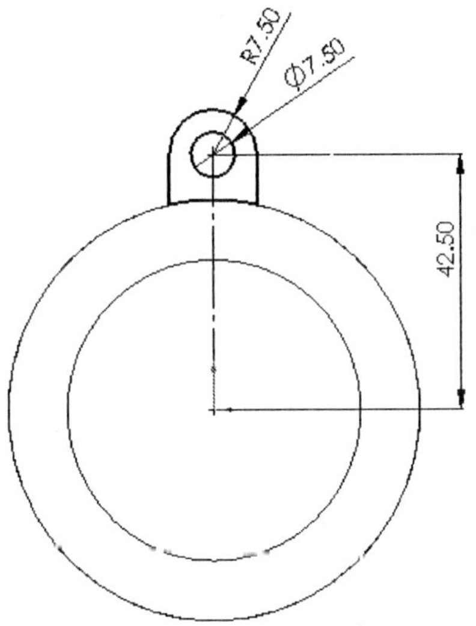

Figure 8-69 *Sketch of the third feature*

4. Exit the sketching environment and extrude the sketch to a depth of 5 mm.

Patterning the Third Feature

After creating the third feature of the model, you need to pattern this feature. This feature will be patterned using the **Circular Pattern** tool. Before proceeding further you need to display the temporary axes. The temporary axis of the base feature will be selected as central axis to create the circular pattern.

1. Choose **View > Temporary Axes** from the menu bar to display the temporary axes.

2. Invoke the **Circular Pattern PropertyManager** and select the temporary axis of the base feature as the central axis.

3. Select the third feature created earlier from the drawing area if not selected in the **Features to Pattern** selection area. The preview of the pattern feature is displayed in the drawing area.

4. Set the value of the **Number of Instances** spinner to **5** and choose **OK** from the **Circular Pattern PropertyManager**.

5. Choose **View > Temporary Axes** from the menu bar to remove the temporary axes from the current display.

The model after creating the pattern feature is displayed in Figure 8-70.

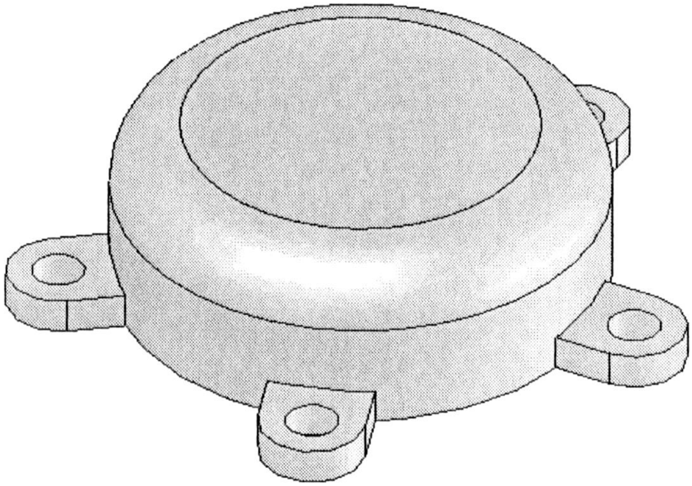

Figure 8-70 Pattern feature added to the model

Editing the Pattern Feature

The pattern created is not the same as required, refer to Figure 8-64. As a result you need to edit the pattern feature.

1. Select **CirPattern1** from the **FeatureManager Design Tree** or select any one of the pattern instance other than the parent instance from the drawing area. Right-click and choose the **Edit Feature** option from the shortcut menu.

   The **CirPattern1 PropertyManager** is displayed. Presently, the number of instances in the pattern feature is 5, but the required number of instances is 3. Therefore, you will edit the number of instances.

2. Set the value of the **Number of Instances** spinner to **3** and choose the **OK** button from the **FeatureManager Design Tree**.

   The model after editing the features is shown in Figure 8-71.

Suppressing the Features

As discussed earlier, sometimes you may need to suppress some features to reduce the complications in the model. The suppressed features are actually not deleted; their display is just turned off. When you suppress a feature, the child features associated with that feature are also suppressed.

1. Select the **Extrude1** feature, which is the third feature of the model, from the **FeatureManager Design Tree**. Right-click and choose the **Suppress** option from the shortcut menu.

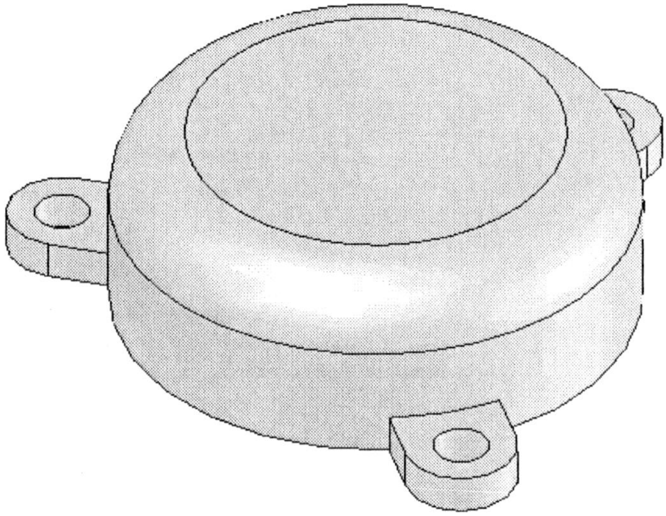

Figure 8-71 The edited pattern feature

The circular pattern feature is the child feature of the extrude feature. Therefore, it is also suppressed. Both the features are not displayed in the drawing area. The **Extrude1** and the **CirPattern1** features are displayed in gray in the **FeatureManager Design Tree**, indicating that both the features are suppressed.

Creating the Cut Feature

The next feature that you are going to create is a cut feature. The sketch for this feature will be created on the top planar face of the base feature.

1. Invoke the **Extrude Cut** tool and select the top planar face of the base feature as the sketching plane.
2. Create the sketch of the cut feature and add the required relations and dimensions to the sketch as shown in Figure 8-72.
3. Exit the sketching environment and specify the end condition as **Through All** from the **Cut-Extrude PropertyManager**.
4. Choose the **OK** button from the **Cut-Extrude PropertyManager**.
5. Now, using the **Linear Pattern** tool create a linear pattern of the cut feature. You can select the dimension 18 as the directional reference. The model after creating the linear pattern is shown in Figure 8-73.

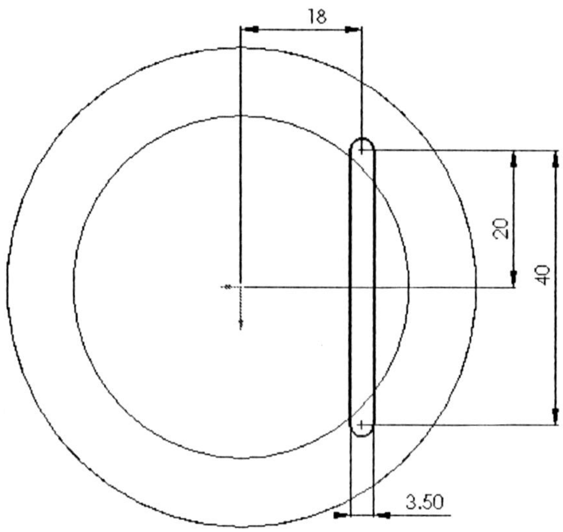

Figure 8-72 Sketch for the cut feature

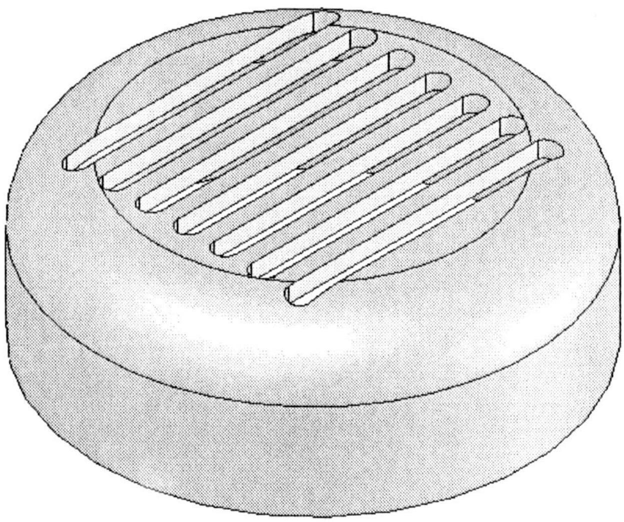

Figure 8-73 Model after patterning the cut feature

6. Create the other features of the model. For dimensions, refer to Figure 8-65. The model after creating the other features is shown in Figure 8-74.

Unsuppressing the Features

After completing the model, you need to unsuppress the features that you suppressed earlier.

1. Press and hold down the CTRL key from the keyboard and select all the suppressed features from the **FeatureManager Design Tree**.

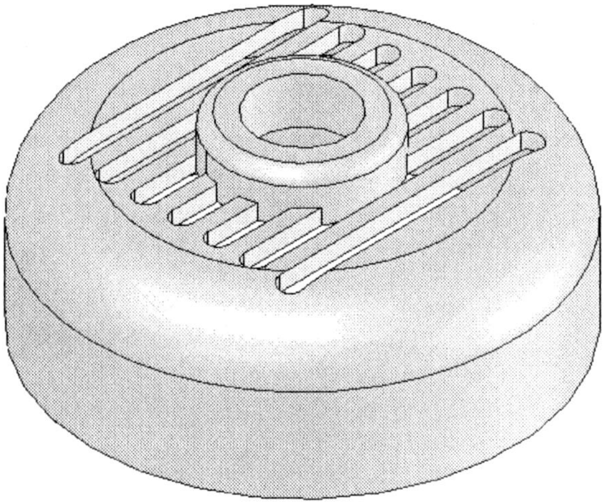

Figure 8-74 *Model after creating other features.*

Note

If you select only the parent suppressed feature and unsuppress it then the child feature will not be unsuppressed. Therefore, you have to select the parent feature and the suppressed child features.

Instead of selecting all the parent and child suppressed features from the ***FeatureManager Design Tree*** *select only the parent feature and choose* ***Edit > Unsuppress with Dependents > All Configurations*** *from the menu bar. You will learn more about the configurations in the later chapters.*

If you unsuppress the child feature then the parent feature will be unsuppressed automatically.

2. Right-click and choose the **Unsuppress** option from the shortcut menu.

   The suppressed features will be restored in the model. The final model after unsuppressing the features is shown in Figure 8-75. The **FeatureManager Design Tree** of the model is shown in Figure 8-76.

Saving the Model

1. Choose the **Save** button from the **Standard** toolbar and save the model with the name given below.

   \My Documents\SolidWorks\c08\c08-tut03.sldprt

2. Choose **File > Close** from the menu bar to close the document.

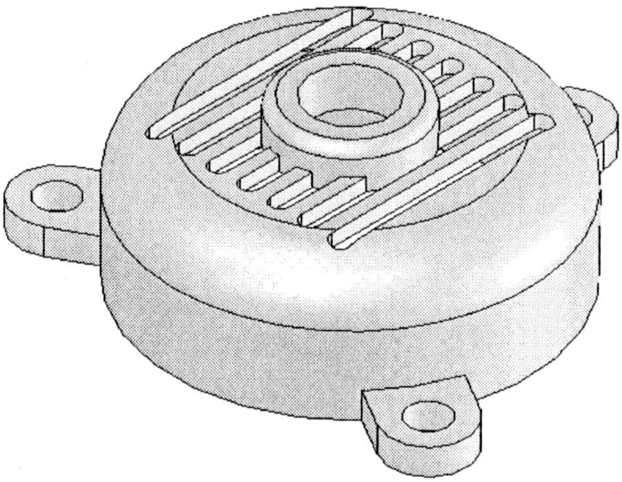

Figure 8-75 The final model

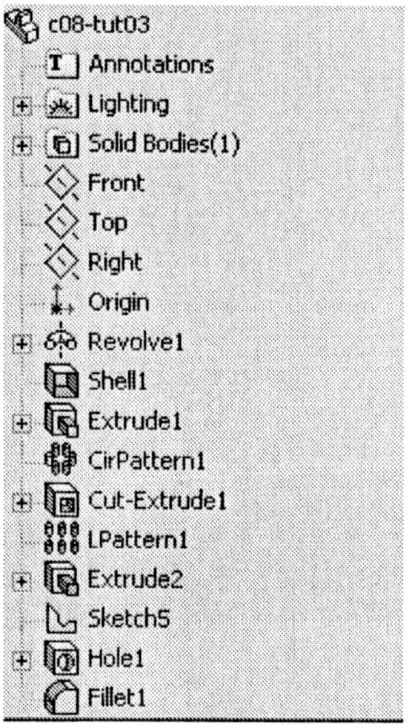

Figure 8-76 The ***FeatureManager Design Tree***

SELF-EVALUATION TEST

Answer the following questions and then compare your answers with the answers given at the end of this chapter.

1. You cannot edit the sketch of a sketched feature. (T/F)

2. The **Edit Feature** option is used to edit any feature. (T/F)

3. You cannot rename the feature in the **FeatureManager Design Tree**. (T/F)

4. You cannot edit the sketch plane of the sketch of a sketched feature. (T/F)

5. Using the **Move/ Size Features** option you cannot edit the sketches. (T/F)

6. The __________ dialog box is displayed when you edit a dimension.

7. The process of changing the position of a feature in the **FeatureManager Design Tree** is known as __________.

8. Using __________ **PropertyManager** you can delete the bodies.

9. The __________ **PropertyManager** is used to move or copy the bodies.

10. The __________ dialog box is displayed when there is any error in a feature.

REVIEW QUESTIONS

Answer the following questions.

1. The __________ **PropertyManager** is invoked to delete a body.

2. You can rotate a body using __________ **PropertyManager**.

3. __________ key is used from the keyboard to copy a feature or a sketch.

4. The __________ key is used from the keyboard to cut a feature or sketch.

5. When __________ tool is active then the preview of the feature is displayed in temporary graphics while editing the sketches.

6. The __________ **PropertyManager** is displayed to edit the sketch plane of a sketch.

7. To add the selected feature in a new folder choose **Add to New Folder** from the shortcut menu. (T/F)

8. For reordering the features, select the feature in the **FeatureManager Design Tree** and drag the feature to the required position. (T/F)

9. The **Modify** dialog box is invoked using a single click on the dimension to modify. (T/F)

10. If you want to modify the sketch by dragging the fully or partially defined sketch the **Override Dims on Drag/Move** option should be selected. (T/F)

EXERCISES

Exercise 1

Create the model that is sectioned and shown in Figure 8-77. The other views and dimensions of the model are also given in the same figure. The complete model is shown in Figure 8-78.

(Expected time: 45 min)

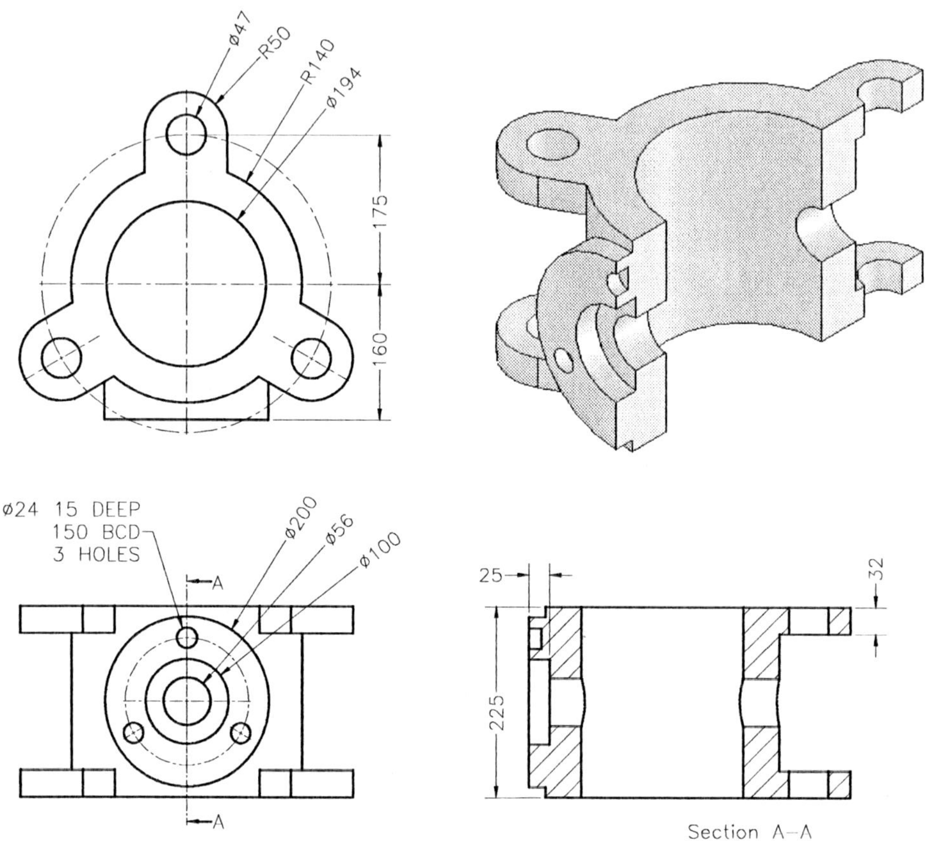

Figure 8-77 *Views and dimensions of the model for Exercise 1*

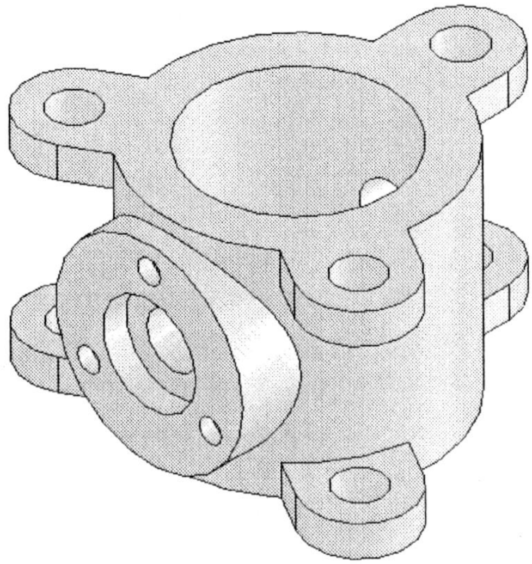

Figure 8-78 Model for Exercise 1

Exercise 2

Create the model shown in Figure 8-79. The dimensions of the model are shown in Figure 8-80.

(Expected time: 30 min)

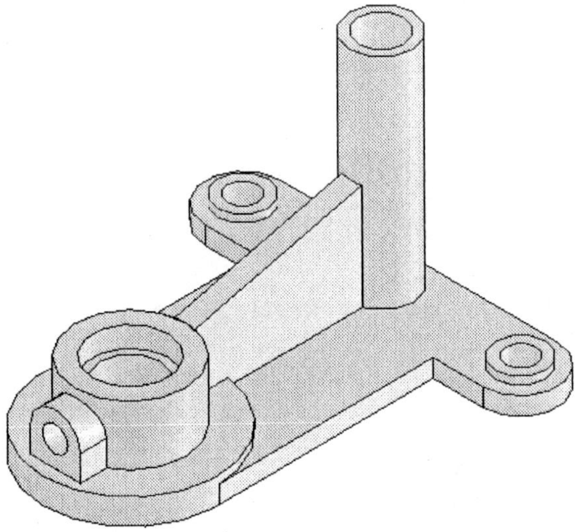

Figure 8-79 Model for Exercise 2

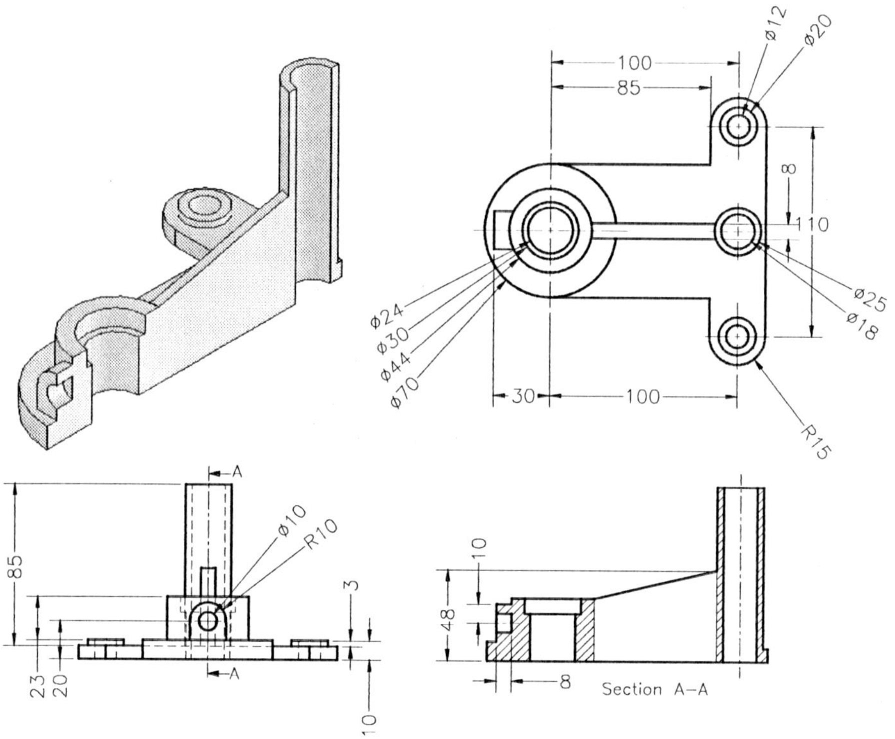

Figure 8-80 Views and dimensions of the model for Exercise 2

Answers to Self-Evaluation Test

1. F, **2.** T, **3.** F, **4.** F, **5.** F, **6. Modify**, **7.** Reordering, **8. Delete Body**, **9. Move/Copy Body**, **10. Rebuild Errors**

Chapter 9

Advanced Modeling Tools-III

Learning Objectives

After completing this chapter you will be able to:

- *Create Sweep Feature.*
- *Create Loft Feature.*
- *Create 3D Sketches.*
- *Edit 3D Sketches.*
- *Create Curves.*
- *Extrude 3D Sketches.*
- *Create Draft Feature.*
- *Create Deform Feature.*

ADVANCED MODELING TOOLS

Some of the advanced modeling tools were discussed in earlier chapters. The remaining advanced modeling tool are discussed in this chapter. The advanced modeling tools that are discussed in this chapter include sweep, loft, draft, curves, 3D sketches, and so on.

Creating the Sweep Feature

CommandManager:	Features > Swept Boss/Base
Menu:	Insert > Boss/Base > Sweep
Toolbar:	Features > Swept Boss/Base

One of the most important advanced modeling tool is the **Swept Boss/Base** tool. This tool is used to extrude a closed profile along an open or closed path. Therefore, to create a sweep feature you need a profile and a path. A profile is the section for the sweep feature and a path is the course taken by the profile while creating the sweep feature. The profile has to be a sketch, but the path can be a sketch, edge, or a curve. You will learn more about how to create the curves later in this chapter. An example of the profile and path for creating the sweep feature is shown in Figure 9-1. Choose the **Swept Boss/Base** button from the **Features CommandManager** to invoke the **Sweep PropertyManager**. The **Sweep PropertyManager** is shown in Figure 9-2.

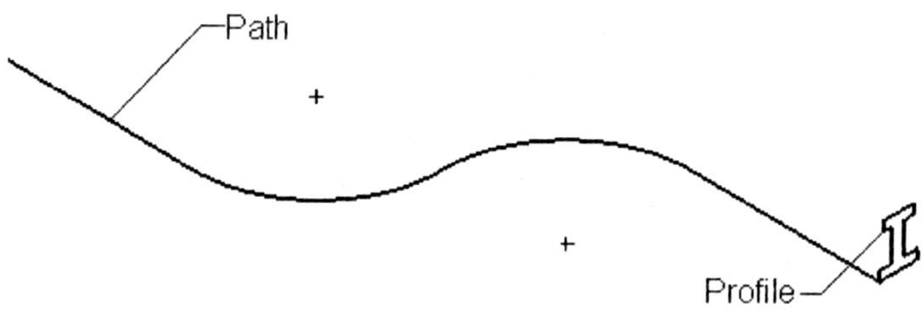

Figure 9-1 *Profile and path to create a sweep feature*

After invoking the **Sweep PropertyManager**, you are prompted to select the sweep profile. Select the sketch that is created as the profile from the drawing area. As soon as you select the sketch, the sketch is highlighted in green and the **Profile** callout is displayed. Now, you are prompted to select the path for the sweep feature. Select the sketch or an edge to be used as the path of the sweep feature. When you select the sketch or an edge, it is highlighted in magenta and the **Path** callout is displayed. The sweep feature is also displayed in temporary graphics in the drawing area. Choose the **OK** button from the **Sweep PropertyManager** to end feature creation. Figure 9-3 shows a sweep feature.

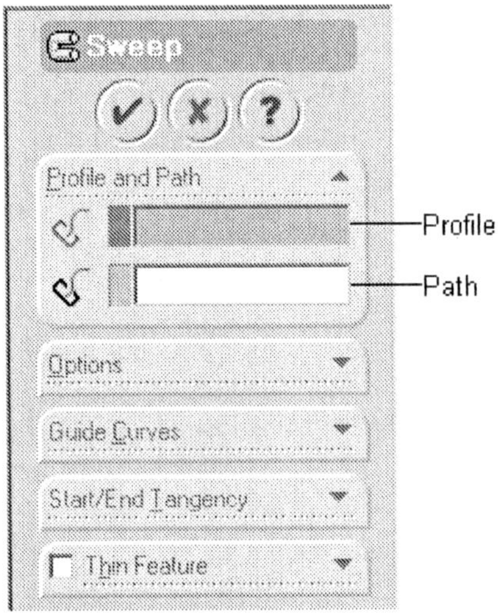

Figure 9-2 The ***Sweep PropertyManager***

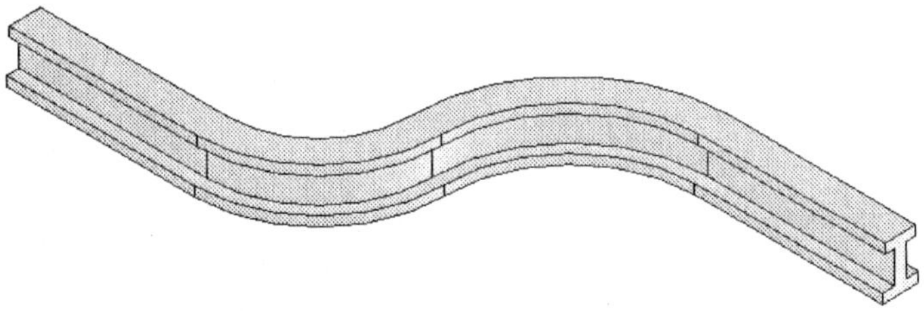

Figure 9-3 Sweep feature

It is not necessary that the sketch created for the profile of the sweep feature intersect the path of the sweep feature. However, the plane on which the profile is drawn should lie at one of the endpoints of the path. Figure 9-4 shows the nonintersecting sketches of profile and path. Figure 9-5 shows the resultant sweep feature. Figure 9-6 shows a sketch of a profile and a closed path. Remember, the plane on which the profile is drawn should intersect the closed path. Figure 9-7 shows the resultant sweep feature.

Tip. *You can also use the* ***Contour Select Tool*** *to select a contour as the section for the sweep feature.*

You can also use a shared sketch as the section of the sweep feature.

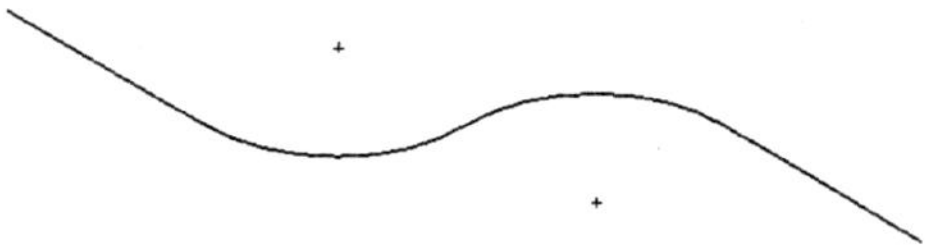

Figure 9-4 *Nonintersecting sketches of profile and path*

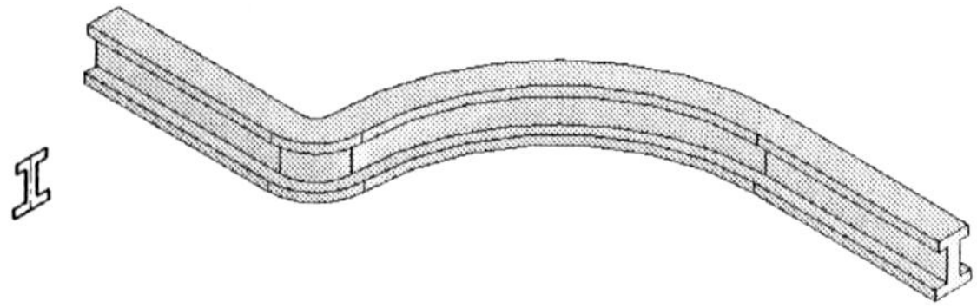

Figure 9-5 *The resultant sweep feature*

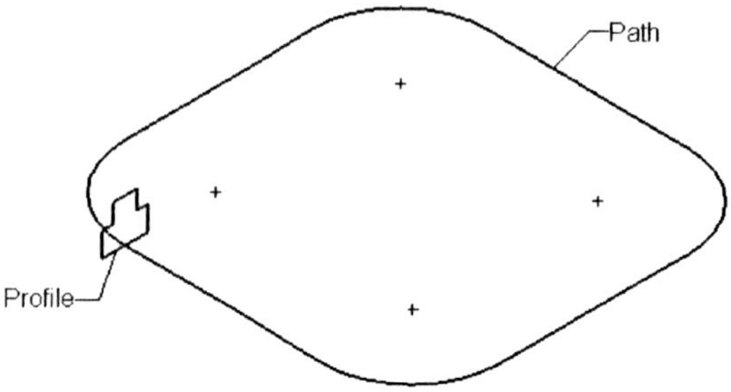

Figure 9-6 *Sketch of a profile and a closed path*

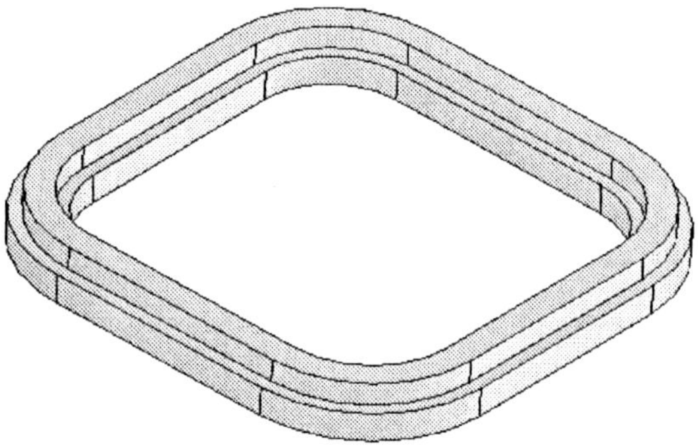

Figure 9-7 *The resultant sweep feature*

The various other options available in the **Sweep PropertyManager** to create advanced sweep features are discussed next.

Sweep Using the Follow Path and Keep Normal Constant Options

When you are creating a sweep feature, by default, the **Follow Path** option is selected in the **Orientation/twist Type** drop-down list available in the **Options** rollout. The **Options** rollout is displayed in Figure 9-8. The **Path alignment type** drop-down list is also displayed. Using this drop-down list you can set the options to create an even sweep feature when the curvature

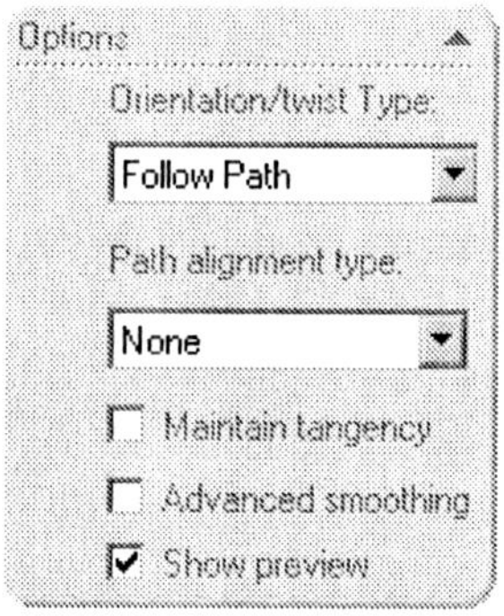

Figure 9-8 *The* ***Options*** *rollout*

along the path fluctuates unevenly. You will learn more about these options in the advanced chapters.

When you create a sweep feature using the **Follow Path** option, the section will follow the path to create the sweep feature. If you select the **Keep normal constant** option from the **Orientation/twist Type** drop-down list, the section will be swept along the path with a normal constraint. The section will not change its orientation along the sweep path. Therefore, the starting face and the end face of the sweep feature will be parallel. Figure 9-9 shows the sketches of the path and profile for the sketch feature. Figure 9-10 shows the sweep feature created using the **Follow Path** option. Figure 9-11 shows the sweep feature created using the **Keep normal constant** option. The other options available in the **Orientation/twist Type** drop-down list are discussed later in this chapter.

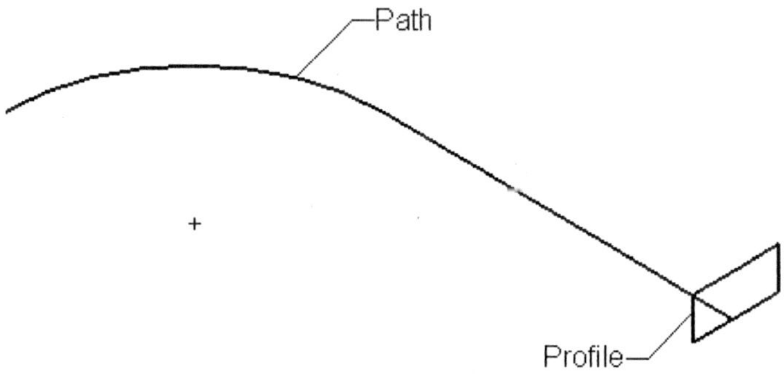

Figure 9-9 *Sketches for the sweep feature*

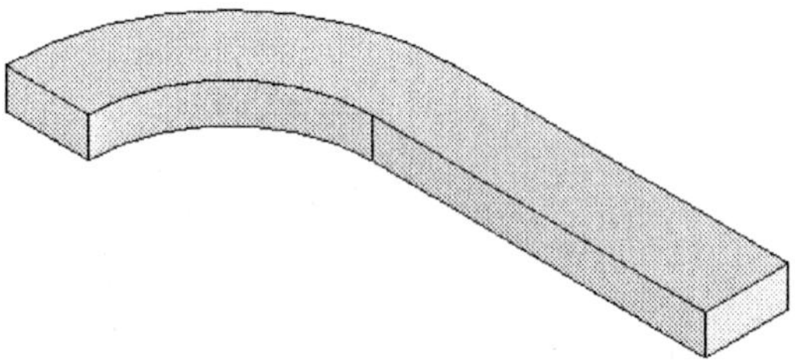

Figure 9-10 *Sweep feature with the **Follow Path** option selected from the **Orientation/twist Type** drop-down list*

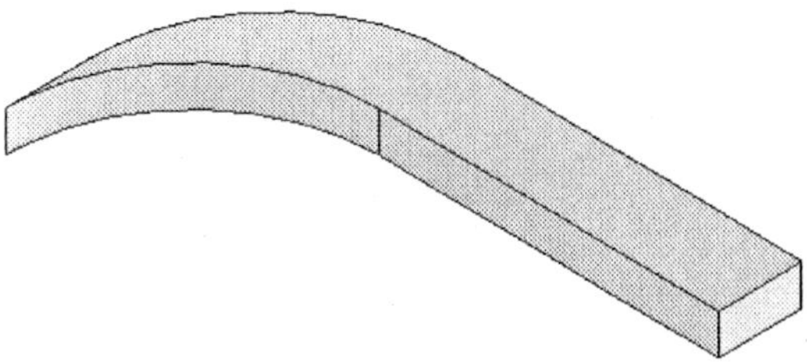

Figure 9-11 *Sweep feature with the **Keep normal constant** option selected from the **Orientation/twist Type** drop-down list*

Tip. *When you select a model edge as the sweep path, the **Tangent Propagation** check box is displayed in the **Options** rollout. If this check box is selected, the edges tangent to the selected edge are selected automatically as the path of the sweep feature.*

Maintain Tangency

The **Maintain Tangency** check box is available in the **Options** rollout. This option is used when the sweep section has tangent entities and you want the profile to maintain its tangency along the path while sweeping.

Advanced smoothing

The **Advanced smoothing** check box is available in the **Options** rollout. This option is used if the sweep section has circular or elliptical arcs and you want a smooth surface to be created in the sweep feature.

Show preview

The **Show preview** check box available in the **Options** rollout is used to display the preview of the sweep feature in the drawing area. This check box is selected by default. If you clear this check box, the preview of the sweep feature will not be displayed in the drawing area.

Merge result

The **Merge result** check box is available only when you have at least one feature in the current document. This check box is selected by default. If you clear this check box, it will result in creating the sweep feature as a separate body.

Align with end faces

The **Align with end faces** check box is available in the **Options** rollout only when at least one feature has already been created in the current document. When this option is selected, the sweep feature is extended or trimmed to align with end faces. Figure 9-12 shows the profile and path for creating the sweep feature. Figure 9-13 shows the resultant sweep feature created with the **Align with end faces** check box cleared. Figure 9-14 shows the resultant sweep feature created with the **Align with end faces** check box selected.

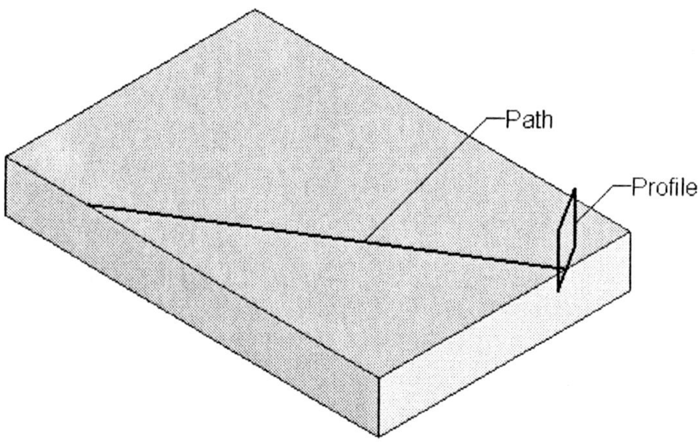

***Figure 9-12** Sketches for the sweep feature*

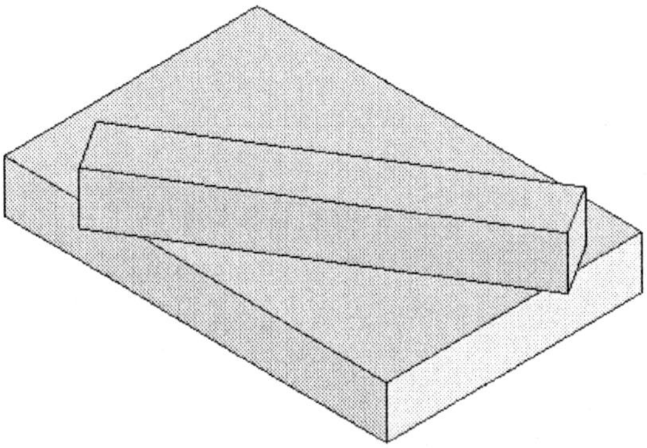

Figure 9-13 *Resultant sweep feature with the* ***Align with end faces*** *check box cleared*

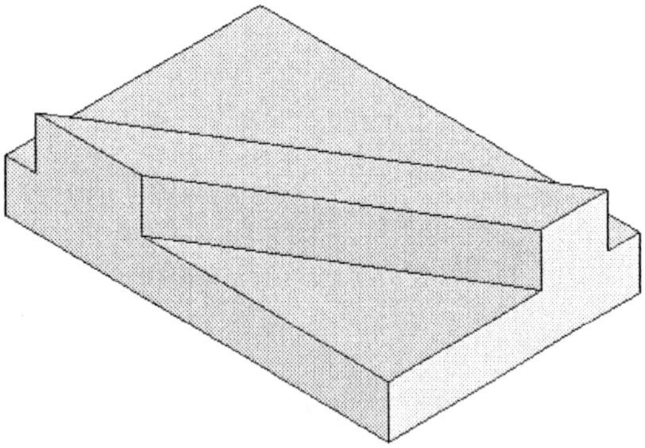

Figure 9-14 *Resultant sweep feature with the* ***Align with end faces*** *check box selected*

Note

If the sweep feature does not merge, you need to reduce the size of the profile.

Sweep with Guide Curves

The sweep using guide curves is the most important option in the advanced modeling tools. In this sweep feature, the section of the sweep profile varies according to the guide curves along the sweep path. To create this type of feature, you need a profile, path, and the guide curves. After creating the sketch of the profile, path, and guide curve apply the coincident relation between the guide curves and the profile. You need to make sure that the guide curves intersect the profile. Now, invoke the **Sweep PropertyManager**. Select the profile and the path; the preview of the sweep feature is displayed in the drawing area. Click on the black arrow on the right of the **Guide Curves** rollout to invoke this rollout. The **Guide Curves** rollout is shown in Figure 9-15.

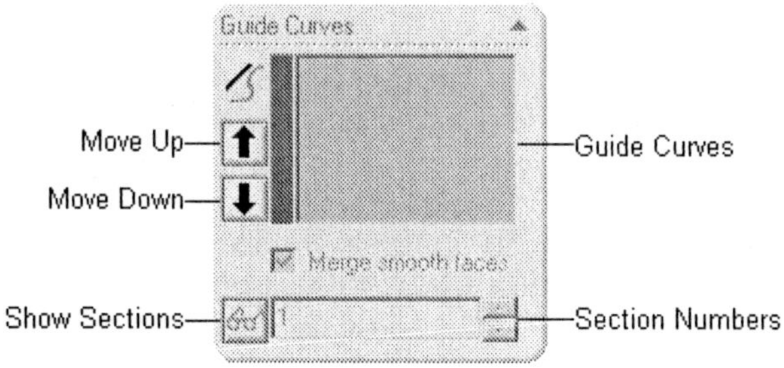

Figure 9-15 *The* ***Guide Curves*** *rollout*

Select the sketch of the guide curve; the selected guide curve is displayed in purple and a **Guide Curve** callout is also displayed attached to the guide curve. The preview of the sweep feature is also displayed in the drawing area. Choose the **OK** button from the **Sweep PropertyManager**. Figure 9-16 shows the sketch for the sweep feature with guide curve. Figure 9-17 shows the resultant sweep feature creation.

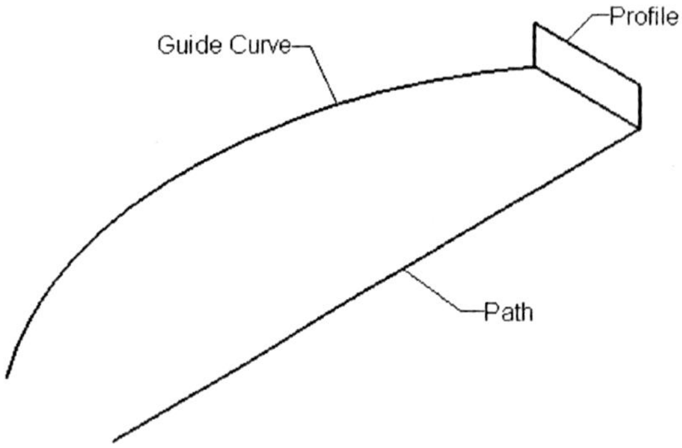

Figure 9-16 Sketches for the sweep feature with guide curves

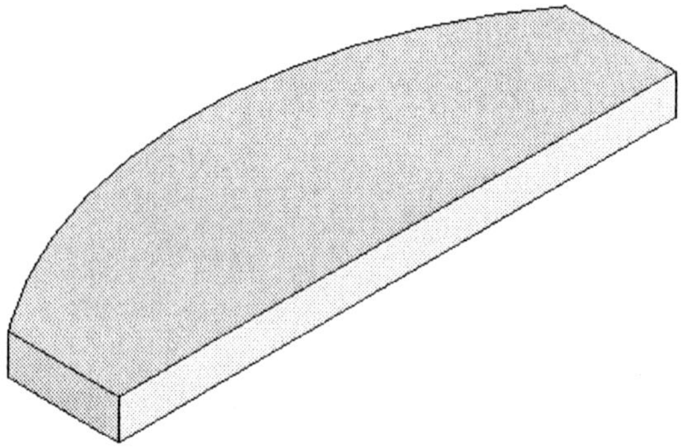

Figure 9-17 Resultant sweep feature

In the previous case the path of the sweep feature was a straight line and the guide curve was an arc. In the next case, an arc is selected as the path of the sweep feature and a straight line is selected as the guide curve. Figure 9-18 shows the sketches for the sweep feature. Figure 9-19 shows the resultant sweep feature.

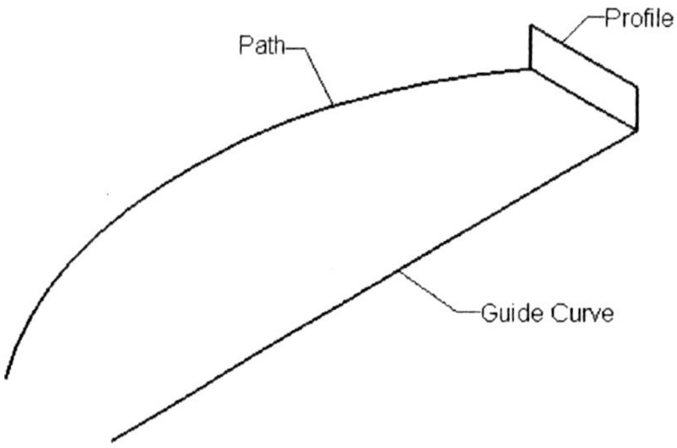

Figure 9-18 Sketches for the sweep feature

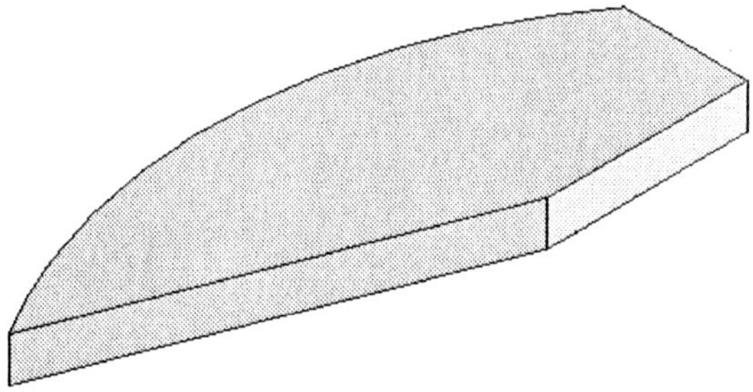

Figure 9-19 Resultant sweep feature

Move Up and Move Down

The **Move Up** button and the **Move Down** button available on the left of the **Guide Curves** selection area are used to change the sequence of the selected guide curve.

Merge smooth faces

The **Merge smooth faces** check box available in the **Guide Curves** rollout is selected by default. This option is used to merge all the smooth faces together, resulting in a smooth sweep feature. After creating the sweep feature, when you edit it and clear the **Merge smooth faces** check box, the **Sweep Preview Warning** dialog box is displayed as shown in Figure 9-20. In this dialog box you are prompted that the feature you are creating may fail because of change in smooth face option. Choose the **Yes** button from this dialog box if you want to accept the change option. When you create a sweep feature with guide curves and this option cleared, the resulting feature do not merge the smooth faces together.

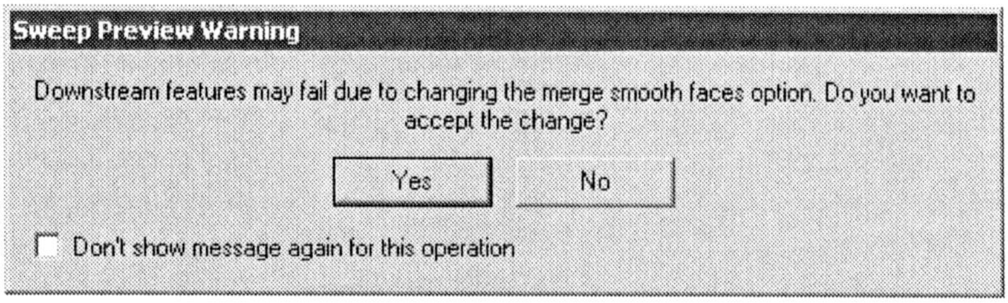

Figure 9-20 *The* ***Sweep Preview Warning*** *dialog box*

This results in a sweep feature with noncontinuous curvature surface. Figure 9-21 shows a sweep feature created with the **Merge smooth faces** check box selected. Figure 9-22 shows the same sweep feature with the **Merge smooth faces** check box cleared.

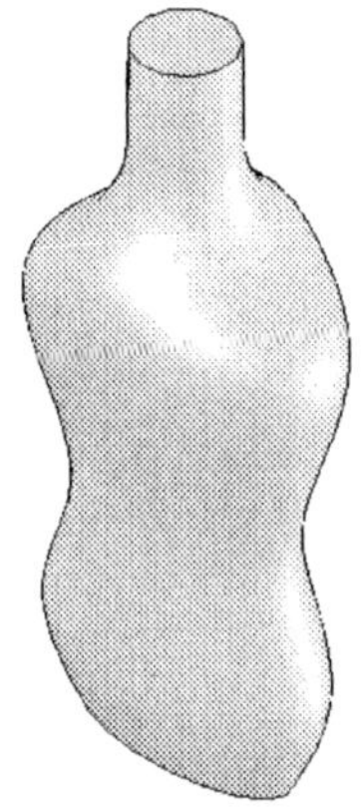

Figure 9-21 *Sweep feature with the* ***Merge smooth faces*** *check box selected*

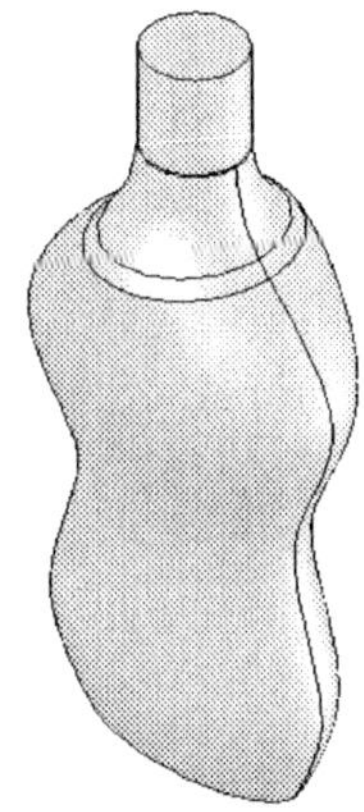

Figure 9-22 *Sweep feature with the* ***Merge smooth faces*** *check box cleared*

Note

Remember that if you create the sweep feature with the ***Merge smooth faces*** *check box cleared, the resultant feature will be generated faster and the adjacent faces and edges are easily merged. Also the lines and arcs in the guide curve match accurately while creating the sweep feature.*

Show Sections

The **Show Sections** button available in the **Guide Curves** rollout is used to display the intermediate sections while creating the sweep feature with guide curves. To display the intermediate profiles or sections along the sweep path, choose the **Show Section** button from the **Guide Curves** rollout. The **Section Number** spinner is invoked. Using this spinner you can view the sections of the profile along the sweep path. The maximum value of spinner goes up to the number of sections that fit inside the sweep feature. Figure 9-23 shows a section being displayed using the **Show Section** option and constrained by the guide curves.

Sweep Feature Using the Follow path and 1st guide curve Option

When you create a sweep feature with guide curve using the **Follow path and 1st guide curve**

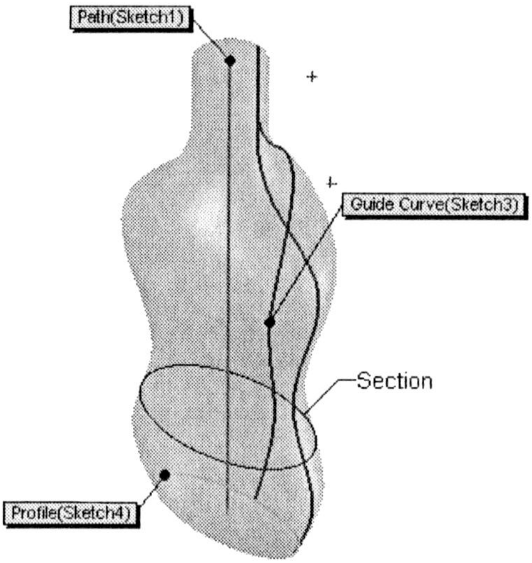

Figure 9-23** Section being displayed using the **Show Section

option from the **Options** rollout, the profile follows the path and the 1st guide curve to create the feature. For creating a sweep feature using this option invoke the **Sweep PropertyManager** and select the profile, path, and guide curve(s). By default, the **Follow path** option is selected in the **Orientation/twist Type** drop-down list of the **Options** rollout. Select the **Follow path and 1st guide curve** option from the **Orientation/twist Type** drop-down list. Choose the **OK** button from the **Sweep PropertyManager** to end feature creation.

Sweep Feature Using the Follow 1st and 2nd guide curves Option

Using this option, the profile of the sweep feature follows the 1st and 2nd guide curves to create the resultant sweep feature. For creating this type of sweep feature select the **Follow 1st and 2nd guide curves** option from the **Orientation/twist Type** drop-down list from the **Options** rollout. Choose the **OK** button from the **Sweep PropertyManager** to end feature creation.

Figure 9-24 shows the sketches of the profile, path, and guide curves for creating a sweep feature. Figure 9-25 shows the sweep feature created using the **Follow path and 1st guide curve** option. Figure 9-26 shows the sweep feature created using the **Follow 1st and 2nd guide curves** option.

Start/End Tangency

The **Start/End Tangency** rollout available in the **Sweep PropertyManager** is used to define the tangency conditions on the start and the end of the feature. Invoke the **Start/End Tangency** rollout, as shown in Figure 9-27. The various options available in the **Start/End Tangency** rollout are discussed next.

Start tangency type

The **Start tangency type** drop-down list is used to specify the options to define the tangency at the start of the sweep feature. The various options available in this drop-down list are discussed next.

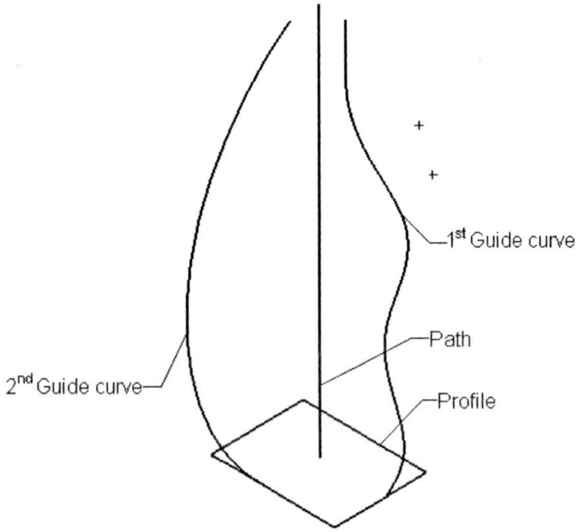

Figure 9-24 *Sketches of profile, path, and guide curves*

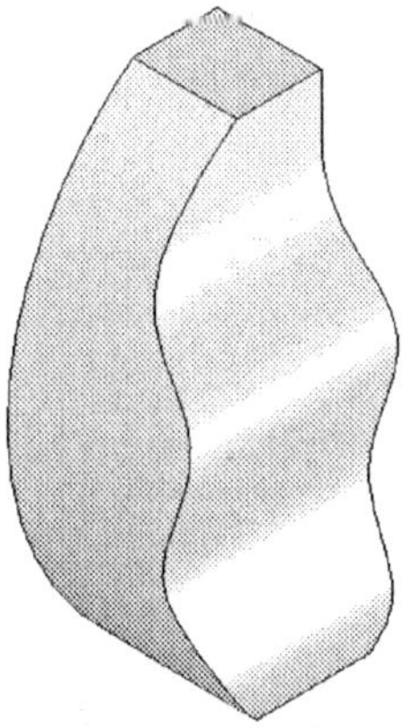

Figure 9-25 *Sweep feature using the **Follow path and 1st guide curve** option*

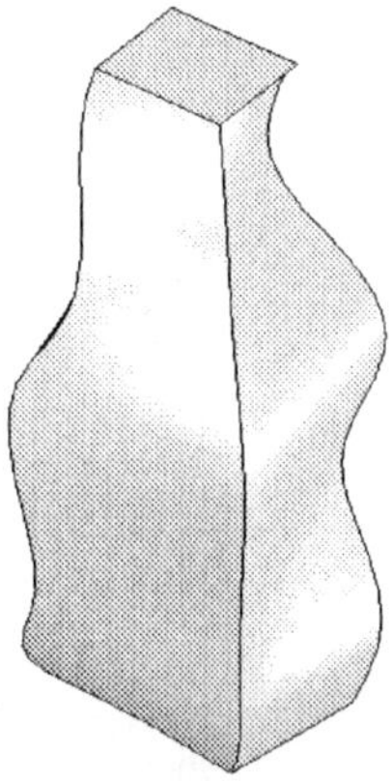

Figure 9-26 *Sweep feature using the **Follow 1st and 2nd guide curves** option*

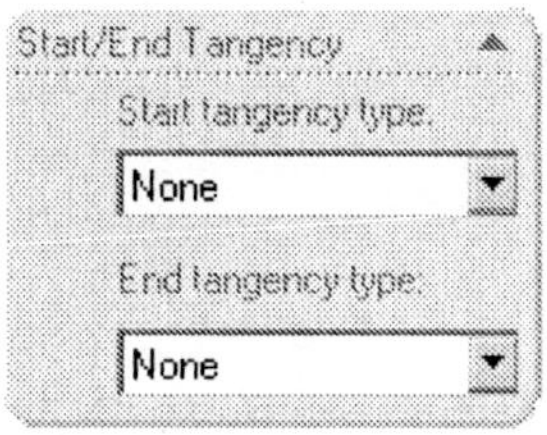

Figure 9-27 *The **Start/End Tangency** rollout*

None

The **None** option is selected by default and is used to create a sweep feature without applying any start tangency.

Path Tangent

The **Path Tangent** option is used to maintain the sweep feature normal to the path at the start.

Direction Vector

When you use the **Direction Vector** option, the starting of the sweep feature will be tangent to a virtual normal created from the selected entity. When you select this option the **Direction Vector** selection area is also displayed and you need to select a linear edge, axis, planar face, or plane.

All Faces

The **All Faces** option is used to sweep the feature tangent to the adjoining faces of the existing geometry at the start.

The options available in the **End tangency type** drop-down list are the same as those discussed above except the **All Faces** option. The only difference is that the options in this drop-down list are applied to the end of the sweep feature. Figure 9-28 shows the sketches and the references to create the sweep feature with start and end tangency using the **Direction Vector** option. Figure 9-29 shows the resultant sweep feature.

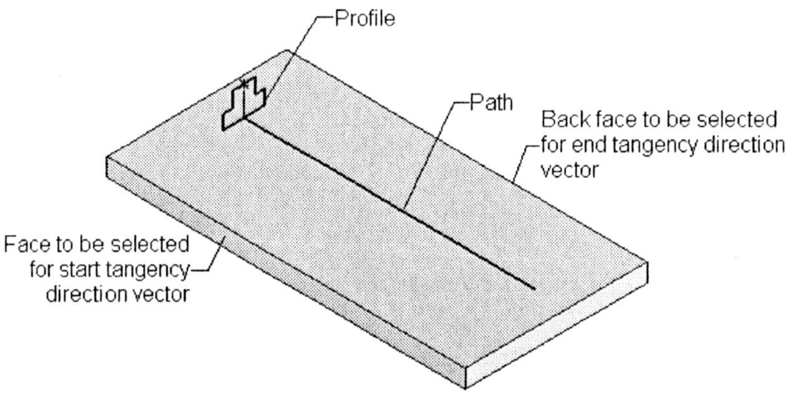

*Figure 9-28 Profile, path, and references for tangency using the **Direction Vector** option*

Creating a Thin Sweep Feature

You can also create a thin sweep feature by specifying the thickness using the **Thin Feature** rollout. This rollout is invoked by selecting the check box provided at the left of the **Thin Feature** rollout. The **Thin Feature** rollout is shown in Figure 9-30. The options available in this rollout are the same as those discussed in the earlier chapters in which extruding and revolving of thin features was discussed. Figure 9-31 shows a thin sweep feature.

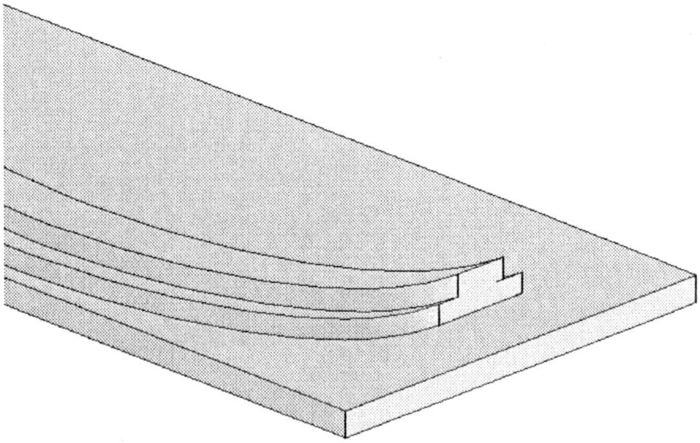

***Figure 9-29** Resultant sweep feature*

***Figure 9-30** The **Thin Feature** rollout*

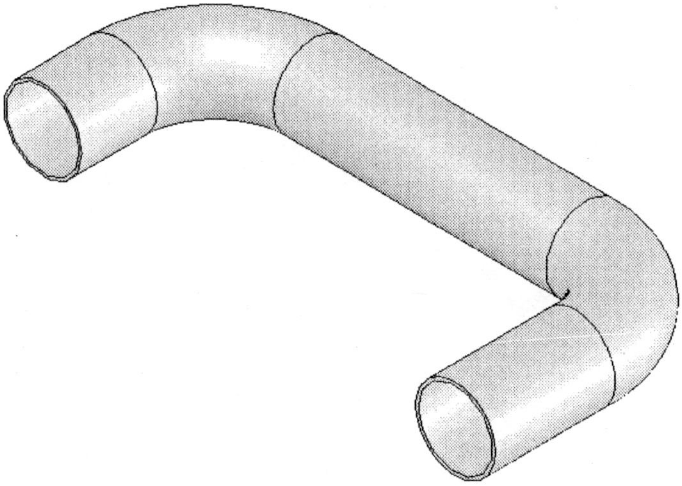

***Figure 9-31** Thin sweep feature*

Creating Cut Sweep Features

CommandManager:	Features > Swept Cut	*(Customize to Add)*
Menu:	Insert > Cut > Sweep	
Toolbar:	Features > Swept Cut	*(Customize to Add)*

You can also remove material from an existing feature or model by creating a cut-sweep feature. To create a cut-sweep feature, choose the **Swept Cut** button from the **Features CommandManager** after customizing it to invoke the **Cut-Sweep PropertyManager** as shown in Figure 9-32. The options available in the **Cut-Sweep PropertyManager** are the same as those discussed in the **Sweep PropertyManager** with the only difference being that it is meant for cut operation. Figure 9-33 shows the profile and the path. Figure 9-34 shows the resultant cut-sweep feature created using the **Swept Cut** tool.

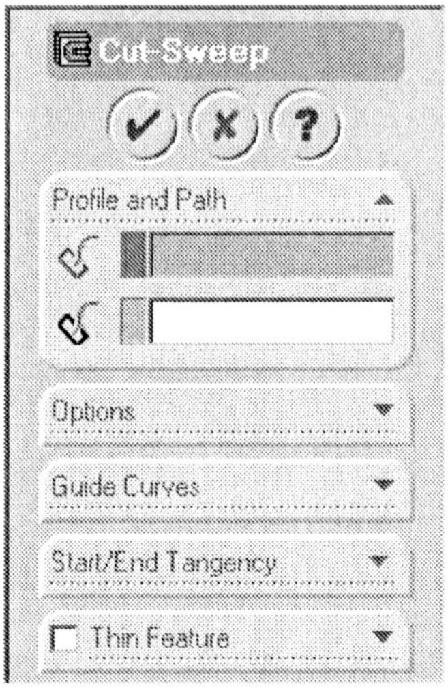

Figure 9-32 *The* ***Cut-Sweep PropertyManager***

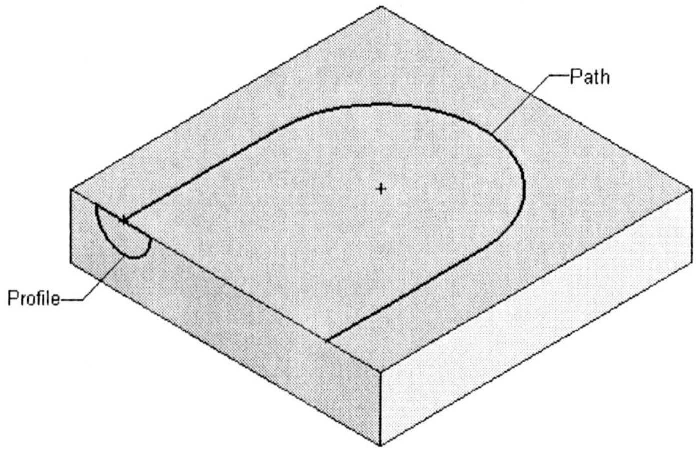

Figure 9-33 *Sketches for the cut sweep feature*

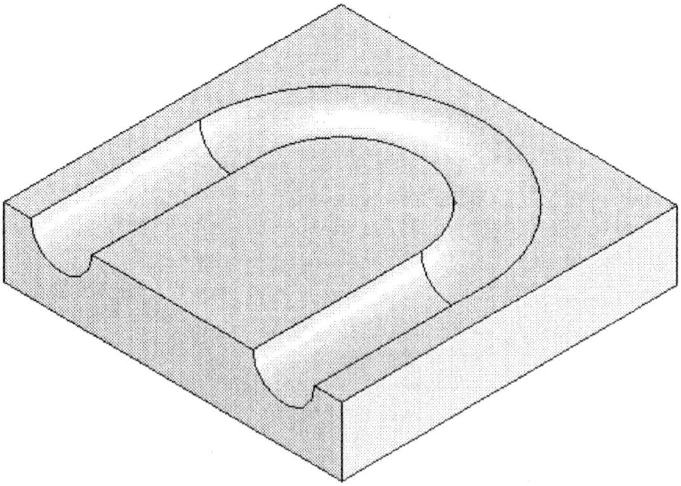

Figure 9-34 *Resultant cut sweep feature*

Creating the Loft Feature

CommandManager:	Features > Lofted Boss/Base
Menu:	Insert > Boss/Base > Loft
Toolbar:	Features > Lofted Boss/Base

The lofted features are created by blending more than one similar or dissimilar sections together to get a free form type of shape. These similar or dissimilar sections may or may not be parallel to each other. The sections for solid lofts should be closed sketches.

To create a loft feature invoke the **Loft PropertyManager** by choosing the **Lofted Boss/Base** button from the **Features CommandManager**. The **Loft PropertyManager** is shown in Figure 9-35.

After creating the sketches when you invoke the **Loft PropertyManager**, you are prompted to select at least two profiles. Select the profiles from the drawing area. As you select the profiles, the preview of the loft feature along with a connector is displayed in the drawing area. Choose the **OK** button from the **Loft PropertyManager** to end feature creation. Figure 9-36 shows the preview of the loft feature along with the connector. Figure 9-37 shows the resultant loft feature.

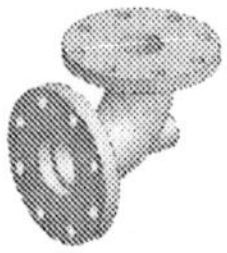

Tip. *You can also use the* ***Contour Select Tool*** *to select a contour that will be used as the profile to create the loft feature. You can also use a shared sketch as a profile that is currently in use by some other sketched feature.*

The loft feature can be reshaped using the handles of the connector. The handles that appear in magenta are the endpoints of the connector. Press and hold down the left mouse button on a handle. The handle appears in red. Drag the cursor to specify a new location and release

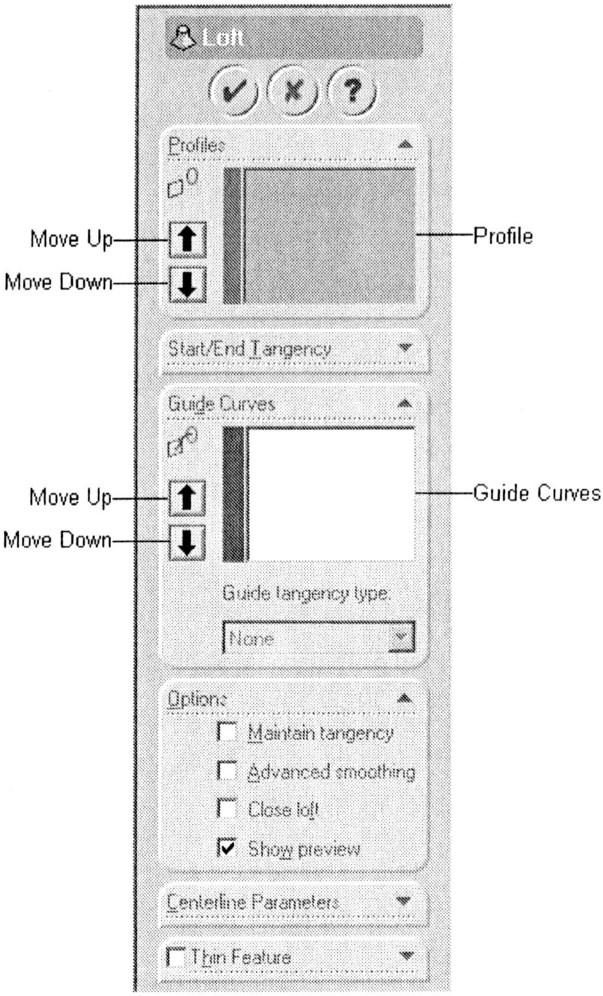

*Figure 9-35 The **Loft PropertyManager***

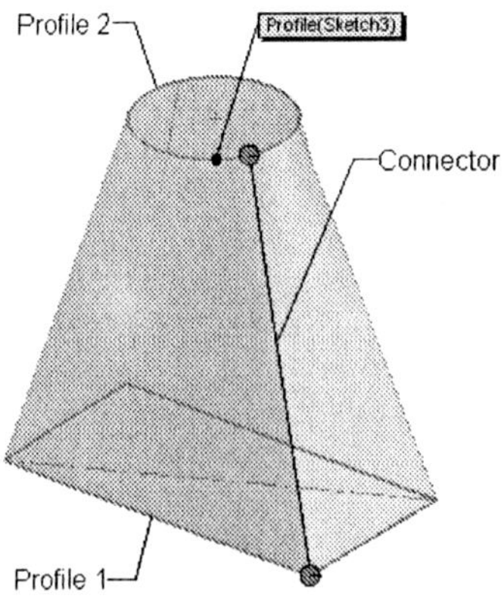

Figure 9-36 Three sketches for the loft feature

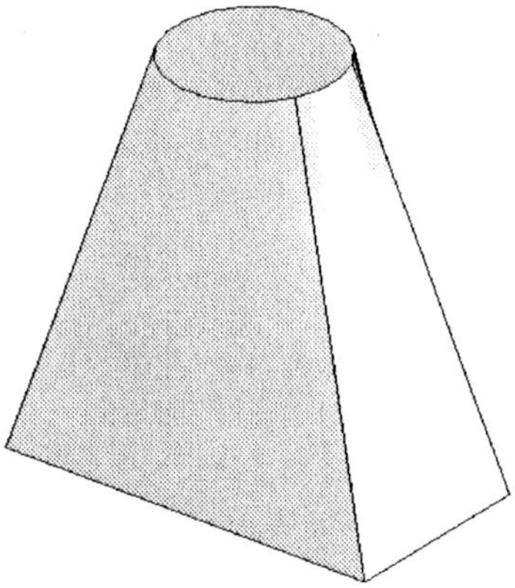

Figure 9-37 Resultant loft feature

the left mouse button to place the connector handle at a new location. The process of controlling the loft shape using the connectors is known as **Loft Synchronization**. Figure 9-38 shows the preview of the loft feature after modifying the location of the handle of the connector. Figure 9-39 shows the resultant loft feature.

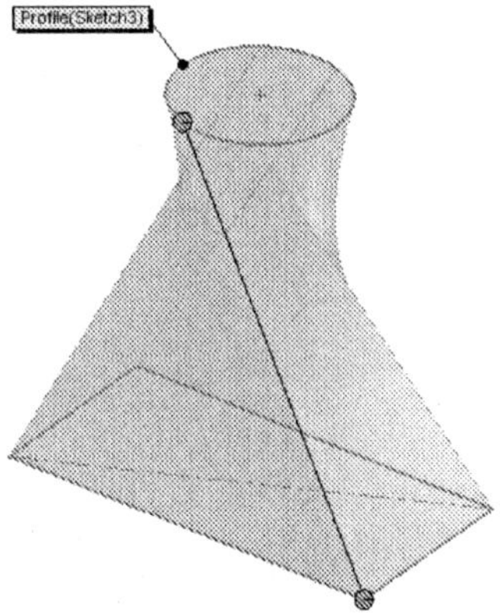

Figure 9-38 *Preview of the loft feature after modifying the location of the connector*

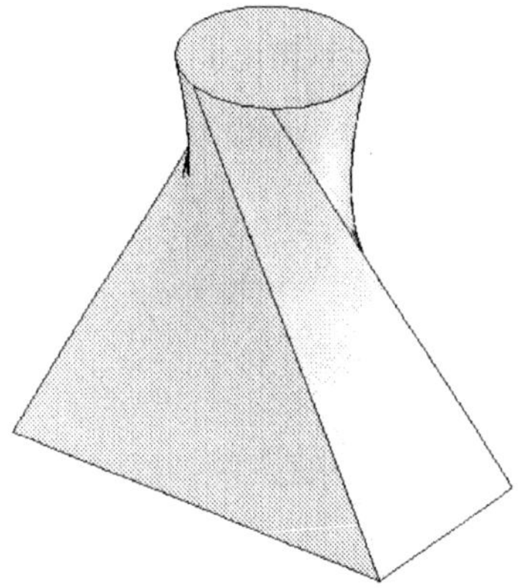

Figure 9-39 *Resultant loft feature*

Note

The reshaping or twisting done using the default connector available while creating the loft feature is global twisting. Global twisting means if you change the location of one connecting point of the profile, then the other connecting points of the profile automatically change their respective positions with respect to the modified connecting point.

In global twisting of a nontangent profile the handle of the connectors moves only from vertex to vertex.

You can also add additional connectors to manipulate the loft feature. A connector can only be added to a profile that has vertices. To add a connector right-click in the drawing area to invoke the shortcut menu. Choose the **Add Connector** option from the shortcut menu. A connector will be added to the loft feature. Using this option you can also add more connectors. The maximum number of connectors depends on the number of vertices of the profile. To view or to add the connectors to the loft feature select the **Show All Connectors** option from the shortcut menu. After adding the connectors you can modify the connectors by dragging the handles of the connectors. Figure 9-40 shows the preview of the loft feature after modifying all the additional connectors. Figure 9-41 shows the resultant loft feature.

Note

The reshaping or twisting done using the additional connectors is known as local twisting. This is because the twisting done using one connector do not affect the other connecting points of the profile. Therefore you can independently modify all the connectors simply by dragging the handles along the profile.

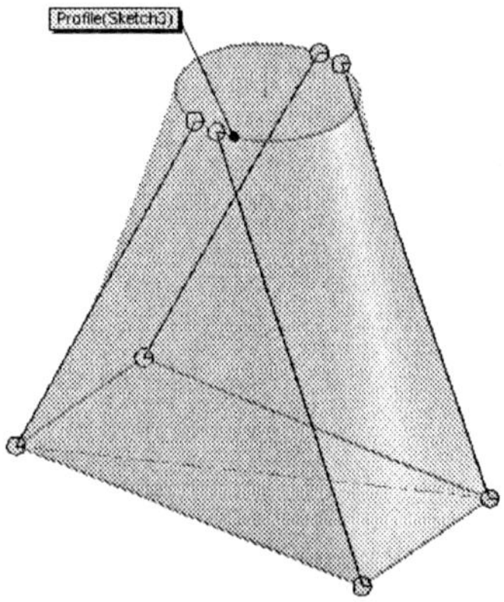

Figure 9-40 Preview after adding all connectors

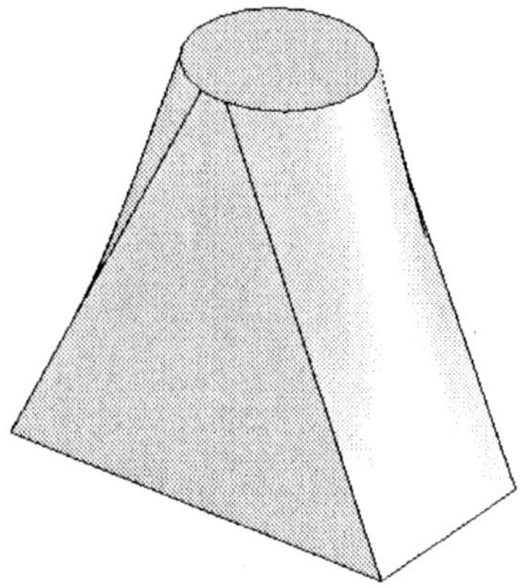

Figure 9-41 Resultant loft feature

Tip. *Invoke the shortcut menu and choose* ***Hide All Connectors*** *to hide all the connectors. By using this option, the connectors are not displayed in the preview but the settings done using these connectors remain as it is in the loft feature. To hide a specific connector select the connector and then invoke the shortcut menu. Then, choose the* ***Hide Connector*** *option.*

If you choose the ***Reset Connectors*** *option from the shortcut menu then all the connectors and the settings related to the connectors are deleted from the memory of the feature and the default connector is displayed with default connection points between the profiles of the loft feature.*

Start/End Tangency

The **Start/End Tangency** rollout available in the **Loft PropertyManager** is used to define the tangency at the start and end sections of the loft feature. Invoke this rollout to define the tangency. By default, the **None** option is selected in the **Start tangency type** drop-down list and the **End tangency type** drop-down list. This means that tangency is not applied to the loft feature. The other options available in these drop-down lists are discussed next.

Normal to Profile

The **Normal to Profile** option is used to define the tangency normal to the profile. When you choose this option, you are provided with a spinner to specify the length of tangent. A **Reverse Direction** button is also provided to flip the direction of tangent. The **Start/End Tangency** rollout with **Normal to Profile** option selected in the **Start tangency type** and the **End tangency type** drop-down lists is shown in Figure 9-42.

Direction Vector

The **Direction Vector** option is used to define the tangency at the start and at the end of the loft feature by defining a direction vector. When you invoke this option you are provided with the **Direction Vector** selection area and the spinners to define the length of tangents. You need to select the direction vectors to specify the tangent at the start and at the end of the loft feature. You can also specify the length of the tangents using the spinners provided in the **Start/End Tangency** rollout. The **Start/End Tangency** rollout with the **Direction Vector** option selected is displayed in Figure 9-43.

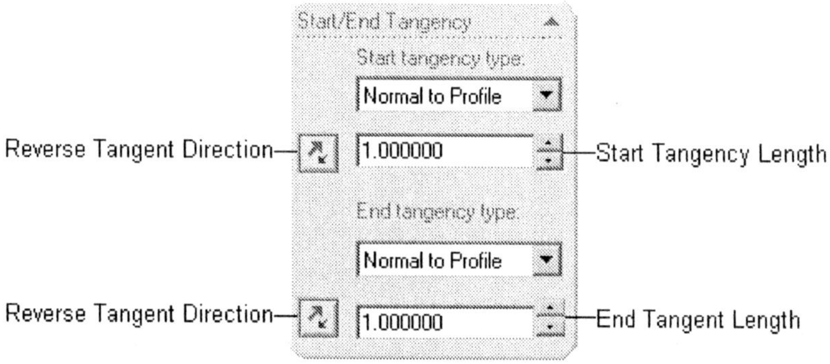

***Figure 9-42** The **Start/End Tangency** rollout with the **Normal to Profile** option selected*

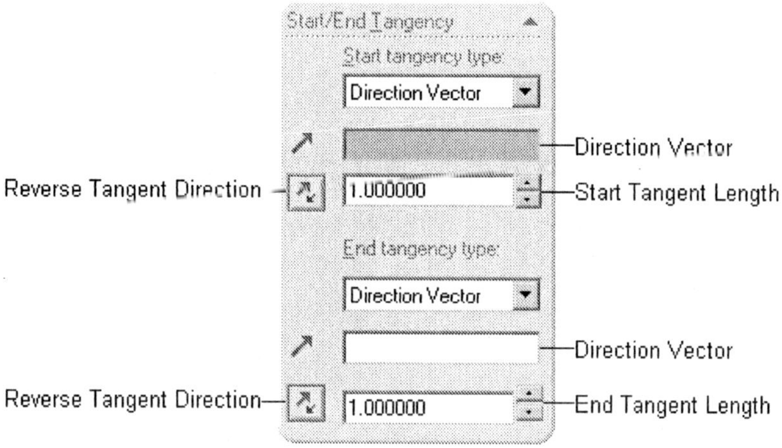

***Figure 9-43** The **Start/End Tangency** rollout with the **Direction Vector** option selected*

Figure 9-44 shows the sections for the loft feature. Figure 9-45 shows the initial preview of the loft feature. Figure 9-46 shows the preview of the loft feature with tangent at the start of the loft. Figure 9-47 shows the tangent at the start and at the end of the loft. Figure 9-48 shows the final loft feature.

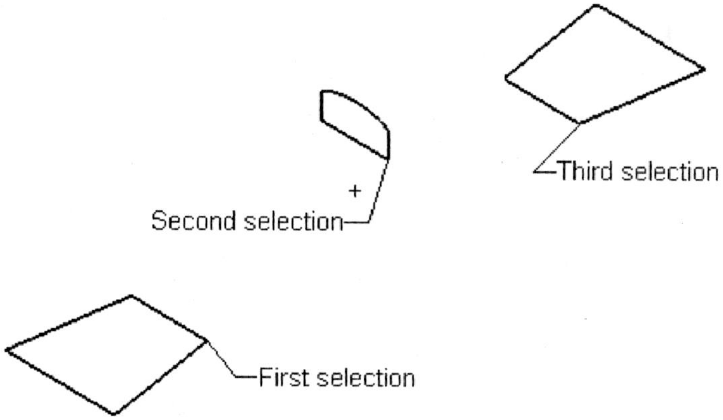

***Figure 9-44** Sections, selection points, and sequence of selection*

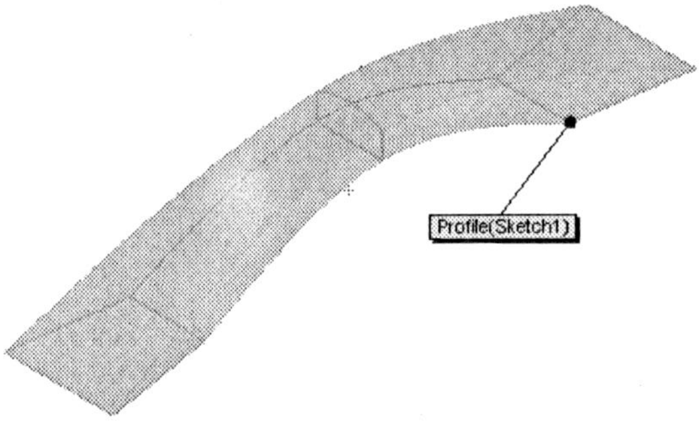

Figure 9-45 Preview of the loft feature

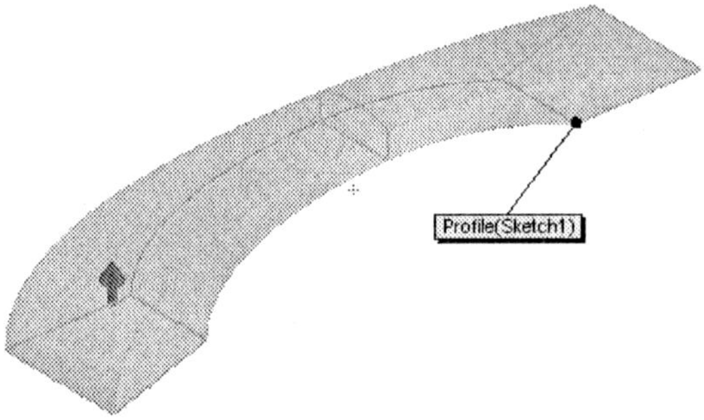

Figure 9-46 Tangent applied at the start

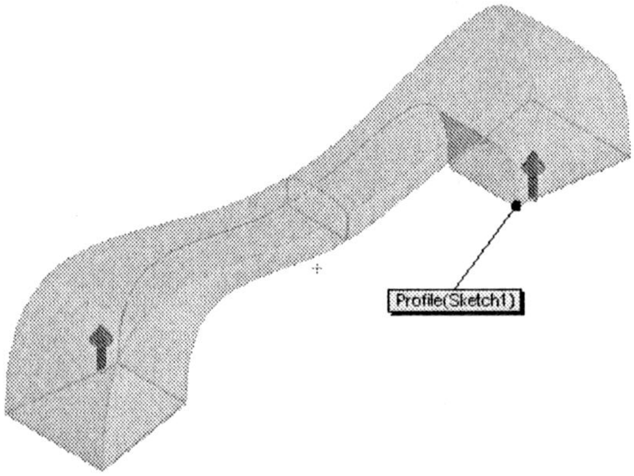

Figure 9-47 Tangent applied at start and end

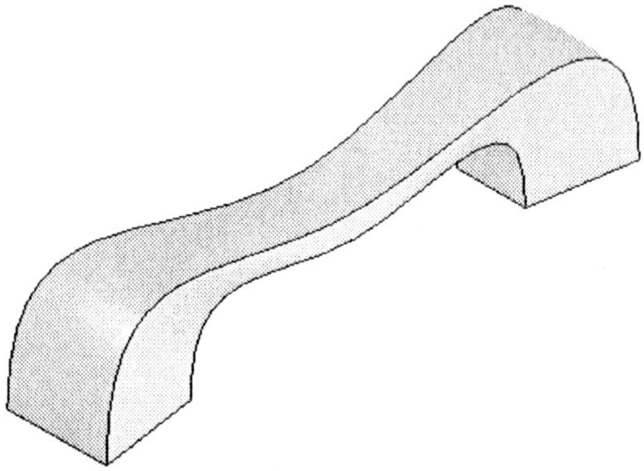

Figure 9-48 The final loft feature

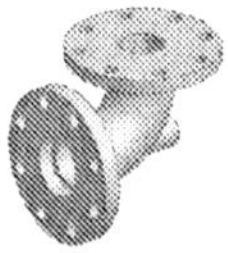

Tip. *You can also define the tangent length dynamically by dragging the red arrows provided at the start and at the end of the loft feature. As you drag the arrow the preview of the tangent length and the tangent value in the spinner modifies automatically.*

Guide Curves

You can also define the guide curve between the profiles of the loft feature to define the path of transition of the loft feature. The sketches created for the guide curve must have a **Pierce** relation with the sketches that define the loft section. The **Guide Curves** rollout is displayed in Figure 9-49.

Figure 9-50 shows the profiles and the guide curves for creating a loft feature with guide curves. Figure 9-51 shows the resultant loft feature created using the guide curves.

Options

The **Options** rollout of the **Loft PropertyManager** is provided with many options to improve the creation of the loft feature. All the options available in this rollout are the same as those discussed earlier while discussing the sweep option. An additional option available with this rollout is discussed next.

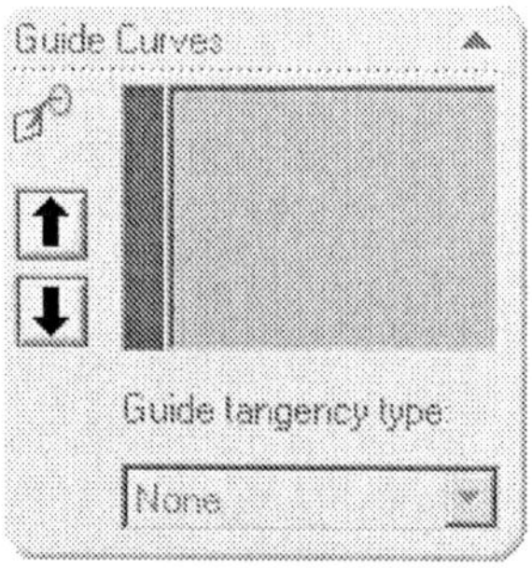

Figure 9-49 *The **Guide Curves** rollout*

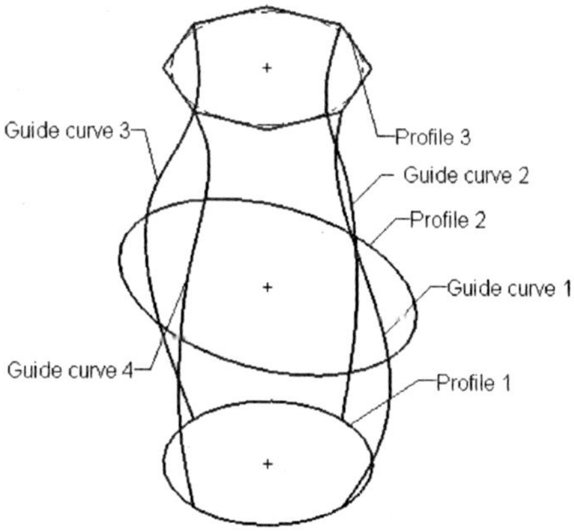

Figure 9-50 *Profiles and guide curves*

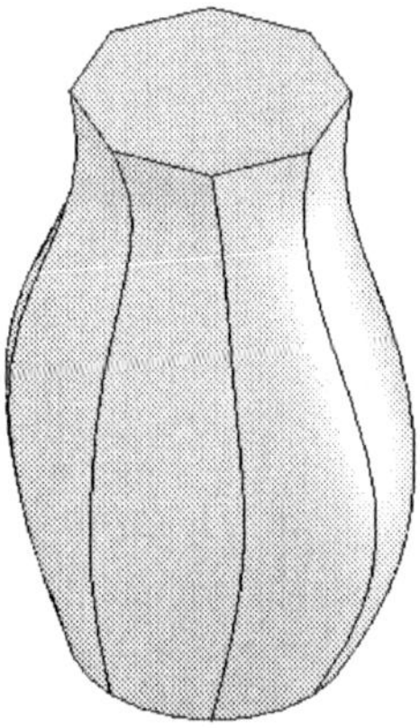

Figure 9-51 *Resultant loft feature*

Close Loft

A closed loft feature is one in which the end and the start of the loft feature are joined together. The **Close loft** check box is selected to create a close loft feature. Figure 9-52 shows a loft feature created with the **Close loft** check box cleared. Figure 9-53 shows the loft feature created with the **Close loft** check box selected.

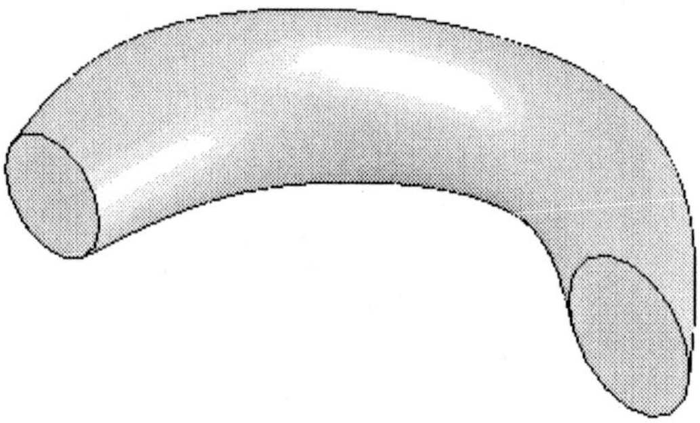

Figure 9-52 *Loft feature with the **Close Loft** check box cleared*

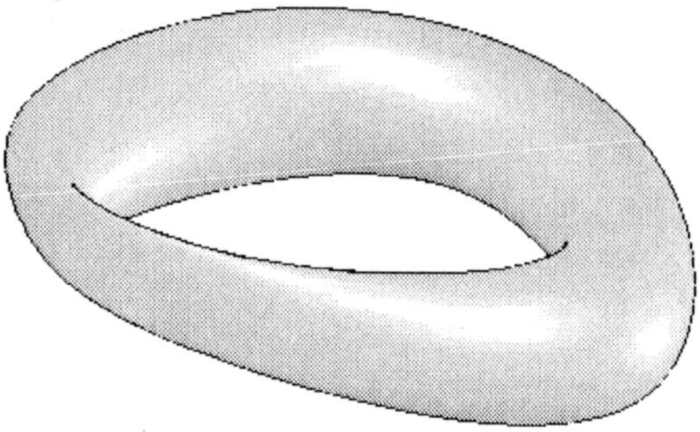

Figure 9-53 *Loft feature with the **Close loft** check box selected*

Centerline Parameters

The **Centerline Parameters** rollout is used to create a loft feature by blending two or more than two sections along a specified path. The path that specifies the transition is called centerline. The **Centerline Parameters** rollout is displayed in Figure 9-54. The various options available in this rollout are discussed next.

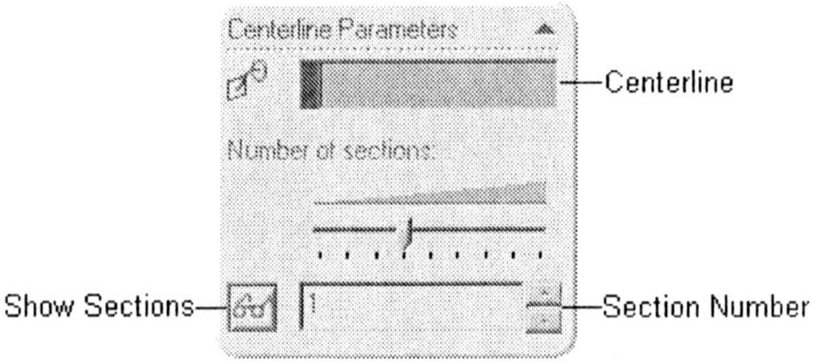

***Figure 9-54** The **Centerline Parameters** rollout*

Centerline

After invoking the **Centerline Parameters** rollout you need to define the centerline. Therefore, select the sketch that defines the centerline for the loft feature. The name of the sketch will be displayed in the **Centerline** selection area.

Number of sections

The **Number of sections** slider bar provided in the **Centerline Parameters** rollout is used to define the number of intermediate sections. These intermediate sections define the accuracy and the smoothness of the loft feature.

The **Show Sections** button and the **Section Number** spinner available in the **Centerline Parameters** rollout are used to display the intermediate sections as discussed earlier. Figure 9-55 shows the sketches of the profiles and the centerline used to create the loft feature. Figure 9-56 shows the resultant loft feature.

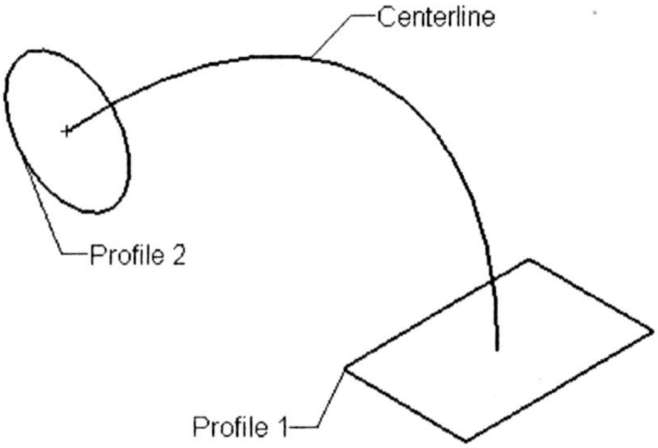

***Figure 9-55** Profiles and centerline*

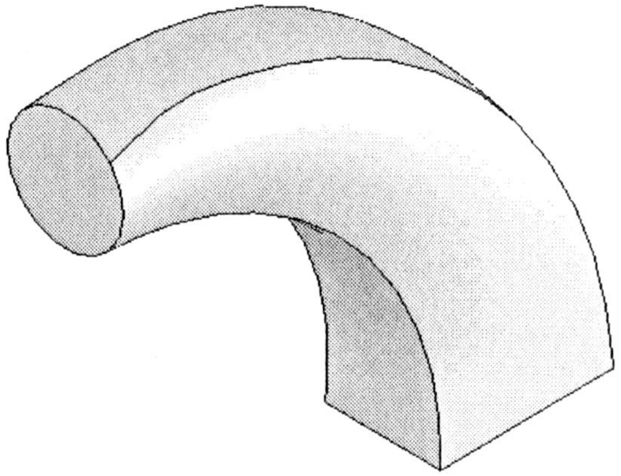

***Figure 9-56** The resultant loft feature*

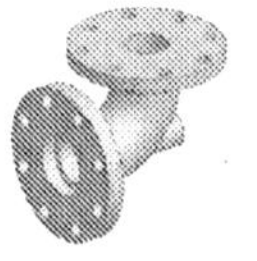

Tip. *If you need to create a smooth loft feature between a circle and a polygon, it is a good practice to split the circle into a number of arcs. The number of arcs that forms the circle should be the same as the number of sides of the polygon. Now, when you create the loft feature the number of connecting points in both the sections are predefined, which results in a smoother loft feature.*

You can also create a thin loft feature by defining the thin parameters using the **Thin Feature** rollout. The options available in this rollout are the same as those discussed in the earlier chapters in which extruding and revolving of thin features was discussed. Figure 9-57 shows a thin loft feature created using the **Thin Features** rollout available in the **Loft PropertyManager**.

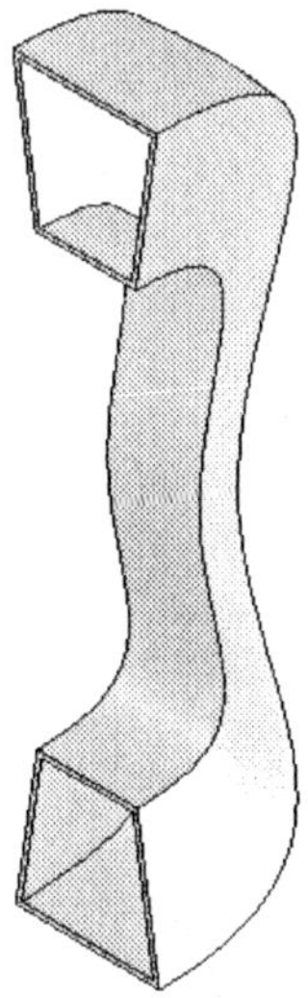

Figure 9-57 *A thin loft feature*

Add a Section in the Loft Feature

After creating a loft feature you can also add a section in the loft feature. To add a section in the loft feature select any one face except the end faces of the loft feature and right-click to invoke the shortcut menu. Choose the **Add Loft Section** option from the shortcut menu. The **Add Loft Section PropertyManager** is displayed. You are provided with a plane that can be moved or rotated dynamically. To move the plane, place the cursor on the arrows provided on the plane. The arrows will turn red in color. Press and hold down the left mouse button and drag the cursor to move the plane. To rotate the plane, select the plane on the edges and drag the cursor. Set the position of the plane by dynamically moving and rotating the plane to define the position where you need to add the sketch. Figures 9-58 and 9-59 show the plane being moved and rotated. After specifying the position of the plane choose the **OK** button from the **Add Loft Section PropertyManager**. Figure 9-60 shows the section added in the model. This is a closed section created using a spline.

Expand the **Loft1** feature and select the sketch that is added using the **Add Loft Section** tool. Right-click and choose the **Edit Sketch** option from the shortcut menu. Edit the sketch as shown in Figure 9-61 and rebuild the part. Figure 9-62 shows the resultant modified loft feature.

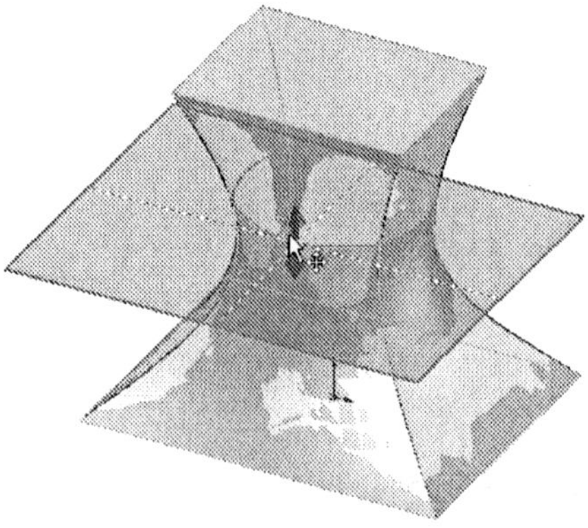

Figure 9-58 Moving the plane

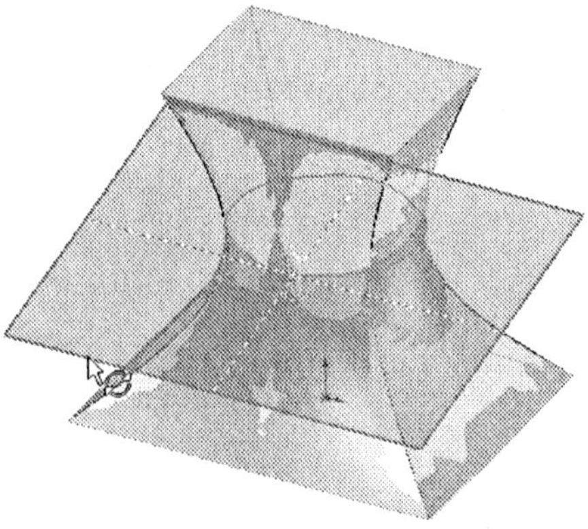

Figure 9-59 Rotating the plane

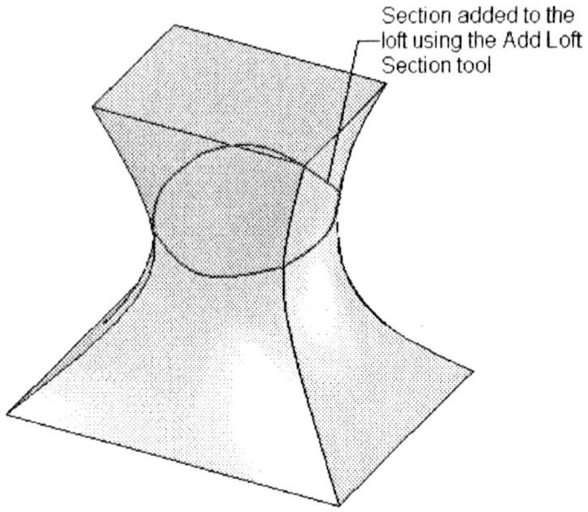

Figure 9-60 Resultant section added to loft feature

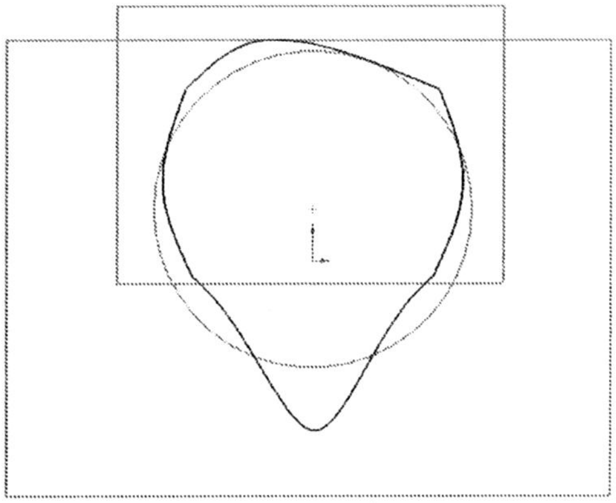

Figure 9-61 Modified loft section

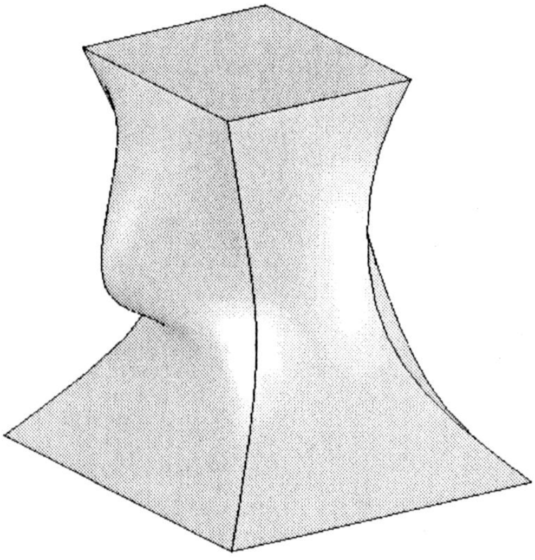

Figure 9-62 Resultant loft feature

The **Cut Loft PropertyManager** is used to create a cut loft feature. This option is invoked by choosing **Insert > Cut > Loft** from the menu bar.

Creating 3D Sketches

CommandManager:	Sketch > 3D Sketch
Menu:	Insert > 3D Sketch
Toolbar:	Sketch > 3D Sketch

In previous chapters you have learned how to create 2D sketches in the sketching environment. In this chapter you will learn how to create 3D sketches. The 3D sketches are mostly used to create 3D paths for the sweep features, 3D curves, and so on. Figure 9-63 shows a chair frame created by sweeping a profile along a 3D path.

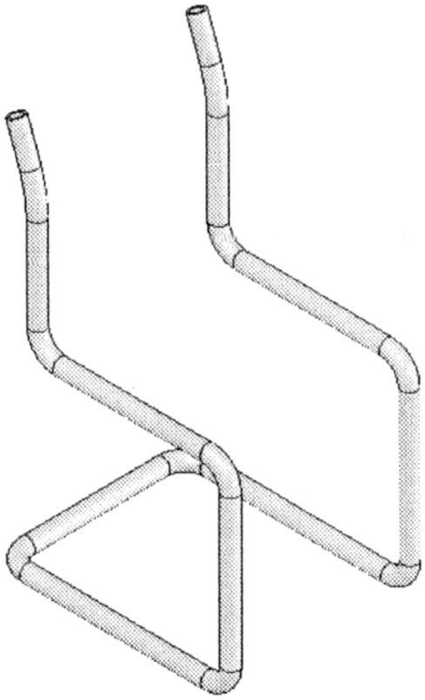

Figure 9-63 *A chair frame created by sweeping a profile along a 3D path*

To create a 3D sketch choose the **3D Sketch** button from the **Sketch CommandManager**. When you choose this option, the 3D sketching environment is invoked and the origin is displayed in red color. For creating a 3D sketch, you do not need to select a sketching plane. When you invoke the 3D sketching environment some of the sketching tools are activated in the **Sketch CommandManager**. You can use these tools in the 3D sketching environment. The sketching tools that can be used in the 3D sketching environment are discussed next.

Line

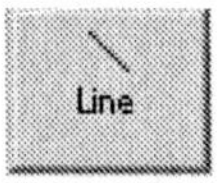

For creating lines in the 3D sketching environment, it is better to orient the view to isometric. Now, choose the **Line** button from the **Sketch CommandManager** to invoke the **Line** tool. The select cursor will be replaced by the line cursor with **XY** displayed at the bottom of the line cursor. This means that the sketch will be created in the XY plane by default. You can toggle between the default planes using the TAB key from the keyboard. Move the cursor to the origin to start the sketch or to the location from where you want to start the sketching. Press and hold down the left mouse button at this location and you are provided with a space handle. The space handle includes a coordinate system in the current plane.

If you toggle the plane using the TAB key then the coordinate system will change with respect to the current plane. Now, drag the cursor to a location where you want to define the endpoint of the line. The length of the line is displayed above the line cursor. Release the left mouse button at this location. Once you complete sketching the first line, move the cursor to the endpoint of the line drawn earlier. When the line cursor turns yellow in color, toggle the plane using the TAB key. Drag the cursor to create another line. To end drawing line right-click to invoke the shortcut menu and select the **Select** option. Figures 9-64 through 9-66 show sketching in different planes in the 3D sketching environment. Figure 9-67 shows an example of a 3D sketch.

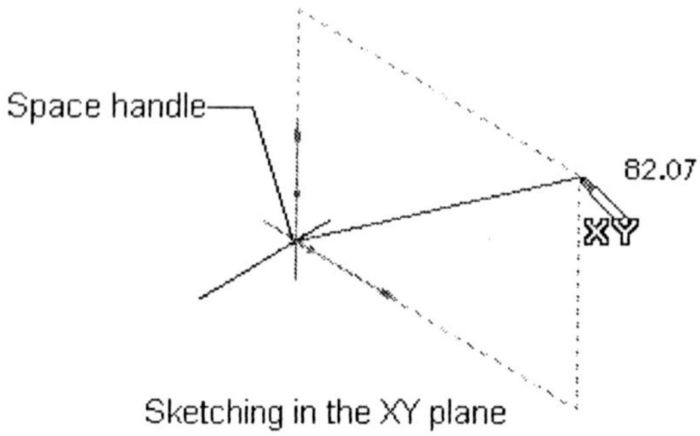

Figure 9-64 Sketching in the XY plane

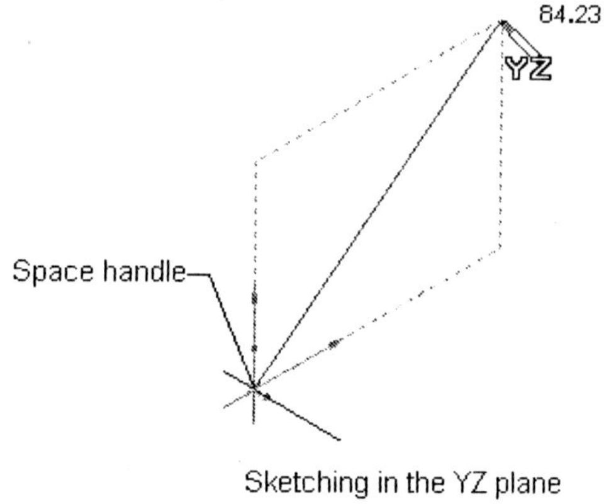

Figure 9-65 Sketching in the YZ plane

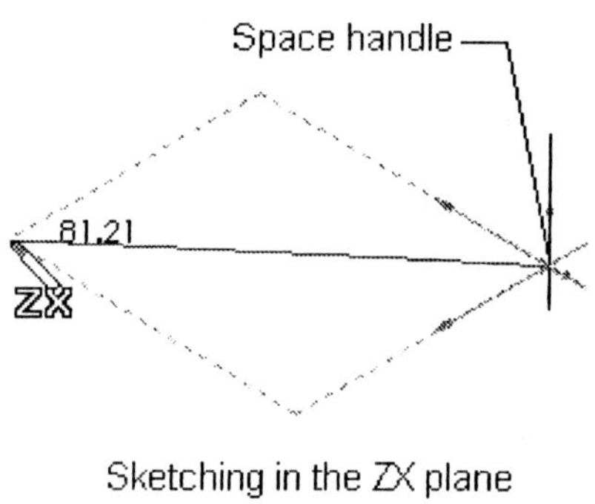

Figure 9-66 Sketching in the ZX plane

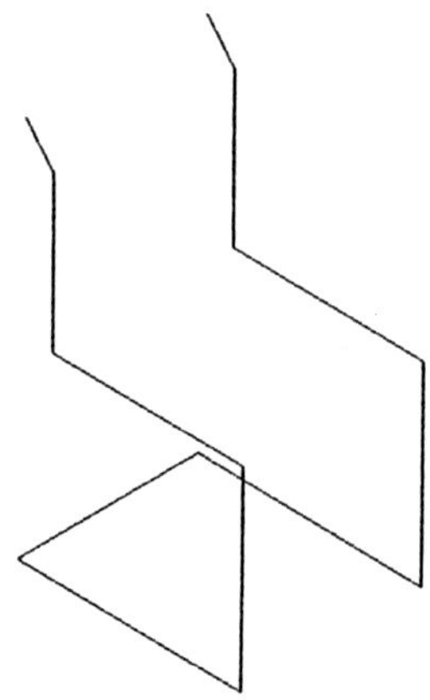

*Figure 9-67 3D sketch created using the **Line** tool*

Spline

You can also draw a spline in the 3D sketching environment. To draw a 3D spline choose the **Spline** button from the **Sketch CommandManager**. The select cursor will be replaced by the spline cursor. Move the cursor to the desired location from where you want to start the sketching. Specify the start point of the spline; the space handle is displayed. You can toggle between the default planes using the TAB key.

Move the cursor where you want to specify the second point of spline and specify the second point of spline. You can follow the same procedure to continue drawing spline. After creating the spline, right-click and select the **Select** or **End Spline** option from the shortcut menu to end spline creation.

Point

You can also draw points in the 3D sketching environment. To draw a 3D point, first you have to invoke the 3D sketching environment and it is better to orient the view to isometric. Choose the **Point** button from the **Sketch CommandManager** to invoke the **Point** tool. The select cursor will be replaced by the point cursor. Use the left mouse button to draw points.

Centerline

You can also draw centerlines in the 3D sketching environment. Invoke the 3D sketching environment and orient the view to isometric view. Choose the **Centerline** button from the **Sketch CommandManager** to draw a centerline. The select cursor will be replaced by the line cursor. The procedure of drawing the centerline is the same as that discussed for drawing the lines.

The dimensioning of 3D sketches is the same as the dimensioning of 2D sketches.

Editing the 3D Sketches

You can perform a number of editing operations on 3D sketches. These editing operations include **Convert Entities**, **Sketch Chamfer**, **Trim Entities**, **Fit Spline**, **Sketch Fillet**, **Extend Entities**, **Construction Geometry**, and **Split Entities**. All these tools were discussed in the earlier chapters.

Creating Curves

You can also create various types of curves in SolidWorks. These curves are mostly used to create complex shapes generally using the **Swept Boss/Base** and **Lofted Boss/Base** tools. The types of curves that can be created in SolidWorks are discussed next.

Creating Projection Curve

CommandManager:	Curves > Project Curve	*(Customize to Add)*
Menu:	Insert > Curve > Projected	
Toolbar:	Curves > Project Curve	

This option allows you to project a single closed/open sketched entity on one or more than one planar or curved faces. You can also project a sketched entity on another sketched entity to create a 3D curve. To create a projected curve, you first need to create at least two sketches or a sketch and at least one feature and then choose the **Project Curve** button from the **Curves CommandManager** after customizing. When you choose the **Project Curve** button, the **Projected Curve PropertyManager** is displayed as shown in Figure 9-68. The confirmation corner is also displayed in the drawing area. The two different options to create projected curves are discussed next.

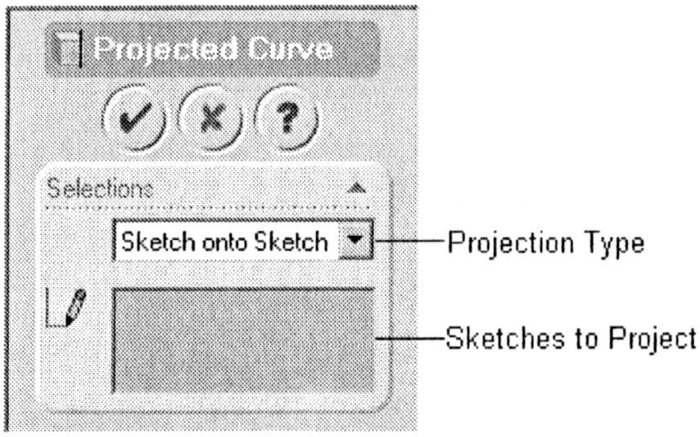

Figure 9-68 The Projected Curve PropertyManager

Sketch onto Sketch

When you invoke the **Projected Curve PropertyManager**, the **Sketch onto Sketch** option is selected by default in the **Projection Type** drop-down list. You are prompted to select two sketches to project onto one another. Select two sketches that do not lie on the same plane to project them onto one another. When you select the sketches, the names of the sketches are displayed in the **Sketches to Project** selection area. The preview of the projected curve is also displayed in the drawing area. Choose the **OK** button from the **Projected Curve PropertyManager** or choose **OK** from the confirmation corner. Figure 9-69 shows the two sketches selected to create a projected curve. Figure 9-70 shows the preview of the projected curve. Figure 9-71 show the resultant projected curve.

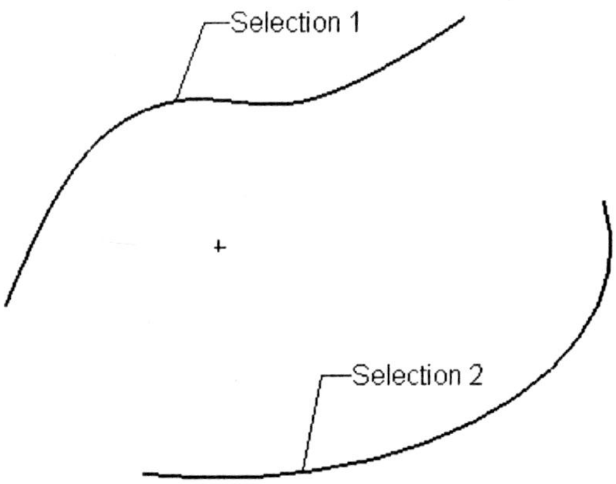

Figure 9-69 Sketches to be selected

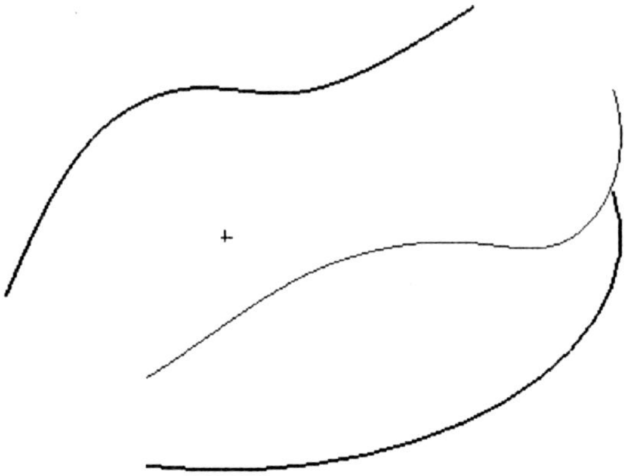

Figure 9-70 Preview of the projected curve

Sketches onto Face(s)

The **Sketches onto Face(s)** option is available in the **Projection Type** drop-down list. This option is used to project a sketch on a planar or curved face(s). When you choose this option, the **Sketch to Project** and the **Projection Faces** selection area are displayed in the **Selections** rollout. You are also provided with a **Reverse Projection** check box The **Projected Curve PropertyManager** with the **Sketches onto Face(s)** option selected is displayed in Figure 9-72. After you select this option, select a sketch to project and the face(s) on which to project. Now, select the sketch from the drawing area and the face or

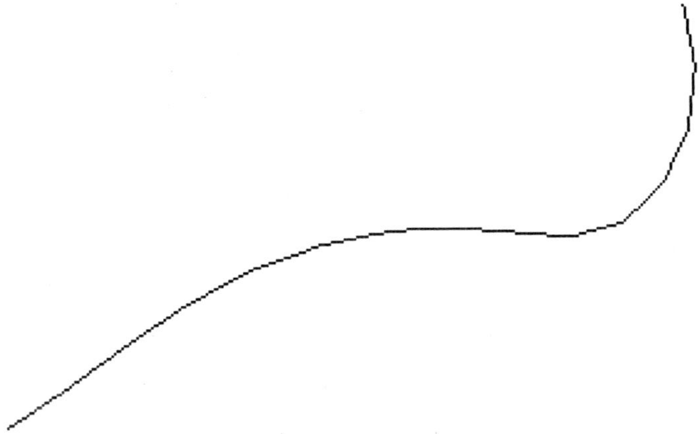

Figure 9-71 *Resultant projected curve*

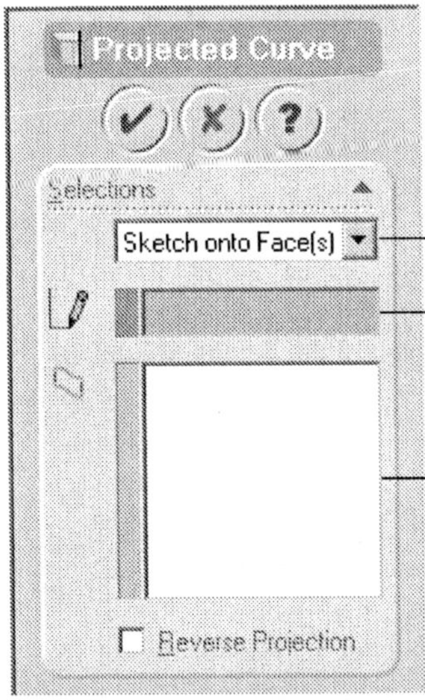

Figure 9-72 *The* ***Projected Curve PropertyManager*** *with the* ***Sketches onto Face(s)*** *option selected*

faces on which you want to project the sketch. The selected sketch is highlighted in green and the selected face is highlighted in magenta. You are also provided with an arrow in the drawing area. This arrow is used to reverse the direction of projection. You can also reverse the direction of projection using the **Reverse Projection** check box. Choose the **OK** button from the **Projected Curve PropertyManager** or choose **OK** from the confirmation corner. Figure 9-73 shows the sketch to be selected for projection and the face to be selected on which the sketch will be projected. Figure 9-74 shows the resultant projected sketch.

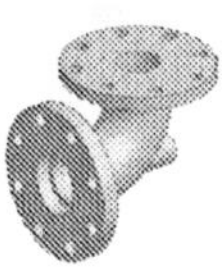

Tip. *You can also use the* ***Contour Select Tool*** *to select the contours for creating a projected curve or a split line. You will learn more about split line later in this chapter.*

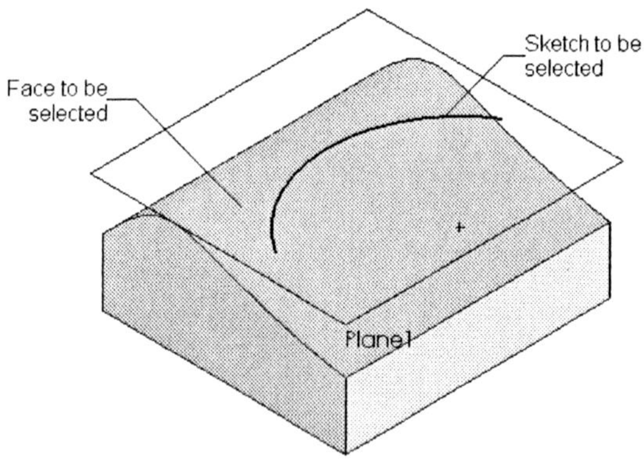

Figure 9-73 Sketch and the face to be selected

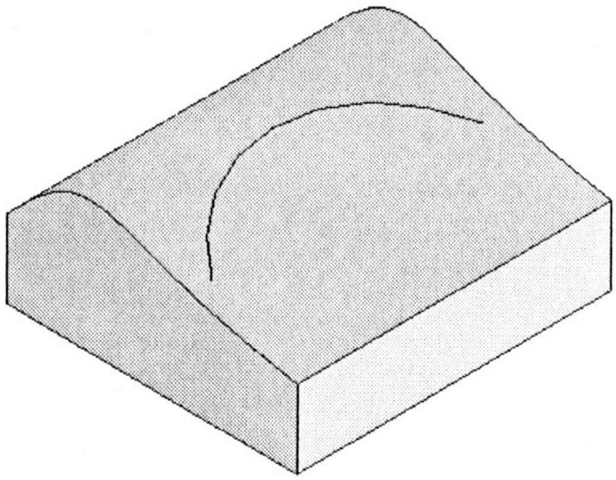

Figure 9-74 Resultant projected curve

Creating Split Lines

CommandManager:	Curves > Split Line	*(Customize to Add)*
Menu:	Insert > Curve > Split Line	
Toolbar:	Curves > Split Line	

The **Split Line** tool is generally used to project a sketch on a planar or a curved face and in turn it splits or divides the single face into two or more than two faces. To create a split line and to divide the faces, you need to invoke the **Split Line PropertyManager**. Choose the **Split Line** button from the **Curves CommandManager** after customizing it to invoke the **Split Line PropertyManager** as shown in Figure 9-75. There are two methods of creating a split line, which are discussed next.

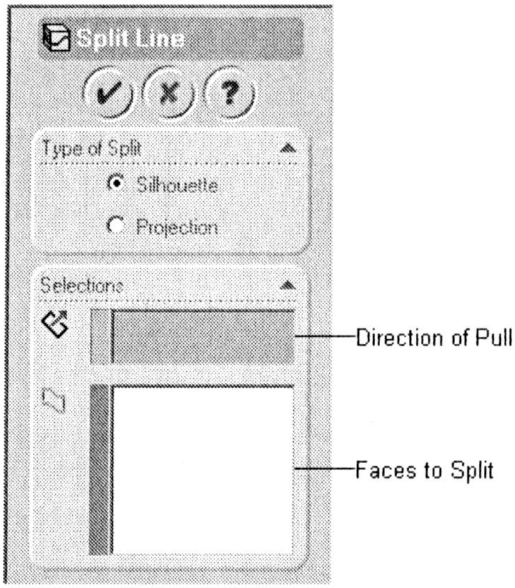

*Figure 9-75 The **Split Line PropertyManager***

Silhouette

Using this option you can split a curved face by creating a silhouette line at the intersection of the projection of direction entity and the curved face. When you invoke the **Split Line PropertyManager**, the **Silhouette** option available in the **Type of Split** rollout is selected by default. Now, change the type of split or select the direction of pull and faces to split. You need to select an edge, a line, or a plane that defines the direction of pull. Select the direction of pull; the selected entity will be highlighted in magenta color. Now, select the curved face; the selected face will be highlighted in green color. Choose the **OK** button from the **Split Line PropertyManager** or choose the **OK** option from the confirmation corner. The curved face will be divided in two or more faces. Consider a case in which you need to split a circular face using this option. Plane will be selected to define the direction of pull. Figure 9-76 shows the plane and the face to be selected. Figure 9-77 shows the resultant split line created to split the selected face.

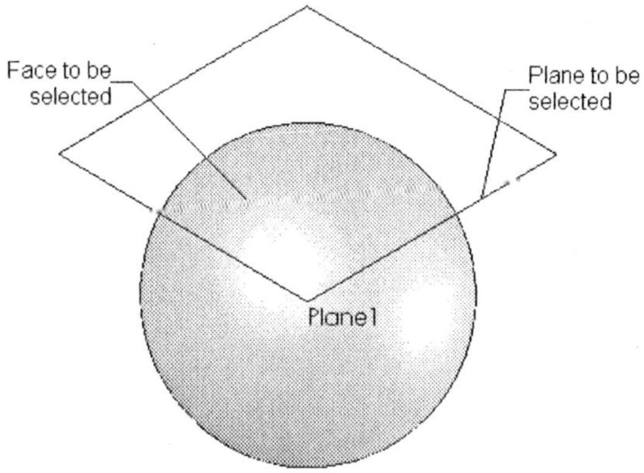

Figure 9-76 Plane and the face to be selected

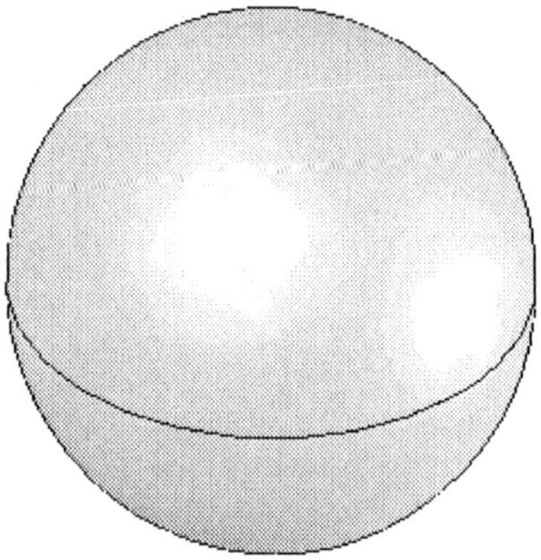

Figure 9-77 Resultant split line

Projection

Using this option you can project a sketched entity onto a planar or a curved face to create a split line on that face. The split line split the selected face on which the sketch is projected. To use this option invoke the **Split Line PropertyManager** and choose the **Projection** option from the **Type of Split** rollout. The **Split Line PropertyManager** with the **Projection** option selected is shown in Figure 9-78. You are prompted to change the type of split or select the sketch to project, direction, and faces to split. Select the sketch and the selected sketch will be displayed in magenta. Now, you need to select the face to split. Select the face; the selected face will be displayed in green. Choose the **OK** button from the **Split Line PropertyManager**. The selected face or faces will split into two or more than two faces depending upon the sketch that is used to project. Figure 9-79 shows the sketch and face to be selected. Figure 9-80 shows the resultant split line created to split the selected face.

The other options available in the **Selections** rollout of the **Split Line PropertyManager** are discussed next.

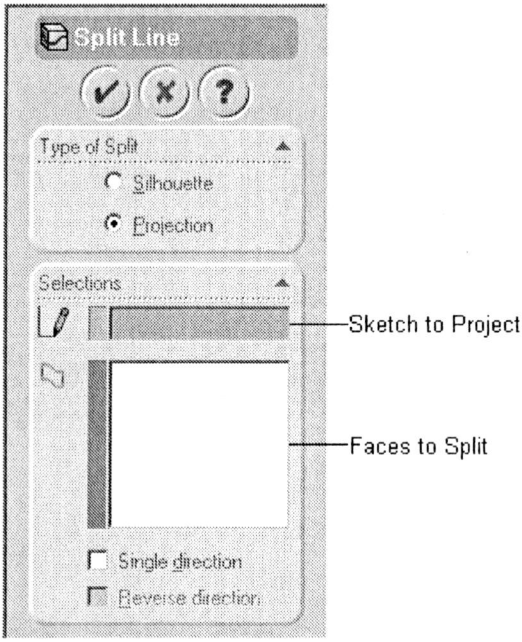

***Figure 9-78** The **Split Line PropertyManager** with the **Projection** option selected*

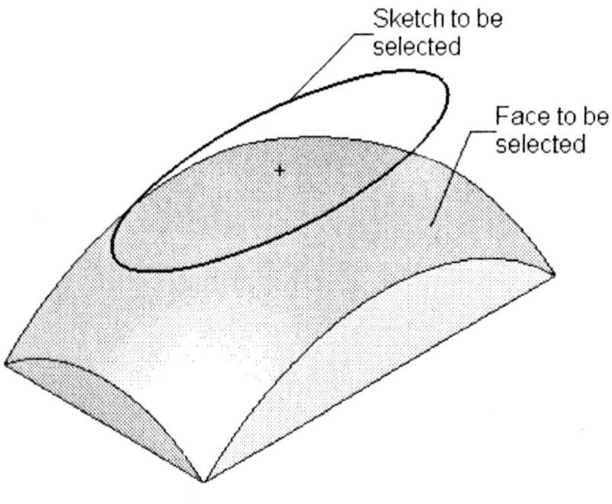

***Figure 9-79** Sketch and the face to be selected*

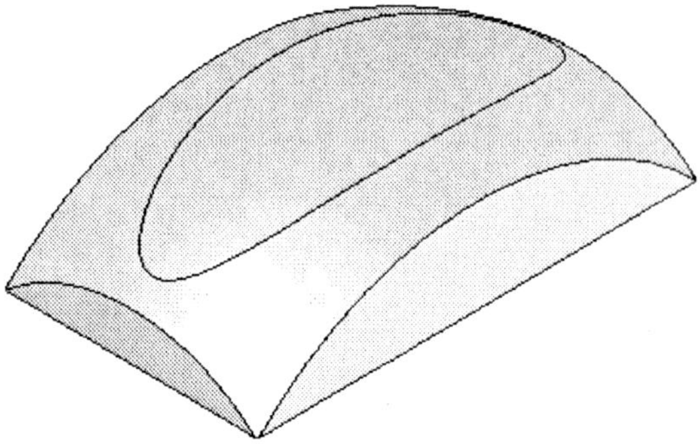

***Figure 9-80** Resultant split line*

Single direction
While creating a split on a cylindrical face, if the sketching plane on which the sketch is created lies within the model, the split line will be created on two sides of the cylindrical face. The **Single direction** check box available in the **Selections** rollout is used to create the split line only in one direction.

Reverse direction
The **Reverse direction** check box is available only if the **Single direction** check box is selected. The **Reverse direction** check box is used to reverse the direction in which

the split line should be created. Figure 9-81 shows the split line created on both sides of the model. Figure 9-82 shows the split line created on single side of the model.

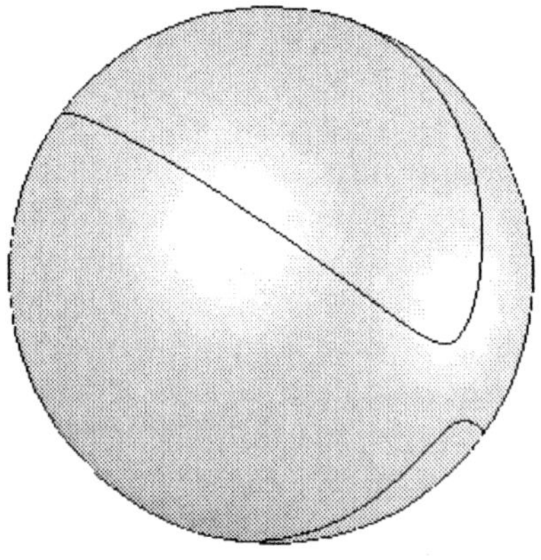

Figure 9-81 Split line created on both sides

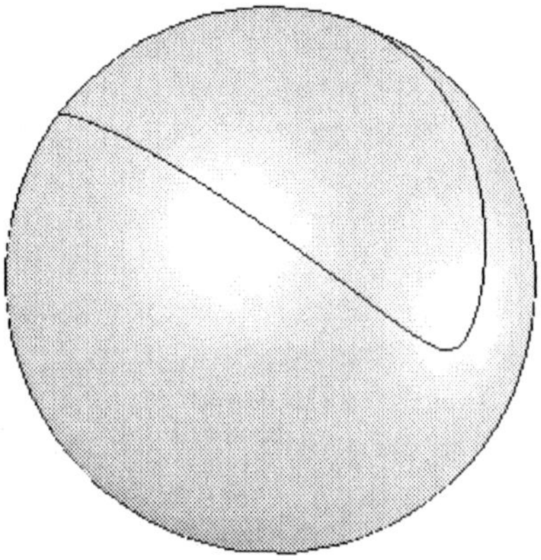

Figure 9-82 Split line created on single side

Creating Composite Curve

CommandManager:	Curves > Composite Curve
Menu:	Insert > Curve > Composite
Toolbar:	Curves > Composite Curve

The **Composite Curve** option is used to create a curve by combining 2D or 3D curves, sketched entities, and part edges into a single curve. You need to make sure that the selected entities form a continuous chain. The composite curve is mainly used while creating a sweep or a loft feature. To create a composite curve you need to invoke the **Composite Curve PropertyManager**. Choose the **Composite Curve** button from the **Curves CommandManager** to invoke the **Composite Curve PropertyManager**. The **Composite Curve PropertyManager** is shown in Figure 9-83.

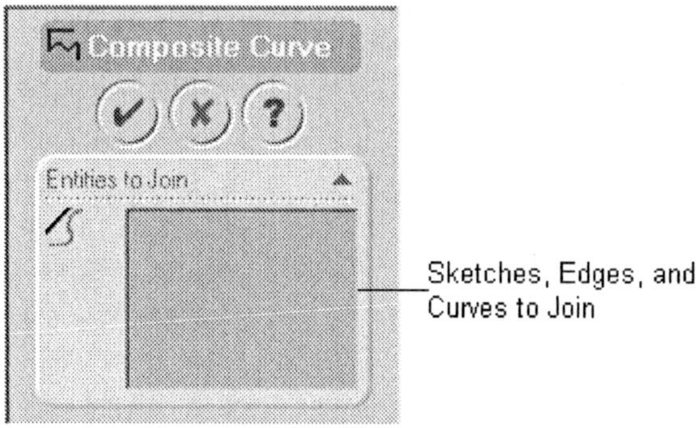

Figure 9-83** The **Composite Curve PropertyManager

When you invoke the **Composite Curve PropertyManager** you are prompted to select a continuous set of sketches, edges, and/or curves. Select the continuous edges, curves, or sketched entities to create a composite curve. Choose the **OK** button to end curve creation.

Creating Curve Through XYZ Points

CommandManager:	Curves > Curve Through XYZ Points
Menu:	Insert > Curve > Curve Through XYZ Points
Toolbar:	Curves > Curve Through XYZ Points

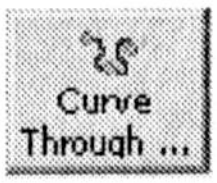

The **Curve Through XYZ Points** option is used to create a curve by specifying the coordinate points. To create a curve using this option, choose the **Curve Through XYZ Points** button from the **Curves CommandManager**. The **Curve File** dialog box is displayed as shown in Figure 9-84. Double-click in the cell under the **X** column to enter the X coordinate for the start point for the curve. Similarly, double-click in the **Y** cell and **Z** cell to enter the respective Y and Z coordinates of the start point of the curve. Double-click in the cell below the first cell to enter the coordinates for the second point for creating the curve. Similarly, specify the coordinates of the other points of the curve, see Figure 9-85. When you enter the coordinates of the points the preview of the curve is displayed in the drawing area. Choose the **OK** button from the **Curve File** dialog box to complete feature creation, see Figure 9-86.

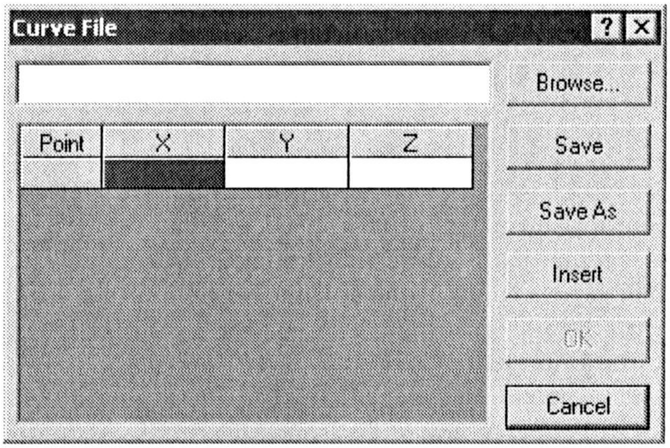

***Figure 9-84** The **Curve File** dialog box*

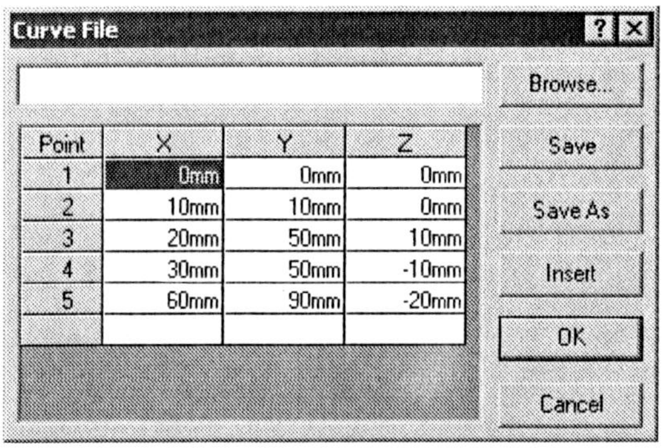

***Figure 9-85** Coordinates entered in the **Curve File** dialog box*

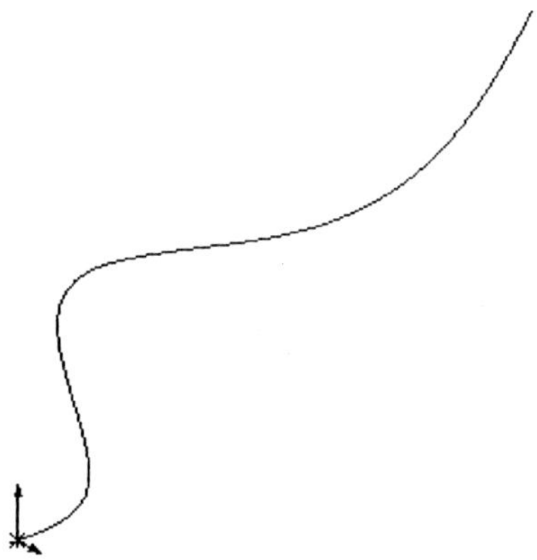

***Figure 9-86** Resultant 3D curve*

> **Tip**. *You can select a row and using the DELETE key from the keyboard you can delete the entire row. Using the SHIFT key select the entire row and using the* ***Insert*** *button, you can add a row between two rows.*

You can also save the current set of coordinates using the **Save** button available in the **Curve File** dialog box. When you choose this button, the **Save As** dialog box is displayed. Browse the folder in which you need to save the coordinates and enter the name of the file in the **File name** edit box and choose the **Save** button. The curve file is saved with extension *.sldcrv*. Using the **Save As** button you can save the current set of coordinates with some other name.

Using the **Browse** button, you can open an existing curve file. When you choose the **Browse** button, the **Open** dialog box will be displayed. You can browse the previously saved curve file to specify the coordinate points. You can also write the coordinates in a text (notepad) file and save it. In the **Open** dialog box, choose the **Text Files (*.txt)** option from the **Files of type** drop-down list and browse the text file to specify the coordinates.

Creating Curve Through Reference Points

CommandManager:	Curves > Curve Through Reference Points
Menu:	Insert > Curve > Curve Through Reference Points
Toolbar:	Curves > Curve Through Reference Points

The **Curve Through Reference Points** option enables you to create a curve by selecting the sketched points, vertices, origin, endpoints, or center points. To create a curve through reference points, you have to invoke the **Curve Through Reference Points PropertyManager**. Choose the **Curve Through Reference Points** button from the **Curves CommandManager** to invoke the **Curve Through Reference Points PropertyManager** as shown in Figure 9-87.

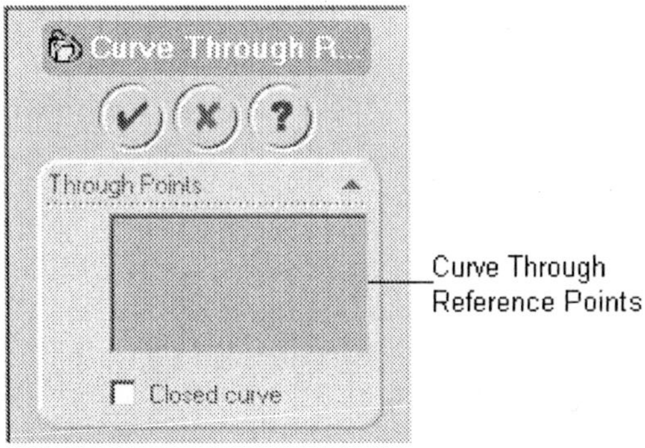

Figure 9-87 The ***Curve Through Reference Points PropertyManager***

When you invoke this tool, you are prompted to select vertices to define the through points for the curve. Select the points to define the curve. As you define the points, the preview of the resultant curve created using the selected points is displayed in the drawing area. After specifying all the points, choose the **OK** button. You can select the **Closed curve** check box to create a closed curve. Figure 9-88 shows the vertices to be selected to create the curve through reference points. Figure 9-89 shows the resultant curve.

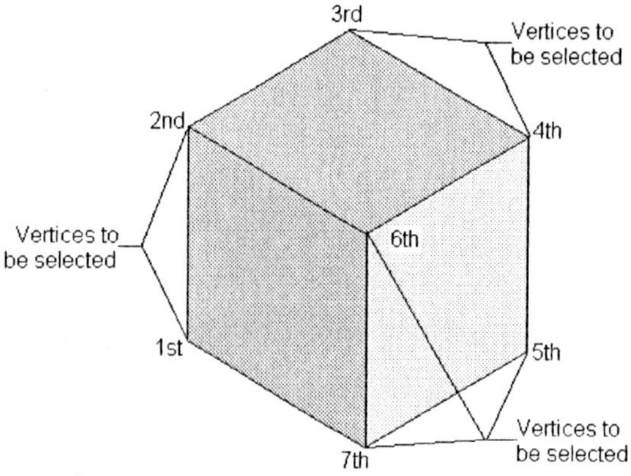

Figure 9-88 *Vertices to be selected*

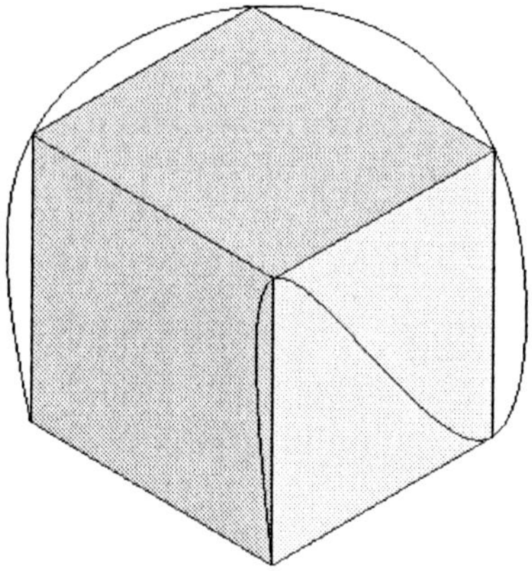

Figure 9-89 *Resultant 3D curve*

Creating Helical/Spiral Curve

CommandManager:	Curves > Helix and Spiral
Menu:	Insert > Curve > Helix/Spiral
Toolbar:	Curves > Helix and Spiral

The **Helix and Spiral** tool is used to create a helical curve or a spiral curve. The helical or spiral curve is generally used as the sweep path to create springs, threads, spiral coils, and so on. Figure 9-90 shows a spring created by sweeping a circular profile along a helical path. Figure 9-91 shows a spiral coil created by sweeping a rectangular profile along a spiral path.

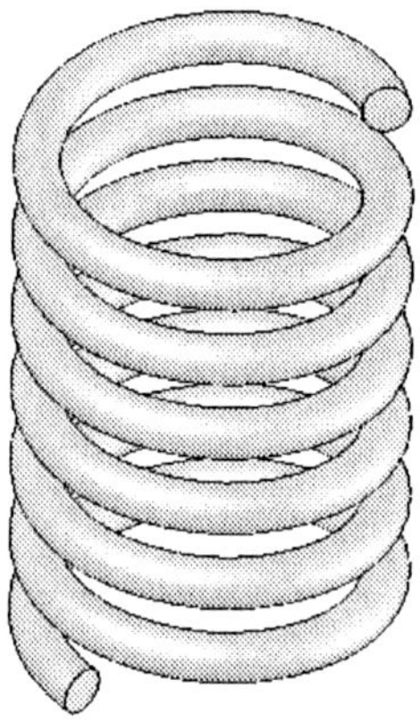

Figure 9-90 *Spring*

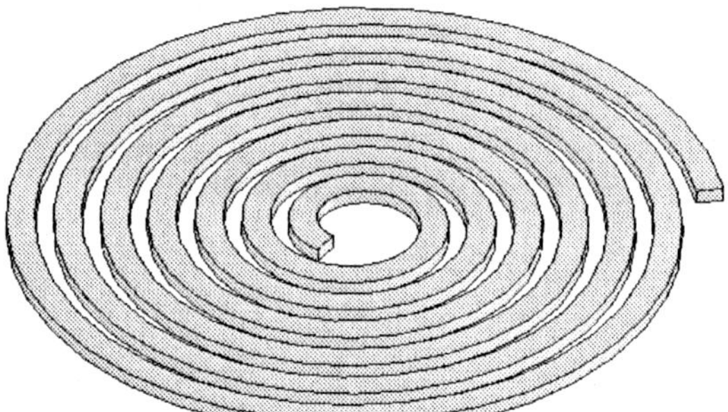

Figure 9-91 *Spiral coil*

To create a helix choose the **Helix and Spiral** button from the **Curves CommandManager**. The **Helix/Spiral PropertyManager** is displayed and you are prompted to select a plane, a planar face, or an edge on which to sketch a circle to define the helix cross-section or a sketch that contains a single circle. The circle will define the diameter of the spring. If you are creating a spiral, the sketch will define the starting diameter of the spiral curve. After you select a plane the **Helix/Spiral PropertyManager** disappears and you are in the sketching environment to draw the cross-section sketch for the helix or spiral. After creating the sketch exit the sketching environment. The **Helix/Spiral PropertyManager** will appear again as shown in Figure 9-92.

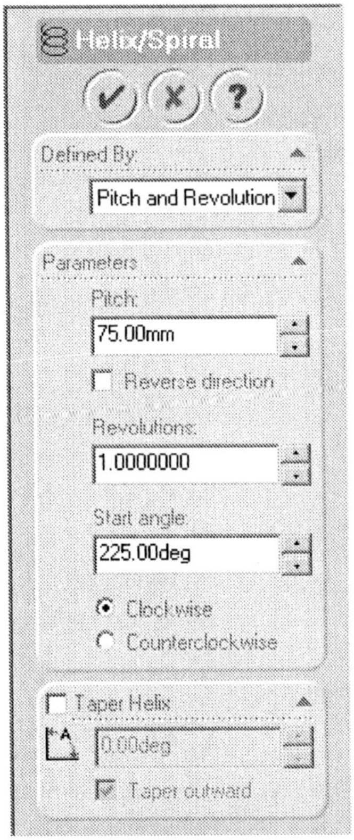

Figure 9-92** The **Helix/Spiral PropertyManager

Two arrows are displayed, one emerging from the center of the circle, which defines the direction of helix. The second arrow is tangent to the circle, which defines the rotation of the helix, that is, whether the helix is clockwise or counterclockwise. The preview of the helix curve is displayed in the drawing area with the default values. There are various methods to specify the parameters of the helical curve. These methods are discussed next.

Pitch and Revolution

The **Pitch and Revolution** option is selected by default in the **Type** drop-down list of the **Defined By** rollout. Using this option, you can specify the pitch of the helical curve and the number of revolutions. When this option is selected, the **Pitch** spinner and the **Revolution** spinner are available in the **Parameters** rollout to define the value of pitch and number of revolutions. You can also select the **Reverse direction** check box available in the **Parameters** rollout to reverse the direction of helix creation. Using the **Start angle** spinner you can specify the start angle of the helical curve. The **Clockwise** and **Counterclockwise** radio buttons available in this rollout are used to define the direction of rotation of helix.

Height and Revolution

The **Height and Revolution** option available in the **Type** drop-down list is used to define the parameters of the helix curve in the form of the total helix height and the number of revolutions. When you choose this option, the **Height** and **Revolutions** spinners are displayed in the **Parameters** rollout along with other options to specify the required parameters.

Height and Pitch

The **Height and Pitch** option available in the **Type** drop-down list is used to define the parameters of the helix curve in terms of the height and the pitch of the helix. When you select this option, the **Height** and **Pitch** spinners are displayed in the **Parameters** rollout along with other options to specify the required parameters.

When you specify the parameters to create the helix curve, the preview in the drawing area modifies automatically. Figure 9-93 shows a helix curve.

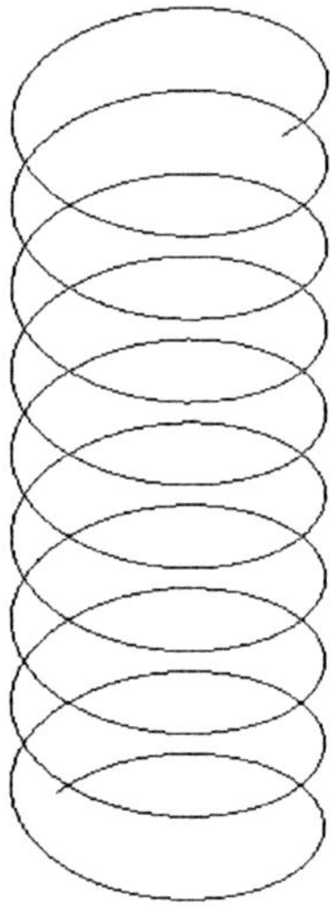

Figure 9-93 *Helix Curve*

You can also create a tapered helix using the **Taper Helix** rollout The procedure of creating a taper helical curve is discussed next.

Taper Helix

Invoke the **Taper Helix** rollout by selecting the **Taper Helix** check box available in the **Taper Helix** rollout. The **Taper Angle** spinner and the **Taper outward** check box are enabled. Using the **Taper Angle** spinner you can specify the value of the angle between the central axis of the helix and the periphery of the helix. The **Taper outward** check box is selected by default and is used to create an outward taper. When you specify the parameters to create a tapered helical curve, the preview of the helical curve updates automatically in the drawing area. Figure 9-94 shows a tapered helical curve. Figure 9-95 shows a tapered helical curve created with the **Taper outward** check box selected.

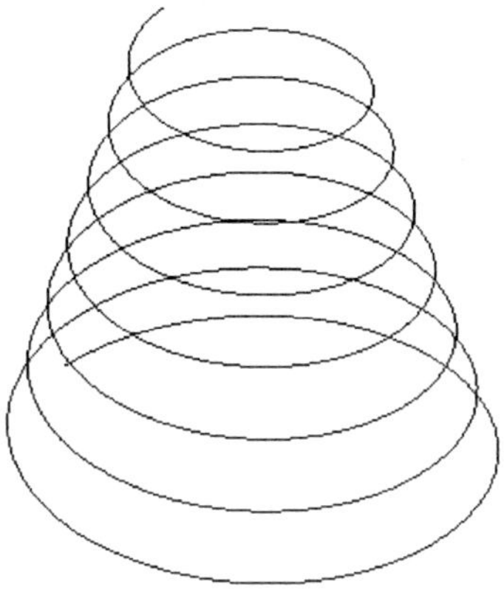

Figure 9-94 *Tapered helical curve*

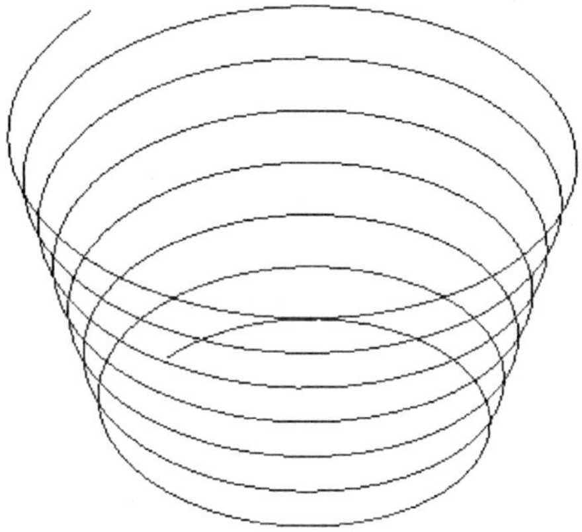

Figure 9-95 *Tapered helical curve with the* ***Taper outwards*** *option selected*

By default, the helical curve is created in the clockwise direction. Therefore, the **Clockwise** radio button is selected in the **Parameters** rollout of the **Helix/Spiral PropertyManager**. If you need to create the helical curve in the counterclockwise direction, you need to select the **Counterclockwise** radio button. After setting all the parameters, choose the **OK** button from the **Helix/Spiral PropertyManager**.

Creating a Spiral Curve

For creating a spiral curve select the **Spiral** option from the **Type** drop-down list available in the **Defined By** rollout. The preview of the spiral curve will be displayed in the drawing area. You can define the pitch and the number of revolutions in the **Pitch** and **Revolutions** spinners, respectively. The other options are the same in the **Parameters** rollout, as were discussed earlier. After specifying all the required parameters, choose the **OK** button. Figure 9-96 shows a spiral curve.

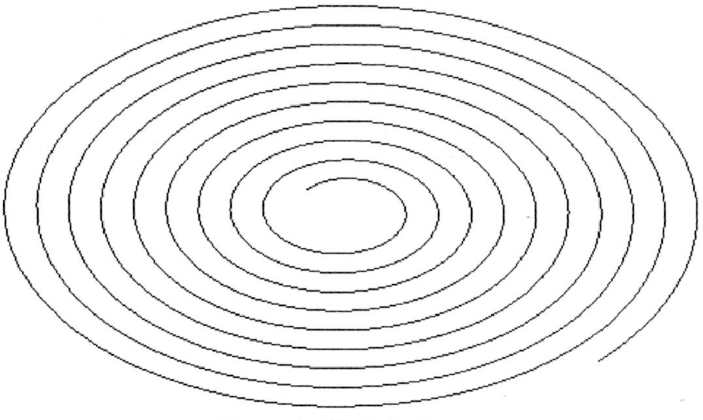

Figure 9-97 *Spiral curve*

Extruding a 3D Sketch

One of the major enhancements of SolidWorks 2004 is the introduction of extruding a 3D sketch. A 3D sketch is always extruded along a direction vector. The direction vector could be a line, edge, planar face, or a plane. To extrude a 3D sketch invoke the **Extrude PropertyManager** and then select the 3D sketch that you need to extrude. You also need to select a direction vector along which the sketch will be extruded. Set the value of the depth of the extrude feature and choose the **OK** button from the **Extrude PropertyManager**. Figure 9-97 shows a 3D sketch and the direction vector along which it will be extruded. Figure 9-98 shows the resultant extrude feature.

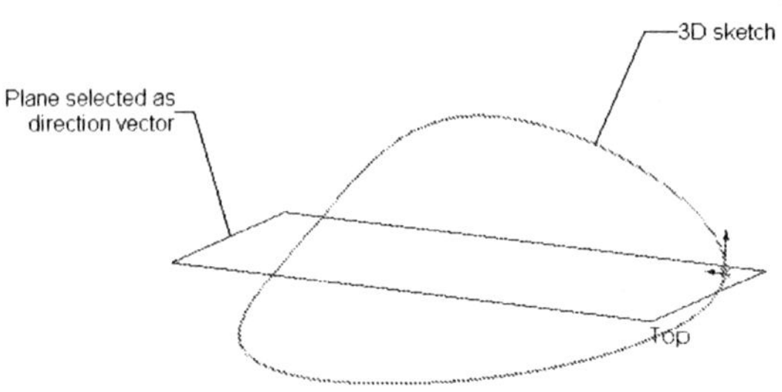

Figure 9-97 3D sketch and the direction vector

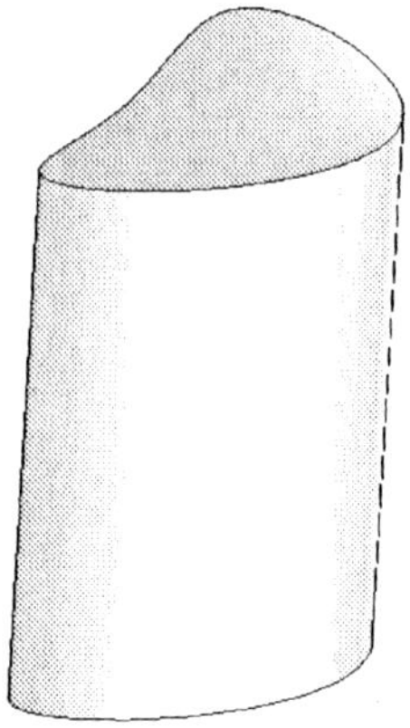

Figure 9-98 Resultant extrude feature

Creating Draft Features

CommandManager:	Features > Draft
Menu:	Insert > Features > Draft
Toolbar:	Features > Draft

The **Draft** tool is used to add taper to the selected faces of the model. One of the main application of creating the draft feature is in models that are to be moulded or casted so that it is easier to remove them from the mould or die. To create a draft feature you need to invoke the **Draft PropertyManager**. Choose the **Draft** button from the **Features CommandManager** to invoke the **Draft PropertyManager**. The **Draft PropertyManager** is shown in Figure 9-99.

After invoking this tool, you are prompted to select a neutral plane and the faces to draft. You will notice that the **Neutral Plane** option is selected by default in the **Type of Draft** drop-down list in the **Type of Draft** rollout. Therefore, you need to select a neutral plane for creating the draft feature. Select a planar face or a plane that acts as a neutral plane. The selected entity will appear in magenta with the **Neutral Plane** callout. A **Reverse Direction** arrow is also displayed in the drawing area. The selection mode in the **Faces to Draft** selection area of the **Faces to Draft** rollout is activated. Select the faces to apply the draft. The selected faces will be displayed in green with the **Draft Face** callout. Now, set the value of the draft angle in the **Draft Angle** spinner available in the **Draft Angle** rollout. Choose the **OK** button from the **Draft PropertyManager** or choose **OK** from the confirmation corner.

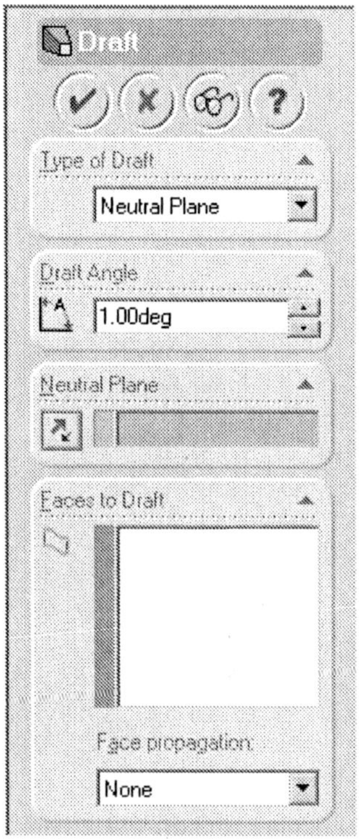

Figure 9-99 *The* ***Draft PropertyManager***

Figure 9-100 shows the neutral face and the faces to be selected to add the draft. Figure 9-101 shows the resultant draft feature. The other options in the **Draft PropertyManager** are discussed next.

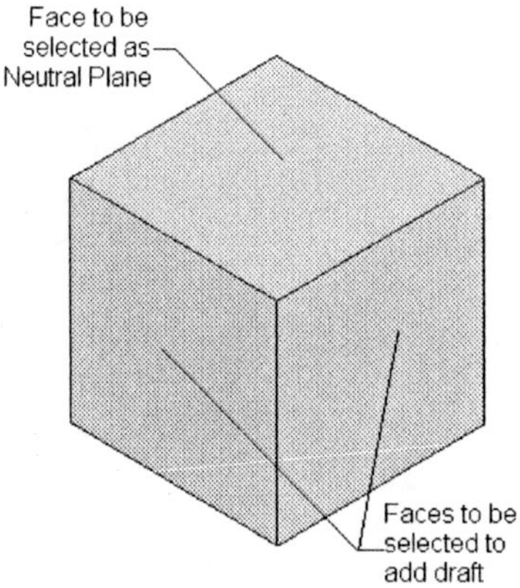

Figure 9-100 *Faces to be selected*

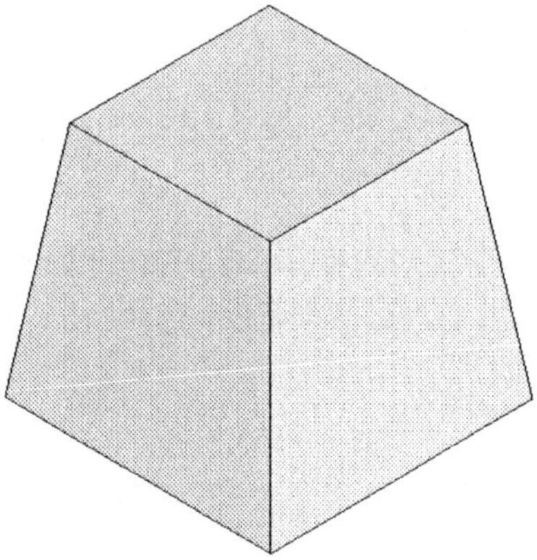

Figure 9-101 *Resultant draft feature*

Reverse Direction

The **Reverse Direction** button available on the left of the **Neutral Plane** selection area in the **Neutral Plane** rollout is used to reverse the direction of draft creation. You can also reverse the direction of draft creation by selecting the **Reverse Direction** arrow from the drawing area.

Face propagation

The options available in the **Face propagation** drop-down list in the **Faces to Draft** rollout are used to extend the draft feature to the other faces. The options available in this drop-down list are discussed next.

None

The **None** option is selected by default. This option is used when you do not need to apply any type of face propagation.

Along Tangent

The **Along Tangent** option is used to apply the draft to the faces tangent to the selected face.

All Faces

The **All Faces** option is used to apply the draft to all the faces attached to the neutral plane or face.

Inner Faces

This option is used to draft all the faces inside the model that are attached to the neutral plane or face.

Outer Faces

This option is used to draft all the outside faces of the model that are attached to the neutral plane or face.

The remaining advanced draft options are discussed in advanced chapters. These chapter are available in part 2 of this book.

Creating the Deform Feature

New

CommandManager:	Features > Deform	*(Customize to Add)*
Menu:	Insert > Features > Deform	
Toolbar:	Features > Deform	*(Customize to Add)*

One of the greatest enhancements in SolidWorks 2004 is the introduction of **the Deform** tool. This is a free-style tool and is used to manipulate the shape of a particular portion of the model. You can also manipulate the shape of the entire model using this tool. To create a deform feature choose the **Deform** button from the **Features CommandManager** after customizing it. The **Deform PropertyManager** is invoked as shown in Figure 9-102. There are generally two methods of creating a deform feature. Both the methods of creating the deform feature are discussed next.

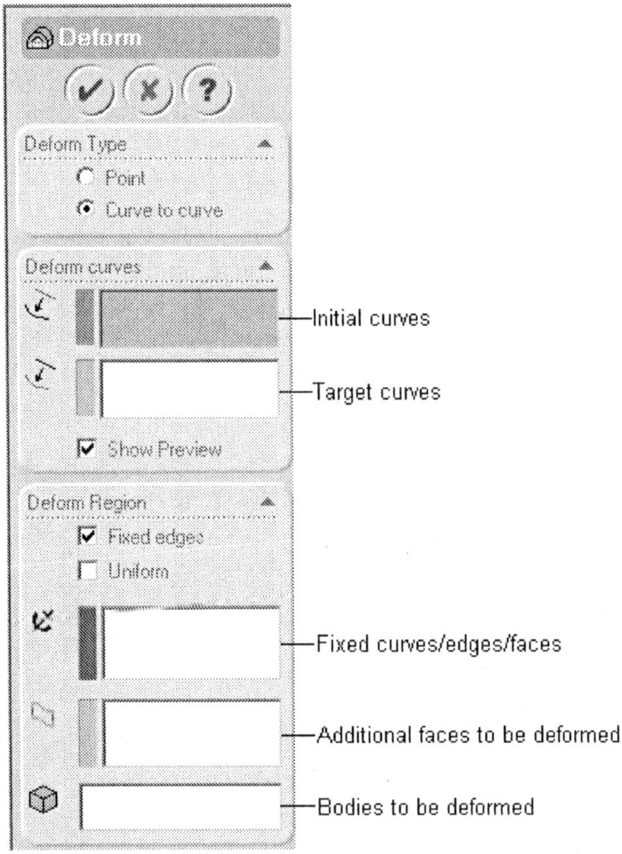

*Figure 9-102 The **Deform PropertyManager***

Creating the Deform Feature Using the Curve to curve Option

To create a deform feature using the **Curve to curve** option, invoke the **Deform PropertyManager**; the **Curve to curve** radio button is selected by default. The various options available to create the deform feature using this option are discussed next.

Deform curves

The **Deform curves** rollout is used to define the curves that are used to deform the shape of the model. When you invoke this tool you are prompted to select the geometry to define the deformation and fixed geometry for the deform operation. First you need to select the initial curve. The initial curve will define the reference from where the deformation will start. Select an edge or a chain of edges or curves that will define the initial curve. Next, you need to select the target curves. Click once in the **Target curves** selection area and select the target curve, which can be a sketch, an edge, or a curve. The profile of the target curve will define the shape of the deform feature.

Deform Region

The **Deform Region** rollout is used to define the options to deform a specific area of the model. By default the entire model is deformed when you specify the initial curves and the target curves. The **Fixed edges** check box is selected by default. Therefore, the **Fixed curves/edges/faces** selection area is invoked. Using this selection area you can select the curves, edges, or faces that you need to keep fixed while creating the deform feature. The **Additional faces to be deformed** selection area is used to specify the additional faces that you need to deform. If you clear the **Fixed edges** check box, then the **Fixed curves/ edges/faces** selection area and the **Additional faces to be deformed** selection area disappears.

The **Uniform** check box available in the **Deform Region** rollout is used to uniformly deform the model. If you select the **Uniform** check box from the **Deform Region** rollout the **Deform radius** spinner is displayed. The **Bodies to be deformed** selection area is used to define the bodies that you need to deform. This option is used in a multibody model.

Shape Options

The options available in this rollout are used to define the intensity of stiffness, shape, accuracy, and intensity of weight along the fixed reference or moving reference.

Note

*The weight slider is available only in the **Shape Options** rollout, if the **Uniform** check box available in the **Deform Region** rollout is cleared.*

Figure 9-103 shows the respective references to be selected and Figure 9-104 shows the resultant preview of the deform feature.

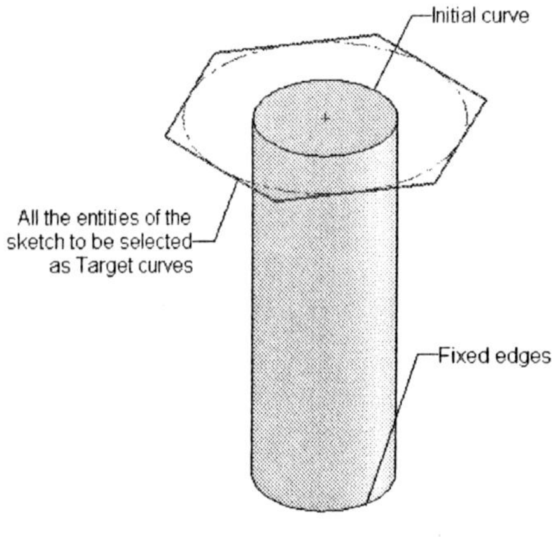

Figure 9-103 References to be selected

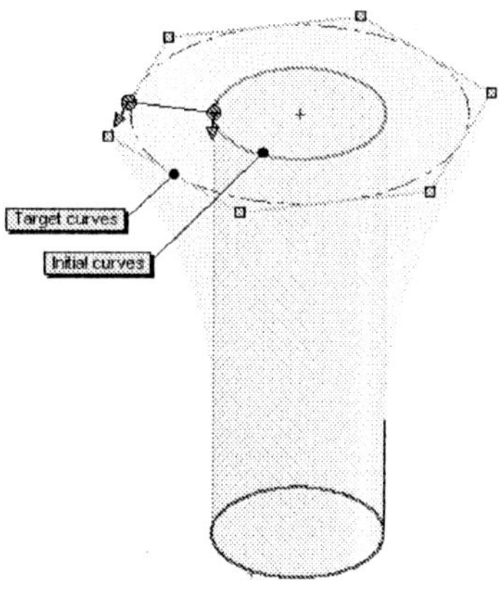

Figure 9-104 Resultant preview of the deform feature

When you select the initial curve and the target curve a connector is displayed connecting both the curves. Using the handles provided on the sides of the connector you can position the connector. To define additional connectors invoke the shortcut menu and choose the **Add Connector** option from the shortcut menu. After setting all the parameters choose the **OK** button from the **Deform PropertyManager**. Figure 9-105 shows the resultant deform feature.

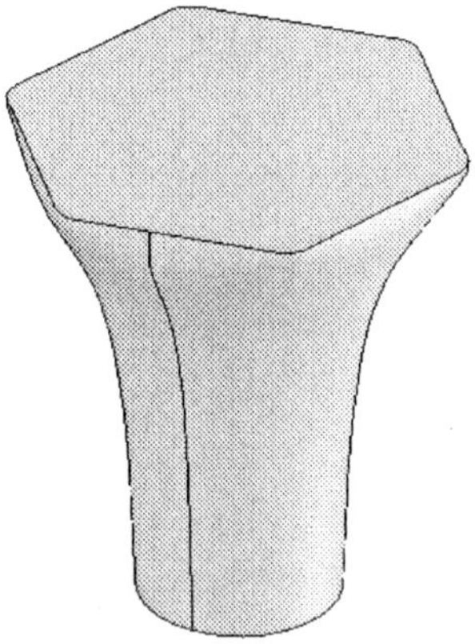

Figure 9-104 Resultant deform feature

Creating the Deform Feature Using the Point Option

To create the deform feature using the **Point** option, select the **Point** radio button from the **Deform Type** rollout as shown in Figure 9-106. You are prompted to select geometry to define the region, deformation point, and fixed geometry for the deform operation. You need to select a point on the model face, vertex, or edge from where you need to deform the model. As soon as you select any of the above-mentioned entities from the model, the model will be

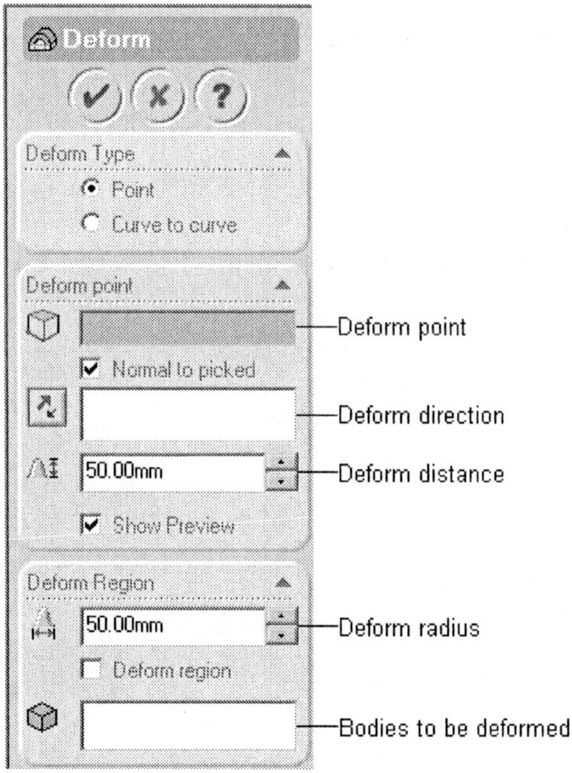

Figure 9-106 Deform PropertyManager

deformed at the selected point using the default value of deform distance and deform radius. However, you can set the value of the deform distance and the deform radius using the **Deform distance** and **Deform radius** spinners from the **Deform point** and **Deform Region** rollouts respectively. Using the **Reverse deform direction** button from the **Deform point** rollout you can reverse the direction of the deform feature.

You can also define a deform direction to create a deform feature. To specify a vector direction to create a deform feature click once in the **Deform direction** selection area and select the direction in which you need to create the deform feature. By default, the entire model is deformed using this tool. You can also specify a particular portion that you need to deform. To specify a region to deform select the **Deform region** check box. The options available to specify a region for deformation are the same as discussed earlier.

Figure 9-107 shows the point where you need to select the face of the model. Figure 9-108 shows the resultant deform feature. Figure 9-109 shows the deform feature created using the **Point** option and the **Deform region** check box selected. Figure 9-110 shows the deform feature with the **Deform region** check box selected and the front face of the base feature selected as the additional faces to deform.

You can also define the deform axis if you are deforming the entire model. The axis of deformation is specified in the **Deform axis** selection area available in the **Shape Options** rollout. Other options available are the same as discussed earlier.

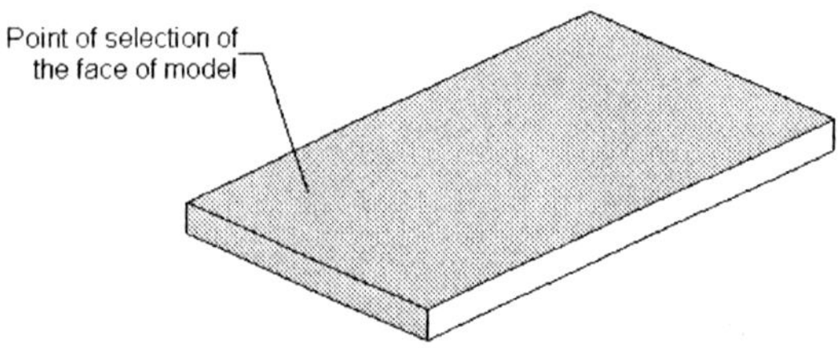

Figure 9-107 *References to be selected*

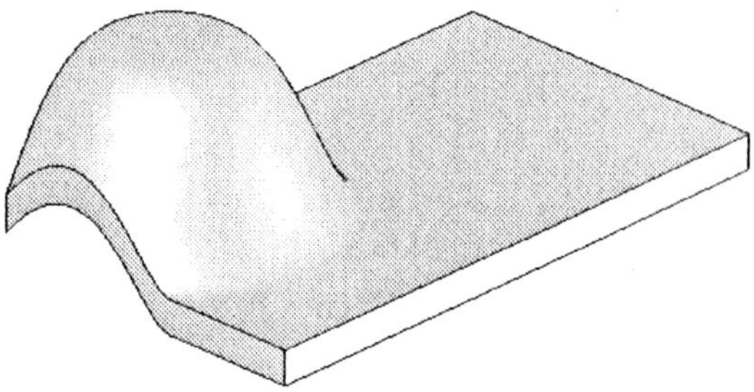

Figure 9-108 *Resultant preview of deform feature*

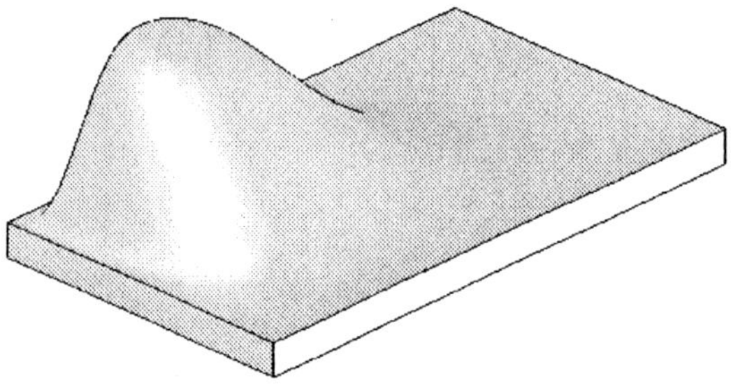

Figure 9-109 *Deform feature with the **Point** option and the **Deform region** check box selected*

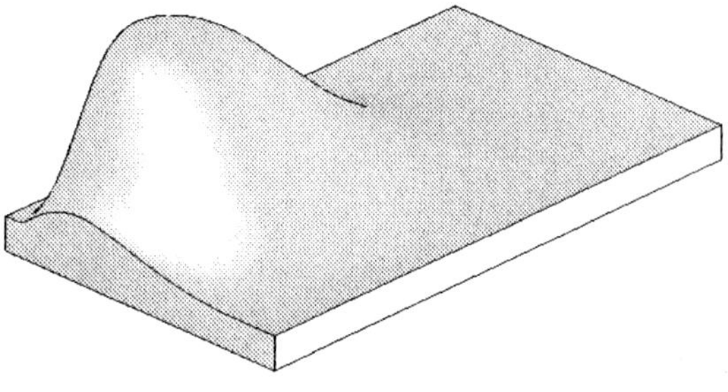

Figure 9-110 *Deform feature with front face selected as the additional face to deform*

Note

If the ***Ignore warning*** *check box available in the* ***Shape Options*** *rollout of the* ***Deform PropertyManagar*** *is selected, the warning message about error in feature creation will not be displayed and it will create the deform feature even if it has some errors.*

TUTORIALS

Tutorial 1

In this tutorial you will create the model shown in Figure 9-111. The dimensions of the model are shown in the same figure. **(Expected time: 45 min)**

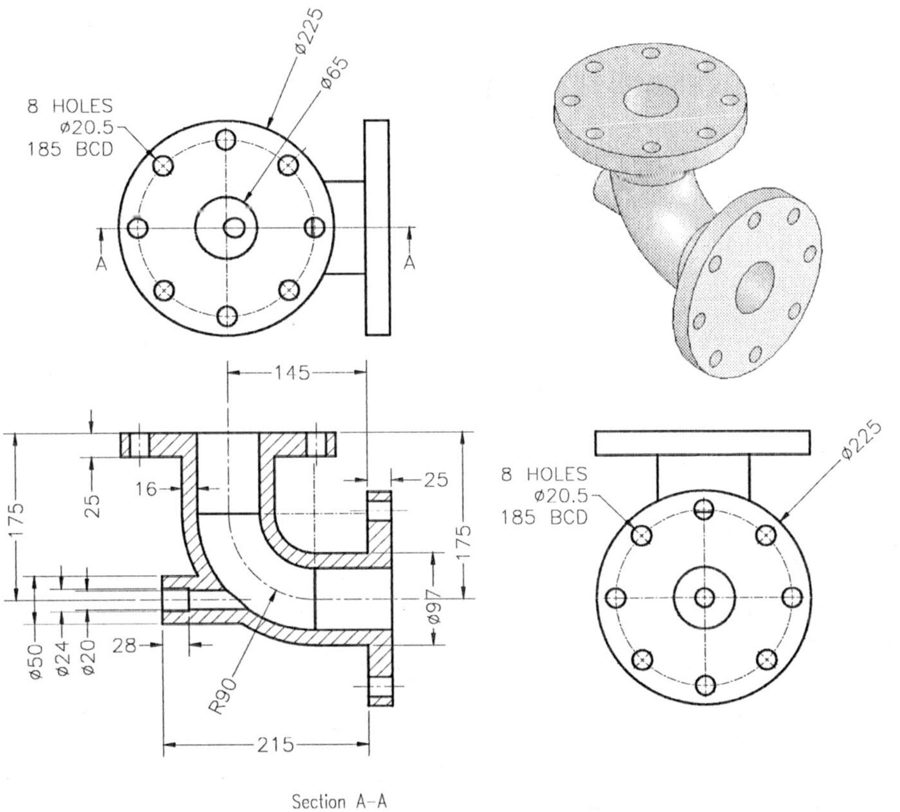

Figure 9-111 *Views and dimensions of the model for Tutorial 1*

The steps to be followed to complete this tutorial are listed next.

a. The base feature of the model is a sweep feature. First, you need to create the path for the sweep feature on the front plane. Next, you need to create a plane normal to this path. Select the newly created plane as the sketching plane and create the profile for the sweep feature. You will create a thin sweep feature because the base feature of the model is a hollow feature, refer to Figures 9-112 through 9-114.

b. Create the extrude features on both ends of the sweep feature, refer to Figure 9-115.
c. Create a plane at an offset distance from the right face of the model. Create the circular feature by extruding the feature using the **Up To Next** option.
d. Create the hole using the **Simple Hole** tool, refer to Figure 9-115.
e. Create the pattern of the hole feature, refer to Figure 9-115.
f. Create the counterbore hole using the cut revolve option, refer to Figure 9-115.

Creating the Path for the Sweep Feature

As discussed earlier, the base feature of the model is a sweep feature. To create the sweep feature, you first need to create the path of the sweep feature. This path will be created on the **Front** plane.

1. Start a new SolidWorks Part document using the **New SolidWorks Document** dialog box.

2. Draw the sketch of the path for the sweep feature on the **Front** plane and add the required relations and dimensions to the sketch as shown in Figure 9-112. Exit the sketching environment and change the view to isometric.

Creating the Profile for the Sweep Feature

After creating the path for the sweep feature, you will create the profile for the sweep feature. For creating the profile, first you need to create a reference plane normal to the path. The newly created plane will be selected as the sketching plane for creating the profile for the sweep feature.

1. Invoke the **Plane PropertyManager** and using the **Normal to Curve** option, create a plane normal to the path as shown in Figure 9-113.

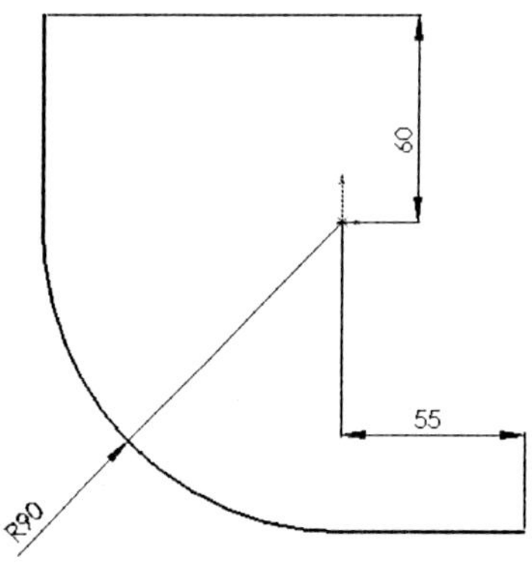

Figure 9-112 *Sketch of the path*

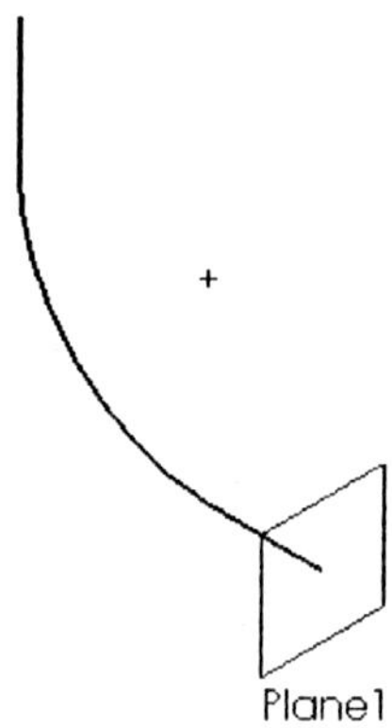

Figure 9-113 *Plane created normal to path*

2. Invoke the sketching environment by selecting the newly created plane as the sketching plane.

3. Create the sketch of the profile of the sweep feature using the **Circle** tool. The center of the circle is at the origin and the diameter of the circle is 97 mm.

4. After creating the profile for the sweep feature exit the sketching environment.

Creating the Sweep Feature

The sweep feature that you are going to create is a thin sweep feature. You will use the **Thin Feature** rollout to specify the parameters of the thin feature.

1. Choose the **Swept Boss/Base** button from the **Features CommandManager**. The **Sweep PropertyManager** is invoked and you are prompted to select the sweep profile.

2. Select the profile for the sweep feature. The selected profile will be highlighted in green and the **Profile** callout will also be displayed.

   After selecting the profile, you are prompted to select the sweep path.

3. Select the path for the sweep feature. The selected path will be highlighted in magenta and the **Path** callout is also displayed. The preview of the sweep feature is also displayed in the drawing area.

4. Select the **Thin Feature** check box available on the left of the **Thin Feature** rollout. The **Thin Feature** rollout is invoked.

5. Set the value of thickness in the **Thickness** spinner to **16**.

   Since the wall thickness added to the model is reverse to the required direction, therefore, you need to reverse the direction of creation of thin feature.

6. Choose the **Reverse Direction** button from the **Thin Features** rollout. Choose the **OK** button from the **Sweep PropertyManager** or choose **OK** from the confirmation corner.

   The base feature created by sweeping a profile along a path is shown in Figure 9-114.

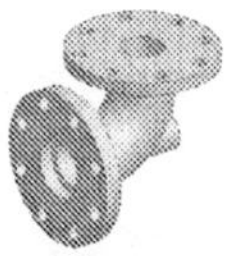

Tip. *Instead of creating a thin sweep feature, you can also create a solid sweep feature and then add a shell feature to hollow the base feature.*

Creating the Remaining Features

1. Create the remaining join features of the model using the extrude option.

2. Using the **Simple Hole PropertyManager** create the holes and pattern them using the **Circular Pattern** tool. Create the counterbore hole using the **Hole Wizard** or using the revolve cut option.

   The final solid model of Tutorial 1 is shown in Figure 9-115. The model tree of the model is shown in Figure 9-116.

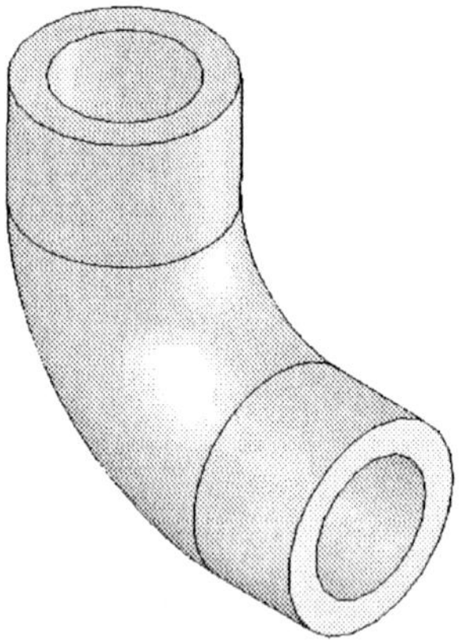

Figure 9-114 *Base feature of the model*

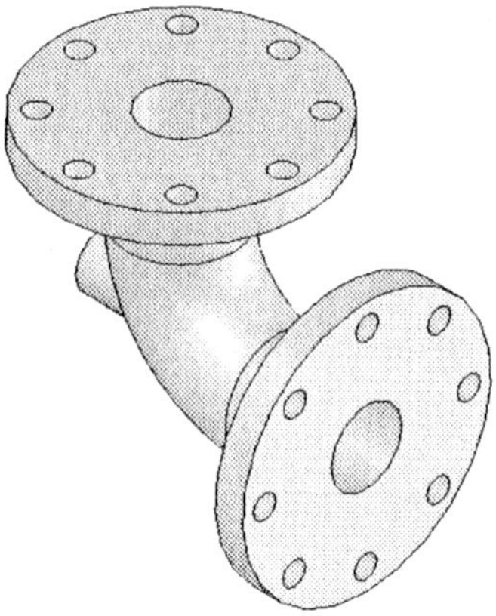

Figure 9-115 *Final model of Tutorial 1*

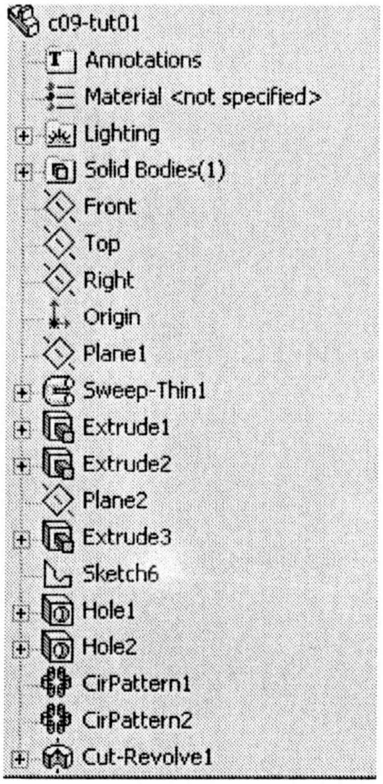

Figure 9-116 *The* ***FeatureManager Design Tree*** *for the model*

Saving the Model

1. Choose the **Save** button from the **Standard** toolbar and save the model with the name given below

   \My Documents\SolidWorks\c09\c09-tut01.sldprt

2. Choose **File > Close** from the menu bar to close the file.

Tutorial 2

In this tutorial you will create the chair frame shown in Figure 9-117. The dimensions of the chair frame are shown in Figure 9-118. **(Expected time: 30 min)**

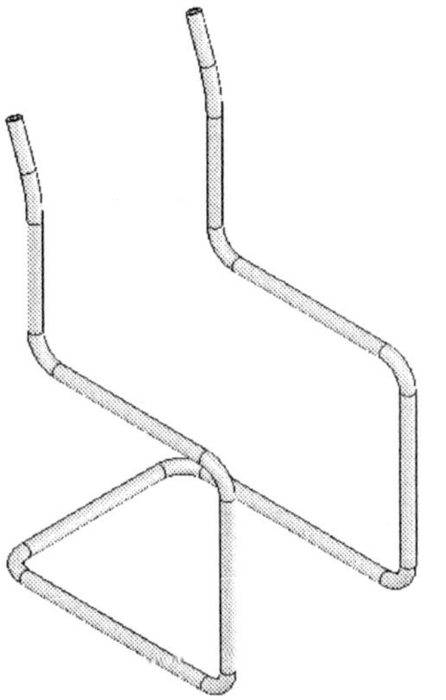

Figure 9-117 *Model of Tutorial 2*

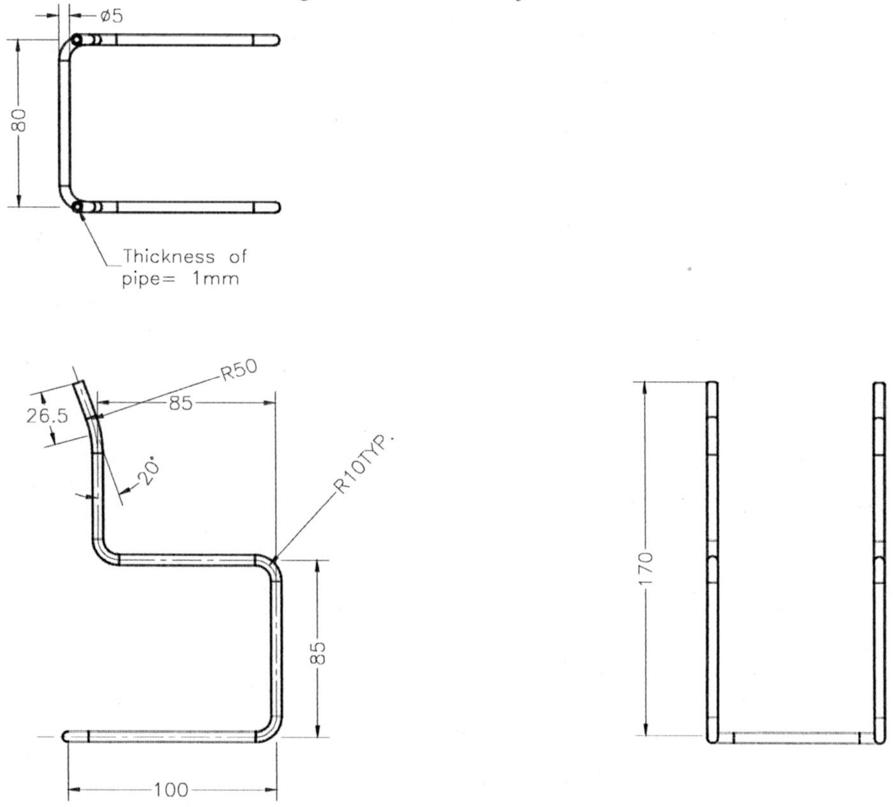

Figure 9-118 *Views and dimensions of the model for Tutorial 2*

The steps to be followed to complete this tutorial are listed next.

a. The chair frame is created by sweeping a profile along a 3D path. The 3D path will be created in the 3D sketching environment. Therefore, you need to invoke the 3D sketching environment and then create the sketch of the 3D path. You will create only the left half of the 3D path in the 3D sketching environment, refer to Figure 9-119.
b. Create a plane normal to the 3D path. Selecting the newly created plane as the sketching plane, create the sketch of the profile.
c. Sweep the profile along the 3D path using the **Thin Feature** option, refer to Figure 9-120.
d. Mirror the sweep feature using the **Front** plane, refer to Figure 9-121.

Creating the Path of Sweep Feature Using 3D Sketching Environment

It is evident from Figure 9-112 that the model is created by sweeping a profile along a 3D path. Therefore, you need to create a path of the sweep feature in the 3D sketching environment.

1. Start a new SolidWorks part document.

2. Create a plane at an offset distance of 40 mm from the **Front** plane.

3. Change the current view to isometric.

4. Choose **Insert > 3D Sketch** from the menu bar to invoke the 3D sketching environment. You can also use the **3D Sketch** button from the **Sketch CommandManager** to invoke the 3D sketching environment.

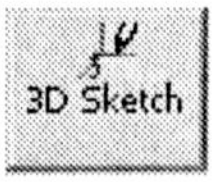

The 3D sketching environment is invoked and the sketch origin is displayed in red. You are also provided with the confirmation corner on the top right of the drawing area. The sketching tools that can be used in the 3D sketching environment are also invoked.

5. Invoke the **Line** tool. The select cursor is replaced by the line cursor. The **XY** that is displayed below the line cursor suggests that by default it will sketch in the **XY** plane.

The first line you need to create is in the **ZX** plane. Therefore, you will toggle the plane before you start creating the sketch.

6. Choose the TAB key from the keyboard twice to switch to **ZX** plane.

7. Move the line cursor to the origin. When the cursor turns yellow in color, press and hold down the left mouse button. A space handle will be displayed at the origin that indicates the direction of **X**, **Y**, and **Z**.

8. Drag the cursor in the direction where the **Z** symbol is displayed below the cursor. This indicates that you are creating the line in the **Z** direction. Release the left mouse button when the value of the length of line displays a value close to **40**.

9. Move the cursor to the endpoint of the previous line and when the cursor turns yellow in color, press and hold down the left mouse button.

10. Drag the cursor in the direction where the **X** symbol is displayed below the cursor, which indicates that the line is being created in the **X** direction. Release the left mouse button when the length of the line above the line cursor shows a value close to **100**.

11. Move the cursor to the endpoint of the previous line. When the cursor snaps to the endpoint, press and hold down the left mouse button.

12. Press the TAB key to switch to the **XY** plane. Drag the cursor in the direction where the **Y** symbol will be displayed below the cursor, suggesting that the line is being created in the **Y** direction.

13. Release the left mouse button when the value above the cursor shows a value close to **85**.

14. Similarly, create the remaining sketch and add the required relations and dimensions to the sketch. The final sketch is displayed in Figure 9-119.

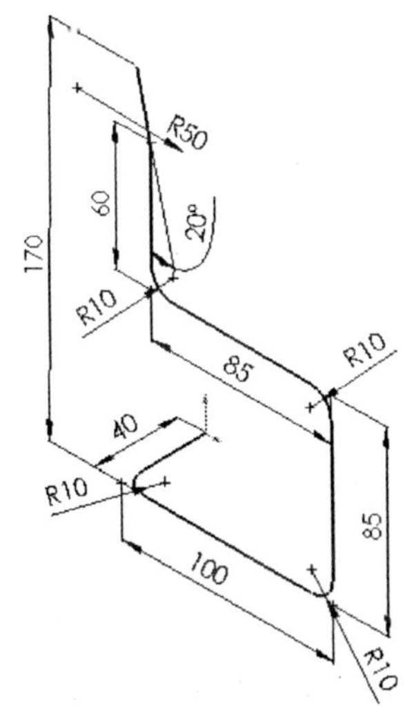

***Figure 9-119** Sketch of the 3D path*

15. Add a **Coincident** relation between the upper point of the 3D sketch and **Plane1**.

16. Exit the 3D sketching environment.

Creating the Profile of the Sweep Feature

As discussed earlier, the sweep feature will be created by sweeping the profile along a 3D path. You have created the path and now you need to create the profile for the sweep feature.

1. Select the **Front** plane as the sketching plane and invoke the sketching environment.

   Since the **Front** plane is normal to the 3D path, therefore, you do not need to create a reference plane. If any one of the default plane is not normal to the path, then you need to create a reference plane normal to the path.

2. Create a circle of diameter 5 mm, that will be used as the sketch of the profile of sweep feature. Refer to Figure 9-118.

3. Exit the sketching environment.

Sweeping the Profile Along the 3D Path

After creating the 3D path and the profile for the sweep feature, you need to sweep the profile along the 3D path using the **Swept Boss/Base** tool.

1. Choose the **Swept Boss/Base** button from the **Features CommandManager**. The **Sweep PropertyManager** is invoked.

2. Make sure that the sketch of the profile is selected. The selected profile will be displayed green in color and the profile callout is also displayed.

   You are prompted to select the path for the sweep feature.

3. Select the path for the sweep feature. The path will be displayed in magenta color and the path callout is also displayed. The preview of the sweep feature is also displayed in the drawing area.

   As evident from Figure 9-118, the frame of the chair is made from a hollow pipe. Therefore, you need to create a thin sweep feature to create a hollow chair frame.

4. Invoke the **Thin Feature** rollout and set the value of the **Thickness** spinner to **1**.

5. Choose the **Reverse Direction** button from the **Thin Features** rollout to reverse the direction of thin feature creation.

6. Choose the **OK** button from the **Sweep PropertyManager** to end feature creation.

   The model after creating the sweep feature is displayed in Figure 9-120.

7. Using the **Mirror** tool, mirror the sweep feature along the **Front** plane. The model after mirroring is displayed in Figure 9-121. The **FeatureManager Design Tree** of the model is shown in Figure 9-122.

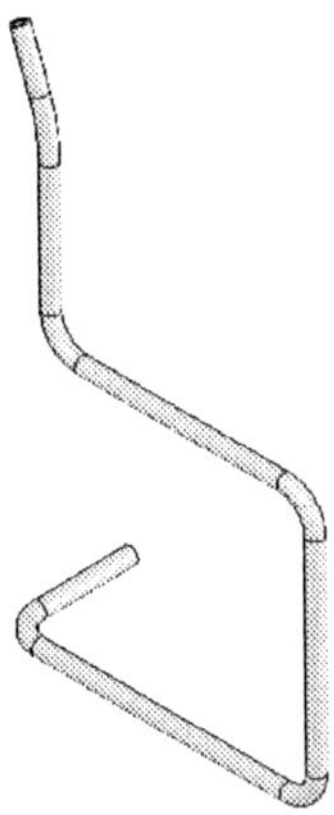

Figure 9-120 Sweep feature created by sweeping a profile along a 3D path

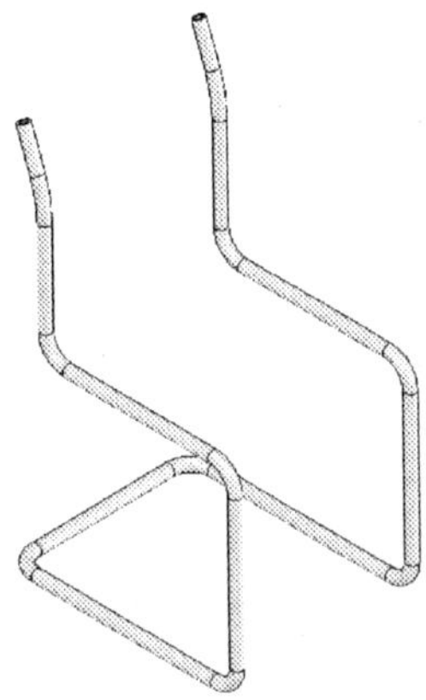

Figure 9-121 The final model

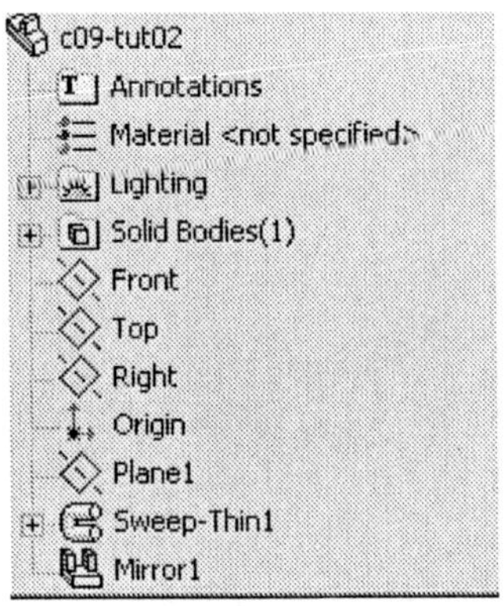

Figure 9-122 The ***FeatureManager Design Tree*** *for Tutorial 2*

Saving the Model

1. Choose the **Save** button from the **Standard** toolbar and save the model with the name given below.

   \My Documents\SolidWorks\c09\c09-tut02.sldprt

2. Close the document.

Tutorial 3

In this tutorial you will create the spring shown in Figure 9-123. The dimensions of the spring are shown in Figure 9-124. **(Expected time: 45 min)**

The steps to be followed to complete this tutorial are listed next.

a. First you need to create the helical path of the spring, refer to Figure 9-125.

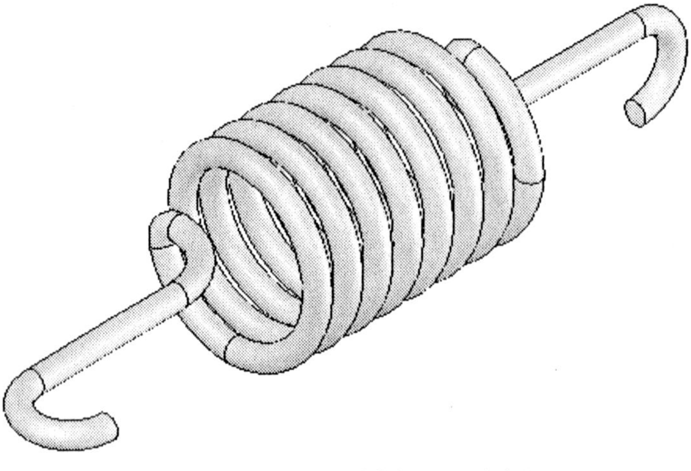

Figure 9-123 Model for Tutorial 3

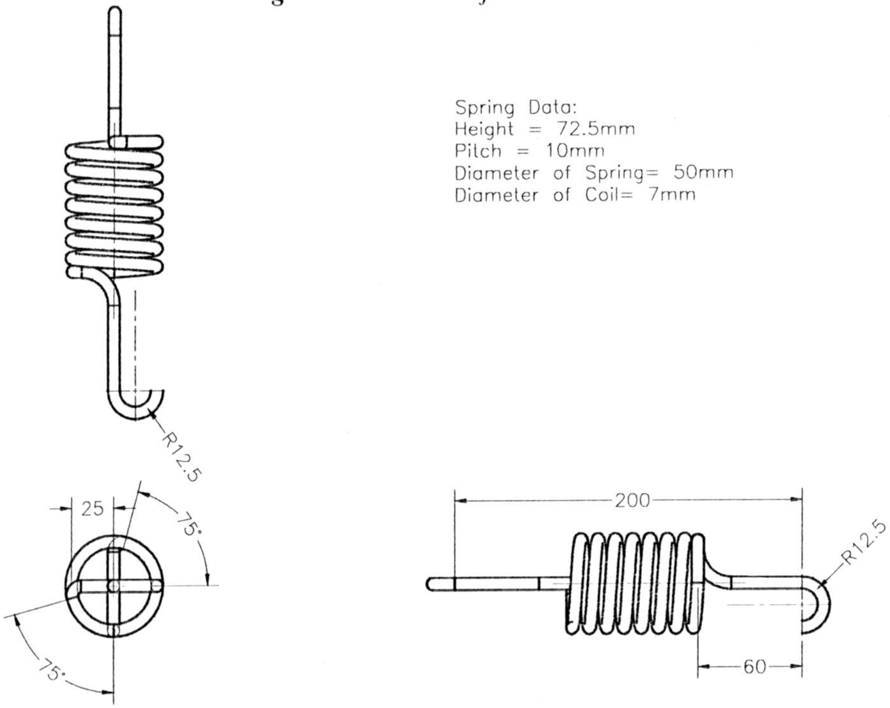

Figure 9-124 Views and dimensions of the model for Tutorial 3

b. Create the side clips of the spring, refer to Figure 9-128.
c. Combine the end clips and the helical curve to create a single curve using the **Composite Curve** option, refer to Figure 9-129.
d. Create the profile on the plane normal to the curve and sweep the profile along the curve, refer to Figures 9-130 and 9-131.

Creating the Helical Curve

For creating this model first you need to create the helical curve of the spring. For creating the helical curve first you need to create a circular sketch. This sketch will define the diameter of the spring.

1. Start a new SolidWorks Part document.

2. Choose the **Helix and Spiral** button from the **Curves CommandManager** and select the **Front** plane as the sketching plane.

3. Draw a circle of diameter 50. Change the view to isometric view and exit the sketching environment.

   The **Helix/Spiral Curve PropertyManager** is displayed and the preview of the helical curve with the default values is displayed in the drawing area.

4. Select the **Height and Pitch** option from the **Type** drop-down list available in the **Defined By** rollout.

5. Set the value of the **Height** spinner to **72.5** and the value of the **Pitch** spinner to **10**.

   You will observe that the preview of the helical curve is updated automatically when you modify the values in the spinners.

6. Set the value of the **Starting angle** spinner to **0** and choose the **OK** button from the **Helix/Spiral PropertyManager**.

   The helical curve created using the above steps is shown in Figure 9-125.

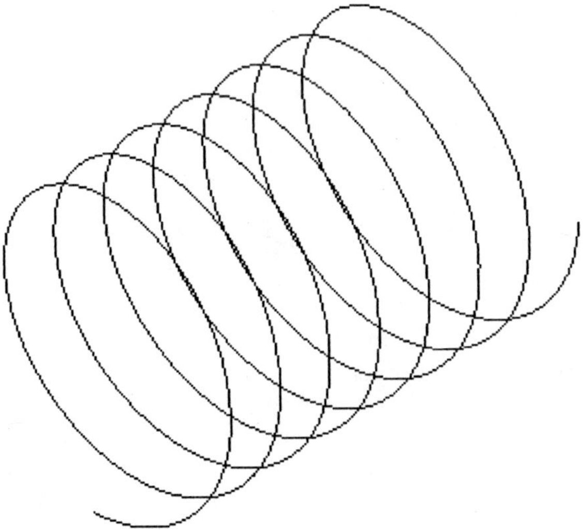

Figure 9-125 *Helical curve*

Creating the Sketch of the End Clips of the Spring

After creating the helical curve, you need to create the sketch that defines the path of the end clips. There are two end clips in this spring, and each end clip is created using two sketches. First, you will create the right end-clip and then the left end-clip.

1. Select the **Front** plane as the sketching plane and invoke the sketching environment.

   The first sketch of the right end-clip consists of two arcs. The first arc will be created using the **Centerpoint Arc** tool and the second arc will be created using the **3 Point Arc** tool.

2. Choose the **Centerpoint Arc** button from the **Sketch CommandManager** and specify the centerpoint of the arc at the origin.

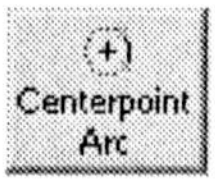

3. Move the cursor toward the right and specify the start point of the arc when the radius of the reference circle shows a value close to **25**.

4. Move the cursor in the counterclockwise direction. Specify the endpoint of the arc when the value of the angle above the cursor shows a value close to **75**.

5. Choose the **3 Point Arc** button from the **Sketch CommandManager**. Specify the start point of the arc on the upper endpoint of the previous arc. Create the arc as shown in Figure 9-126.

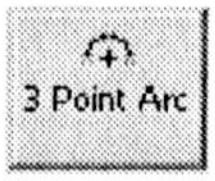

6. Add the **Pierce** relation between the lower endpoint of the first arc and the helical curve. Add the other relations and the dimensions to fully define the sketch. The fully defined sketch is shown in Figure 9-126.

7. Exit the sketching environment.

   After creating the first sketch of the right end-clip, you need to create the second sketch of the right end-clip. The second sketch of the right end-clip will be created on the **Right** plane.

8. Select the **Right** plane and invoke the sketching environment.

9. Create the sketch as shown in Figure 9-127. You need to apply the **Pierce** relation between the upper arc and the previous sketch and the left end point of left arc of the current sketch.

10. Add a **Tangent** relation between the left arc of the current sketch and the upper arc of the sketch created previously. Add the required relations and dimensions to the current sketch. The sketch after applying all the relations and dimensions is shown in Figure 9-127.

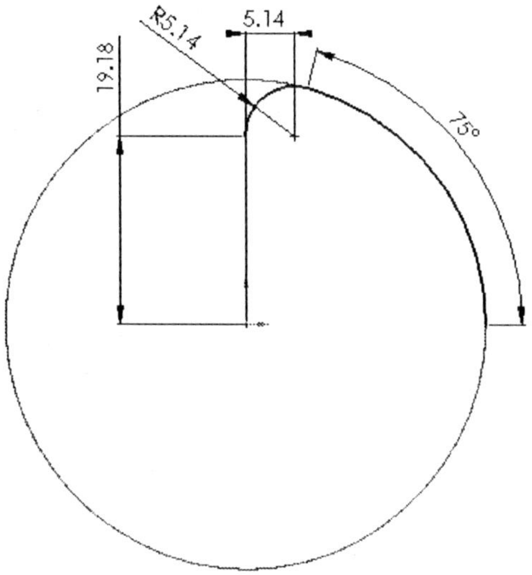

Figure 9-126 First sketch of right end-clip

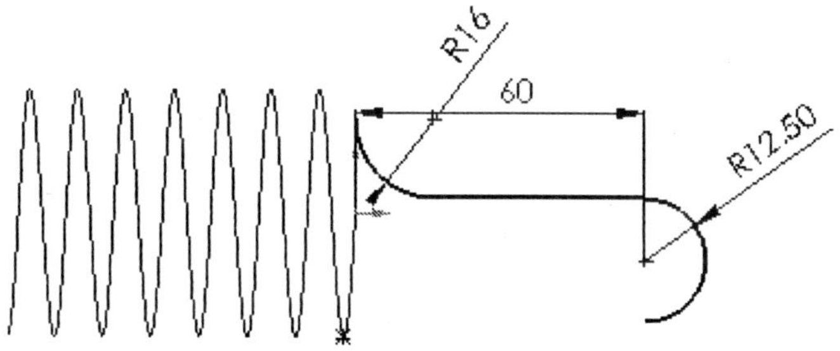

Figure 9-127 Second sketch of right end-clip

Similarly, create the sketch of the left end-clip. Figure 9-128 shows the path of the spring after creating the helical curve, right end-clip, and the left end-clip.

Creating the Composite Curve

After creating all the required sketches and the helical curve, you need to combine them so that they form a single curve. This is done because the path of the sweep feature has to be a single curve or sketch.

1. Choose the **Composite Curve** button from the **Curves CommandManager** or choose **Insert > Curve > Composite** from the menu bar. The **Composite Curve PropertyManager** is displayed and the confirmation corner is also available.

2. Select both the end-clips and the helical curve from the drawing area or from the **FeatureManager Design Tree**.

3. Choose the **OK** button from the **Composite Curve PropertyManager** or choose **OK** from the confirmation corner.

   The composite curve is created. The composite curve created is displayed in Figure 9-129.

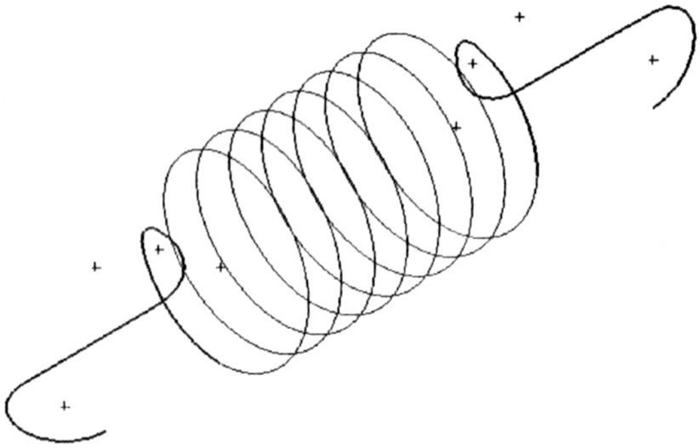

Figure 9-128 Path for the spring

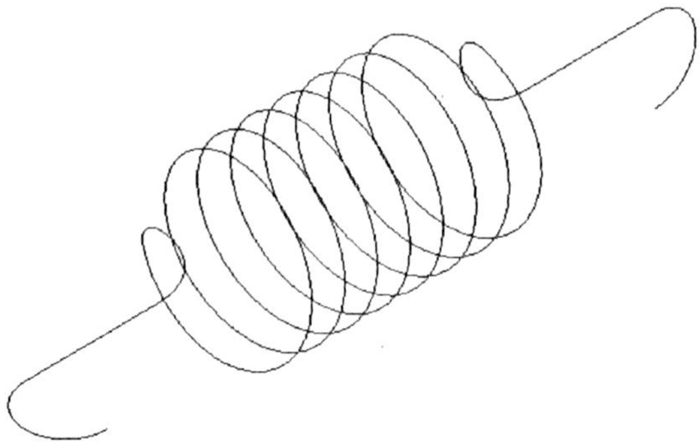

Figure 9-129 Composite curve

Creating the Profile for the Sweep Feature

Now, you need to create the profile for the sweep feature. The profile of the sweep feature will be created on a plane normal to the curve on the endpoint of the right end-clip.

1. Create a plane normal to the path and at the endpoint of the right end-clip.

2. Create the profile of the sweep feature and add the **Pierce** relation between the center of the circle and the composite curve. Add the required dimension to the sketch.

3. Exit the sketching environment. Figure 9-130 shows the profile and path for the sweep feature.

Creating the Sweep Feature

You will create a sweep feature to complete the creation of spring.

1. Invoke the **Sweep PropertyManager** and you are prompted to select the sweep profile.

2. Select the sweep profile from the drawing area.

   You are prompted to select the sweep path.

3. Select the path and choose the **OK** button from the **Sweep PropertyManager**.

   The spring created after sweeping the profile along the path is shown in Figure 9-131.

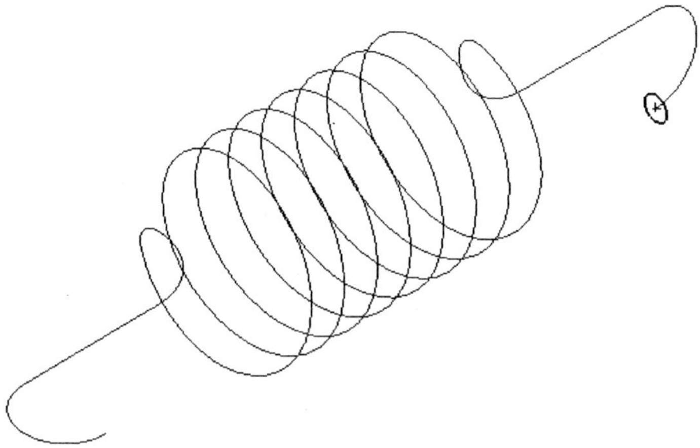

Figure 9-130 Profile and path of the sweep feature

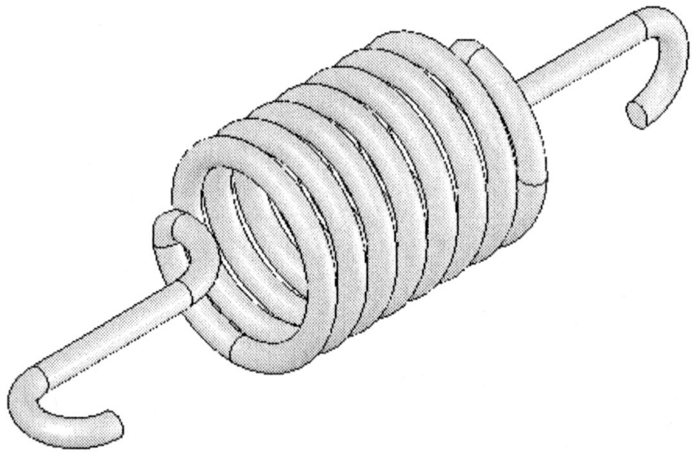

Figure 9-131 Final model

The **FeatureManager Design Tree** of the spring is shown in Figure 9-132.

Saving the Model

1. Choose the **Save** button from the **Standard** toolbar and save the model with the name given below.

   \My Documents\SolidWorks\c09\c09-tut03.sldprt

2. Close the document.

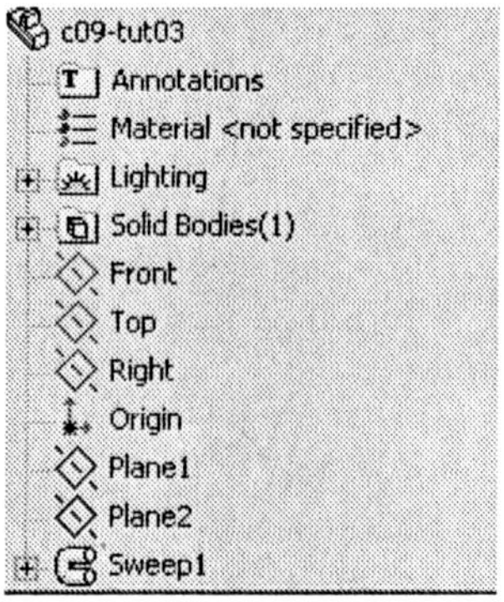

Figure 9-132 FeatureManager Design Tree

SELF-EVALUATION TEST

Answer the following questions and then compare your answers with the answers given at the end of this chapter.

1. You need a profile and a path to create a sweep feature. (T/F)

2. At least two sections are required to create a loft feature. (T/F)

3. You cannot sweep a closed profile along a closed path. (T/F)

4. You cannot create a thin sweep feature. (T/F)

5. You can also create a loft feature using open sections. (T/F)

6. The __________ **PropertyManager** is used to create a loft feature.

7. The __________ option is used to create a single curve by joining the continuous chain of existing sketches, edges, or curves.

8. You have to apply __________ relation between the sketch and the guide curve while sweeping a profile along a path using guide curves.

9. __________ option is used to create a curve by defining coordinates.

10. __________ **PropertyManager** is invoked to create a cut-sweep feature.

REVIEW QUESTIONS

Answer the following questions.

1. Using the __________ rollout you can define the tangency at the start and end in the sweep feature.

2. The __________ rollout is used to create a thin loft feature.

3. You need to invoke __________ dialog box to create a spiral curve.

4. The __________ **PropertyManager** is invoked to create a curve projected on a surface.

5. The __________ dialog box is used to specify the coordinates to create the curve.

6. Which button is selected to invoke the 3D sketching environment?

   (a) **2D Sketch** (b) **3D Sketching Environment**
   (c) **3D Sketch** (d) **Sketch**

7. Which rollout available in the **Sweep PropertyManager** is used to define the tangency?

   (a) **Start/End Tangency** (b) **Tangency**
   (c) **Options** (d) None of these

8. Which button available in the **Features PropertyManager** is used to invoke the **Draft PropertyManager**?

   (a) **Draft** (b) **Taper Angle**
   (c) **Draft Feature** (c) **Draft Angle**

9. In which rollout can you define the pull direction while creating a draft feature using the parting line option?

   (a) **Pull Direction** (b) **Direction of Pull**
   (c) **Reference Direction** (d) **Options**

10. Which button available in the **Guide Curves** rollout is used to display the sections while creating the sweep feature with guide curves?

   (a) **Preview Sections** (b) **Show Sections**
   (c) **Sections** (d) **Preview**

EXERCISES

Exercise 1

Create the model of the Upper Housing shown in Figure 9-133. The dimensions of the model are shown in Figure 9-134. **(Expected time: 1 hr)**

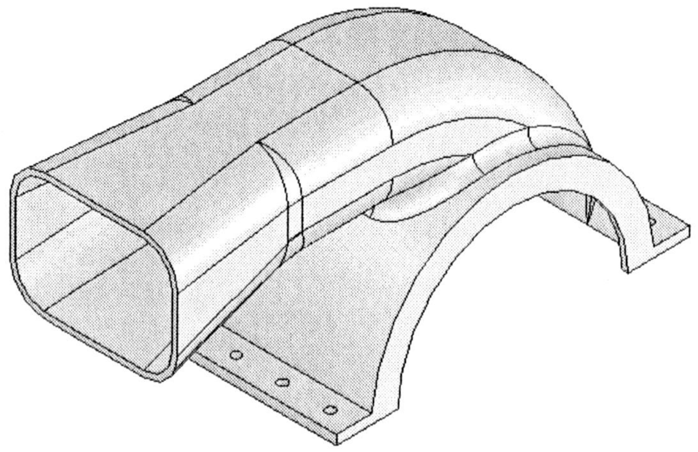

Figure 9-133 Model of the Upper Housing

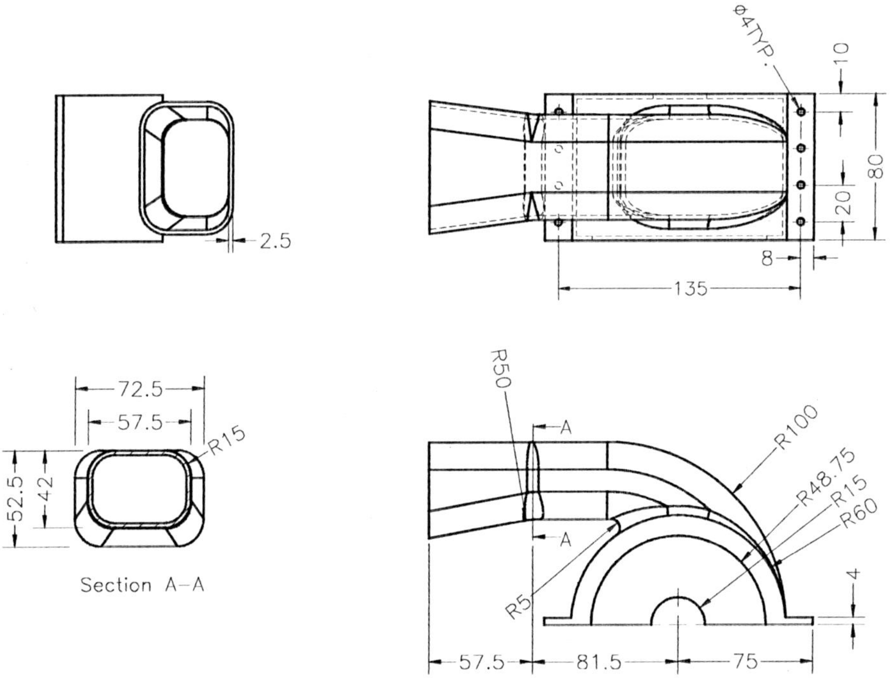

Figure 9-134 Dimensions of the Upper Housing

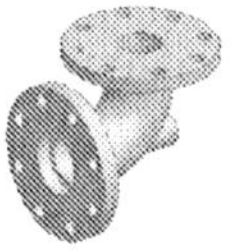

Tip. *This model is divided in three major parts. The first part is the base and is created by extruding the sketch to a distance of 80 mm using the **Mid Plane** option.*

The second part of this model is the right portion of the discharge venturi. This portion is created using the sweep feature. The path of the sweep feature will be created on the right plane. You need to create a plane at the left endpoint of the path and normal to the path.

The third part of this model is the left portion of the discharge venturi. This will be created using the loft feature. You need to create the first section of the loft feature on the planar face of the sweep feature created earlier. The second section will be created on a plane at an offset distance from the planar face of the sweep feature created earlier. Create a loft feature using the two sections created earlier. The other features needed to complete the model are fillets, hole, circular pattern, and so on.

Exercise 2

Create the model shown in Figure 9-135. The dimensions of the model are shown in Figure 9-136. **(Expected time: 1 hr)**

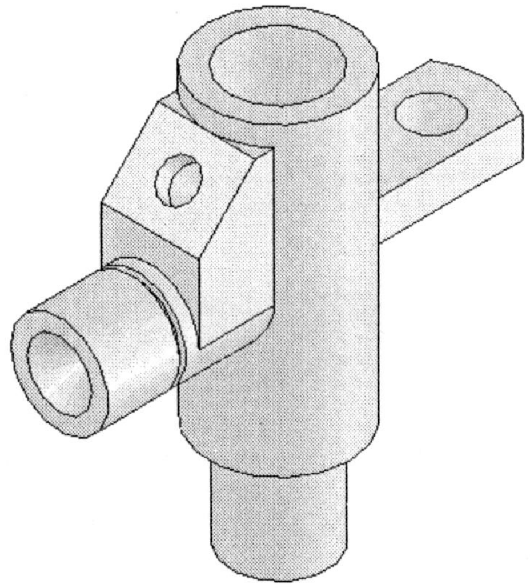

Figure 9-135 *Model for Exercise 2*

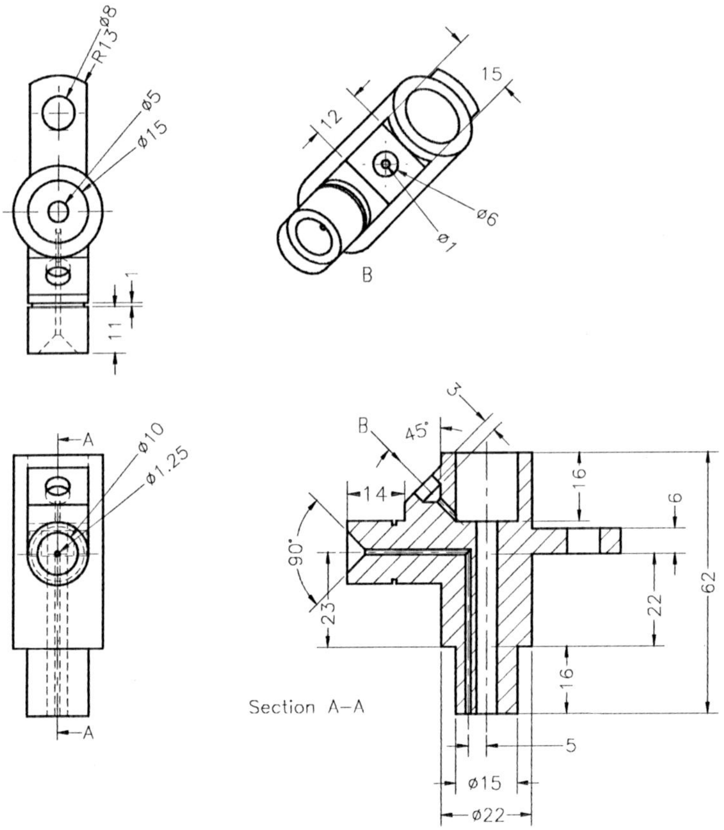

***Figure 9-136** Views and dimensions of the model for Exercise 2*

Answers to Self-Evaluation Test

1. T, **2.** T, **3.** F, **4.** F, **5.** T, **6. Loft, 7. Composite Curve, 8. Coincident, 9. Curve Through Free Points, 10. Cut-Sweep**

Chapter 10

Assembly Modeling-I

Learning Objectives

After completing this chapter you will be able to:

- *Create Bottom-up Assemblies.*
- *Add Mates to Assemblies.*
- *Create Top-down Assemblies.*
- *Move Individual Components.*
- *Rotate Individual Components.*

ASSEMBLY MODELING

An assembly design is defined as a design consisting of two or more components assembled together at their respective work positions. The components are assembled together in the **Assembly** mode using parametric relations. In SolidWorks, these relations are called mates. These mates allow you to constrain the degrees of freedom of the components on their respective work positions. To proceed to the **Assembly** mode of SolidWorks, invoke the **New SolidWorks Document** dialog box and choose the **Assembly** button as shown in Figure 10-1. Choose the **OK** button to create a new assembly document.

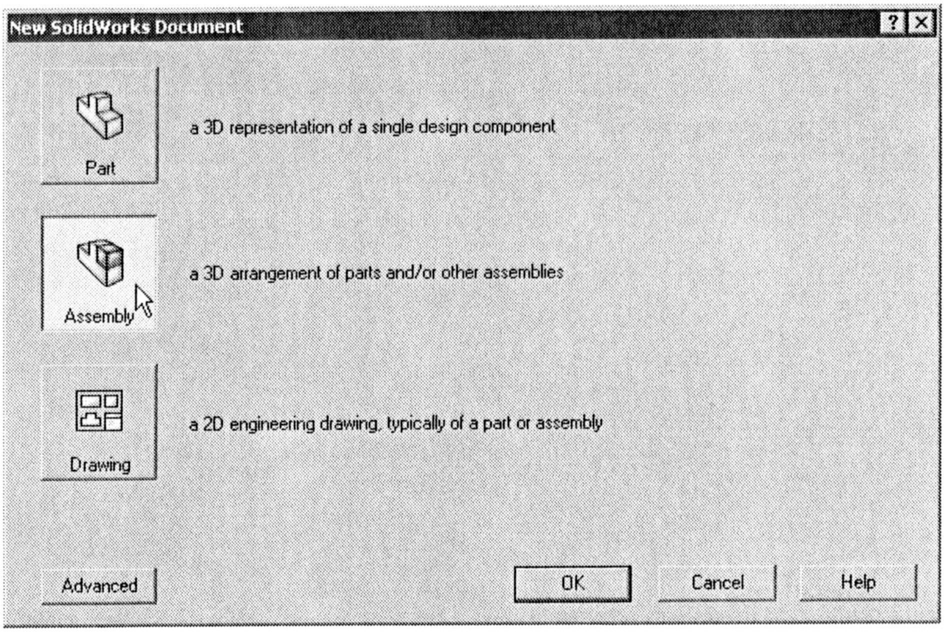

***Figure 10-1** The **New SolidWorks Document** dialog box*

A new SolidWorks document is created in the **Assembly** mode and the **Insert Component PropertyManager** is invoked. The screen display of SolidWorks in the **Assembly** mode with the **Insert Component PropertyManager** is shown in Figure 10-2. The **Assembly CommandManager** is also invoked as displayed in Figure 10-2.

Types of Assembly Design Approach

In SolidWorks, the assemblies are created using two types of design approaches. The first design approach is the bottom-up approach, and the second design approach is the top-down approach. Both these design approaches are discussed next.

Bottom-up Assembly Design Approach

The bottom-up assembly design approach is the traditional and the most widely preferred approach of assembly design. In this assembly design approach, all the components are created as separate part documents and these parts are placed and referenced in the assembly as external components. In this type of approach, the components are created in the **Part** mode

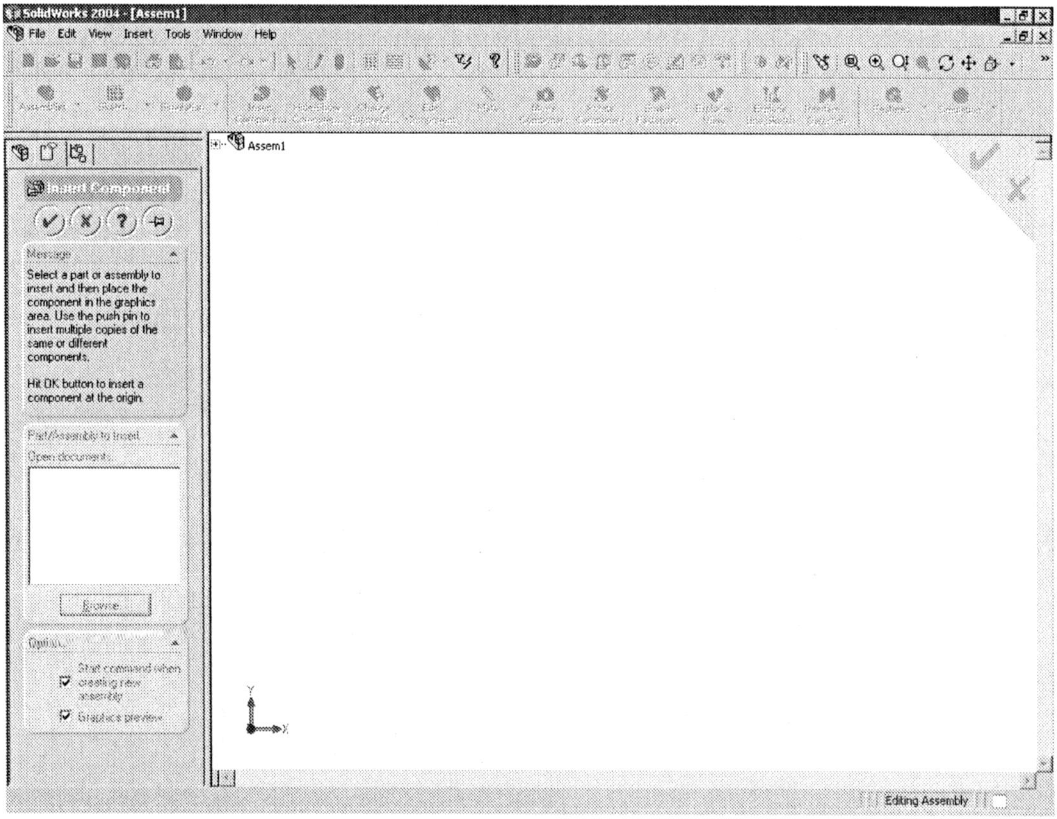

Figure 10-2 *Screen display of SolidWorks in the* ***Assembly*** *mode*

as *.sldprt* documents. After creating all the components of the assembly, you will open an assembly document (*.sldasm*) and insert all the components using the tools provided in the **Assembly** mode. After inserting the components, they are assembled using the assembly mates.

The main advantage of this assembly design approach is that as the components are designed individually, you can maintain the relationships between the features easily. Hence, this approach allows you to pay more attention and to focus more on the individual components. This approach is preferred while handling large assemblies.

Top-down Assembly Design Approach

In the top-down assembly design approach, all the components are created in the same assembly document. Therefore, the top-down assembly design approach is entirely different from the bottom-up design approach. In this approach you will start your work in the assembly document and the geometry of one part helps you to define the geometry of the other.

Note

You can also create an assembly with a combination of the bottom-up and the top-down assembly approaches.

CREATING BOTTOM-UP ASSEMBLIES

As mentioned earlier, the bottom-up assemblies are the assemblies in which the components are created as separate part documents in the **Part** mode. After creating the components, they are inserted in the assembly and then assembled using the assembly mates. For starting an assembly design with this approach, you first need to insert the components in the assembly. It is recommended that the first component should be placed at the origin of the assembly document. By doing this the default planes of the assembly and the part will coincide and the component will be in the same orientation as it was in the **Part** mode. When you place the first component in the assembly, that component is fixed at its placement position. The techniques used to place the components in the assembly file are discussed next.

Placing Components in the Assembly File

In SolidWorks, there are various options to place the components in the assembly. These options are discussed next.

Placing the Components Using the Insert Component PropertyManager

CommandManager: Assemblies > Insert Components
Menu: Insert > Component > Existing Part/Assembly
Toolbar: Assembly > Insert Components

One of the major enhancements in the **Assembly** mode of SolidWorks 2004 is the introduction of the **Insert Component PropertyManager**. When you start a new SolidWorks document in the **Assembly** mode, the **Insert Component PropertyManager** is displayed as shown in Figure 10-3. The **Message** rollout available in the **Insert Component PropertyManager** prompts you to select a part or assembly and then place the component in the display area. When you choose the **Browse** button available in the **Part/Assembly to Insert** rollout, the **Open** dialog box is displayed; browse the location where the component is saved. Select the component and choose the **Open** button. The cursor is replaced by the component cursor and graphic preview of the component is also displayed. The name of the selected component is displayed in the **Insert Component** selection area of the **Part/Assembly to Insert** rollout. It is recommended that the origin of the first component should be aligned with the assembly origin. To place the component origin on the assembly origin choose the **OK** button from the **Insert Component PropertyManager**.

To place the second component, invoke the **Insert Components PropertyManager** and choose the **Browse** button from the **Part/Assembly to Insert** rollout. Select the component from the **Open** dialog box. The cursor is replaced by the component cursor and the preview of the component is also displayed in the drawing area. Select a point anywhere in the drawing area to place the second component.

When you select a component to insert in the assembly, the **Thumbnail Preview** rollout is displayed in the **Insert Components PropertyManager**. You can invoke this rollout to view the thumbnail preview of the selected component.

The **Start command when creating new assembly** check box available in the **Options** rollout

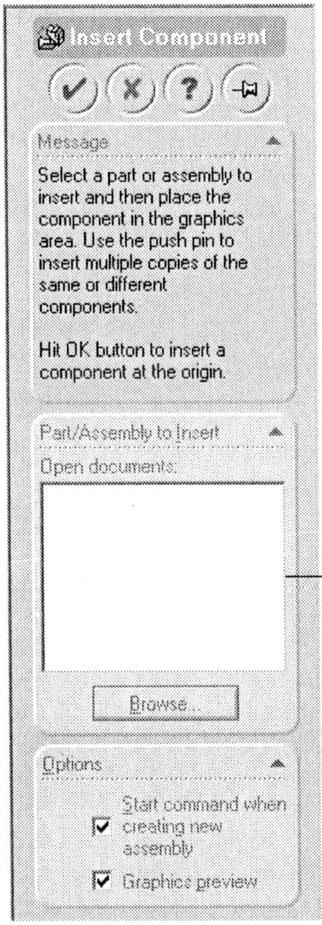

Figure 10-3** The **Insert Component PropertyManager

is selected by default and is used to invoke the **Insert Component PropertyManager** automatically when creating a new SolidWorks assembly document. The **Graphics preview** check box available in the **Options** rollout is selected by default and is used to display the graphics preview of the component selected to insert.

Tip. *If you need to place multiple instances of a component, then choose the **Keep Visible** button from the **Insert Component PropertyManager** and select the placement points in the drawing area to place the multiple instances of the selected component. If the **Keep Visible** button is selected from the **Insert Components PropertyManager**, then you can select different components one after the other to place them in the assembly document.*

*If some part documents are opened in the SolidWorks window and you create an assembly document, then those parts are displayed in the **Insert Component** selection area. You can directly select the name of the parts from the **Insert Component** selection area and place them in the drawing area.*

Tip. *When you place the first component in the assembly, the component is displayed in the* ***FeatureManager Design Tree****. The nomenclature of the name of the first component is displayed as* ***(f) Name of Component <1>****. In this nomenclature* ***(f)*** *implies that the component is fixed. You cannot move a fixed component. You will learn more about fixed and floating components later in this chapter. After this the name of the component is displayed. After the name of the component,* ***<1>*** *symbol is displayed. This implies the serial number of the same component in the entire assembly.*

The ***(-)*** *symbol implies the component is floating and is underdefined. You need to apply the required mates to that component to fully define the component. You will learn more about assembly mates later in this chapter.*

The ***(+)*** *symbol implies that the component is overdefined.*

If no symbol appears before the name of the component, the component is fully defined.

Creating an Assembly From the Part Document

Menu:	File > Make Assembly from Part
Toolbar:	Standard > Make Assembly from Part/Assembly

You can also create an assembly document from the part document. If the part document of the base component of the assembly is opened in SolidWorks, then choose the **Make Assembly from Part/Assembly** button from the **Standard** toolbar or choose **File > Make Assembly from Part** from the menu bar. You can also use CTRL+A from keyboard as the shortcut key. If the **New SolidWorks Document** dialog box is invoked, choose the **OK** button from this dialog box. An assembly document is created and the **Insert Component PropertyManager** is invoked. Choose the **OK** button from this **PropertyManager** to place the component at the origin.

Note

If the last time you invoked the ***New SolidWorks Document*** *dialog box in the novice mode, then you just need to select the* ***OK*** *button from the* ***New SolidWorks Document*** *dialog box if it is displayed while creating an assembly from a part document. If you have used the* ***New SolidWorks Document*** *in the advanced mode last time, then you first need to select the assembly template and then choose the* ***OK*** *button. When you choose the* ***Make Assembly from Part/Assembly*** *button and if a* ***SolidWorks*** *warning box is displayed, then choose* ***No*** *from the warning box. The* ***New SolidWorks Document*** *dialog box is displayed in advanced mode.*

Placing the Components Using the Opened Document Window

Another most widely used method of adding the components in the assembly is the use of existing opened document windows. For example, the assembly that you are going to create consists of three components. Open the part documents of all the parts in the SolidWorks window and create a new assembly document. Close the **Insert Component PropertyManager**. Now, choose **Window > Tile Horizontally** or **Tile Vertically** from the menu bar. All the SolidWorks document windows will be tiled vertically or horizontally, depending on the option

> **Tip**. *Only the information about the mates is stored in the assembly file. The feature information of the parts is stored in the individual part files. Therefore, the size of the assembly file is small.*
>
> *It is recommended that all the parts of an assembly be saved in the same folder in which the assembly is saved. This is because if the location of the parts is changed, the component will not be displayed in the assembly and assembly file will show errors.*

selected. You first need to place the first component in the assembly and this needs to be placed at the origin of the assembly. If the origin is not displayed by default, then choose **View > Origins** from the menu bar. Move the select cursor on the name of the first component to be placed in the **FeatureManager Design Tree** of the window of that component. Press and hold down the left mouse button at this location. Drag the cursor to the assembly origin in the assembly window. When the coincident symbol appears below the component cursor, place the component in the assembly. Similarly, place the other components in the assembly. Figure 10-4 displays placing first component from an existing opened document window to the assembly document window. If another existing assembly document is opened, you can also drag and drop the part from that assembly document.

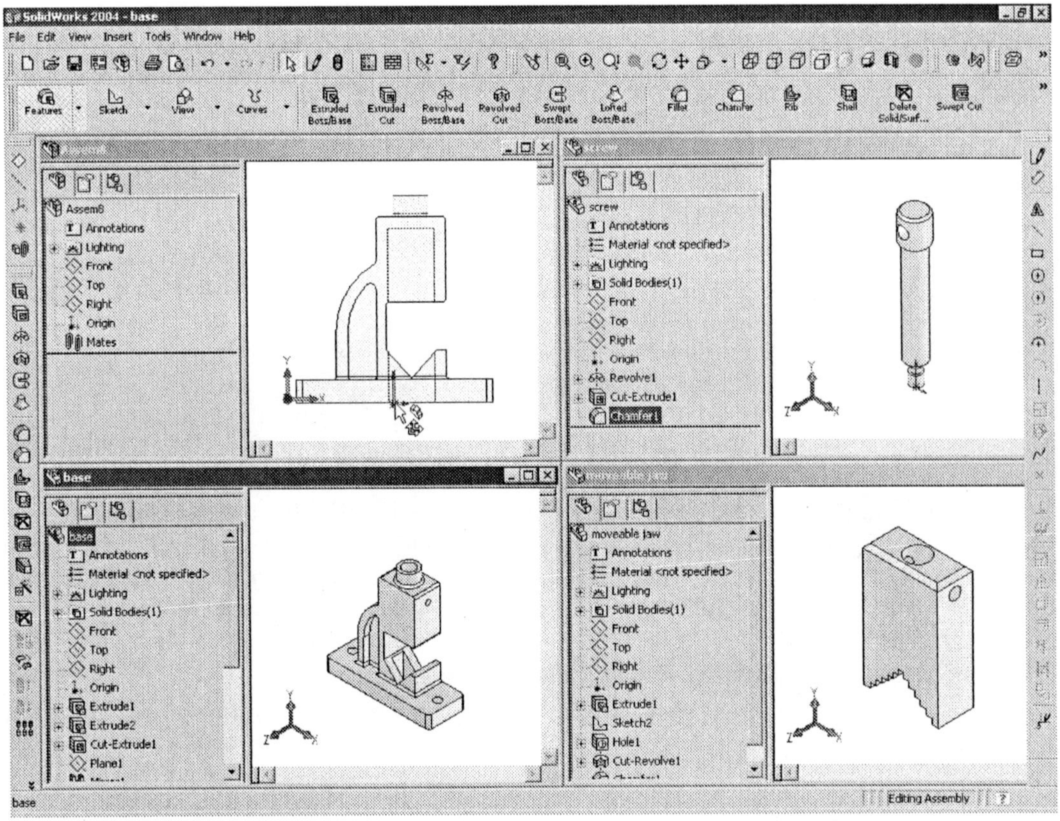

Figure 10-4 *Placing a component in the assembly file from an existing window*

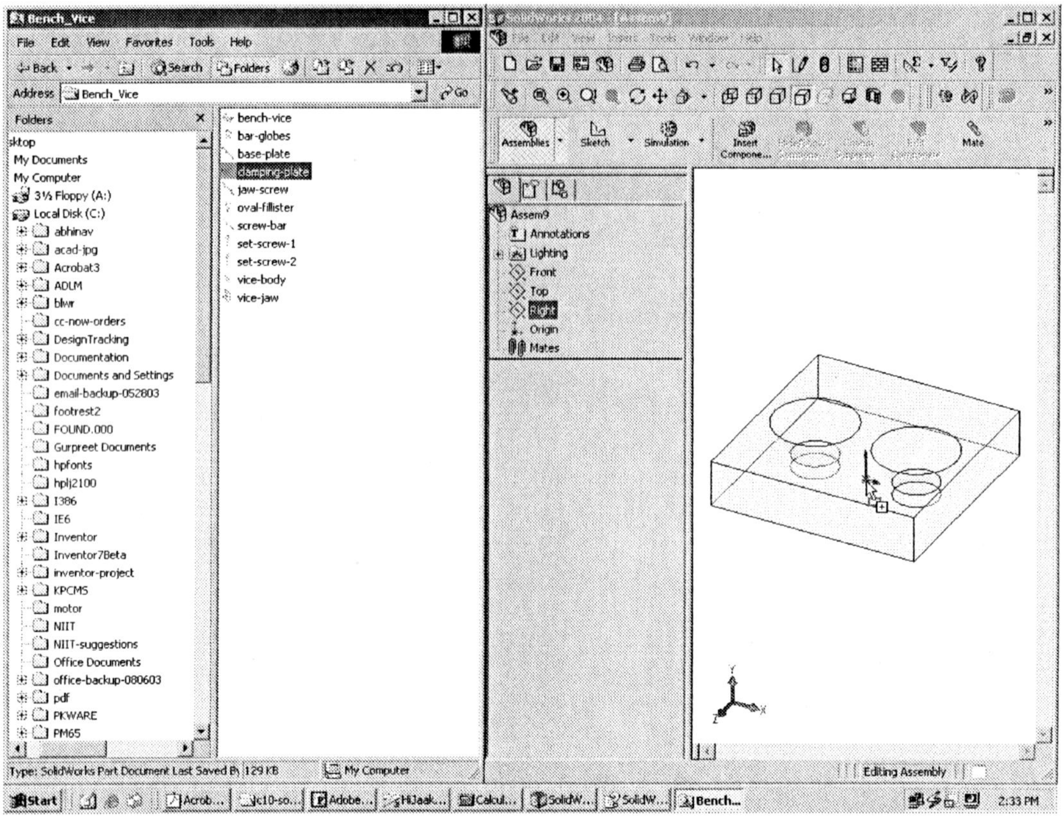

Figure 10-5 Dropping the part from Windows Explorer to the assembly window

Placing the Components by Dragging From Windows Explorer

You can also place the components in the assembly file by dragging them from Windows Explorer. Open Windows Explorer and browse the location where the part documents are saved. Tile the Windows Explorer window and the SolidWorks window such that you can view both the windows. Move the cursor on the icon of the part document in Windows Explorer. Press and hold down the left mouse button and drag the cursor to the assembly document window. Drop the part in the assembly document. Similarly, place the other parts by dragging from Windows Explorer and dropping them in the assembly document. When you place the components using this method, the part origin will not coincide with the assembly origin. You need to manually add the constraint to coincide the origin of the part with the assembly origin. Figure 10-5 shows the part being dropped in the assembly window.

Placing the Components From Internet Explorer

You can also place the components from Internet Explorer. For placing the components from Internet Explorer, you need Internet Explorer 4.0 or later. Browse the location of the SolidWorks part file link on the Web. Drag the hyperlink and drop it in the drawing area of the assembly document. The **Save As** dialog box will be displayed; save the part document at the desired location.

Placing the Additional Instances of Existing Component in the Assembly

Sometimes, you need more than one instance of the component to be placed in the assembly document. Press and hold down the CTRL key from the keyboard. Select the component in the assembly file and drag the cursor to the location where you want to place the instance of the selected component. Release the left mouse button to drop the new instance of the component. Similarly, you can place as many instances of the existing component as you want by following the above-mentioned procedure.

Placing the Component Using the Feature Palette Window

You can also place a component in the assembly file from the **Feature Palette** window. If you have saved components in the **Feature Palette** window, you can place that component by dragging and dropping from this window. **Feature Palette** window is invoked by choosing **Tools > Feature Palette** from the menu bar. You will learn more about the **Feature Palette** window in advanced chapters in part 2 of this book.

Assembling the Components

After placing the components in the assembly document, you need to assemble them. By assembling the components, you will constrain the degree of freedom of the components. As mentioned earlier, the components are assembled using the mates. Mates help you to precisely place and position the component with respect to the other components and the surroundings in the assembly. You can also define the linear and rotatory movement of the component with respect to the other components. In addition, you can create a dynamic mechanism and check the stability of the mechanism by precisely defining the combination of mates. There are two methods of adding mates to the assembly. The first method is using the **Mate PropertyManager**. The second and the most widely used method of adding the mates to the assembly is **Smart Mates**. Both methods are discussed next.

Assembling Components Using the Mate PropertyManager

Enhanced

CommandManager:	Assemblies > Mate
Menu:	Insert > Mate
Toolbar:	Assembly > Mate

In SolidWorks, the mates can be applied using the **Mate PropertyManager**. Choose the **Mate** button available in the **Assemblies CommandManager** or choose **Insert > Mate** from the menu bar. The **Mate PropertyManager** is invoked as shown in Figure 10-6.

Select a planar face, curved face, axis, or point on the first component and then select the entity from the second component. The selected entities will be highlighted in green. The names of the selected entities are displayed in the **Entities to Mate** selection area of the **Mate Selections** rollout. The **Mate** pop-up toolbar is invoked as shown in Figure 10-7. The most suitable mates to be applied to the current selection set are displayed in the **Mate** pop-up menu toolbar and in the **Standard Mates** rollout of the **Mate PropertyManager**. The most appropriate mate is selected by default. The preview of the assembly using most appropriate mate is displayed in the drawing area. You can also select the mates from the given list of the

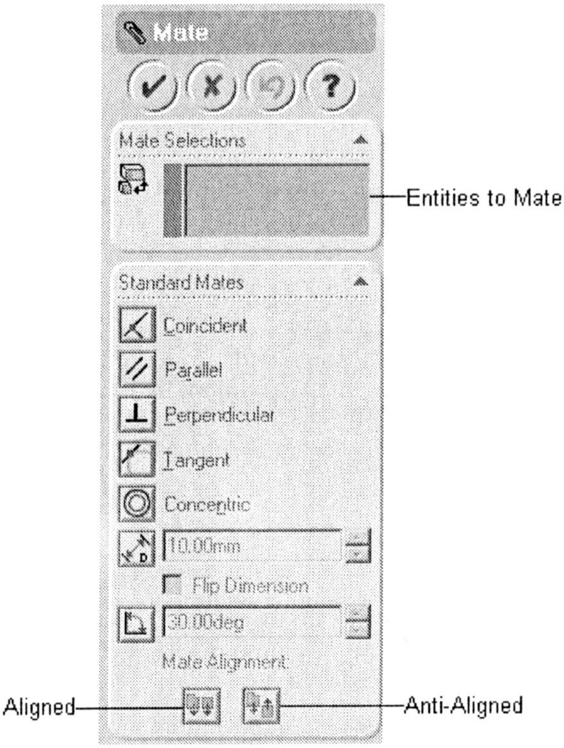

*Figure 10-6 The **Mate PropertyManager***

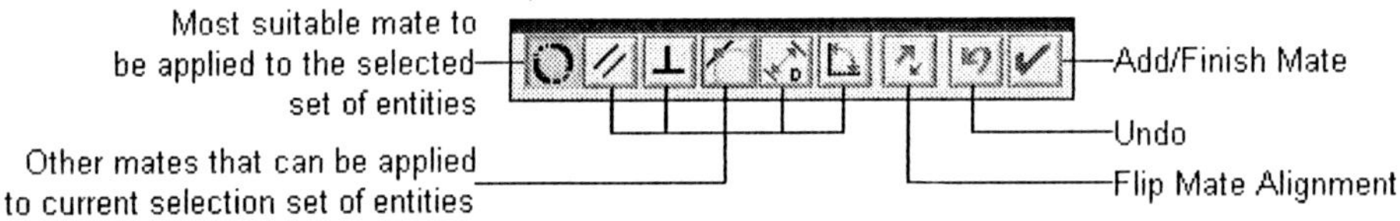

*Figure 10-7 The **Mate** pop-up toolbar*

mates that can be applied to the current selection set. As you select some other mate from the **Mate** pop-up toolbar the preview of the assembly is displayed using the newly selected mate. Now, choose the **Add/Finish Mate** button from the **Mate** pop-up toolbar. The **Mate PropertyManager** is still displayed, and you can add other mates to the assembly. After adding all the mates choose the **OK** button from the **Mate PropertyManager**. The various types of mates that can be applied are discussed next.

Coincident

The **Coincident** mate is generally applied to make two planar faces coplanar. But you can also apply the **Coincident** mate to other entities. The details of the geometries on which the **Coincident** mate can be applied are shown in Figure 10-8.

When you choose the **Coincident** button from the **Mate** pop-up toolbar, the preview of the model is displayed according to the current selection of the mate and the model is assembled in aligned or anti-aligned direction, depending on the current orientation of the model. Depending upon the preview of the assembled component the **Aligned** or

ASSEMBLY MATE COMBINATIONS (USING COINCIDENT MATE)		Second Component									
		Cone	Cylinder	Line	Point	Sphere	Circular/ Arc Edge	Extrusion	Surface	Plane	Cam
First Component	Cylinder	✗	✗	✓	✓	✗	✓	✗	✗	✗	✗
	Sphere	✗	✗	✗	✓	✗	✗	✗	✗	✗	✗
	Cone	✓	✗	✗	✓	✗	✓	✗	✗	✗	✗
	Circular/ Arc Edge	✗	✓	✗	✗	✗	✓	✗	✗	✓	✗
	Line	✗	✓	✓	✓	✗	✗	✗	✗	✓	✗
	Point	✗	✓	✓	✓	✓	✗	✓	✓	✓	✓
	Extrusion	✗	✗	✗	✓	✗	✗	✗	✗	✗	✗
	Surface	✗	✗	✗	✓	✗	✗	✗	✗	✗	✗
	Plane	✗	✗	✓	✓	✗	✓	✗	✗	✓	✗

Figure 10-8 *Table displaying the combinations for applying the* ***Coincident*** *mate*

Anti-Aligned button is selected in the **Standard Mates** rollout. You can choose the alignment of the assembled component by choosing the **Flip Mate Alignment** button from the **Mate** pop-up toolbar or switch between **Aligned** and **Anti-Aligned** buttons in the **Mate PropertyManager.** Figure 10-9 displays the faces to be selected to apply the coincident mate. Figure 10-10 shows the resultant mate applied with the default option, that is, **Anti-Aligned** selected. Figure 10-11 shows the coincident mate applied with the **Aligned** button selected.

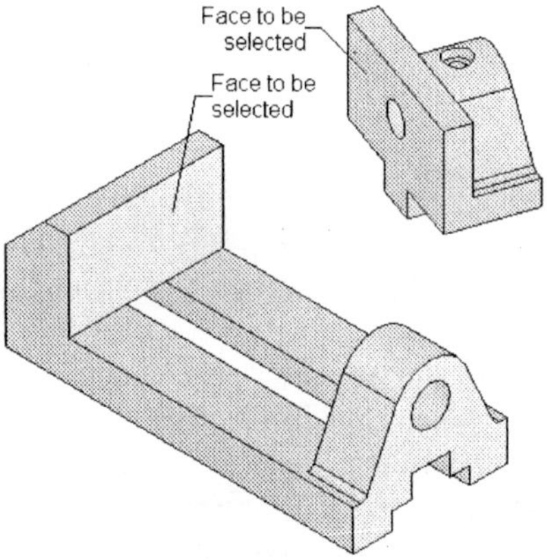

Figure 10-9 *Faces to be selected*

Concentric

The **Concentric** mate is generally used to align the central axis of one component with the central axis of the other. You generally need to select the circular faces to apply the concentric mate. You can also apply the **Concentric** mate between a point and a circular face or edge. The other combinations of applying the **Concentric** mate are displayed in the table given in Figure 10-12. For applying a **Concentric** mate,

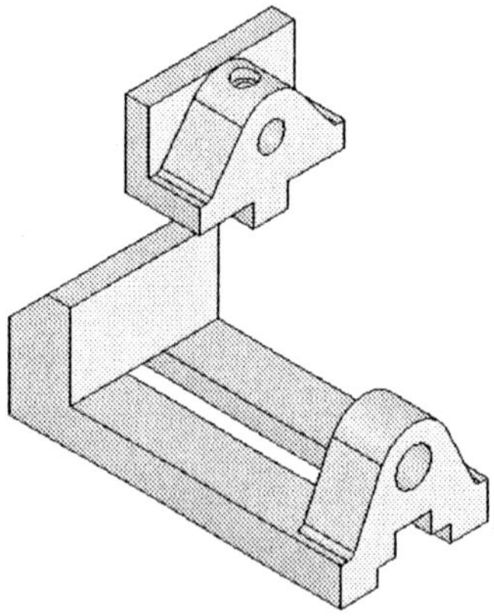

Figure 10-10 Coincident mate applied with the ***Anti-Aligned*** *button selected*

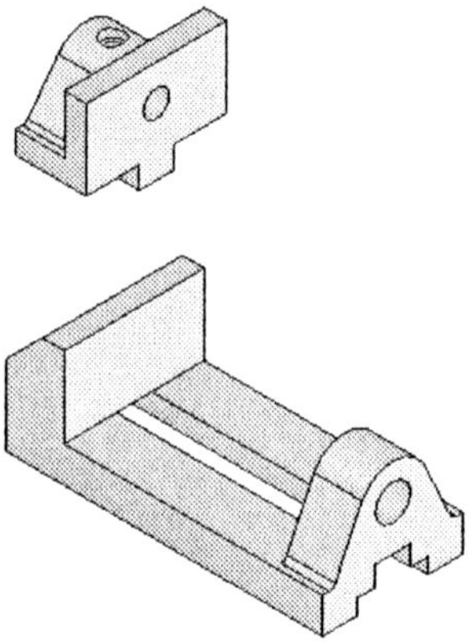

Figure 10-11 Coincident mate applied with the ***Aligned*** *button selected*

ASSEMBLY MATE COMBINATIONS (USING CONCENTRIC MATE)		Second Component									
		Cone	Cylinder	Line	Point	Sphere	Circular/ Arc Edge	Extrusion	Surface	Plane	Cam
First Component	Cylinder	✓	✓	✓	✓	✓	✓	✗	✗	✗	✗
	Sphere	✗	✓	✓	✓	✓	✗	✗	✗	✗	✗
	Cone	✓	✓	✓	✓	✗	✓	✗	✗	✗	✗
	Circular/ Arc Edge	✗	✓	✓	✗	✗	✓	✗	✗	✗	✗
	Line	✓	✓	✗	✗	✓	✓	✗	✗	✗	✗
	Point	✓	✓	✗	✗	✓	✗	✗	✗	✗	✗
	Extrusion	✗	✗	✗	✗	✗	✗	✗	✗	✗	✗
	Surface	✗	✗	✗	✗	✗	✗	✗	✗	✗	✗
	Plane	✗	✗	✗	✗	✗	✗	✗	✗	✗	✗

Figure 10-12 Table displaying the combinations for applying the ***Concentric*** *mate*

invoke the **Mate PropertyManager**. Select two entities from two different components. The names of the selected entities will be displayed in the **Entities to Mate** selection area. The **Concentric** button will be chosen in the **Mate** pop-up toolbar. If the **Concentric** button is not chosen by default, you need to manually select this button. The preview of the models being assembled after applying this mate is displayed in the graphics area. You can toggle between the **Aligned** and **Anti-Aligned** buttons according to the requirement. Choose the **Add/Finish Mate** button from the **Mate** pop-up toolbar.

Figure 10-13 shows the faces to be selected to apply the concentric mate. Figure 10-14 shows the concentric mate applied with the **Aligned** radio button selected. Figure 10-15 shows the concentric mate applied with the **Anti-Aligned** radio button selected.

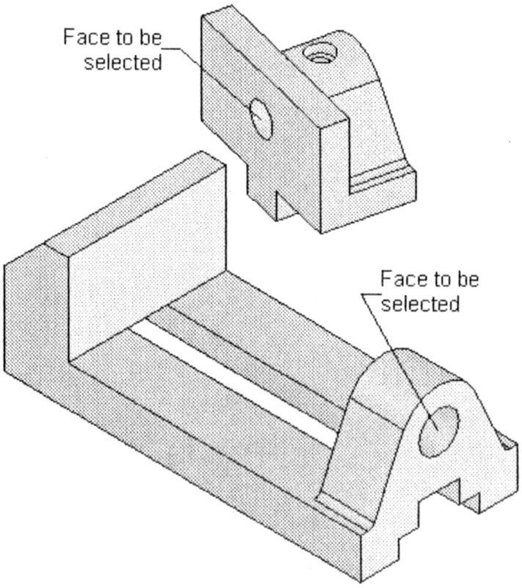

Figure 10-13 *Faces to be selected*

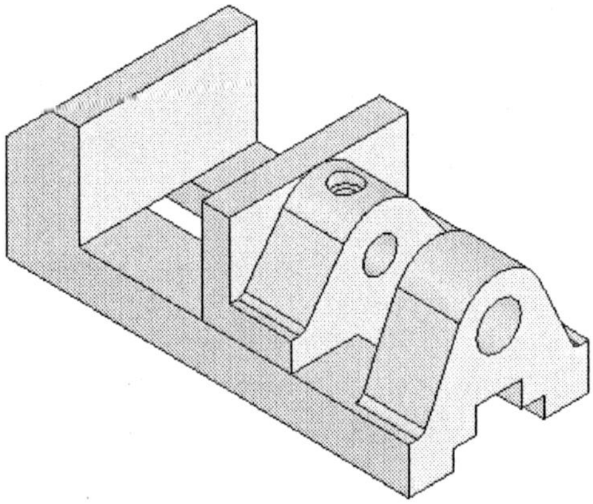

Figure 10-14 *Concentric mate applied with the* ***Aligned*** *button selected*

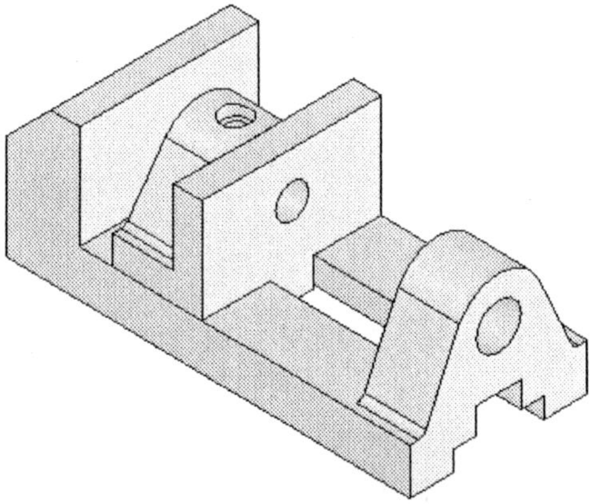

Figure 10-15 *Concentric mate applied with the* ***Anti-Aligned*** *button selected*

Distance

The **Distance** button is chosen to apply the **Distance** mate between the components. To apply the **Distance** mate, invoke the **Mate PropertyManager** and select the entities from both the components. Choose the **Distance** button from the **Mate** pop-up toolbar. The **Distance** spinner is displayed in the **Mate** pop-up toolbar. Set the value of the distance in the **Distance** spinner. The preview of the assembly updates automatically after you set the value of the distance. Using the **Flip Dimension** button available at the left of the **Distance** spinner you can specify a negative distance value. If needed, toggle between the **Aligned** button and the **Anti-Aligned** button. Figure 10-16 shows the combinations of components to apply the distance mate. Figure 10-17 shows the faces to be selected. Figure 10-18 shows the **Distance** mate applied between two components. You can also toggle between the **Aligned** and **Anti-Aligned** options.

ASSEMBLY MATE COMBINATIONS (USING DISTANCE MATE)		Second Component									
		Cone	Cylinder	Line	Point	Sphere	Plane	Extrusion	Surface	Circular/ Arc Edge	Cam
First Component	Cylinder	✗	✓	✓	✓	✗	✓	✗	✗	✗	✗
	Sphere	✗	✗	✓	✓	✓	✗	✗	✗	✗	✗
	Cone	✓	✗	✗	✗	✗	✗	✗	✗	✗	✗
	Plane	✗	✓	✓	✓	✓	✓	✗	✗	✗	✗
	Line	✗	✓	✓	✓	✓	✓	✗	✗	✗	✗
	Point	✗	✓	✓	✓	✓	✓	✗	✗	✗	✗
	Extrusion	✗	✗	✗	✗	✗	✗	✗	✗	✗	✗
	Surface	✗	✗	✗	✗	✗	✗	✗	✗	✗	✗
	Circular/ Arc Edge	✗	✗	✗	✗	✗	✗	✗	✗	✗	✗

Figure 10-16 *Table displaying the combinations for applying the* ***Distance*** *mate*

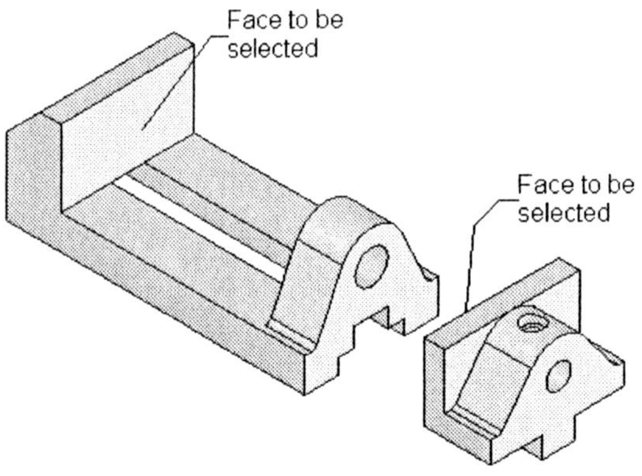

Figure 10-17 *Selecting the faces to apply the* ***Distance*** *mate*

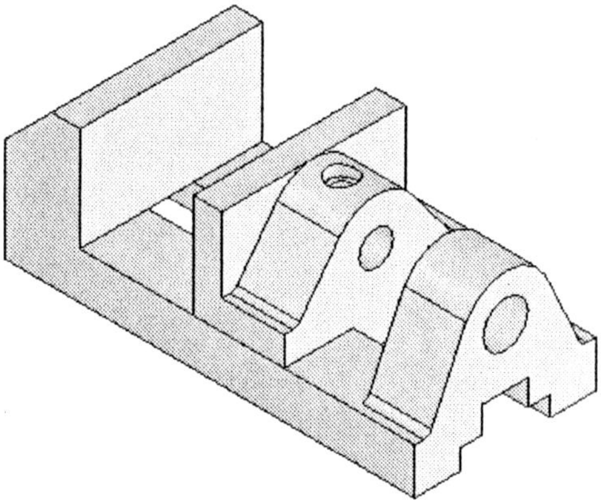

Figure 10-18 ***Distance*** *mate applied to the selected faces*

Angle

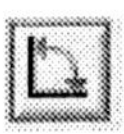

The **Angle** button is used to apply the **Angle** mate between two components. The **Angle** mate is used to specify the angular position between the selected plane, planar face, or edges of two components. To apply this mate, invoke the **Mate PropertyManager** and select the entities from two components. Choose the **Angle** button from the **Mate** pop-up toolbar. The **SolidWorks** warning box is displayed, which informs you about the minimum and the maximum angle value that you can specify to the current selection set. Choose the **OK** button from this message window. The preview of the models will be modified with the default value of the angle. The **Angle** spinner is invoked and you can set the value of the angle in this spinner. You can also change the angle direction using the **Flip Dimension** button provided on the right of the angle spinner. You can also toggle between the **Aligned** button and the **Anti-Aligned** button. Choose the **Add/Finish**

Mate button from the **Mate** pop-up toolbar. Figure 10-19 shows the table of combinations for applying the angle mate.

ASSEMBLY MATE COMBINATIONS (USING ANGLE MATE)		Second Component									
		Cylinder	Extrusion	Line	Plane	Sphere	Circular/ Arc Edge	Cone	Surface	Point	Cam
First Component	Cylinder	✓	✓	✓	X	X	X	X	X	X	X
	Extrusion	✓	✓	✓	X	X	X	X	X	X	X
	Line	✓	✓	✓	X	X	X	X	X	X	X
	Plane	X	X	X	✓	X	X	X	X	X	X
	Sphere	X	X	X	X	X	X	X	X	X	X
	Circular/ Arc Edge	X	X	X	X	X	X	X	X	X	X
	Cone	✓	✓	✓	X	X	X	✓	X	X	X
	Point	X	X	X	X	X	X	X	X	X	X
	Surface	X	X	X	X	X	X	X	X	X	X

***Figure 10-19** Table displaying the combinations for applying the **Angle** mate*

Figure 10-20 shows the faces to be selected to apply the angular mate. Figure 10-21 shows the assembly after applying the **Angular** mate with the angle value of 90-degree.

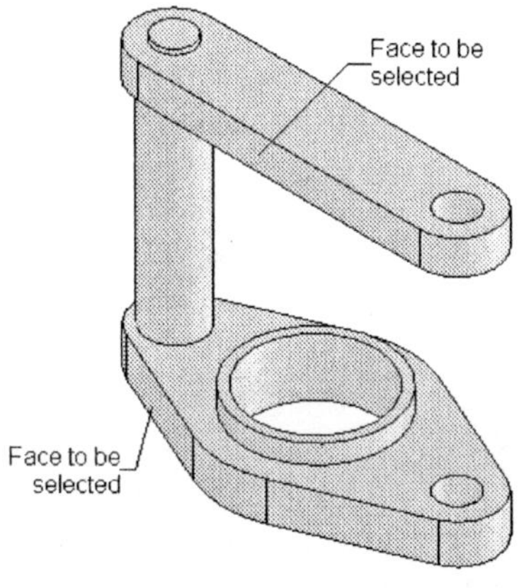

***Figure 10-20** Faces selected to apply the **Angle** mate*

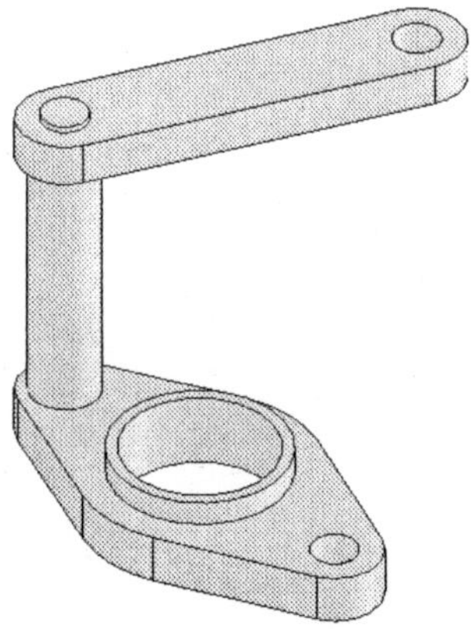

***Figure 10-21** Assembly after applying the **Angle** mate*

Parallel

The **Parallel** button available in the **Mate** pop-up toolbar is used to apply the **Parallel** mate between the two components. To apply the **Parallel** mate, invoke the **Mate PropertyManager** and select the two entities from the two components. Choose the **Parallel** button from the **Mate** pop-up menu to apply the mate constraint. You can also toggle between the **Align** button and the **Anti-Aligned** button. Choose the **Add/Finish Mate** button from the **Mate** pop-up toolbar. Figure 10-22 shows the combination of components on which you can apply the **Parallel** mate. Figure 10-23 shows the entities to be selected to apply the **Parallel** mate. Figure 10-24 shows the assembly after applying the parallel mate.

ASSEMBLY MATE COMBINATIONS (USING PARALLEL MATE)		Second Component									
		Cylinder	Extrusion	Line	Plane	Sphere	Circular/ Arc Edge	Cone	Surface	Point	Cam
First Component	Cylinder	✓	✓	✓	X	X	X	X	X	X	X
	Extrusion	✓	✓	✓	X	X	X	X	X	X	X
	Line	✓	✓	✓	✓	X	X	X	X	X	X
	Plane	X	X	✓	✓	X	X	X	X	X	X
	Circular/ Arc Edge	X	X	X	X	X	X	X	X	X	X
	Sphere	X	X	X	X	X	X	X	X	X	X
	Cone	✓	✓	✓	X	X	X	✓	X	X	X
	Surface	X	X	X	X	X	X	X	X	X	X

Figure 10-22 *Table displaying the combinations for applying the* ***Parallel*** *mate*

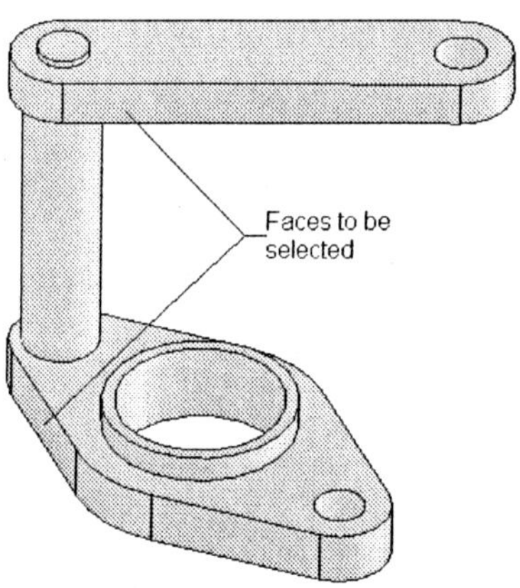

Figure 10-23 *Faces selected to apply the* ***Parallel*** *mate*

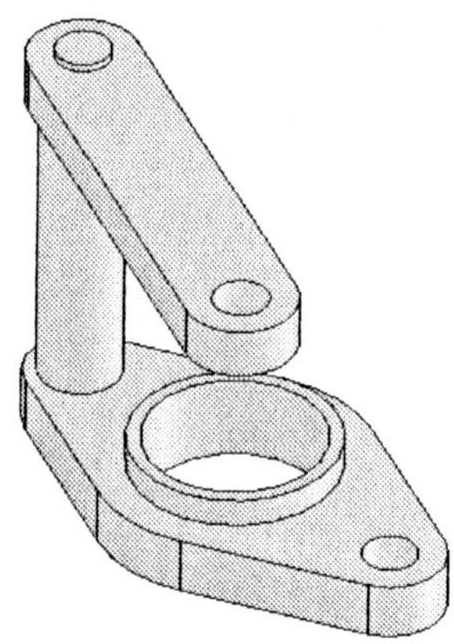

Figure 10-24 *Assembly after applying the* ***Parallel*** *mate*

Perpendicular

The **Perpendicular** button available in the **Mate Settings** rollout is used to apply the **Perpendicular** mate between two components. Invoke the **Mate PropertyManager** and select two entities from two components. Choose the **Perpendicular** button from the **Mate** pop-up toolbar. You can also toggle between the **Aligned** button and the **Anti-Aligned** button. Choose the **Add/Finish Mate** button from the **Mate** pop-up toolbar. Figure 10-25 shows the table displaying the combinations for applying the **Perpendicular** mate. Figure 10-26 shows the entities to be selected. Figure 10-27 shows the **Perpendicular** mate applied to the assembly.

ASSEMBLY MATE COMBINATIONS (USING PERPENDICULAR MATE)		Second Component									
		Cylinder	Extrusion	Line	Plane	Sphere	Circular/ Arc Edge	Cone	Surface	Point	Cam
First Component	Cylinder	✓	✓	✓	X	X	X	X	X	X	X
	Extrusion	✓	✓	✓	X	X	X	X	X	X	X
	Line	✓	✓	✓	✓	X	X	X	X	X	X
	Plane	X	X	✓	✓	X	X	X	X	X	X
	Circular/ Arc Edge	X	X	X	X	X	X	X	X	X	X
	Sphere	X	X	X	X	X	X	X	X	X	X
	Cone	✓	✓	✓	X	X	X	✓	X	X	X
	Surface	X	X	X	X	X	X	X	X	X	X

Figure 10-25 *Table displaying the combinations for applying the* ***Perpendicular*** *mate*

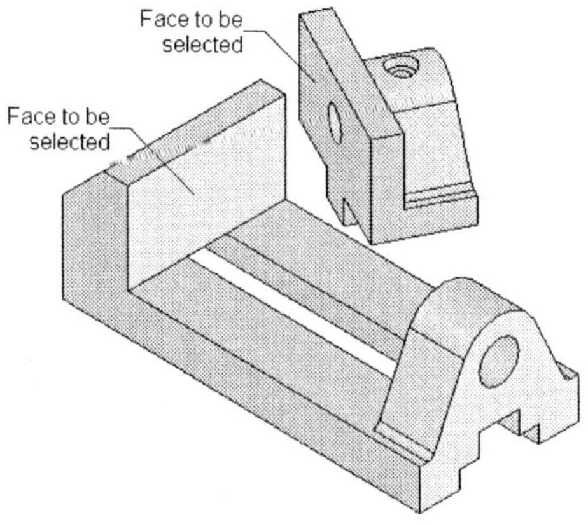

Figure 10-26 *Faces selected to apply the* ***Perpendicular*** *mate*

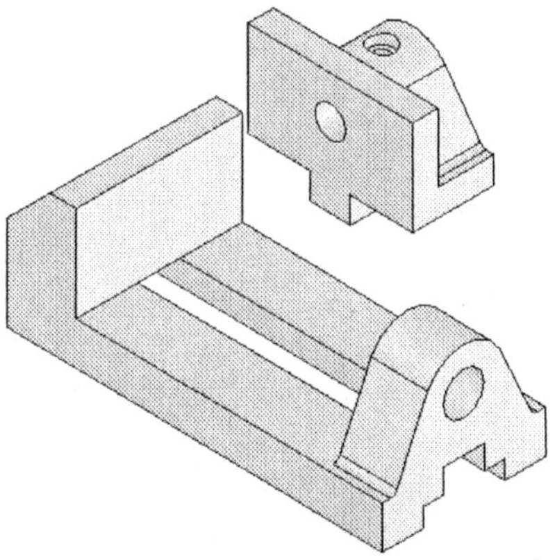

Figure 10-27 *Assembly after applying the* ***Perpendicular*** *mate*

Tangent

The **Tangent** button available in the **Mate** pop-up toolbar is used to apply the **Tangent** mate between the two components. Figure 10-28 shows the table displaying the combinations for applying the tangent mate. Figure 10-29 shows the entities to be selected to apply the tangent mate. Figure 10-30 shows the tangent mate applied to the assembly.

ASSEMBLY MATE COMBINATIONS (USING TANGENT MATE)		Second Component								
		Cone	Cylinder	Line	Point	Sphere	Plane	Surface	Cam	Extrusion
First Component	Cylinder	✓	✓	✓	X	✓	✓	✓	✓	✓
	Sphere	✓	✓	✓	X	✓	✓	X	X	X
	Cone	X	X	X	X	✓	✓	X	X	X
	Plane	X	✓	X	X	✓	X	✓	✓	✓
	Line	X	✓	✓	✓	✓	✓	✓	✓	✓
	Extrusion	X	✓	X	X	X	✓	X	X	X
	Surface	X	✓	X	X	X	✓	X	X	X
	Cam	X	X	X	X	X	X	X	X	X

***Figure 10-28** Table displaying the combinations for applying the **Tangent** mate*

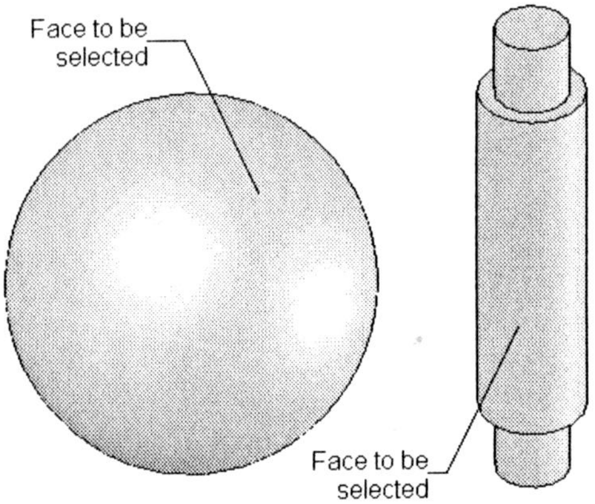

***Figure 10-29** Entities to be selected to apply the **Tangent** mate*

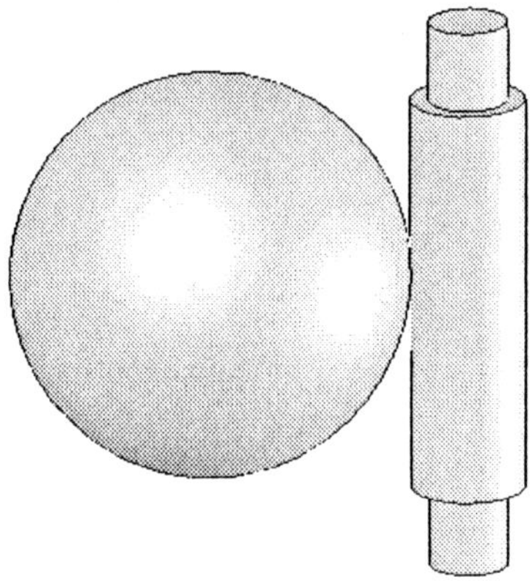

***Figure 10-30** Assembly after applying the **Tangent** mate*

You will learn about some advanced assembly mates in Chapter 11.

The options that are available in the **Mate PropertyManager** are discussed next.

Mate

The **Mate** rollout is used to display the mates that are applied between the selected entities.

Options

The **Add to new folder** check box available in the **Options** rollout is used to add the currently applied mate in a folder. This folder is placed in the **Mates** mategroup in the **FeatureManager Design Tree**. You can also drag and drop the other mates from the **Mate** mategroup in the newly created folder. The **Show popup dialog** check box available in the **Options** rollout is selected by default and is used to display the **Mate** pop-up toolbar. The **Show preview** check box is selected by default and is used to display the dynamic preview of the assembly as you apply the mates to the components. The **Use for positioning only** check box is used only to define the position of the component by applying the mate. This mate is displayed in the **Mate** rollout but when you exit the **Mate PropertyManager**, this mate is not displayed in the **Mate** mategroup in the **FeatureManager Design Tree**.

Assembling Components Using Smart Mates

Smart Mates is the most attractive feature of the assembly design environment of SolidWorks. The **Smart Mates** technology is the patent of SolidWorks Corporation. This also speeds up the design process in the assembly environment of SolidWorks. To add smart mates to the components that have been placed in the assembly, choose the **Move Component** button from the **Assembly CommandManager**. The **Move Component PropertyManager** is displayed. Now, choose the **SmartMates** button available in the **Move** rollout. The **Move Component PropertyManager** is replaced by the **SmartMates PropertyManager** as shown in Figure 10-31.

***Figure 10-31** The SmartMates PropertyManager*

The select cursor is replaced by the move cursor. You need to double-click the entity of the first component to add a mate. The component will be displayed in transparent. The cursor will be replaced by the smart mates cursor. Press and hold down the left mouse button on the selected entity and drag the cursor to the entity with which you want to mate the previously selected entity. The symbol of the constraint that can be applied between the two entities will be displayed below the smart mates cursor. You can use the TAB key to toggle between the aligned and the anti-aligned option while applying **Smart Mates**. When the symbol of the mate is displayed below the cursor, release the left mouse button. The **Mate** pop-up toolbar is displayed; choose the **Add/Finish Mate** button from this toolbar. Figure 10-32 shows the face to be selected to apply smart mate. Figure 10-33 shows the component being dragged.

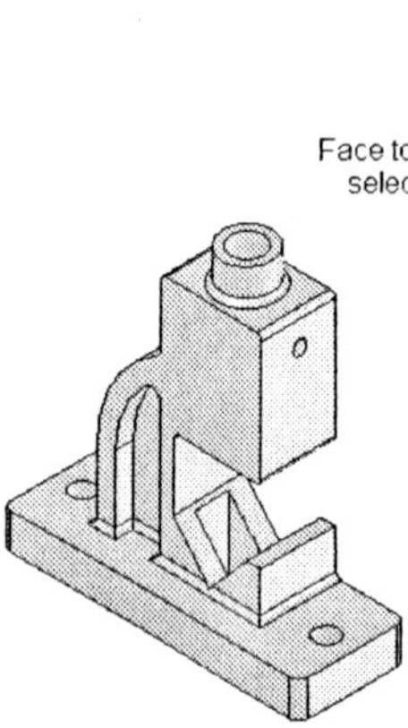

Figure 10-32 *Face to be selected to apply smart mate*

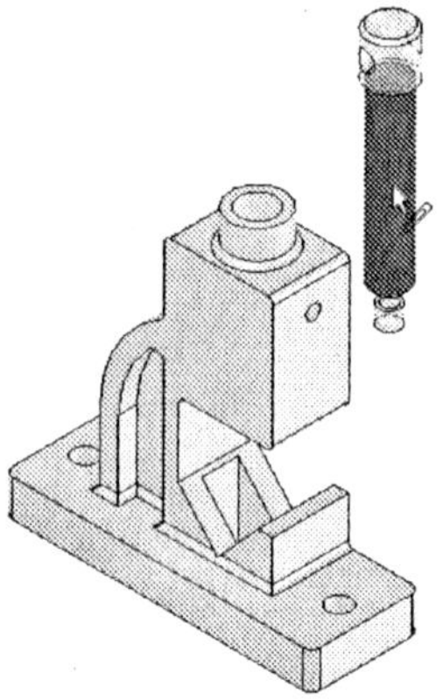

Figure 10-33 *Component being dragged to apply smart mate*

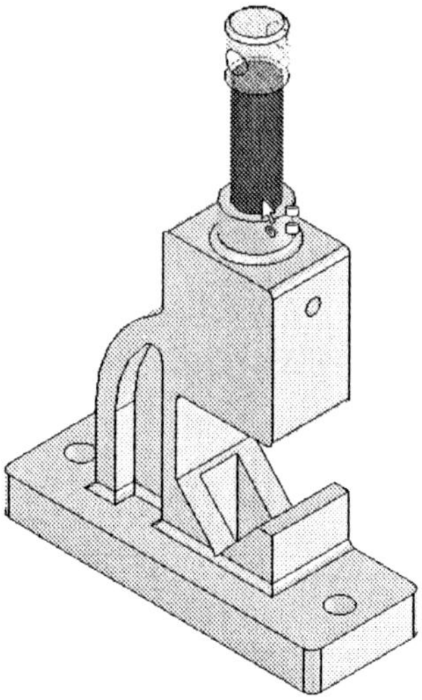

Figure 10-34 *Concentric symbol appears below the smart mates cursor*

Figure 10-34 shows that the concentric symbol appears below the smart mates cursor when the cursor is placed near a circular face of the other component.

Figure 10-35 shows a planar face selected to apply smart mate. You can use the TAB key to toggle between the aligned and anti-aligned mates. Figure 10-36 shows that the coincident symbol appears below the cursor. Figure 10-37 shows the assembly after applying the **Coincident** mate using smart mate.

Note

To apply smart mates without invoking the ***SmartMates*** *tool, press and hold down the ALT key from keyboard and select the entity of the first component and drag the component to the entity of the other component with which you need to apply the mate. As you release the left mouse button, the* ***Mate*** *pop-up toolbar is displayed. Choose the* ***Add/Finish Mate*** *button from this toolbar.*

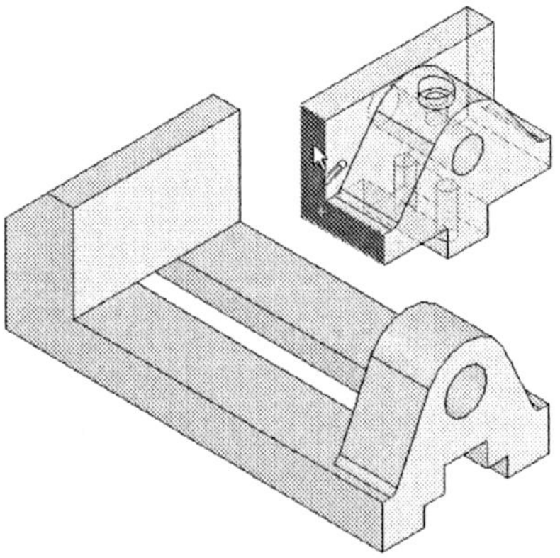

Figure 10-35 *Planar face to be selected to apply a mate using smart mates*

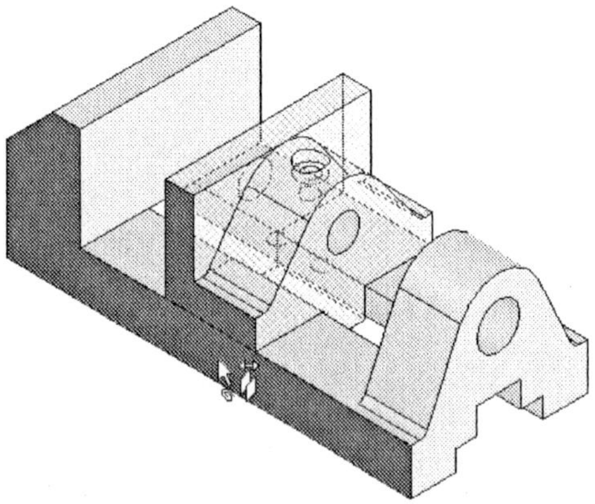

Figure 10-36 *Coincident symbol appears after dragging the component near another planar face*

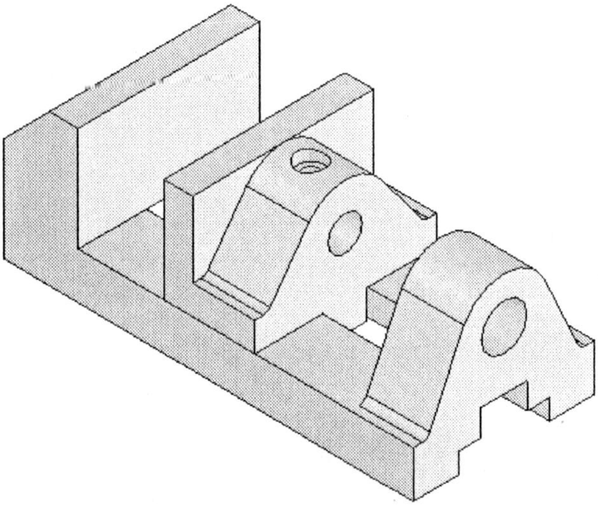

Figure 10-37* *Coincident *mate applied to the assembly using smart mates*

Note

When you drag a component for applying a smart mate, the selected entity of the first component will snap all the corresponding entities of the second component. You can press the ALT key from the keyboard to exit the snap. To again enter the snap mode press the ALT key from the key board.

Tip. *You can also add smart mates without dragging the component. To add a smart mate without dragging, invoke the* ***Smart Mates*** *tool. Double-click the entity from the first component. The component will be displayed in transparent. Now, select the entity from the second component. The preview with the most appropriate mate is displayed and the* ***Mate*** *pop-up toolbar is invoked.*

Geometry-Based Mates

In the assembly design environment of SolidWorks, you can also add geometry-based mates. Geometry-based mates are also a type of smart mates, and are applied while you are placing a component in the assembly environment. Consider a case in which the first component is already placed in the assembly environment. Now, open the part document of the second component. Choose **Window > Tile Horizontal** or **Tile Vertical** from the menu bar.

Suppose, you need to insert the revolved feature of the second component in the circular slot of the first component. Also, the same time you also need to align the larger bottom face of the second component with the upper face of the first component. Press and hold down the left mouse button on the edge of the second component as shown in Figure 10-38. Drag the cursor to the assembly window near the upper edge of the circular slot of the first component. The second component mated with the first component will be displayed in temporary graphics as shown in Figure 10-39. The coincident symbol is also displayed below the smart mates cursor. You can also toggle the direction of placement of the component. Figure 10-40 shows

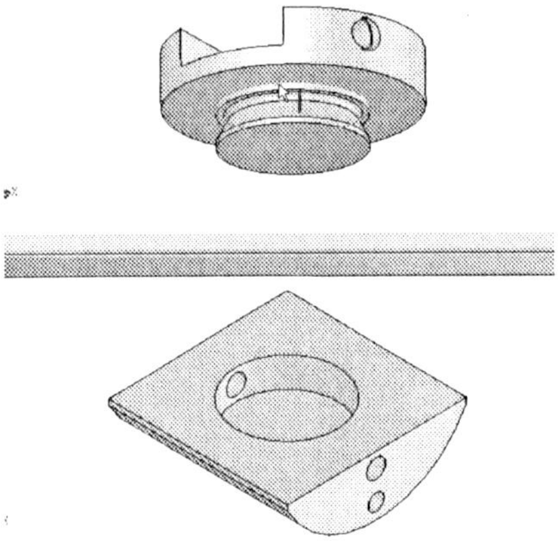

Figure 10-38 Edge of the second component to be selected.

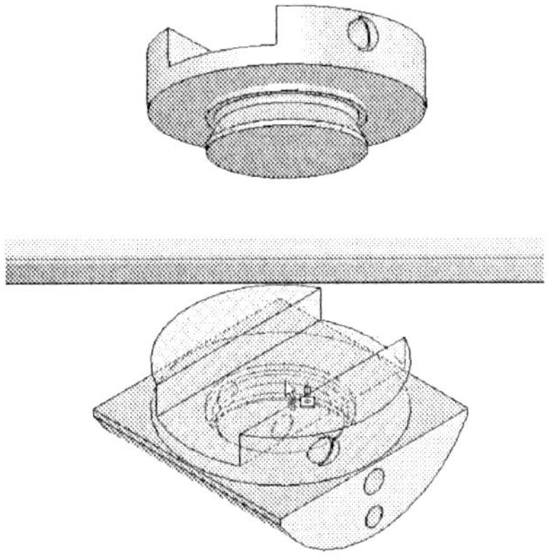

Figure 10-39 Component being dragged into the assembly window for applying geometry-based mates

that the direction of placement is flipped using the TAB key. Again press the TAB key to return to the default direction. Release the left mouse button to place the component. If you expand the **Mates** option from the **FeatureManager Design Tree**, you will notice that two mates are applied to the assembly. One is the coincident mate and the second is the concentric mate. Figure 10-41 shows the assembly after adding the geometry-based smart mates.

Feature-Based Mates

In the assembly design environment of SolidWorks, you can also add feature-based mates. For adding feature-based mates, one of the feature of the first component must be a circular base or boss feature and the second component must have a hole or a circular cut feature. The

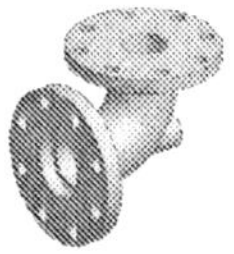

Tip. *You can add the geometry-based mates between two linear edges, two planar faces, two vertices, two conical faces, two axes, between an axis and a conical face, and between two circular edges.*

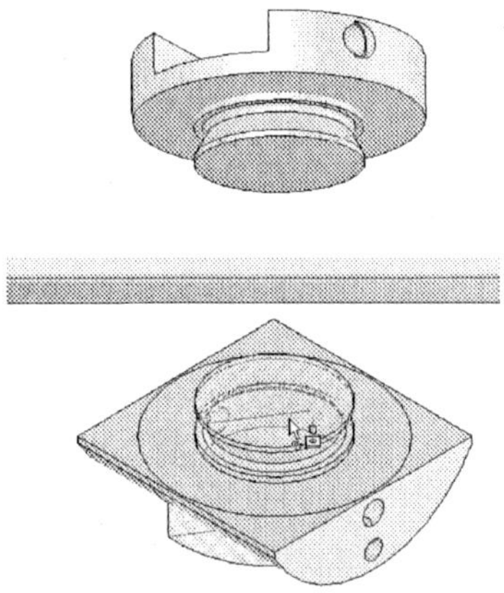

Figure 10-40 *The placement direction of component is flipped using the TAB key*

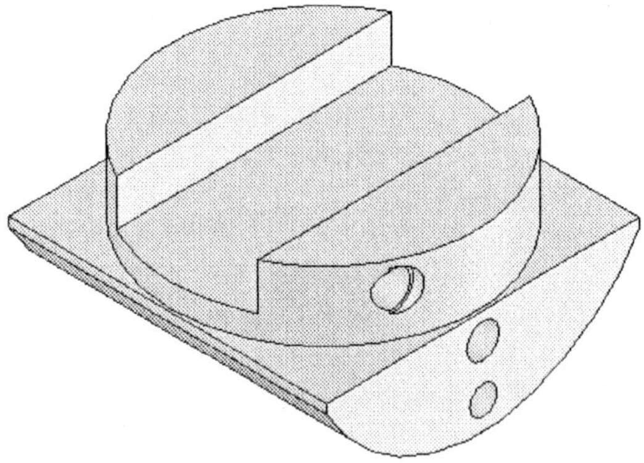

Figure 10-41 *Assembly after adding a geometry-based mate*

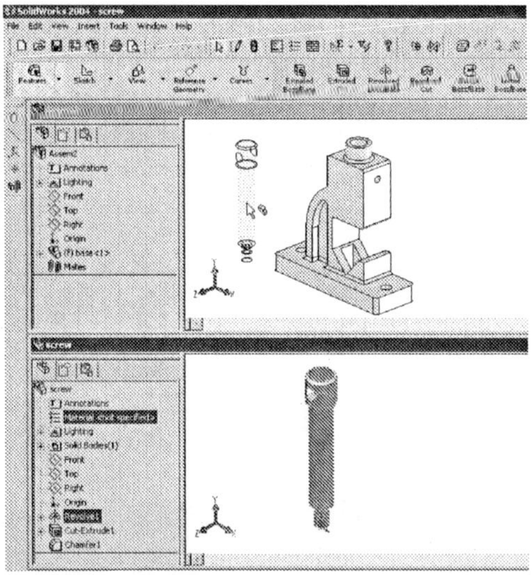

Figure 10-42 *Component being dragged by selecting the revolved feature*

feature can be an extruded or a revolved feature. In the assembly document one of the feature must be placed earlier. Open the part document of the component to be placed using feature-based mates. Tile both the document windows horizontally or vertically. In the **FeatureManager Design Tree** of the part document select the extruded or revolved feature and drag it to the assembly window. Place the cursor at a location where you need to place the component. The mate symbol will be displayed below the cursor. The preview of the part mates with the second component will be displayed in temporary graphics. You can also change the alignment or direction of placement using the TAB key. Release the left mouse button to drop the component. Figure 10-42 shows the component being dragged by selecting the revolved feature from the **FeatureManager Design Tree** of part document. Figure 10-43 shows the preview of the component being inserted into the hole of the second component using feature-based mates. Figure 10-44 shows the component assembled using feature-based mates.

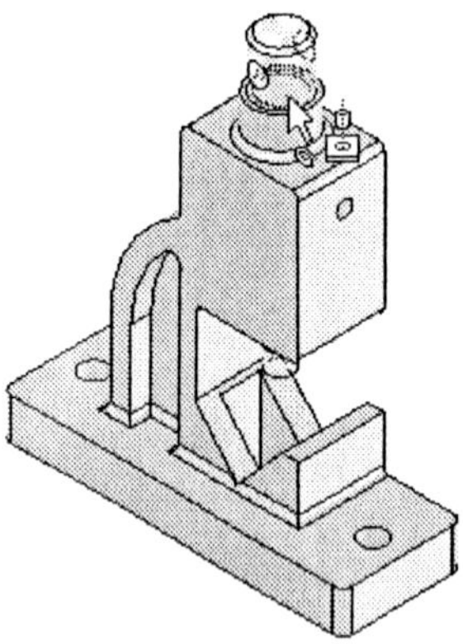

Figure 10-43 *Preview of the component being inserted using feature-based mates*

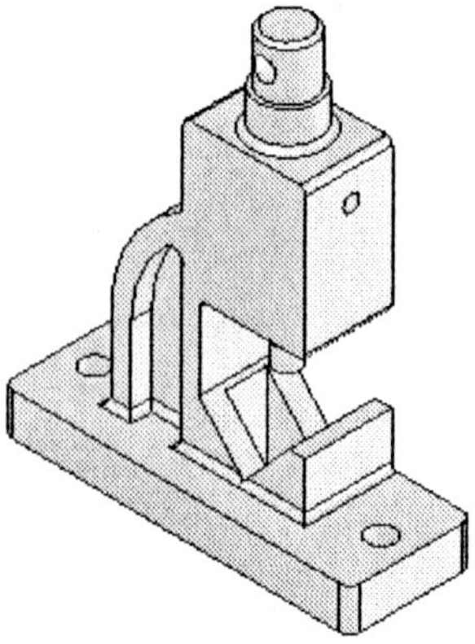

Figure 10-44 *Component assembled using feature-based mates*

Tip. *Feature-based mates are applied only to the components having cylindrical or conical features. You cannot add feature-based mates using features other than cylindrical or conical geometry. If you are adding feature-based mates to a component having a conical face, the second component must have a conical face. You cannot add a feature-based mate if the geometry of the feature of one component is cylindrical and that of the second component is conical.*

If you are adding feature-based mates using the features having conical geometry, there must be a planar face adjacent to the conical face of both features.

Pattern-Based Mates

Pattern-based mates are used to assemble the components that have a circular pattern created on the circular feature. The best example of this type of component is the flange or a shaft coupling. Remember that to create pattern-based mates, all the components that will be assembled must have a circular pattern on the mating faces. To create pattern-based mates, select the outer edge of the second component and drag it to the circular edge of the first component that is already placed in the assembly document. The preview of the component assembled with the first component will be displayed. Using the TAB key you can switch the part with respect to the pattern instances. Release the left mouse button to drop the part. Figure 10-45 shows the component being dragged to the assembly document window. Figure 10-46 shows the preview of the component being assembled. Figure 10-47 shows the component assembled using pattern-based mates.

Assembling Components Using Mate Reference

CommandManager:	Reference Geometry > Mate Reference	*(Customize to Add)*
Menu:	Insert > Component > New Part	
Toolbar:	Reference Geometry > Mate Reference	*(Customize to Add)*

In SolidWorks you can define the mate reference to the **Part** in the part mode or in the **Assembly** mode. The mate references allow you to define the mating references such as planar surfaces, axis, edges, and so on before assembling the component.

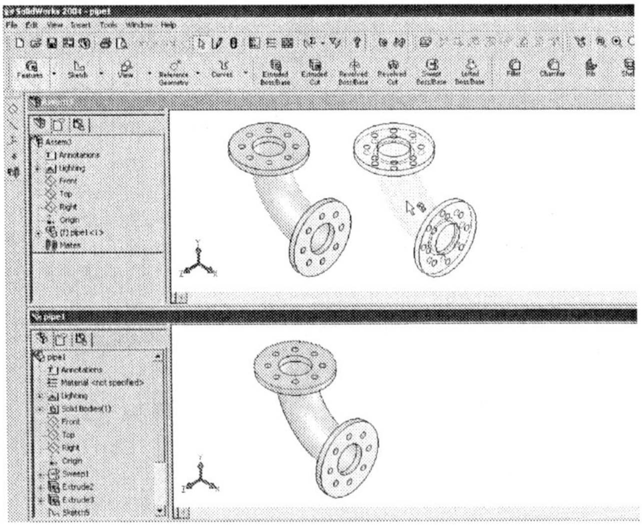

***Figure 10-45** Component being dragged into the assembly document*

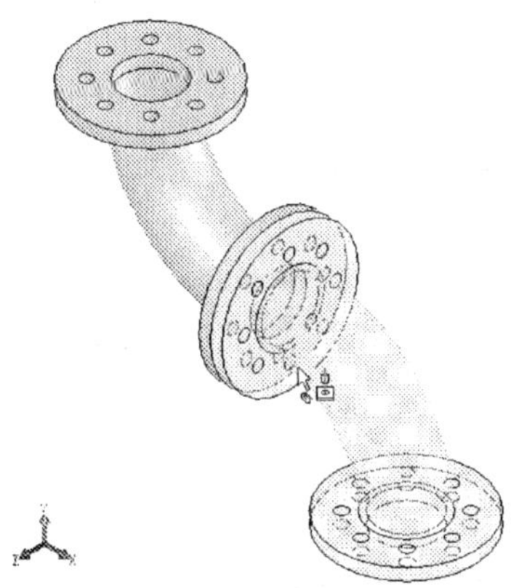

***Figure 10-46** Preview of the component being assembled using pattern-based mates*

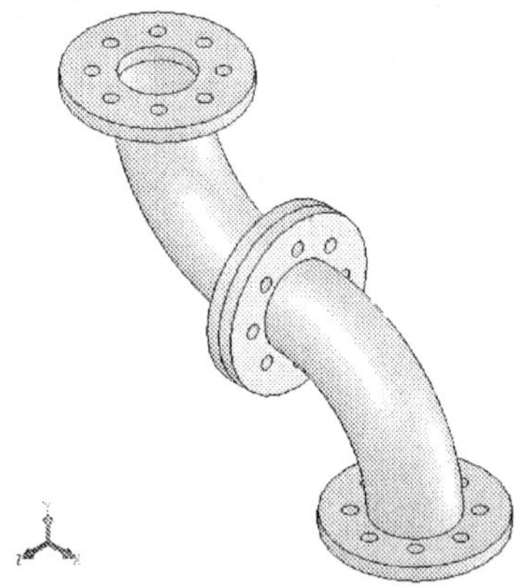

***Figure 10-47** Component assembled using pattern-based mates*

If is recommended that the mate reference should be added to the component in the **Part** mode. The mate references are created using the **Mate Reference PropertyManager**. The **Mate Reference PropertyManager** is invoked by choosing the **Mate Reference** button from the **Reference Geometry Command** after customizing it. The **Mate Reference PropertyManager** is shown in Figure 10-48.

The **Mate Reference Name** edit box is used to define the name of the mate reference. The **Primary Reference Entity** rollout is used to define the primary mate reference. The **Mate Reference Type** drop-down list is used to define the type of mate. The **Mate Reference Alignment** drop-down list is used to define the type of alignment.

The **Secondary Reference Entity** rollout is used to define the secondary mate reference. The **Tertiary Reference Entity** rollout is used to define the tertiary mate reference.

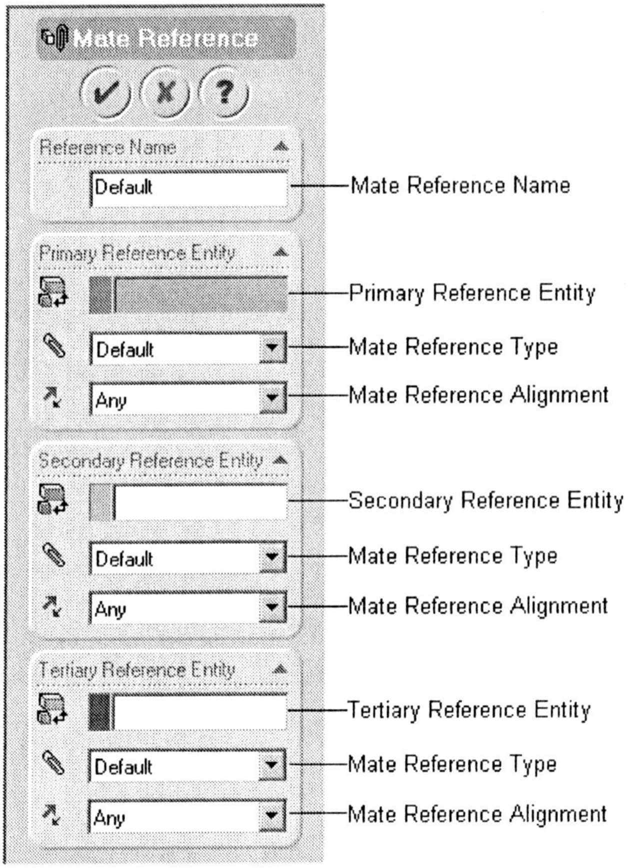

*Figure 10-48 The **Mate Reference PropertyManager***

To assemble a component using the mate reference, you need to define the mate reference to both the components. The names of the mate reference should be the same for both the components. After defining mate references to both the components, you need to place the first component coincident with the origin in the assembly document. Now, drag the second component and you will notice that the second component is aligned to the references that were defined as mate references. Therefore, you do not need to apply mates in the assembly environment.

CREATING TOP-DOWN ASSEMBLIES

As mentioned earlier, top-down assemblies are the assemblies in which all the components are created in the same assembly file. However, to create the components, you require an environment in which you can draw the sketches or the sketched features and the environment where you can convert the sketches into features. In other words, you can say that to create the components in the assembly file you need a sketcher environment as well as a part modeling environment in the assembly file. In SolidWorks, you can invoke the sketcher environment and the part modeling environment in the assembly document itself. The basic procedure to create the components in the assembly or in other words to create the top-down assembly is discussed next.

Creating the Component in the Top-Down Assembly

CommandManager:	Assemblies > New Part	*(Customize to Add)*
Menu:	Insert > Component > New Part	
Toolbar:	Assembly > New Part	*(Customize to Add)*

Before creating the first component in the top-down assembly, you first need to save the assembly document. After creating a new assembly document, choose the **Save** button from the **Standard** toolbar to save the assembly file. It is recommended that you create a new folder and save the assembly file and the other referenced file in the same folder.

Now, choose the **New Part** button from the **Assemblies CommandManager** after customizing it. The **Save As** dialog box is displayed and you need to specify the name and location to save the new component. Enter the name of the new component in the **File name** edit box and choose the **Save** button. The select cursor will be replaced by the new component cursor. Now, you need to place the new component in the assembly document. Using the left mouse button place the component on the **Front** assembly plane, which is displayed in the **FeatureManager Design Tree**. The plane on which you place the component is selected as the sketching plane and the sketcher environment is invoked automatically. You will notice that the **Edit Part** button available in the **Assemblies CommandManager** is selected. This means that the part modeling environment is invoked in the assembly document. You can draw the sketch of the base feature in the current sketching environment or you can also exit the sketching environment and select any other sketching plane to create the sketch. After creating the sketch of the base feature, exit the sketching environment. Now, as the **Features CommandManager** is not available by default, you need to invoke it and use the tools available in the **Features CommandManager** to convert the sketch into a model. Similarly, create the remaining features in the model.

After creating all the features, choose the **Edit Part** button from the **Assembly** toolbar to exit the part modeling environment. The newly created component will have an **Inplace** mate with the default assembly plane on which it was placed earlier. Therefore, the newly created component is fixed. Using the procedure mentioned above, create other components.

Whenever you create a component in a top-down assembly, the component is fixed using the **Inplace** mate. You can also delete this mate by selecting the **Inplace** mate by expanding the **Mates** option available in the **FeatureManager Design Tree**. After selecting the mate use the DELETE key to delete the mate. Now, this component is floating, and you can move this component. You can also assemble this component according to your requirement. You will learn more about fixed and floating components later in this chapter.

Note

As discussed earlier, when you place the first component in the assembly in bottom-up assembly design approach, that component is fixed by default. Therefore, you cannot apply any mates to a fixed component. If you need to add some mates to the fixed component, you first need to float the component. Select the component from the drawing area or from the ***FeatureManager Design Tree****. Right-click to invoke the shortcut menu and choose the* ***Float*** *option from the shortcut menu. The components placed after the first component are floating components by default. If you need to fix a floating component, select the component and invoke the shortcut menu. Choose the* ***Fix*** *option from the shortcut menu.*

MOVING THE INDIVIDUAL COMPONENTS

SolidWorks allows you to move the individual unconstrained components in the assembly document without affecting the position and location of the other components. There are three methods of moving an individual component. Two methods of moving the individual component are discussed next. The third method is discussed later in this chapter.

Moving the Individual Components by Dragging

One of the major enhancements in the **Assembly** mode of SolidWorks 2004 is that you can move the component placed in the assembly without invoking any tool. To move an individual component, simply select the component and drag the cursor to move the component. Release the left mouse button to place the component at the desired location.

Moving the Individual Components Using the Move Component Tool

CommandManager:	Assemblies > Move Component
Toolbar:	Assembly > Move Component

You can also move an individual component using the **Move Component** tool. Choose the **Move Component** button from the **Assemblies CommandManager**. The **Move Component PropertyManager** is invoked. You will notice that the **Free Drag** option is selected in the **Move** drop-down list in the **Move** rollout. You are prompted to select a component and drag to move it. The select cursor is replaced by the move cursor; select the component and drag the cursor to move the component. Release the left mouse button to move the component to the desired location. The other options available in the **Move** drop-down list to move the component are discussed next.

Along Assembly XYZ

Using the **Along Assembly XYZ** option from the **Move** drop-down list, you can move the component dynamically along the X, Y, and Z axes of the assembly document. Choose the **Along Assembly XYZ** option from the **Move** drop-down list in the **Move** rollout. An assembly coordinate system in magenta color is displayed in the drawing area and you are prompted to select a component and drag parallel to an assembly axis to move along that axis. Select the component and drag to move it along any one of the assembly axis.

Along Entity

The **Along Entity** option available in the **Move** drop-down list is used to move the component along the direction of the selected entity. When you invoke this option, the **Selected item** selection area is displayed. You are prompted to select an entity to drag along, then select a component and drag to move it. Select an entity to define the direction in which you need to move the component. The name of the selected entity is displayed in the **Selected item** selection area. Now, select and drag the component to move it along the selected entity.

By Delta XYZ

The **By Delta XYZ** option available in the **Move** drop-down list is used to move the selected component to a given distance in a specified direction. When you select this option, the **Delta X**, **Delta Y**, and **Delta Z** spinners are invoked and you are prompted to select a component and enter the distance to move in the PropertyManager. An assembly coordinate system is also displayed in magenta color. Select the component to move and specify the distance in the respective direction spinners in X, Y, or Z direction. Choose the **Apply** button to move the component.

To XYZ Position

The **To XYZ Position** option is used to specify the coordinates of the origin of the part where the component will be placed after moving. When you select this option, the **X Coordinate**, **Y Coordinate**, and the **Z Coordinate** spinners are invoked and you are prompted to select a component and enter the XYZ coordinates for the part's origin. An assembly coordinate system is also displayed in magenta color. Select the component and enter the respective coordinates of the part origin in the spinners and choose the **Apply** button to move the component.

ROTATING THE INDIVIDUAL COMPONENTS

SolidWorks allows you to rotate an individual unconstrained component in the assembly document without affecting the position and location of the other components. The **Rotate Component** tool is used to rotate the component. There are three methods of rotating an individual component. Two methods of rotating an individual component are discussed next. The third method is discussed later in this chapter.

Rotating the Individual Components by Dragging

In SolidWorks 2004, you can rotate the component placed in the assembly without invoking any tool. To rotate an individual component, simply select the component using the right mouse button and drag the cursor to rotate the component. Release the right mouse button after attaining the desired orientation of the individual component.

Rotating Individual Components Using the Rotate Component Tool

CommandManager:	Assemblies > Rotate Component
Toolbar:	Assembly > Rotate Component

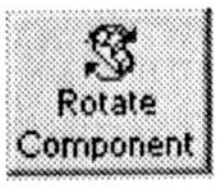

You can also rotate an individual component using the **Rotate Component** tool. To rotate an individual component using this tool choose the **Rotate Component** button from the **Assemblies CommandManager**. The **Rotate Component PropertyManager** is invoked. You will notice that the **Free Drag** option is selected in the **Rotate** drop-down list in the **Rotate** rollout. Therefore, you are prompted to select a component and drag to rotate it. The select cursor is replaced by the rotate cursor; select the component and drag the cursor to rotate the component. The other options available in the **Rotate** drop-down list to rotate the component are discussed next.

Along Entity

The **Along Entity** option available in the **Rotate** drop-down list is used to rotate the component with respect to a selected entity. The selected entity is defined as the rotational axis. When you invoke this option, the **Selected item** selection area is displayed and you are prompted to select an axis entity to rotate about. Select an entity to define the rotational axis. The name of the selected entity is displayed in the **Selected item** selection area. Now, select and drag the component around the selected axis.

By Delta XYZ

The **By Delta XYZ** option available in the **Rotate** drop-down list is used to rotate the selected component to a given incremental angle along the specified axis. When you select this option, the **Delta X**, **Delta Y**, and **Delta Z** spinners are invoked and you are prompted to select a component and enter the desired rotation in the PropertyManager. Select the component to rotate and specify the rotation angle in the respective spinners in the direction in which you need to rotate the component. Choose the **Apply** button to rotate the component.

Tip: *You can toggle between the **Move PropertyManager**, **Rotate PropertyManager**, and the **SmartMates PropertyManager**. If you invoke the **Move PropertyManager**, invoke the **Rotate** rollout to switch to the **Rotate PropertyManager**. If you select the **SmartMates** button, then the system will switch you to the **SmartMates PropertyManager**. From the **Rotate** or **SmartMates PropertyManager**, if you invoke the **Move** rollout, the system will switch you to the **Move PropertyManager***

Moving and Rotating an Individual Component Using the Triad

One of the major enhancements in the **Assembly** mode of SolidWorks 2004 is the introduction of Triad. Using the triad, you can move an individual component. To move an individual component using the triad, first you need to select the component and then right-click to invoke the shortcut menu. Choose the **Move with Triad** option from the shortcut menu. The triad will be displayed on the component.

Using this triad you can move and rotate a component. The various components of the triad are displayed in Figure 10-49.

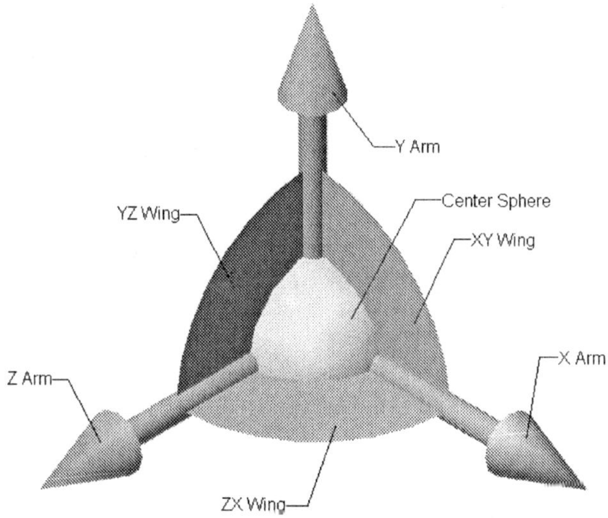

Figure 10-49 *Triad*

Press and hold down the left mouse button on the X arm, the select cursor is replaced by the move/rotate cursor. Use the left mouse button to drag the selected component in the X direction. If you use the right mouse button and drag the cursor, the selected component will rotate about the X axis. Select the Y arm and drag the cursor to move/rotate the component in the Y direction. Select the Z arm and drag the cursor to move/rotate the component in the Z direction.

You can also move a component in the XY plane. To move the component in the XY plane select the XY wing of the triad and drag the cursor to move the component. To move the component in the YZ plane, select the YZ wing of the triad and drag the cursor to move the component. Similarly, by selecting the ZX wing and dragging the cursor, you can move the component in the ZX plane.

If you select the center sphere and invoke the shortcut menu, various options are available, which are discussed next.

Show Translate XYZ Box

The **Show Translate XYZ Box** option available in the shortcut menu is used to display the **Translate XYZ** box. Using this box you can specify the value of the X, Y, and Z coordinates of the destination where you need to place the selected component. Specify the X, Y, and Z coordinates in the respective edit boxes and then choose **OK** from the box. When you move the component by dragging, the values of the X, Y, and Z spinners change automatically.

Show Translate Delta XYZ Box

The **Show Translate Delta XYZ Box** option available in the shortcut menu is used to display the **Translate Delta XYZ** box. Using this box you can specify the incremental value by which you need to move the selected component in the X, Y, or Z direction. Set the incremental value in the X, Y, or Z edit box and then choose **OK** from the box.

Show Rotate Delta XYZ Box

The **Show Rotate Delta XYZ Box** option available in the shortcut menu is used to display the **Rotate Delta XYZ** box. Using this box, you can specify the incremental value by which the selected component will rotate in the X, Y, or Z direction. Set the incremental value in the X, Y, or Z edit box and then choose **OK** from the box.

Align to Assembly

The **Align to Assembly** option available in the shortcut menu is used to align the triad with respect to the current component origin.

TUTORIALS

Tutorial 1

In this tutorial you will create all the components of the Bench Vice and then assemble them. The bench vice assembly is shown in Figure 10-50. The dimensions of the various components are given in Figures 10-51 through 10-54. **(Expected time: 2 hrs 45 min)**

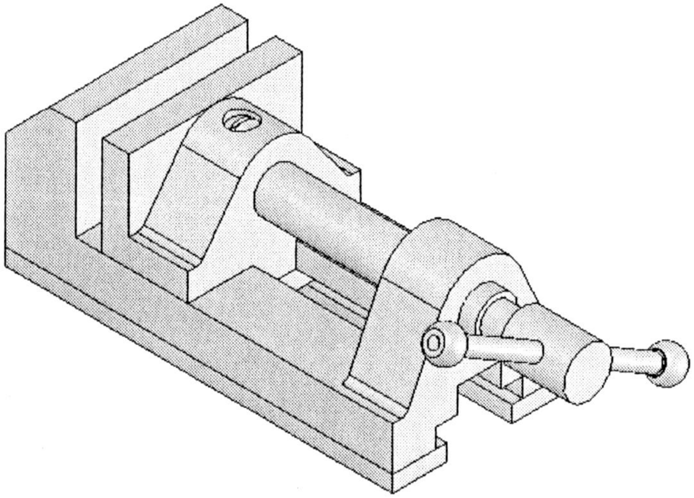

Figure 10-50 *Bench Vice assembly*

The steps to be followed to complete the assembly of this tutorial are listed below.

a. Create all the components in individual part documents and save them. The part documents will be saved in */My Documents/SolidWorks/c10/Bench Vice* folder.

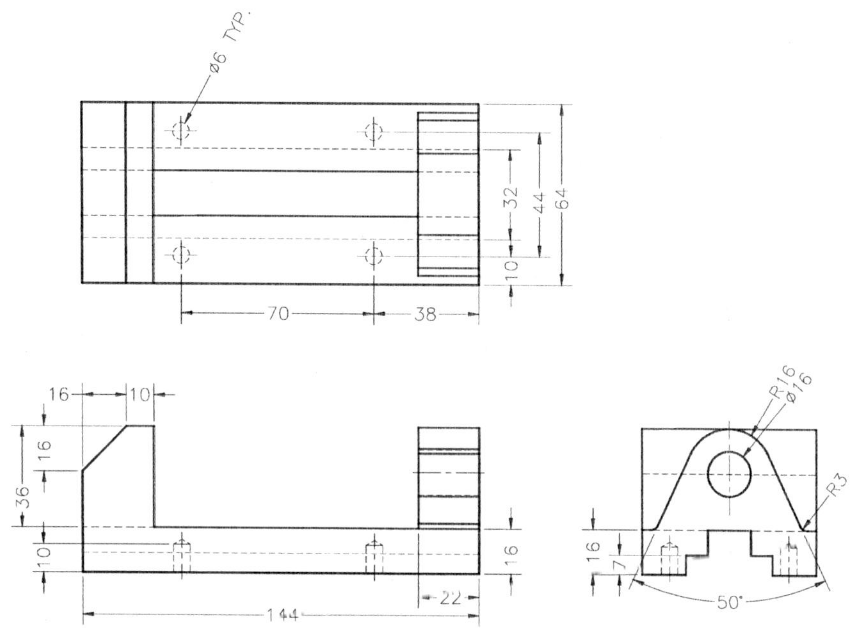

Figure 10-51 Views and dimensions of the Vice Body

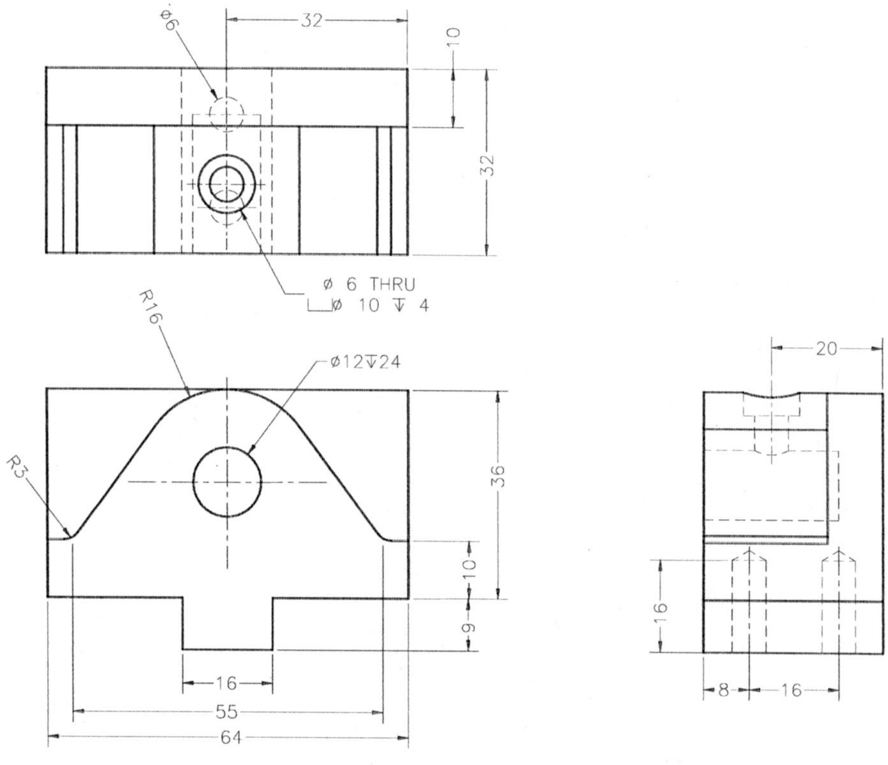

Figure 10-52 Views and dimensions of Vice Jaw

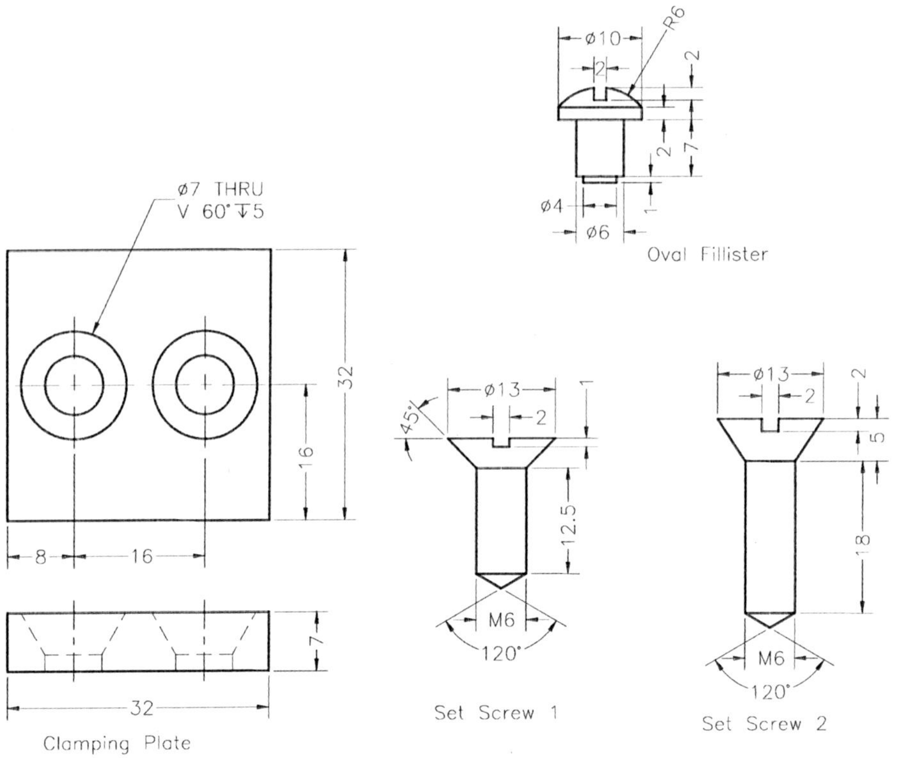

Figure 10-53 *Views and dimensions of Clamping Plate, Oval Fillister, Set Screw 1, and Set Screw 2*

b. Open Vice Body and Vice Jaw part documents and define the mate references in both the part documents.
c. Create a new assembly document and open all the part documents. Place the first component, which is Vice Body, by dragging and dropping from the part document window. Now, drag and drop the Vice Jaw in the assembly document. It will automatically assemble with the Vice Jaw because the mate references are already defined in both the part documents, refer to Figures 10-57 and 10-58.
d. Drag and drop the jaw screw in the assembly document. Apply the required mates, refer to Figures 10-60 through 10-65.
e. Next, analyze the assembly for degrees of freedom of the components.
f. After analyzing the assembly, apply the required mates to constrain all the degrees of freedom, refer to Figures 10-66 and 10-67.
g. Next, assemble the Clamping Plate, refer to Figure 10-69 through Figure 10-71.
h. Next, assemble the Oval Fillister using the feature-based mates, refer to Figure 10-72.
i. Similarly, assemble the other components.

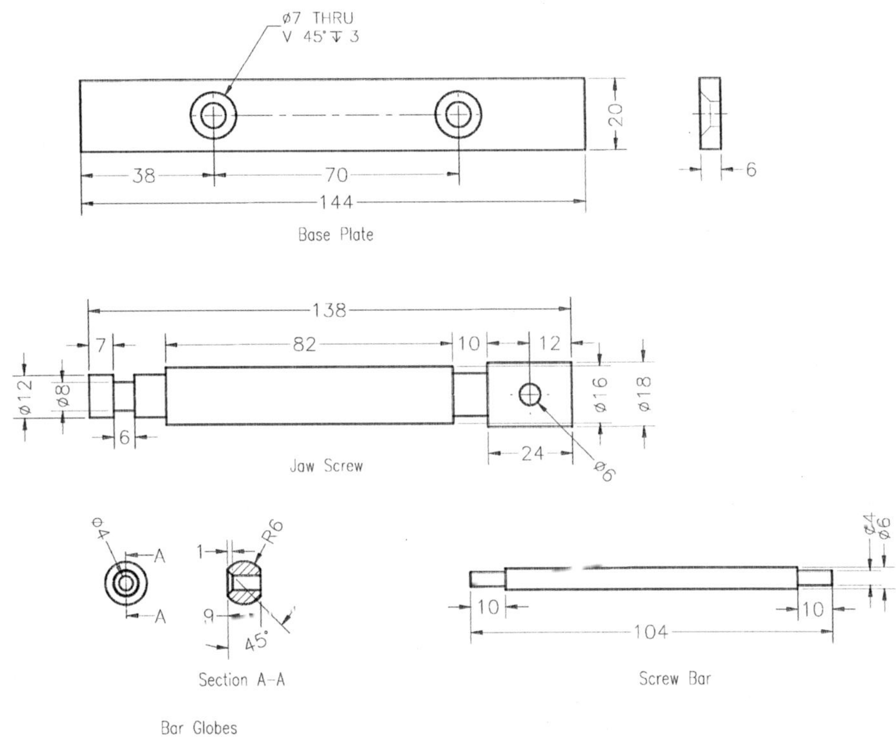

Figure 10-54 *Views and dimensions of Base Plate, Jaw Screw, Screw Bar, and Bar Globes*

Creating the Components

1. Create all the components of the Bench Vice assembly as separate part documents. Specify the names of the documents as shown in Figures 10-51 through 10-54. The files should be saved in the folder */My Documents/SolidWorks/c10/Bench Vice*. Make sure that the *Bench Vice* is your current folder.

Creating the Mate References

In this tutorial you will assemble the first two components of the assembly using the mate references. For assembling the components using the mate references, first you need to create the mate reference. Therefore, you need to open the part documents in which you will add the mate references.

1. Invoke the **Open** dialog box from the **Standard** toolbar.

2. Double-click on the Vice Body.

   The vice-body part document is opened in the SolidWorks window.

3. Choose **Insert > Reference Geometry > Mate Reference** from the menu bar.

The **Mate Reference PropertyManager** is invoked. The selection mode in the **Primary Reference Entity** selection area is active.

4. Select the planar face shown in Figure 10-55 of the model as the primary reference. The selected planar face will be highlighted in green.

5. Select the **Coincident** option from the **Mate Reference Type** drop-down list in the **Primary Reference Entity** rollout.

   The selection mode in the **Secondary Reference Entity** selection area is active.

6. Select the planar face shown in Figure 10-55 of the model as the secondary reference. The selected face will be highlighted in magenta.

7. Select the **Coincident** option from the **Mate Reference Type** drop-down list in the **Secondary Reference Entity** rollout.

   The selection mode in the **Tertiary Reference Entity** selection area is active.

8. Select the planar face of the model shown in Figure 10-55 as the tertiary reference. The selected face will be highlighted in purple.

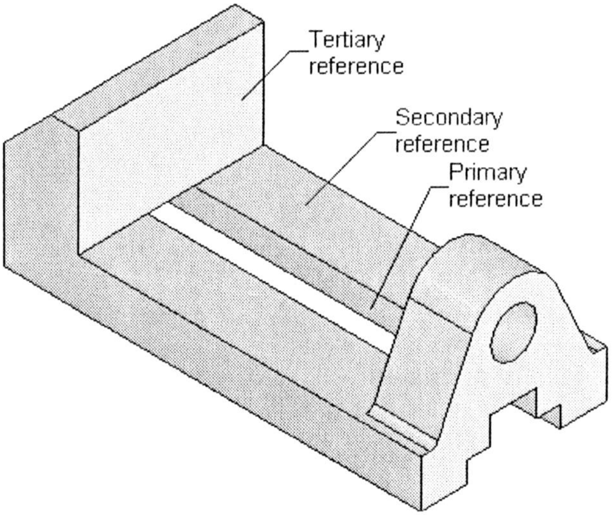

Figure 10-55 *Faces to be selected as mate references*

9. Select the **Parallel** option from the **Mate Reference Type** drop-down list in the **Tertiary Reference Entity** rollout.

10. Enter **Vice Mate Reference** as the name of mate reference in the **Mate Reference Name** edit box available in the **Reference Name** rollout.

11. Choose the **OK** button from the **Mate Reference PropertyManager**.

12. Similarly, create the mate reference in the Vice Jaw part document. The faces to be selected as reference are displayed in Figure 10-56. The name of the mate reference in the **Reference Name** rollout should be same in both the part documents.

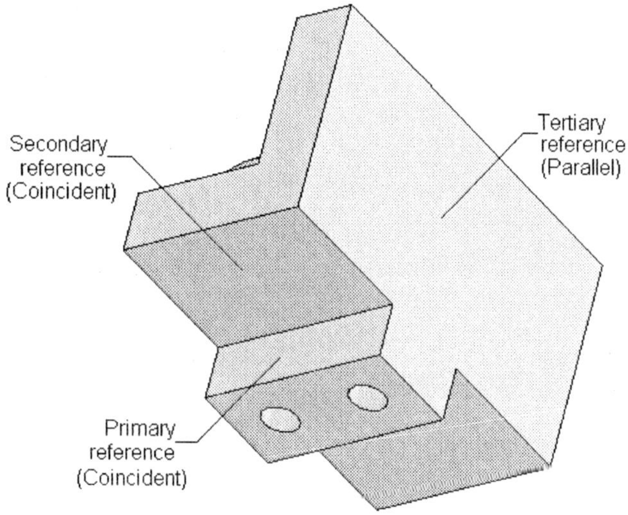

Figure 10-56 *Faces to be selected as mate references*

13. Close all the part documents except *Vice Body.sldpart* and *Vice Jaw.sldprt* if they are opened.

Assembling First Two Components of the Assembly

After creating the mate references in the part documents, you need to assemble the components. For assembling the components you need to start a new SolidWorks assembly document.

1. Start a new SolidWorks assembly document.

   The **Insert Component PropertyManager** is invoked automatically and the preview of the Vice Body is displayed along with the component cursor. Make sure that the Vice Body is selected in the **Insert Component** selection area.

2. Choose the **Keep Visible** button from the **Insert Component PropertyManager** because after placing this component, you also need to place the second component that is displayed in the **Open documents** selection area.

   It is recommended that the first component of the assembly should be placed at the assembly origin.

3. Choose the **OK** button from the **Insert Component PropertyManager** to place the first component coincident to the origin.

   As the **Keep Visible** button is selected, the preview of the Vice Body is again displayed with the component cursor in the drawing area.

4. Change the current view to isometric.

   Next, you need to place the second component in the assembly. As discussed earlier, the second component of the assembly, which is the Vice Jaw, will be assembled with the vice body using the mate references.

5. Select the Vice Jaw from the **Insert Component** selection area. The preview of the Vice Jaw will be available in the drawing area.

   When you move the cursor close to the vice body in the assembly document, the preview of the Vice Jaw assembled after applying the mates with the vice body is displayed in the assembly document.

6. Place the component at the required location. The mates specified in the mate references will be applied between the vice jaw and the vice body. Choose the **Keep Visible** button and exit the **Mate PropertyManager**.

   Figure 10-57 shows the second component being placed in the assembly document. Figure 10-58 shows the vice jaw assembled with the vice body.

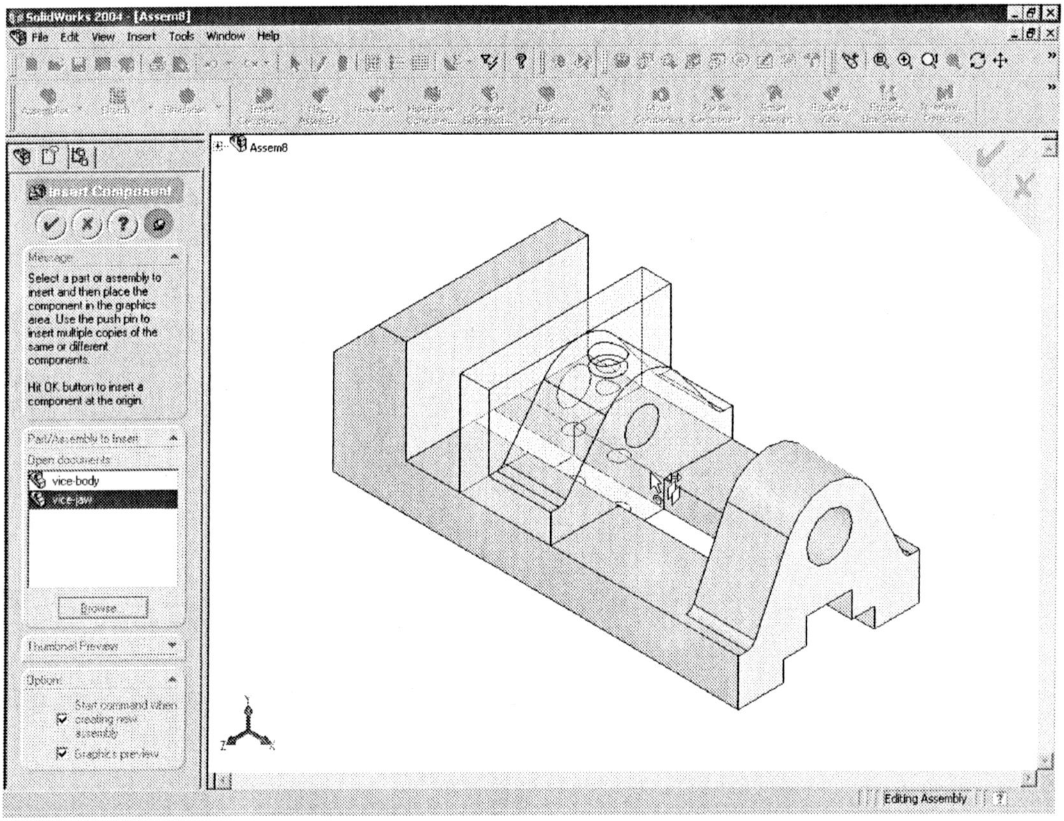

Figure 10-57 *Second component being dragged*

Tip. *The mates that are defined in the mate reference are applied to the components when you place the components in the assembly. You can view the mates applied to both the components by expanding the* **Mates** *option from the* **FeatureManager Design Tree** *of the assembly document.*

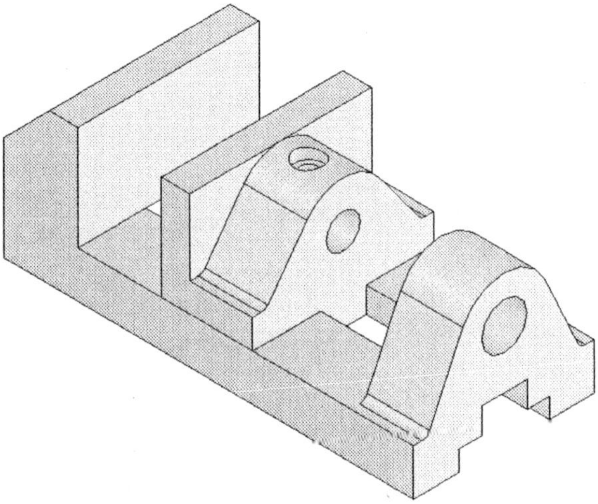

Figure 10-58 *Vice jaw assembled with the Vice Body*

It is recommended that you close all the part documents.

8. Close the part document windows of all the parts that are placed in assembly document.

Assembling the Jaw screw

Now, you need to place the Jaw Screw in the assembly document.

1. Choose the **Insert Component** button from the **Assemblies CommandManager**.

2. Choose the **Browse** button from the **Part/Assembly to Insert** rollout. The **Open** dialog box is displayed.

3. Double-click on the Jaw Screw to open the part document of Jaw Screw.

   The preview of the Jaw Screw is displayed in the drawing area.

4. Click anywhere in the drawing area to place the Jaw Screw.

   Figure 10-59 shows the Jaw Screw placed arbitrarily in the assembly document. Next, you need to add the assembly mates to assemble the components placed in the assembly.

5. Press and hold down the ALT key from the keyboard. Select the face for the location shown in Figure 10-60.

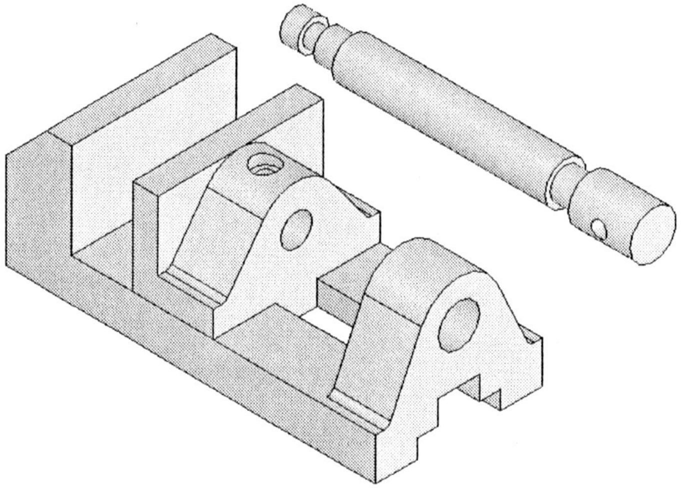

Figure 10-59 *Jaw Screw placed in the assembly document*

6. Drag the cursor, the Jaw Screw is displayed in transparent.

   The select cursor is replaced by the smart mates cursor.

7. Hold the Jaw Screw close to the bottom end and drag it to the hole in the Vice Body as shown in Figure 10-61. The symbol of concentric mate will be displayed below the cursor.

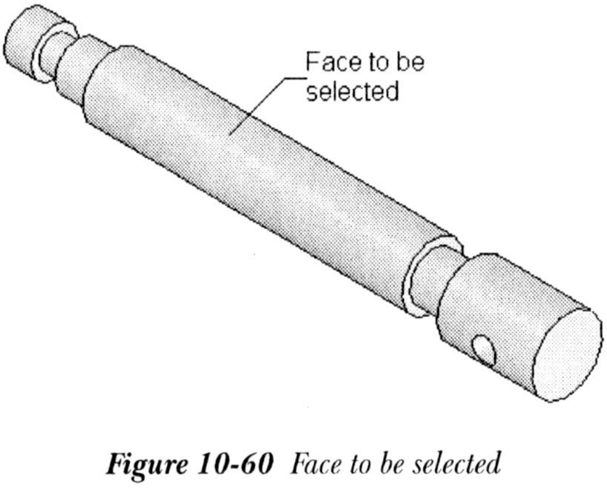

Figure 10-60 *Face to be selected*

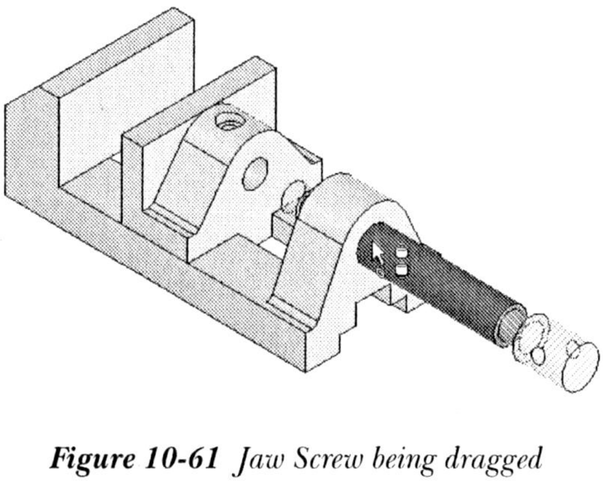

Figure 10-61 *Jaw Screw being dragged*

8. Release the left mouse button at this location. The **Mates** pop-up toolbar is displayed and the **Concentric** button is selected by default, which suggests that the **Concentric** mate is the most appropriate mate to be applied. Choose the **Add/Finish Mate** button from the **Mates** pop-up toolbar. Figure 10-62 shows the Jaw Screw assembled to the Vice Body.

   Next, you need to apply the coincident mate between the planar faces of the Jaw Screw and the Vice Jaw.

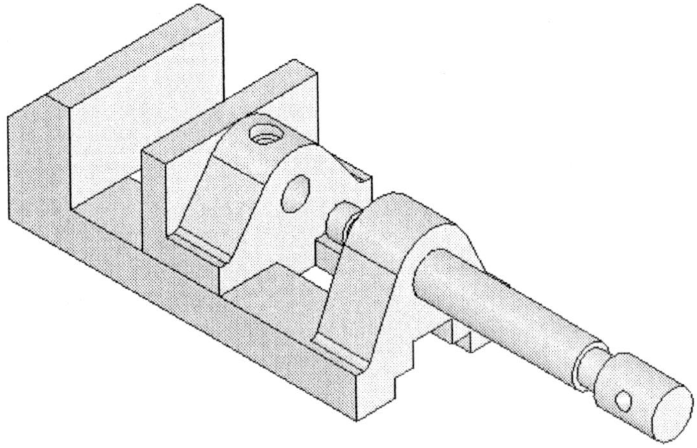

Figure 10-62 ***Concentric*** *mate applied between the Jaw Screw and the Vice Body*

9. Choose the **Mate** button from the **Assemblies CommandManager** to invoke the **Mate PropertyManager**.

10. Rotate the assembly view and select the face of the Vice Jaw as shown in Figure 10-63. Next, select the face of the Jaw Screw as shown in Figure 10-64.

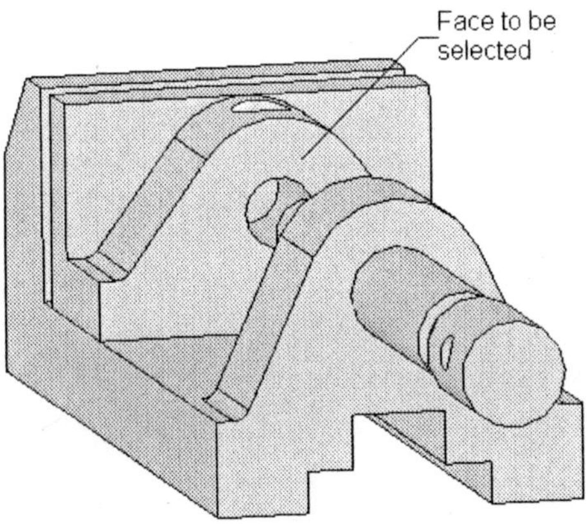

Figure 10-63 *Face to be selected*

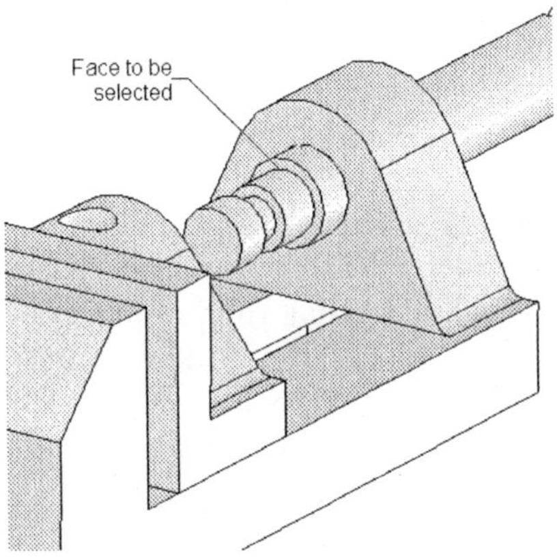

Figure 10-64 *Face to be selected*

As soon as you select the faces, the **Mates** pop-up toolbar is invoked and the preview of the assembly with coincident mate is displayed in the drawing area.

11. Choose the **Add/Finish Mate** button from the **Mate** pop-up toolbar. The assembly after adding the **Coincident** mate is displayed in Figure 10-65.

In real world, there are two types of assemblies. First one is the fully defined assembly in which all the degrees of freedom of all the components are restricted. The other type of

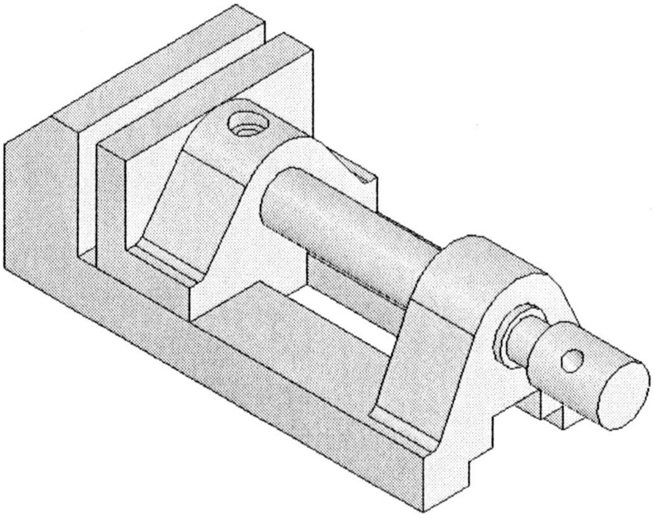

Figure 10-65 *Assembly after applying the* ***Coincident*** *mate to the Jaw Screw*

assemblies are those in which some degrees of freedom of the components are left free so that they can be moved or rotated. These type of assemblies are used for mechanism, which you will learn in the next chapter.

After adding the **Coincident** mate, you need to move the assembly to analyze the degree of freedom of components of the assembly. After analyzing the components of an assembly you will add the mates to constrain that degree of freedom.

12. Select the circular face of the Jaw Screw using the left mouse button and drag the cursor.

13. Select the circular face of the Jaw Screw using the right mouse button and drag the cursor.

    You will notice that the Jaw Screw is rotating on its axis and it is moving along the X axis. Also, it is forcing the Vice Jaw to move along the X axis. Originally, this degree of freedom of Vice Jaw and Jaw Screw needs to be left free so that the assembly can function. But in this chapter you will restrict this degree of freedom so as to create a fully defined assembly.

14. Select the two faces, one of the Vice Body and other of the Jaw Screw as shown in Figures 10-66 and 10-67.

15. Choose the **Distance** button from the **Standard Mates** rollout and set the **Distance** spinner value to **10**. As you choose the **Distance** button the **PropertyManager** changes to the **Distance 1 PropertyManager**.

16. Choose the **OK** button from the **Distance 1 PropertyManager**. The **Mate PropertyManager** is displayed again.

17. Expand the **FeatureManager Design Tree** displayed in the drawing area and select the

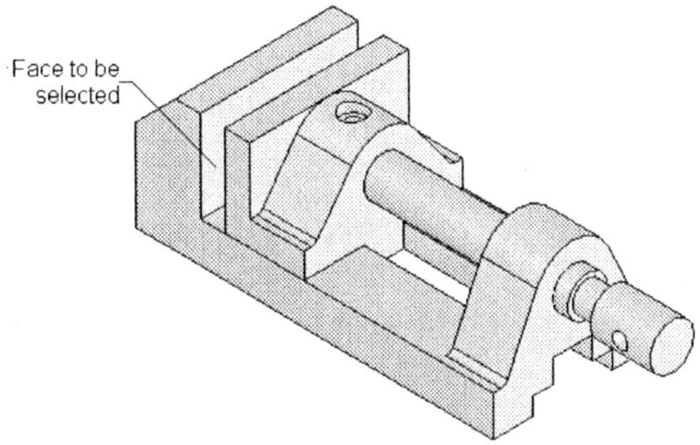

Figure 10-66 *Face to be selected*

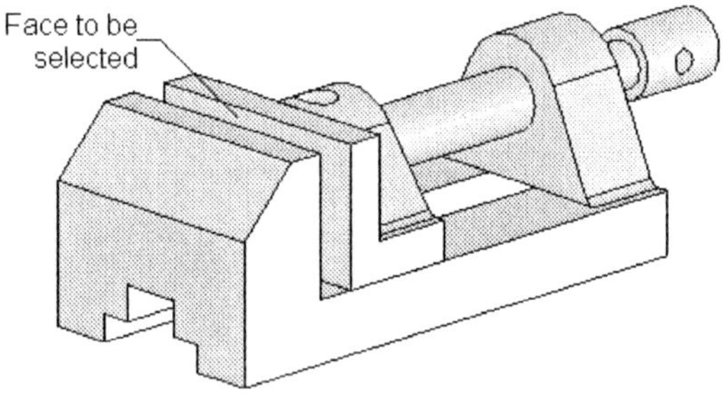

Figure 10-67 *Face to be selected*

Top assembly plane. Now, expand the **Jaw Screw** from the **FeatureManager Design Tree** and select the **Top** plane. The **Mates** pop-up toolbar is displayed.

18. Choose the **Angle** button from the **Mate** pop-up toolbar. The **SolidWorks** warning box is displayed and it informs you about the minimum and the maximum angle value that you can specify to the current selection set.

19. Choose the **OK** button from the **SolidWorks** warning box and set the value of the **Angle** spinner to **45**. Choose the **Add/Finish Mate** button from the **Mates** pop-up toolbar.

20. Choose the **OK** button from the **Mate PropertyManager** to exit it.

Assembling the Clamping Plate

Now, you need to assemble the Clamping Plate with the assembly.

1. Invoke the **Insert Component PropertyManager** and choose the **Browse** button to invoke the **Open** dialog box. Open and place the Clamping Plate in the drawing area.

2. Rotate the assembly such that the bottom face of the assembly is displayed as shown in Figure 10-68.

3. Select the Clamping Plate using the right mouse button and drag the cursor to rotate the Clamping Plate as shown in Figure 10-69.

4. Apply the **Concentric** mate between the two cylindrical faces of the clamping plate and the two holes of the Vice Jaw, refer to Figure 10-70. You may have to move the Clamping Plate after applying the first mate.

   Move the Clamping Plate by dragging. Now, apply the **Coincident** mate between the faces of the Clamping Plate and the Vice Jaw as shown in Figure 10-71.

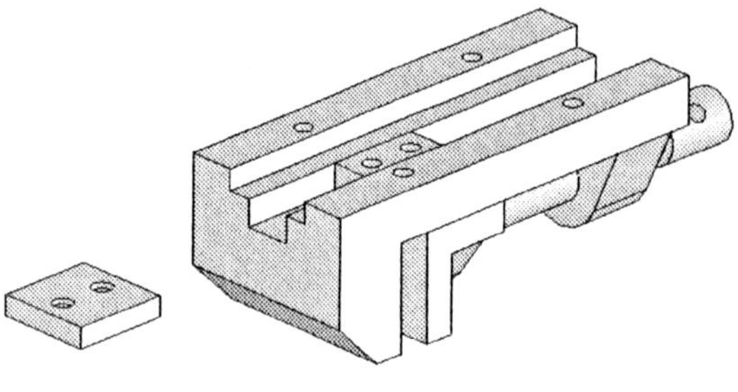

Figure 10-68 Rotated assembly

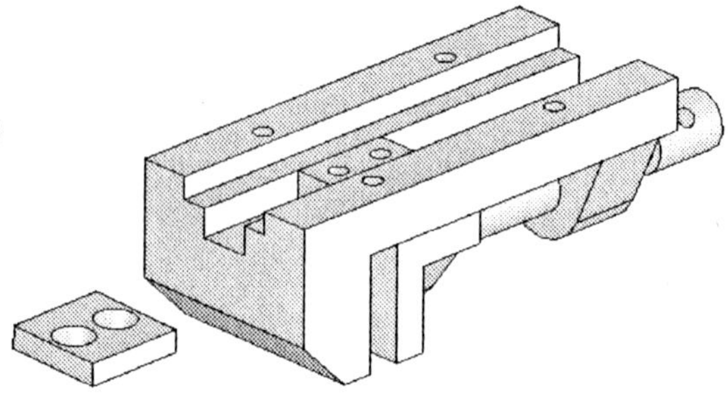

Figure 10-69 Clamping Plate after rotating

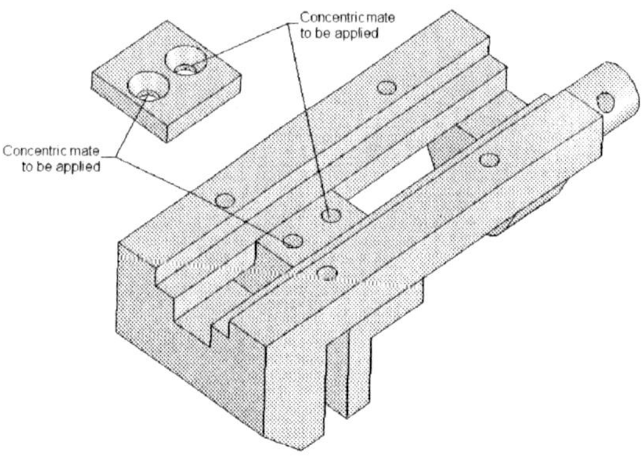

Figure 10-70 Faces to be selected to apply mate

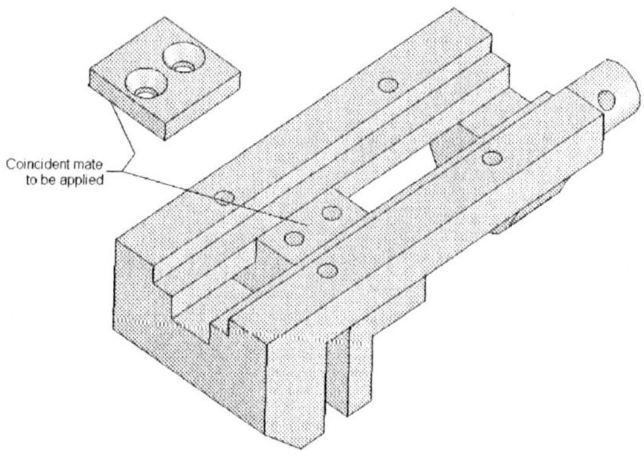

Figure 10-71 Faces to be selected to apply mate

Similarly, assemble the Screw Bar, Support Plates, and Bar Globes. The assembly after assembling all these components is shown in Figure 10-72.

Assembling the Remaining Components

Next, you need to assemble the Oval Fillister, Set Screw 1, and Set Screw 2. These fasteners will be assembled using feature-based mates.

1. Open the part documents of Oval Fillisters, Set Screw 1, and Set Screw 2.

2. Choose **Window > Tile Horizontally** from the menu bar to rearrange the windows.

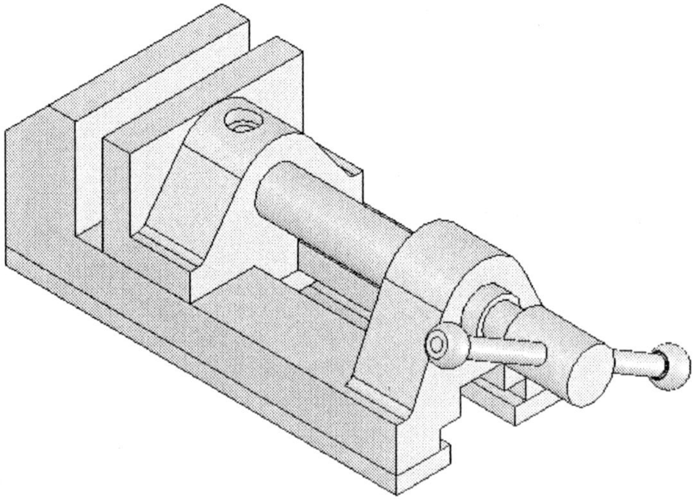

Figure 10-72 *Assembly after assembling the Vice Body, Vice Jaw, Jaw Screw, Screw Bar, Clamping Plate, Base Plate, and Bar Globes*

3. Select the **Revolve1** feature from the **FeatureManager Design Tree** of the Oval Fillister part document. Drag the cursor to place the component in the assembly as shown in Figure 10-73.

4. Drop the component when the coincident symbol is displayed below the cursor.

   Similarly, assemble Set Screw 1 and Set screw 2 using feature-based mates. Use the TAB key from the keyboard to reverse the direction. The rotated view of the assembly after assembling Set Screw 1 and Set Screw 2 is shown in Figure 10-74.

5. Choose the **Save** button to save the assembly with the name *Bench Vice* in the folder */My Documents/SolidWorks/Bench Vice*.

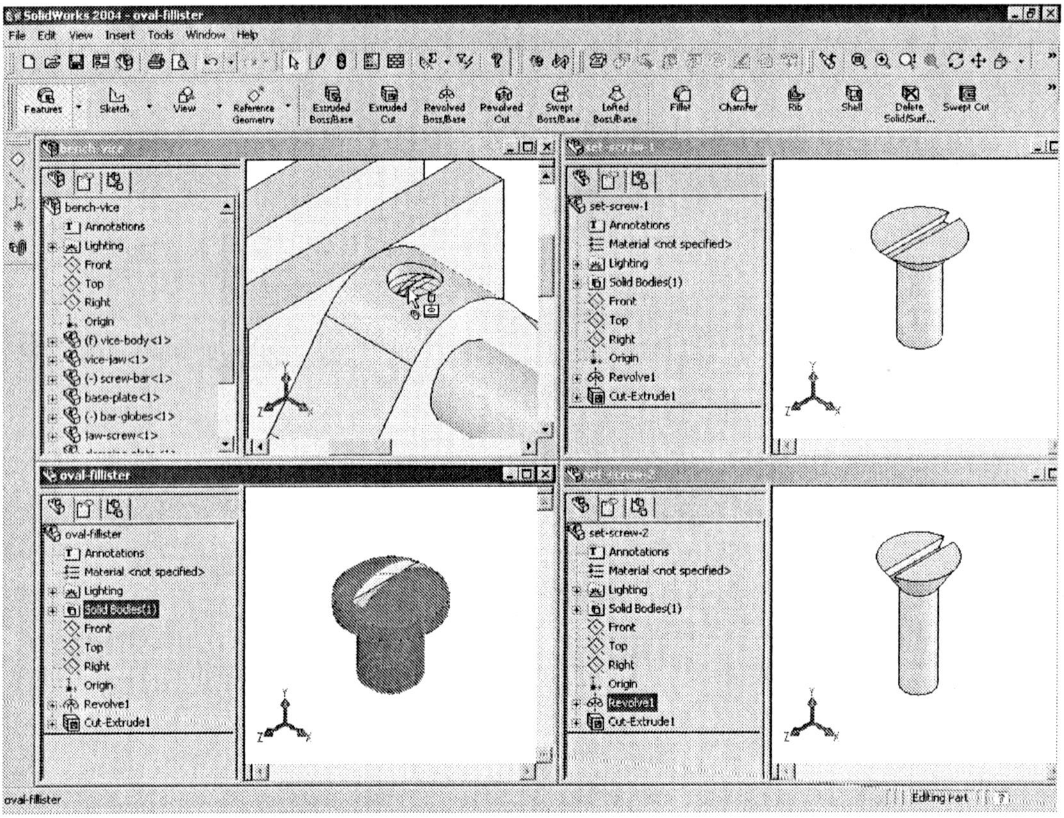

Figure 10-73 Dragging the Oval Fillisters in the assembly using feature-based mates

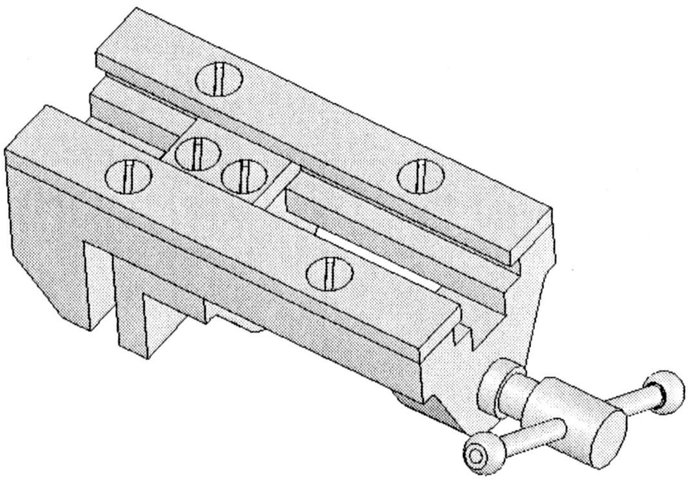

Figure 10-74 Viewing the final assembly from the bottom

Tutorial 2

In this tutorial you will create all the components of the Pipe Vice and then assemble them. The Pipe Vice assembly is shown in Figure 10-75. The dimensions of the various components are given in Figures 10-76 and 10-77. **(Expected time: 2 hrs 45 min)**

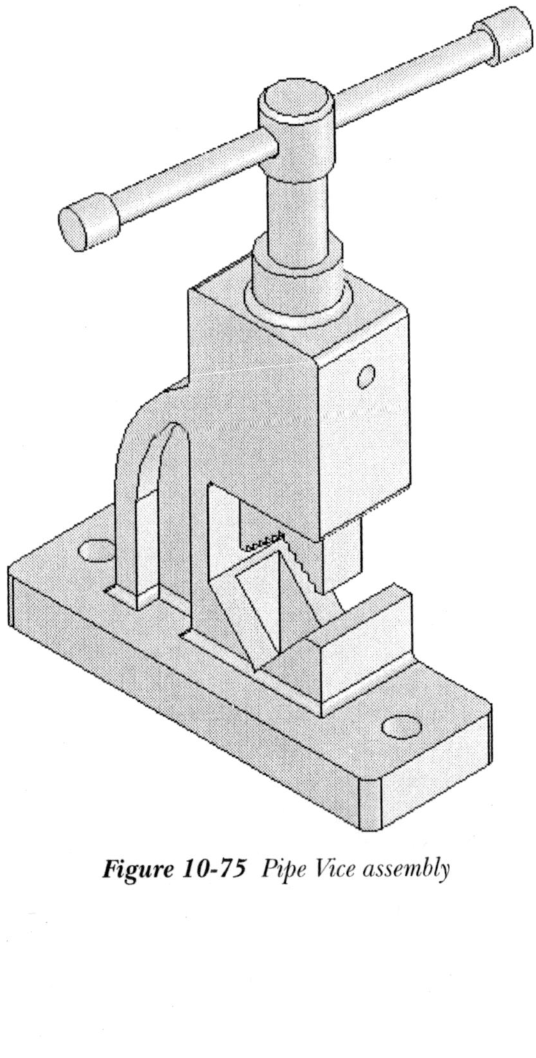

Figure 10-75 *Pipe Vice assembly*

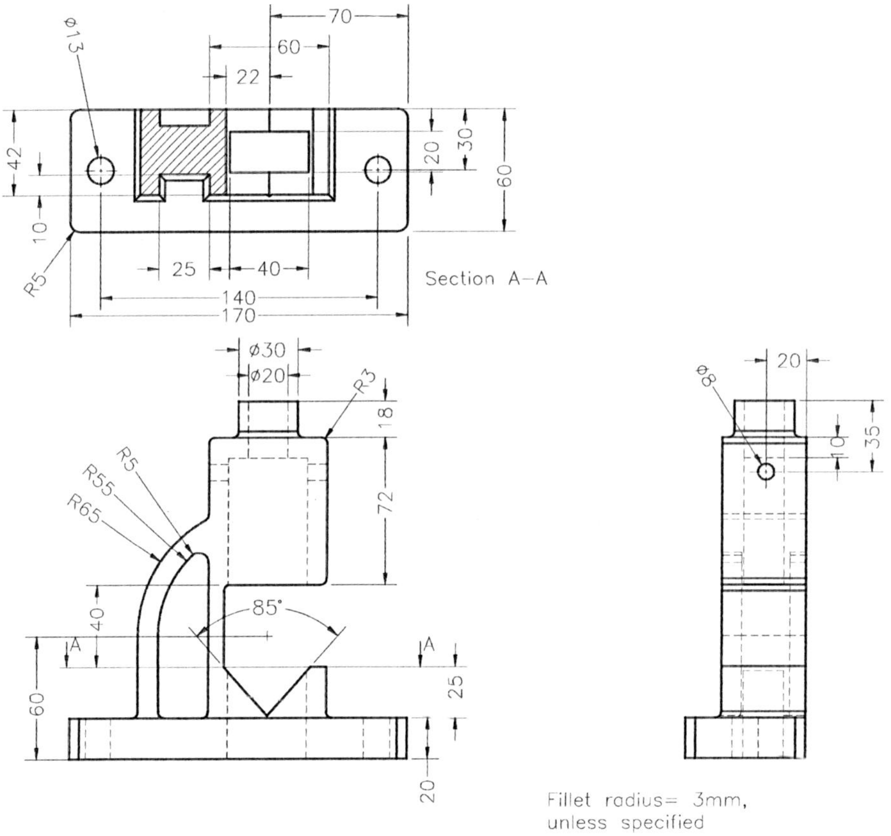

Figure 10-76 Views and dimensions of the base

The steps to be followed to complete the assembly of this tutorial are listed below.

You need to create all the components of the Pipe Vice assembly as separate part documents. After creating the parts you will assemble the parts in the assembly document. Therefore, in this tutorial you are using the bottom-up approach for creating the assembly.

a. Create all the components in individual part documents and save them. The part documents will be saved in */My Documents/SolidWorks/c10/Pipe Vice*.
b. Place the base at the origin of the assembly.
c. Place the Moveable Jaw and the Screw in the assembly. Apply the mates between Moveable Jaw and the Screw, refer to Figures 10-78 through 10-80.
d. Now, assemble the Screw with the Base.
e. Place the other components in the assembly and apply the required mates to the assembly, refer to Figure 10-81.

Creating the Components

1. Create all the components of the Pipe Vice assembly as separate part documents. Specify the names of the files as shown in Figures 10-76 and 10-77. The documents should be

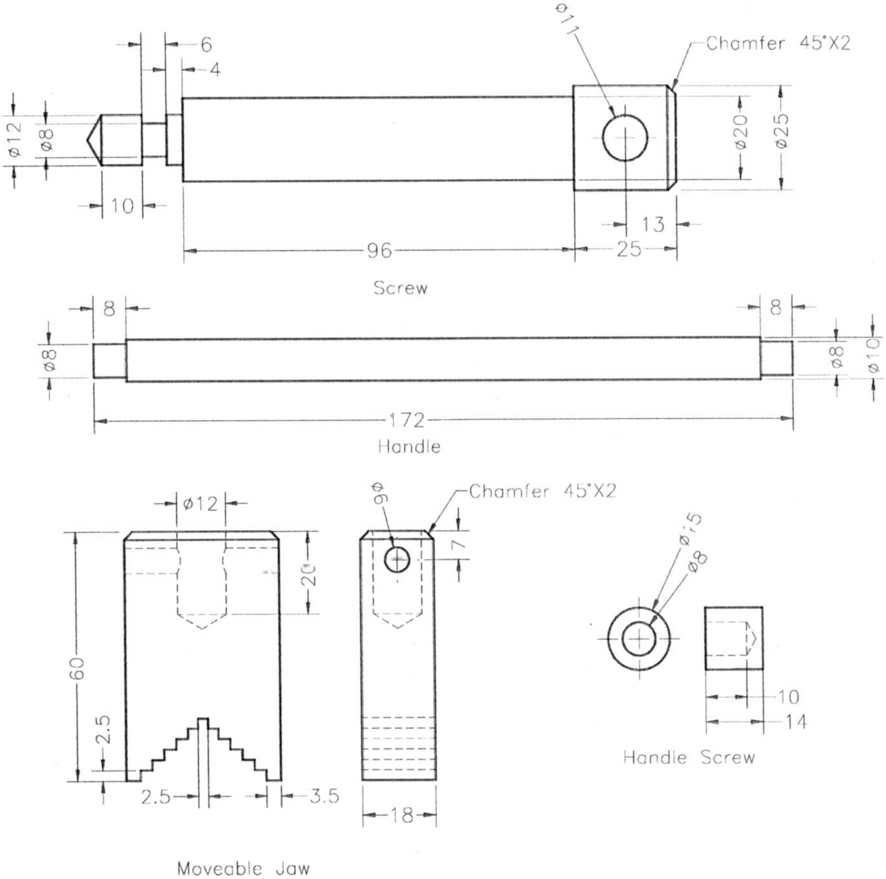

***Figure 10-77** Views and dimensions of the Screw, Handle, Moveable Jaw, and Handle Screw*

saved in the folder */My Documents/SolidWorks/c10/Pipe Vice*.

Inserting the First Component in the Assembly

After creating all the components of the pipe vice assembly, you need to start a new SolidWorks assembly document.

1. Start a new SolidWorks assembly document, the **Insert Component PropertyManager** is invoked by default.

2. Choose the **Browse** button from the **Part/Assembly to Insert** rollout. The **Open** dialog box is displayed. Double-click on Base.

3. Choose the **OK** button from the **Insert Component PropertyManager** to place the Base origin coincident to the origin of the assembly document.

4. Change the view orientation to isometric.

Inserting and Assembling the Moveable Jaw and the Screw

After placing the first component in the assembly document, you need to place the Moveable Jaw and the Screw in the assembly document. After placing these components you will apply the required mates.

1. Choose the **Insert Components** button from the **Assemblies CommandManager**. Choose the **Keep Visible** button from the **PropertyManager** and invoke the **Open** dialog box by choosing the **Browse** button from the **Part/Assembly to Insert** rollout.

2. Double-click on Moveable Jaw. Place the component anywhere in the assembly document such that it does not interfere with existing component.

3. Similarly, place the Screw in the assembly document and choose the **OK** button from the **Insert Component PropertyManager**. Figure 10-78 shows the Moveable Jaw, Screw, and the Base placed in the assembly document.

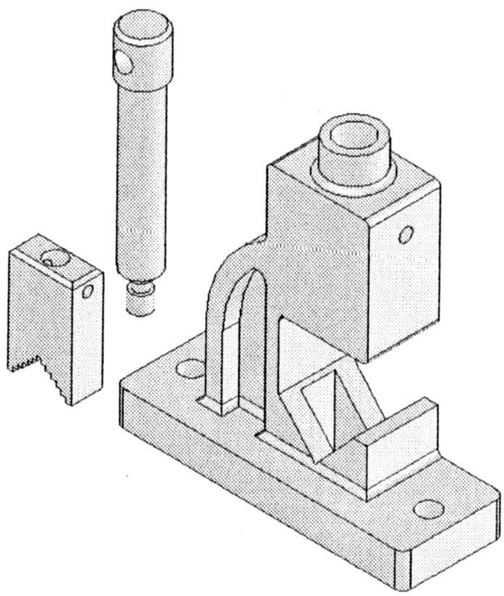

Figure 10-78 *The Moveable Jaw, Screw, and the Base placed in the assembly document*

First, you need to assemble the Screw with the Moveable Jaw. Therefore, for assembling the Screw with the Moveable Jaw, you need to fix the Moveable Jaw.

4. Select the Moveable Jaw from the drawing area or from the **FeatureManager Design Tree**. Right-click to invoke the shortcut menu.

5. Choose the **Fix** option from the shortcut menu. The Moveable Jaw is fixed and you cannot move or rotate the Moveable Jaw.

6. Invoke the **Move PropertyManager** and choose the **SmartMates** button from the **Move**

rollout. Double-click the lowermost cylindrical face of the Screw. The Screw will be displayed in transparent.

7. Drag the cursor to the hole located on the top of the Moveable Jaw. Release the left mouse button as soon as the concentric symbol is displayed below the cursor. Choose the **Add/Finish Mate** button from the **Mates** pop-up toolbar.
8. Select the Screw and move it up so that it is not inside the Moveable Jaw.
9. Right-click in the drawing area and choose **Clear Selections** to clear the current selection.
10. Rotate the assembly and double-click the lower flat face of the Screw; the Screw will be displayed in transparent.
11. Again, rotate the model and select the top planar face of the Moveable Jaw. A coincident relation is applied between the two selected faces. Choose the **Add/Finish Mate** button from the **Mates** pop-up toolbar.
12. Choose the **OK** button from the **SmartMates PropertyManager**. Figure 10-79 shows the Screw after applying the mates.

    Next, you need to assemble the Screw and the Moveable Jaw with the Base.

13. Select the Moveable Jaw and invoke the shortcut menu. Choose the **Float** option from the shortcut menu.

    Now, the Moveable Jaw and the Screw assembled to it can be moved.

14. Using smart mates, add a concentric mate between the cylindrical face of the screw and the hole created at the top face of the Base.
15. Invoke the **Mate PropertyManager** and select the front planar face of the Moveable Jaw and the front planar face of the Base.
16. Choose the **Parallel** button from the **Mates** pop-up toolbar and choose the **Add/Finish Mate** button to add a parallel mate between the selected faces.
17. Next, select the faces as shown in Figure 10-80.
18. Choose the **Distance** button from the **Mates** pop-up toolbar. Set the value of the **Distance** spinner to **35**.
19. Choose the **Add/Finish Mate** button from the **Mates** pop-up toolbar and then choose the **OK** button from the **Mate PropertyManager**.

    Similarly, assemble the other components of Pipe Vice. Figure 10-81 shows the final Pipe Vice assembly.

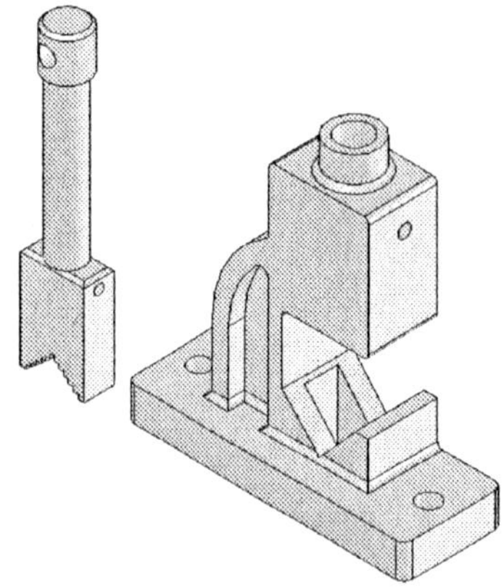

Figure 10-79 *The Screw assembled with the Moveable Jaw*

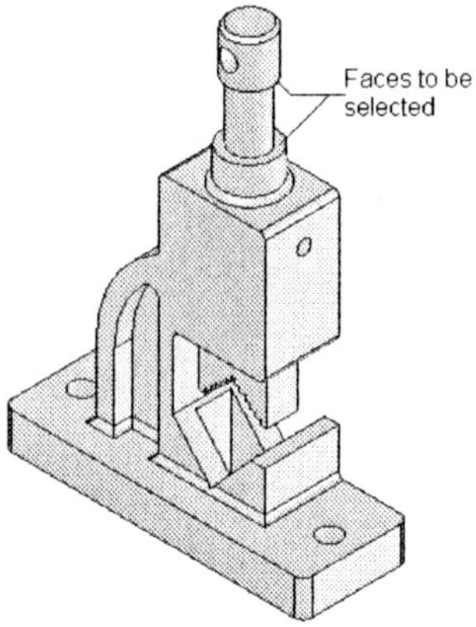

Figure 10-80 *Faces to be selected*

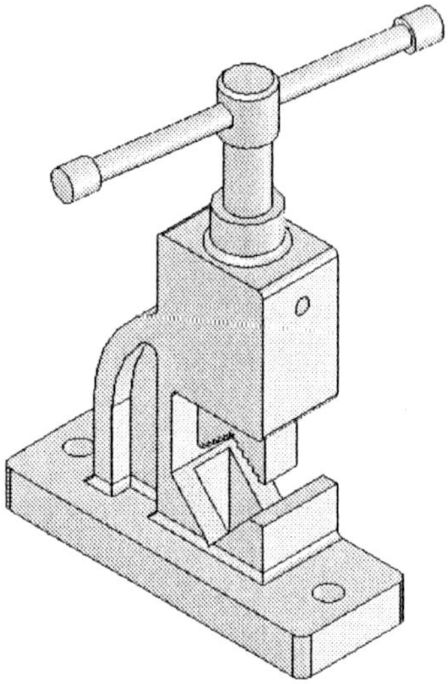

Figure 10-81 *Final Pipe Vice assembly*

20. Choose the **Save** button to save the assembly document in the */My Documents/SolidWorks/Pipe Vice* folder.

SELF-EVALUATION TEST

Answer the following questions and then compare your answers with the answers given at the end of this chapter.

1. The bottom-up assembly design approach is the traditional and the most widely preferred approach of assembly design. (T/F)

2. In the top-down assembly design approach, all the components are created in the same assembly document. (T/F)

3. The **Coincident** mate is generally applied to make two planar faces coplanar. (T/F)

4. The most suitable mates that can be applied to the current selection set are displayed in the **Mate Selections** rollout of the **Mate PropertyManager**. (T/F)

5. Feature-based mates are applied only to the components having cylindrical features. (T/F)

6. Pattern-based mates are used to assemble the components that have a circular pattern created on the circular feature. (T/F)

7. Choose the __________ button from the **Assemblies CommandManager** to invoke the **Rotate Component PropertyManager**.

8. The __________ option available in the **Rotate** drop-down list is used to rotate the selected component to a given incremental angle along the specified axis.

9. The __________ mate is generally used to align the central axis of one component with the central axis of the other.

10. The __________ option available in the **Move** drop-down list is used to move the component along the direction of the selected entity.

REVIEW QUESTIONS

Answer the following questions.

1. The names of the selected entities are displayed in the __________ selection area of the **Mate PropertyManager**.

2. Choose __________ from the menu bar to place a component in the assembly document.

3. Using the __________ option from the **Move** drop-down list, you can move the component dynamically along the X, Y, and Z axes of assembly document.

4. The __________ button available in the **Standard Mates** rollout is used to make the two selected entities normal to each other.

5. The __________ option is used to specify the coordinates of the origin of the part where the component will be placed after moving.

6. If you are adding feature-based mates using the features having conical geometry, then there must be a __________ face adjacent to the conical face of both the features.

7. The most widely used method of adding the mates to components in the assembly in SolidWorks is

   (a) **Smart Mates** (b) **Mate PropertyManager**
   (c) By dragging from part document (d) None of these

8. Which button is used to make the **Mate PropertyManager** available after applying a mate to the selected entities?

(a) **Help** (b) **OK**
(c) **Keep Visible** (d) **Cancel**

9. Which option available in the **Rotate** drop-down list is used to rotate the component with respect to the selected entity?

(a) **Along Entity** (b) **Selected Edge**
(c) **Reference Entity** (d) None of these

10. Which option is used to specify the coordinates of the origin of the part where the component will be placed after moving?

(a) **To XYZ Position** (b) **Reference Position**
(b) **Along Entity** (c) None of these

EXERCISE

Exercise 1

Create the Plummer Block assembly as shown in Figure 10-82. The dimensions of the components of this assembly are shown in Figures 10-83 through 10-85.

(Expected time: 1 hr)

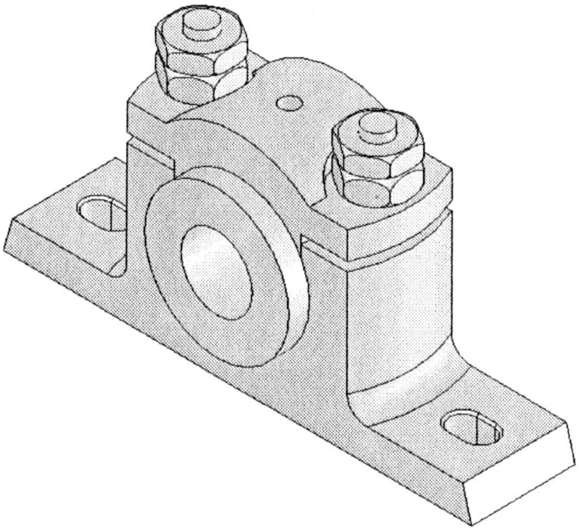

Figure 10-82 *Plummer Block assembly*

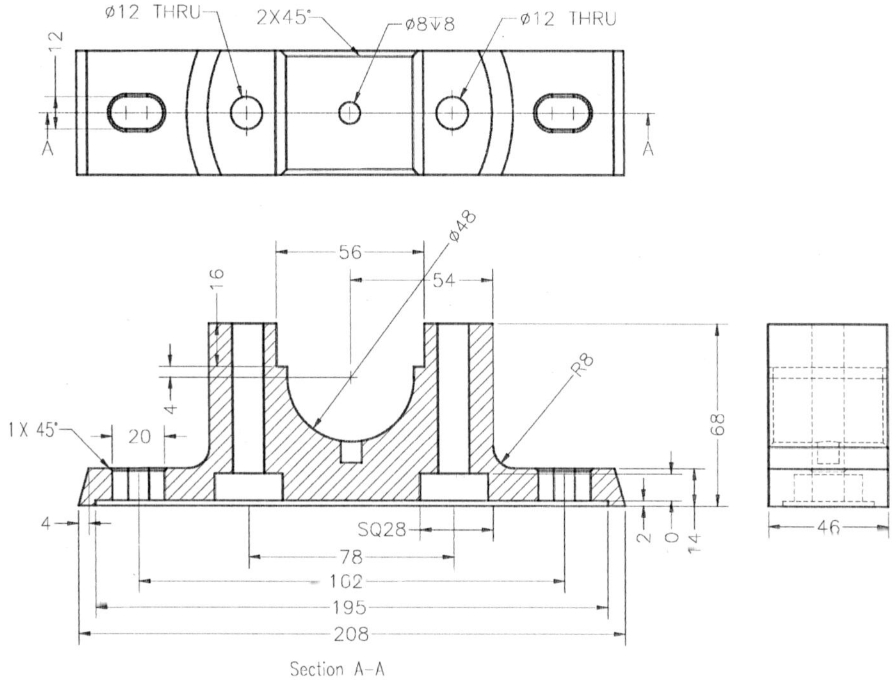

Figure 10-83 *Views and dimensions of Casting*

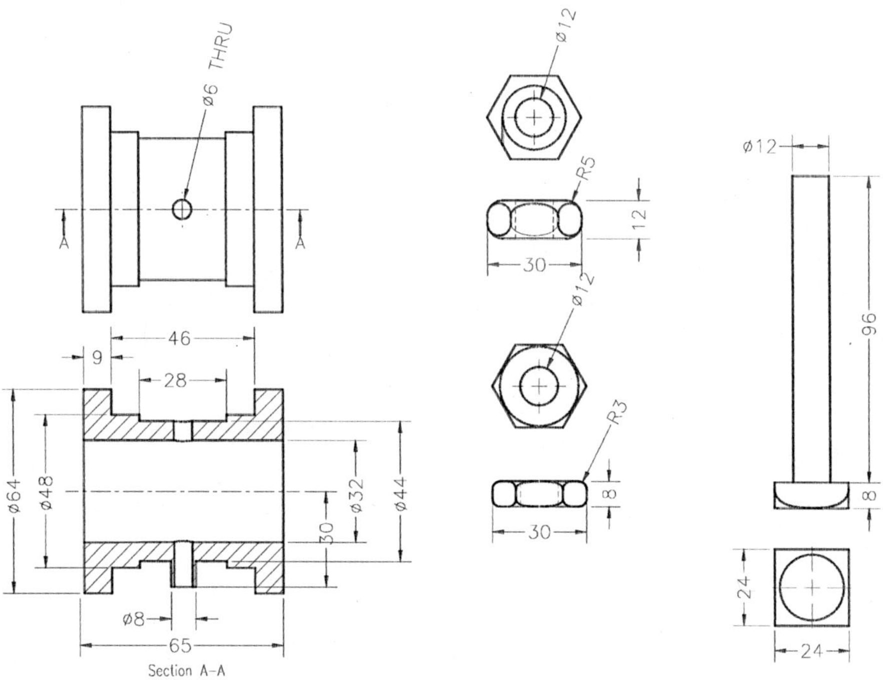

Figure 10-84 *Views and dimensions of Brasses, Nut, Lock Nut, and Bolt*

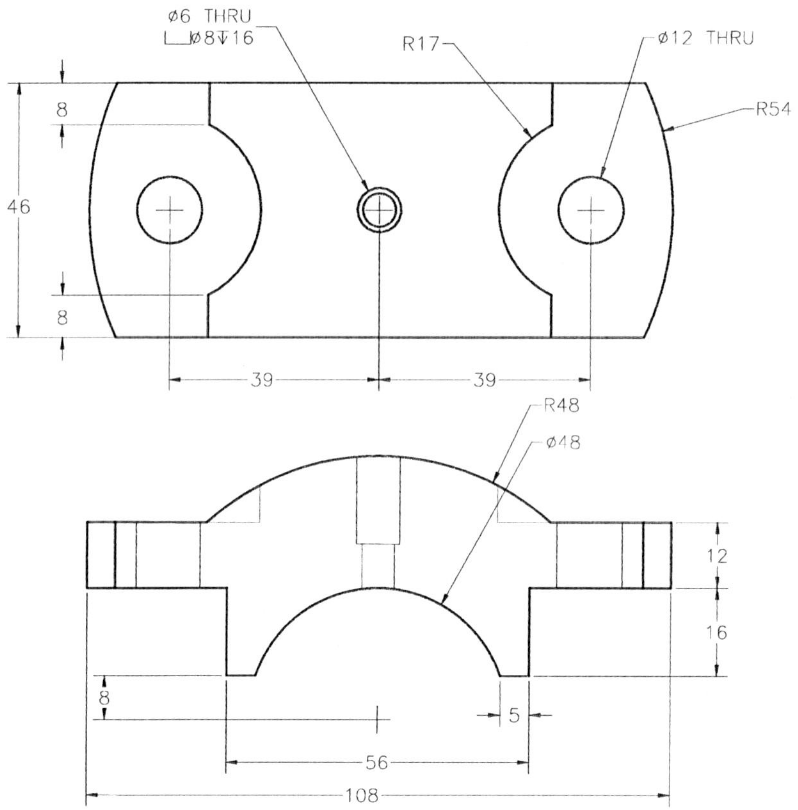

Figure 10-85 *Views and dimensions of Cap*

Answers to Self-Evaluation Test

1. T, **2.** T, **3.** T, **4.** T, **5.** T, **6.** T, **7. Rotate Component**, **8. By Delta XYZ**, **9. Coincident**, **10. Along Entity**

Chapter 11

Assembly Modeling-II

Learning Objectives

After completing this chapter you will be able to:

- *Apply Advanced Mates.*
- *Create Subassemblies.*
- *Delete Components and Subassemblies.*
- *Edit Assembly Mates.*
- *Replace Mate Entities.*
- *Edit Components and Subassemblies.*
- *Dissolve Subassemblies.*
- *Replace Components in Assemblies.*
- *Create Patterns of Components in the Assembly.*
- *Create Mirrored Components.*
- *Hide and Suppress Components in Assemblies.*
- *Change the Transparency Condition of Assembly.*
- *Check Interference in the Assembly.*
- *Create Assemblies for Mechanism.*
- *Detect Collision while the Assembly is in Motion.*
- *Create the Exploded State of the Assembly.*
- *Create the Explode Line Sketch.*

ADVANCED ASSEMBLY MATES

In the previous chapter, you learned how to place the components in an assembly document and how to apply assembly mates to the components. In this chapter you will learn how to apply advanced mates to the components. To apply the advanced mates to the components invoke the **Mate PropertyManager** and invoke the **Advanced Mates** rollout available in this **PropertyManager**. The **Advanced Mates** rollout is displayed in Figure 11-1.

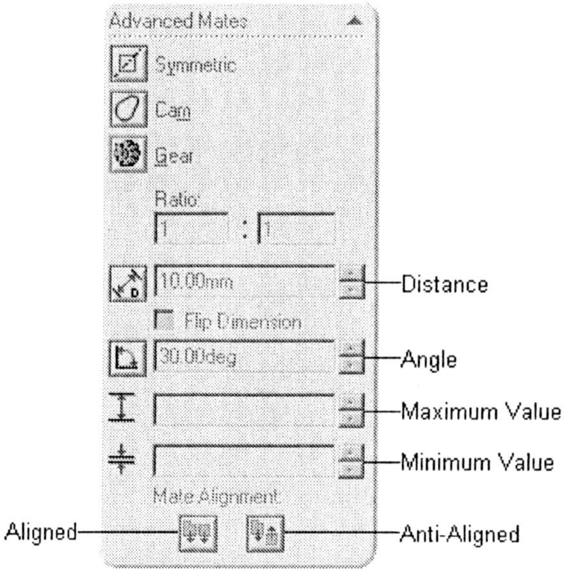

***Figure 11-1** The **Advanced Mates** rollout*

The advanced mates that you will learn in this chapter are the **Symmetric** mate, **Cam** mate, **Gear** mate, and **Limit** mate. All these mates are discussed next.

Applying the Symmetric Mate

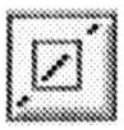

The symmetric mate is applied to create a symmetric relation between two components. To apply the **Symmetric** mate between two components choose the **Symmetric** button available in the **Advanced Mates** rollout. When you choose this button, the **Symmetry Plane** selection area is displayed in the **Mate Selections** rollout. Select two similar entities from two components to which you need to apply the symmetric mate. The similar entities that you need to select could be edges, vertices, planar faces, and so on. Next, you need to select a plane or a planar face as a symmetry plane. Click once in the **Symmetry Plane** selection area to invoke the selection mode and select the symmetric plane. The **Symmetry Plane** callout is attached to the symmetry plane. Choose the **OK** button from the **Symmetric1 PropertyManager**. Figure 11-2 shows the entities and the plane to be selected to apply the **Symmetric** mate. Figure 11-3 shows the **Symmetric** mate applied to the assembly.

Applying the Cam Mate

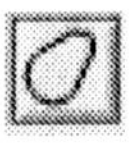

You can add a cam mate between a face that is formed by closed and continuous extruded faces from a single profile and a face of the follower. The face of the follower

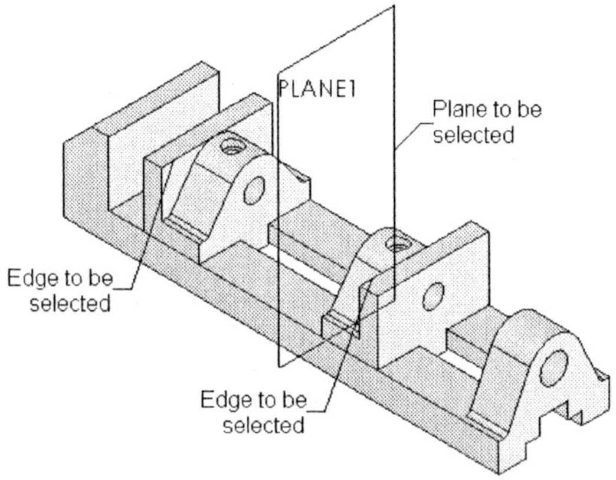

Figure 11-2 Entities and the plane to be selected to apply the symmetric mate

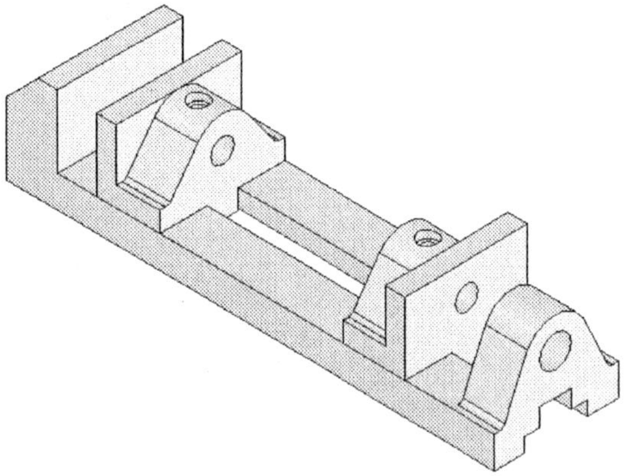

Figure 11-3 Assembly after applying the symmetric mate

could be a cylindrical or a planar face or it could be a vertex. To apply a cam mate you need to select a chain of continuous tangent faces that will act as surface of the cam. After this, you need to select a curved face, a planar face, or a vertex from the follower, that will remain in contact with the selected chain of tangent faces. To apply a cam mate invoke the **Advanced Mates** rollout of the **Mate PropertyManager**. Choose the **Cam Follower** button from this rollout. The **Cam Follower** selection area is displayed in the **Mate Selections** rollout. Select the chain of continuous faces that will act as the cam surface. The selected faces are highlighted in green. Now, click once in the **Cam Follower** selection area and select the face from the second component that will act as the surface of the follower. The follower will touch the surface of Cam and the **Cam Follower** callout is attached to the selected face. Choose the **OK** button from the **CamMate Tangent1 PropertyManager**. Now, if you rotate the cam, the follower will attain the respective linear motion, that is, following the cam surface. Figure 11-4 shows the faces to be selected and Figure 11-5 shows the preview of the assembly after applying the **Cam** mate.

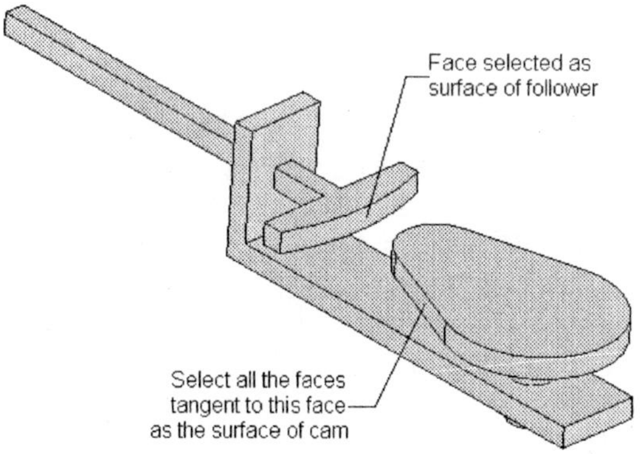

Figure 11-4 Faces selected to apply cam mate

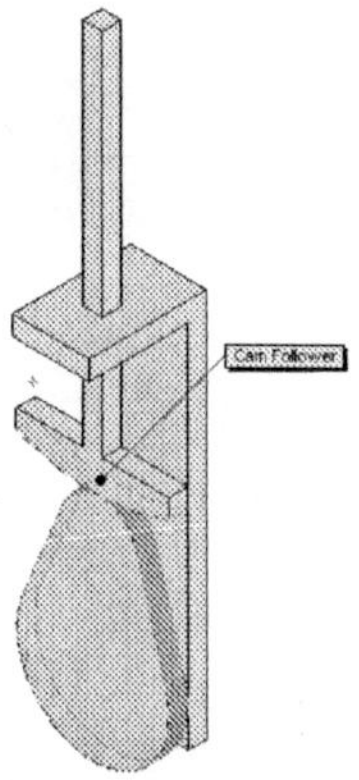

Figure 11-5 Preview of the assembly with cam mate

Note

For better understanding of this mate make sure that the front faces of the cam and follower are coplanar. Also, you need to make sure that proper degree of freedom of the components of the cam and follower assembly should be kept free.

Applying Gear Mate

One of the major enhancements in the **Assembly** mode of SolidWorks 2004 is the introduction of gear mate. The gear mate allows you to rotate two components such as gears with respect to each other about a selected axis. You can also define a specific gear ratio between the selected components. To apply a gear mate between two components invoke the **Advanced Mates** rollout from the **Mate PropertyManager**. Choose the **Gearmate** button from the **Advanced Mates** rollout, the **Ratio** edit boxes and the **Reverse** check box are also invoked. Next you need to select the references along which the two components will rotate. These references could be cylindrical faces, conical faces, axes, circular edges, or linear edges. Select the required references, the names of references are displayed in the **Entities to Mate** selection area. The **Teeth/Diameter** callouts are also displayed attached to selected references. You can set the ratio of rotation using the **Ratio** edit boxes or using the **Teeth/Diameter** callouts. The **Reverse** check box is selected to change the direction of rotation of one component with respect to the other. Choose the **OK** button from the **GearMate1 PropertyManager**. After adding this mate rotate any one of the component and you will observe that the other component is rotating with the specified gear ratio. Generally, this mate is used to display the rotation of the gears and direction. Figure 11-6 shows the cylindrical faces that you need to select to apply gear mate. Figure 11-7 shows the **Teeth/Diameter** callouts displayed after selecting the respective faces.

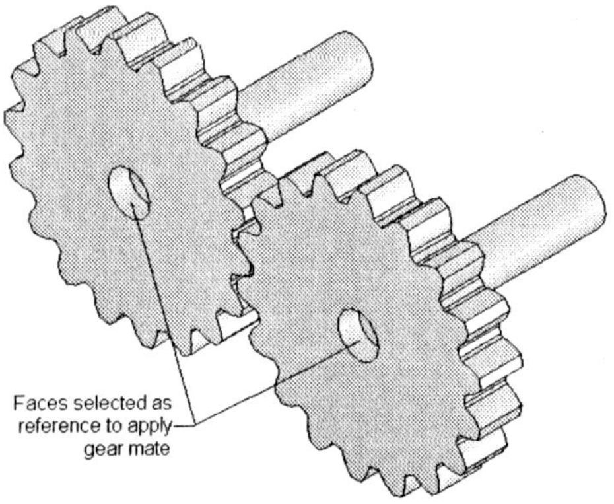

Figure 11-6 *Faces to be selected to apply gear mate*

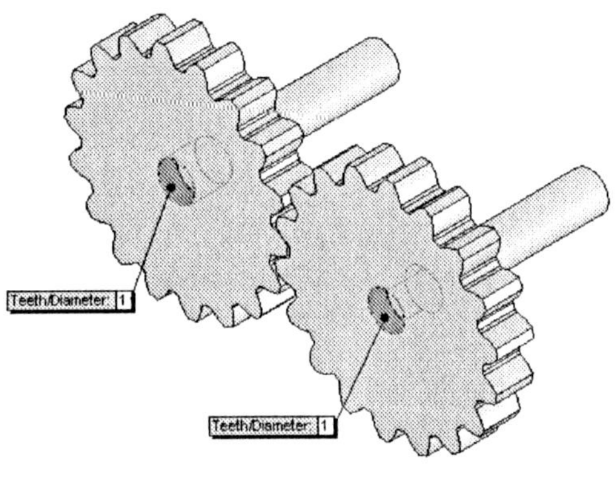

Figure 11-7 ***Teeth/Diameter*** *callouts attached to the selected faces*

Applying Limit Mate

The introduction of the limit mate is another major enhancement in the **Assembly** mode of SolidWorks 2004. Limit mate is used to apply a distance mate or angle mate to the selected components, but the difference between simple distance or angle mate and limit mate is that in limit mate you can move the component within the specific range of values. To apply a limit mate invoke the **Advanced Mates** rollout from the **Mate PropertyManager** and then apply the distance or the angle mate as required. Next, you need to specify the maximum and minimum values of the limit mate in the **Maximum Value** and the **Minimum Value** spinners respectively. Choose the **OK** button from the **PropertyManager**.

CREATING SUBASSEMBLIES

In the previous chapter, you learned to place the components in the assembly document and apply the assembly mates to the components. In this chapter you will also learn how to create the subassemblies and place the subassemblies in the main assembly.

For placing the subassemblies in the main assembly, there are two approaches that are followed in the assembly design environment. These two approaches are discussed next.

Bottom-up Subassembly Design

In the **Bottom-up Subassembly** design approach, the subassembly is created in the assembly environment and then it is saved as an assembly file. When you need to place the subassembly in the main assembly, open the main assembly document and invoke the **Insert Component PropertyManager**. When you choose the **Browse** button, the **Open** dialog box is displayed; choose **Assembly** (**.asm, *.sldasm*) from the **Files of Type** drop-down list. All the assemblies saved in the current location will be displayed in the selection area. Select the subassembly from the selection area and choose the **Open** button from the **Open** dialog box. Select a point in the drawing area to place the subassembly. Now, using the assembly mates, assemble the subassembly with the main assembly. Figure 11-8 shows a subassembly of piston and articulated rod. Figure 11-9 shows the main assembly of the piston and master rod. Figure 11-10 shows the main assembly after assembling the subassembly with the main assembly.

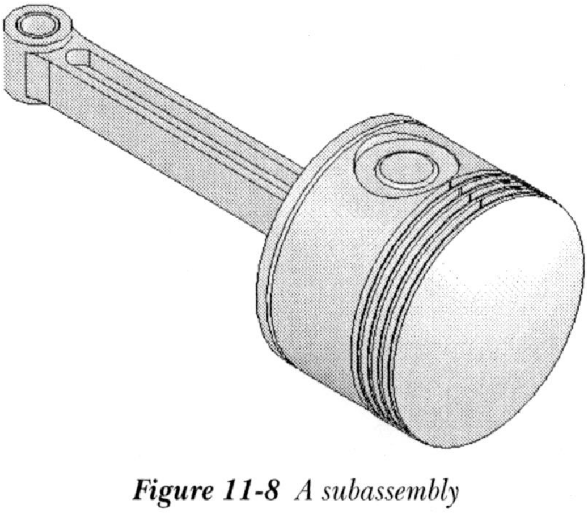

Figure 11-8 *A subassembly*

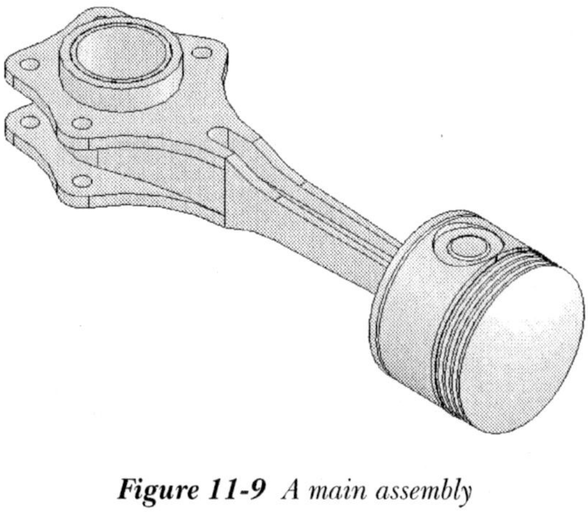

Figure 11-9 *A main assembly*

Top-down Subassembly Design

The **Top-down Subassembly** design approach is the most flexible subassembly design approach. In this, you create a new subassembly in the main assembly document. This approach is generally used in conceptual design or while managing a large assembly. To create a new subassembly in the assembly document, choose **Insert > Component > New Assembly** from the menu bar. The **Save As** dialog box is displayed; save the subassembly in the current location. You can drag and place the components in the new subassembly from the **FeatureManager Design Tree**. You will learn more about this approach while discussing the editing of assemblies and subassemblies.

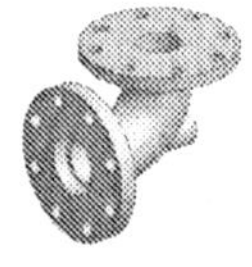

Tip. *You can also place a subassembly in the main assembly using the drag and drop method that was discussed in the previous chapter.*

When you place a subassembly in the main assembly, an assembly icon will be displayed with the name of the subassembly in the ***FeatureManager Design Tree****. If you expand the subassembly in the* ***FeatureManager Design Tree****, all the parts assembled in the subassembly will be displayed.*

You can also create a subassembly of the components that are already placed in an assembly file. Press and hold down the CTRL key and select the components either from the drawing area or from the ***FeatureManager Design Tree****. Invoke the shortcut menu and choose* ***Form New Sub-assembly Here*** *from the menu bar or choose* ***Assembly from [Selected] Components*** *from the menu bar. The* ***Save As*** *dialog box is displayed; you need to enter the name of the subassembly and save it in the current location. You will observe that the selected components will be combined and an assembly icon will be displayed in the* ***FeatureManager Design Tree****.*

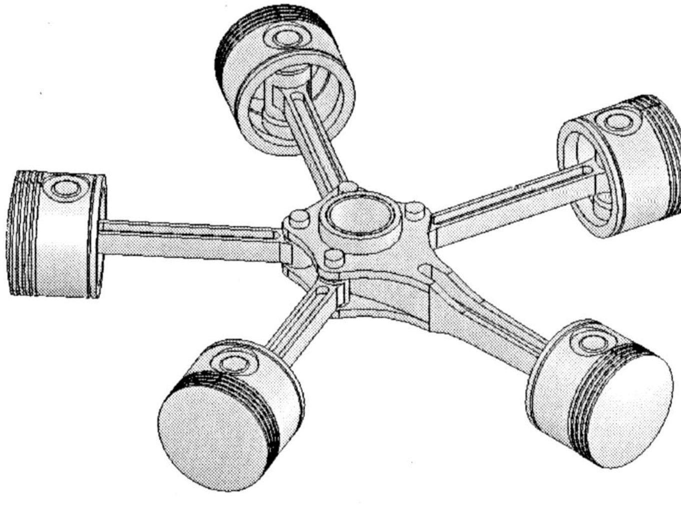

Figure 11-10 *Main assembly after assembling the subassemblies*

DELETING COMPONENTS AND SUBASSEMBLIES

After creating the assembly, at a certain stage of your design cycle you may need to delete any component or the subassembly. To delete a component of the assembly, select the component either from the drawing area or from the **FeatureManager Design Tree**. Invoke the shortcut menu; choose the **Delete** option from the shortcut menu. You can also delete the selected component by pressing the DELETE key from the keyboard. When you delete a component, the **Confirm Delete** dialog box is displayed. The name of the component and the items dependent on that component are displayed in this dialog box. Choose the **Yes** button from the **Confirm Delete** dialog box.

If you need to delete the subassembly, select the subassembly from the **FeatureManager Design Tree** and press the DELETE key. The **Confirm Delete** dialog box is displayed, choose the **YES** button from this dialog box. You should note that when you delete the subassembly, all the components of the subassembly are deleted.

EDITING ASSEMBLY MATES

Generally, after creating the assembly or during the process of assembling the components, you need to edit the assembly mates. The editing operations that can be performed on the assembly mates are modifying the type of assembly mate, modifying the angle and offset values, changing the component to which the mate was applied, and so on. In SolidWorks, for editing the mates, you first need to expand the **Mates** option available at the bottom of the **FeatureManager Design Tree**. Now, select the mate that you need to modify and invoke the shortcut menu. Choose the **Edit Feature** option from the shortcut menu. The **Mate PropertyManager** will be displayed. The name of the **Mate PropertyManager** will depend on the name and sequence of the mate applied. Figure 11-11 shows a partial view of the **Mate PropertyManager** to edit the **Coincident** mate. You can edit the entities to mate, type of mate, value of offset, value of angle, and so on using this **PropertyManager**.

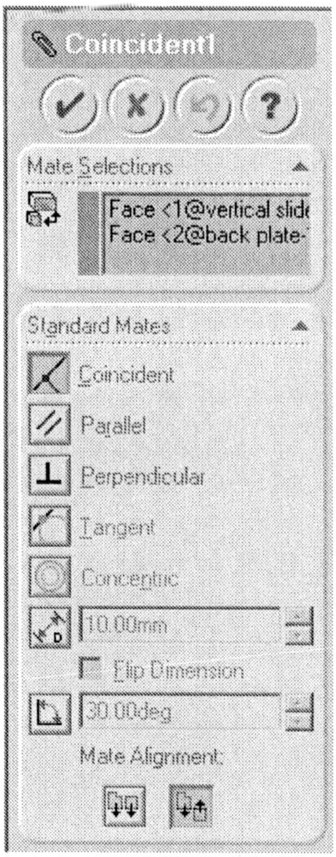

Figure 11-11 *Partial view of the **Mate PropertyManager** to edit the **Coincident** mate*

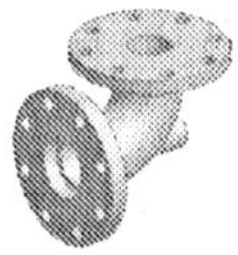

Tip. *When you move the cursor on a mate in the* ***FeatureManager Design Tree****, the entities used in the mate are highlighted in red color in the drawing area. When you select the mate from the* ***FeatureManager Design Tree****, the entities used in the selected mate will be highlighted in green color in the drawing area.*

Replacing the Mated Entities

CommandManager:	Assemblies > Replace Mate Entities	*(Customize to Add)*
Toolbar:	Assembly > Replace Mate Entities	*(Customize to Add)*

As discussed, you can edit the mate entities using the **Mate PropertyManager**. The mate entities can also be modified using the **Mated Entities PropertyManager**. Select the **Mate** option from the **FeatureManager Design Tree** and invoke the shortcut menu. Choose the **Replace Mate Entities** option from the shortcut menu. The **Mated Entities PropertyManager** is displayed as shown in Figure 11-12. You can customize the **Assemblies CommandManager** and then select the

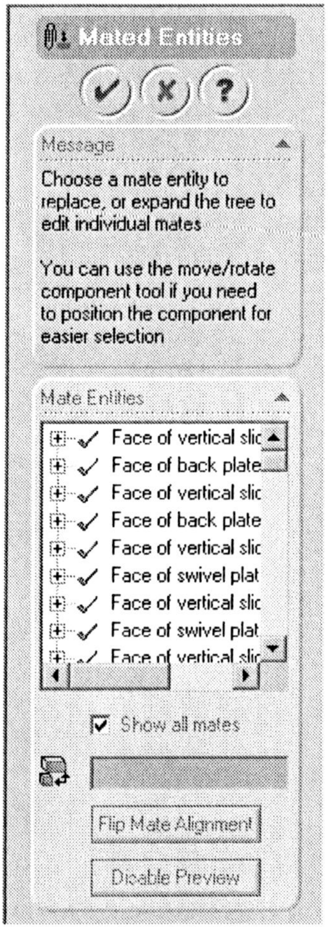

Figure 11-12 The ***Mated Entities PropertyManager***

specific mate whose entities you need to replace and then choose the **Replace Mate Entities** button from the **Assemblies CommandManager**. When you invoke the **Mated Entities PropertyManager**, you are prompted to select the entity to be replaced. Select the entity from the **Mate Entities** selection area. You can also expand the entity tree to edit an individual mate. After selecting, the selected face will be highlighted in green color in the drawing area. The name of the selected entity will be displayed in the **Replacement Mate Entity** selection area. You are prompted to select the entity to be mated. Select the entity that will replace the previously selected entity. If the selected entity overdefines the mate, or the mating is not possible between the entities then the **SolidWorks** warning box will be displayed. You will be informed about the possible cause of error.

The **Flip Mate Alignment** button is used to flip the direction of mate. The **Disable Preview** button is used to disable the preview of the assembly after replacing the mate entity. The **Show All** check box is used to display all the mated entities.

Tip. *Select the **Mate** option from the **FeatureManager Design Tree**, and invoke the shortcut menu. Choose the **Parent/Child Relationship** option from the shortcut menu. The **Parent Child Relationship** dialog box will be displayed. You can display the child and parent relationship of any component placed in the assembly using the **Parent Child Relationship** dialog box.*

EDITING THE COMPONENTS

CommandManager:	Assemblies > Edit Component
Toolbar:	Assembly > Edit Component

After placing and mating the components in the assembly document, at some stage of your design cycle you may need to edit the components. The editing of components includes editing the features, editing the sketches, and editing the sketch planes. For editing the components, you first need to select the component and invoke the part modeling environment in the assembly document. Select the component from the drawing area or from the **FeatureManager Design Tree**. Now, choose the **Edit Component** button from the **Assembly CommandManager**. The part modeling environment will be invoked and the entire assembly except the component to be edited is displayed in transparent. If the assembly is not displayed in transparent then choose the **Assembly Transparency** button displayed in the **Assemblies CommandManager**. When you choose this button a flyout is invoked, the options available in this flyout are used to set the transparency settings. If you select the **Opaque** option from the flyout, then all the components of the assembly are set to opaque. If you select the **Maintain Transparency** option from the flyout, then the default transparency settings of the individual components are retained. If you select the **Force Transparency** option from the flyout, then all the components of the assembly except the component being edited are set as transparent. The name of the component to be edited will be displayed in magenta in the **FeatureManager Design Tree**. Select the desired feature to edit, and invoke the respective **PropertyManager**. The **PropertyManager** relative to that feature will be displayed and you can easily edit the parameters of the feature. You can also add new features to the component. This type of edit is technically termed as **Editing in the Context of Assembly**. After editing the component, again choose the **Edit Component** button from the **Assembly CommandManager** to return to the assembly environment.

Tip. *You can also modify the dimensions of a component assembled or placed in the assembly by double-clicking the feature of that component. All the dimensions of that feature will be displayed in the drawing area. Invoke the* ***Modify*** *dialog box by double-clicking the dimension to modify. Enter the new dimension in the* ***Modify*** *dialog box and choose the ENTER key from the keyboard. Since the dimension is modified, but the geometry of the feature has not changed, therefore, you need to rebuild the assembly. Choose the* ***Rebuild*** *button from the* ***Standard*** *toolbar to rebuild the entire assembly. You can also rebuild the assembly using CTRL+B key from the keyboard.*

While in the part editing mode in the assembly document, you can use the ***Move/Size Features*** *tool to edit the features dynamically using editing handles.*

Note

If you need to edit components separately in their part documents, select the component and invoke the shortcut menu. Select the ***Open Part*** *option from the shortcut menu. The part document of the selected component will be opened. You can edit the component individually in the part document. After editing the component, save the part and close the part document and return to the assembly document. The* ***SolidWorks 2004*** *dialog box will be displayed as shown in Figure 11-13. This dialog box prompts you that models contained within the assembly have changed. Would you like to rebuild the assembly now? Choose the* ***Yes*** *button from this dialog box.*

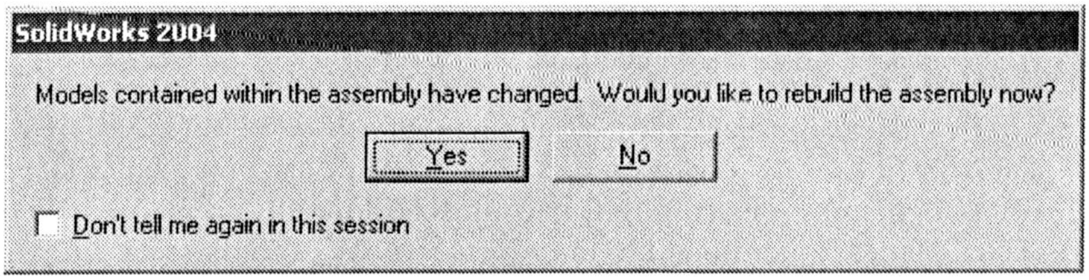

Figure 11-13 *The* ***SolidWorks 2003*** *dialog box*

EDITING THE SUBASSEMBLIES

CommandManager:	Assemblies > Edit Component
Toolbar:	Assembly > Edit Component

To edit the subassemblies, select the subassembly from the **FeatureManager Design Tree** and choose the **Edit Component** button from the **Assembly CommandManager**. You can set the transparency setting using the **Assembly Transparency** button available in the **Assemblies CommandManager**. You can add components in the subassembly, modify the mates, and replace the components while in the editing mode. After editing the subassembly, choose the **Edit Component** button to exit the editing mode.

Note

*For editing a component of the subassembly, select the component from the drawing area and invoke the shortcut menu. Choose the **Edit Part** option from the shortcut menu. The part editing mode will be invoked in the assembly document.*

*To select a component of the subassembly from the **FeatureManager Design Tree**, you need to expand the subassembly. All the components assembled in that subassembly will be displayed when you expand the subassembly.*

DISSOLVING THE SUBASSEMBLY

Dissolving the subassembly means the components of the subassembly become the components of the current assembly. When you dissolve a subassembly, the subassembly is removed from the main assembly and the components of the subassembly become the components of the main assembly. To dissolve a subassembly, select the subassembly from the **FeatureManager Design Tree** and invoke the shortcut menu. Select the **Dissolve Sub-assembly** option from the shortcut menu. The subassembly will be removed from the **FeatureManager Design Tree** and the components of the subassembly will be displayed as the components of the main assembly in the **FeatureManager Design Tree**.

REPLACING THE COMPONENTS

CommandManager:	Assemblies > Replace Components	*(Customize to Add)*
Toolbar:	Assembly > Replace Components	*(Customize to Add)*

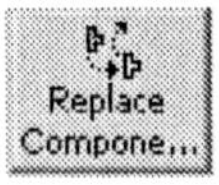

Sometimes, in the assembly design cycle you may need to replace a component of the assembly with some other component. To replace a component, select the component to be replaced and invoke the shortcut menu. Expand the shortcut menu and select the **Replace Components** option from the shortcut menu. The **Replace PropertyManager** will be displayed as shown in Figure 11-14. You can also invoke the **Replace PropertyManager** by choosing the **Replace Components** button from the **Assemblies CommandManager** after customizing it.

When you invoke the **Replace PropertyManager**, you are prompted to select components to be replaced. If the component is already selected before invoking this tool, then that component will be selected automatically. The name of the selected component is displayed in the **Components to be Replaced** selection area. Next, you need to specify the replacement component. Choose the **Browse** button from the **Selection** rollout; the **Open** dialog box will be displayed. Select the replacement component and choose the **Open** button from the **Open** dialog box. The name and location of the replacement component is displayed in the **Replacement Component** area.

The **Re-attach mates** check box is selected by default. This option is used to invoke the **Mate Entities PropertyManager** after you choose the **OK** button from the **Replace PropertyManager**. Using the **Mate Entities PropertyManager**, you can replace the mate entities.

If this check box is cleared, the **Rebuild Errors** dialog box is displayed after you exit the **Replace PropertyManager**. The **Rebuild Errors** dialog box informs you the name of the

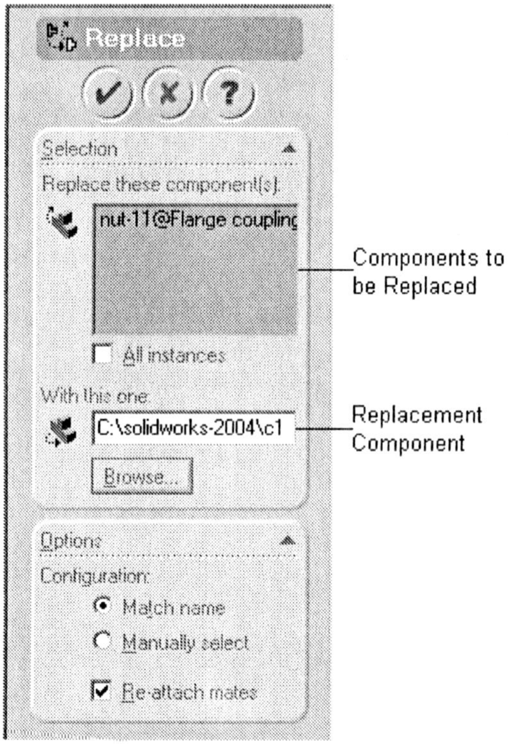

Figure 11-14** The **Replace PropertyManager

mates that contain errors. Therefore, you need to redefine the mates. For redefining the mates, expand the **Mates** option from the **FeatureManager Design Tree**. Select the mate that contains symbol on the left. Invoke the shortcut menu and select the **Edit Definition** option. The **Mate PropertyManager** will be displayed and you can edit the mate entities.

The **All instances** check box available in the **Selection** rollout is used to replace all the instances of the selected component. The options available in the **Configuration** area are used to define the selection procedure of the configurations. You will learn more about configurations later.

Figure 11-15 shows the assembly in which the bolt is to be replaced by a pin. Figure 11-16 shows the faces of the pin to be used as mate entities after replacing the component. Figure 11-17 shows the isometric view of the assembly after bolts are replaced by pins.

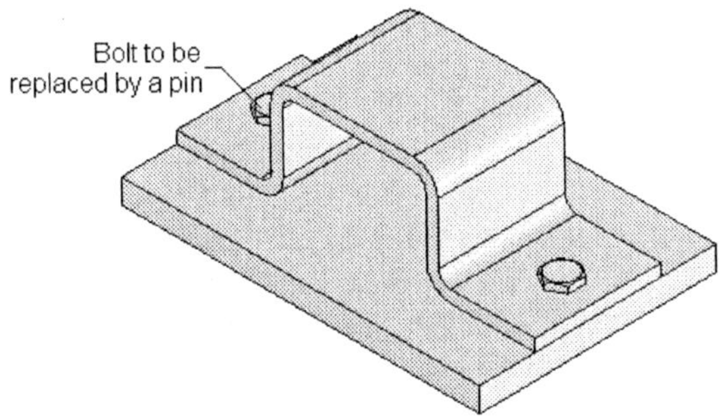

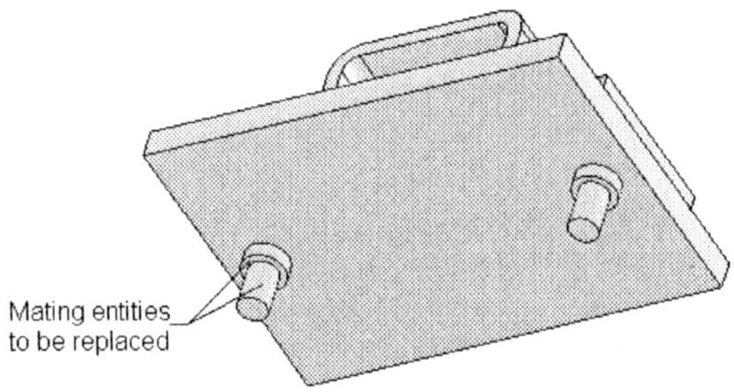

Figure 11-15 *Bolt to be replaced by a pin*

Figure 11-16 *Mating entities to be replaced*

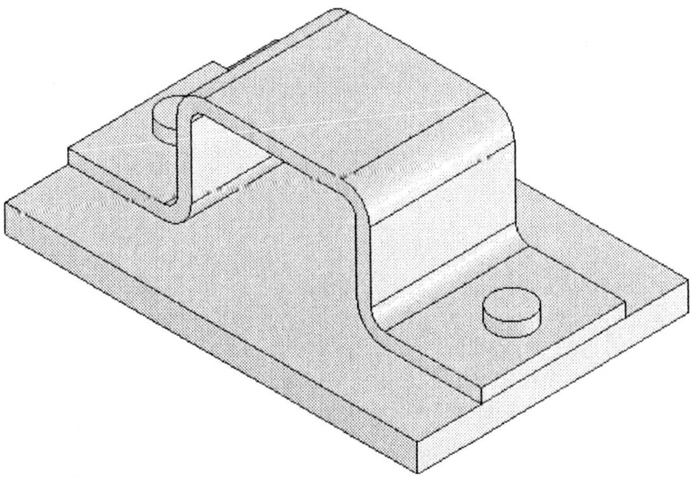

Figure 11-17 *Bolt replaced by pin in the assembly*

Tip. *If symbol is displayed on the* ***Mates*** *option in the* ***FeatureManager Design Tree****, this means that this mate group has some errors. Expand the mate group; the mate that displays or sign will have some errors. Select that mate and invoke the shortcut menu. Select* ***Whats Wrong?*** *from the shortcut menu to display the* ***Rebuild Errors*** *dialog box. The possible cause of error will be displayed in the* ***Rebuild Errors*** *dialog box.*

If you select the ***Mate Diagnostic*** *option from the shortcut menu, the* ***Diagnostics PropertyManager*** *will be displayed. The* ***Diagnose*** *button is used to display the entities that are the cause of errors in the mate. Choose the* ***Analyze*** *button; the name of the mate and the entity will be displayed in the* ***Analyze Problem*** *rollout.*

Select the ***Mates*** *mategroup with error symbol from the* ***FeatureManager Design Tree*** *and invoke the shortcut menu. You can choose the* ***Whats Wrong?*** *or* ***Mate Diagnostics*** *option to analyze all the mates having errors.*

CREATING PATTERNS OF THE COMPONENTS IN AN ASSEMBLY

Menu: Insert > Component Pattern > Linear Pattern/ Circular Pattern/ Feature Driven

While working in the assembly design environment of SolidWorks, you may need to assemble more than one instance of the component about a specified arrangement. Consider the case of a flange coupling where you have to assemble eight instances of the nut and bolt to fasten the coupling. Therefore, you need to make the instances of the nut and bolt and then assemble all the eight nuts and the eight bolts manually. However, this is a very tedious and time-consuming precess. Therefore, to reduce the time in the assembly design cycle, SolidWorks has provided a tool to create the patterns of the components. There are three types of component patterns provided in SolidWorks. These are discussed next.

Feature-driven Pattern

The feature driven pattern is used to pattern the instances of the components using an existing pattern feature. For creating a derived pattern, choose **Insert > Component Pattern > Feature Driven** from the menu bar. The **Feature Driven PropertyManager** is displayed as shown in Figure 11-18.

*Figure 11-18 The **Feature Driven PropertyManager***

You are prompted to select the driving feature pattern and select components to pattern. Select the components to pattern from the drawing area. The selected components will be displayed in green. Click once in the **Driving Feature** selection area and select a pattern feature that will drive the component pattern. The preview of the resultant pattern is displayed in the drawing area. Choose the **OK** button from the **Feature Driven PropertyManager**.

Figure 11-19 shows the components to be selected to pattern and an instance of an existing pattern feature to be selected. Figure 11-20 shows the assembly after creating the derived pattern feature.

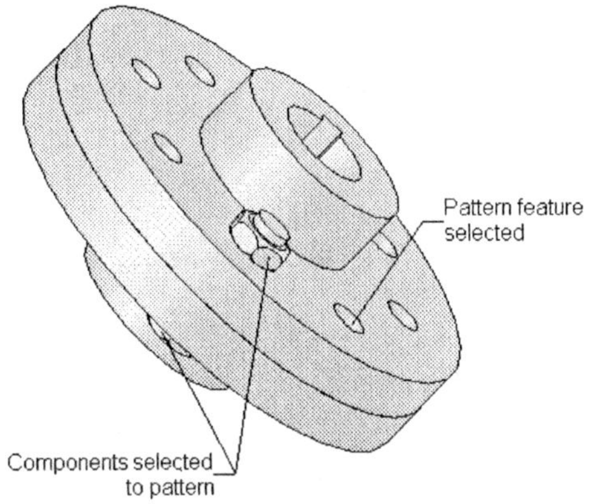

Figure 11-19 *Components and the instance of the pattern feature to be selected*

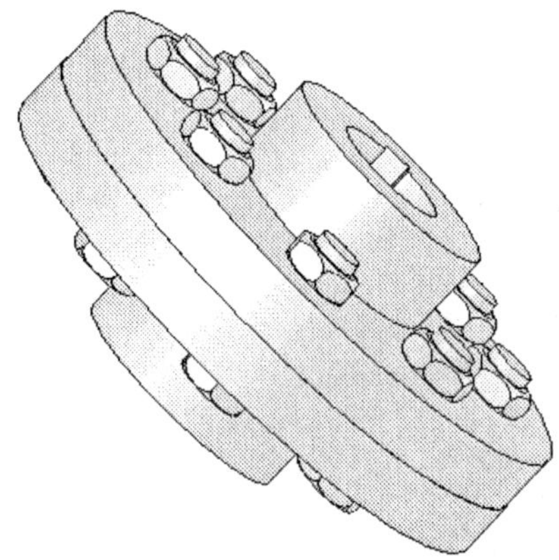

Figure 11-20 *Assembly after creating the derived component pattern*

If you need to skip some of the instances of the pattern, then after creating the component pattern select the derived pattern feature from the **FeatureManager Design Tree** and invoke the shortcut menu. Select the **Edit Feature** option from the shortcut menu, the resultant **PropertyManager** is displayed. Click once in the **Instances to Skip** selection area. Filled dots in magenta are displayed with the preview of the instances of pattern. Select the dot on the instance that you need to skip, the selected instances disappear from the display and the color of the dot changes to red. To retain the instances again select the dot. After selecting the instances to skip, choose the **OK** button from the **PropertyManager**.

Note

If you change the number of instances of the pattern feature using which the derived component pattern is created, the number of instances of the component pattern are automatically modified. This displays the associative and parametric nature of the derived component pattern.

After editing the number of entities in the pattern feature, you may have to choose the ***Rebuild*** *button.*

Remember that if the feature that was used to assemble the seed component is deleted while reducing the number of features in the pattern, you will have to modify the mates. This is because the original feature on which the mates were applied does not exist any more and so the mates give an error.

Local Pattern

You can also create the patterns of the components individually even if there is no pattern feature. This type of component pattern is known as local pattern. You can create two types of local pattern: first is the linear pattern and the other is the circular pattern. Both the types of local patterns are discussed next.

Tip. *After creating the pattern you will observe that the* ***DerivedCirPattern1*** *feature will be displayed in the* ***FeatureManager Design Tree****. If the derived pattern creates a rectangular pattern of the components, the name of the feature will be* ***DerivedLPattern1****. The number at the last of the feature displays the sequence number of the derived pattern feature.*

When expand this feature, all the instances of the patterned component are displayed. Select the instance to be skipped from the ***FeatureManager Design Tree*** *and press the DELETE key. The* ***Confirmation*** *dialog box will be displayed; choose the* ***Yes*** *button from the dialog box.*

If you need to restore the skipped pattern instance, then select the derived pattern from the ***FeatureManager Design Tree*** *and invoke the shortcut menu. Choose the* ***Edit Feature*** *option from the shortcut menu. Using the options available in the* ***PropertyManager*** *you can restore the skipped instance.*

Linear Pattern

For creating the local linear pattern, choose **Insert > Component Pattern > Linear Pattern** from the menu bar. The **Linear Pattern PropertyManager** is displayed, and you are prompted to select an edge or axis for direction reference, and select components to pattern. Select the direction reference and the component to pattern from the drawing area. Other options available in this **PropertyManager** are the same as those discussed while creating linear pattern of the features, faces, and bodies in Chapter 7.

Circular Pattern

For creating the local circular pattern, choose **Insert > Component Pattern > Circular Pattern** from the menu bar. The **Circular Pattern PropertyManager** is displayed, and you are prompted to select an edge or axis for direction reference, and select components to pattern. Select the direction reference and the component to pattern from the drawing area. Other options available in this **PropertyManager** are the same as those discussed while creating circular pattern of the features, faces, and bodies in Chapter 7.

Tip. *If one instance of the component pattern is modified or edited, the other instances of the component will also be modified.*

You can also create the component pattern of a component pattern feature.

MIRRORING THE COMPONENTS

Menu: Insert > Mirror Components

In the **Assembly** mode of SolidWorks, you can also mirror a component to place the new instance of the component in the assembly document. For mirroring the component, choose **Insert > Mirror Components** from the menu bar. The **Mirror Components PropertyManager** is displayed as shown in Figure 11-21.

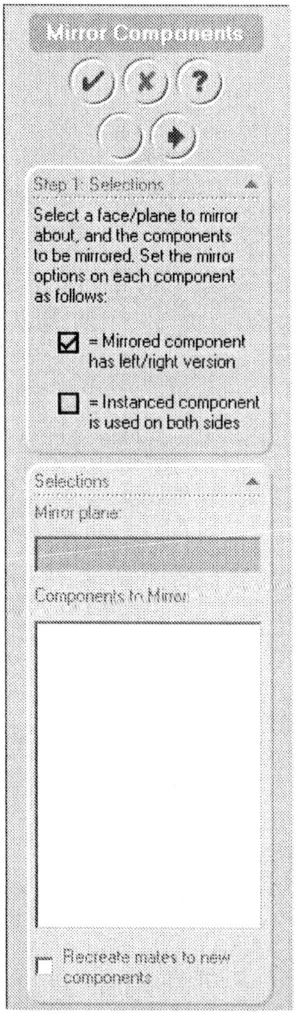

Figure 11-21 The ***Mirror Components PropertyManager***

Select the planar face or plane that will act as mirror plane as shown in Figure 11-22. Now, select the component to mirror; you will observe that the name of the component with a check box is displayed in the **Components to Mirror** area. Leave this check box cleared and choose the **Next** button from the **Mirror Components PropertyManager**. The preview of the mirrored component is displayed as shown in Figure 11-23. The **Orientation** rollout is displayed and using the options available in this rollout you can change the orientation of the mirrored component. Choose the **OK** button from the **Mirror Components PropertyManager**. The mirrored instance of the selected component is displayed in the drawing area as shown in Figure 11-24.

If you select the check box available with the name of the selected component in the **Components to Mirror** area, a new component that is mirrored along the selected mirror plane and saved in a specified location is created. Select the check box available along the name of the selected component and choose the **Next** button. The **Filenames** rollout is

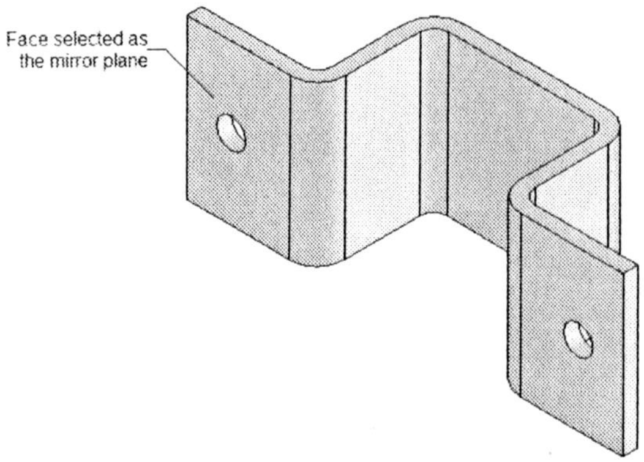

Figure 11-22 *Face to be selected as mirror plane*

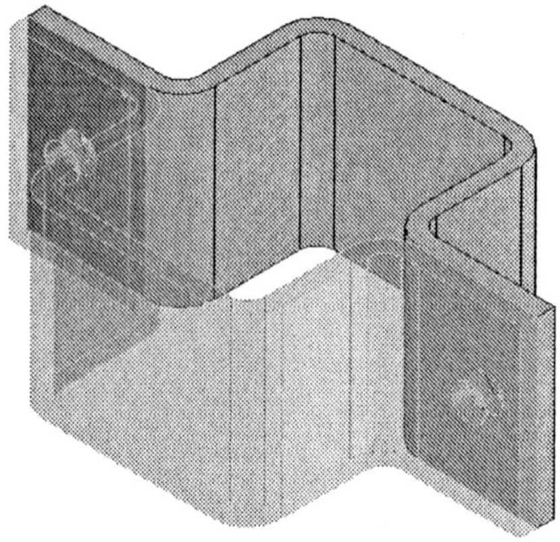

Figure 11-23 *Preview of the mirrored component*

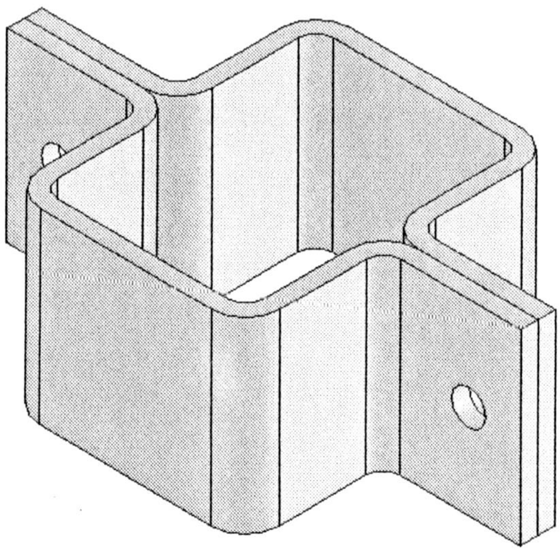

Figure 11-24 *Instance created by mirroring the component*

displayed; choose the **Browse** button available in the **Filenames** rollout. This button is used to specify the name and location where the file will be saved. Using the **Place files in one folder** check box, you can save all the mirrored instances in one folder. Using the **Add** drop-down list, you can add a suffix or a prefix to the file name.

SIMPLIFYING THE ASSEMBLIES USING THE VISIBILITY OPTIONS

When you are assembling the components, whether it be a large assembly or a small assembly, you may need to simplify the assembly using the visibility options. By simplifying, you can hide the components at any stage of the design cycle. You can also set the transparency of any component for simplifying the assembly. You can also suppress and unsuppress the components at any stage of the design cycle. The various methods of simplifying the assembly are discussed next.

Hiding the Components

CommandManager:	Assemblies > Hide/Show Components
Menu:	Edit > Hide > This Configuration
Toolbar:	Assembly > Hide/Show Components

If you need to hide the component placed in the assembly, choose the component from the drawing area or from the **FeatureManager Design Tree**. Using the CTRL key you can select more than one component to hide. Right-click to invoke the shortcut menu and select the **Hide** option from the shortcut menu. The component will disappear from the drawing area. The icon of the component is displayed in transparent in the **FeatureManager Design Tree**. To unhide the hidden component, select the icon of the component from the **FeatureManager Design Tree** and invoke the shortcut menu. Show the **Show** option from the shortcut menu. The hidden component will be redisplayed in the drawing area. You can also use the **Hide/Show Components** button from the **Assembly CommandManager** to hide or show the components.

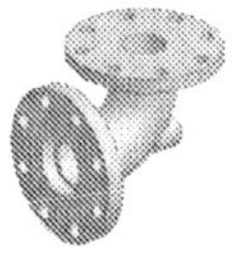

Tip. *You will observe that the **Lightweight** check box is provided in the **Open** dialog box. If you select this check box before opening the assembly file, the assembly will be opened with lightweight components.*

*A lightweight component is a component in which the feature information is available in the part document and only the graphical representation of the component is displayed in the assembly document. Therefore, the assembly environment becomes light. An icon of a lightweight component is displayed as a feather attached to the component icon in the **FeatureManager Design Tree**.*

*To get the feature information of the lightweight component, you need to resolve the component to normal state. Therefore, select the component from the drawing area or from the **FeatureManager Design Tree** and invoke the shortcut menu. Select the **Set to Resolved** option from the shortcut menu. If you need to set a resolved component to a lightweight component, select the component and invoke the shortcut menu. Select **Set to Lightweight** from the shortcut menu.*

Suppressing the Components

CommandManager:	Assemblies > Change Suppression State
Menu:	Edit > Suppress > This Configuration
Toolbar:	Assembly > Change Suppression State

You can also suppress the components placed in the assembly to simplify the assembly representation. To suppress a component, select the component from the drawing area or from the **FeatureManager Design Tree**. Invoke the shortcut menu and choose **Suppress**. The component will not be displayed in the assembly document and the icon of the suppressed component will be displayed in grey in the **FeatureManager Design Tree**. To unsuppress the suppressed component, select the component to be resolved from the **FeatureManager Design Tree** and invoke the shortcut menu. Choose the **Set to Resolve** option from the shortcut menu.

You can also suppress the component using the **Change Suppression State** button from the **Assembly CommandManager**. Using the **Change Suppression State** button, you can suppress the selected component as well as set the selected component to lightweight. When you select the component and choose the **Change Suppression State** button from the **Assembly CommandManager**, a flyout is displayed. This flyout has three options. The first option is the **Suppress** option, the second option is the **Lightweight** option, and the third option is the **Resolve** option. The **Resolve** option is used to set the suppressed or lightweight components to resolve state.

Changing the Transparency Conditions

In SolidWorks you can change the transparency of the components or selected faces to simplify the assembly. First, you will learn how to change the transparency of the components. Select the component to change its transparency and invoke the shortcut menu. Select the **Properties** option from the **Component** area of the shortcut menu. The **Component Properties** dialog box will be displayed. Choose the **Color** button available in this dialog box. The **Assembly Instance Color** dialog box is displayed as shown in Figure 11-25.

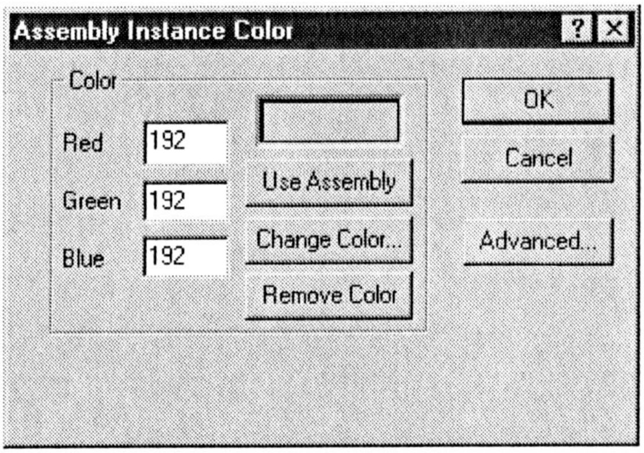

***Figure 11-25** The **Assembly Instance Color** dialog box*

Choose the **Advanced** button from this dialog box. The **Advanced Properties** dialog box is displayed as shown in Figure 11-26. You can set the transparency of the component from the **Transparency** slider. You can also set the various other advanced color settings such as **Ambient**, **Diffuse**, **Specularity**, **Shininess**, and **Emission** from this dialog box.

You can also change the transparency of a selected face. To set the transparency of a face, select the face and invoke the shortcut menu and choose the **Properties** option from the **Face** area of the shortcut menu. The **Entity Property** dialog box is displayed. Choose the **Advanced** button from this dialog box to invoke the **Advanced Properties** dialog box. Using this dialog box you can set the transparency of the selected face.

Note

You will learn about assembly envelope and advanced selection of component using assembly envelope in the advanced chapters in part 2 of this book.

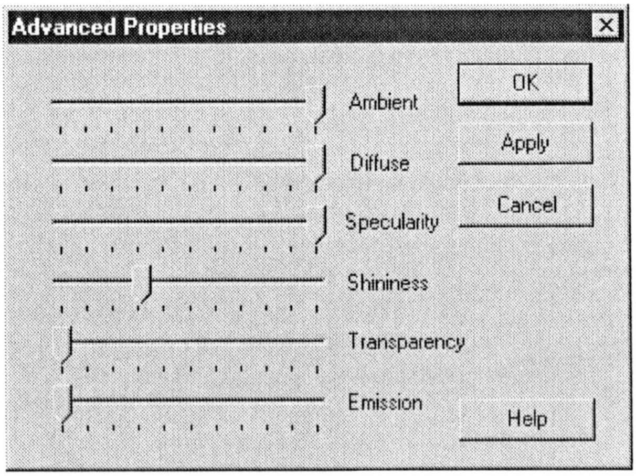

***Figure 11-26** The **Advanced Properties** dialog box*

Tip. *The face that is made transparent cannot be selected by clicking. You need to select some other faces and then choose the **Select Other** option from the shortcut menu. Right-click until the desired face is highlighted. Once the desired face is highlighted, left-click to select it. Now, right-click anywhere in the drawing area to display the shortcut menu.*

CHECKING THE INTERFERENCES IN THE ASSEMBLY

CommandManager:	Assemblies > Interference Detection
Menu:	Tools > Interference Detection
Toolbar:	Assembly > Interference Detection

After creating the assembly design, the first and the most essential step is to check the interference between the components of the assembly. If there is interference between the components, the components may not assemble properly after they are out from the machine shop or tool room. Therefore, before sending the part file and assembly for the detailing and drafting purpose, it is essential to check the interference. To check the interference, choose the **Interference Detection** button from the **Assemblies CommandManager**. The **Interference Detection PropertyManager** is displayed as shown in Figure 11-27.

The name of the current assembly is displayed in the **Components to Check** selection area. You can also check the interference between two or more than two components. To select the components, first clear the current selection set and select the components from the assembly. Choose the **OK** button from this **PropertyManager** to check the interference. If there is any interference between the components in the assembly, it will be displayed in the assembly and all the interferences are also displayed in the **Interference Results** selection area. The **Component 1** and **Component 2** selection areas displays the name of the components in between which the first interference is detected. Select the interference from the **Interference Results** selection area to view the names of the interfering components in the **Component 1** and **Component 2** selection areas. Also, as you select the interference, the components are

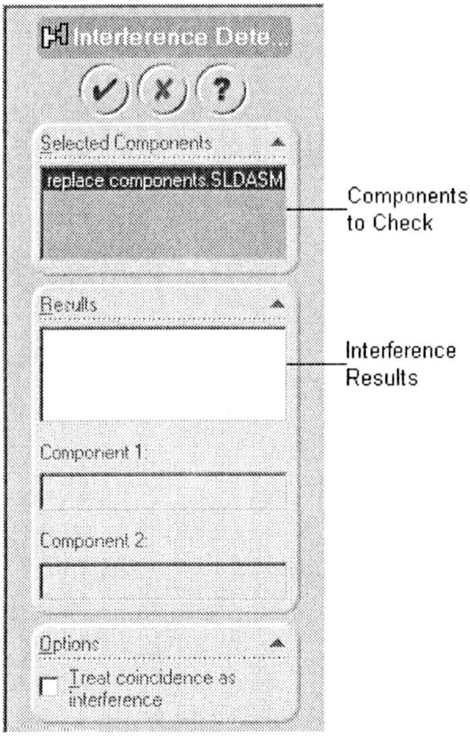

***Figure 11-27** The **Interference Volumes** dialog box*

highlighted in the assembly and the amount of interfering component is displayed in the assembly. The **Treat coincidence as interference** check box available in the **Options** rollout is used to display the coincident mates as interference. After analyzing the assembly you can edit or modify the part.

CREATING THE ASSEMBLIES FOR MECHANISM

As mentioned earlier, there are two types of assemblies. The first one is a fully defined assembly in which the relative movement of all the components is contained. The second type of assembly is that in which the components are not fully defined and some degree of freedom is kept unconstrained. As a result they can move in certain direction with respect to the surroundings of the assembly. This flexibility in turn helps you to create mechanisms and then you can move the assembly to check the mechanism that you have designed. Consider the case of a Bench Vice in which you are assembling the Vice Jaw with the Vice Body. For this assembly to work, the linear movement of the Vice Jaw when placed on the Vice Body should be free. Therefore, while creating this assembly for mechanism, you should not apply the mates for constraining the linear motion of the Vice Jaw with respect to the Vice Body. Figure 11-28 shows the degree of freedom that has to be free to create an assembly for motion.

After creating the assembly for mechanism by defining minimum mates, invoke the **Move Component** tool. Select one of the face of the component that you need to move and drag the cursor to move the assembly. While moving the assembly for mechanism design, there are

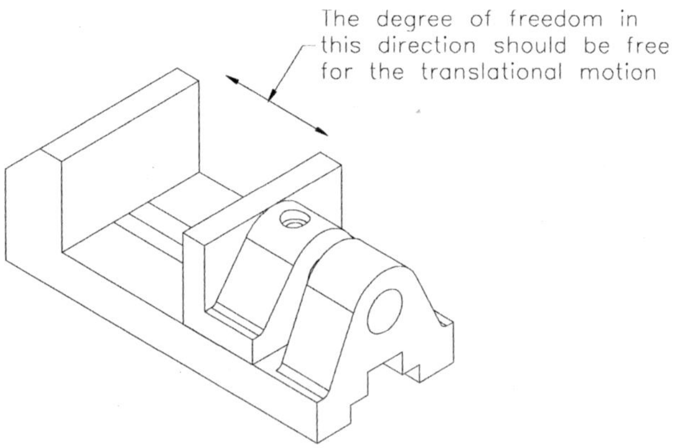

Figure 11-28 Direction in which the degree of freedom should be free

some options available for analyzing the assembly. These options of analyzing the assembly are discussed next.

Analyzing the Collisions Using the Collision Detection Tool

In SolidWorks, you can also analyze if there is any collision between the components of the assembly while the assembly is in motion. To analyze the collision when the assembly is in motion, invoke the **Move Component PropertyManager** and invoke the **Options** rollout and select the **Collision Detection** radio button as shown in Figure 11-29. The **Check between** area is displayed; the options available in this area are used to specify the components between which the collision will be detected. The options available in the **Check between** area are discussed next.

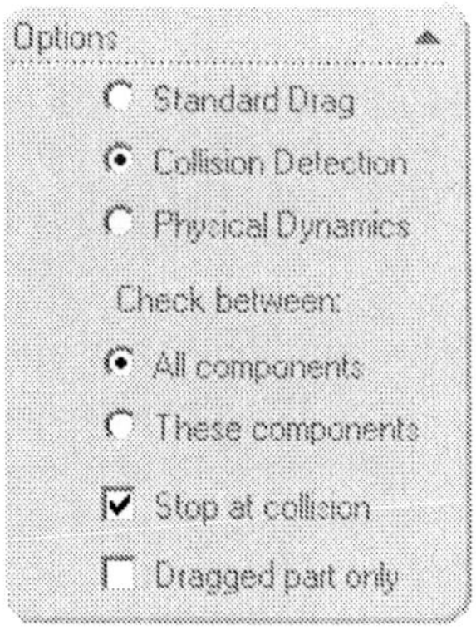

***Figure 11-29** The **Options** rollout of the*
Move Component PropertyManager

All components

The **All components** radio button is selected to check the collision between all the components when the assembly is in motion.

These components

The **These components** radio button is selected to check the collision only between the selected components when the assembly is in motion. When you select the **These components** radio button, the **Components for Collision Check** selection area and the **Resume Drag** button are displayed in the **Options** rollout as shown in Figure 11-30. After selecting the components between which the collision has to be detected, the name of the components is displayed in the **Components for Collision Check** selection area. After selecting the components, choose the **Resume Drag** button from the **Options** rollout and drag the cursor to move the assembly.

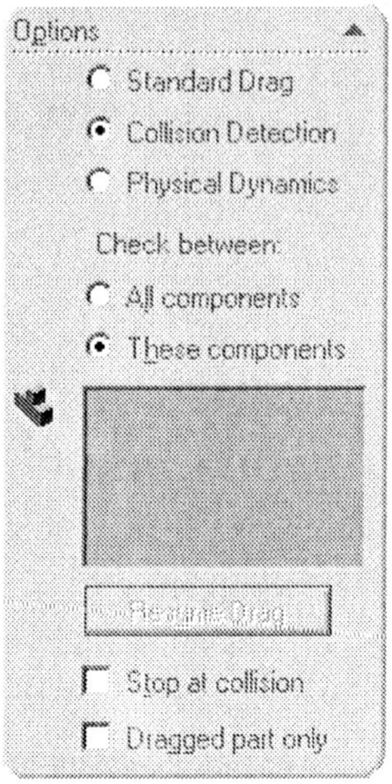

*Figure 11-30 The **Options** rollout of the **Move PropertyManager** with the **These components** radio button selected from the **Check between** area*

Stop at collision

The **Stop at collision** check box is selected to stop the motion of the assembly when one of the component collides with another component during the assembly motion.

Dragged part only

The **Dragged part only** check box is selected when you need to detect the collision only between the components that you selected to move. If you do not select this check box, the components that you selected to move and any other component that moves because of mates with the selected components are detected for collision.

After setting the options in the **Options** rollout, drag the assembly either using the **Move Component** tool or using the **Rotate Component** tool. You can also use the **Rotate Component** tool to move the assembly. To use the **Rotate Component** tool invoke the **Rotate** rollout and drag the assembly to move. If a component of the assembly collides with another component while the assembly is in motion, the faces of the components that collide with each other will be displayed in green color. If the **Stop at collision** check box is selected, the motion of the assembly will be stopped when one of the components collides with another component of the assembly.

Consider the assembly shown in Figure 11-31. In this assembly you need to move the slider in the given direction. Invoke the **Move PropertyManager** and select the **Collision Detection** radio button from the **Options** rollout. Drag the slider to move in the given direction. Figure 11-32 shows that the slider collides with the extrusion feature created in the vertical column of the base component. The faces of the component that collides are displayed in green color. If the **Stop at collision** check box is selected, you cannot move the component further after it collides with one of the components of the assembly.

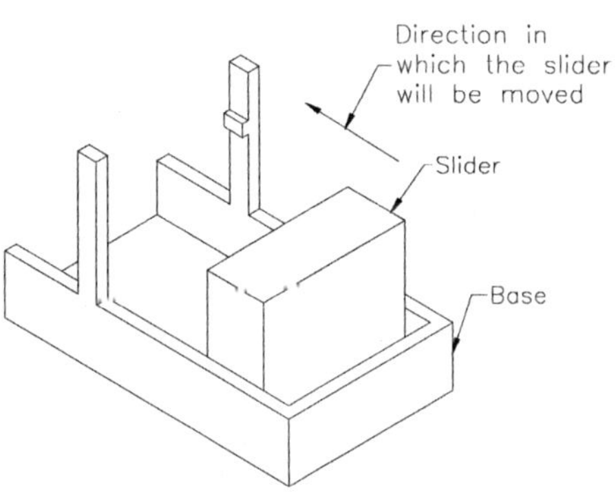

Figure 11-31 *Direction in which the slider will be moved inside the base*

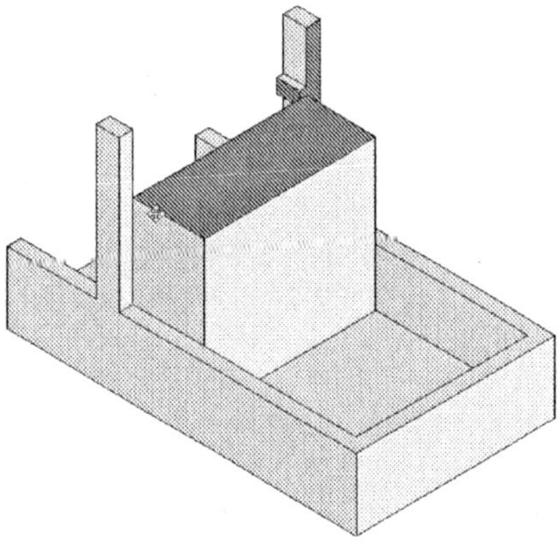

Figure 11-32 *The faces of the components highlighted in green after collision*

Once the collision is detected in the assembly, you can edit and modify the components that collide during the assembly motion.

CREATING THE EXPLODED STATE OF THE ASSEMBLY

CommandManager:	Assemblies > Exploded View
Menu:	Insert > Exploded View
Toolbar:	Assembly > Exploded View

In SolidWorks, you can also create an exploded state of the assembly. The exploded state of the assembly is created using the **Assembly Exploder** dialog box. To invoke this dialog box, choose the **Exploded View** button from the **Assemblies CommandManager**. You can also invoke this dialog box by first invoking the **ConfigurationManager**. Select the **Default** option from the **ConfigurationManager** and invoke the shortcut menu. Choose the **New Explode View** option from the shortcut menu. The **Assembly Exploder** dialog box will be displayed as shown in Figure 11-33.

There are two options to create an exploded state of the assembly. The first option is to create the exploded state automatically. To create an automatic explode state, choose the **Auto Explode** button from the **Assembly Exploder** dialog box. The components of the assembly

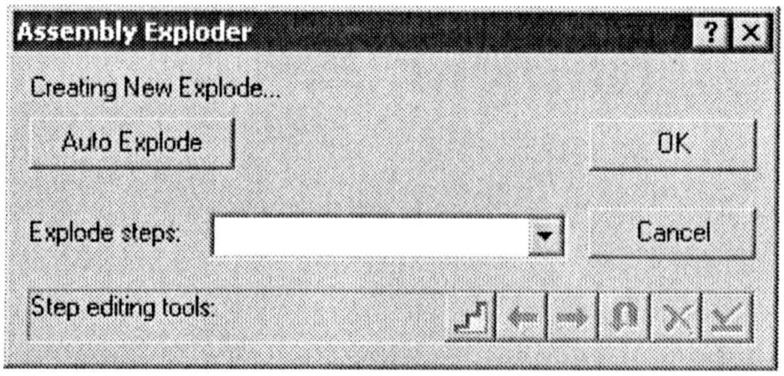

Figure 11-33 *The* ***Assembly Exploder*** *dialog box*

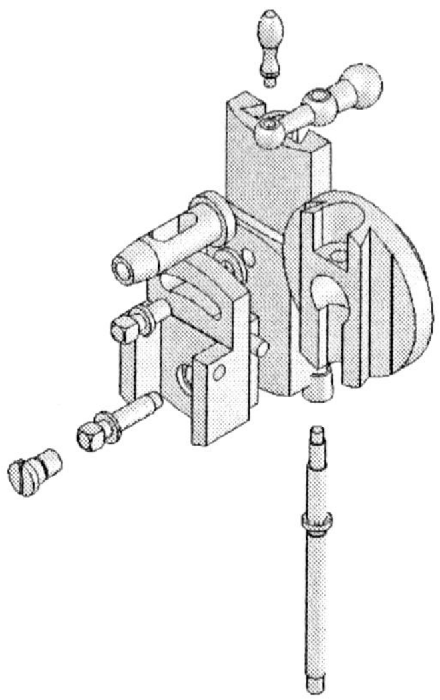

Figure 11-34 *An assembly exploded using the* ***Auto Explode*** *tool*

will be exploded and will be placed arbitrarily in the assembly document. Figure 11-34 shows an automatic exploded view of an assembly.

Creating a Systematic Explode State

To create a systematic explode state of the assembly, choose the **New** button from the **Assembly Exploder** dialog box. The **Assembly Exploder** dialog box will expand and some more options are displayed as shown in Figure 11-35.

As you expand this dialog box, you are prompted to create a new explode step. The selection is active in the **Direction to explode along** selection area. Therefore, you need to select the direction in which you need to explode the component. Select an edge, face, plane, axis, or a sketch to define the direction along which the components will be exploded. After selecting the explode direction, the selection mode in the **Components to explode** selection area is activated. Select the components to be exploded in the current direction. Specify the explode distance using the **Distance** spinner. You can also reverse the direction of explode using the **Reverse direction** check box. The **Explode related components together** check box is selected

Figure 11-35 *The expanded **Assembly Exploder** dialog box*

to explode all the components that are related with the selected component using the assembly mates.

After exploding the components in a particular direction with same distance, you need to explode the other components. Therefore, you need to create another explode step. Choose the **New** button from the **Assembly Exploder** dialog box. Now, using the same procedure, explode the other set of components. You can create as many explode steps as you want while exploding the assembly.

After creating all the explode steps choose the **OK** button from the **Assembly Exploder** dialog box to complete the explode state creation. If you need to switch back to the unexplode state, expand the **Default** option from the **ConfigurationManager** and select **ExplView1**. Invoke the shortcut menu and choose **Collapse** from the shortcut menu. You can also edit each explode step by expanding **ExplView1** and select the explode step to be modified. Invoke the shortcut menu and choose the **Edit Definition** option from the shortcut menu. The **Assembly Exploder** dialog box is displayed and you can modify the parameters of the explode step. Figure 11-36 shows the systematic exploded view of an assembly.

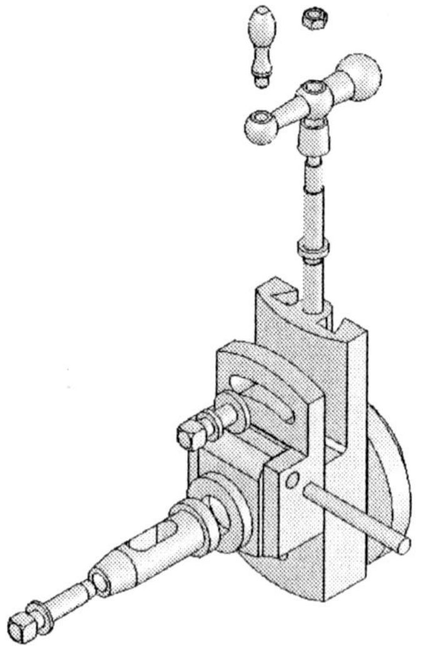

Figure 11-36 The systematic exploded state of an assembly

Creating the Explode Line Sketch

CommandManager:	Assemblies > Explode Line Sketch
Menu:	Insert > Explode Line Sketch
Toolbar:	Assembly > Explode Line Sketch

Explode lines are the parametric axes that display the direction of explosion of the components in the exploded state. Figure 11-37 shows an exploded assembly with explode lines.

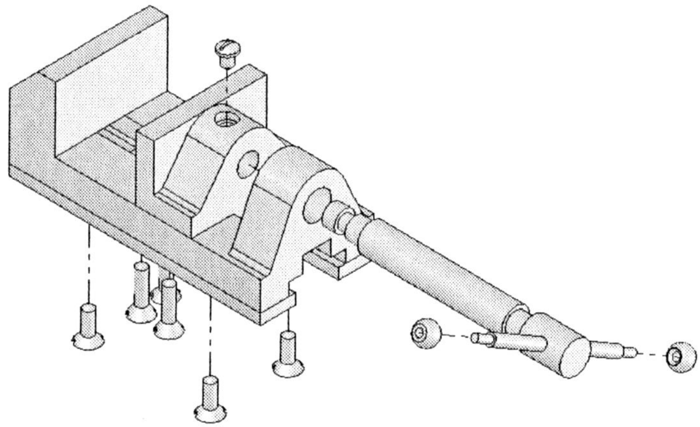

Figure 11-37 Explode line sketch created on an exploded assembly

To create an explode line sketch, choose the **Explode Line Sketch** button from the **Assembly CommandManager**. The sketcher environment in invoked and the **Explode Sketch** toolbar is invoked. The **Route Line PropertyManager** is also displayed as shown in Figure 11-38.

Figure 11-38 *The* ***RouteLine PropertyManager***

You are prompted to select a cylindrical face, planar face, vertex, point, arc, or line entities. One by one select the cylindrical faces of the two components to create an explode line between them. For example, to create an explode line between the Oval Fillister and the Vice Jaw, select the cylindrical face of the Oval Fillister that goes inside the Vice Jaw. Now, select the cylindrical hole of the Vice Jaw. The preview of the explode line appears. Choose **OK** to create the explode line. The names of the selected faces are displayed in the **Items To Connect** selection area. As you select an entity to create an explode line, an arrow is also displayed with the line. You can use that arrow to reverse the direction of explode line creation. You can also select the **Reverse** check box from the **Options** rollout of the **Route Line PropertyManager** to reverse the direction of the explode line creation. After creating the explode line in all the components, choose the **OK** button from the **Route Line PropertyManager** and exit the sketching environment.

TUTORIALS

Tutorial 1

In this tutorial you will create the assembly shown in Figure 11-39. This assembly will be created in two parts, one will be the subassembly and the second will be the main assembly. You will also create the exploded state of the assembly and then create the explode line sketch. The exploded state of the assembly is displayed in Figure 11-40. The views and dimensions of all the components of this assembly are displayed in Figures 11-41 through 11-44. **(Expected time: 3 hrs)**

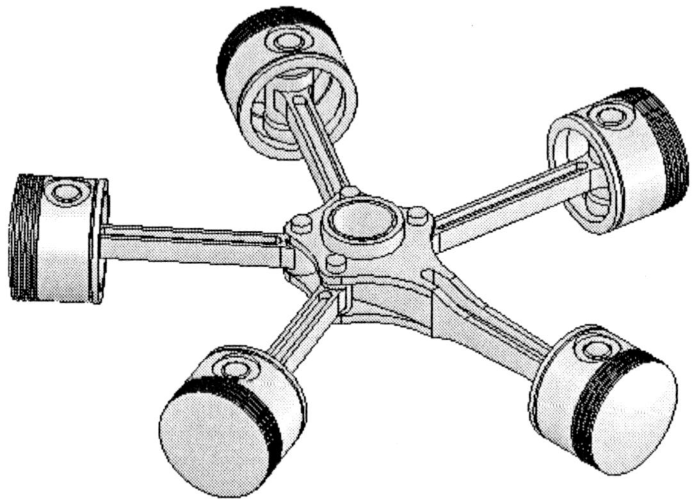

Figure 11-39 *The radial engine subassembly*

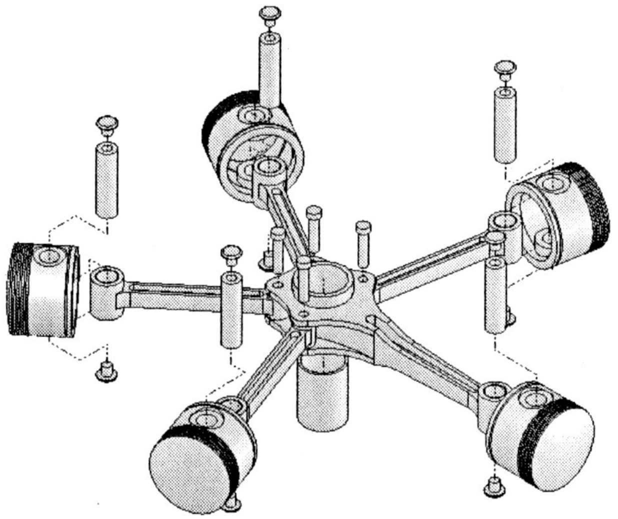

Figure 11-40 *Exploded view of the assembly*

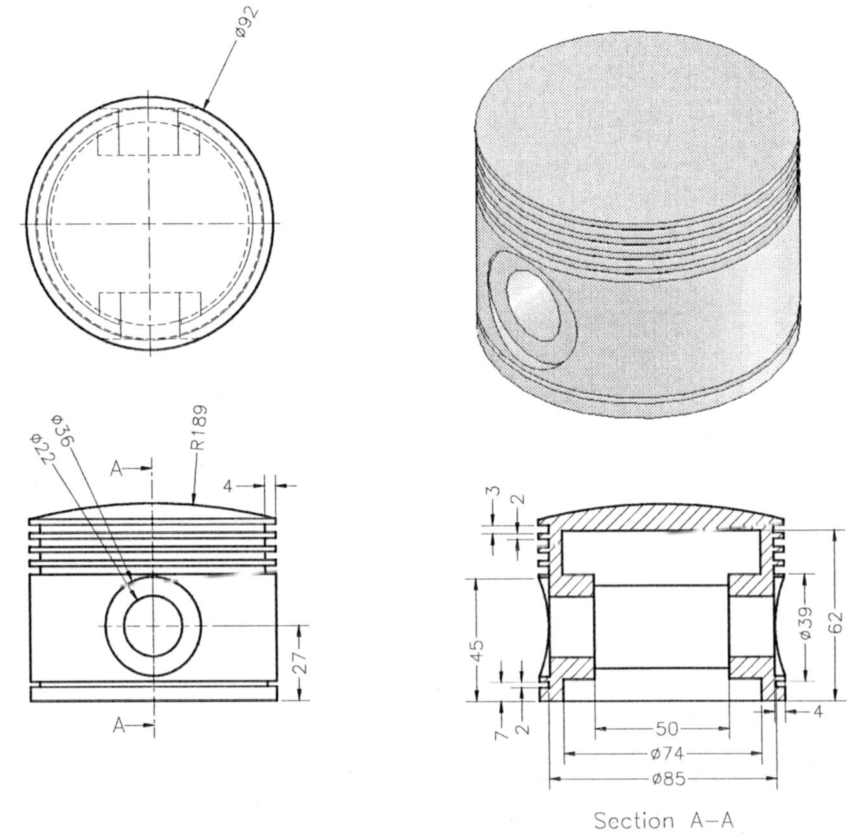

Figure 11-41 *Views and dimensions of the Piston*

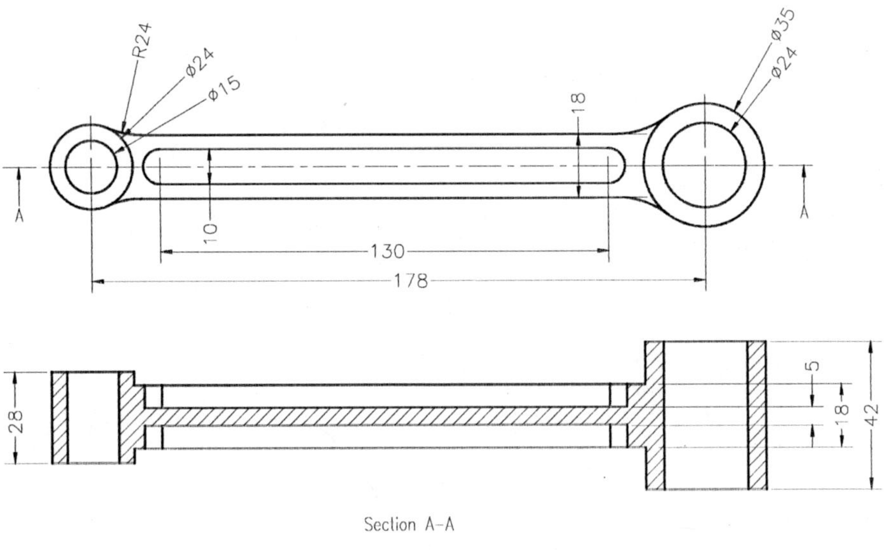

Figure 11-42 *Views and dimensions of the Articulated rod*

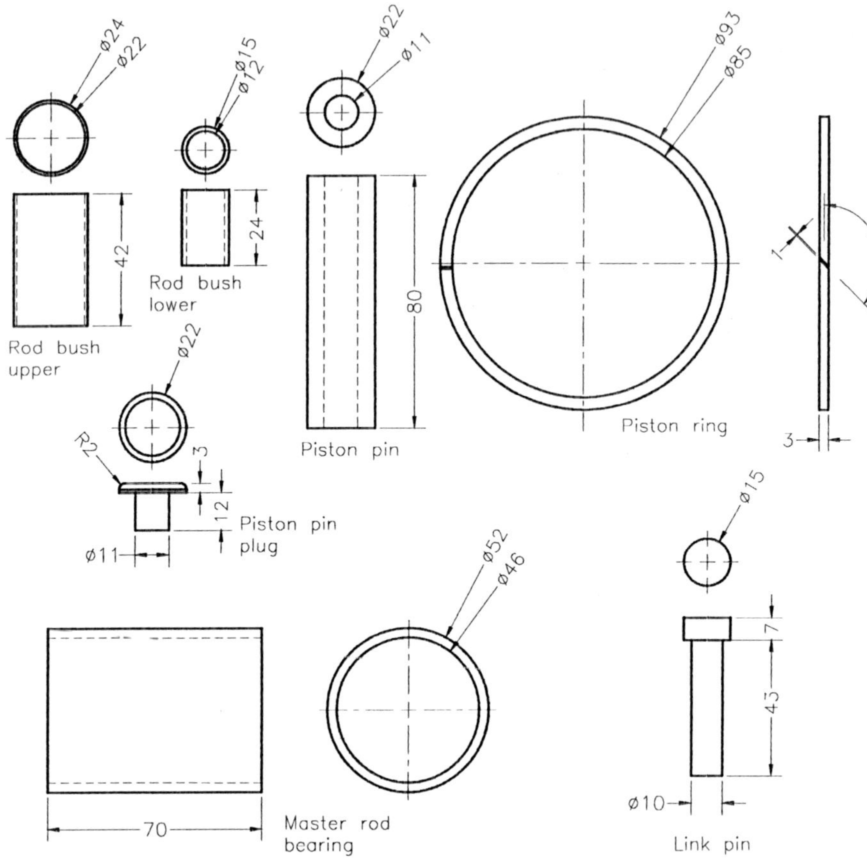

Figure 11-43 Views and dimensions of other components

The steps to be followed to complete this tutorial are listed next.

Since this assembly is a large assembly, therefore, you need to divide this assembly in two steps. One will be the subassembly and the second will be the main assembly. First, you need to create the subassembly, which consists of Articulation Rod, Piston, Piston Rings, Piston Pin, Rod Bush Upper, Rod Bush Lower, and Piston Pin Plug. After creating this subassembly you will create the main assembly in which you will assemble the Master Rod with the Piston, Piston Rings, Piston Pin, Rod Bush Upper, and Piston Pin Plug. After assembling the components in the main assembly you will assemble the subassembly with the main assembly.

a. Create all the components of the assembly in the **Part** mode and save the components in a folder named *Radial Engine Subassembly*.
b. Start a new assembly document and assemble the components to create a subassembly.

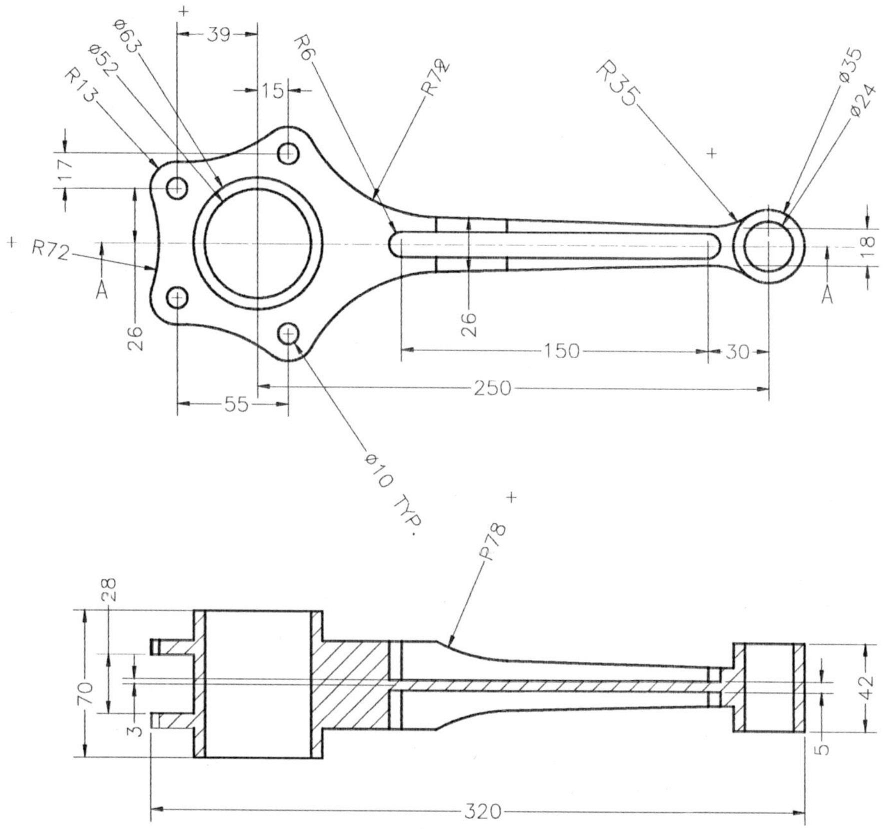

Section A–A

Figure 11-44 *Views and dimensions of Master rod*

c. Start a new assembly document and assemble the components of the main assembly.
d. Assemble the subassembly in the main assembly.
e. Create the exploded view of the assembly and then create the explode line sketch.

Creating the Components

1. Create a folder with the name **Radial Engine Subassembly** in the */My Documents/ SolidWorks/ c11* folder. Create all the components in the individual part documents and save them in this folder.

Note

When you create the Master Rod, remember that the holes on the left of the Master Rod should be created using the sketch-driven pattern. This is done because while assembling the Link Pin you will create the derived pattern of the Link Pin using the sketch-driven pattern feature.

Creating the Subassembly

As discussed earlier, you will first create the subassembly. After that, you will assemble the subassembly with the main assembly.

1. Start a new SolidWorks assembly document and exit the **Insert Component PropertyManager**. Now, save it with the name **Piston Articulation Rod Subassembly** in the same folder in which the parts are created.

2. First place the Articulated Rod at the origin of the assembly and then place the other components such as Piston, Piston Pin, Piston Pin Plug, Rod Bush Upper, and Rod Bush Lower in the assembly document.

3. Apply the required mates to assemble these components. Figure 11-45 shows the sequence in which you need to assemble the components. The exploded view and the explode line sketch is only given for your reference, you will learn more about exploded view and explode line sketch later. The assembly after assembling the Articulated Rod, Piston, Piston Pin, Piston Pin Plug, Rod Bush Upper, and Rod Bush Lower is shown in Figure 11-46.

   It is clear from the assembly that you need to assemble four instances of Piston Ring. You will only assemble one instance of the Piston Ring at the uppermost groove of the ring. After assembling this instance, you will create a local linear pattern of the Piston Ring.

4. Insert the Piston Ring in the assembly document and assemble the Piston Ring at the uppermost groove of the Piston using the assembly mates, see Figure 11-46. In this figure, the color of Piston Ring is changed by selecting it and choosing the **Component Properties** option from the shortcut menu. From the **Component Properties** dialog box, choose the **Color** button to change the color.

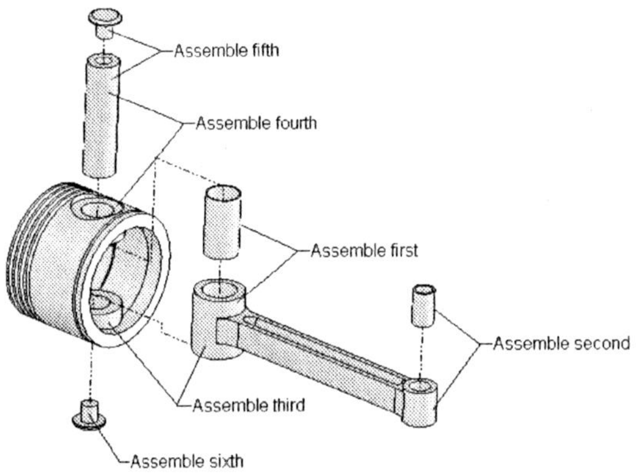

Figure 11-45 *Assembly of Articulated Rod, Piston, Piston Pin, Piston Pin Plug, Rod Bush Upper, and Rod Bush Lower*

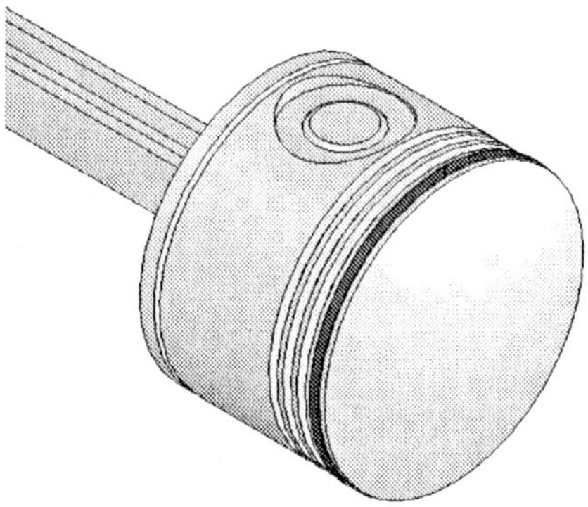

Figure 11-46 *First instance of the Piston Ring assembled with the Piston*

Next, you need to create the local linear pattern of the Piston Ring.

5. Choose **Insert > Component Pattern > Linear Pattern** from the menu bar. The **Linear Pattern PropertyManager** is displayed.

6. Select any one of the horizontal edge of the Articulated Rod to define the direction of pattern creation.

7. Click once in the **Components to Pattern** selection area and select the Piston Ring from the drawing area.

   The preview of the linear pattern with default settings is displayed in the drawing area.

8. Select the **Reverse Direction** button to reverse the direction of pattern creation.

9. Set the value of the **Spacing** spinner to **5** and set the value of the **Number of Instances** spinner to **4**.

10. Choose the **OK** button from the **Linear Pattern PropertyManager.**

    The subassembly after patterning the Piston Ring is shown in Figure 11-47

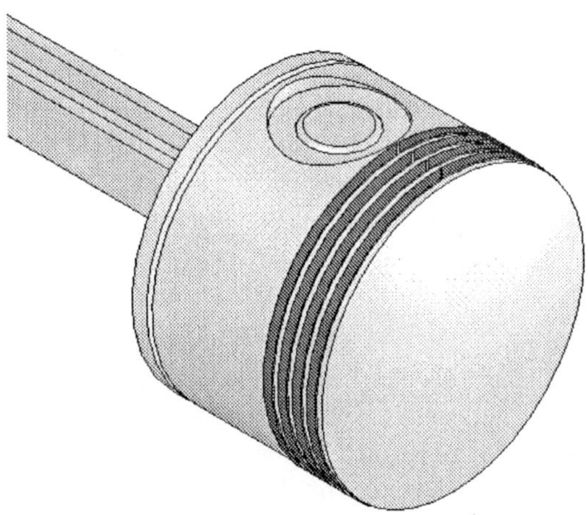

Figure 11-47 *Subassembly after patterning the Piston ring*

The subassembly is completed. Save and close the assembly document.

Creating the Main Assembly

Next, you will create the main assembly. After creating the main assembly you will assemble the subassembly with the main assembly.

1. Start a new SolidWorks assembly document and exit the **Insert Component**

PropertyManager. Now, save it with the name **Radial Engine Subassembly** in the same folder in which the parts are created.

2. First, place the Master rod at the origin of the assembly and then place the Piston, Piston Pin, Piston Pin Plug, Piston Ring, Rod Bush Upper, and Master Rod Bearing in the current assembly document.

3. Using the assembly mates and local linear pattern option assemble all the components of the main assembly.

   The components after assembling in the main assembly are displayed in Figure 11-48.

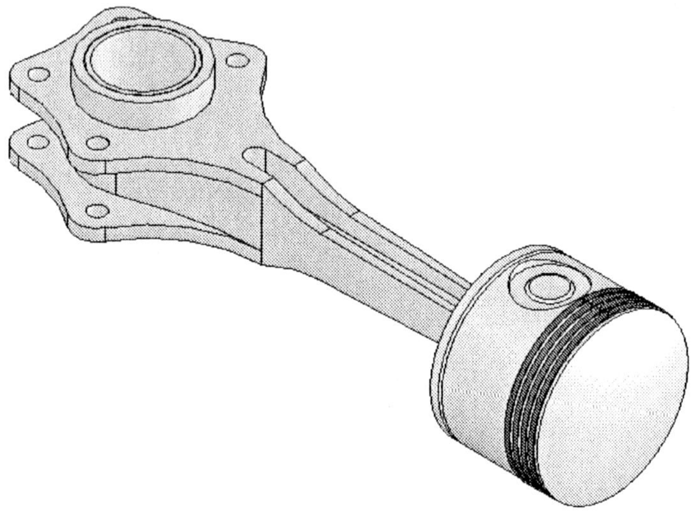

Figure 11-48 *Components assembled in the Main assembly.*

Assembling the Subassembly With the Main Assembly

Now, you will place the subassembly in the main assembly and then assemble them.

1. Choose the **Insert Components** button from the **Assemblies CommandManager**. The **Insert Component PropertyManager** is displayed.

2. Choose the **Browse** button available in the **Part/Assembly to Insert** rollout. The **Open** dialog box is displayed.

3. Choose **Assembly** (**.asm, *.sldasm*) from the **Files of type** drop-down list.

4. Double-click on *Piston Articulated Rod Subassembly* and place the subassembly in the main assembly. Figure 11-49 shows the subassembly and the main assembly placed together.

5. Using the assembly mates assemble the subassembly with the main assembly. Refer to Figure 11-50, which shows you the assembly structure that will help you in assembling the instances of the subassembly.

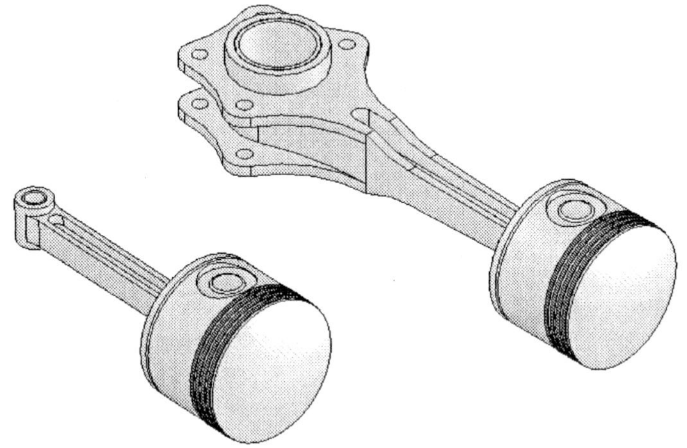

Figure 11-49 Subassembly and the main assembly placed together

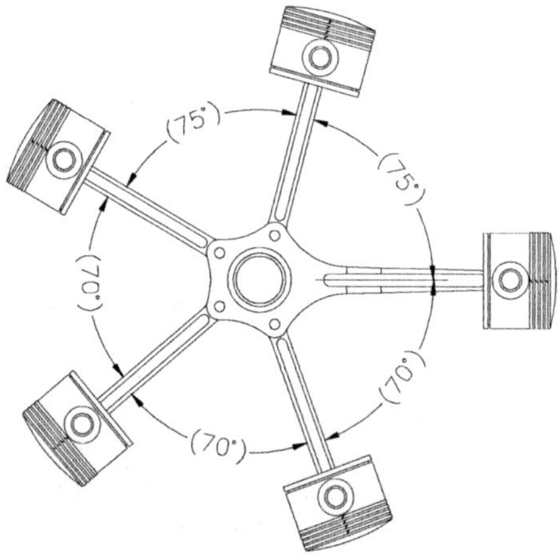

Figure 11-50 Assembly structure

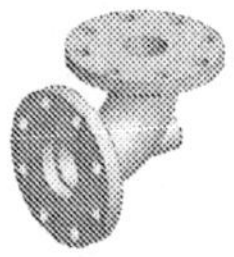

Tip. *You can create more than one instance of the subassembly by holding down the CTRL key; select and drag the subassembly from the* **FeatureManager Design Tree**. *Release the left mouse button to place the assembly in the current assembly document.*

Figure 11-51 shows all the instances of the subassembly assembled with the main assembly.

Assembling the Link Pin

After assembling the subassembly with the main assembly you need to assemble the Link Pin with the main assembly.

1. Place the Link Pin in the current assembly document and using the assembly mates, assemble the Link Pin with the main assembly. Figure 11-52 shows the first instance of Link Pin assembled with the main assembly.

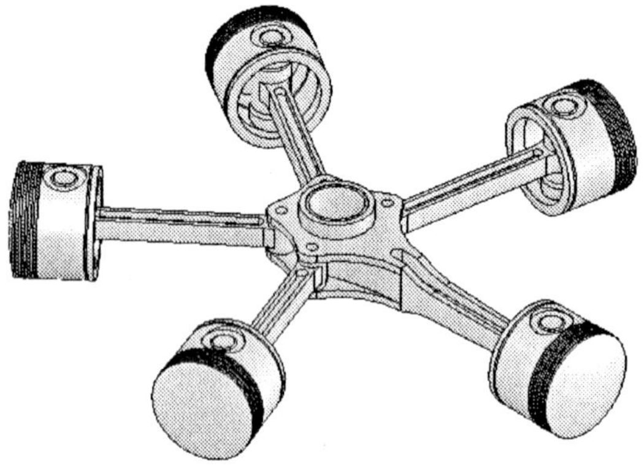

Figure 11-51 *Subassembly assembled to the main assembly*

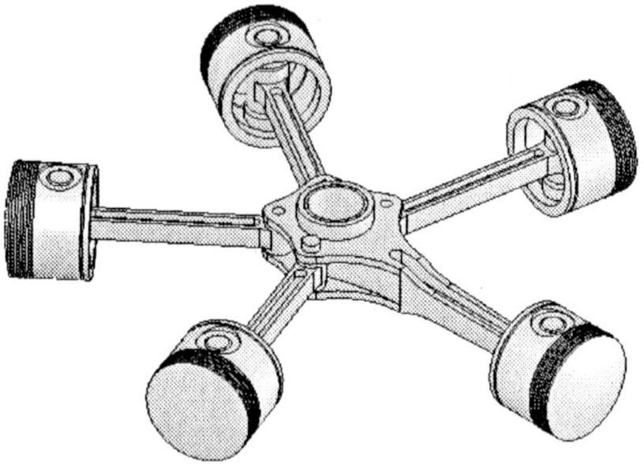

Figure 11-52 *First instance of the Link pin assembled with the main assembly*

As discussed earlier, the other instances of the Link Pin will be assembled using the sketch-driven pattern feature of the holes created on the left of the master rod.

2. Choose **Insert > Component Pattern > Feature Driven** from the menu bar. The **Feature Driven PropertyManager** is displayed.

3. Select the Link Pin from the main assembly. The name of the Link Pin is displayed in the **Components to Pattern** selection area.

4. Click once in the **Driving Feature** selection area to activate the selection mode in this area.

5. Select any one of the hole instance from the master rod. The name of the sketch pattern feature will be displayed in the **Driving Feature** selection area and the preview of the resultant pattern is also displayed.

6. Choose the **OK** button from the **Feature Driven PropertyManager**.

Figure 11-53 displays the final assembly.

Exploding the Assembly

After creating the assembly you need to explode the assembly using the **Assembly Exploder** tool.

1. Choose the **Exploded View** button from the **Assemblies CommandManager**.

The **Assembly Exploder** dialog box is displayed.

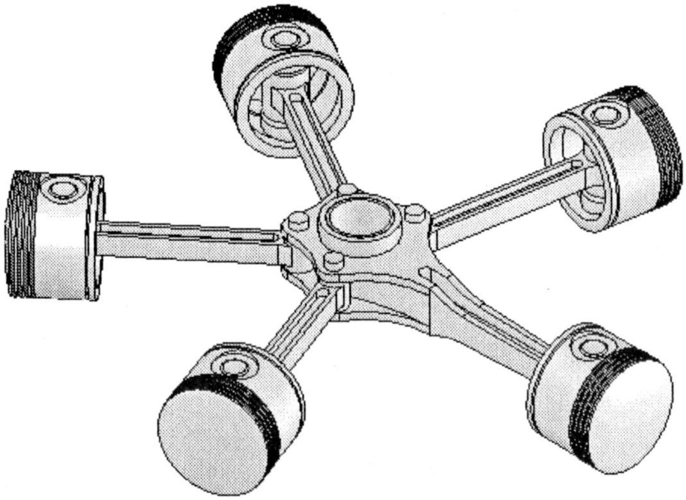

***Figure 11-53** The final assembly*

2. Choose the **New** button from the **Assembly Exploder** dialog box. 

3. Select any vertical edge from the assembly to define the direction in which the components will be exploded.

4. Select the **Component part only** radio button on the right of the **Components to explode** selection area.

5. Select all the Piston Pin Plugs from the assembly.

6. Set the value of the **Distance** spinner to **170** and choose the **Apply** button from the **Assembly Exploder** dialog box. 

   All the instances of the Piston Pin Plug are exploded.

7. Choose the **New** button from the **Assembly Exploder** dialog box to create a new assembly explode step. 

8. Select any vertical edge from the assembly to define the direction of explode and select the **Component part only** radio button that is available on the right of the **Components to explode** selection area.

9. Select all the instances of the Piston Pin from the assembly.

10. Set the value of the **Distance** spinner to **140** and choose the **Apply** button to end the explode step creation. 

11. Similarly, explode the other components of the assembly. Make sure that every time you explode the components of the assembly, the **Component part only** radio button should be selected. The assembly after exploding the components is shown in Figure 11-54.

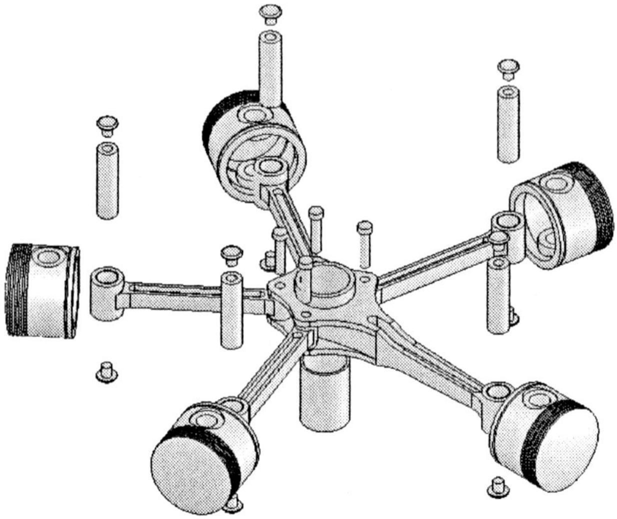

Figure 11-54 *Final exploded assembly*

Creating the Explode Line Sketch

After exploding the assembly, you need to create the explode line sketch of the exploded state of the assembly.

1. Choose the **Explode Line Sketch** button from the **Assemblies CommandManager**. The **Route Line PropertyManager** is displayed and you are prompted to select a cylindrical face, planar face, vertex, point, arc, or line entities.

2. Choose the **Keep Visible** button from the **Route Line PropertyManager** if it is not chosen automatically to keep it visible on the screen.

3. Select the cylindrical face as shown in Figure 11-55 as the first selection. The name of the selected face will be displayed in the **Items To Connect** selection area. The preview of the explode line sketch is displayed at the center of the selected face.

4. Refer to Figure 11-55 and select the other cylindrical faces to create the explode line sketch.

5. After selecting all the faces, choose the **OK** button. An exploded line will be created.

Similarly, create explode lines between other parts of the exploded assembly. Figure 11-56 shows the assembly after creating the explode line sketch.

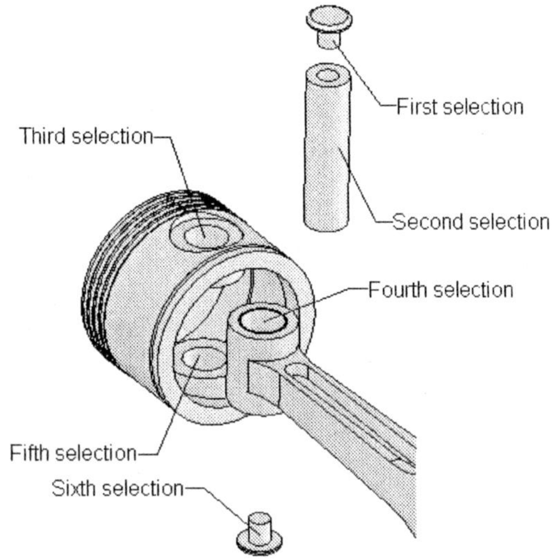

Figure 11-55 *Faces to be selected to create explode line sketch*

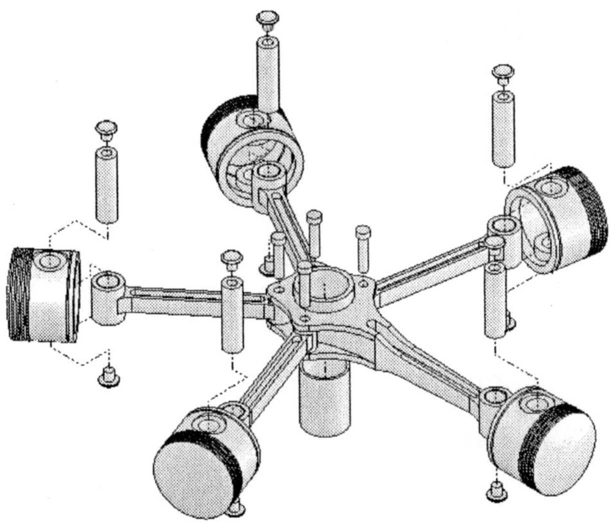

Figure 11-56 *Explode line sketch created to the exploded state of the assembly*

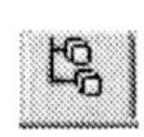

6. Invoke the **ConfigurationManager** and expand the **Default** option. Select **ExplView1** and invoke the shortcut menu. Choose the **Collapse** option from the shortcut menu to switch back to the collapse state of the assembly.

7. Save and close the assembly document.

Tutorial 2

In this tutorial you will modify the assembly created in Tutorial 1 (Bench Vice) of Chapter 10. You will modify the design of the components of the assembly and then suppress some mates that enable the assembly to move along a particular degree of freedom. After that you will check the assembly for collision detection when the assembly is in motion and then you will modify the assembly and check the interference. **(Expected time: 1 hr)**

The steps that will be followed to complete this tutorial are listed next.

a. Save the Bench Vice assembly folder in *c11* directory and then open the Bench Vice assembly.
b. Modify the design of the components within the context of assembly.
c. Suppress the mate to enable the Vice Jaw to move along the slide ways of the Vice Body.
d. Check the new assembly design for collision detection when the assembly is in motion. Modify the design if there is any collision between the components.
e. Check the interference in the modified assembly.

Opening the Bench Vice Assembly

The assembly created in Tutorial 1 of Chapter 10 is the Bench Vice assembly. You need to copy and save that assembly in the current folder of Chapter 11.

1. Copy the folder in which the Bench vice assembly is saved and then paste it in the *c11* folder.

2. Invoke the **Open** dialog box and browse the Bench Vice assembly document. Double-click on it to open the assembly document.

Modifying the Design of the Components of Bench Vice Assembly

Because of some alteration in the design of some of the components of this assembly, you need to modify the components. You will modify and edit the components in the context of the assembly.

Before you start modifying the components, it is recommended that you hide some of the components to simplify the assembly. Hiding the components will simplify the assembly and facilitate in selection while editing and modifying the components.

1. Press and hold down the CTRL key from the keyboard and select the Clamping Plate, Base Plate, all four Set Screw 1, all four Set Screw 2, Oval Fillister, Screw Bar, and Jaw Screw from the **FeatureManager Design Tree**.

2. Invoke the shortcut menu and select the **Hide** option from the shortcut menu.

The selected components are hidden and are not displayed in the drawing area of the assembly document.

The design alteration includes creating a through slot on the right face of the Vice Jaw. For modifying the design of the Vice Jaw, first you need to enable the part editing environment.

3. Select the Vice Jaw from the assembly and then choose the **Edit Component** button from the **Assemblies CommandManager**. The part modeling environment is invoked in the assembly document.

4. Make sure that the Vice Body is transparent. If it is not, then choose the **Assembly Transparency** button from the **Assemblies CommandManager**. A flyout is displayed. Select the **Force Transparency** option from the flyout. The Vice Body is displayed in transparent as shown in Figure 11-57.

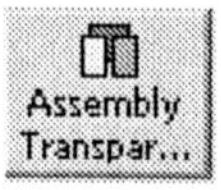

5. Select the right face of the Vice Jaw and invoke the sketching environment.

6. Create the sketch of the slot as shown in Figure 11-58.

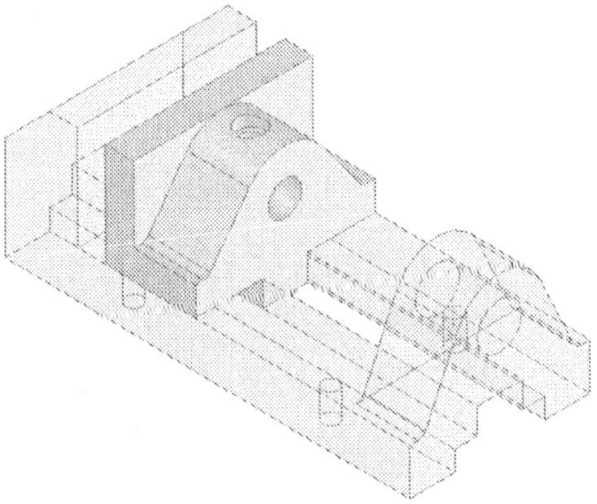

Figure 11-57 *Vice Jaw in the part edit mode in the assembly document*

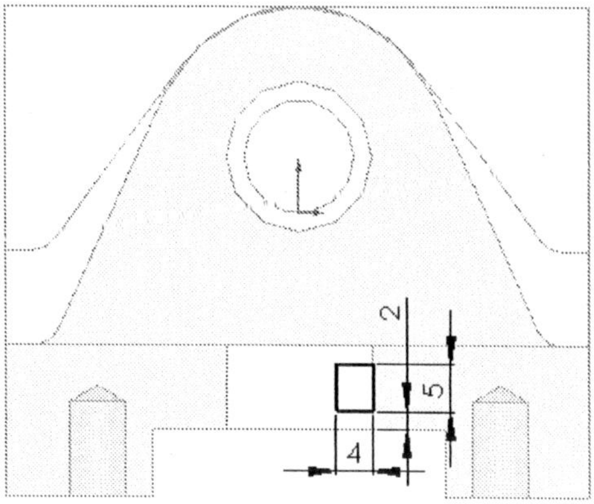

Figure 11-58 *Sketch of the slot*

7. Invoke the **Cut-Extrude PropertyManager**.

8. Create the cut feature using the **Through All** option.

9. Choose the **Edit Component** button to exit the part editing environment.

Figure 11-59 shows the assembly after modifying the design of the Vice Jaw.

10. Similarly modify the design of the Vice Body. You need to create a blind extruded boss feature up to 60 mm depth on the right face of the component. You need to reverse the direction of feature creation.

The sketch of the feature is shown in Figure 11-60. Figure 11-61 shows the assembly after exiting the part editing environment.

11. Choose the **Save** button from the **Standard** toolbar. The **SolidWorks** information box is displayed and you are informed that some models referenced in the document are modified and they must be saved. Choose **Yes** to save the referenced models also.

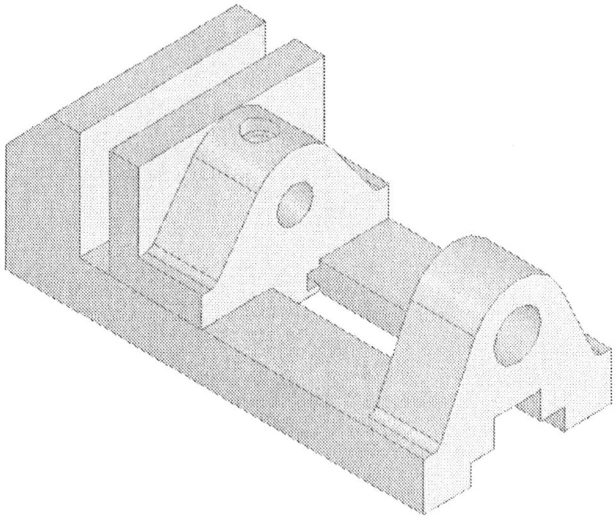

Figure 11-59 Modified Vice Jaw

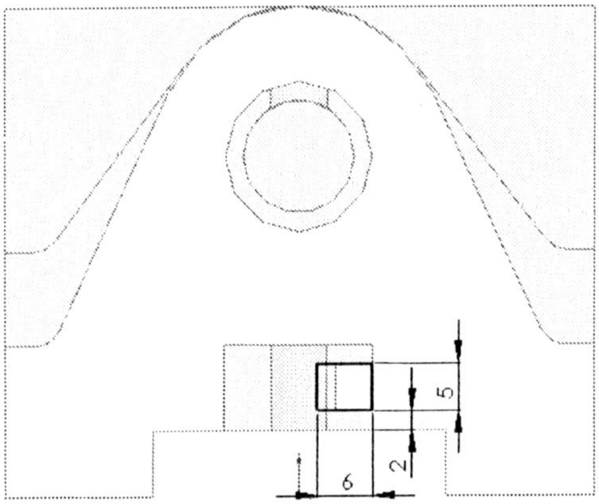

Figure 11-60 Sketch of the extruded boss feature

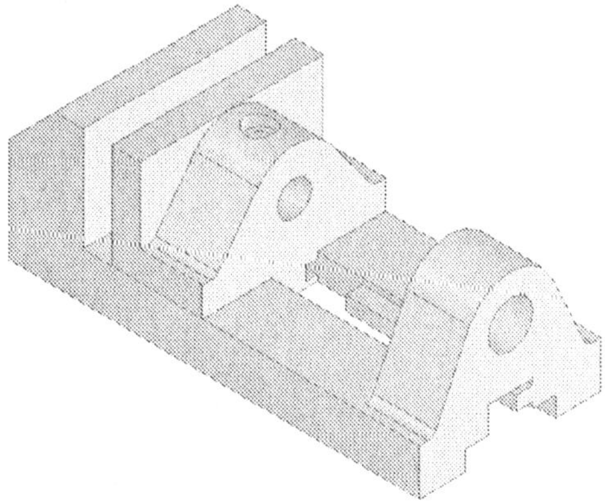

Figure 11-61 Modified assembly

Suppressing the Mate to Make the Movement of Vice Jaw Free in a Specified Direction

To analyze the movement of the Bench vice assembly, you need to make the movement of the Vice Jaw free in the X direction. By doing so the Vice Jaw will slide on the sideways of the Vice Body.

1. Expand the **Mates** mategroup from the **FeatureManager Design Tree** and select the **Distance1** mate. The planar faces of the Vice Jaw and the Vice Body to which this mate is applied are highlighted in green.

2. Invoke the shortcut menu. Choose the **Suppress** option from the shortcut menu.

   Now, the degree of freedom in the X direction is free.

3. Choose the **Move Component** button from the **Assemblies CommandManager** and select a horizontal edge of the Vice Jaw. When you drag the cursor, you will observe that you can move the Vice Jaw in the X direction.

4. Drag the Vice Jaw back to its original position and choose the **OK** button from the **Move Component PropertyManager**.

Analyzing the Collision Between the Components When the Assembly is in Motion

Next, you will analyze the collision between the components of the assembly when the assembly is in motion.

1. Again, choose the **Move Component** button from the **Assemblies CommandManager**. Select the **Collision Detection** radio button from the **Options** rollout.

2. Select Vice Jaw and drag the cursor to move it in the direction as shown in Figure 11-62.

3. When you move the Vice Jaw in the specified direction, you will observe that the right face of the Vice Jaw and the newly created extrusion feature of the Vice Body are highlighted in green as shown in Figure 11-63.

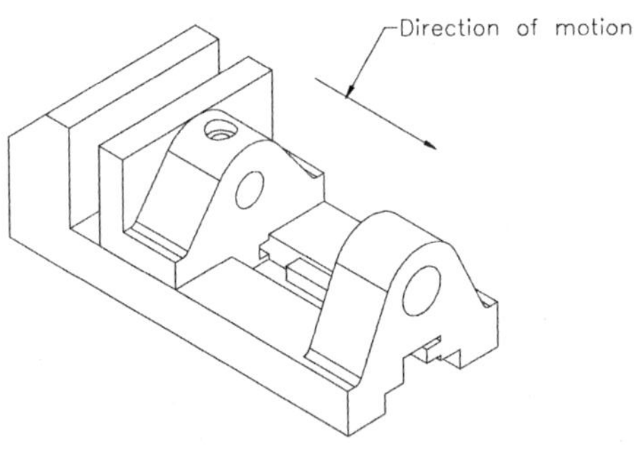

Figure 11-62 *Direction in which the Vice Jaw will be moved*

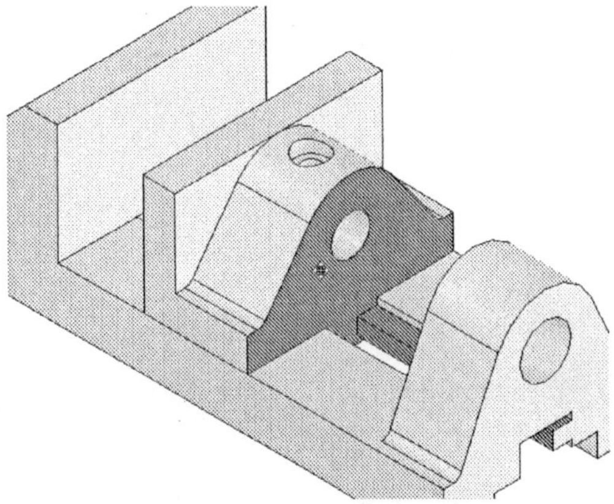

Figure 11-63 *Faces of the Vice Jaw and the Vice Body highlighted in green*

This indicates that the Vice Jaw collides with the Vice Body. Leave the assembly at this location.

4. Choose the **OK** button from the **Move Component PropertyManager**.

   Since the collision is detected in the assembly, you need to modify the design of one of the components. In this case, you will modify the dimensions of the extruded boss feature.

5. Double-click the newly created extrusion feature of the Vice Body. The dimensions of the newly created feature will be displayed.

6. Double-click the dimension that has a of value **6**; the **Modify** dialog box is displayed; set the value of the dimension to **4** and press the ENTER key.

7. Press CTRL+B from the keyboard to rebuild the entire assembly.

8. Choose the **Interference Detection** button from the **Assemblies CommandManager**. The **Interference Detection PropertyManager** is displayed. Choose the **OK** button from the **Interference Detection PropertyManager**.

9. You will observe that **0 Interference** is displayed in the **Interference Results** selection area in the **Results** rollout of this **PropertyManager**.

10. Choose the **Cancel** button from the **Interference Detection PropertyManager**.

    Next, you need to show all the components of this assembly.

11. Press and hold the CTRL key and select the hidden components from the **FeatureManager Design Tree** and invoke the shortcut menu. Choose the **Show** option from the shortcut menu.

12. Save the assembly document and all the referenced part documents.

SELF-EVALUATION TEST

Answer the following questions and then compare your answers with the answers given at the end of this chapter.

1. You can create subassemblies in the assembly environment of SolidWorks. (T/F)

2. You cannot create a subassembly of the components that are already placed in an assembly document. (T/F)

3. When you move the cursor on a mate in the **FeatureManager Design Tree**, the entities used in the mate are highlighted in red color in the drawing area. (T/F)

4. You cannot edit the assembly mates. (T/F)

5. While in the part editing mode in the assembly document, you can use the **Move/Size Features** tool to edit the features dynamically using editing handles. (T/F)

6. The component patterns created individually without the use of any existing pattern feature are known as __________ patterns.

7. The component patterns created using an existing pattern feature are known as __________ patterns.

8. In ________ component, the feature information is available in the part document and only the graphical representation of the component is displayed in the assembly document.

9. After selecting the component, choose the ________ option from the shortcut menu to change the transparency condition of the selected component.

10. To create the explode line sketch, choose the ________ button from the **Assemblies CommandManager**.

REVIEW QUESTIONS

Answer the following questions.

1. Which option is used to open a component separately in the part document?

   (a) **Modify** (b) **Edit**
   (c) **Open Part** (d) None of these

2. Which option is used to define whether a component collides with another component of the assembly or not?

   (a) **Collision Detection** (b) **Interference Detection**
   (c) **Mass Properties** (d) None of these

3. Which check box is selected in the **Open** dialog box to open an assembly with lightweight parts?

   (a) **Lightweight** (b) **Open Lightweight**
   (c) **Lightweight parts** (d) **Lightweight assembly**

4. Which button available in the **Assembly CommandManager** is used to suppress a component?

   (a) **Change Suppression State** (b) **Suppress**
   (c) **Hide/Show Component** (d) **Move Component**

5. Which button is selected from the **Assemblies CommandManager** to create an exploded view?

   (a) **Exploded View** (b) **Assembly Exploder**
   (c) **Mate** (d) None of these

6. The ________ radio button is used to create a local linear pattern.

7. To show the hidden component, select the icon of the component from the **FeatureManager Design Tree** and invoke the shortcut menu. Choose the ________ option from the shortcut menu.

8. The __________ option is used to pattern the instances of the components using an existing pattern feature.

9. The explode state of the assembly is created using the __________ dialog box.

10. The __________ check box is selected to stop the motion of the assembly when one of the component collides with another component when the assembly is in motion.

EXERCISE

Exercise 1

Create the assembly as shown in Figure 11-64. Make sure that when you create this assembly, the back plate should be fixed and the entire assembly can move in the Y direction with respect to the back plate. Keep the rotational degree of freedom of the screw rod free so that it can also rotate on its axis. After creating the assembly you need to explode the assembly and create the explode line sketch. The exploded view of the assembly with explode line sketch is shown in Figure 11-65. The dimensions of the model are given in Figures 11-66 thought 11-70. **(Expected time: 4 hrs)**

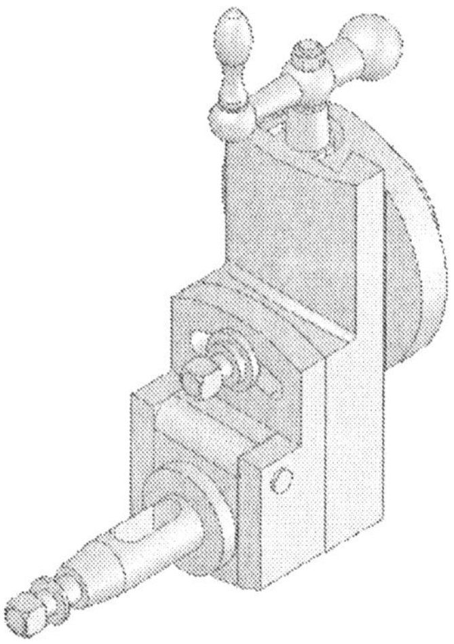

Figure 11-64 *Shaper tool holder assembly*

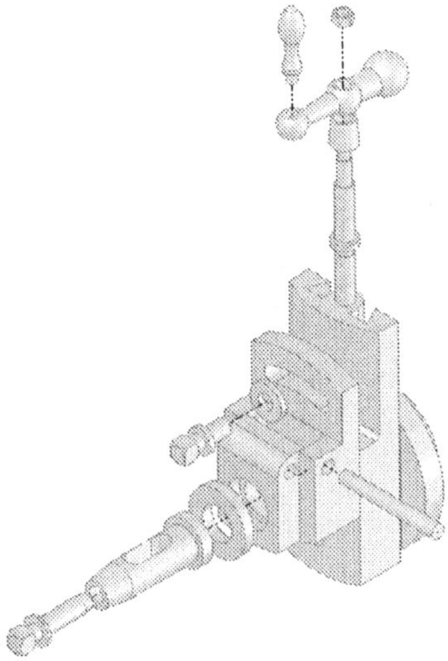

Figure 11-65 *Exploded view of the Shaper tool holder assembly with explode lines*

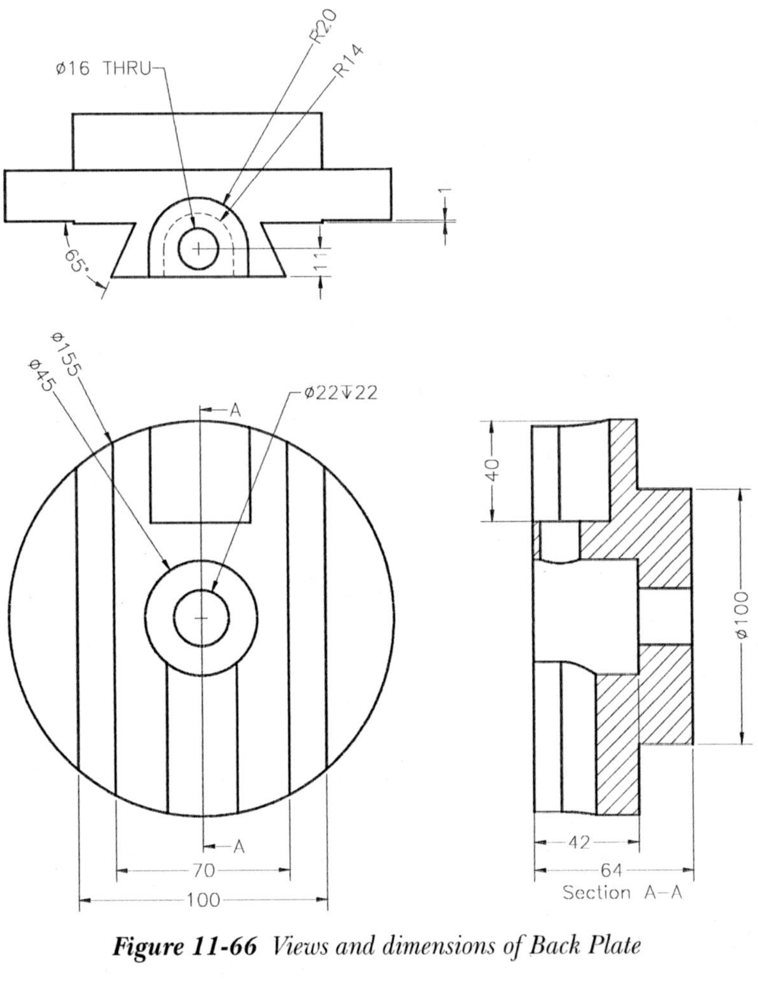

Figure 11-66 *Views and dimensions of Back Plate*

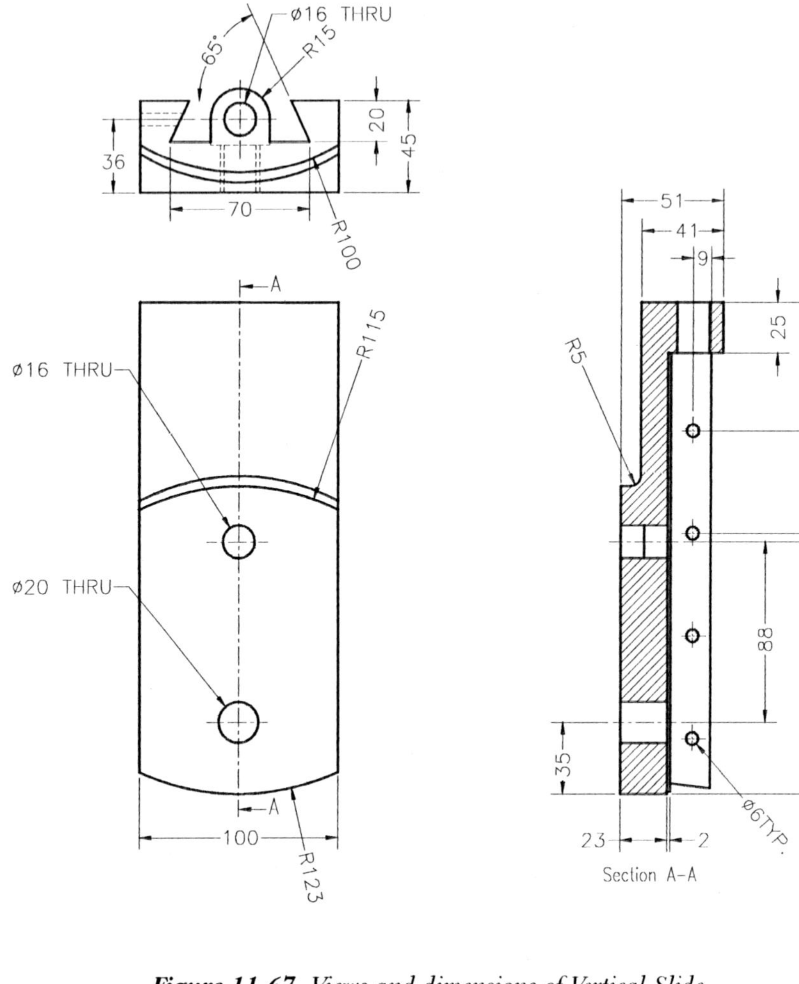

Figure 11-67 Views and dimensions of Vertical Slide

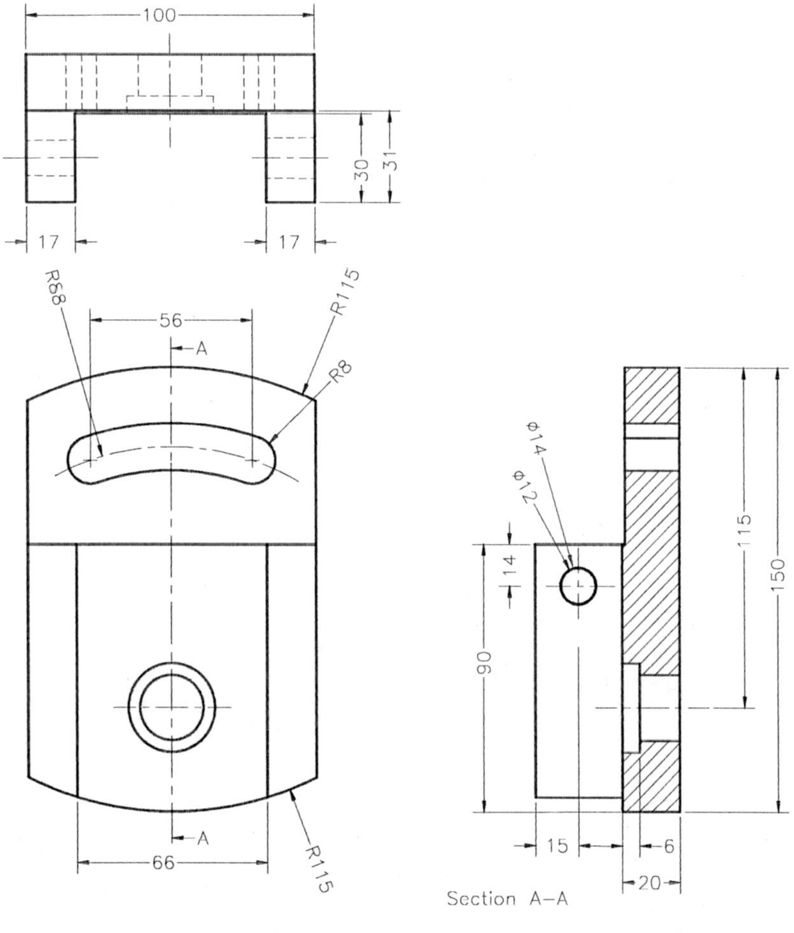

Figure 11-68 *Views and dimensions of Swivel Plate*

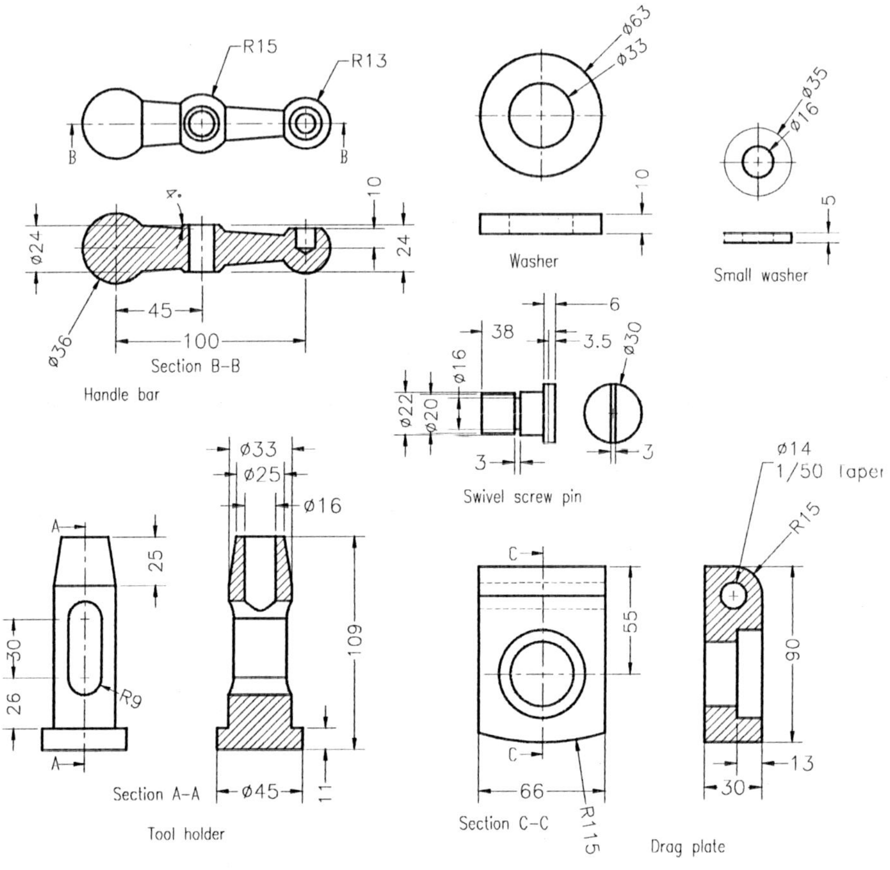

Figure 11-69 Views and dimensions of the components

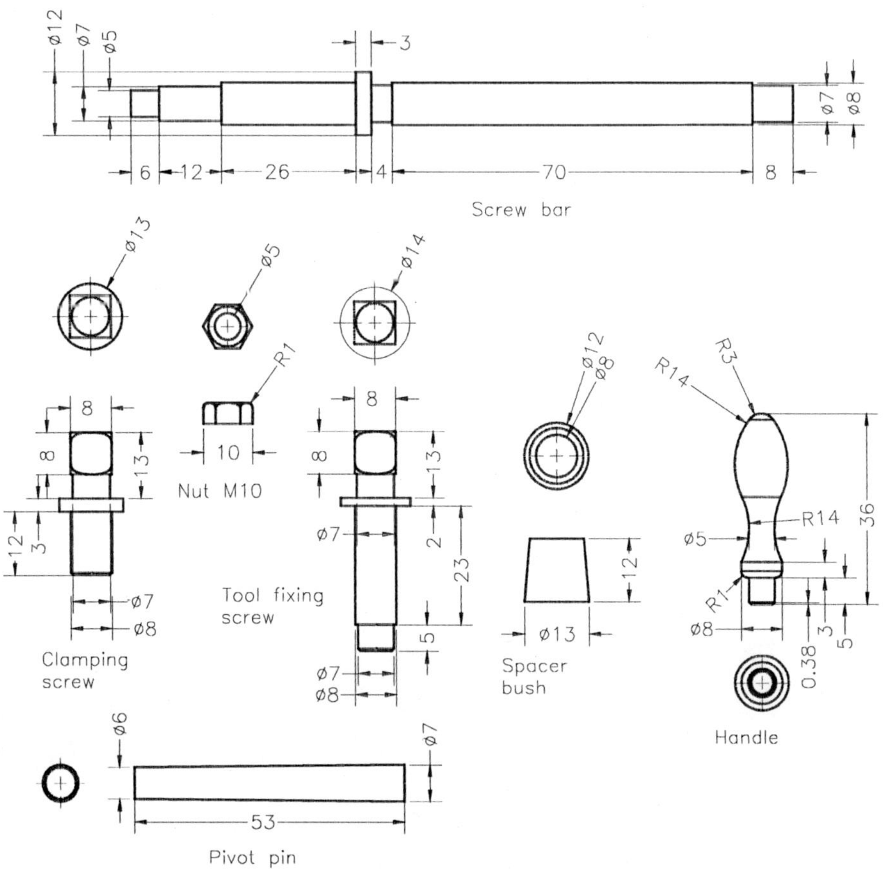

Figure 11-70 Views and dimensions of the components

Answers to Self-Evaluation Test

1. T, **2.** F, **3.** T, **4.** F, **5.** F, **6. Local, 7. Derived, 8. Lightweight, 9. Component Properties, 10. Explode Line Sketch**

Chapter 12

Working With Drawing Views-I

Learning Objectives

After completing this chapter you will be able to:

- *Generate standard three views.*
- *Generate Model Views.*
- *Generate Relative Views.*
- *Generate Predefined Views.*
- *Generate Projected Views.*
- *Generate Section Views.*
- *Generate Aligned Section Views.*
- *Generate Broken-out Section Views.*
- *Generate Auxiliary Views.*
- *Generate Detail Views.*
- *Generate Crop Views.*
- *Generate Broken Views.*
- *Generate Alternate Position Views.*
- *Generate the View of an Assembly in Exploded State.*
- *Work with Interactive Drafting.*
- *Edit the Drawing Views.*
- *Change the Scale of the Drawing Views.*
- *Delete Drawing Views*
- *Modify the Hatch Pattern of the Section Views.*

THE DRAWING MODE

After you have created the solid models of the parts, or assemblies, you need to generate two-dimensional (2D) drawing views. The 2D drawing views are the lifeline of all the manufacturing systems because at the shop floor or machine floor, the machinist mostly needs the 2D drawing for manufacturing. SolidWorks has provided a specialized environment, known as the **Drawing** mode, that has all the tools that are required to generate and modify the drawing views, and add dimensions and annotations to them. In other words, you can get the final shop floor drawing using this mode of SolidWorks. You can also sketch the 2D drawings in the **Drawing** mode of SolidWorks using the sketching tools provided in this mode.

In other words, there are two types of drafting methods available in SolidWorks: Generative drafting and Interactive drafting. Generative drafting is a technique of generating the drawing views using a solid model or an assembly. Interactive drafting is a technique in which you use the sketching tools to sketch a drawing view in the **Drawing** mode. In this chapter, you will learn about generating the drawing views of parts or assemblies.

One of the major advantage of working in SolidWorks is that this software is bidirectionally associative in nature. This property ensures that if the modifications are made in a model in the **Part** mode, the same modification will be reflected in the **Assembly** mode and the **Drawing** mode, and vice versa.

STARTING A DRAWING DOCUMENT

To generate the drawing views, you need to start a new drawing document. In SolidWorks 2004, there are two methods of starting a drawing document. You can use the **New SolidWorks Document** dialog box or use the option available in the part of assembly document to start a drawing document. Both these methods are discussed next.

Starting a New Drawing Document Using the New SolidWorks Document Dialog box

To start a new drawing document for generating the drawing views, invoke the **New SolidWorks Document** dialog box. Choose the **Drawing** button, as shown in Figure 12-1, and choose the **OK** button.

A new drawing document is started and the **Model View PropertyManager** is invoked automatically. The appearance of this **PropertyManager** will depend on whether any part or assembly document was opened or not when you started the new drawing document. Figure 12-2 shows the initial screen of the drawing document with the **Model View PropertyManager**.

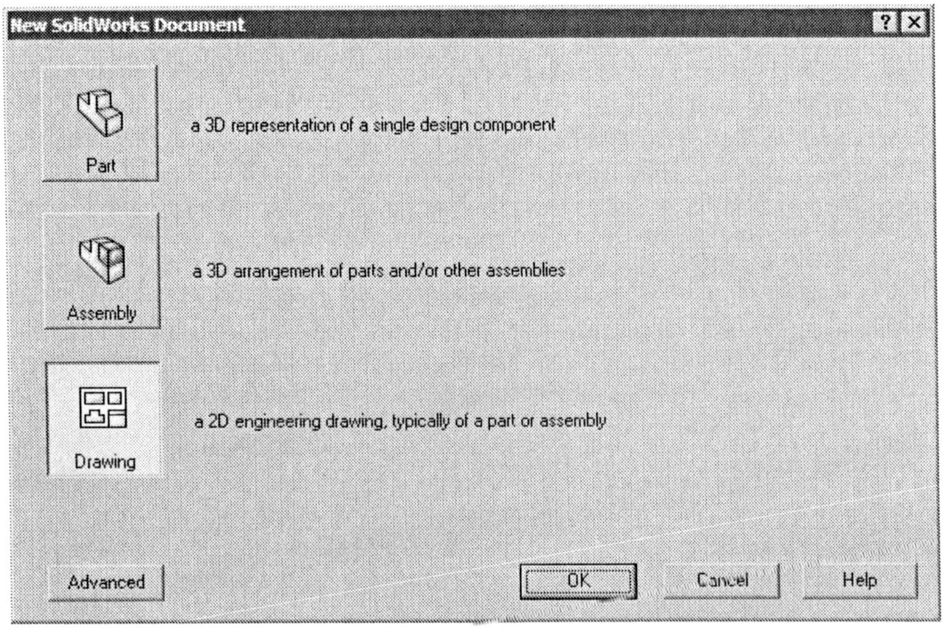

*Figure 12-1 The **New SolidWorks Document** dialog box*

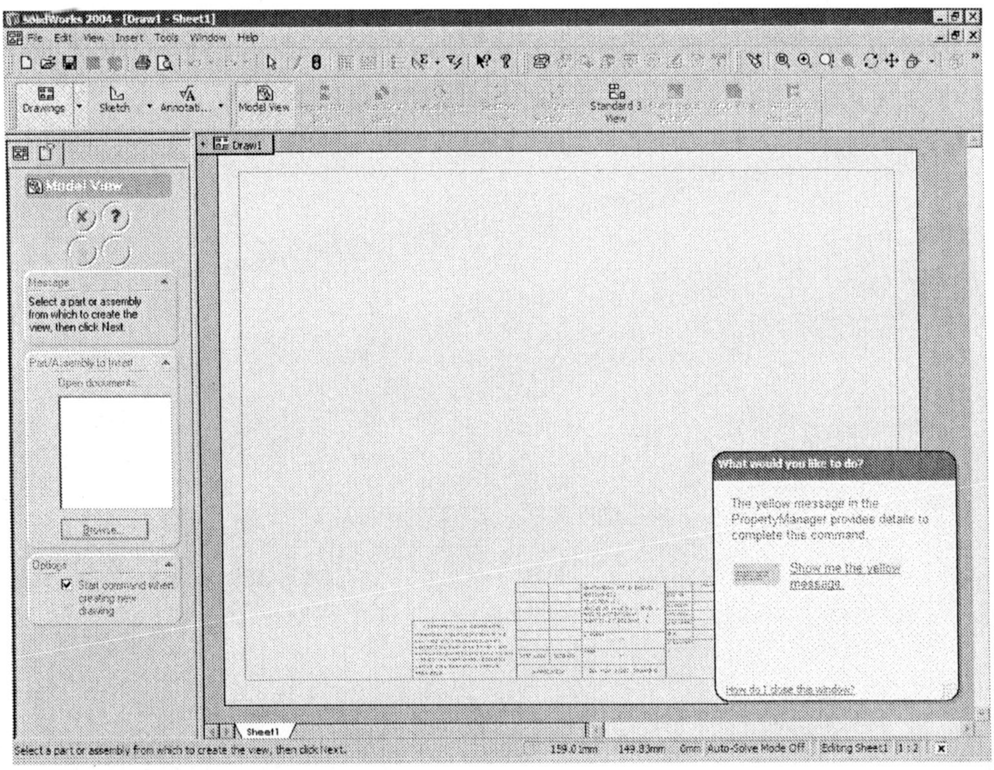

*Figure 12-2 The initial screen of the drawing document with the **Model View PropertyManager***

Starting a New Drawing Document From Within the Part/Assembly Document

This is the second method of starting a new drawing document. This method is recommended when the part or the assembly document for which you want to generate the drawing views is opened. In this case, choose the **Make Drawing from Part/Assembly** button from the **Standard** toolbar of the part or the assembly document. If the **SolidWorks** warning box is displayed, choose the **OK** button. A new drawing document will be started and the **Sheet Format/Size** dialog box is displayed. Note that if you are using the advanced version of the **New SolidWorks Document** dialog box for starting new files, it will be automatically displayed when you choose the **Make Drawing from Part/Assembly** button. Select the available drawing template file from one of the tabs of this dialog box. A new drawing document will be opened. The **Sheet Format/Size** dialog box will be displayed only if you selected the drawing template from the **Templates** tab. Figure 12-3 shows the new drawing document with the **Sheet Format/Size** dialog box.

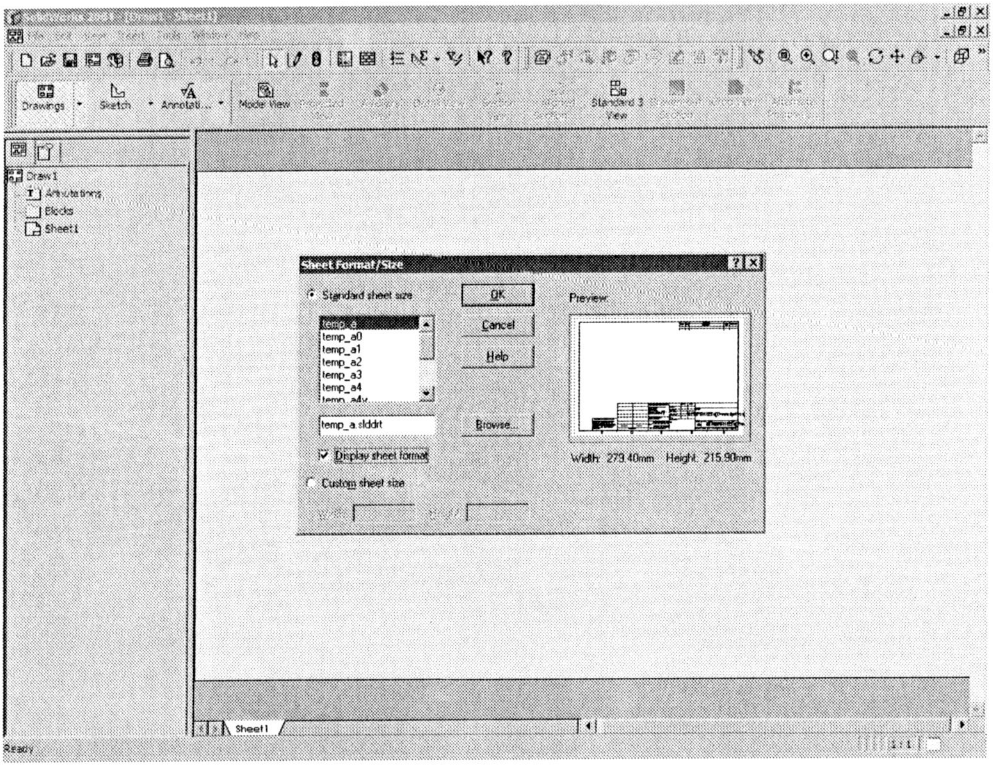

Figure 12-3 *Starting a new drawing document from within the part/assembly document*

You can select the format and the size of the required sheet from the **Sheet Format/Size** dialog box. When you choose **OK** from this dialog box, the **Model View PropertyManager** is displayed and you can generate the drawing views of the selected part or assembly.

Tip. *If you choose the* ***Cancel*** *button from the* ***Sheet Format/Size*** *dialog box, a blank custom sheet of size 431.80 mm x 279.40 mm will be inserted in the drawing document.*

TYPE OF VIEWS

In SolidWorks, you can generate nine type of views. Generally, you first need to generate a standard view such as the top view or the front view and then use this view to derive the remaining views. After generating a standard view, you can generate or derive the following views from the standard view.

Model View

The model views are generally used to create the base view in the drawing sheet. You can generate orthogonal views such as the front, top, left, and so on as the model view. You can also generate isometric, trimetric, or dimetric views as the model view.

Projected View

The projected view is generated by taking an existing view as the parent view. This view is generated by projecting the lines normal to the parent view. The resultant view will be an orthographic view.

Section View

A section view is generated by chopping a part of an existing view using a plane and then viewing the parent view from a direction normal to the section plane. In SolidWorks, the section plane is defined using one or more sketched line segments.

Aligned Section View

An aligned section view is used to section the features that are created at a certain angle to the main section planes. Align sections straighten these features by revolving them about an axis that is normal to the view plane. Remember that the axis about which the feature is straightened should lie on the cutting planes.

Auxiliary View

An auxiliary view is generated by projecting the lines normal to a specified edge of an existing view.

Detail View

A detail view is used to display the details of a portion of an existing view. You can select the portion whose detailing has to be shown in the parent view. The portion that you have selected will be magnified and will be placed as a separate view. You can control the magnification of the detail view.

Broken View

A broken view is used to display a component by removing a portion of it from between, keeping the ends of the drawing view intact. This type of view is used to display the components whose length to width ratio is very high. This means that either the length is very large as compared to the width or the width is very large as compared to the length. The broken view will break the view along the horizontal or vertical direction such that the drawing view fits the area you require.

Broken-out Section View

A broken-out section view is used to remove a part of the existing view and display the area of the model or the assembly that lies behind the removed portion. This type of view is generated using a closed sketch that is associated with the parent view.

Crop View

A crop view is used to crop an existing view enclosed in a closed sketch associated to that view. The portion of the view that lies inside the associated sketch is retained and the remaining portion is removed.

Alternate Position View

The alternate position view is used to create a view in which you can show the maximum and minimum range of motion of the assembly. The main position is displayed in the drawing view in continuous lines and the alternate position of the assembly is shown in the same view in dashed lines (phantom lines).

GENERATING STANDARD DRAWING VIEWS

A standard view is generally the first view that you generate in the current drawing sheet. There are a number of methods of generating the standard drawing views. All these methods are discussed next.

Generating Model Views

Command Manager:	Drawing > Model View
Menu:	Insert > Drawing View > Model
Toolbar:	Drawing > Model View

As mentioned earlier, the model views can be used to generate the base view in the drawing sheet. Whenever you start a new drawing document, the **Model View PropertyManager** is automatically invoked.

If you started the new drawing document from within the part or the assembly document, the part or the assembly is automatically selected and you can place the view. However, if you started the new drawing document using the **New SolidWorks Document** dialog box, a message will be displayed in the **Model View PropertyManager** and you will be prompted to select a part or assembly to generate the drawing view. If there are any part or assembly document opened, it will be displayed in the selection area of the **Part/Assembly to Insert** rollout. You can preview the part or the assembly document by opening the **Thumbnail Preview** rollout, as shown in Figure 12-4.

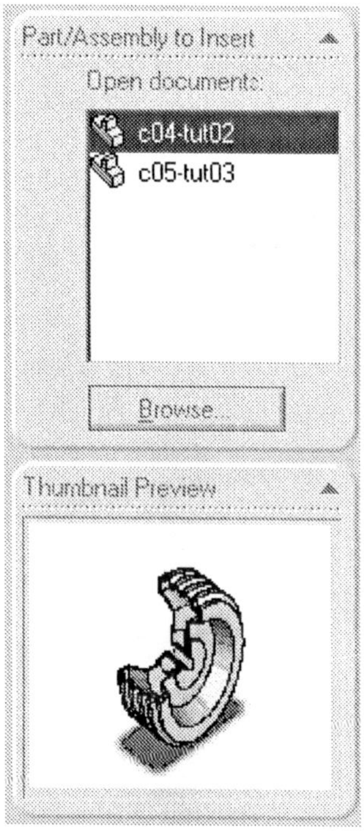

Figure 12-4 *Selecting the document to generate the model views*

You can also choose the **Browse** button and use the **Open** dialog box to select the document. After selecting the required part or assembly document, double-click on it in the **Part/Assembly to Insert** rollout. You can also choose the **Next** button available above the **Message** rollout in the **Model View PropertyManager**.

The **Model View PropertyManager** is automatically modified and now provides the options related to generating the standard views, see Figure 12-5. The various options available in this **PropertyManager** are discussed next.

Orientation Rollout

The list box available in the **Orientation** rollout is used to specify the orientation of the view. Selecting the **Preview** check box allows you to preview the drawing view before it is placed.

Note

Even after placing the model view, you can change its orientation. To do this, place the view and make sure that it is still selected. This can be confirmed by the box that appears around the model view. If the box is displayed in green with grip point, the view is selected. Now, double-click on the required view orientation from the ***Orientation*** *rolllout in the* ***Model View PropertyManager****. The orientation of the view is automatically modified.*

Options Rollout

The **Auto-start projected view** check box available in this rollout is used to automatically invoke the **Projected View** tool to generate the projected view immediately after you place

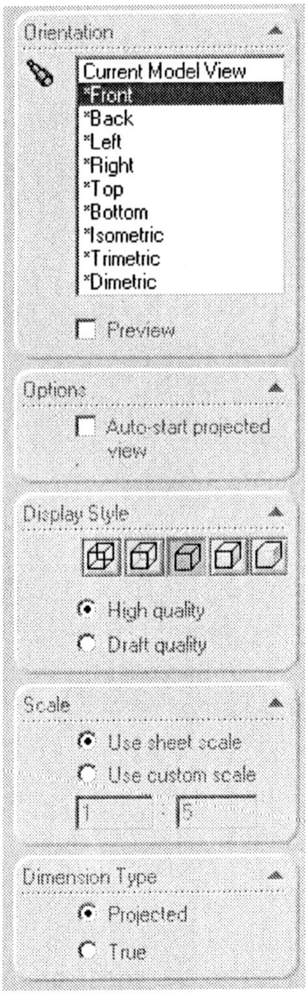

***Figure 12-5** Partial view of the **Model View PropertyManager** after selecting the document to generate the drawing views*

the model view. The model view that you generated will be automatically taken as the parent view to generate the projected view.

Display Style Rollout

The options available in this rollout are used to specify the display style for the model view. These display styles are similar to those available in the **View** toolbar to display the part or assemblies in the part or assembly documents.

Scale Rollout

Whenever you select a template to generate the drawing views, a default scale that is used to generate the drawing views is automatically defined in it. This is the reason the **Use sheet scale** radio button is selected in the **Scale** rollout. If you want to define a custom scale for the model view, select the **Use custom scale** radio button and specify the scale factor in the edit boxes available below this radio button.

Dimension Type Rollout

The radio buttons available in this rollout are used to specify whether the model view will have true dimensions or projected dimensions. The true dimensions are the exact model dimensions that were specified while creating the model. The projected dimensions are reduced dimensions that are used in case of isometric, dimetric, or trimetric view. Generally, the value of the projected dimension is about 81.6% of the value of true dimension.

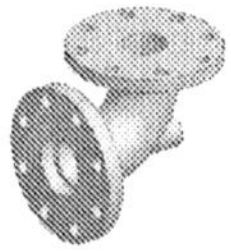

Tip. *The drawing views will be generated depending on the default projection type of the current sheet. If the sheet is configured for first angle projection, the drawing views will be generated according to that. If the drawing sheet is configured for third angle projection, the views will be generated accordingly.*

To change the projection type of the current sheet, right-click on ***Sheet Format1*** *in the* ***FeatureManager Design Tree*** *and choose* ***Properties*** *from the shortcut menu to display the* ***Sheet Properties*** *dialog box. In this dialog box, set the required projection type using the* ***Type of projection*** *area.*

Generating the Three Standard Views

CommandManager:	Drawing > Standard 3 View
Menu:	Insert > Drawing View > Standard 3 View
Toolbar:	Drawing > Standard 3 View

Using the **Standard 3 View** option, you can generate three default orthographic views of the specified part or assembly. To create the three standard views, choose the **Standard 3 View** button from the **Drawing CommandManager**. The **Standard 3 View PropertyManager** will be displayed. If there is any part or assembly document opened in the current session of SolidWorks, it will be displayed in the list box available in the **Part/Assembly to Insert** rollout, as shown in Figure 12-6.

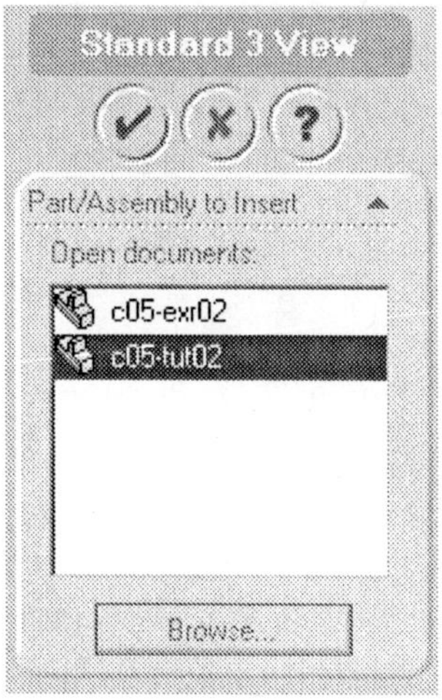

Figure 12-6 Standard 3 View PropertyManager

You can select the document from this list box. If there are no documents opened, you can use the **Browse** button to select the part or assembly document. As soon as you select a document, three standard views are generated based on the default scale of the current sheet. Figure 12-7 shows the three standard views of a model generated in third angle projection using the **Standard 3 View** tool.

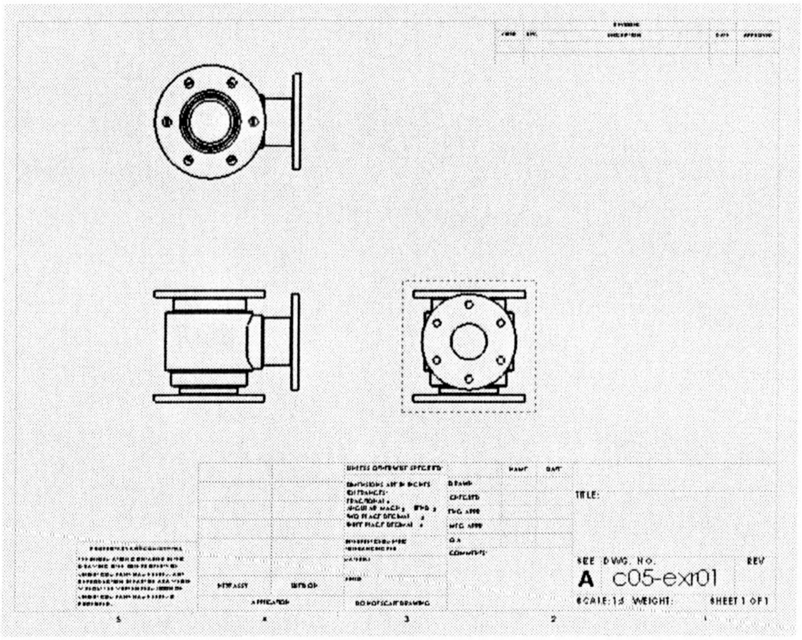

Figure 12-7 *Three standard views generated using the* ***Standard 3 View*** *tool*

Tip. *If the generated view overlaps the title block, you need to move it. To move a view, move the cursor over it. The bounding box of the view is displayed in dashed lines of red color. At this point, left-click to select the view. Now, move the cursor to the boundary of the selected view; the cursor is replaced by the move cursor. Press and hold down the left mouse button and drag the cursor to move the view. Remember that if you move the front view, all the views are moved.*

You need to make sure that the dynamic view activation option is turned on in the ***System Options*** *dialog box. To confirm it, choose* ***Tools > Options*** *from the menu bar to display the* ***System Options - General*** *dialog box. Select the* ***Drawing*** *option to display the settings related to the* ***Drawing*** *mode. Select the* ***Dynamic drawing view activation*** *check box, if it is not selected.*

Note

You will observe that the name of the part document whose drawing views are generated is displayed in the ***DWG NO.*** *text box of the title block. The size of the sheet is also displayed at the lower right corner of the title block. If these parameters are not displayed, try changing the sheet format. You will also learn to generate all the information such as draw by, check by, and so on in advance chapters in part 2 of this book.*

Generating Standard Views Using the Relative View Tool

CommandManager:	Drawing > Relative View *(Customize to Add)*
Menu:	Insert > Drawing View > Relative To Model
Toolbar:	Drawing > Relative View *(Customize to Add)*

The **Relative View** tool is used to generate an orthographic view. The orientation of the view is defined by selecting two reference planes or the planar faces of the model. This option is very useful if you need the orientation of the parent view other than the default orientations.

To create a relative view, open the part or the assembly document and tile them vertically or horizontally with the drawing document. Invoke the **Relative View** tool to display the **Relative View PropertyManager**. You are prompted to select a planar face of the model. Click in the window of part or assembly document. The **Relative View PropertyManager** is displayed in the part or the assembly document, as shown in Figure 12-8.

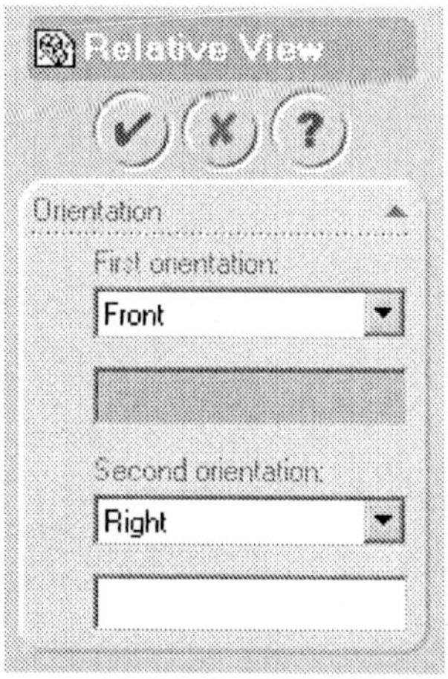

Figure 12-8 Relative View PropertyManager

Select the orientation for the first plane or planar face from the drop-down list available in the **First orientation** area. Then select the plane or planar face of the model to be oriented in that direction. For example, you can select the **Top** option from this drop-down list and then select the top planar face of the model.

As soon as you specify the first reference, the selection area in the **Second orientation** area is highlighted. Select the orientation for the second reference from the drop-down list and then select a plane or planar face. After selecting the references, choose the **OK** button from the **Relative View PropertyManager**. You will return to the drawing document and the resultant view will be attached to the cursor. Place the view at the required location. Figure 12-9 shows the faces of the model selected to generate a standard view and Figure 12-10 shows the resultant view.

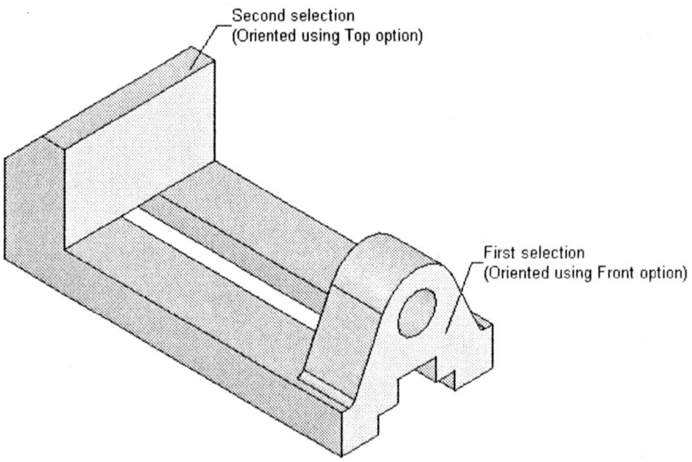

Figure 12-9 *Faces to be selected*

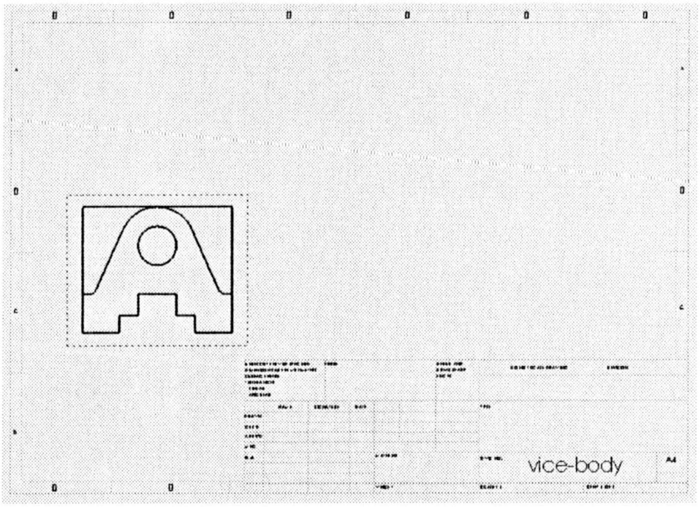

Figure 12-10 *Resultant view*

Generating Standard Views Using the Predefined View Tool

Toolbar:	Drawing > Predefined View	*(Customize to Add)*
Menu:	Insert > Drawing View > Predefined	
Toolbar:	Drawing > Predefined View	*(Customize to Add)*

The **Predefined View** tool is used to create empty views with the predefined orientation. After creating all the predefined views, populate them by dragging the part from the other window by holding the name of the part document in the **FeatureManager Design Tree**. All the predefined views will be populated. This option is basically used to add the empty views in the drawing document and then save it as a drawing template. Next time when you open a new drawing document using that template, you just need to drag and drop the part from other window, all the predefined views will be populated. To create predefined views, choose the **Predefined View** button from the **Drawing CommandManager**. An empty view is attached to the cursor. Specify a point in the drawing document to place the predefined view. The view will be placed in the drawing document and the **Predefined View PropertyManager** is displayed, as shown in Figure 12-11.

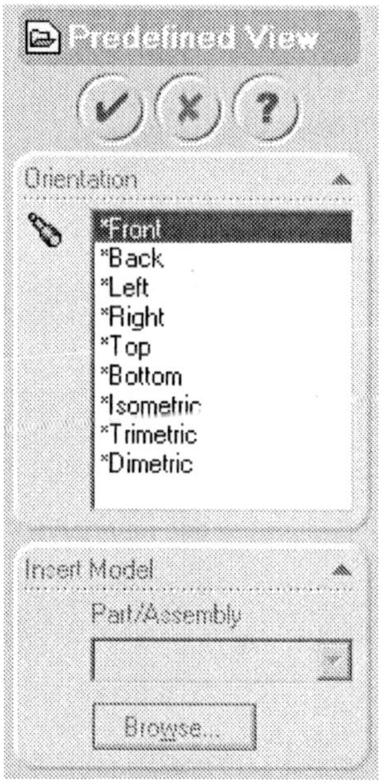

Figure 12-11 *The* ***Predefined View PropertyManager***

Select the view orientation from the **View Orientation** area of the **View Orientation** rollout and choose the **OK** button from the **Predefined View PropertyManager**.

To add the next predefined view, you need to define the alignment option while placing the view. To create additional predefined views, invoke the **Predefined View** tool and place the view in the drawing sheet. The bounding box of the view is displayed. Right-click inside the bounding box and choose **Alignment > Align Horizontal by Center/Align Vertical by Center**. Now, select the previous predefined view to align the corresponding view. Similarly, add other predefined views using this tool.

After creating all the predefined views, open a part or an assembly document and tile the documents windows horizontally or vertically. Now, drag the component from the part document and drop in the drawing document; all the predefined views will be populated. You can also populate individual predefined views by selecting them and choosing the **Browse** button from the **Insert Model** rollout of the **Predefined View PropertyManager**.

Figure 12-12 shows the selected predefined views with the orientation in which the views are created. Figure 12-13 shows the drawing document after populating the drawing views.

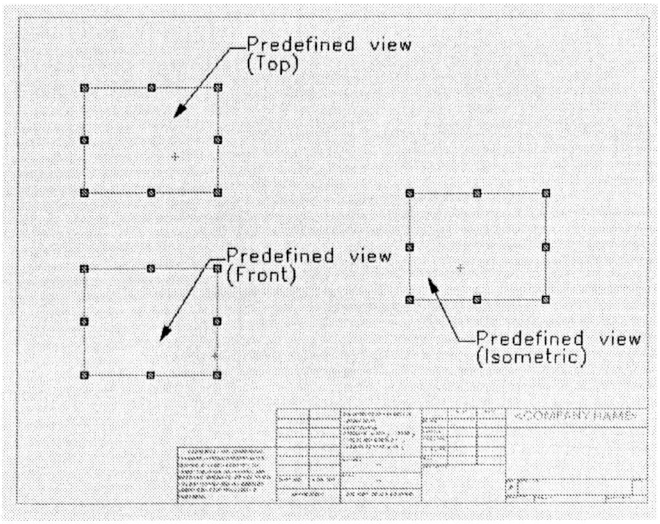

Figure 12-12 *Various predefined views*

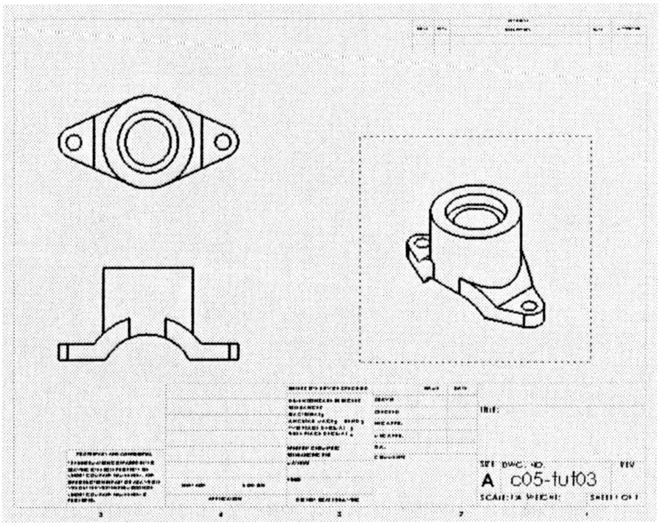

Figure 12-13 *Views created after populating the predefined views*

Note

When you generate the views in the ***Drawing*** *mode of SolidWorks, the views are scaled automatically depending on the size of the sheet.*

In case of predefined views, if the drawing contains more than one predefined view, the views will be scaled automatically.

If only one predefined view is placed in the drawing document, the view will be scaled with respect to the ***Custom Scale*** *value, if specified. Otherwise, the view will be scaled with the default scale factor of the drawing sheet. You can change the view scale using the* ***Sheet Properties*** *dialog box. You will learn more about scaling the views in the next chapter.*

GENERATING THE DERIVED VIEWS

All the views that are generated from a view already placed in the drawing document are known as derived views. The various types of derived views that are generated from the standard views are as follows.

1. Projected view
2. Section view
3. Aligned Section view
4. Broken-out Section view
5. Auxiliary view
6. Detail view
7. Crop view
8. Broken view
9. Alternate Position view

The methods of generating the various derived drawing views are discussed next.

Generating Projected Views

CommandManager:	Drawing > Projected View
Menu:	Insert > Drawing View > Projected
Toolbar:	Drawing > Projected View

As mentioned earlier, the projected views are generated by projecting the normal lines from an existing view. To generate a projected view, choose the **Projected View** button from the **Drawing CommandManager**. The **Projected View PropertyManager** is displayed and you are prompted to select a drawing view from which you need to project the normal lines. The select cursor is replaced by the view cursor. Select the parent view and move the cursor vertically to generate the top view or the bottom view, or move the cursor horizontally to create right or left view. Specify a point on the drawing sheet to place the view. For generating more than one projected views, choose the **Keep Visible** button to pin the **Projected View PropertyManager**. Figure 12-14 shows the front view generated from the top view.

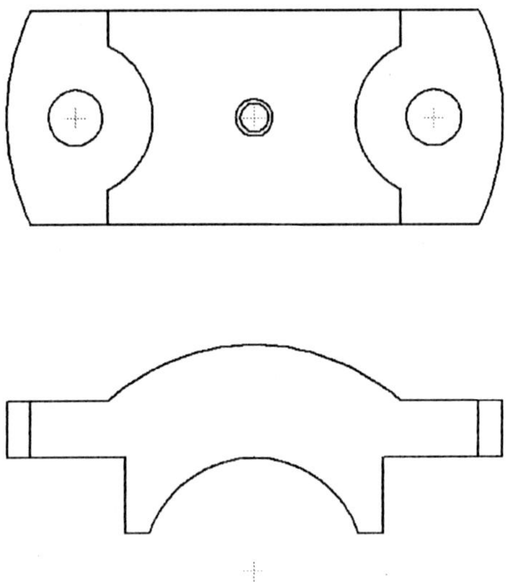

Figure 12-14 Front view generated from the top view

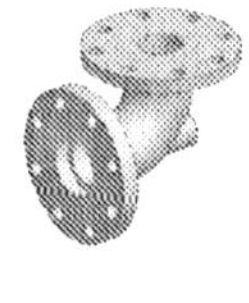

Tip. *When you generate a projected drawing view, the drawing view is aligned to the parent view. To place the projected view that is not in alignment with the parent view, press and hold down the CTRL key before placing the view. Now move the cursor to the desired location and place the view.*

All the standard views and derived views such as projected views, section view, detailed view, and so on are linked to their parent view by a Parent-Child relationship. If you select the child view, the bounding box of the parent view will also be displayed.

Select the child view and invoke the shortcut menu and choose the ***Jump to Parent View*** *option from the shortcut menu. The parent view will be selected automatically.*

Generating Section Views

CommandManager:	Drawing > Section View
Menu:	Insert > Drawing View > Section
Toolbar:	Drawing > Section View

As mentioned earlier, section views are generated by chopping a portion of an existing view using a cutting plane (defined by sketched lines) and then viewing the parent view from a direction normal to the cutting plane.

In SolidWorks, you can use the **Section View** tool to create a full section view, as shown in Figure 12-15, or a half section view, as shown in Figure 12-16. A full section view is defined using a single line segment but a half section view is defined using three line segments. Note that the section plane for a full section view can be defined after invoking the **Section View** tool. But if you need to generate a half section view, you will have to draw the line segments to

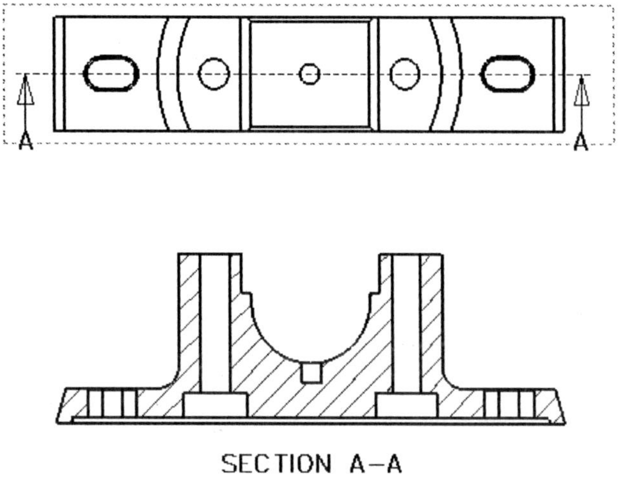

Figure 12-15 *Full section view*

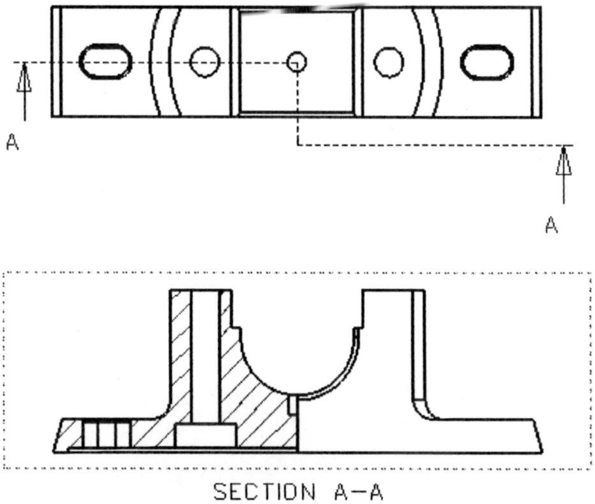

Figure 12-16 *Half section view*

define the section plane before invoking the **Section View** tool. To do this, select the drawing view that you want to use as the parent view and then choose the **Sketch** button from the **CommandManager**. The sketching environment will be invoked. You can use the inferencing lines to draw the lines for the section plane.

To create a full section view, activate the view in which you need to create the section line. The view symbol will be displayed below the cursor and the bounding box of the view is displayed. Now, choose the **Section View** button from the **Drawing CommandManager**. The **Section View PropertyManager** is displayed and you are prompted to sketch a line to continue view creation. Note that if the **SolidWorks** warning box is displayed, which warns you that no drawing view is selected, you need to turn on the option of dynamic view activation. You can zoom using the **Zoom** tool to increase the display area of the activated view. Refer to the Tip on Page 10 to know more about how to turn on this option.

After activating the view, draw a line that will define the section plane. As soon as you specify the endpoint of the section line, section view is defined and is attached to the cursor. Also, the other options in the **Section View PropertyManager** are displayed.

If you want to generate half section view, select all the line segments that define the section plane and then invoke the **Section View** tool. The section view is defined and is attached to the cursor.

Move the cursor and specify a point on the drawing sheet to place the section view. The name and the scale factor of the drawing view is displayed below the section view. The **Section View PropertyManager** is still active, as shown in Figure 12-17.

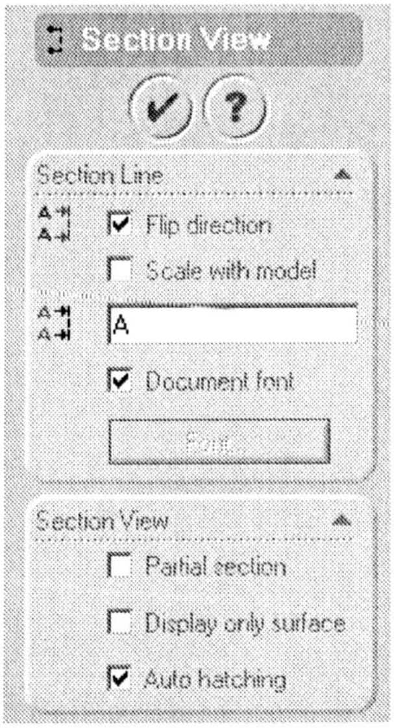

Figure 12-17 *Partial view of the* ***Section View PropertyManager***

You can use the **Flip direction** check box to flip the direction of the section view. The view will be automatically modified in the drawing sheet. The **Scale with model** check box is used to scale the drawing view if the model is scaled in the part document. You will learn more about scaling the model in the next chapter.

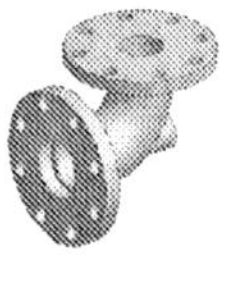

Tip. *When you create a section view and move the cursor to place the section view, you will observe that the view is aligned to the direction of arrows on the section line. If you need to remove this alignment to place the section view, press and hold down the CTRL key and move the view to the desired location. Select a point in the drawing sheet to place the view.*

Note

The default hatch pattern in the section view depends on the material assigned to the model. Also, by default the spacing of the hatch is not what is required. Therefore, you may need to increase the spacing of the hatch pattern. You will learn more about editing the hatch pattern later in this chapter.

There are some other options available in the **Section View PropertyManager** using which you can create partial section view and the surface section view. The options are discussed next

Creating the Partial Section View

If the section line does not cut through the model, the **SolidWorks** information box will be displayed. This dialog box prompts you that the section line does not completely cut through the bounding box of the model in this view. Do you want this to be a partial section cut? For creating the partial section view, choose the **Yes** button from this dialog box. If you choose the **No** button from this dialog box, then the complete section view will be created. Figure 12-18 shows a partial section view generated from the top view.

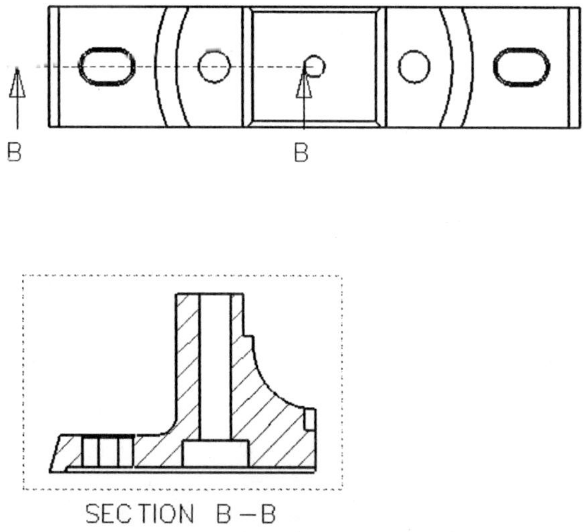

Figure 12-18 *A partial section view*

Creating the Surface Section View

A surface section view is the one in which only the sectioned surface is displayed in the section view. All the other edges or faces are not displayed in the surface section view. To create a surface section view, you first need to create the section view and then select the **Display only surface** check box from the **Section View PropertyManager**. Figure 12-19 shows a surface section view.

Tip. *Sometimes while generating the sectioned views, the view is upside down even if you have set the projection type to third angle. In such cases, you need to flip the direction of the section line from the* ***Section View PropertyManager****.*

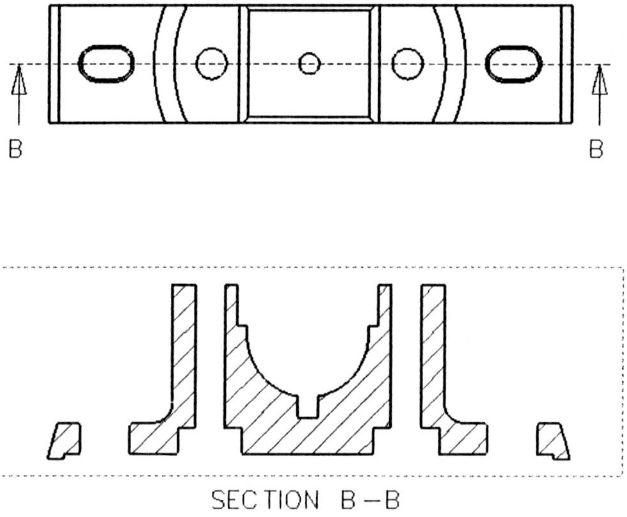

Figure 12-19 *A surface section view*

Generating the Section View of an Assembly

According to the drawing standards, when you create the section view of an assembly, some components such as fasteners, shafts, keys, and so on should not be sectioned. Therefore, when you create the section view of an assembly, the **Section View** dialog box is displayed, as shown in Figure 12-20.

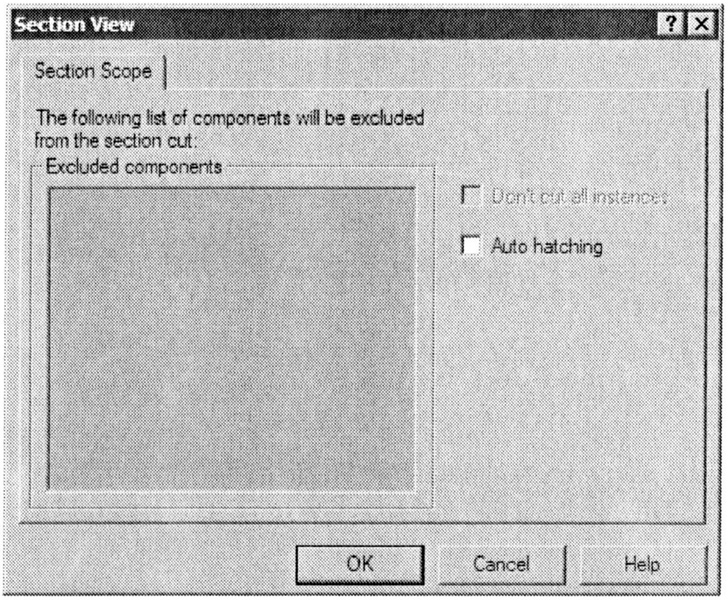

Figure 12-20 *The* ***Section View*** *dialog box*

This dialog box allows you to select the components that will be excluded from the section cut. You can also select the component from the parent view. But if the components are not visible in the parent view, you can invoke the **FeatureManager Design Tree** and expand the parent drawing view. Then expand the assembly tree view to display all the components of the assembly. Now, select the components that you do not want to be sectioned. The name of the selected component is displayed in the **Excluded components** selection area. The **Auto hatching** check box is used to automatically define the hatch patterns. You can even change the hatch pattern. The method of changing the hatch pattern is discussed later.

If you have more than one instance of the component in the assembly, and you need to exclude all the instances of the component from the section view, select the component from the drawing sheet and select the name of the component from the **Exclude components** selection area. Select the **Don't cut all instances** check box from the **Section Scope** dialog box. All the instances of the selected component will be excluded from the section view. Figure 12-21 shows an assembly section view with fasteners excluded from the cut.

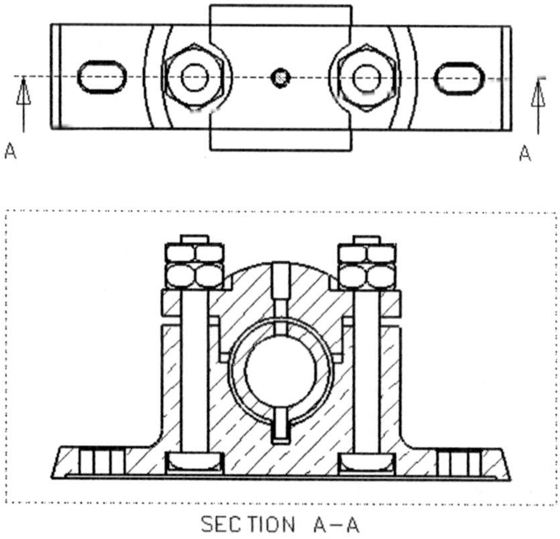

Figure 12-21 *Section view of an assembly with some of the components excluded from the cut*

Tip. *You can add or remove the components that are sectioned by right-clicking the drawing view and choosing **Properties** from the shortcut menu. Now, choose the **Section Scope** tab and add or remove the components.*

Generating Aligned Section Views

CommandManager:	Drawing > Aligned Section View
Menu:	Insert > Drawing View > Aligned Section
Toolbar:	Drawing > Aligned Section View

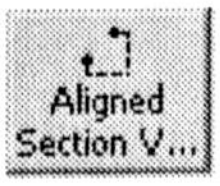

This tool is used to generate a section view of the component in which at least one of the feature is at an angle. In the aligned section view, the sectioned portion revolves about an axis normal to the viewing plane such that it is straightened. For example, refer to Figure 12-22.

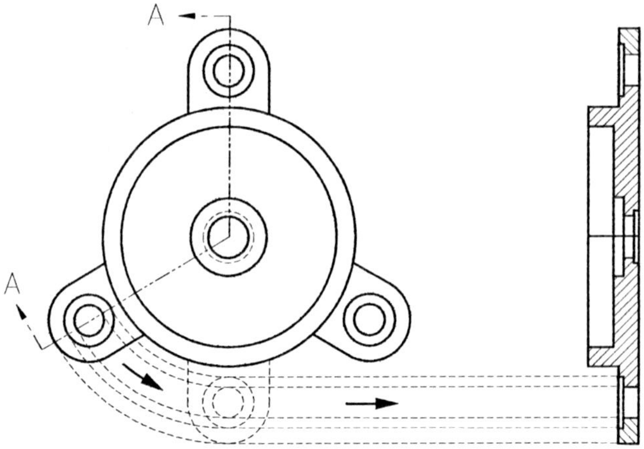

Figure 12-22 *Aligned section view*

This figure explains the concept of an aligned section view of a model. Notice that the inclined feature that is sectioned in this view is straightened. As a result, the section view is longer than the parent view. To create the aligned section view, activate the view. Choose the **Aligned Section View** button from the **Drawing CommandManager**. Draw the sketch that defines the section plane. The aligned section view will be attached to the cursor; place the view at an appropriate location in the drawing sheet. Note that the resultant view will be projected normal to the line that is drawn in the end in the section sketch. Therefore, to get the aligned section view similar to that shown in Figure 12-22, the inclined line in the section sketch should be drawn first and then the vertical line. Figure 12-23 shows the aligned section view in which the vertical line in the section sketch is drawn first. This is the reason the section view is projected normal to the inclined line that is drawn last. On the other hand, Figure 12-24 shows the view in which the inclined line is drawn first.

Note

You can also create a sketch associated to a view. This sketch can be selected as the section plane for generating the section view. To create an associated sketch, activate the view and draw the sketch that defines the section plane using the ***Line*** *tool.*

If you create a sketch to define the section plane for the aligned section view before invoking the ***Aligned Section View*** *tool, the view will be projected normal to the line that you select last. However, if you select the sketch to define the section plane by dragging a window around it, the view will be projected normal to the line that was drawn last.*

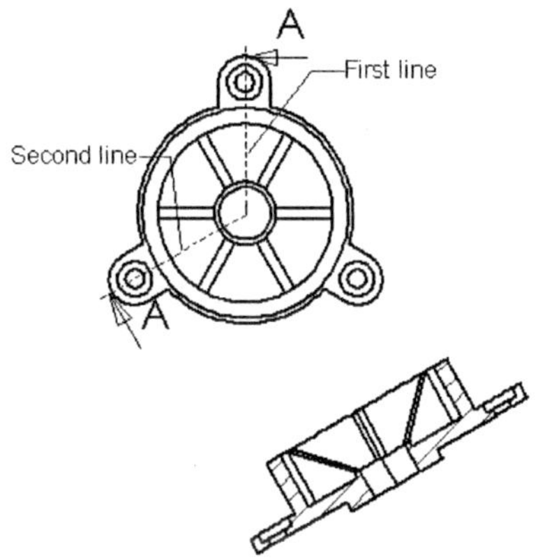

Figure 12-23 Aligned section view

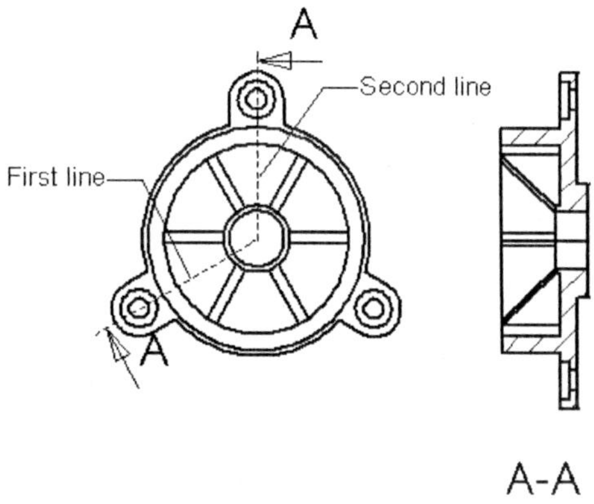

Figure 12-24 Aligned section view

Generating Broken-out Section Views

CommandManager:	Drawing > Broken-out Section
Menu:	Insert > Drawing View > Broken-out Section
Toolbar:	Drawing > Broken-out Section

This tool is used to create a broken-out section view. A broken-out section view is used to remove a part of the existing view and display the area of the model or the assembly behind the removed portion. This type of view is generated using a closed sketch that is associated with the parent view. To create a broken-out section view, activate the view on which you need to create the broken-out section view. Choose the **Broken-out Section** button from the **Drawing CommandManager**. The **Broken-out Section PropertyManager** is displayed and it prompts you to create a closed spline to continue section creation. The cursor will be replaced by the spline cursor. Create a closed sketch using the spline cursor. If you do not want a spline profile, select a closed profile before choosing the **Broken-out Section** button. Figure 12-25 shows an associated sketch created for creating a broken-out section view.

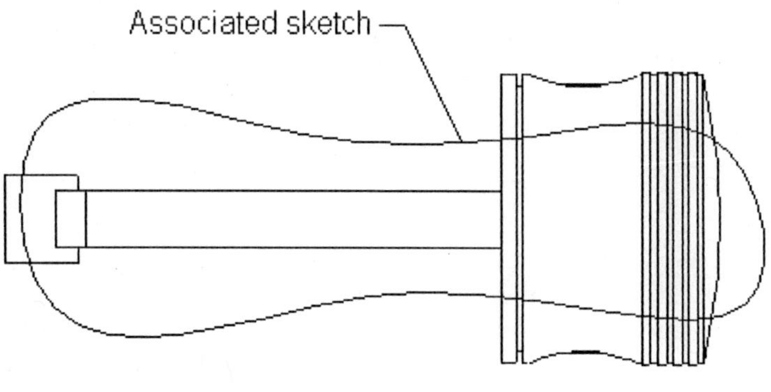

Figure 12-25 Sketch for creating a broken-out section view

When you create a closed sketch, the options are displayed in the **Broken-out Section PropertyManager**, as shown in Figure 12-26, and you are prompted to specify the depth of the broken-out section.

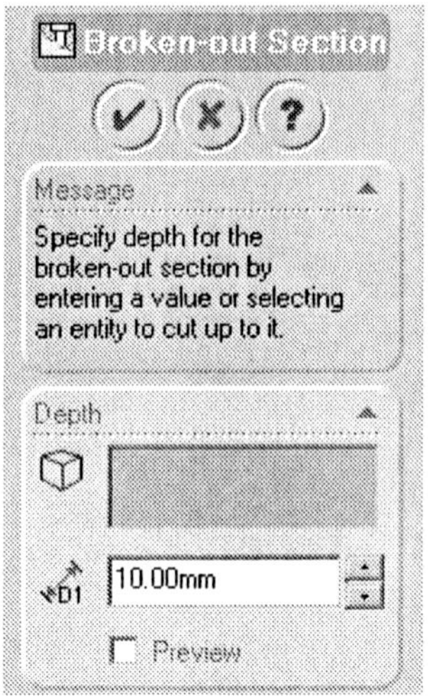

Figure 12-26** The **Broken-out Section PropertyManager

Select the **Preview** check box to preview the broken-out section view. The **Auto hatching** check box is used to define the hatch pattern automatically to the section drawing view of the assembly. Note that if you are creating the broken-out section view of a part, then the **Auto hatching** check box is not available in the **Broken-out Section PropertyManager**. Figure 12-27 shows the preview of the broken-out section view.

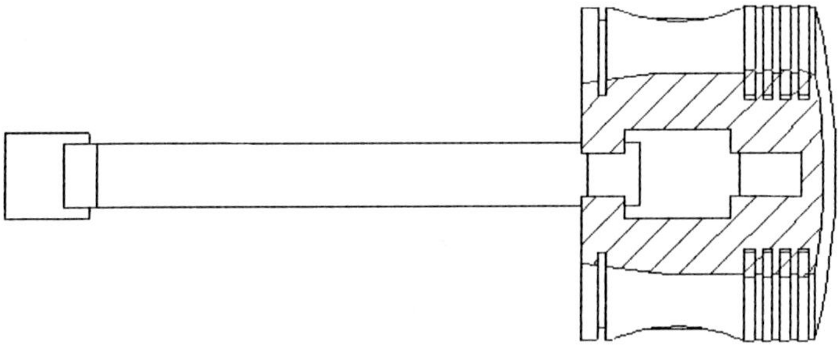

***Figure 12-27** Preview of the broken-out section view*

Set the value of the depth of the broken-out section in the **Depth** spinner. The preview of the section will be modified dynamically in the drawing view. After setting the value of the depth of the broken-out section, choose the **OK** button from the **Broken-out Section PropertyManager**. Figure 12-28 shows a broken-out section view with a different depth value.

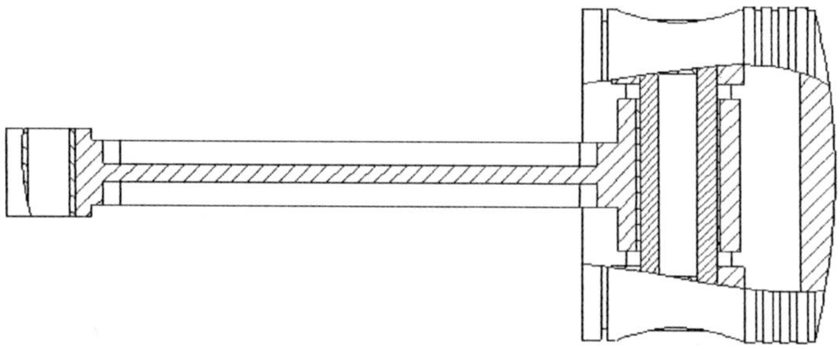

Figure 12-28 Broken-out section view

Generating Auxiliary Views

CommandManager:	Drawing > Auxiliary View
Menu:	Insert > Drawing View > Auxiliary
Toolbar:	Drawing > Auxiliary View

An auxiliary view is a drawing view that is generated by projecting the lines normal to a specified edge of an existing view. SolidWorks also allows you to create a line segment associated to the view that you can use to generate the auxiliary view. But note that the associated line segment needs to be created before invoking this tool.

To create an auxiliary view, choose the **Auxiliary View** button from the **Drawing CommandManager**. The **Auxiliary View PropertyManager** is displayed and you are prompted to select a reference edge to continue. Select the edge or the associated sketch that will be used to generate the auxiliary view. A view will be attached to the cursor and some options are displayed in the **Auxiliary View PropertyManager**, as shown in Figure 12-29. Also, you are prompted to specify the location to place the view.

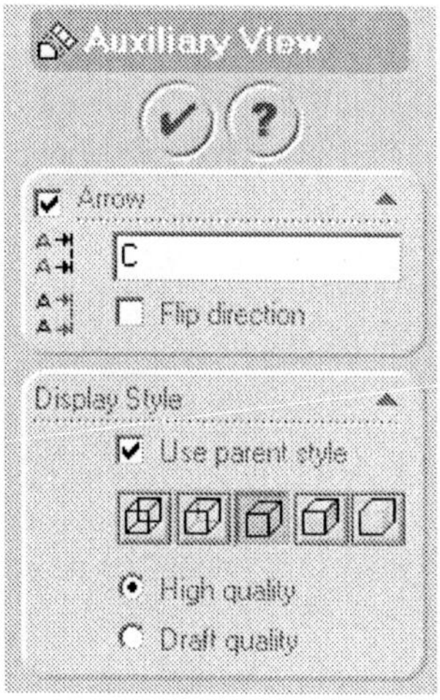

*Figure 12-29 Partial view of the **Auxiliary View PropertyManager***

The check box available in the **Arrow** rollout is used to display the arrow of the viewing direction in the drawing views. The name of the auxiliary view is specified in the **Label** edit box. Using the **Flip Direction** check box you can flip the viewing direction for creating the auxiliary view. Figure 12-30 shows the reference edge to be selected to create the auxiliary view.

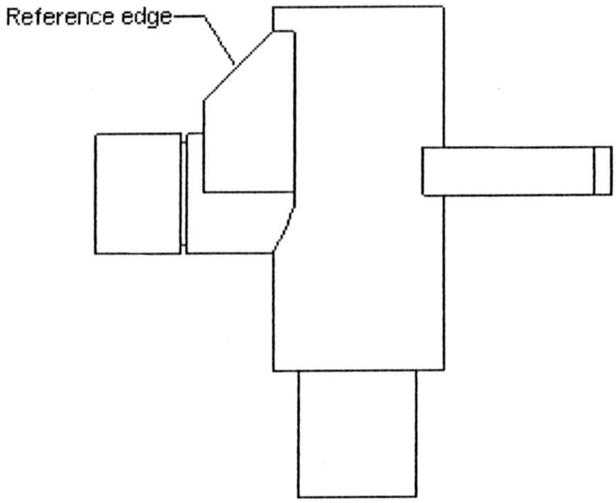

Figure 12-30 *Reference edge to be selected to create the auxiliary view*

Figure 12-31 shows the auxiliary view created with the default viewing direction. Figure 12-32 shows the auxiliary view created with the **Flip Direction** check box selected.

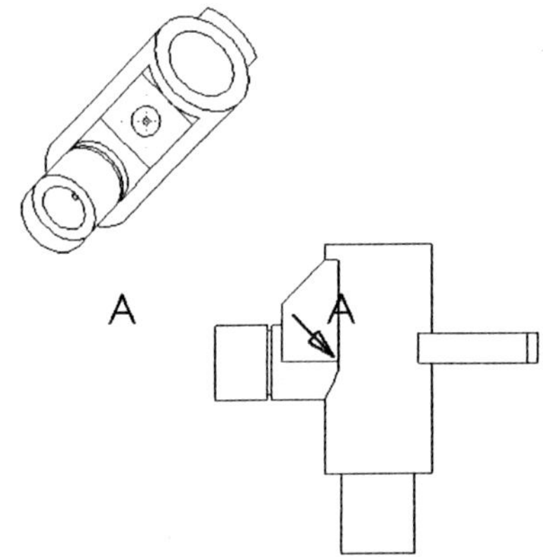

Figure 12-31 *Auxiliary view created with the* ***Flip Direction*** *check box cleared*

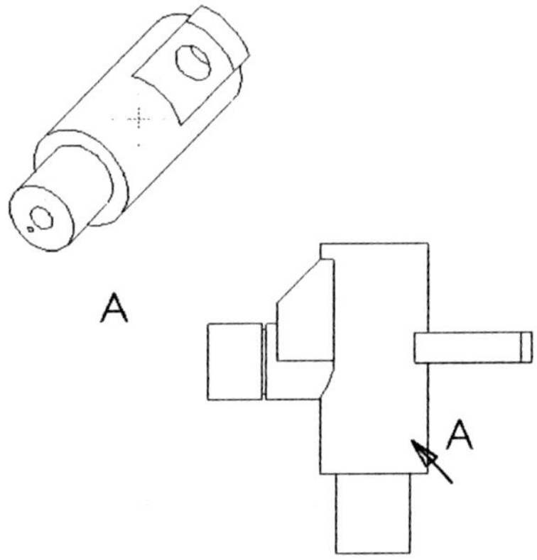

Figure 12-32 *Auxiliary view created with the* ***Flip Direction*** *check box selected*

Generating Detail Views

CommandManager:	Drawing > Detail View
Menu:	Insert > Drawing View > Detail
Toolbar:	Drawing > Detail View

A detail view is used to display the details of a portion of an existing view. You can select the portion whose detailing has to be shown in the parent view. The portion that you select will be magnified and placed as a separate view. You can control the magnification of the detail view. To create a detail view, you first need to activate the view from which you will generate the detail view. Choose the **Detail View** button from the **Drawing CommandManager**. The **Detail View PropertyManager** is displayed and you are prompted to sketch a circle to continue view creation. The cursor is replaced by a circle cursor. Create the circle on the portion of the view that is to be displayed in the detail view. If you want to use a profile other than the circle, you need to create it associated to the sketch before invoking the **Detail View** tool.

As soon as you draw the circle, the detail view is attached to the cursor and the options are displayed in the **Detail View PropertyManager**, as shown in Figure 12-33. You are also prompted to select a location for the new view. Specify a point on the drawing sheet to place the view. The options available in the **Detail View PropertyManager** are discussed next.

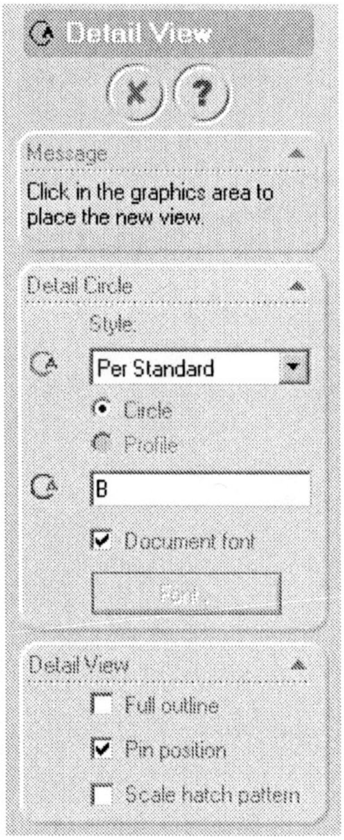

*Figure 12-33 The **Detail View PropertyManager***

Detail Circle

This rollout is used to define the options to display the circle of the detail view. Using the options available in the rollout, you can also apply the leader to the detail view. The options available in this rollout are discussed next.

Style

The **Style** area has the **Style** drop-down list to specify the style of the closed profile. By default, the **Circle** radio button is selected below the **Style** drop-down list. Therefore, the portion of the parent view that is shown in the detail view is highlighted in circle. If you have created a closed profile for defining the portion to be shown in the detail view, select the **Profile** radio button. The options available in the **Style** drop-down list are discussed next.

Per Standard. The **Per Standard** option is used to create the detail view as per default standards.

Broken Circle. The **Broken Circle** option is used to display the area of the parent view to be displayed in the detailed view in a broken circle.

With Leader. The **With Leader** option is used to add the leader to the callout of the detail view.

No Leader. The **No Leader** option is used to remove the leader from the callout of the detail view.

Connected. This option is used to create a line that connects the detail view with the closed profile in the parent view.

Detail View

This rollout is used to set the parameters of the detail view. The various options available in this rollout are discussed next.

Full outline

The **Full outline** check box is used to display the complete outline of the closed profile in the detail view.

Pin position

The **Pin position** check box is used to pin the position of the detail view.

Scale hatch pattern

The **Scale hatch pattern** check box is used to scale the hatch pattern with respect to the scale factor of the detail view when you create a detail view of a section view.

Figure 12-34 shows the detail view created using the **Detail View** tool.

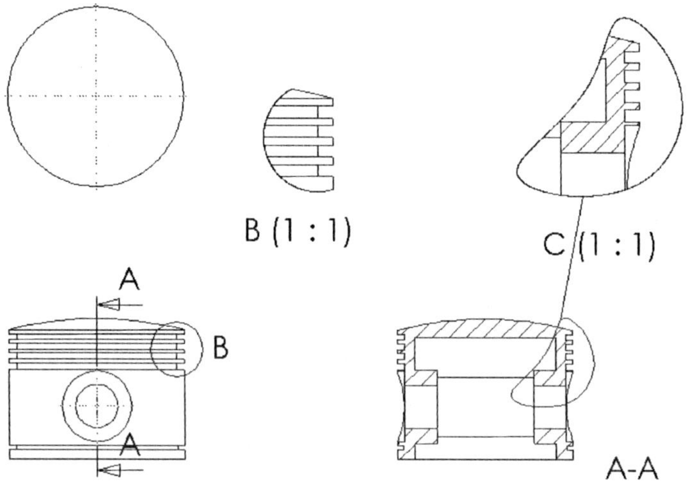

Figure 12-34 *Detail views generated using the existing views*

Tip. *When you create a detail view, by default the detail view is scaled as 1:1. You can define the scale factor in the* ***System Options*** *dialog box so that whenever you create a detail view, it will be created with the scaling factor provided by you. To specify the scale factor for the detail view, invoke the* ***System Options*** *dialog box and select the* ***Drawings*** *option from the left of this dialog box. Set the value of the scale factor of the detail view in the* ***Detail view scaling*** *edit box and choose the* ***OK*** *button. Hence forth, the detail view will be created of the scale factor defined in the* ***System Options*** *dialog box.*

Cropping Drawing Views

CommandManager:	Drawing > Crop View
Menu:	Insert > Drawing View > Crop
Toolbar:	Drawing > Crop View

This tool is used to crop an existing view using a closed sketch associated to that view. The portion of the view that lies inside the associated sketch is retained and the remaining portion is removed. To crop the view, you first need to create a closed profile associated to the view that defines the area of the view that will be displayed. Remember that the area of the view outside this closed profile will not be displayed when you crop the view. Select the closed profile and choose the **Crop View** button from the **Drawing CommandManager**. Note that this button will not be available until you select the associated closed sketch. Figure 12-35 shows the closed profile used to crop the view and Figure 12-36 shows the cropped view.

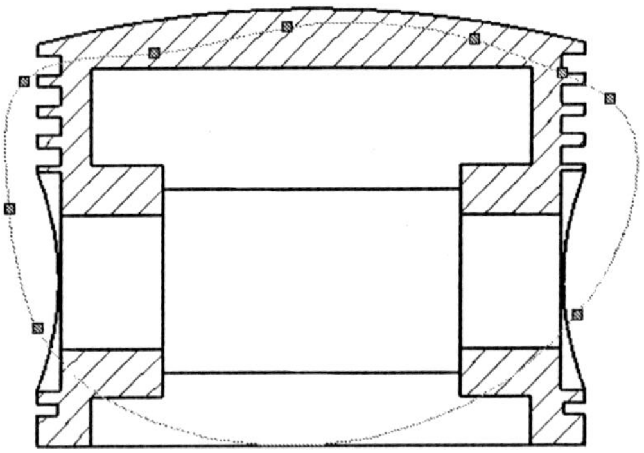

Figure 12-35 Closed profile to crop the view

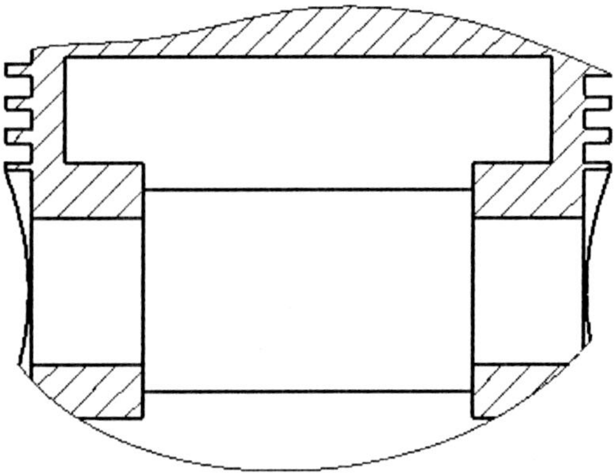

Figure 12-36 Resultant crop view

Tip. *You can remove the cropping of view by selecting the crop view and invoking the shortcut menu. Choose **Crop View > Remove Crop** from the shortcut menu. The initial view will be displayed in the drawing sheet.*

*If you need to edit the closed profile of the crop view, select the crop view and invoke the shortcut menu. Choose **Crop View > Edit Crop** from the shortcut menu. The sketch of the closed profile and the complete view is displayed in the drawing sheet. Edit the closed profile and choose the **Rebuild** button from the **Standard** toolbar or use CTRL+B from the keyboard.*

Generating Broken View

A broken view is used to display a component by removing a portion of it from between, keeping the ends of the drawing view intact. This type of view is used for displaying the components in which the length to width ratio is very high. This means that either the length is very large as compared to the width or the width is very large as compared to the length. The broken view will break the view along the horizontal or vertical direction such that the drawing view fits the area you require. To create a broken view, you first need to define the break line. Select the view you need to break and choose **Insert > Horizontal Break/Vertical Break**, depending on the direction in which you need to break the component. Two break lines will be displayed on the selected view, as shown in Figure 12-37.

Tip. *If you suppress the features of a model whose drawing views are generated, the suppressed features will not be displayed in the drawing views. As you unsuppress the feature, it will be displayed in the drawing views.*

When you hide or suppress the components of an assembly, the hidden or suppressed components are not displayed in the drawing views.

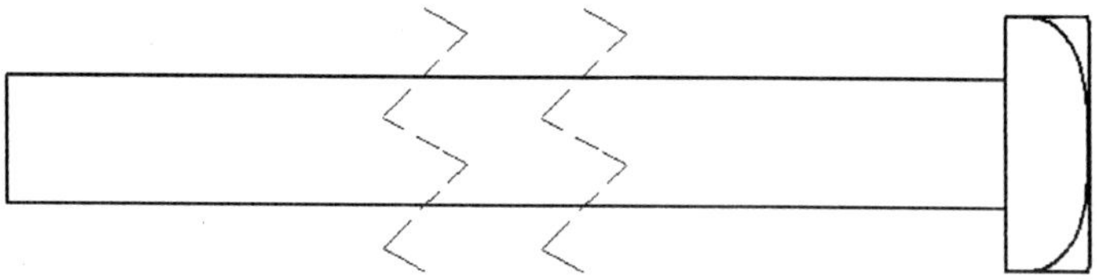

Figure 12-37 Break lines added to the view

After adding the break lines, you need to move the break lines to define the gap in the broken view. Select the break lines and move them away from each other, as shown in Figure 12-38. Now, select the view and invoke the shortcut menu. Choose the **Break View** option from the shortcut menu. The broken view will be created, as shown in Figure 12-39.

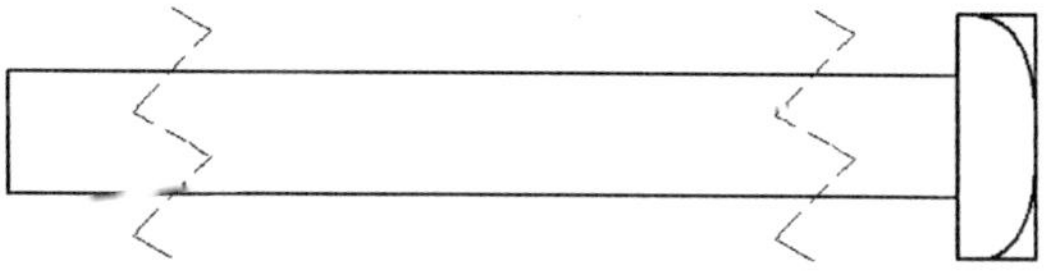

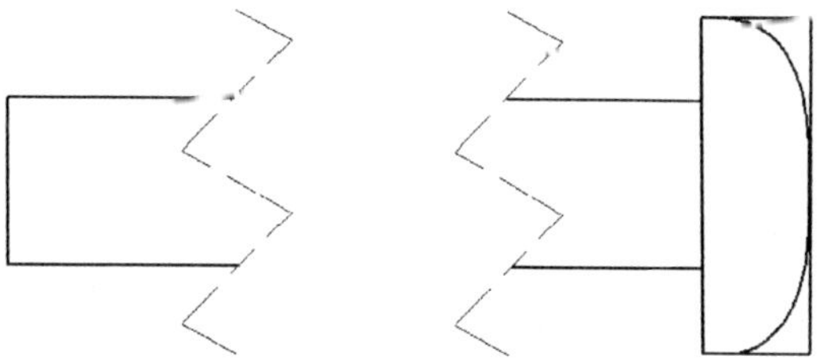

Figure 12-38 Extended gap between break lines *Figure 12-39 Resultant broken view*

You can also break an isometric view; the procedure of breaking an isometric view or any 3D view is the same as discussed earlier. Figure 12-40 shows a broken isometric view.

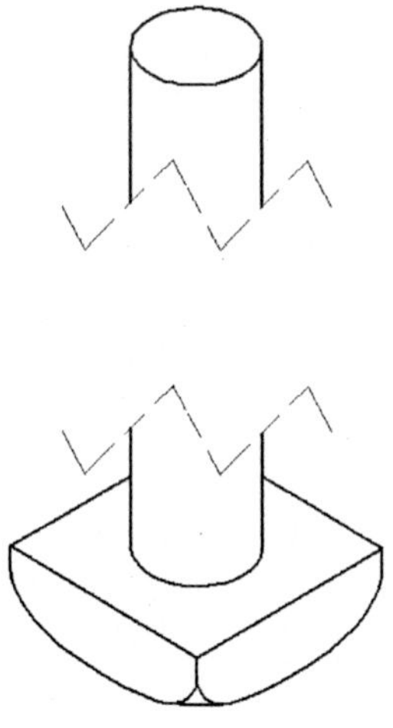

Figure 12-40 A broken isometric view

Note

Select the break line to increase or decrease the gap between the broken view. As you move the break line, the broken view is modified dynamically.

If you generate a projected view from a broken view, the resultant projected view is also a broken view.

If you break a 3D view placed horizontally, the two parts of the view as a result of the ***Broken View*** *tool will lose their alignment.*

Tip. *You can change the style of the break line by selecting the break line and invoking the shortcut menu. The various break line styles available are straight cut, curve cut, zig zag cut, and small zig zag cut.*

To unbreak the broken view, select the view and invoke the shortcut menu. Choose the ***Un-Break View*** *option from the shortcut menu.*

Select the view and invoke the shortcut menu. Choose the ***Break View*** *option to again break the view.*

If you select the break lines and press the DELETE key from the keyboard, the broken view will be replaced by the parent view.

Alternate Position View

CommandManager:	Drawing > Alternate Position View
Menu:	Insert > Drawing View > Alternate Position
Toolbar:	Drawing > Alternate Position View

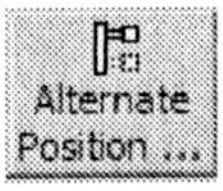

The alternate position view is used to create a view in which you can show the maximum and minimum range of motion of an assembly. The main position is displayed with continuous lines in the drawing view and the alternate position of the assembly is shown in the same view with dashed (phantom) lines. To create an alternate position view, you first need to activate and select the view of the assembly drawing on which you need to create the alternate position view. Choose the **Alternate Position View** button from the **Drawing CommandManager**. The **Alternate Position PropertyManager** is displayed as shown in Figure 12-41.

The **Alternate Position PropertyManager** prompts you to select a new configuration, click OK or enter and define the new configuration parameters. If you have not created any configurations, the **New Configuration** radio button will be automatically selected to create a new configuration. Enter the name of the configuration in the edit box given below. Choose the **OK** button from the **Alternate Position PropertyManager** to create a new configuration. Choose **OK** from the **Tangent Edge Display** dialog box, if displayed.

The assembly document is opened and the **Move Component PropertyManager** is displayed in the assembly document. The **Move Component PropertyManager** prompts you to move

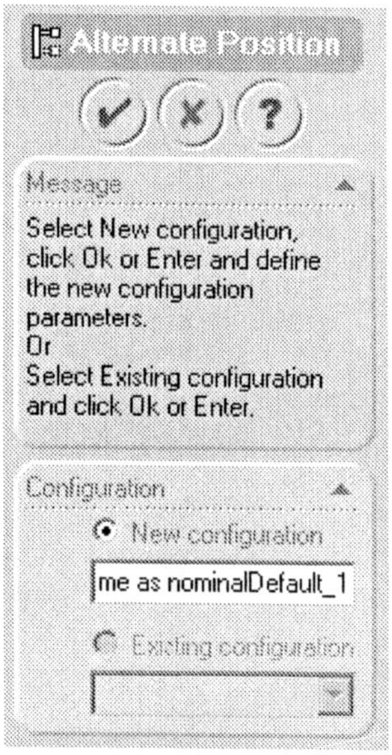

Figure 12-41 *The* ***Alternate Position PropertyManager***

the desired components to the position to be shown in the alternate view. Note that the component or components that you need to move to show in the alternate view should have that particular degree of freedom free to move. These components should not be fully defined in the assembly. Select and drag the cursor to move the components to the desired location. After defining the alternate position of the components, choose the **OK** button from the **Move Component PropertyManager**. You will return to the drawing document automatically. The alternate position of the components that are moved will be displayed in phantom lines in the drawing view, see Figure 12-42.

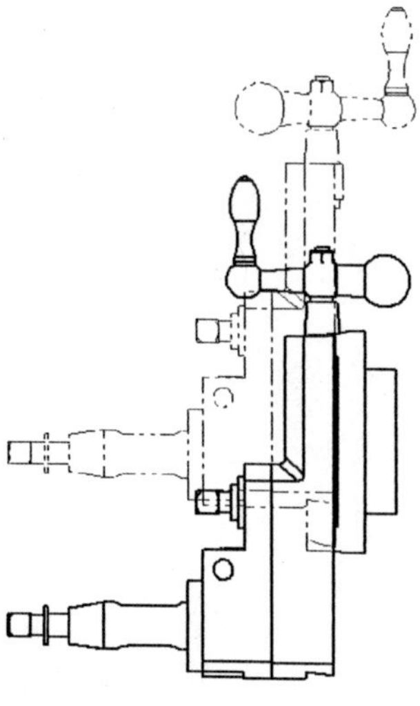

Figure 12-42 *Alternate position view*

You can also create the alternate position view of an isometric view or of any 3D view. The procedure of creating the alternate position view of a 3D view is the same as that discussed earlier. Figure 12-43 shows the alternate position view of an isometric view.

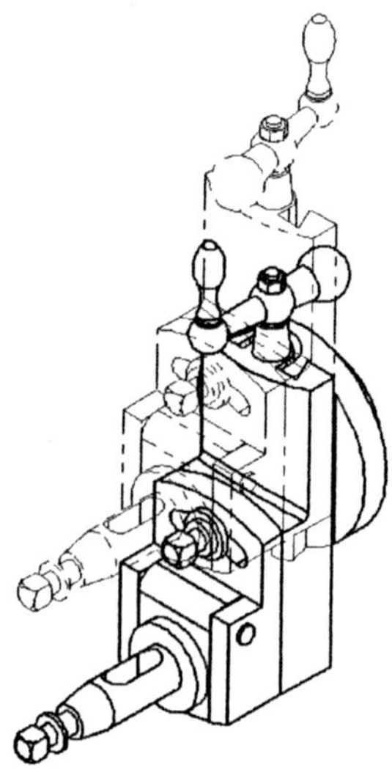

Figure 12-43 *Alternate view of an isometric view.*

Note

When you create an alternate view of an assembly, a new configuration is created inside the assembly document with the same name that is specified to the configuration while creating the alternate position view. Open the assembly document and invoke the ***ConfigurationManager****; you will observe that a new configuration is created along with the default configuration. By default, the newly created configuration is selected. Therefore, the assembly is displayed with the components moved to their extreme positions. To switch back to the default configuration, select* ***Default*** *from the* ***ConfigurationManager*** *and invoke the shortcut menu. Choose the* ***Show Configuration*** *option from the shortcut menu. You will observe that the assembly will be displayed with the moved components back at their original positions. If the* ***Show Configuration*** *option is not available in the shortcut menu, the assembly is at the default configuration.*

Creating the Drawing View of the Exploded State of the Assembly

You can create the drawing view of the exploded state of the assembly. To generate the view of the exploded state of an assembly, you need to have an exploded state defined in the assembly document. Generate the isometric view of the assembly on the drawing sheet. Select the view and invoke the shortcut menu. Choose the **Properties** option from the shortcut menu. The **View Properties** tab of the **Drawing View Properties** dialog box is displayed. Select the **Show in exploded state** check box from the **Configuration information** area and choose the **OK** button from the **Drawing View Properties** dialog box. Figure 12-44 shows the drawing view of the exploded state of assembly with explode lines.

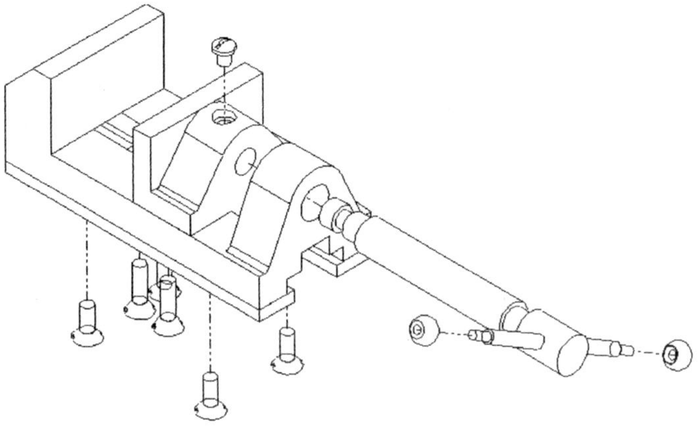

Figure 12-44 *Drawing view of the exploded state with explode lines*

Tip. *If the assembly in the assembly document is in the exploded state and you drag and drop the assembly to generate the drawing views, all the views of the assembly will be generated in the exploded state.*

If you need to switch to the collapse state in the drawing view, select the view and clear the ***Show in exploded state*** *check box from the* ***View Properties*** *tab of the* ***Drawing View Properties*** *dialog box.*

WORKING WITH INTERACTIVE DRAFTING IN SolidWorks

As mentioned earlier, you can also sketch the 2D drawings in the drawing document of SolidWorks. In technical terms, sketching 2D drawings is known as interactive drafting. Before starting the drawing, it is recommended that you insert an empty view. To create an empty view, choose **Insert > Drawing View > Empty** from the menu bar. An empty view is attached to the cursor. Select a point at the desired location in the drawing document to place an empty view. Now, select the empty view to activate and use the tools in the **Sketch CommandManager** to sketch the view.

Because the sketch that you draw is associated to the empty view, the sketched entities will also move if you move the empty view by selecting and dragging it.

EDITING AND MODIFYING THE DRAWING VIEWS

In SolidWorks, you can perform various kinds of editing operations and modifications on the drawing views. For example, you can change the orientation of the view, you can change the view scale, or you can delete the view. All these operations are discussed next.

Changing the View Orientation

You can change the orientation of the views generated using the **Model View** or the **Predefined View** option. To change the orientation, select the view; the **Model View PropertyManager** is displayed in both the cases. Double-click on the view orientation that you want to be made current from the **View Orientation** rollout. The orientation of the selected view will be modified. Choose the **OK** button from the **Model View PropertyManager**.

Remember that all the derived views will also change their orientation when you change the orientation of the parent view.

Changing the Scale of the Drawing Views

In SolidWorks, you can also change the scale of the drawing views. To change the scale of the drawing views, select the drawing view and then select the **Use custom scale** radio button from the **Scale** rollout. Set the new scale of the drawing view in the edit boxes available below this radio button. You can also change the scale of the derived views. However, remember that the scale of the parent view will not be changed if you change the scale of a derived view.

Deleting the Drawing Views

The unwanted views are deleted from the drawing sheet using the **FeatureManager Design Tree** or directly from the drawing sheet. Select the view to be deleted from the **FeatureManager Design Tree** and invoke the shortcut menu. Choose the **Delete** option from the shortcut menu. The **Confirm Delete** dialog box will be displayed. Choose the **Yes** button from this dialog box. You can also delete a view by selecting it directly from the drawing sheet and pressing the DELETE key from the keyboard. The **Confirm Delete** dialog box is displayed; choose the **Yes** button from this dialog box. Remember that when you delete a parent view, the projected views are not deleted. However, if you delete the parent view from which a section view or a detail view is generated, the name of the dependent view is also displayed in the **Confirm Delete** dialog box. If you choose **Yes**, the dependent views are also deleted.

Rotating a Drawing View

SolidWorks allows you to rotate a drawing view in the 2D plane. To rotate a drawing view, select the view and choose the Rotate View button from the View toolbar. The **Rotate Drawing View** dialog box is displayed, as shown in Figure 12-45. You can enter the value or rotation angle in this dialog box or you can also dynamically rotate the drawing view by dragging the mouse. If you select the **Dependent views update to change in orientation** check box, the views dependent on the rotated view will also change their orientation.

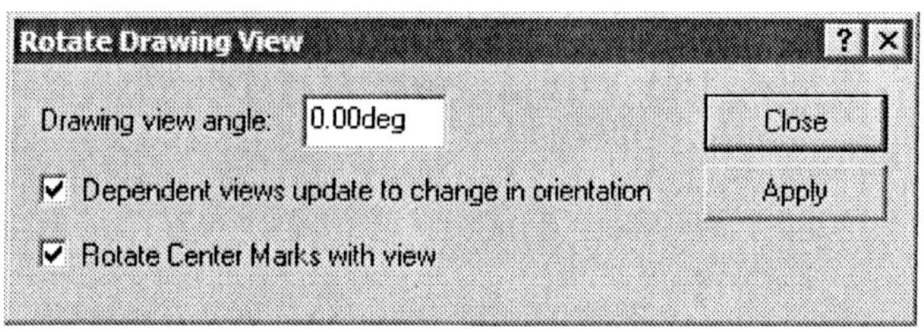

Figure 12-45 Rotate Drawing View *dialog box*

Tip. *You can also copy and paste the drawing view in the drawing sheet. Select the view to copy and press CTRL+C from the keyboard. Now, select anywhere on the drawing sheet to select the sheet and press CTRL+V from the keyboard to paste the drawing view.*

MODIFY THE HATCH PATTERN OF THE SECTION VIEW

As discussed earlier, when you generate a section view of an assembly or a component, a hatch pattern is applied to the component or components. This hatch pattern is based on the material assigned to the components in the part document. If you need to modify the default hatch pattern, select the hatch pattern from the section view and invoke the shortcut menu. Choose the **Crosshatch Properties** option, the **Area Hatch/Fill** dialog box is displayed, as shown in Figure 12-46. The various options available in this dialog box are discussed next.

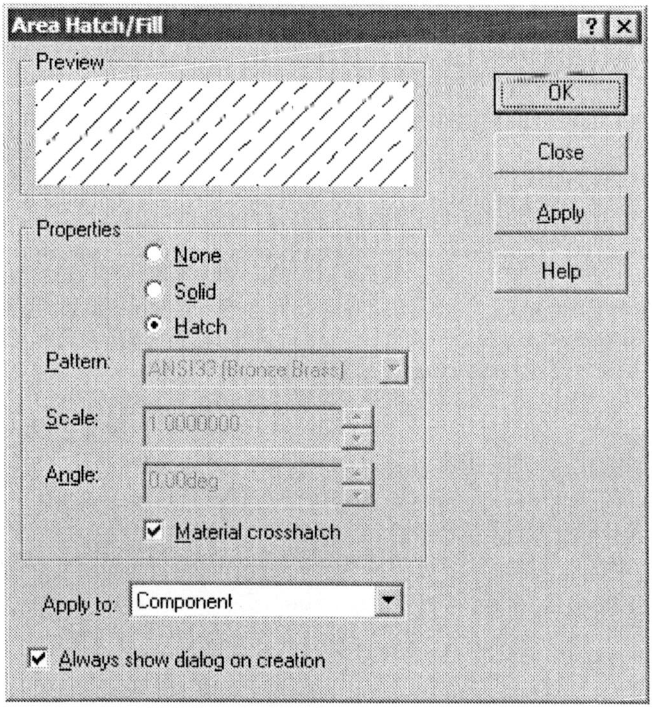

***Figure 12-46** The **Area Hatch/Fill** dialog box*

Preview

The **Preview** area displays the preview of the hatch pattern with the current setting.

Properties

The **Properties** area is used to define the type of hatch pattern and the properties of the hatch pattern. Some of the options in this area are not available by default. This is because by default, material-dependent hatch pattern is applied to the component. If you want to make the other options also available, clear the **Material crosshatch** check box. The various options available in this area are discussed next.

None

The **None** radio button is selected if you do not need to apply any hatch pattern in the section view.

Solid

The **Solid** radio button is used to apply the solid filled hatch pattern to the section view. By default, black color is applied as solid filled hatch pattern.

Hatch

The **Hatch** radio button is selected to apply the standard hatch patterns to the section view. When you select this button, some options available in this dialog box are invoked to define the properties of the hatch pattern. The options available to define the properties of the hatch pattern are discussed next.

Pattern

The **Pattern** drop-down list is used to define the style of the standard hatch pattern you need to apply to the section view. The preview of the hatch pattern selected from this drop-down list is displayed in the **Preview** area of the **Area Hatch/Fill** dialog box.

Scale

The **Scale** spinner is used to specify the scale factor of the standard hatch pattern selected from the **Pattern** drop-down list. When you change the scale factor using this spinner, the preview displayed in the **Preview** area updates dynamically.

Angle

The **Angle** spinner is used to define the angle to the selected hatch pattern.

Material crosshatch

The **Material crosshatch** check box is selected to apply hatch pattern based on the material assigned to the model. Clear this check box to change the type of hatch pattern.

Apply to

The **Apply to** drop-down list is used to define whether you need to apply this hatch pattern to the selected component, region, body, or to the entire view. Note that some of these options are available only while modifying the hatch pattern of an assembly section view.

Note

*The **Always show dialog on creation** check box is used to invoke this dialog box when you manually apply hatch patterns to a closed area. This will be discussed in advanced chapters in part 2 of this book.*

TUTORIALS

Tutorial 1

In this tutorial, you will generate the front view, top view, right view, aligned section view, detail view, and isometric view of the model created in Tutorial 2 of Chapter 7. Use the Standard A4 Landscape sheet format for generating the views. **(Expected time: 30 min)**

The steps to be followed to complete this tutorial are listed next.

a. Copy the part document of Tutorial 2 of Chapter 7 in the folder of the current chapter.
b. Open the copied part document and start a new drawing document from within the part document.
c. Select the standard A4 landscape sheet format and generate the parent view using the **Model View** tool, refer to Figure 12-47.
d. Generate the projected views using the **Projected View** tool, refer to Figure 12-47.
e. Generate the aligned section view using the **Aligned Section View** tool, refer to Figures 12-50 through 12-53.
f. Generate the detail view and the isometric view, refer to Figure 12-54.
g. Save and close the drawing document.

Copying and Opening the Part Document

1. Create a folder with the name *c12* in the *SolidWorks* directory and copy *c07-tut02.sldprt* from the */My Document/SolidWorks/c07* folder to this folder.

2. Start SolidWorks and open the part document of Tutorial 2 of Chapter 7 that you copied in the folder of the current chapter.

Starting the New Drawing Document

As mentioned earlier, one of the latest enhancements of SolidWorks 2004 is its ability to let you start a new drawing document from within the part document. This way, the model in the part document is automatically selected and you can generate its drawing views.

1. Choose the **Make Drawing from Part/Assembly** button from the **Standard** toolbar. 

   If you were using the advanced form of the **New SolidWorks Document** dialog box to start new documents, this dialog box is displayed.

2. Select the drawing template from the **Template** tab of this dialog box and choose **OK**.

   Remember that if you are not using the advanced form of the **New SolidWorks Document** dialog box, this dialog box is not displayed. Instead, a new drawing document is started directly and the **Sheet Format/Size** dialog box is displayed.

3. Select the **A4 - Landscape** sheet from the list box available in this dialog box and choose the **OK** button.

The new drawing document is started with standard A4 sheet and the **Model View** tool is invoked automatically. The model of Tutorial 2 of Chapter 7 is by default selected for generating the drawing views and a view is attached to the cursor.

Generating the Parent View and the Projected Views

Before you proceed with generating the drawing views, you need to confirm whether the projection type for the current sheet is set to third angle or not.

1. Expand the **FeatureManager Design Tree** that is displayed in the drawing area. Select **Sheet1** and then right-click on it. Choose the **Properties** option from the shortcut menu to display the **Sheet Properties** dialog box.

   Make sure that you select this option before right-clicking. If this option is not selected first, the shortcut menu that will be displayed will not have the **Properties** option.

2. Select the **Third angle** radio button from the **Type of projection** area, if it is not already selected. Close this dialog box.

3. Select the **Front** option from the list box available in the **Orientation** rollout of the **Model View PropertyManager**, if not already selected. Also, select the **Preview** check box. The preview of the front view of the model is displayed.

4. Select the **Auto-start projected view** check box in the **Options** rollout. Move the cursor to the middle left of the drawing sheet just above the title block and specify a point at this location to place the front view, refer to Figure 12-47.

   As soon as you specify the location of the view, the front view will be generated and placed at this location. This view is generated at 1:1 scale. Note that because you selected the option to start projected views immediately after generating the front view, the **Projected View PropertyManager** is invoked and the preview of the projected view is attached to the cursor. This view is being generated taking the front view as the parent view.

5. Move the cursor above the front view and specify a point to place the top view, refer to Figure 12-47. The top view of the model is generated and the preview of another projected view with the front view as the base view is attached to the cursor.

6. Move the cursor to the right of the front view and place the right view, refer to Figure 12-47. Choose **OK** from the **Projected View PropertyManager** to exit this tool.

Note

If the origins are displayed in the drawing views, you can turn them off using the ***View*** *menu.*

Tip. *If there are cylindrical or circular features in the model, then while generating the drawing views, center marks are automatically created on those features in the drawing views.*

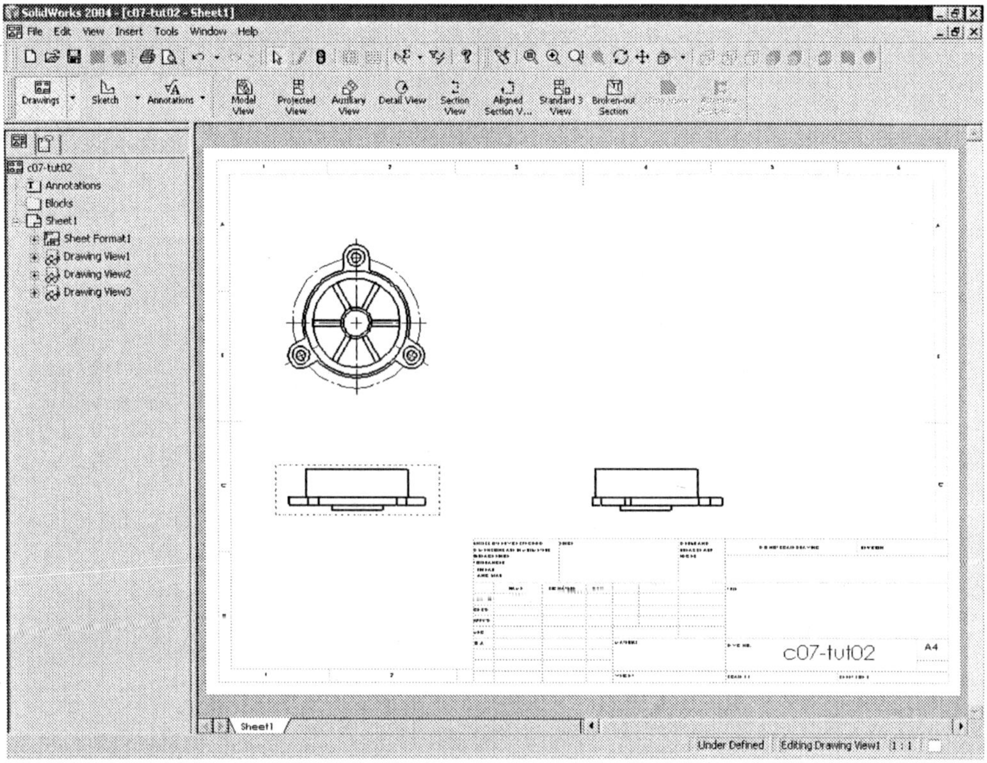

Figure 12-47 Drawing sheet after generating the front, top and right views

Generating the Aligned Section View

Next, you need to generate the aligned section view. The line segments that will be used to generate this view will be drawn before invoking the **Aligned Section View** tool. Remember that irrespective of the line that is drawn last, the view is projected normal to the line selected last. This is the reason you can draw any line first. In this tutorial, you will select the inclined line first and the vertical line later.

1. Click on the top view to activate the view.

2. Choose the **Sketch** button from the **CommandManager** to invoke the sketching tools. Using the **Line** tool, draw the sketch and apply the relations and dimensions to the sketch, as shown in Figure 12-48.

3. Select the dimension and invoke the shortcut menu. Choose the **Hide** option from the shortcut menu.

   Next, you need to select the lines to generate the aligned section view. Note that the vertical line should be selected last to generate the view normal to this line. But because this vertical line coincides with the center marks, it will be difficult to select this line. You will use the **Select Other** option to select this line.

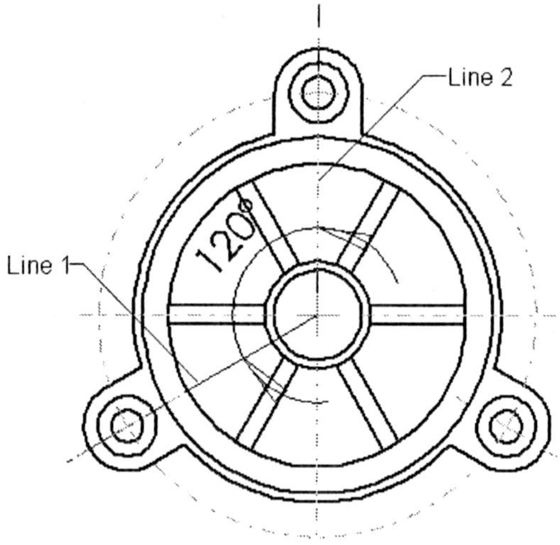

Figure 12-48 Sketch to be used as section sketch for the aligned section view

4. Select the inclined line. Make sure you do not select any segment of the center mark.

5. Now, press and hold the CTRL key down and move the cursor to the vertical line. Right-click on the vertical line and choose the **Select Other** option from the shortcut menu. The select other icon is attached to the cursor.

6. Continue right-clicking on the drawing sheet until the vertical line is not highlighted. When the vertical line is highlighted, left-click. It will be selected. Also, because the CTRL key was pressed, the inclined line will still be in the current selection set.

7. Now, choose the **Aligned Section View** button from the **Drawing CommandManager**. The **Section View PropertyManager** is displayed and the aligned section view will be attached to the cursor.

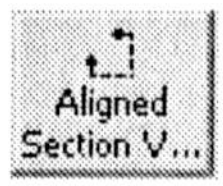

   The view generated is normal to the vertical line. If the direction of viewing the aligned section view is reversed, select the **Flip direction** check box from the **Section Line** rollout of the **Section View PropertyManager**.

8. Move the cursor to the right of the top view and place the aligned section view. Click anywhere on the screen to close the **PropertyManager**. The sheet after generating the aligned section view is shown in Figure 12-49.

Modifying the Hatch Pattern of the Aligned Section View

The gap between the hatching lines in the aligned section view is large. Therefore, you need to modify the spacing.

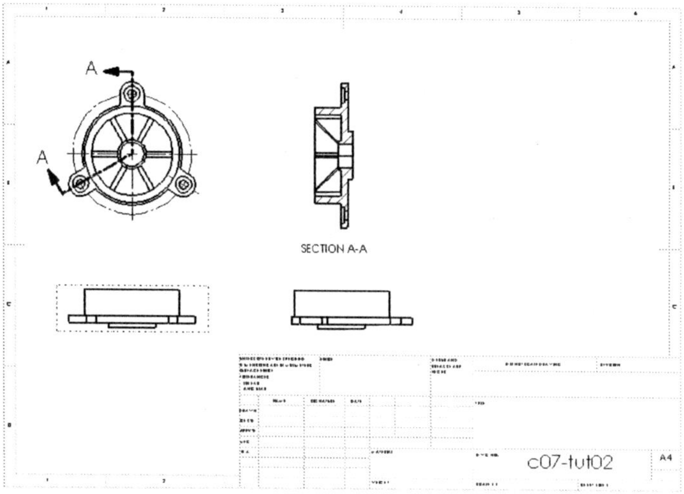

Figure 12-49 Sheet after generating the aligned section view

1. Select the hatch pattern from one of the sections in the aligned section view and right-click to invoke the shortcut menu. Choose **Crosshatch Properties** to display the **Area Hatch/Fill** dialog box.

2. Clear the **Material crosshatch** check box. The **Pattern**, **Scale**, and **Angle** options are available. Set the value of the **Scale** spinner to **2** and select **View** from the **Apply to** drop-down list. Choose **OK** to close the dialog box.

Generating the Detail View

Next, you need to generate the detail view of the right circular feature of the model. Before invoking the **Detail View** tool to generate the detail view, you need to activate a view from which you will drive the detail view.

1. Activate the top view.

2. Choose the **Detail View** button from the **Drawing CommandManager**. The **Detail View PropertyManager** is displayed and you are prompted to sketch a circle to continue view creation.

The cursor will be replaced by the circle cursor.

3. Create a small circle on the right circular feature of the model in the top view, refer to Figure 12-50. As you create the circle, the detail view is attached to the cursor.

4. Place the view on the right side of the drawing sheet above the title block, refer to Figure 12-50.

5. Set the value of the scale factor of the detail view to **3:1** and choose the **OK** button from the **Detail View PropertyManager**.

   You may need to move the drawing view and its label so that the view is not overlapping the title block. Figure 12-50 shows the detail view derived from the top view.

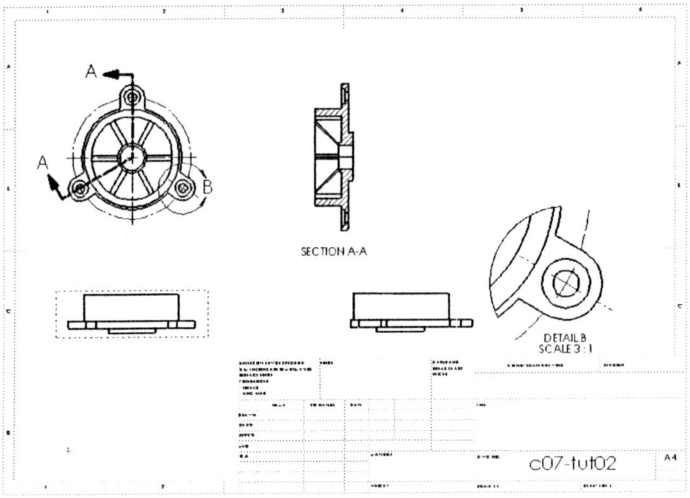

Figure 12-50 *The detail view derived from the top view*

Generating the Isometric View

The last view that you need to generate is the isometric view. You will generate the isometric view using the **Model View** tool.

1. Choose the **Model View** button from the **Drawings CommandManager**. The **Model View PropertyManager** is displayed.

   The name of the part document of Tutorial 2 is displayed in the **Open documents** selection area of the **Part/Assembly to Insert** rollout.

2. Double-click on the name of the document of Tutorial 2 to proceed further. You can also choose the **Next** button in the **Model View PropertyManager**.

3. The **Isometric** option is selected from the **Orientation** rollout. Select the **Preview** check box. The preview of the isometric view is attached to the cursor.

4. Place the view close to the top right corner of the drawing sheet. If the **SolidWorks** information window is displayed, choose **Yes** from this window.

5. Click anywhere on the sheet to close the **PropertyManager**. Figure 12-51 shows the final drawing sheet after generating all the drawing views.

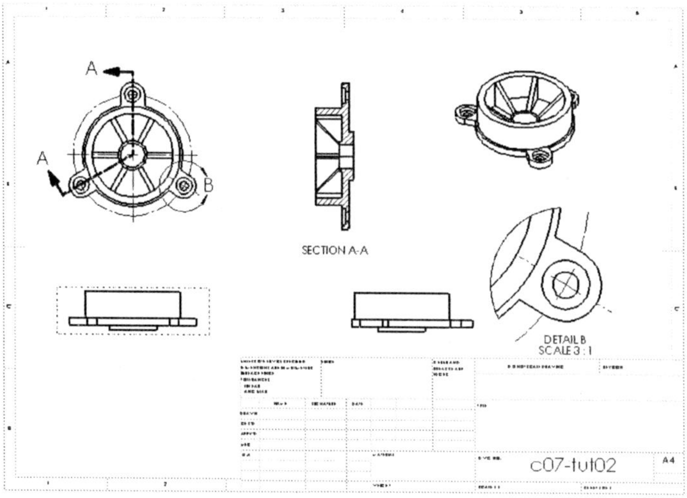

Figure 12-51 Drawing sheet after generating all the views

Saving the Drawing

1. Choose the **Save** button from the **Standard** toolbar and save the drawing document with the name given below.

   \My Documents\SolidWorks\c12\c12-tut01.slddrw

2. Choose **File > Close** to close this document. Also, close the part document of Tutorial 2.

Tutorial 2

In this tutorial, you will generate the drawing view of the Bench Vice assembly created in Chapter 10. You will generate the top view, sectioned front view, right view, and isometric view of the assembly in the exploded state. **(Expected time: 45 min)**

The steps to be followed to complete this tutorial are listed next.

a. Copy the folder of the Bench Vice assembly from Chapter 10 to the *c12* folder.
b. Create the exploded state of the Bench vice assembly, refer to Figure 12-52.
c. Start a new drawing document from within the assembly document using A4 landscape sheet format. Generate the top view using the **Model View** tool, refer to Figure 12-53.
d. Generate the section view using the **Section View** tool, refer to Figure 12-54.
e. Generate the right view using the **Projected View** tool, refer to Figure 12-55.
f. Generate the isometric view and change the state of the isometric view to the exploded state, refer to Figure 12-56.
g. Save and close the drawing and assembly documents.

Copying the Folder of the Bench Vice Assembly

First, you need to copy the folder of the Bench Vice assembly to *c12* folder.

1. Copy the folder of the Bench Vice assembly from the */My Document/SolidWorks/c10* folder to the *c12* folder.

Creating the Exploded View of the Assembly

Before proceeding further to generate the drawing views of the assembly, you need to create the exploded state of the assembly in the **Assembly** mode.

1. Open the Bench Vice assembly and create the exploded state and the explode lines, as shown in Figure 12-52.

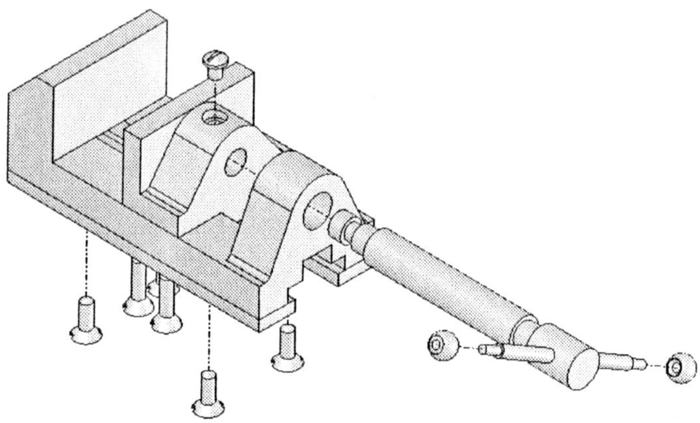

Figure 12-52 *Exploded view of the assembly with explode lines*

It is recommended that whenever you create an exploded state of an assembly, you must revert to the collapse state. If you save the assembly in the exploded state, whenever you generate the drawing views of the assembly, it will generate views with the exploded state.

2. Right-click **Bench Vice Configutarion(s) > Default** in the **ConfigurationManager** and choose **Collapse** to unexplode the assembly.

3. Save the assembly.

Starting a New Drawing Document From Within the Assembly Document

As mentioned earlier, you can also start a drawing document from within the assembly document.

1. Choose the **Make Drawing from Part/Assembly** button from the **Standard** toolbar. 

A new SolidWorks drawing document is started and the **Sheet Format/Size** dialog box is displayed.

2. Select the **A4 - Landscape** sheet and choose the **OK** button to close this dialog box.

   The **Model View PropertyManager** is automatically invoked and you are prompted to select a named view and place the drawing view.

Generating the Top View

1. Using the **FeatureManager Design Tree**, make sure that the projection type for the current sheet is set to third angle. Select the **Top** option from the list box available in the **Orientation** rollout of the **Model View PropertyManager**.

2. Select the **Use custom scale** radio button from the **Scale** rollout.

3. Set the value of the scale factor to **1:2** and then select the **Preview** check box from the **Orientation** rollout. The preview of the top view of assembly is displayed.

4. Make sure the **Auto-start projected view** check box in the **Options** rollout is cleared.

5. Place the view close to the top left corner of the drawing sheet, refer to Figure 12-53. Click anywhere on the sheet to exit the **PropertyManager**.

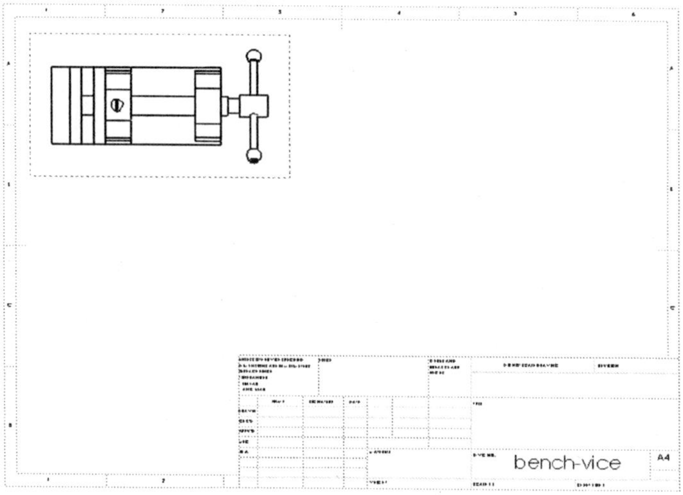

*Figure 12-53 Top view generated using the **Model View** tool*

Creating the Sectioned Front View

Next, you need to generate the sectioned front view that is derived from the top view.

1. Activate the top view and choose the **Section View** button from the **Drawings CommandManager**.

The **Section View PropertyManager** is displayed and you are prompted to sketch a line to continue view creation. The cursor will be replaced by the line cursor.

2. Draw a horizontal line such that it passes through the center of the Bench vice assembly.

   As soon as you specify the endpoint of the line, the **Section View** dialog box is displayed. This dialog box is used to exclude the components from the section cut.

3. Click on the + sign located on the left of the **FeatureManager Design Tree** and continue expanding it until Bench Vice assembly name is not displayed. Next, expand the assembly.

4. Select Screw Bar, Bar Globes, Jaw Screw, Oval Fillister, Set Screw1, and Set Screw2. Now, activate the **Section View** dialog box and select the **Don't cut all instances** check box individually for all the selected items.

5. Select the **Auto hatching** check box and choose the **OK** button from the **Section View** dialog box.

   The preview of the section view is displayed in the drawing sheet as you move the cursor up and down.

   If the direction of viewing of section view is not what is required, flip the direction by selecting the **Flip direction** check box from the **Section Line** rollout.

6. Place the section view below the top view. Click anywhere on the sheet to exit the **PropertyManager**.

7. Modify the hatch scale for the components. Figure 12-54 shows the section view generated using the **Section View** tool after modifying the hatch scale.

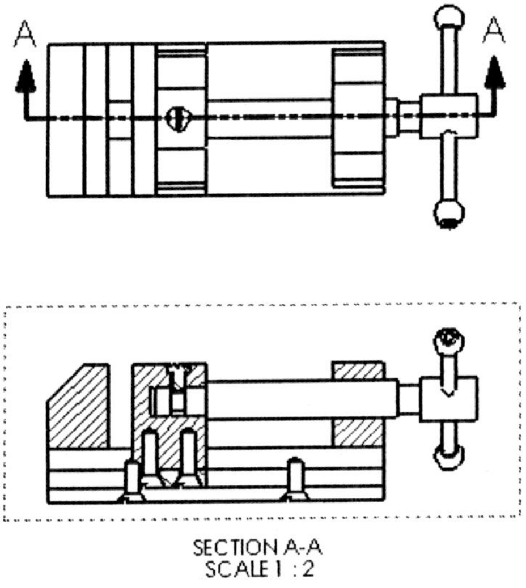

Figure 12-54 *Section view generated using the* ***Section View*** *tool*

Generating the Right-Side View

The next view that you need to generate is the right-side view derived from the sectioned front view and will be generated using the **Projected View** tool.

1. Select the sectioned front view and invoke the **Projected View** tool. The **Projected View PropertyManager** is displayed and a projected view is attached to the cursor.

2. Move the cursor to the right of the sectioned front view and place the view on the right of the sectioned front view.

3. Choose **OK** from the **PropertyManager**. The sheet after generating the projected view is shown in Figure 12-55.

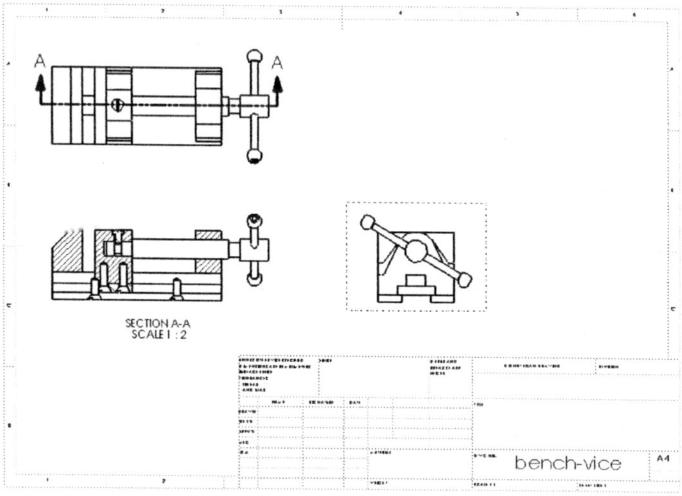

***Figure 12-55** Right-side view generated using the **Projected View** tool*

Creating the Isometric View of the Assembly in the Exploded State

The last view to be generated is the isometric view of the assembly in the exploded state.

1. Using the **Model View** tool, generate the isometric view and place the view close to the upper right corner of the drawing sheet. Choose **Yes** from the **SolidWorks** information box.

2. Set the scale factor of the drawing view to **1:2**.

3. Right-click on the view to invoke the shortcut menu. Choose the **Properties** option from the shortcut menu. The **Drawing View Properties** dialog box is displayed.

4. Select the **Show in exploded state** check box from the **View Properties** tab of the **Drawing View Properties** dialog box and choose the **OK** button.

5. Move the views to place all of them in the drawing sheet. Figure 12-56 shows the final drawing sheet after generating all the drawing views.

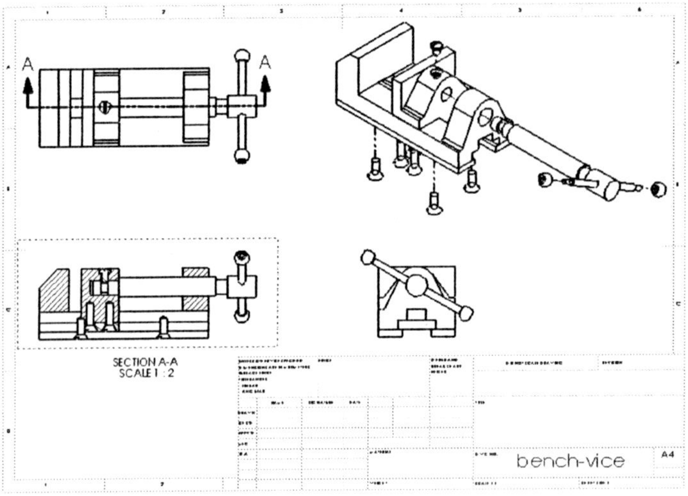

Figure 12-56 Drawing sheet after generating all the views

Saving the Drawing

Next, you need to save the drawing document.

1. Choose the **Save** button from the **Standard** toolbar and save the drawing document with the name given below.

   \My Documents\SolidWorks\c12*c12-tut02.slddrw*

2. Close the drawing and assembly documents.

SELF-EVALUATION TEST

Answer the following questions and then compare your answers with the answers given at the end of this chapter.

1. The **Standard sheet size** radio button is selected by default in the **Sheet Format/Size** dialog box. (T/F)

2. The **Display sheet format** check box is selected if you want to use the empty sheet, without any margin lines or title block. (T/F)

3. The **Relative View** tool is used to generate an orthographic view by defining the orientation of the view using reference planes or planar faces of the model. (T/F)

4. An auxiliary view is a drawing view that is generated by projecting the lines normal to a specified edge of an existing view. (T/F)

5. You cannot change the style of the break line in a broken view. (T/F)

6. In technical terms, creating a 2D drawing in the drawing document is known as __________.

7. To start a new drawing document from within the part document, choose the __________ button from the **Standard** toolbar.

8. The __________ check box available in the **Detail View** rollout of the **Detail View PropertyManager** is used to display the complete outline of the closed profile in the detail view.

9. To change the scale of the drawing views, select the drawing view and select the __________ radio button from the **Scale** rollout.

10. To rotate a drawing view, select the view and choose the __________ button from the **View** toolbar.

REVIEW QUESTIONS

Answer the following questions.

1. Choose the __________ button from the **Drawings CommandManager** to create an alternate position view.

2. The __________ dialog box is used to modify the hatch pattern of a section view.

3. The __________ check box needs to be cleared to modify the scale of the hatch pattern.

4. A __________ view is a section view in which only the sectioned surface is displayed in the section view.

5. The __________ dialog box is displayed to confirm the deletion of the views.

6. The views that are generated from a view already placed in the drawing sheet are known as

   (a) Child views (b) Derived views
   (c) Predefined views (d) Empty views

7. By default, the detail view boundary is in which shape?

   (a) Circle (b) Ellipse
   (c) Rectangle (d) None

8. In which edit box the name of the auxiliary view is specified?

(a) **Label** (b) **Arrow**
(c) **Name** (d) **Detail view label**

9. From which rollout can you select the view orientation in the **Named View PropertyManager**?

(a) **View Orientation** (b) **Define View**
(c) **Specify View** (d) **Scale View**

EXERCISE

Exercise 1

In this exercise you will generate the front view, section right view, isometric view, and the alternate position view on the isometric view of Exercise 1 of Chapter 11. You need to scale the parent view to the scale factor of **1:3**. The views that you need to generate are shown in Figure 12-57. **(Expected time: 30 min)**

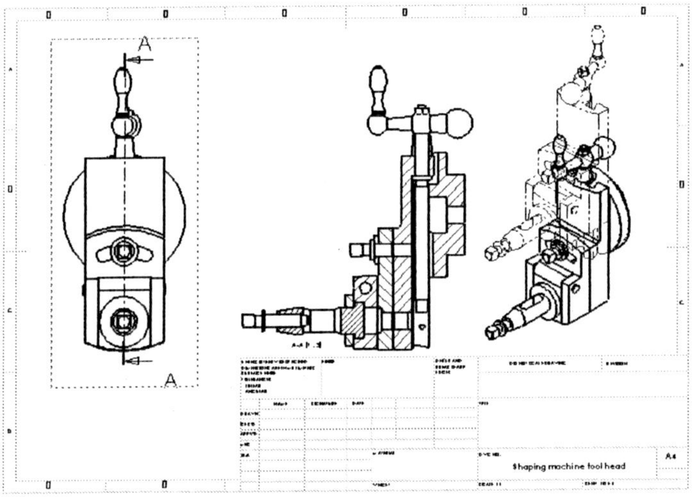

Figure 12-57 *Views of Exercise 1*

Answers to Self-Evaluation Test

1. T, **2.** T, **3.** T, **4.** T, **5.** F, **6.** Interactive drafting, **7. Make Drawing from Model/Assembly, 8. Full outline, 9. Use custom scale, 10. Rotate View**

Chapter 13

Working With Drawing Views-II

Learning Objectives

After completing this chapter you will be able to:

- *Add Annotations in the Drawing Views.*
- *Add Reference Dimensions.*
- *Add Notes to the Drawing Views.*
- *Add Surface Finish Symbols to the Drawing Views.*
- *Add Datum Feature Symbols to the Drawing Views.*
- *Add Geometric Tolerance to the Drawing Views.*
- *Add Datum Target Symbols to the Drawing Views.*
- *Add Center Marks and Center Lines to the Drawing Views.*
- *Add Hole Callouts to the Drawing Views.*
- *Add Cosmetic Threads to the Drawing Views.*
- *Add Weld Symbols to the Drawing Views.*
- *Add Multi-jog Leader to the Drawing Views.*
- *Add Dowel Pin Symbol to the Drawing Views.*
- *Edit the Annotations.*
- *Add BOM*
- *Add Balloons to the Assembly.*
- *Add New Sheets in the Drawing Document.*
- *Edit Sheet Format.*
- *Create User-Defined Sheet Format.*

ADDING ANNOTATIONS TO THE DRAWING VIEWS

After generating the drawing views, you need to generate the dimensions in the drawing views and add the other annotations such as notes, surface finish symbols, geometric tolerance, and so on. There are two types of annotations that can be displayed in the drawing views. The first type of annotations are the generative annotations. These annotations are added while creating the part in the **Part** mode. For example, the dimensions that you add to the sketch and features of the part. The second type of annotations are added manually to the geometry of drawing views such as reference dimensions, notes, surface finish symbols, and so on. Both these types of annotations are discussed next.

Generating Annotations Using the Model Items Tool

CommandManager:	Annotations > Model Items
Menu:	Insert > Model Items
Toolbar:	Annotation > Model Items

The **Model Items** tool is used to generate the annotations that were added while creating the model in the **Part** mode. To invoke this tool, choose the **Model Items** button from the **Drawing Annotation CommandManager**. The **Insert Model Items** dialog box is displayed as shown in Figure 13-1. The various options available in this dialog box are discussed next.

Annotations

The **Annotations** area of this dialog box is used to select the annotations that you need to generate in the drawing views. Using the options available in the **Annotation** area you can generate the cosmetic thread, datums, datum targets, dimensions, geometric tolerances, notes, surface finish, and weld symbols. Select the type of annotation to be generated from this area. With this release of SolidWorks, the **Model Items** tool is enhanced. In the **Annotations** area of the **Insert Model Items** dialog box you are provided with the **Marked for drawing** check box. This check box is selected by default and is used to generate dimensions that are marked to be imported in the drawing view. If you clear this check box, then you can import all the dimensions applied while creating the part.

Reference geometry

The **Reference geometry** area is used to generate the reference geometries that were used in creating the part. You can generate axes, curves, planes, surfaces, and so on. Select the check box of the type of reference geometry that you need to generate in the drawing views.

Import from

The options available in the **Import from** area are used to define the options from where the annotation should be imported. The options available in this area are discussed next.

Entire model

The **Entire model** radio button is selected to import the annotations from the entire model when a feature of a model is selected from the **FeatureManager Design Tree** in

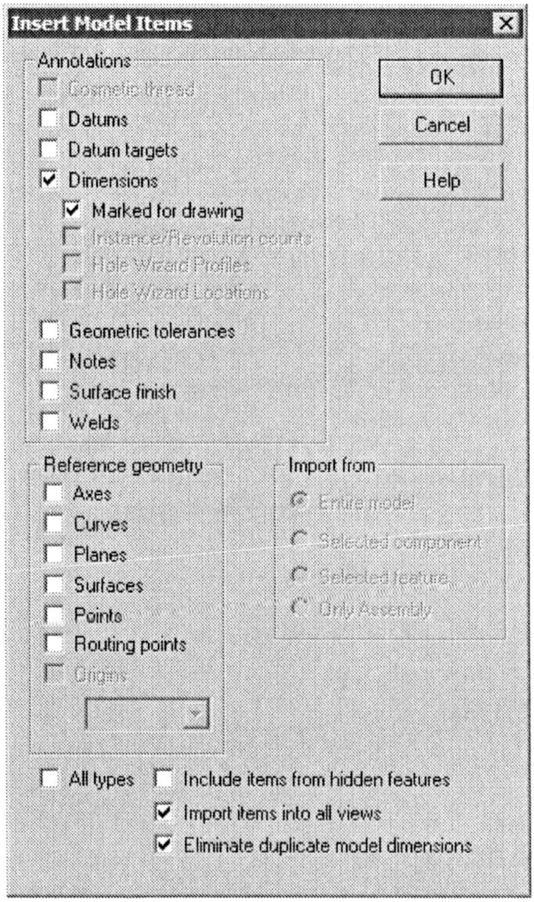

Figure 13-1 *The* ***Insert Model Items*** *dialog box*

the **Drawing** mode. In case of assemblies, the annotations from the entire assembly are imported even if you select a single component of the assembly.

Selected component

The **Selected component** radio button is active if you select a component of the assembly from the **FeatureManager Design Tree**. This radio button is selected to import the annotations from only the selected component.

Selected feature

The **Selected feature** radio button is active if you select a feature of the component from the **FeatureManager Design Tree**. This radio button is selected to import the annotations from only the selected feature or features.

Only Assembly

The **Only Assembly** radio button is active when you generate the drawing views of an assembly. This radio button is selected to import the annotations that are applied to the assembly in the **Assembly** mode such as offset distance and so on.

All types

The **All types** check box is selected if you need to display all the annotations applied to the part or assembly.

Include items from hidden features

The **Include items from hidden features** check box is selected to display the annotation that belongs to a hidden feature of the model. By default, this check box is cleared. It is recommended that you clear this check box because it will eliminate the display of unwanted annotations.

Import items into all views

The **Import items into all views** check box is selected to display the annotations in all the views. The annotations are automatically distributed in all the drawing views. If you need to display the annotations in a selected view or views only, clear this check box. When you clear this check box, the **Import to views** selection area is displayed below the **Help** button in the **Insert Model Items** dialog box. You can select the views in which you need to display the annotations. You can also preselect the views in which you need to display the annotations before invoking the **Insert Model Items** dialog box.

Eliminate duplicate model dimensions

The **Eliminate duplicate model dimensions** check box is selected to prevent the display of the same annotation more than once.

After setting all the parameters in the **Insert Model Items** dialog box, choose the **OK** button to display the annotations.

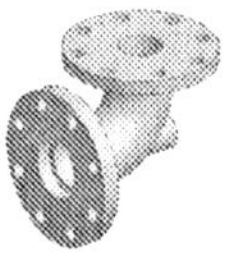

Tip. *When you generate the annotations, mostly the annotations overlap one over the other. Therefore, you may need to move the annotations after generating. To move an annotation, move the cursor on the annotation; a red rectangle is displayed. Press and hold down the left mouse button and drag the cursor to place the annotation at the desired location. Release the left mouse button when the cursor is placed at the desired location.*

The dimensions generated while generating the annotations are the same dimensions that were used to create the model. These dimensions are linked to the model because of the bidirectional associativity in SolidWorks.

To modify the dimension, double-click on it; the ***Modify*** *dialog box is displayed. Using this dialog box, you can modify the value of the dimension. After modifying the dimensions rebuild the drawing views using the* ***Rebuild*** *button. The same change in the dimension will also be made in the original model. If the model is used in any assembly, the changes will also be reflected in the assembly.*

If you need to delete a particular generated annotation, select the annotation from the drawing sheet and press the DELETE key from the keyboard. The annotation deleted from the drawing view will not be deleted from the original model.

Adding Reference Annotations

In SolidWorks, you can add reference annotations in the drawing views. The reference annotations include reference dimensions, notes, surface finish symbols, datum feature symbol, geometric tolerance, and so on. The method of adding reference annotations is discussed next.

Adding Reference Dimensions

CommandManager:	Annotations > Smart Dimension
Menu:	Tools > Dimensions > Smart
Toolbar:	Annotations > Smart Dimension

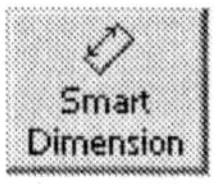

In the **Drawing** mode of SolidWorks, you can use the **Smart Dimension** tool to add reference dimensions to the drawing views. This tool is similar to the **Smart Dimension** tool discussed in the sketching environment of SolidWorks. You can also use the **Dimension/Relations CommandManager** to add chamfer dimensions. The technique of adding chamfer dimensions to the drawing views in the **Drawing** mode of SolidWorks is discussed next.

Chamfer Dimension

The **Chamfer Dimension** tool is used to add the reference dimension to the chamfers in the drawing view. To add a chamfer dimension, choose **Chamfer Dimension** from the **Dimension/Relations CommandManager**. The cursor is replaced by the chamfer dimension cursor. Now, select the inclined chamfered edge and then select a horizontal or vertical edge. The chamfer dimension will be attached to the cursor. Select a point on the sheet to place the dimension. Figure 13-2 shows a chamfer dimension created using the **Chamfer Dimension** tool.

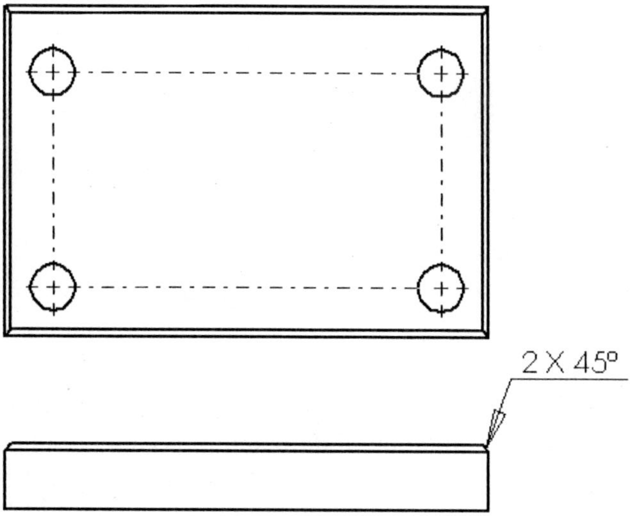

***Figure 13-2** Chamfer dimension added using the **Chamfer Dimension** tool*

Adding Notes to the Drawing Views

CommandManager:	Annotations > Note
Menu:	Insert > Annotations > Note
Toolbar:	Annotation > Note

In SolidWorks, you can also add notes to the drawing views in the **Drawing** mode. To add notes to the drawing view, choose the **Note** button from the **Annotations CommandManager**. The partial view of the **Note PropertyManager** is displayed as shown in Figure 13-3.

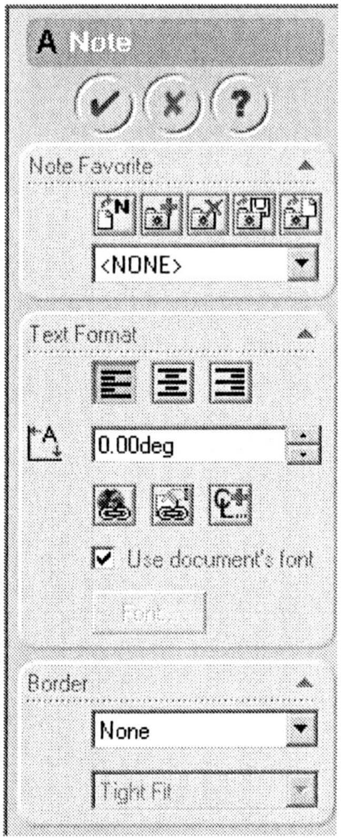

Figure 13-3** Partial view of the **Note PropertyManager

When you invoke the **Note PropertyManager**, a text box is attached to the cursor. When you move the cursor close to a edge in a drawing view, a leader will be displayed attached to the text box. This is because the **Auto Leader** button is chosen in the **Leader** rollout. Place the text box at the desired location. As you place the text box, the text edit box will be invoked in the drawing area and you can enter the text. After entering the text, choose **OK** from the **Note PropertyManager**. The other options available in the **Note PropertyManager** are discussed next.

Note Favorite

With this release of SolidWorks, using the options available in the **Note Favorite** rollout you can save a note as favorite. The options available in this rollout are the same as those discussed in Chapter 3.

Text Format

The **Text Format** rollout is used to set the format of text such as font, size, justification, and rotation of the text. You can also add the symbols and hyperlink to the text using the options available in this rollout.

Border

The options available in the **Border** rollout are used to define the border in which the note text will be displayed. You can use various types of borders from the **Style** drop-down list. The **Size** drop-down list available in this rollout is used to define the size of the border in which the text will be placed.

Leader

The options available in the **Leader** rollout are used to define the style of arrows and leaders that are displayed in the notes.

Adding Surface Finish Symbols to the Drawing Views

CommandManager:	Annotations > Surface Finish
Menu:	Insert > Annotations > Surface Finish Symbol
Toolbar:	Annotation > Surface Finish

You can add the surface finish symbols to the edges or faces in the drawing views using the **Surface Finish** tool. To add the surface finish symbol, choose the **Surface Finish** button from the **Annotations CommandManager**. The **Surface Finish PropertyManager** is displayed as shown in Figure 13-4.

As you invoke this tool, a surface finish symbol is attached to the cursor. The options available in the **Surface Finish PropertyManager** dialog box are discussed next.

Surface Finish Favorite

With this release of SolidWorks, using the options available in the **Surface Finish Favorite** rollout you can save a surface finish as a favorite. The options available in this rollout are the same as those discussed in Chapter 3.

Symbol

The **Symbol** rollout is used to define the type of surface finish symbol that you need to add to the drawing views. The surface finish symbols include the **Basic** surface finish symbol, **Machining Required**, **Machining Prohibited**, **JIS Basic**, **JIS Surface Texture 2**, **JIS Surface Texture 3**, **JIS Surface Texture 4**, **JIS Grind**, **JIS No Machining**, and so on.

Symbol Layout

The options available in the **Symbol Layout** rollout are used to define the parameters of the surface finish symbol.

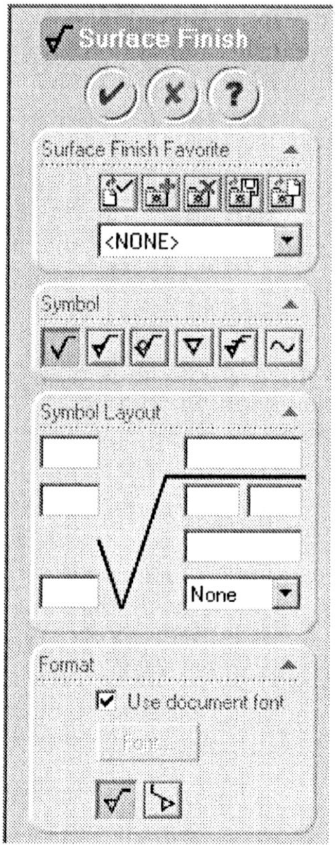

Figure 13-4 Partial view of the ***Surface Finish PropertyManager***

Format

The options available in the **Format** rollout are used to define the font size and the orientation of the surface finish symbol.

The other options available in the **Surface Finish PropertyManager** are the same as those discussed while adding notes.

Adding Datum Feature Symbol to the Drawing Views

CommandManager:	Annotations > Datum Feature
Menu:	Insert > Annotations > Datum Feature Symbol
Toolbar:	Annotation > Datum Feature

The **Datum Feature** tool is used to add a datum feature symbol to an entity in the drawing view. The datum feature symbols are used as datum references while adding the geometric tolerances in the drawing view. To add the datum feature symbol, choose the **Datum Feature** button from the **Annotations CommandManager**. The **Datum Feature PropertyManager** is displayed and a datum feature symbol with default parameters is attached to the cursor. The **Datum Feature PropertyManager** is displayed in Figure 13-5.

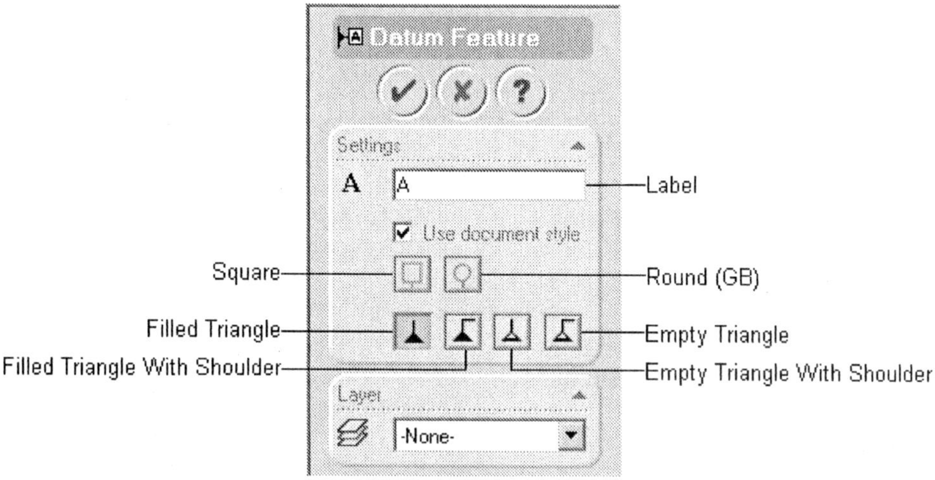

*Figure 13-5 The **Datum Feature PropertyManager***

The options available in the **Datum Feature PropertyManager** are discussed next.

Label

The **Label** edit box in the **Settings** rollout is used to define the label to be used in the datum feature symbol. You can use alphabets or numeric characters as labels.

Square

The **Square** button is used to place the text of the datum feature inside a square. By default, the text of the datum feature is placed using this option. Therefore, using the buttons available below the **Square** button you can set the type of datum feature such as filled triangle, filled triangle with shoulder, empty triangle, empty triangle with shoulder.

Round (GB)

The **Round (GB)** button is used to place the text of the datum feature inside a circle. To place the text inside a circle first you need to clear the **Use document style** check box and then choose the **Round (GB)** button. When you choose this button, additional buttons are displayed below the **Round (GB)** button. These buttons are used to set the style of the datum feature.

After defining all the parameters of the datum feature symbol, specify a point on an existing entity in the drawing sheet and then move the cursor to define the length and the placement of the datum feature symbol. As you place one datum feature symbol, another datum feature symbol is attached to the cursor. Therefore, you can place as many datum feature symbols as you want using the **Datum Feature PropertyManager**. The sequence of the names of datum feature symbols automatically follows the order based on the label as you place multiple datum feature symbols.

Adding Geometric Tolerance to the Drawing Views

Enhanced

CommandManager:	Annotations > Geometric Tolerance
Menu:	Insert > Annotations > Geometric Tolerance
Toolbar:	Annotation > Geometric Tolerance

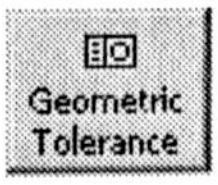

In a shop floor drawing, you need to provide various other parameters along with the dimensions and dimensional tolerance. These parameters can be geometric condition, surface profile, material condition, and so on. All these parameters are defined using the **Geometric Tolerance** tool. To add the geometric tolerance to the drawing views, choose the **Geometric Tolerance** button from the **Annotations CommandManager**. The **Properties** dialog box will be displayed, as shown in Figure 13-6. Also, a geometric tolerance is attached to the cursor.

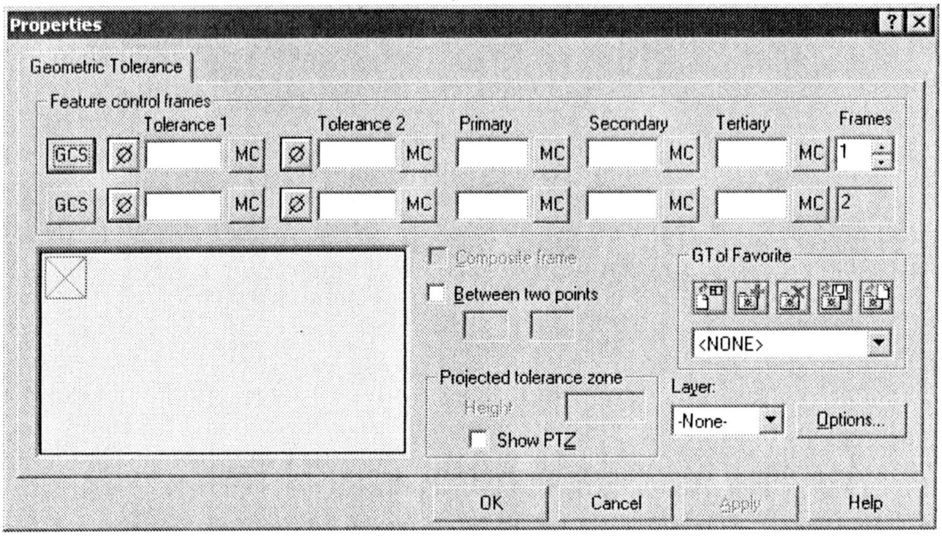

*Figure 13-6 The **Properties** dialog box used to apply geometric tolerance*

The options available in the **Properties** dialog box to add the geometric tolerance to the drawing views are discussed next.

Feature control frames

The **Feature control frames** area is used to define the parameters of the geometric tolerance to be added in the frames. Remember that both the rows in this area are separate frames. The parameters that can be added to these frames are geometric condition symbols, diameter symbol, value of tolerance, material condition, and datum references. The options available in the **Feature control frames** area are discussed next.

GCS

The **GCS** button available in the **Feature control frames** area is used to define the geometric condition symbol. When you choose this button, the **Symbols** dialog box is displayed, which is used to define the geometric condition symbols in the geometric tolerance. The **Symbols** dialog box is shown in

Figure 13-7. You can specify the standard of the geometric condition symbol from the **Symbol library** drop-down list. After selecting the appropriate geometric condition symbol, choose the **OK** button from this dialog box. The preview of the geometric tolerance is displayed in the preview area.

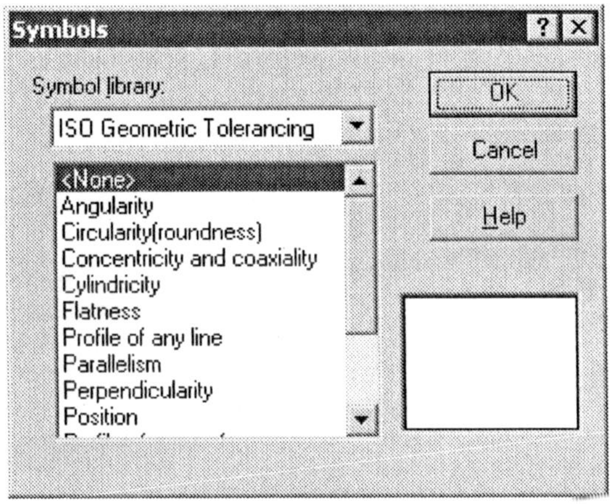

Figure 13-7 *The **Symbols** dialog box used to define the geometric condition symbols*

Tolerance 1

The **Tolerance 1** edit box is used to specify the tolerance value with respect to the geometric condition defined using the **GCS** button. Using the **Diameter** button, you can add a diameter symbol with the tolerance value. Using the **MC** button available next to the **Tolerance 1** edit box, you can specify the material condition in the geometric tolerance. When you choose the **MC** button, the **Symbols** dialog box is displayed, see Figure 13-8. This dialog box is used to define the symbol for the material condition.

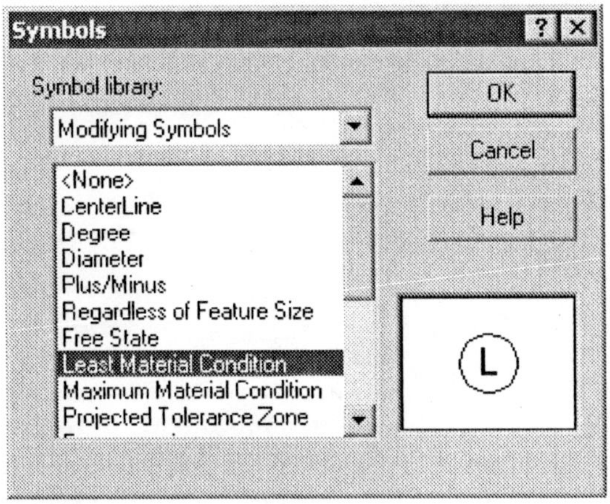

Figure 13-8 *The **Symbols** dialog box used to define the symbol for the material condition*

Tolerance 2

The options available in the **Tolerance 2** edit box are the same as those discussed earlier. This edit box is used to define the second geometric tolerance, if required.

Primary

The **Primary** edit box is used to specify the characters to define the datum reference added to the entities in the drawing view using the **Datum Feature Symbol** tool.

Similarly, you can define the **Secondary** datum reference and the **Tertiary** datum reference.

Frames

The **Frames** spinner is used to increase the frames for applying more complex geometric tolerances.

Projected tolerance zone

The **Projected tolerance zone** area is used to define the height of the projected tolerance. Select the **Show PTZ** check box from the **Projected tolerance zone** area; the **Height** edit box will be displayed. Specify the zone height in this edit box.

Options

The **Options** button is used to define the settings of the leader, arrow, font, and so on. When you choose this button, the **Geometric Tolerance Options** dialog box is displayed. Using this dialog box you can set the parameters for leader, arrow, font, and so on.

Between two points

The **Between two points** check box is selected if the geometric tolerance to be applied is between two points of entities. To apply the geometric tolerance between two points or entities select the **Between two points** check box and specify the tolerance in the edit boxes provided in the **Between two points** area.

GTol Favorite

The options available in the **GTol Favorite** area are used to save the geometric tolerance favorite. Figure 13-9 shows a drawing after adding annotations to some of the entities in the drawing view.

Adding Datum Target Symbols to the Drawing Views

Enhanced

CommandManager:	Annotations > Datum Target (*Customize to Add*)
Menu:	Insert > Annotations > Datum Target
Toolbar:	Annotation > Datum Target (*Customize to Add*)

The **Datum Target** tool is used to add a datum target to the entities in the drawing view. To add the datum targets, choose the **Datum Target** button from the **Annotations CommandManager**. The cursor is replaced by the datum target cursor. The **Datum Target PropertyManager** is displayed as shown in Figure 13-10. Select a model face, an edge, or a line from the view on which you need to add the datum target and then place the datum target. This **PropertyManager** is used to define the properties

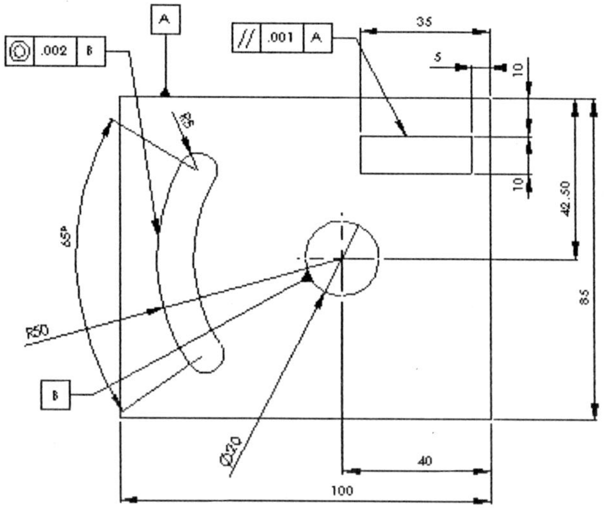

Figure 13-9 A drawing after adding some annotations

of the datum target symbol. Set the parameters of the datum target symbol in the **Settings** rollout of this **PropertyManager**. You can set the target shape as a point, circle, or rectangle; you can also define the diameter of the target area if the shapes of the target are selected as point and circle. If the shape of the target is selected as rectangle then you need to define the width and height of the rectangle. Using the **First reference**, **Second reference**, and the **Third reference** edit boxes you can specify datum references to the datum target symbol. Figure 13-11 shows the datum target symbols added to the entities in the drawing view.

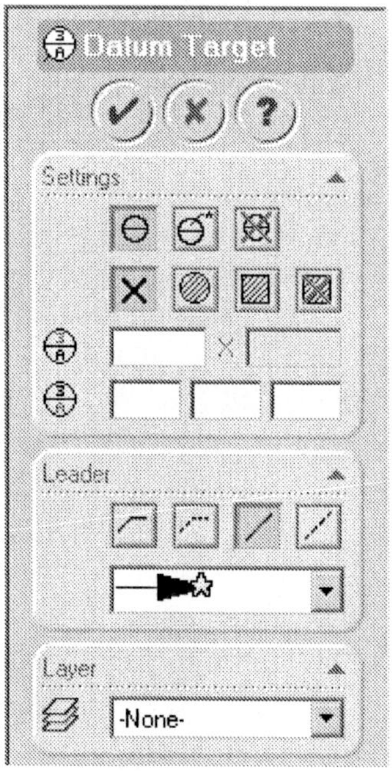

Figure 13-10** The **Datum Target PropertyManager

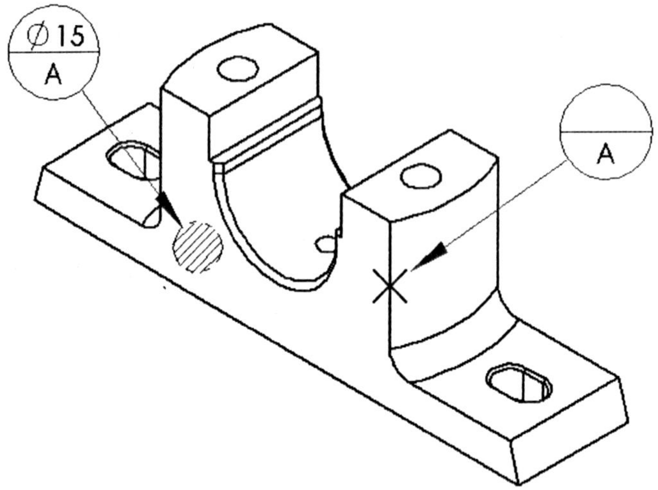

Figure 13-11 Datum target symbols added to the model

Adding Center Marks to the Drawing Views

CommandManager:	Annotations > Center Mark
Menu:	Insert > Annotations > Center Mark
Toolbar:	Annotation > Center Mark

The **Center Mark** tool is used to add center marks to the circular entities. As discussed earlier, center marks are automatically generated when you generate the model. But if the center marks are not generated while generating the drawing view, or if you have sketched a view, you can use this tool to add the center marks to the drawing views. To add the center marks to the drawing views, choose the **Center Mark** button from the **Annotations CommandManager**. The **Center Mark PropertyManager** will be displayed as shown in Figure 13-12.

When you invoke the **Center Mark PropertyManager**, the cursor is replaced by the center mark cursor and you are prompted to select a circular edge or an arc for the center mark insertion. By default, the **Single Center Mark** button is selected in the **Options** rollout of the **Center Mark PropertyManager**. Select the circular edge or arc to add the center mark. Figure 13-13 shows the center marks added using the **Single Center Mark** button selected from the **Options** rollout.

If you add center marks to a hole that is part of a linear or circular pattern, the **Propagate** button appears on the center mark. If you choose this button the center mark is added to all the remaining instances of the pattern.

The center marks in rectangular and circular patterns can also be applied using the **Linear Center Mark** and **Circular Center Mark** buttons. If you select the **Linear Center Mark** button from the **Options** rollout, then you can add the center marks in the linear pattern format. Since the **Connection lines** check box is selected by default, the center marks will be connected

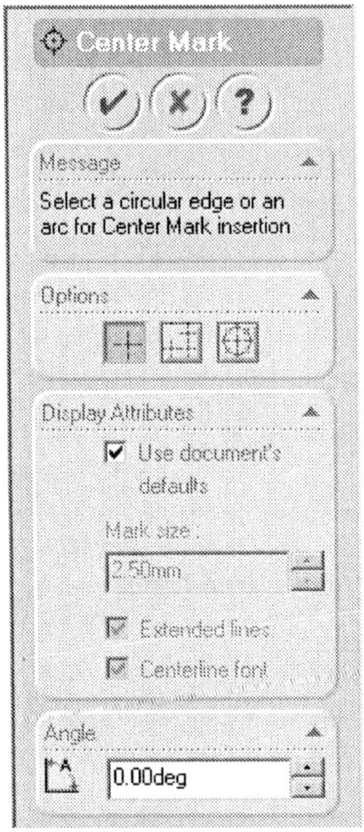

Figure 13-12 *Partial view of the* ***Center Mark PropertyManager***

using the center lines. Figure 13-14 shows the center marks added using the **Linear Center Mark** button with the **Connection lines** check box selected.

Using the **Circular Center Mark** button, you can create the center mark in the circular pattern format. When you select the **Circular Center Mark** button from the **Options** rollout, the **Circular lines** check box, **Radial lines** check box, and the **Base center mark** check box are displayed. The **Circular lines** check box is used to create a circular line passing through the centers of the circles arranged in the form of a circular pattern. The **Radial lines** check box is used to display the radial lines from the center of the pattern to the center of each instance. The **Base center mark** check box is used to display the center mark at the center of the pattern. Figure 13-15 shows the center mark created with the **Base center mark** check box cleared. Figure 13-16 shows the center mark created with the **Base center mark** check box selected. Figure 13-17 shows the center mark created with the **Radial lines** check box selected. Remember that in all these figures, the **Circular lines** check box is selected.

The **Display Attributes** rollout available in the **Center Mark PropertyManager** is used to define the size of the center mark and the extended lines. The **Angle** rollout is used to rotate the center mark at an angle.

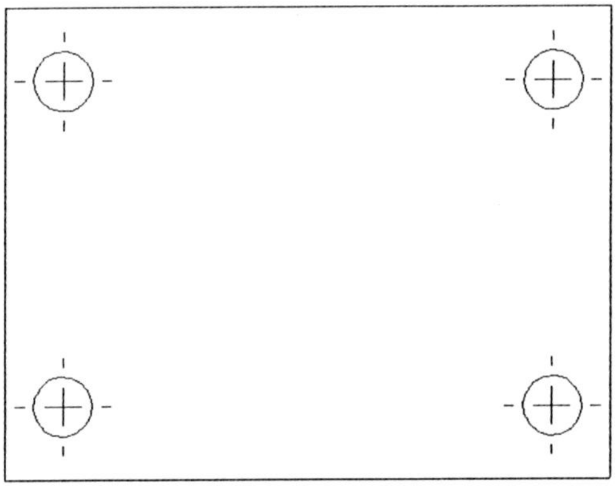

Figure 13-13 *Center marks added using the* ***Single Center Mark*** *option*

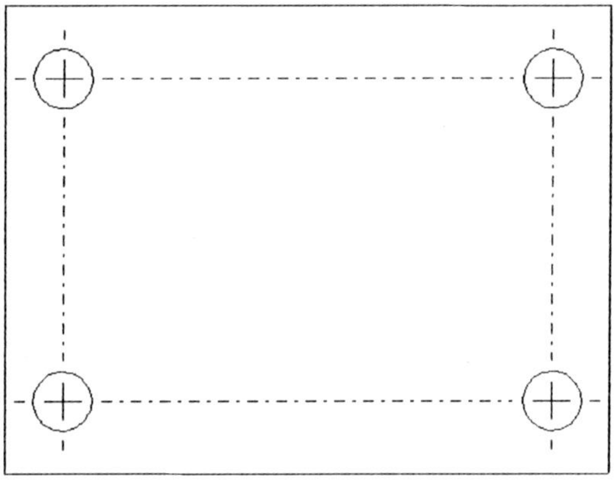

Figure 13-14 *Center marks added using the* ***Linear Center Mark*** *option*

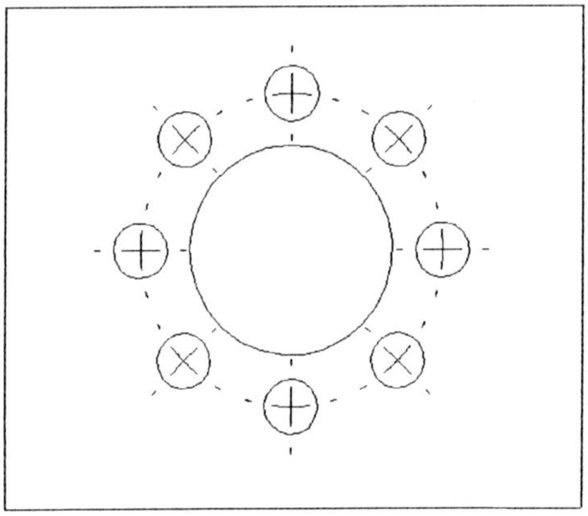

Figure 13-15 *Center marks created with the* ***Base center mark*** *check box cleared*

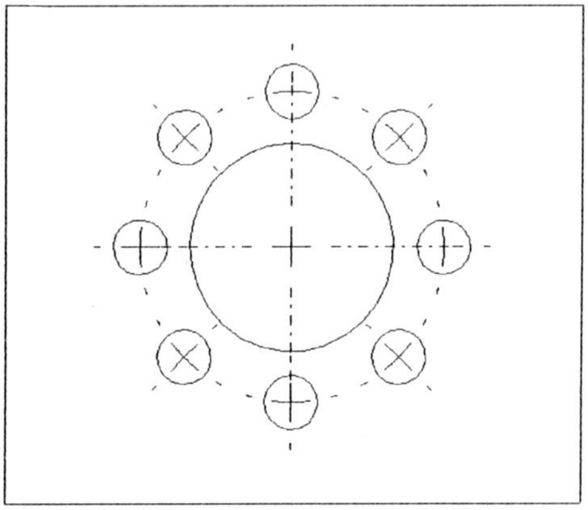

Figure 13-16 *Center marks created with the* ***Base center mark*** *check box selected*

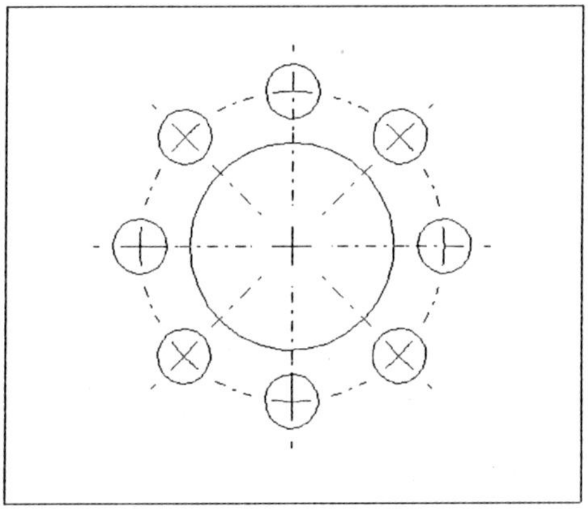

Figure 13-17 *Center marks created with the* ***Radial lines*** *check box selected*

Adding Centerlines to the Drawing Views

CommandManager:	Annotations > Centerline
Menu:	Insert > Annotations > Centerline
Toolbar:	Annotation > Centerline

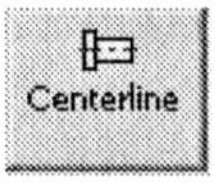

The **Centerline** tool is used to add centerlines to the views by selecting two edges/sketch segments or a single cylindrical/conical face. To add a centerline, choose the **Centerline** button from the **Annotations CommandManager**. The **Centerline PropertyManager** is invoked and it prompts you to select two edges/sketch segments or a single cylindrical/conical face. Select the entity or entities from the view to add the centerline. Figure 13-18 shows the centerlines added to the drawing views by selecting the surface of a cylinder.

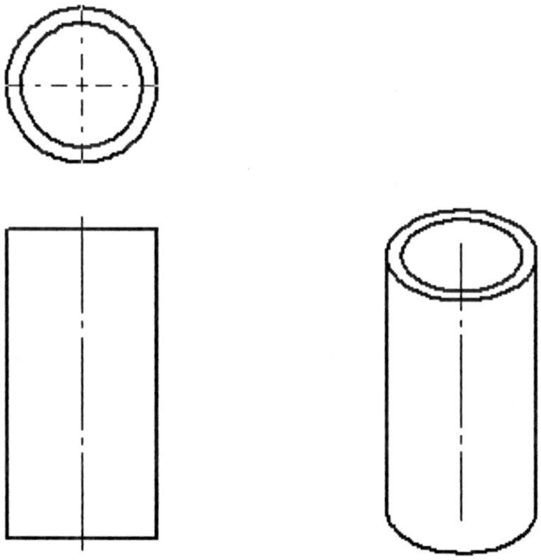

Figure 13-18 *Centerlines added to the front and isometric views using the* ***Centerline*** *tool*

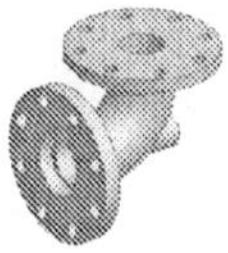

Tip. *You can also set the option for the automatic creation of centerline while generating the drawing views. Choose* ***Tools > Options*** *from the menu bar to invoke the* ***System Options- General*** *dialog box. Select the* ***Document Properties*** *tab from this dialog box. Select the* ***Centerlines*** *check box from the* ***Auto insertion on view creation*** *area and choose* ***OK*** *to close the dialog box.*

You can also directly select a view to add the centerlines.

Adding Hole Callout to the Views

CommandManager:	Annotations > Hole Callout
Menu:	Insert > Annotations > Hole Callout
Toolbar:	Annotation > Hole Callout

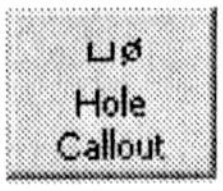

The **Hole Callout** tool is used to generate the hole callouts that are created in the **Part** mode using the **Hole PropertyManager**, the **Hole Wizard** tool, or the cut option. To generate a hole callout, choose the **Hole Callout** button from the **Annotations CommandManager**. The **Dimension PropertyManager** will be displayed and the cursor will be replaced by the hole callout cursor. Select the hole from the drawing views; the hole callout will be attached to the cursor. Pick a point on the drawing sheet to place the hole callout. Figure 13-19 shows a drawing view with hole callouts generated using the **Hole Callout** tool.

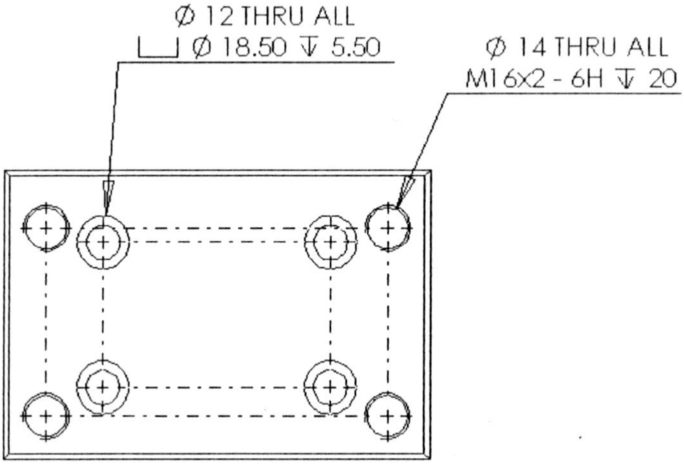

***Figure 13-19** Hole callouts generated using the **Hole Callout** tool*

Adding Cosmetic Threads to the Drawing Views

CommandManager:	Annotations > Cosmetic Thread (*Customize to Add*)
Menu:	Insert > Annotations > Cosmetic Thread
Toolbar:	Annotation > Cosmetic Thread (*Customize to Add*)

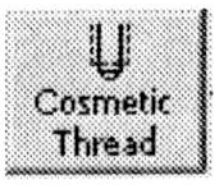

The **Cosmetic Thread** tool is used to add cosmetic threads that will display the thread conventions in the drawing views. To add a cosmetic thread in the drawing view, select the circular edge from the drawing view on which you need to apply the cosmetic thread. Now, choose the **Cosmetic Thread** button from the **Annotations CommandManager**. The **Cosmetic Thread PropertyManager** is displayed as shown in Figure 13-20.

The options available in the **Cosmetic Thread PropertyManager** are discussed next.

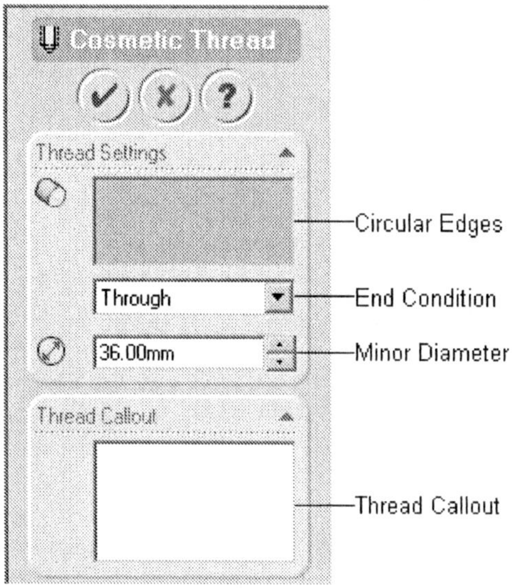

*Figure 13-20 The **Cosmetic Thread PropertyManager***

Thread Settings

The options available in the **Thread Settings** rollout are used to define various parameters of cosmetic thread. When you invoke this tool, you are prompted to select the edges for the threads and set the parameters. Select the required circular edges on which you need to add a cosmetic thread. The name of the selected edge is displayed in the **Circular Edges** selection area. Specify the end condition and set the minor diameter of the thread.

Thread callout

The **Thread Callout** edit box is used to specify the text to be used in the thread callout for the cosmetic thread.

After setting all the parameters, choose the **OK** button from the **Cosmetic Thread PropertyManager**. When you add a cosmetic thread to a generated drawing view, the thread convention will be displayed in all the drawing views. Figure 13-21 shows the cosmetic thread added to the drawing views.

Note

The thread conventions once added in the drawing view can be deleted only from the part document. You cannot delete the thread convention from the drawing document. To delete the thread convention in the part document, expand the ***Hole*** *or* ***Cut*** *feature in the* ***FeatureManager Design Tree****. Now, select the thread convention and delete it.*

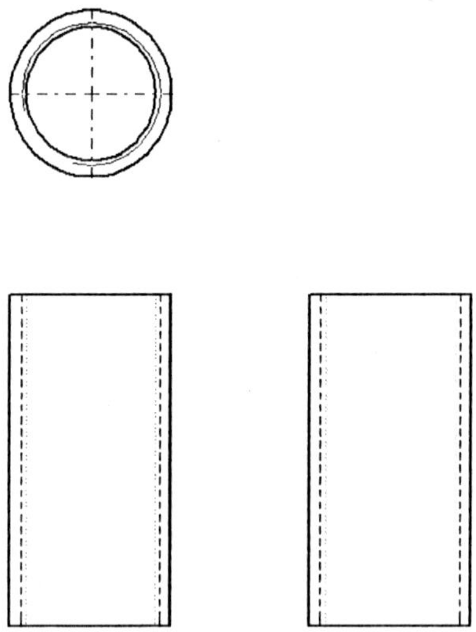

Figure 13-21 *Cosmetic threads added to the drawing views*

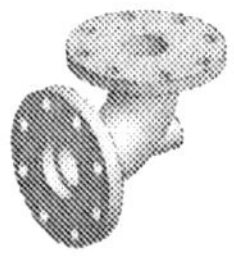

Tip. *You can view the cosmetic threads in the part document also if cosmetic threads are added to the drawing views. You can use any display mode to view the cosmetic threads in the model.*

Adding the Multi-jog Leader to the Drawing Views

CommandManager:	Annotations > Multi-jog Leader (*Customize to Add*)
Menu:	Insert > Annotations > Multi-jog Leader
Toolbar:	Annotation > Multi-jog Leader (*Customize to Add*)

Using the **Multi-jog Leader** tool you can add a multi-jog leader line to the drawing views. A multi-jog leader is basically a leader in which you can add multiple jog lines. To add a multi-jog leader, choose the **Multi-jog Leader** button from the **Annotations CommandManager**. The cursor will be replaced by the multi-jog line cursor. Select a point on the sheet or on an entity from where you need to start the leader. Now, specify the points on the drawing sheet to specify the jogs and then select the second entity where the end of leader will be placed. A multi-jog leader will be created. Double-click the sheet to end multi-jog leader creation. Figure 13-22 shows a multi-jog leader added to the drawing view.

Adding the Dowel Pin Symbols to the Drawing Views

CommandManager:	Annotations > Dowel Pin Symbol (*Customize to Add*)
Menu:	Insert > Annotations > Dowel Pin Symbol
Toolbar:	Annotation > Dowel Pin Symbol (*Customize to Add*)

The **Dowel Pin Symbol** tool is used to add the dowel pin symbol to the holes in the drawing views. The dowel pin symbol is used to confirm the size of the selected hole. To create a dowel pin symbol, select a hole or circular edge from the drawing

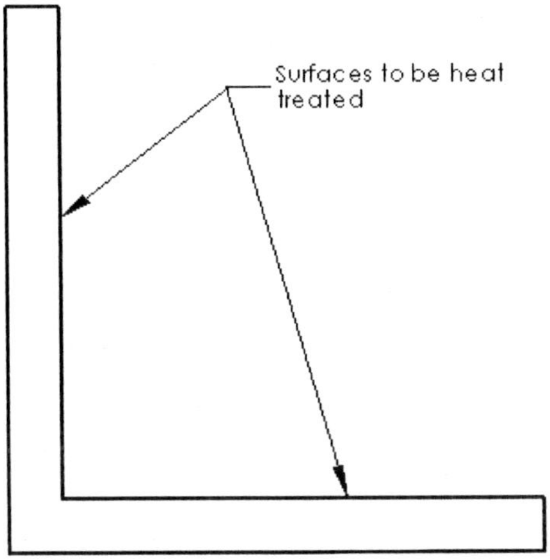

Figure 13-22 *Multi-jog leader added to the drawing view*

view and choose the **Dowel Pin Symbol** button from the **Annotations CommandManager**. The **Dowel Pin Symbol PropertyManager** is displayed. Using the **Flip symbol** check box in the **Display Attributes** rollout, you can flip the direction of the dowel pin symbol.

EDITING THE ANNOTATIONS

You can edit the annotation added to the drawing views by selecting them or by double-clicking them to display their respective **PropertyManager** or dialog box. Using the **PropertyManager** or dialog box you can edit the parameters of the selected annotation.

ADDING THE BOM IN THE DRAWING

CommandManager:	Annotations > Bill of Materials
Menu:	Insert > Tables > Bill of Materials
Toolbar:	Annotation > Bill of Materials

The BOM (Bill of Materials) is a table that displays the list of the components used in the assembly. This table can also be used to provide information related to the number of components in an assembly, their names, quantity, and other related information required to assemble the components in an assembly. Remember that the sequence of the parts in the BOM depends on the sequence in which they were inserted in the assembly document. The BOM placed in the drawing document is parametric in nature. Therefore, if you add or delete a part from the assembly in the assembly document, the change will be reflected in the BOM in the drawing document.

To insert a BOM in the drawing file having the drawing views of an assembly, select any one view from the drawing document and choose the **Bill of Materials** button from the **Annotations CommandManager**. The **Bill Of Materials PropertyManager** is invoked. Its partial view is displayed in Figure 13-23.

The options available in the **Bill Of Materials PropertyManager** are discussed next.

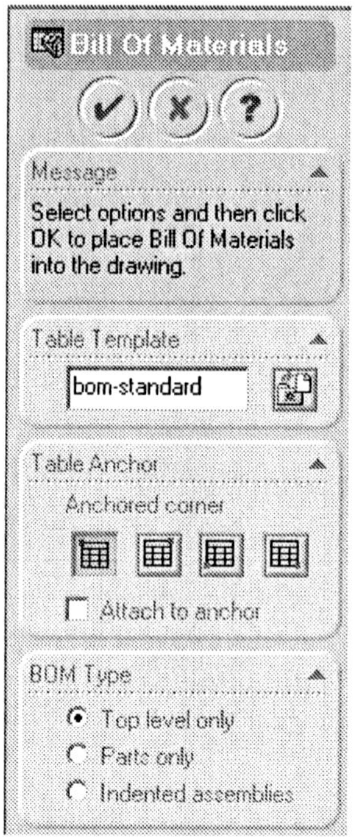

Figure 13-23 The partial view of the ***Bill Of Materials PropertyManager***

Table Template

The **Table Template** rollout is used to define the template to be used in BOM. By default, the **bom-standard** template is selected. If you choose the **Open table template for Bill of Materials** button, the **Select BOM Template** dialog box is displayed. Using this dialog box you can choose any of the default templates available in SolidWorks.

Table Anchor

The buttons available in the **Table Anchor** rollout are used to select the corner of the BOM using which it will be attached to the drawing sheet. If the **Attach to anchor** check box is selected, the table will be automatically attached to the anchor point of the drawing sheet.

BOM Type

The **BOM Type** rollout is used to specify the level of the assembly. The options available in this rollout are discussed next.

Top level only

The **Top level only** radio button is used to list only the parts and the subassemblies in BOM. The parts of the subassemblies are not listed in the BOM.

Parts only

If you select the **Parts only** radio button then subassemblies are not listed in the BOM. The components of the subassemblies are listed as individual components in the BOM.

Indented assemblies

The **Indented assemblies** radio button is selected to list the components and the subassemblies. The components of the subassemblies are listed as indented below their respective subassemblies. These components are not listed with their respective item number.

Configurations

The **Configurations** rollout is used to specify the configuration for creating the BOM. By default the **Default** configuration is selected. You will learn more about configurations in the advanced chapters in part 2 of this book.

Keep Missing Items

After creating the BOM, if a component is deleted from the assembly, the component will also be deleted from the BOM. If you need to retain the deleted component in the BOM, select the **Keep Missing Items** check box. The **Keep Missing Items** rollout is also invoked. If you select the **Strikeout** check box available in the **Keep Missing Items** rollout, then the name of the missing components will be displayed in the BOM with a strikeout.

Zero Quantity Display

The options available in the **Zero Quantity Display** rollout are used to set the option to display the component using -, 0, blank in the BOM if that component is not displayed in a particular configuration. You will learn more about configurations in the advanced chapters.

Item Numbers

You can specify a numeric value from where the sequence of the components starts in a BOM. You can set this value in the **Start at** edit box in this rollout.

After setting all the parameters in the **Bill Of Materials PropertyManager** choose the **OK** button. A BOM will be attached to the cursor. Specify a point on the drawing sheet to place the BOM. Figure 13-24 shows a BOM added to a drawing sheet.

ADDING BALLOONS TO THE DRAWING VIEWS

After adding the BOM, you need to add the balloons to the components in the drawing views. Until the previous release, the balloons were added to drawing views using the **Balloon** tool. From this release, you can also add automatic balloons using the **AutoBalloon** tool. Both these methods of adding balloons to the components in the drawing views are discussed next.

Adding Balloons Using the Balloon Tool

CommandManager:	Annotations > Balloon
Menu:	Insert > Annotations > Balloon
Toolbar:	Annotation > Balloon

The **Balloon** tool is used to manually add the balloons to the components of the assembly in the drawing view. The naming of balloons depends on the sequence of the parts in the BOM. To add balloons to the drawing view, choose the **Balloon**

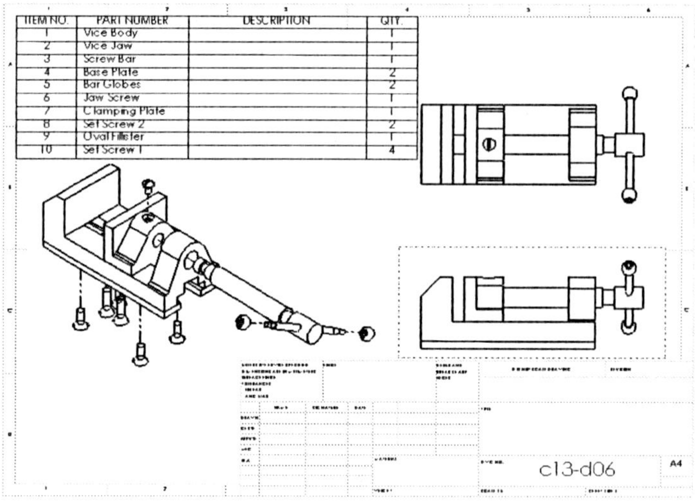

Figure 13-24 *BOM added to the drawing sheet*

button from the **Annotations CommandManager** the **Balloon PropertyManager** is displayed. The partial view of the **Balloon PropertyManager** is displayed in Figure 13-25.

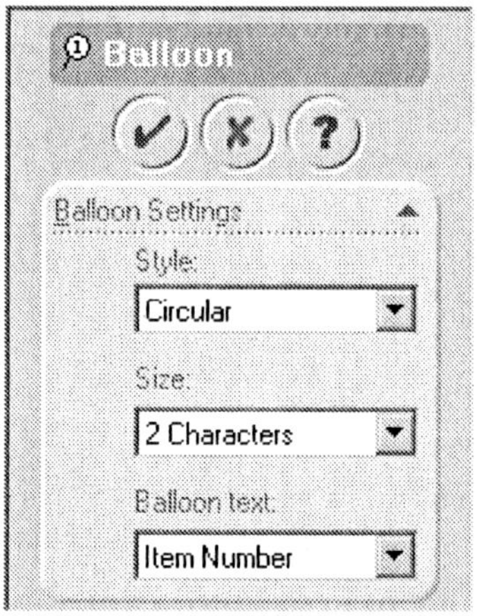

Figure 13-25 *Partial view of the* ***Balloon PropertyManager***

After invoking the **Balloon PropertyManager**, you are prompted to select one or more locations to place balloons. Select the components from the assembly drawing view to add the balloons. If you select the face of a component, the balloon will have a filled circle at the attachment point. However, if you select an edge of the component, the balloon will have a closed filled arrow. After adding balloons to all the components, choose the **OK** button from the **Balloon PropertyManager**. Figure 13-26 shows the drawing sheet in which balloons are added.

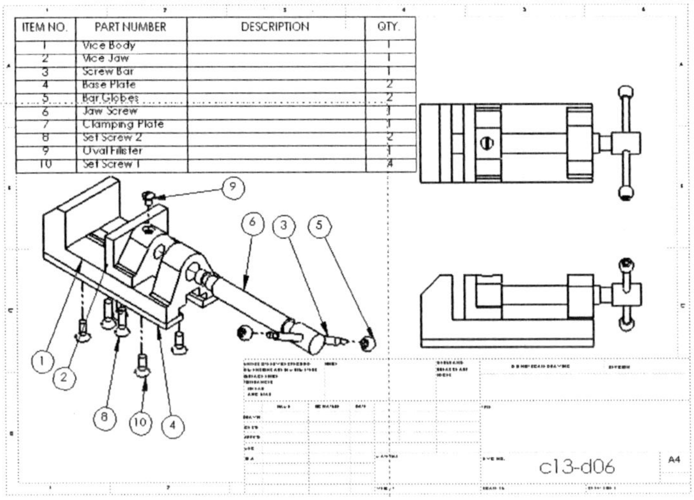

Figure 13-26 Balloons added to the assembly drawing view

Adding Balloons Using the AutoBalloon Tool

CommandManager: Annotations > AutoBalloon
Menu: Insert > Annotations > Auto Balloon
Toolbar: Annotation > AutoBalloon

One of the major enhancements in the **Drawing** mode of SolidWorks is the introduction of the **AutoBalloon** tool. As discussed earlier, you need to manually select the components of the assembly on which you need to add the balloons. This tool enables you to add balloons automatically. To add balloons automatically select the drawing view in which you need to add the balloons and choose the **AutoBalloon** button from the **Annotations CommandManager**. Balloons are added to the selected drawing view with the default settings and the **Auto Balloon PropertyManager** is also displayed as shown in Figure 13-27. The options available in the **Auto Balloon PropertyManager** are discussed next.

Balloon Layout

The options available in the **Balloon Layout** rollout are used to set the layout of the balloons when they are placed automatically. By default, the balloons are placed in a square form because the **Square** button is chosen in the **Balloon Layout** rollout. You can also arrange the balloons in circular, aligned to top, aligned to bottom, aligned to left, and aligned to right forms. The **Ignore multiple instances** check box is selected by default to eliminate the creation of multiple instances of a balloon.

Balloon Settings

The **Balloon Settings** rollout is used to set the style, size, and type of the text of balloons.

After setting all the parameters choose the **OK** button from the **AutoBalloon PropertyManager**.

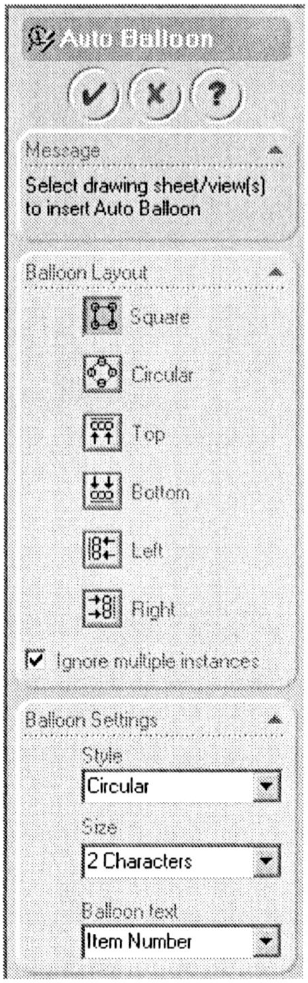

Figure 13-27 Partial view of the ***Auto Balloon PropertyManager***

ADDING NEW SHEETS TO THE DRAWING VIEWS

You can also add new sheets to the drawing document. A multisheet drawing document can be used when you need to generate the drawing views of all the components and the drawing views of the assembly in the same document. You can switch between the sheets easily to refer to the drawings of different parts of an assembly within the same document and you do not need to open the separate drawing documents. To add a sheet to the drawing document, select **Sheet1** from the **FeatureManager Design Tree** and invoke the shortcut menu. Choose the **Add Sheet** option from the shortcut menu to add a sheet to the drawing document or choose **Insert** > **Sheet** from the menu bar. The **Sheet Properties** dialog box is displayed. Using this dialog box you can specify the name, standard sheet format, and scale projection type for the drawing sheet to be added. Choose the **OK** button from this dialog box; a new sheet will be added in the drawing document. Figure 13-28 shows a drawing document with four drawing sheets added and **Sheet 4** as the active sheet.

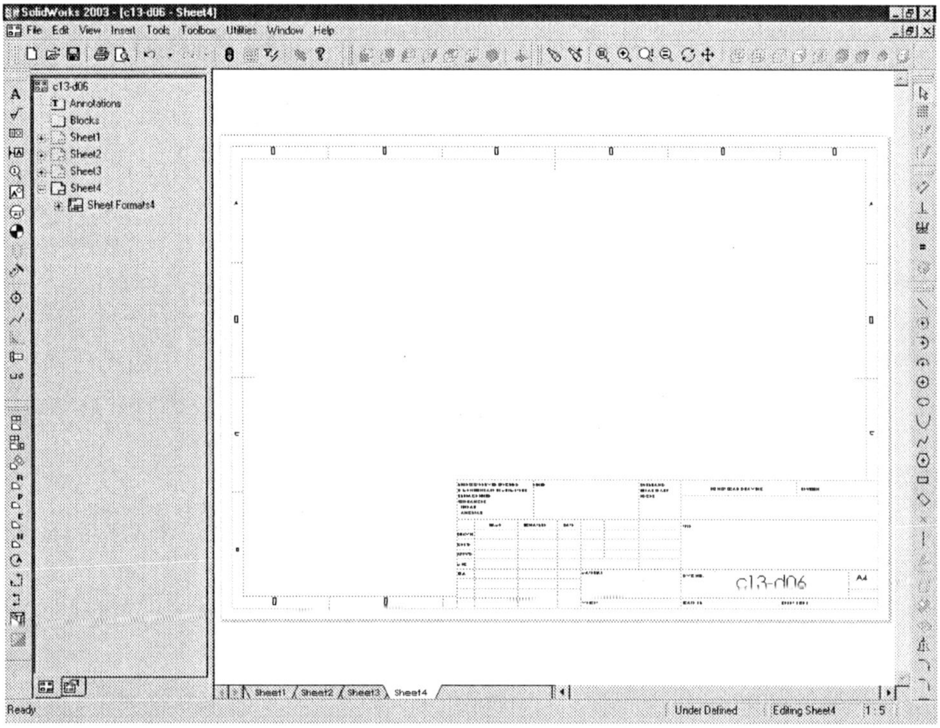

Figure 13-28 Drawing sheets added to the drawing document

To activate a drawing sheet, select the sheet from the **FeatureManager Design Tree** and invoke the shortcut menu. Choose the **Activate** option from the shortcut menu to activate the selected sheet. You can also activate the sheet by selecting the sheet tab from the bottom of the drawing document.

EDITING THE SHEET FORMAT

You can edit the standard sheet format according to your requirement. To edit the standard sheet format, make that sheet active and select it from the **FeatureManager Design Tree**. Right-click on the sheet and invoke the shortcut menu. Choose the **Edit Sheet Format** option from the shortcut menu. All the entities, annotations, and views will disappear from the drawing sheet. Using the sketching tools available in the **Sketch CommandManager**, you can edit the sheet format. After editing the sheet format, select the active sheet and invoke the shortcut menu. Choose **Edit Sheet** from the shortcut menu to switch back to the edit sheet environment.

CREATING A USER-DEFINED SHEET FORMAT

In SolidWorks, you can also create a user-defined sheet format. To create a user-defined sheet format, start a new drawing document with a custom size. On the basis of your design requirements, you can set a size of the sheet in the **Width** and **Height** edit boxes available in the **Sheet Format/Size** dialog box. After defining the sheet, select **Sheet1** from the **FeatureManager Design Tree** and invoke the shortcut menu. Choose the **Edit Sheet Format** option from the shortcut menu. The edit sheet format environment is invoked. Using the standard sketching tools available in the **Sketch** CommandManager, you can create or modify the sheet format. After creating or modifying the sheet format, switch back to the edit sheet environment. Choose **File > Save Sheet Format** from the menu bar. The **Save Sheet Format** dialog box is displayed. Browse the location where you need to save the sheet format. Specify the name of the sheet format and choose the **Save** button from the **Save Sheet Format** dialog box. Next time when you invoke the **Sheet Format/Size** dialog box, you can retrieve the saved sheet format by selecting the **Standard sheet format** radio button. Using the **Browse** button, browse to the location where you have saved the sheet format. Figure 13-29 shows a user-defined sheet format.

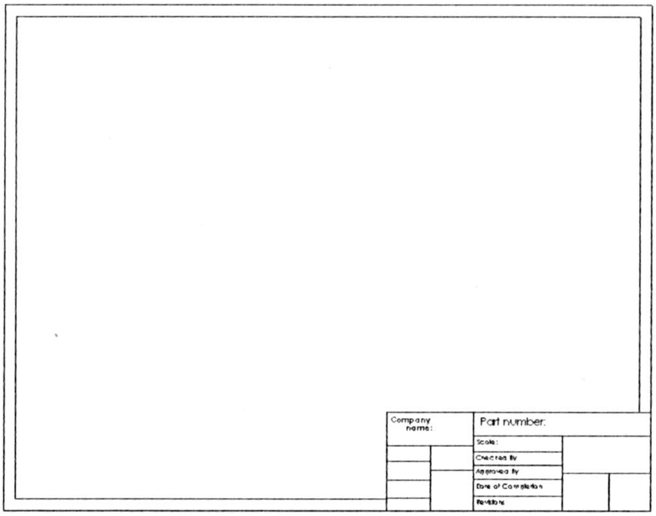

Figure 13-29 *A user-defined sheet format*

TUTORIALS

Tutorial 1

In this tutorial you will open the drawing created in Tutorial 1 of Chapter 12 and generate the dimensions and add the annotations to the drawing. After that you will change the display of the front and the right view to hidden lines visible, and the display of the isometric and aligned section views to shaded mode. **(Expected time: 45 min)**

The steps to be followed to complete this tutorial are listed next.

a. Copy the part and the drawing document in the folder of the current chapter.
b. Configure the font settings and generate the dimensions using the **Model Items** tool.
c. Arrange the dimensions and delete the unwanted dimensions, refer to Figures 13-31 and 13-32.
d. Add the datum symbol and geometric tolerance to the drawing views, refer to Figures 13-32 and 13-34.
e. Change the model display state of the drawing views, refer to Figure 13-35.

Copying the Documents in the Folder of the Current Chapter

Before you proceed, you need to copy the model and the drawing document in the folder of the current chapter.

1. Create a folder with the name *c13* in the *SolidWorks* folder and copy *c07-tut02.sldprt* and *c12-tut01.slddrw* from the */My Document/SolidWorks/c12* folder to this folder.

Opening the Drawing Document

After copying the part and the drawing document, you need to open the drawing document in the SolidWorks window.

1. Invoke the **Open** dialog box and open *c12-tut01.slddrw* document from the folder of the current chapter.

   The drawing document in which you need to add the detailing is displayed in the drawing area, as shown in Figure 13-30.

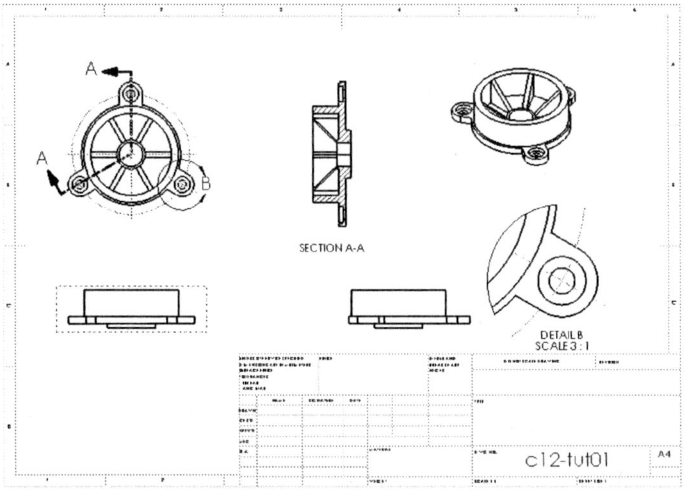

***Figure 13-30** Drawing views generated in Chapter 12*

Applying the Documents Settings

Before you start generating the model dimensions, you need to configure the document settings. These settings will allow the dimensions and other annotations in the current sheet to be viewed properly.

1. Invoke the **Document Properties - Detailing** dialog box and select the **Annotations Font** option from the area on the left.

2. The **Note** option is selected in the **Annotation type** area. Select this option again. The **Choose Font** dialog box will be invoked.

3. Select the **Points** radio button from the **Height** area and set the value of the font size to **9** from the list box.

4. Choose the **OK** button from the **Choose Font** dialog box. Similarly, set the font size for **Dimension, Detail,** and **Section** options to **9**.

5. Now, select the **Arrows** option from the area on the left. The options related to size and shape of the arrows are displayed.

6. Set the value of the **Height** as **1, Width** as **3,** and **Length** as **3** in the **Size** area.

7. In the **Section/View size** area, set the value of the **Height** as **2, Width** as **4,** and **Length** as **8**.

8. Choose the **OK** button from the **Document Properties - Detailing - Arrows** dialog box.

Generating the Dimensions

Next, you need to generate the dimensions using the **Model Items** tool. As discussed earlier, if you do not select any view and generate the dimensions using the **Model Items** tool, all the dimensions will be displayed in all the views. Sometimes the dimensions overlap each other. Therefore, you will select the view in which you need to generate the dimension and then you will invoke the **Model Items** tool.

1. Select the top view and choose the **Model Items** button from the **Annotation CommandManager**. The **Insert Model Items** dialog box is displayed. The name of the selected view is displayed in the **Import to views** selection area.

2. Choose the **OK** button from the **Insert Model Items** dialog box.

The dimensions of the model that can be displayed in the selected view are generated. If a radial dimension is attached to the counterbore hole in the top view, it needs to be deleted because you will later add a hole callout to this counterbore hole.

3. Select the radial dimension and press the DELETE key from keyboard to delete the dimension. Similarly, delete the dimension of diameter **20**.

Note that if SolidWorks 2004 is configured for ANSI system on your machine, the dimension arrowheads are displayed as no filled. You need to change the arrowheads to filled.

4. Drag a window such that all the dimensions are enclosed inside the window. Release the left mouse button to select all the generated dimensions. The **Dimension PropertyManager** is displayed.

5. Select the **Filled Arrow** option from the **Style** drop-down list of the **Arrow** rollout. Choose **OK** to close the **Dimension PropertyManager**.

The arrowheads will be changed to filled arrowheads. Note that the dimensions that are generated are scattered arbitrarily on the drawing sheet. Therefore, you need to arrange the dimensions by moving them to the required locations.

6. Select the dimension one by one and drag them to the desired location, refer to Figure 13-31. You can reverse the direction of arrowheads by clicking on the control point that is displayed on the arrowheads.

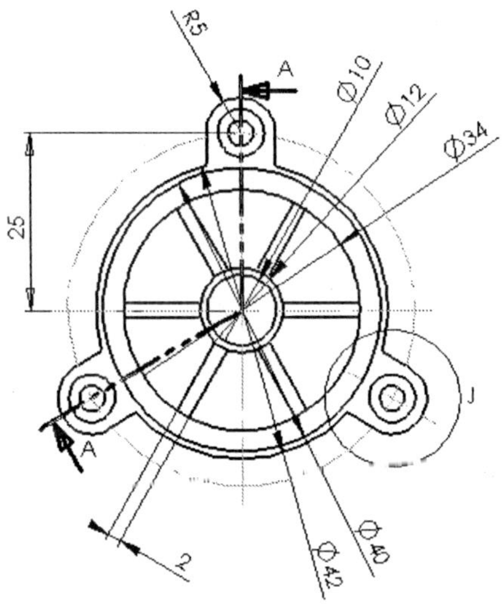

Figure 13-31 *Top view after generating and arranging the dimensions*

7. Select the aligned section view and generate the dimensions using the **Model Items** tool. After placing the dimensions, move them to appropriate places, refer to Figure 13-32. You may need to change the arrowheads to closed filled if they are not filled.

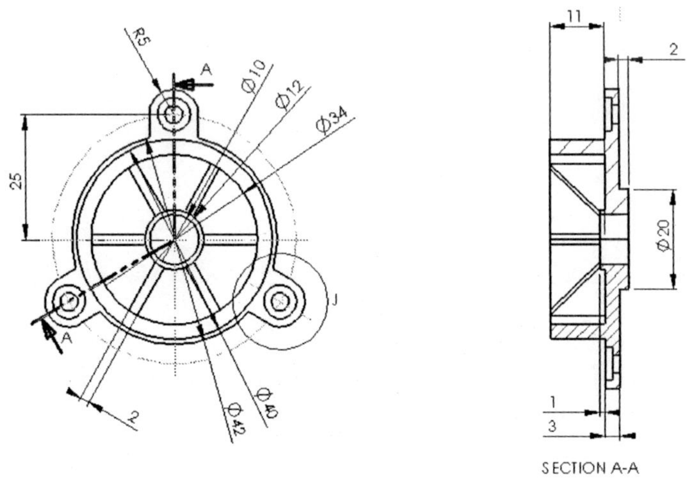

Figure 13-32 *Partial view of the sheet after generating and arranging the dimensions in the aligned section view*

8. Choose the **Hole Callout** button from the **Annotation CommandManager** and select the outer circle of one of the counterbore feature in the front view. The hole callout will be attached to the cursor. Pick a point on the drawing sheet to place the hole callout.

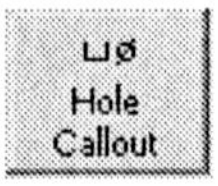

Adding the Datum Feature Symbol to the Drawing View

After generating the dimensions, you need to add the datum feature symbol to the drawing view. Datum feature symbols are used as the datum reference for adding the geometric tolerance to the drawing views.

1. Select the edge of the outer cylindrical feature from the top view and then choose the **Datum Feature** button from the **Annotations CommandManager**. The **Datum Feature PropertyManager** is displayed and a datum callout is attached to the cursor.

2. Place the datum symbol at an appropriate location, refer to Figure 13-33.

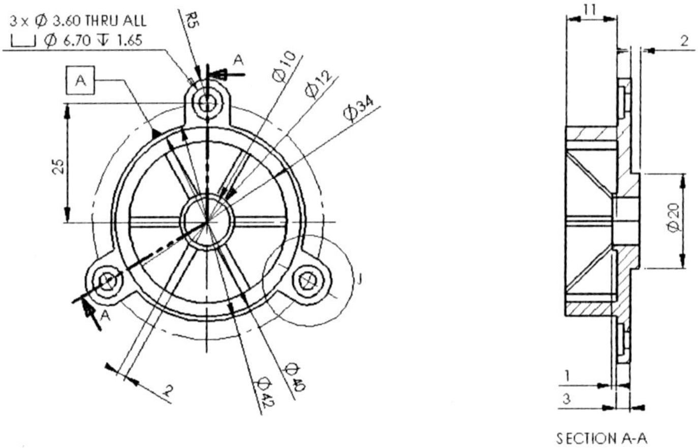

Figure 13-33 Datum feature symbol hole description added to the top view

3. Choose the **OK** button from the **Datum Feature PropertyManager**.

Adding the Geometric Tolerance to the Drawing View

After defining the datum feature symbol, you will add the geometric tolerance to the drawing view.

1. Select the circular edge that has a diameter of 12 from the top view and choose the **Geometric Tolerance** button from the **Annotation CommandManager**.

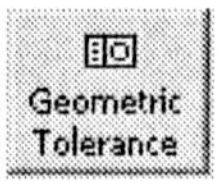

The **Properties** dialog box is displayed. It is used to specify the parameters of the geometric tolerance.

2. Choose the **GCS** button corresponding to the first row in the **Feature control frames** area. The **Symbols** dialog box is displayed.

3. Select the **Concentricity and coaxiality** option from the list box and choose the **OK** button from the **Symbols** dialog box.

4. Set the value of tolerance in the **Tolerance 1** edit box to **0.002**.

5. Specify **A** in the **Primary** edit box to define the primary datum reference.

6. Choose the **OK** button from the **Properties** dialog box. The geometric tolerance will be placed attached to the selected circular edge. You may need to move the geometric tolerance if it overlaps the dimensions. Figure 13-34 shows the drawing view after adding and rearranging the geometric tolerance.

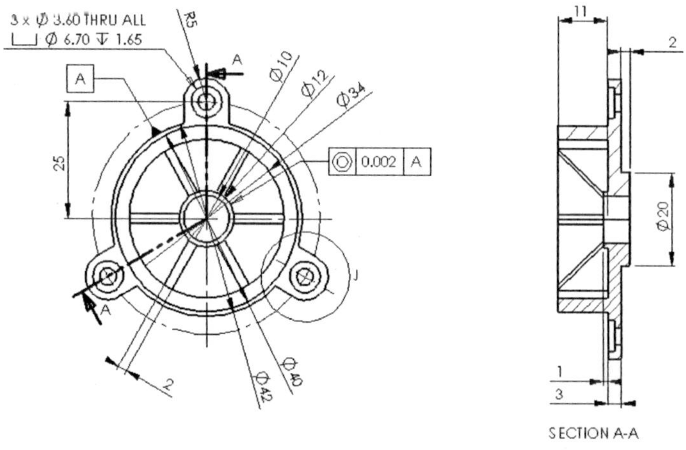

Figure 13-34 Geometric tolerance added to the drawing view

Changing the View Display Options

After adding all the annotations to the drawing views, you need to change the display setting of the drawing views.

1. Press and hold down the CTRL key and select the front view and the right-side view from the drawing sheet.

2. Choose the **Hidden Lines Visible** button from the **View** toolbar. You can also choose this button from the **Display Style** rollout of the **Multiple Views PropertyManager** that is displayed when you select the two views.

The hidden lines will be displayed in the selected drawing views.

3. Now, select the isometric view and the aligned section view from the drawing sheet.

4. Choose the **Shaded With Edges** button from the **View** toolbar or the **Multiple Views PropertyManager**. Figure 13-35 shows the final drawing sheet after changing the display view settings.

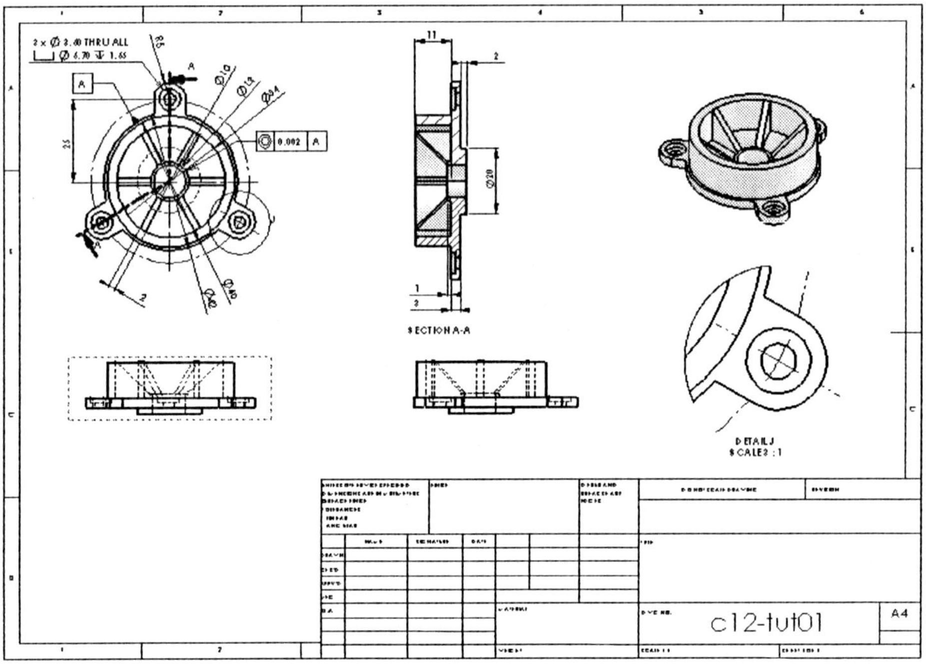

Figure 13-35 *Final drawing sheet*

5. Save and close the drawing document.

Tutorial 2

In this tutorial, you will generate the BOM of the Bench Vice assembly and then add balloons to the isometric view in the exploded state. **(Expected time: 45 min)**

The steps to be followed to complete this tutorial are listed next.

a. Copy the Bench Vice folder that contains parts, assembly, and the drawing document to the folder of the current chapter.
b. Delete the views that are not required in the drawing sheet, refer to Figure 13-37.
c. Move to arrange the views in the drawing sheet, refer to Figure 13-37.
d. Set the anchor on the drawing sheet where the BOM will be attached.

e. Generate the BOM, refer to Figure 13-38.
f. Add balloons to the isometric view, refer to Figure 13-39.

Copying the Bench Vice Assembly Folder to the Current Folder

1. Copy the Bench Vice folder from the */My Document/SolidWorks/c12* folder to the folder of the current chapter.

Opening the Drawing Document

After copying the folder, you need to open the drawing document in the SolidWorks window.

1. Open *c12-tut02.slddrw* document.

The drawing document in which you need to generate the BOM and balloons is displayed in Figure 13-36.

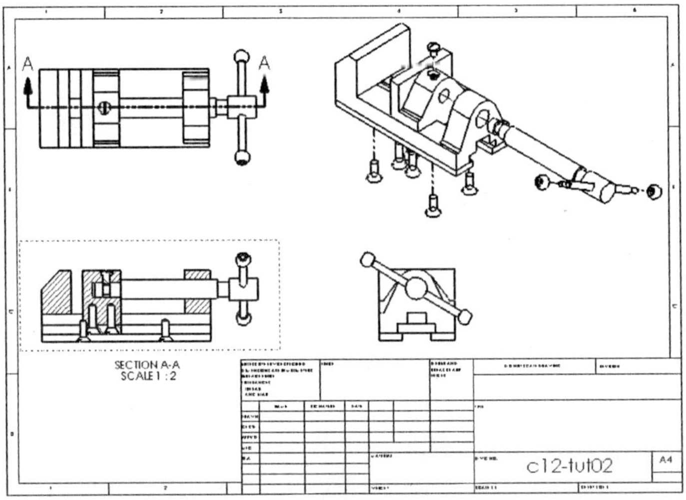

Figure 13-36 *Drawing views generated in Chapter 12*

Deleting the Unwanted View

You need to delete the right view because this view is not required in this tutorial.

1. Select the right-side view and press the DELETE key from the keyboard.

The **Confirm Delete** dialog box will be displayed.

2. Choose the **Yes** button from this dialog box. The view will be deleted from the current drawing sheet.

Moving the Isometric View

You need to move the exploded isometric view because the BOM will be generated and placed at the top right corner of the drawing sheet.

1. Select the isometric view; the border of the view is displayed in green.
2. Move the cursor to the border. The cursor will be replaced by the move cursor.
3. Drag the cursor to move the drawing view. Place it at the required location, refer to Figure 13-37.

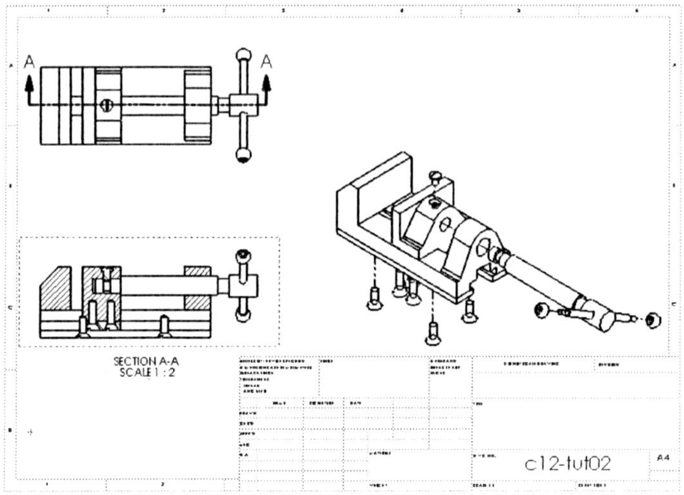

Figure 13-37 *Drawing sheet after moving the drawing view*

Setting the Anchor for the BOM

Before generating the BOM, you need to set the anchor of the BOM. The anchor is a point on the drawing sheet with which one of the corners of the BOM coincides. By default, the anchor is defined at the top left corner of the drawing sheet. But in this tutorial, you need to add the BOM on the top right corner of the drawing sheet. Therefore, you will set anchor before generating the BOM.

1. Expand **Sheet1** from the **FeatureManager Design Tree** and then expand **Sheet Format1**.
2. Select the **Bill of Materials Anchor1** option and right-click to invoke the shortcut menu. Choose the **Set Anchor** option from the shortcut menu. The drawing views will disappear from the sheet.
3. Specify the anchor point on the inner top right corner of the drawing sheet. A point will be placed at the selected location.

After setting the anchor, you need to switch back to the edit sheet environment.

4. Select **Sheet1** from the **FeatureManager Design Tree** and right-click to invoke the shortcut menu. Choose the **Edit Sheet** option from the shortcut menu.

Generating the BOM

Next, you need to generate the BOM. As discussed earlier, the BOM generated in SolidWorks is parametric in nature. If a component is deleted or added in the assembly, the change is automatically reflected in the BOM. But before you generate the BOM, you need to set the text parameters for the BOM.

1. Invoke the **Document Properties - Detailing - Annotation Font** dialog box. Select the **Tables** option from the **Annotation type** area to display the **Choose Font** dialog box.

2. Select the **Points** radio button from the **Height** area and set the value of font size to 9 point. Close both the dialog boxes.

3. Similarly, change the text height of balloons to 12 points.

4. Select the isometric view and choose **Bill of Materials** from the **Annotations CommandManager**. The **Bill Of Materials PropertyManager** is displayed.

5. Select the **Top Right** button from the **Anchored corner** area of the **Table Anchor** rollout. Also, select the **Attach to anchor** check box.

6. Choose the **OK** button from the **Bill Of Materials PropertyManager**.

   The BOM is generated. You will notice that the **Description** column is also displayed in the BOM. But this column is not required. Therefore, you need to delete this column.

7. Move the cursor over the heading **Description** and right-click to display the shortcut menu. Choose **Delete > Column** from the cascading menu. This column will be deleted. The drawing sheet after generating the BOM and deleting the **Description** column is displayed in Figure 13-38.

Note

If the color of the BOM is gray and is not clearly visible, invoke the ***System Options - Colors*** *dialog box. Select the* ***Imported Annotations (Driven)*** *from the* ***System colors*** *list box and change its color to black*

Adding the Balloons to the Components

After generating the BOM, you need to add the balloons to the components. Before proceeding further, make sure that you have changed the font height of balloons to 12 point in the last section.

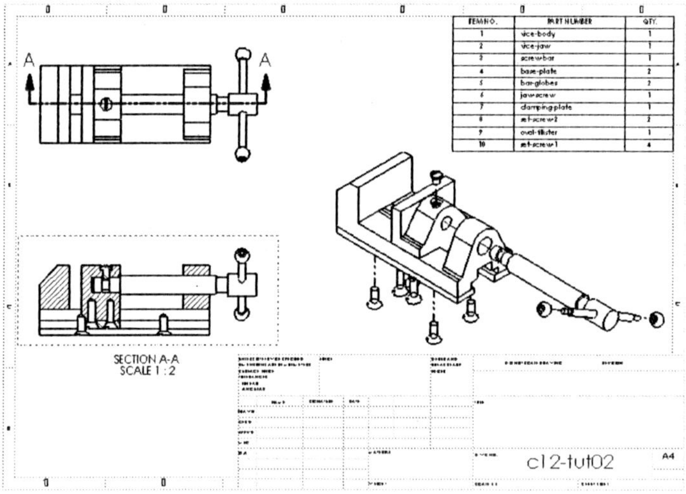

Figure 13-38 *Drawing sheet after generating the BOM*

1. Select the isometric view and choose the **AutoBalloon** button from the **Annotations CommandManager**. Balloons are automatically added to all the components in the isometric view. Also, the **Auto Balloon PropertyManager** is displayed.

   Because the **Ignore multiple instances** check box is selected in the **Balloon Layout** rollout, the multiple instances of any component are ignored.

2. Select **1 Character** from the **Size** drop-down list in the **Balloon Settings** rollout. Choose **OK** to close this **PropertyManager**.

   The balloons are added to all the components, except Clamping Plate. This is because the clamping plate is not visible in the isometric view. You will notice that the balloons are not properly arranged on the sheet. They are placed arbitrarily. Therefore, you need to manually drag each balloon and place them properly.

3. Move the cursor over one of the balloons and when it is highlighted, drag it to place it at a better location, refer to Figure 13-39.

4. Similarly, drag and place the remaining balloons also at proper locations. The final drawing sheet after adding and rearranging balloons is shown in Figure 13-39.

5. Save the drawing and close the document.

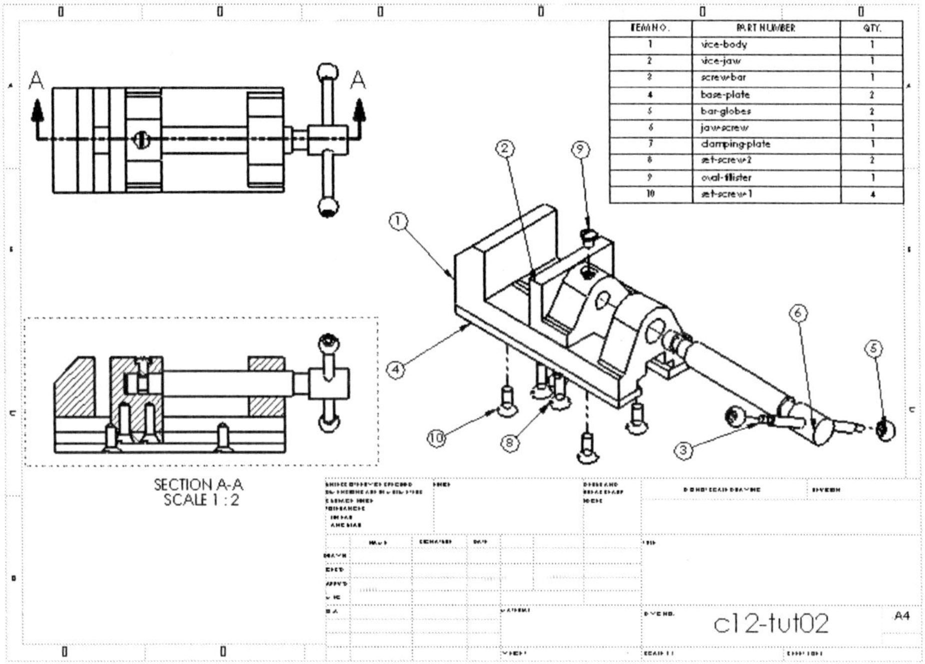

***Figure 13-39** Final drawing sheet after adding balloons*

SELF-EVALUATION TEST

Answer the following questions and then compare your answers with the answers given at the end of this chapter.

1. You cannot add annotations while creating the part in SolidWorks. (T/F)

2. You can add the surface finish symbols to the drawing views. (T/F)

3. The **Projected tolerance zone** area is used to define the quality of the projected tolerance. (T/F)

4. You can set the target shape as a point, circle, or rectangle while adding a datum target. (T/F)

5. You can also set the option for the automatic creation of centerline while generating the drawing views. (T/F)

6. The __________ spinner is used to define the major diameter of the thread.

7. The __________ tool is used to add the balloons to the components of the assembly in the drawing view.

8. You can also create automatic balloons using the __________ tool.

9. Using the __________ rollout available in the **Bill Of Materials PropertyManager** you can specify the template that you need to create the BOM.

10. The __________ tool is used to create a hole callout.

REVIEW QUESTIONS

Answer the following questions.

1. The __________ tool is used to add the cosmetic threads that are used to display the thread conventions in the drawing views.

2. The __________ tool is used to add reference dimensions to the drawing views.

3. You can change the model display setting from hidden lines removed to hidden lines visible, or wireframe, or shaded using the options available in the __________ toolbar.

4. Select the __________ check box from the **Auto insertion on view creation** area to automatically create centerlines while generating the view.

5. The __________ tool is used to create the centerlines in the views.

6. Which **PropertyManager** is invoked to add automatic balloons to the selected drawing view?

   (a) **AutoBalloon** (b) **Balloon**
   (c) **Properties** (d) **Center Mark**

7. Which **PropertyManager** is displayed when you choose the **Cosmetic Thread** button from the **Annotations CommandManager**?

   (a) **Cosmetic Thread Properties** (b) **Cosmetic Thread**
   (c) **Cosmetic Thread Convention** (d) None of these

8. Which **PropertyManager** is used to add center marks to the drawing views?

   (a) **Add Center Mark** (b) **Create Center Mark**
   (c) **Center Mark** (c) **Cosmetic Thread**

9. Which rollout available in the **Cosmetic Thread PropertyManager** is used to define the depth of the cosmetic thread?

   (a) **Thread Settings** (b) **Thread Depth**
   (c) **Cosmetic Thread** (d) None of these

10. Which **PropertyManager** is used to add balloons in the drawing views?

(a) **Add Balloons** (b) **Balloon Properties**
(c) **Balloons** (d) None of these

EXERCISE

Exercise 1

In this exercise you will generate the isometric view in exploded view of the assembly created in Tutorial 1 of Chapter 11 on the standard A4 sheet format. The scale of the view will be 1:5. After generating the view generate the BOM and add balloons to the assembly view shown in Figure 13-40. **(Expected time: 30 min)**

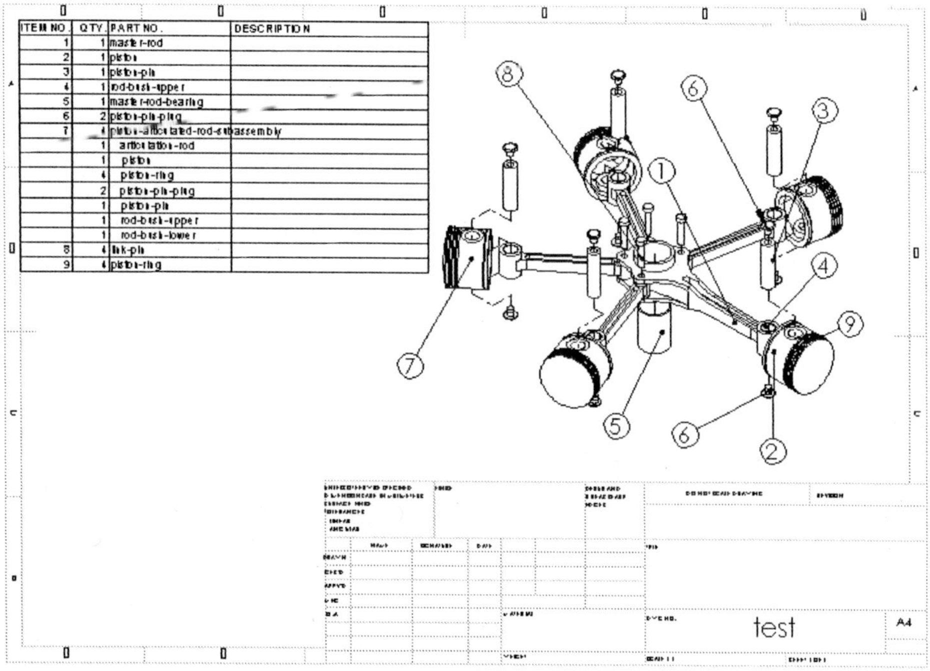

Figure 13-40 *Drawing view for Exercise 1*

Answers to Self-Evaluation Test

1. T, **2.** T, **3.** F, **4.** T, **5.** T, **6. Minor Diameter**, **7. Balloon**, **8. AutoBalloon**, **9. Table Template**, **10. Hole Callout**

Student Project

Student Project 1

Create all the components of the Butterfly Valve shown in Figure 1 and then assemble them. The dimensions for the various components are given in Figure 2 through Figure 9. Assume the missing dimensions for the components.

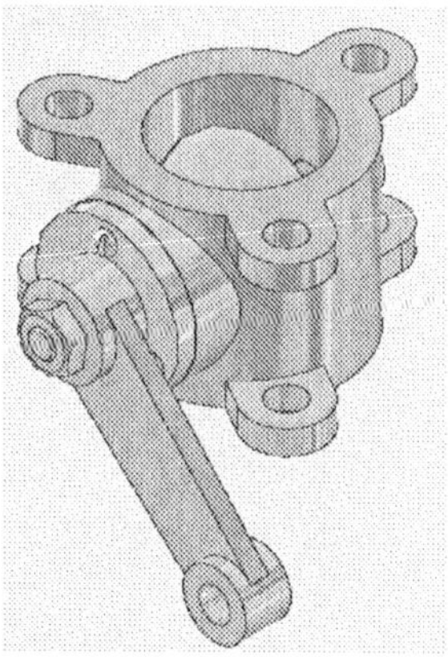

Figure 1 Butterfly Valve assembly

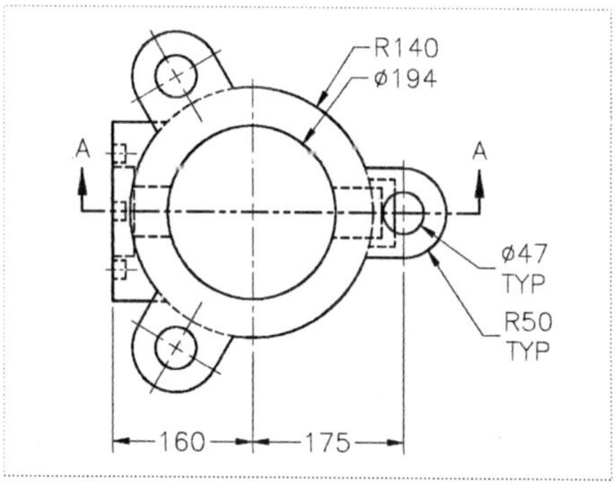

Figure 2 Top view of the Body

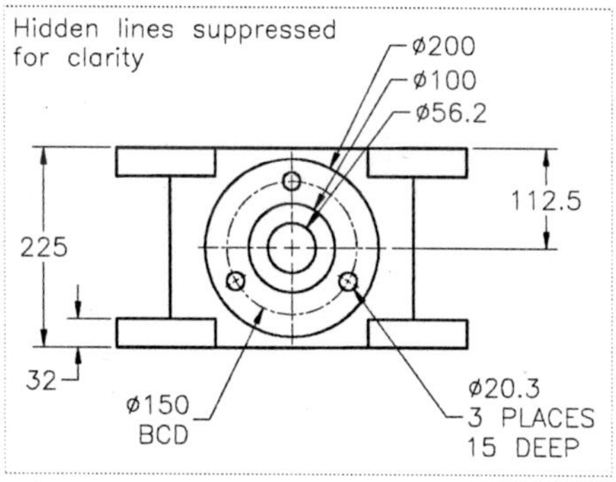

Figure 3 Left side view of the Body

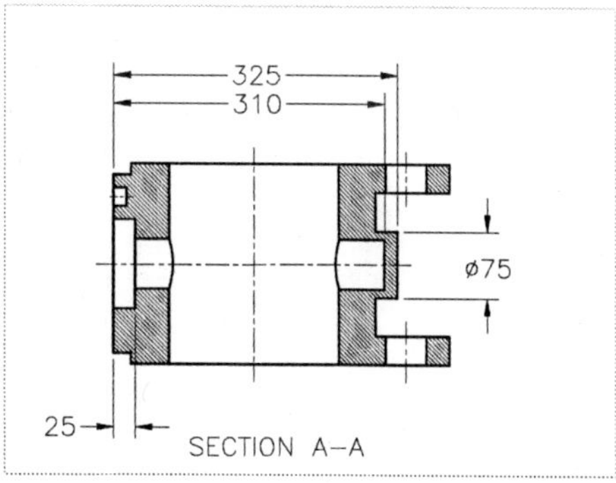

Figure 4 Sectioned front view of the Body

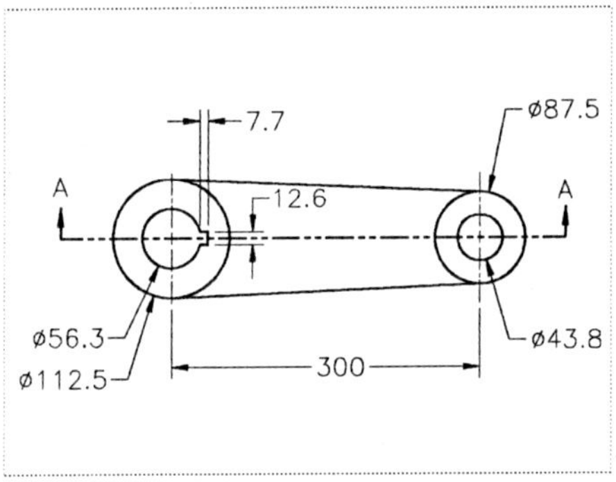

Figure 5 Top view of the Arm

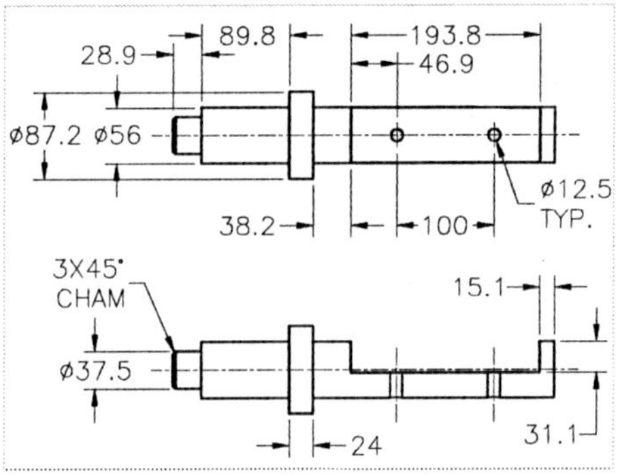

Figure 6 Dimensions of the Shaft

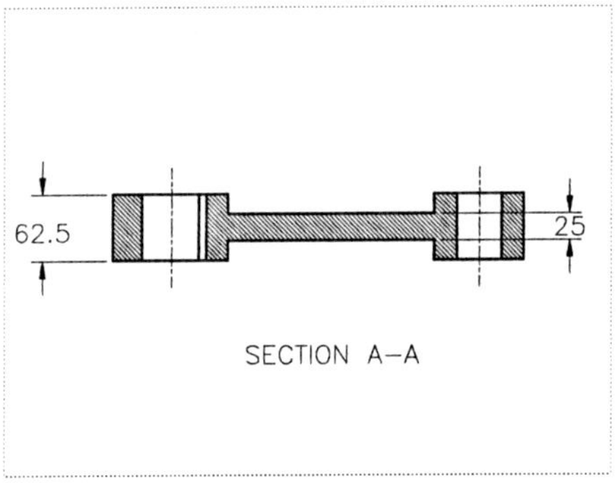

Figure 7 Sectioned front view of the Arm

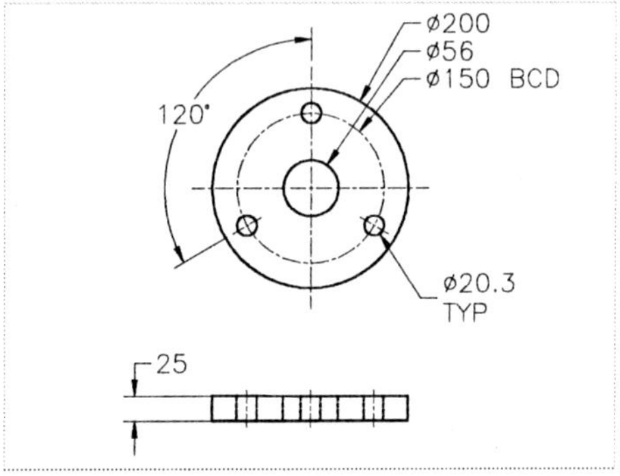

Figure 8 Dimensions of the Retainer

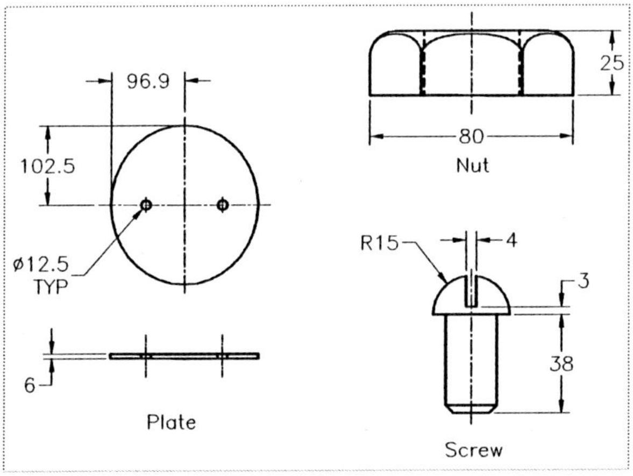

Figure 9 Dimensions of the Plate, the Nut, and the Screw

Student Project 2

Create different components of the Double Bearing assembly and then assemble them as shown in Figure 10. Figure 11 shows the exploded view of the assembly. The dimensions of various components are given in Figure 12 to Figure 16.

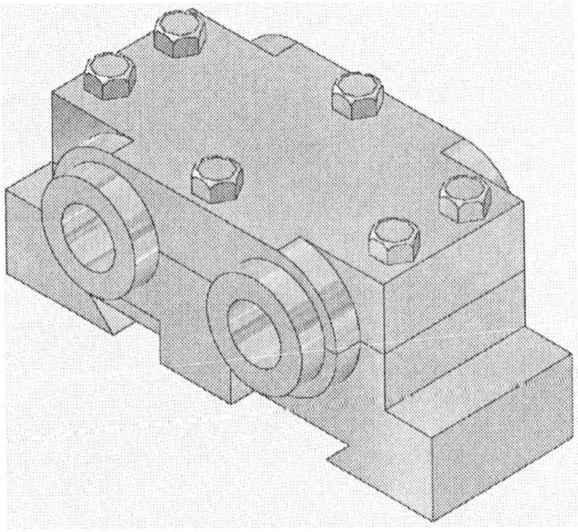

Figure 10 *Double Bearing assembly*

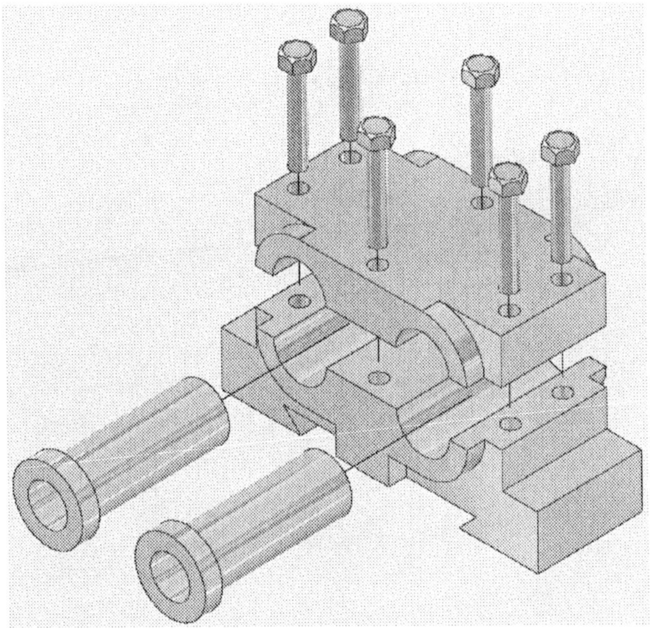

Figure 11 *Exploded view of Double Bearing assembly*

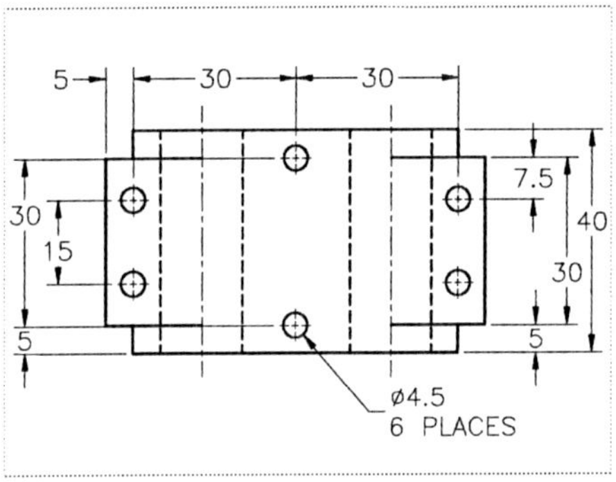

Figure 12 *Top view of the Cap*

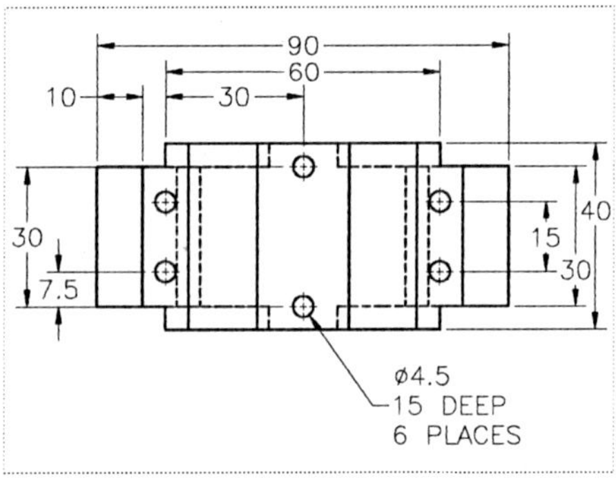

Figure 13 *Top view of the Base*

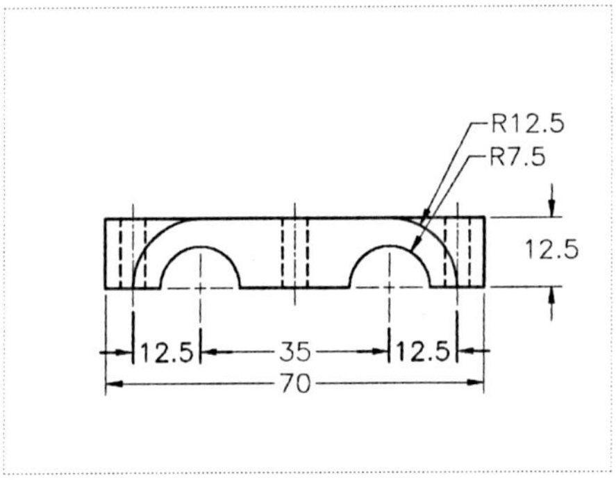

Figure 14 *Front view of the Cap*

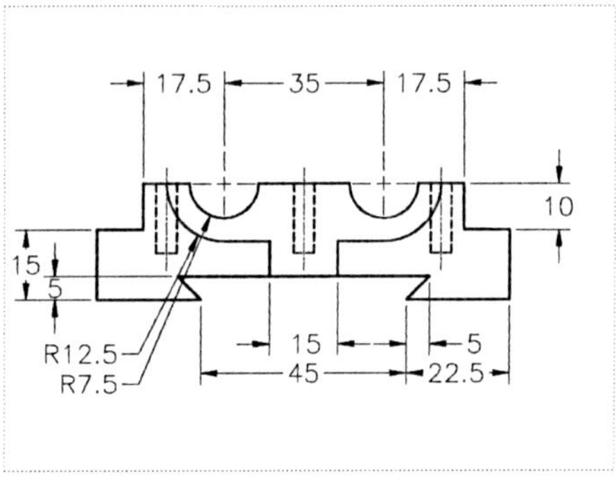

Figure 15 *Front view of the Base*

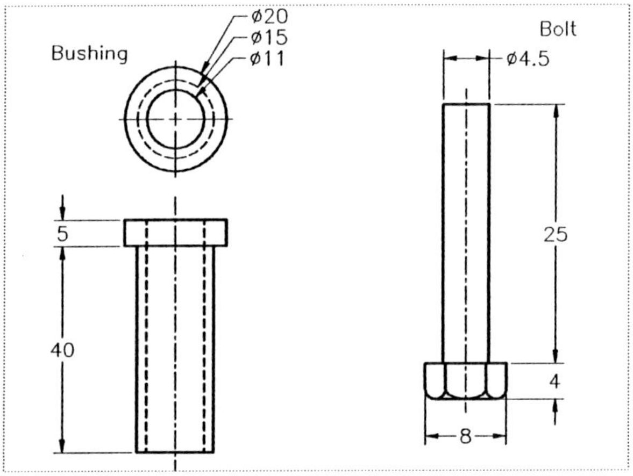

Figure 16 *Dimensions of the Bushing and the Bolt*

Student Project 3

In this tutorial you will create all the components of the Wheel Support assembly and then assemble them together as shown in Figure 17. The exploded view of the assembly is shown in Figure 18. The dimensions of the components are shown in Figure 19, to Figure 23.

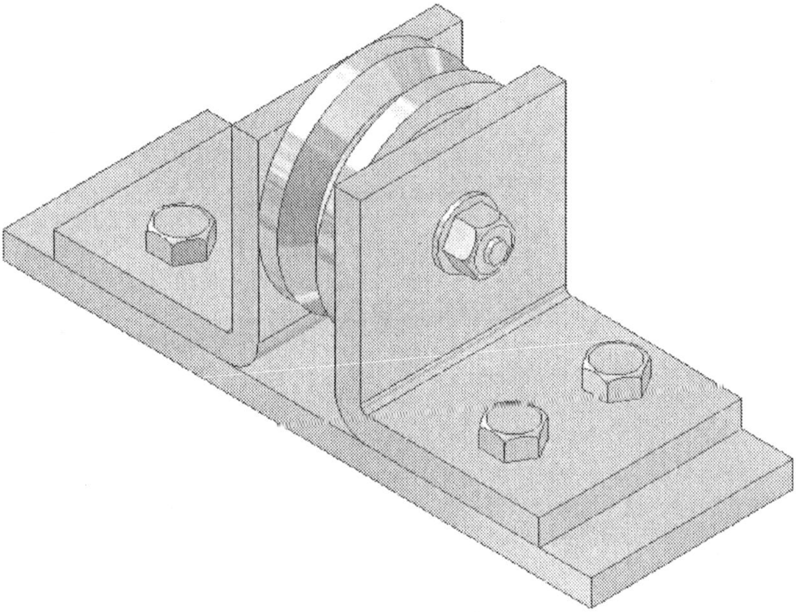

Figure 17 *Wheel Support assembly*

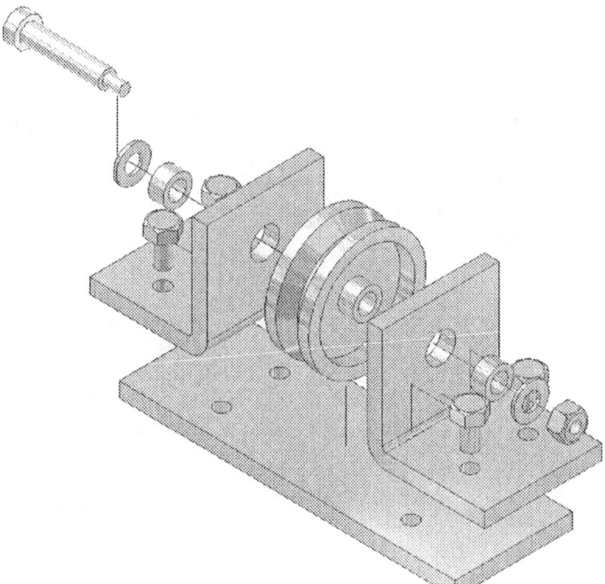

Figure 18 *Exploded view of the Wheel Support assembly*

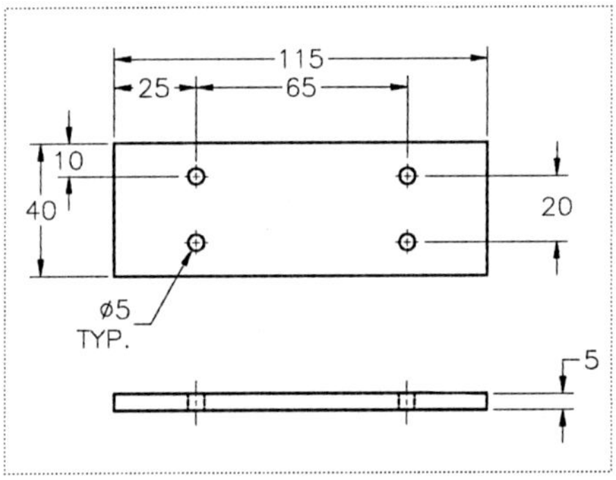

Figure 19 *Dimensions of the Base*

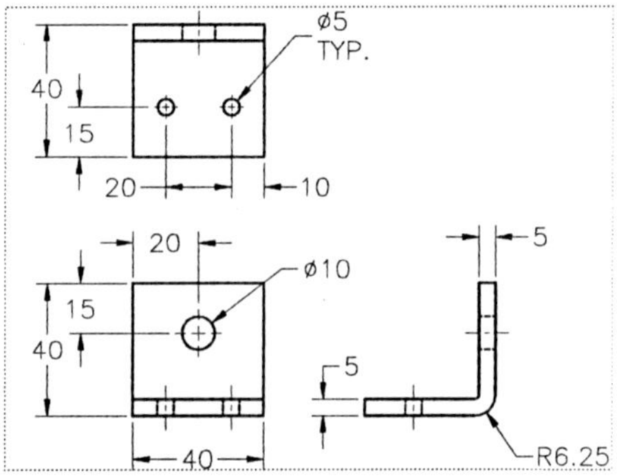

Figure 20 *Dimensions of the Support*

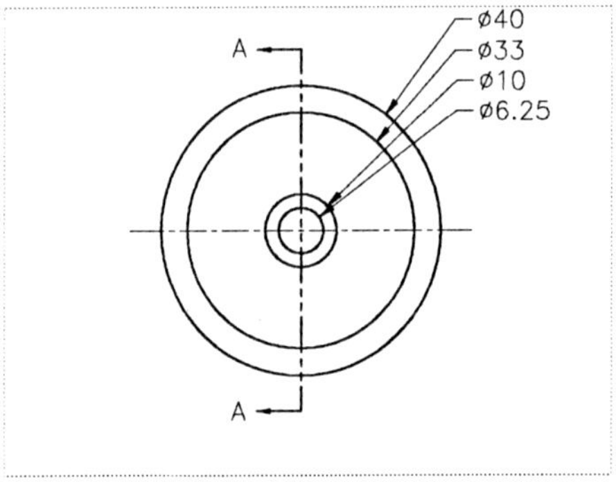

Figure 21 *Front view of the Wheel*

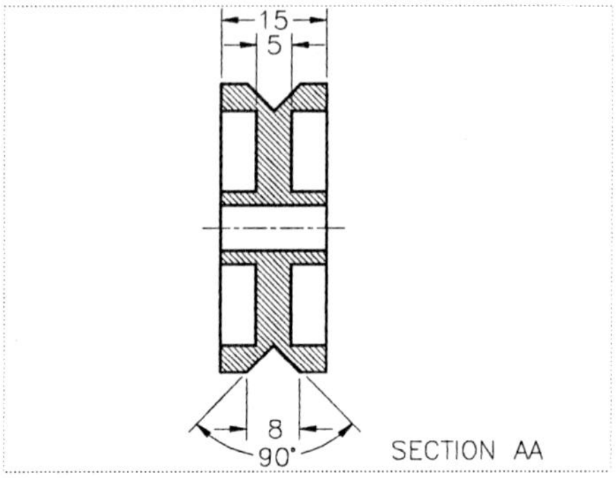

Figure 22 *Sectioned side view of the Wheel*

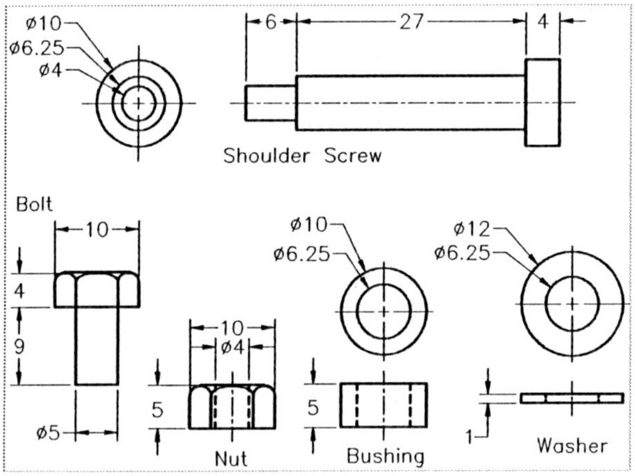

Figure 23 *Dimensions of the Shoulder Screw, Bolt, Nut, Bushing, and Washer*

Index

Symbols

2D Command Line Emulator xxxix
3 Pt Arc tool 1-23
3D Sketch tool 9-25
3DSketch tool 9-25

A

Absorbed Features xl
Add a Section in the Loft Feature 9-23
Add Loft Section option 9-23
Add Loft Section PropertyManager 9-23
Add Relations PropertyManager 3-2
Adding annotations to the Drawing Views 13-2
Adding Center Lines to Views 13-17
Adding Center Marks to the Drawing Views 13-14
Adding Datum Feature Symbol to the Drawing Views 13-8
Adding Datum Target Symbols to the Drawing Views 13-12
Adding Geometric Relations to the Sketch 3-2
Adding Geometric Tolerance to the Drawing Views 13-10
Adding Hole Callout to the Views 13-18
Adding New Sheets to the Drawing Views 13-26
Adding Notes to the Drawing Views 13-6
Adding Relations 3-2
Adding Surface Finish Symbols to the Drawing Views 13-7
Adding the Bom and Balloons In the drawing 13-21
Adding the Cosmetic Threads to the Drawing Views 13-18
Adding the Dowel Pin Symbols to the Drawing Views 13-20
Adding the Multi-jog Leader to the Drawing Views 13-20
Advance Dimensioning Techniques 4-2
Advanced Assembly Mates 11-2
Advanced Mates rollout 11-2
Advanced Modeling Tools 6-2, 9-2
Advanced smoothing option 9-6
Align with end faces option 9-6
Aligned Dimensioning 3-20
Aligned Section View 12-5, 12-21
Aligned Section View tool 12-21
Alternate Position PropertyManager 12-32
Alternate Position View 12-6, 12-32
Alternate Position View tool 12-32
Analyzing Collisions Using Collision Detection tool 11-23
Angle mate 10-14
Angle-distance radio button 2-6
Angular Dimensioning 3-20
Annotations CommandManager xxxv
Applying Gear Mate 11-4
Applying Limit Mate 11-4
Applying the Cam Mate 11-2
Applying the Symmetric Mate 11-2
Arc PropertyManager 1-23
Assemblies CommandManager xxxii
Assembling Components Using Mate Reference 10-24
Assembling Components Using Smart Mates 10-19

Assembling the Components 10-9
Assembly 1-4
Assembly Exploder dialog box 11-27
Assembly from [Selected] Components 11-7
Assembly Instance Color dialog box 11-21
Assembly Mode xx
Assembly mode 10-2
Assembly Modeling 10-2
Assigning Material and Texture to the Models 4-32
Assigning Material to a Model 4-32
Assigning Textures to a Model 4-33
Auto Explode tool 11-25
Auto-Backup Option xli
Autodimension PropertyManager 4-3
Autodimension the Sketches 4-2
Autodimension tool 4-2
Automatic Inferencing Lines 1-13
Automatic Relations 3-7, xxxvi
Auxiliary View 12-5
Auxiliary View PropertyManager 12-25
Auxiliary View tool 12-25
Axis PropertyManager 5-11
Axis tool 5-10

B

Bidirectional Associativity xxxv
Bodies to Keep dialog box 5-28
Bottom-up assembly xx
Bottom-up Assembly Design Approach 10-2
Bottom-up Subassembly Design 11-5
Broken View 12-5, 12-30
Broken-out Section PropertyManager 12-25
Broken-out Section tool 12-23
Broken-out Section View 12-6

C

C-Bored Drilled 6-13
C-Drilled Drilled 6-13
C-Sunk Drilled 6-13
cam mate 11-2
Center Mark PropertyManager 13-14
Center Mark tool 13-14
Centerline Parameters option 9-22
Centerline PropertyManager 13-17
Centerline tool 1-18, 13-17
Centerpoint Arc tool 1-22
Chamfer Dimension tool 13-5
Chamfer PropertyManager 6-34
Chamfer tool 6-33
Chamfering the Sketched Entities 2-5
Change Suppression State tool 11-19
Changing the Scale of the Drawing Views 12-36
Changing the Sketch Plane of the Sketches 8-4
Changing the Transparency Conditions 11-20
Changing the View Orientation 12-36
Checking the Interferences in the Assembly 11-21
Child Features xl
Circle PropertyManager 1-19
Circle tool 1-18
Circular Pattern 11-16
Circular Sketch Step and Repeat dialog box 2-23
Circular Sketch Step and Repeat tool 2-21
Close loft option 9-21
Coincident mate 10-10
Collision Detection xxxix
Color Scheme xli
CommandManager xxv
Composite Curve PropertyManager 9-33
Composite Curve tool 9-33
Concentric mate 10-11
Concept of fully defined sketch 3-27
Concept of the Feature Scope 5-31
Confirm Delete dialog box 8-13
Connectors 9-17
Constant Radius Fillet 6-17
Contour Select Tool 9-2
Coordinate System PropertyManager 5-17
Coordinate System tool 5-16
Copy Confirmation dialog box 8-11
Copying Features Using Drag and Drop 8-11
Cosmetic Thread dialog box 6-15, 13-19
Cosmetic Thread tool 13-18
COSMOSXpress xxxix
counterbore hole 6-4
Counterbored 6-13
Counterdrilled 6-13
countersink hole 6-7
Countersunk 6-13
Creating 3D Sketches 9-25

Creating a Plane at an Angle to an Existing Plane 5-7
Creating a Plane at Some Offset From an Existing Plane 5-8
Creating a Plane Normal to Curve 5-9
Creating a Plane On Surface 5-10
Creating a Plane Parallel to an Existing Plane or Planar face 5-7
Creating a Plane Using Through Lines/Points 5-6
Creating a Reference Axis Using One Line/Edge/Axis 5-11
Creating a Reference Axis Using Two Planes 5-11
Creating a Reference Axis Using Two Points or Vertices 5-12
Creating a Systematic Explode State 11-26
Creating a Thin Sweep Feature 9-12
Creating a User-Defined Sheet Format 13-27
Creating an Assembly From the Part Document 10-6
Creating Base Features by Extruding the Sketches 4-11
Creating Base Features by Revolving the Sketches 4-20
Creating Bottom-Up Assemblies 10-4
Creating Chamfer 6-33
Creating Circular Pattern Feature 7-15
Creating Circular Pattern of the Sketched Entities 2-21
Creating Composite Curve 9-33
Creating Curve-driven Pattern 7-21
Creating Curve Through Free Points 9-34
Creating Curve Through Reference Points 9-35
Creating Curve Through XYZ Points 9-34
Creating Curves 9-27
Creating Cut Features 5-25
Creating Cut Sweep Features 9-14
Creating Draft Features 9-40
Creating Edge Chamfer 6-33
Creating Extruded Cuts 5-25
Creating Fillets 6-15
Creating Folders in the FeatureManager Design Tree 8-21
Creating Helical Curve 9-36
Creating Helical/Spiral Curve 9-36
Creating Linear Pattern Feature 7-8
Creating Linear Pattern of the Sketched Entities 2-16
Creating Mirror Feature 7-2
Creating Multi-thickness Shell 6-39
Creating New Planes 5-5
Creating Patterns 2-15
Creating PATTERNS of the Components in An Assembly 11-14
Creating Projection Curve 9-27
Creating Reference Axis 5-10
Creating Reference Coordinate System 5-16
Creating Reference Points 5-14
Creating Revolved Cuts 5-30
Creating Simple Hole 6-2
Creating Sketch-driven Pattern 7-19
Creating Solid Extruded Features 4-12
Creating Solid Revolved Features 4-21
Creating Split Lines 9-30
Creating Standard Holes Using Hole Wizard 6-3
Creating Subassemblies 11-5
Creating the Component in the Top-Down Assembly 10-27
Creating the Deform Feature 9-42
Creating the Explode Line Sketch 11-28
Creating the Exploded State of the Assembly 11-25
Creating the Loft Feature 9-15
Creating the Partial Section View 12-19
Creating the Shell Feature 6-37
Creating the Spiral Curve 9-39
Creating the Surface Section View 12-19
Creating the Sweep Feature 9-2
Creating the Wrap Feature 6-41
Creating Table-driven Pattern 7-24
Creating the Dome Feature 7-31
Creating the Rib Feature 7-26
Creating the Shape Feature 7-35
Creating Thin Extruded Features 4-17
Creating Thin Revolved Features 4-24
Creating Top-down Assemblies 10-26
Creating Vertex Chamfer 6-35
Crop View 12-6
Crop View tool 12-29
Cropping Drawing Views 12-29
Curvature Continuous in the Face Fillet With Hold 6-29

Curve File dialog box 9-34
Curve Through Free Points tool 9-34
Curve Through Reference Points PropertyManager 9-35
Curve Through Reference Points tool 9-35
Curve Through XYZ Points tool 9-34
Curves CommandManager xxviii
Customize dialog 2-2
Cut, Copy, and Paste Features and Sketches from One Document to the Other 8-11
Cut Loft PropertyManager 9-25
Cut Sweep tool 9-14
Cut-Extrude PropertyManager 5-24
Cut-Revolve PropertyManager 5-31
Cut-Sweep PropertyManager 9-14, 9-15

D

Dangling 3-28
Dassault Systemes xix
Datum Feature Symbol tool 13-8
Datum Target tool 13-12
Default Planes 5-4
Deform PropertyManager 9-43
Deform tool 9-42
Delete Body PropertyManager 8-15
Delete Solid/Surface tool 8-13
Deleting Components and Subassemblies 11-6
Deleting the Bodies 8-13
Deleting the Drawing Views 12-36
Deleting the Features 8-13
Deleting the over defining dimensions 3-30
Deleting the Sketched Entities 1-36
Dependent Features xli
Design Table xxxviii
Detach Segment on Drag 2-30
Detail View 12-5
Detail View PropertyManager 12-27
Detail View tool 12-27
Determining Section Properties of Closed Sketches 4-10
Diagnostics PropertyManager 11-13
Diameter Dimensioning 3-22
Dimension PropertyManager 3-10
Dimension tool 3-10,
Dimension/Relations CommandManager xxvi
Dimensioning of the True Length of an Arc 4-5
Dimensioning Standard and Units xxxiii
dimensioning the sketch 3-8
Direction of Extrusion 5-21
Direction Vector option 9-18
Display Modes of the Model 4-29
Displaying the Section View of the Model 7-37
Dissolve Sub-assembly option 11-11
Dissolving the Subassembly 11-11
Distance mate 10-13
Dowel Pin Symbol tool 13-20
Draft PropertyManager 9-40
Draft Quality HLR/HLV 4-30
Draft tool 9-40
Dragging and Dropping Features from One Document to Other 8-12
Drawing 1-5
Drawing 3 Point Arcs 1-23
Drawing a Construction Circle 1-19
Drawing a Parabolic Curve 1-33
Drawing a Rectangle at an Angle 1-26
Drawing Arcs 1-20
Drawing Centerpoint Arcs 1-22
Drawing Circles 1-18
Drawing Construction Lines or Centerlines 1-18
Drawing Continuous Lines 1-14
Drawing Display Tools 1-34
Drawing Ellipses 1-31
Drawing Elliptical Arcs 1-32
Drawing Individual Lines 1-14
Drawing Lines 1-13
Drawing Mode xxi
Drawing Parallelogram 1-26
Drawing Parallelograms 1-26
Drawing Points 1-30
Drawing Rectangles 1-25
Drawing Splines 1-29
Drawing Tangent or Normal Arcs Using the Line Tool 1-17
Drawing Tangent/Normal Arcs 1-20
Drawing View of the Exploded State of the Assembly 12-34
Drawings CommandManager xxxiii
Dynamically Modifying and Copying the Sketched Entities 2-29
Dynamically Rotating the View of the Model 4-26

E

Edit Circular Step and Repeat 2-25
Edit Component 11-9
Edit Linear Step and Repeat 2-25
Edit Material 4-32
Edit Material tool 4-32
Edit Sketch PropertyManager 5-4
Edit Texture tool 4-33
Editing and Modifying the Drawing Views 12-35
Editing Assembly Mates 11-7
Editing by Double-Clicking the Entities and Feature 8-4
Editing Features and Sketches by Cut, Copy, and Paste 8-10
Editing in the Context of Assembly 11-9
Editing Patterns 2-25
Editing the 3D Sketches 9-27
Editing the Annotations 13-21
Editing the Components 11-9
Editing The Features Of the model 8-2
Editing the Sheet Format 13-27
Editing the sketched entities 2-2
Editing the Sketches of the Sketched Features 8-4
Editing the Sketches Using Move/Size Features 8-9
Editing Using Edit Feature Option 8-2
Editing Using the Move/Size Features 8-6
Ellipse PropertyManager 1-31, 1-33
Ellipse tool 1-31
Equations xxxviii
Error Diagnostics 6-40
Explode Line Sketch 11-28
Exploded View tool 11-25
Extend Entities tool 2-3
Extending the Sketched Entities 2-3
Extrude PropertyManager 4-12
Extruded Boss/Base tool 4-11
Extruded Cut tool 5-25
Extruding a 3D Sketch 9-40

F

Face Fillet 6-27
Face Fillet Using the Hold Line 6-29
Feature Driven PropertyManager 11-14
Feature Palette xxxviii
Feature-Based Mates 10-22
Feature-based Modeling xxxiv
Feature-driven Pattern 11-14
FeatureManager Design Tree xl
Features CommandManager xxvi
Fillet PropertyManager 6-16
Fillet tool 6-15
Fillet With and Without Tangent Propagation 6-19
Filleting the Sketched Entities 2-4
Follow 1st and 2nd guide curves option 9-10
Follow path and 1st guide curve option 9-9
Follow Path option 9-4
Force Transparency 11-9
Form New Sub-assembly Here 11-7
Front plane 5-4
Full Round Fillet 6-30
Full section view 12-17
Fully Defined 3-27

G

Gear mate 11-4
GearMate1 PropertyManager 11-4
Generating Aligned Section Views 12-21
Generating Annotations Using the Model Items Tool 13-2
Generating Auxiliary Views 12-25
Generating Broken View 12-30
Generating Broken-out Section Views 12-23
Generating Detail Views 12-27
Generating Model Views 12-6
Generating Projected Views 12-15
Generating Section Views 12-16
Generating Standard Drawing Views 12-6
Generating Standard Views Using the Relative View Tool 12-11
Generating the Derived Views 12-15
Generating the Section View of an Assembly 12-20
Generating the Three Standard Views 12-9
Generative drafting xxi
Geometric Relations xxxvi
Geometric Tolerance tool 13-10
Geometry-Based Mates 10-22
Getting Started With SolidWorks xxi
Graphic cosmetic thread 6-11

Guide Curves 9-7, 9-20
Guide Curves option 9-20

H

Half section view 12-17
Handling Multiple Bodies in Cut Feature 5-28
Helix and Spiral tool 9-36
Helix Curve dialog 9-37
Helix/Spiral PropertyManager 9-36
Helix/Spiral tool 9-36
Hidden Lines Removed 4-30
Hidden Lines Visible 4-29
Hide/Show Components tool 11-19
Hiding the Bodies 8-15
Hiding the Components 11-19
Hole Callout tool 13-18
Hole Definition dialog box 6-5
Hole Placement dialog box 6-14
Hole PropertyManager 6-3
Hole Wizard tool 6-3

I

Importance of the sketching planes 5-2
Important terms and their definitions xxxiv
Inferencing Lines 1-12
Insert Component PropertyManager 10-5
Insert Components tool 10-4
Insert Model Items dialog box 13-2
Interactive drafting xxi
Interference Detection tool 11-21
Interference Volumes dialog box 11-23
Invalid Solution Found 3-29

K

Keep constrained corners check box 2-5
Keep normal constant option 9-5

L

Learning about Sketcher Terms 1-11
Library Feature xxxviii
Lightweight 3-37, 11-18
Limit mate 11-4
Line Cursor Parameters 1-15
Line PropertyManager 1-16
Line tool 1-13
Linear Diameter Dimensioning 3-23
Linear Pattern 11-16
Linear Sketch Step and Repeat dialog box 2-16
Linear Sketch Step and Repeat tool 2-16
Local Pattern 11-15
Loft PropertyManager 9-15, 9-17
Loft Synchronization 9-17
Lofted Boss/Base tool 9-15

M

Maintain Tangency option 9-6
Maintain Transparency 11-9
Make Assembly from Part 10-6
Mate Diagnostic 11-13
Mate Diagnostics option 11-13
Make Drawing from Part/Assembly tool 12-4
Mate pop-up toolbar 10-11
Mate PropertyManager 10-10
Mate Reference tool 10-24
Mate tool 10-9
Mated Entities PropertyManager 11-9
Materials Editor PropertyManager 4-33
Measure dialog box 4-6
Measure tool 4-6
Measuring Distances 4-6
Measuring Distances and Viewing Section Properties 4-6
Merge results option 9-6
Merge smooth faces option 9-8
Mirror Components PropertyManager 11-17
Mirror Components tool 11-16
Mirror Entities tool 2-9
Mirroring Items While Sketching (Automatic Mirroring 2-9
Mirroring the Components 11-16
Mirroring the Existing Items 2-9
Mirroring the Sketched Entities 2-9
Model View 12-5
Model View PropertyManager 12-6, 12-8
Model View tool 12-6
Modeling Using the Contour Selection Method 5-21
Modify box 3-8
Modify Sketch tool 2-30
Modify the Hatch Pattern of the section view 12-37

Modifying a Parabola 2-29
Modifying a Sketched Arc 2-27
Modifying a Sketched Circle 2-27
Modifying a Sketched Line 2-27
Modifying a Sketched Point 2-29
Modifying a Sketched Polygon 2-28
Modifying a Spline 2-28
Modifying an Ellipse or an Elliptical Arc 2-29
Modifying Sketched Entities 2-27
Modifying the Dimensioning Standards 1-8
Modifying the Linear and Angular Units 1-8
Modifying the Snap and Grid Settings 1-9
Modifying the View Orientation 4-27
Molds CommandManager xxx
Move Component tool 10-28
Move Confirmation dialog box 8-7
Move or Copy Entities tool 2-11
Move or Copy PropertyManager 2-13, 2-14
Move or Copy tool 2-11
Move/Copy Bodies 8-16
Move/Copy Bodies tool 8-16
Move/Copy Body PropertyManager 8-17
Move/Size Features tool 8-6
Moving and Rotating Component Using Triad 10-30
Moving or Copying the Sketched Entities 2-11
Moving, Rotating, and Scaling Sketched Entities 2-11
Moving the Individual Components 10-28
Moving the Individual Components by Dragging 10-28
Multi-jog Line tool 13-20
Multiple Radius Fillet 6-18

N

Named View dialog box 4-30
New Part tool 10-27
New SolidWorks Document dialog box 1-4
New View 4-29
No Solution Found 3-28
Note tool 13-6

O

Offset Entities PropertyManager 2-9
Offset Entities tool 2-7
Offset From Surface 5-19
Offsetting the Sketched Entities 2-7
Point PropertyManager 1-31
Opaque 11-9
Opening an existing file 3-36
Orientation dialog box 4-27, 4-28
Orientation/Twist Control option 9-5
Origin 1-12
Other Boss/Base Options 5-17
Over Defined 3-28

P

Pan 1-36
Pan tool 1-36
Parabola tool 1-33
Parallel mate 10-15
Parallelogram at an angle 1-27
Parallelogram tool 1-26
Parametric Modeling xxxiv
Part 1-4
Part Mode xx
Part Mode CommandManagers xxvi
Partial Ellipse tool 1-32
Pattern-Based Mates 10-24
Perpendicular mate 10-16
Perspective 4-30
Perspective View PropertyManager 4-30
Physical Dynamics xxxix
Physical Simulation xxxix
pipe tap hole 6-12
Placing Components in the Assembly File 10-4
Placing the Components by Dragging From Windows Explorer 10-8
Placing the Components From Internet Explorer 10-8
Placing the Components Using the Insert Component 10-4
Placing the Components Using the Opened Document 10-6
Plane PropertyManager 5-5
Plane tool 5-5
Point PropertyManager 5-15
Point tool 1-30, 5-14
Polygon PropertyManager 1-28
Polygon tool 1-27
Polygons 1-27
Predefined View PropertyManager 12-12
Predefined View tool 12-12

Project Curve tool 9-27
Projected Curve PropertyManager 9-27, 9-29
Projected View 12-5
Projected View tool 12-15
Projection option 9-31
Projection tool 9-27
PropertyManager 1-35
Push-Pin 4-28

R

Radius Dimensioning 3-23
Rebuild Errors dialog box 8-23
Rectangle tool 1-25
Redraw tool 1-36
Reference Geometry 5-2
Reference Geometry CommandManager 5-5, xxix
Reference Planes 5-3
Relative View PropertyManager 12-11
Relative View tool 12-11
Remove Textures 4-34
Renaming the Features 8-21
Reordering the Features 8-18
Replace Components tool 11-11
Replace Mate Entities tool 11-8
Replace PropertyManager 11-12
replacing the Components 11-11
Replacing the Mated Entities 11-8
Reset Standard Views 4-29
Restoring Previous View 4-29
Revolve PropertyManager 4-23
Revolved Boss/Base tool 4-20
Revolved Cut tool 5-30
Right plane 5-4
Right-hand thumb rule 4-20
Rolling Back the Model 8-20
Rotate Component tool 10-30
Rotate or Copy tool 2-14
Rotate View tool 4-26
Rotating a Drawing View 12-36
Rotating or Copying the Sketched Entities 2-14
Rotating the Individual Components 10-29
Rotating the Individual Components by Dragging 10-29
Rotating the View Around a Selected Vertex, Edge, or Face 4-26
Rotating the View Freely in 3D Space 4-26
RouteLine PropertyManager 11-28

S

Scale or Copy tool 2-15
Scaling or Copying the Sketched Entities 2-15
Section Properties dialog box 4-11
Section Properties tool 4-10
Section View 12-5
Section View dialog box 12-20
Section View PropertyManager 12-18
Section View tool 12-16
Seed Feature and Seed Component xxxix
Select Loop 6-32
Select Midpoint 6-32
Select Other 6-31
Select Partial Loop 6-32
Select Tangency 6-33
Select Tool 1-13
Selection Methods 6-31
Set to Lightweight 11-18
Set to Resolved 11-18
Setback Fillets 6-20
Setting up the Document Options 1-7
Shaded 4-30
Shaded With Edges 4-30
Shadows In Shaded Mode 4-30
Sheet Format/Size dialog box 12-4
Sheet Metal CommandManager xxix
Shell tool 6-37
Shell1 PropertyManager 6-39
Show preview option 9-6
Silhouette option 9-31
Simple Hole tool 6-2
Simplifying Assemblies using the Visibility Option 11-18
Sketch Chamfer PropertyManager 2-6
Sketch Chamfer tool 2-5
Sketch CommandManager xxvi
Sketch Dimension or Relation Status 3-29
Sketch Fillet PropertyManager 2-5
Sketch Fillet tool 2-4
Sketch onto Sketch option 9-28
Sketch Plane PropertyManager 8-4, 8-5
Sketch Setting 2-30
Sketch Text PropertyManager 2-26
Sketches onto Faces option 9-28

Sketching environment 1-6
sketching environment 1-2
SmartMates PropertyManager 10-18
Spline tool 1-29
Split Entities tool 2-30
Split Line PropertyManager 9-30
Split Line tool 9-30
Splitting the Sketched Entity 2-30
Assembling Components Using the Mate PropertyManager 10-9
Standard 3 View PropertyManager 12-9
Standard 3 View tool 12-9
standard drilled hole 6-9
Standard View flyout 4-29
Standard Views CommandManager xxxi
Start/End Tangency 9-10, 9-18
Start/End Tangency option 9-10
Starting a Drawing Document 12-2
Starting a new Document in SolidWorks 2004 1-3
Suppress tool 8-14
Suppressing the Components 11-19
Suppressing the Features 8-14
Surface Finish tool 13-7
Surfaces CommandManager xxix
Sweep 9-2
Sweep Feature Using the Follow Path and 1st Guide 9-9
Sweep PropertyManager 9-2, 9-3
Sweep tool 9-2
Sweep Using the Follow Path and Keep Normal Constant 9-4
Sweep with Guide Curves 9-7
Swept Boss/Base tool 9-2
Swept Cut tool 9-14
Symmetric mate 11-2
Symmetric1 PropertyManager 11-2
System Requirements xxi

T

Tangent Arc tool 1-20
Tangent mate 10-17
Tangent Propagation option 9-4
tapped hole 6-10
Text tool 2-26
Texture PropertyManager 4-33
Texture Tree 4-34
The Drawing Mode 12-2
The Mate Reference PropertyManager 10-27
Thin Feature rollout 4-18
Through All 5-17
Tip of the Day 1-3, 1-5
Tools CommandManager xxxiii
Top plane 5-4
Top-down assembly xx
Top-down Assembly Design Approach 10-3
Top-down Subassembly Design 11-5
Triad 10-30
Trim Entities tool 2-2
Trimming the Sketched Entities 2-2
Type of Views 12-5
Types of Assembly Design Approach 10-2

U

Under Defined 3-28
Unsuppress tool 8-15
Unsuppress with Dependents tool 8-15
Unsuppressing Features With Dependents 8-15
Unsuppressing the Features 8-15
Up To Body 5-21
Up To Next 5-18
Up To Surface 5-19
Up To Vertex 5-18
Update Standard Views 4-29
Using the Trim Option to Extend a Sketched Entity 2-2

V

Variable Radius Fillet 6-24
View CommandManager xxx
Viewing and Examining Relations 3-35

W

Welcome to SolidWorks 2004 window 1-4
What's Wrong? Functionality 8-22
What's Wrong? functionality xxxix
Windows functionality xxxvi
Wireframe 4-29
Working with Interactive Drafting in SolidWorks 12-35
Wrap tool 6-41
Wrap1 PropertyManager 6-43

Writing Text in the Sketching Environment 2-26

Z

Zoom In/Out 1-35
Zoom to Area 1-35
Zoom to Fit 1-35
Zoom to Selection 1-35

CADCIM Technologies

CADCIM Design Center

CADCIM Design Center offers a broad array of engineering services that will increase your profits by saving you time and money involved in engineering a product. Our expertise as a professional Computer Aided Design and Manufacturing (CAD/CAM) service provider ranges throughout the product development cycle.

Benefits to the Customers of CADCIM Design Center

From concept and initial design optimization to detailed modeling, prototyping, and analysis, CADCIM Design Center provides professional engineering services to its customers that will benefit you in a number of ways.

- *Affordable and cost-effective services at 25 to 50 percent lower cost than that charged by our competitors.*
- *Design team with thousands of man-hours of design experience to meet challenges involving different levels of expertise and resources.*
- *Quality output with precision and accuracy and assurances on the reliability of the services.*
- *Eliminate unexpected costs by ensuring design verification before manufacturing.*
- *Innovative solutions that are purely based on the in-depth understanding of the technology and the challenge of the problem.*
- *An approach that seamlessly integrates our experts into project-specific multidisciplinary teams, ensuring the right blend of know-how.*
- *Anticipating and meeting the design, modeling, and engineering needs of the customer within the set deadlines.*

Services Offered by CADCIM Design Center

We combine our skills and experience with state-of-the-art Computer Aided Design (CAD) software to create a strong outsourcing resource for your design requirements. To cater to these design requirements, the CADCIM Design Center provides the following services.

- *Surface and Solid Modeling*
- *2D Drafting and Detailing*
- *Reverse Engineering*
- *Assembly Creation and Management*
- *Technical Documentation*
- *Platform Migration*
- *Core Cavity Solutions*
- *FEA/FEM*
- *Rendering and functional animations of products and processes*

CADCIM Training

To excel in the design field, you need to update yourself to the latest technology and make sure you have the right technical skills to get the job done. With our CAD/CAM training programs, you have an assurance of highly professional training.

Benefits to the Students of CADCIM Training

- *Rates lower than that of most of the training companies.*
- *Training conducted by faculty members who have vast experience in related field and have been authoring books on the related softwares for more than a decade.*

- *World-class training material, which is used by a number of universities and colleges across the globe.*
- *Courses specially customized to meet the customer requirements.*

Courses Offered

We offer the following courses that are specially designed to meet your requirement.

- *Introduction to Parametric Modeling*
  Part, Assembly, Detailing and Drafting modules using one of the following softwares.

- Pro/E Wildfire	*- SolidWorks*	*- Solid Edge*
- Pro/E 2001	*- Autodesk Inventor*	*- Mechanical Desktop*

- *Advance course in Parametric Modeling*
  Advance Part, Assembly, Detailing and Drafting, Sheet Metal, Weldments, Mold Design, Animation, Photo-Rendering, and Surfacing using one of the above-mentioned softwares.
- *Introduction to CAD*
- *2D drafting and 3D modeling using AutoCAD*

For Universities and Colleges

We have the qualifications and experience to **substitute** *or* **assist** *the faculty to teach courses in Computer Aided Design (CAD), Solid Modeling, and related classes.*

CADCIM Publishing

CADCIM has published several textbooks on Computer Aided Design (CAD). These textbooks are considered one of the best in the market. Some of our textbooks have been translated and published in Russia, China, Philippines, and Italy. The following are some of the features of our textbooks.

- *Free online technical support by contacting* **techsupport@cadcim.com**.
- *Student projects are available for free download from our Web site* **www.cadcim.com**.
- *Additional information in the form of tips and tricks for each textbook can be downloaded for free from our Web site.*
- *Real world mechanical engineering projects are used as tutorials and exercises in the textbooks so that the reader can relate the tutorials with the real world products in the mechanical engineering industry.*
- *Step-by-step explanation of the procedure to create models.*
- *Explanation of commands and tools followed by the tutorials, which helps the readers to first understand the commands and then use them to execute the design problem.*

Free Teaching Resources for Faculty

We are the only publisher that provides these resources free to the faculty.

- *Powerpoint presentations for each chapter, and part, assembly, and drawing files of all tutorials and exercises in the textbooks.*
- *Instructor's guide with solutions of all review questions and exercises.*
- *Course outlines and student projects.*
- *Free online technical support by contacting techsupport@cadcim.com.*

For Faculty/Business Professionals

If you are a faculty member or a business professional, we can provide the following services at competitive prices 25 to 50 percent lower than that charged by our competitors.

- *Update your existing CAD-related textbooks or training material to the latest release of software.*
- *Do comprehensive technical editing and copyediting of your CAD-related publications, textbooks, and training material.*
- *Do page layout of your manuscript, insert artwork, and design the cover pages.*
- *Prepare customized business presentations using state-of-the-art tools.*